THE PHYSICS AND APPLICATIONS OF HIGH BRIGHTNESS ELECTRON BEAMS

THE SCIENCE AND CULTURE SERIES — PHYSICS

1. Perspectives for New Detectors in Future Supercolliders, 1991
2. Data Structures for Particle Physics Experiments: Evolution or Revolution?, 1991
3. Image Processing for Future High-Energy Physics Detectors, 1992
4. GaAs Detectors and Electronics for High-Energy Physics, 1992
5. Supercolliders and Superdetectors, 1993
6. Properties of SUSY Particles, 1993
7. From Superstrings to Supergravity, 1994
8. Probing the Nuclear Paradigm with Heavy Ion Reactions, 1994
9. Quantum-Like Models and Coherent Effects, 1995
10. Quantum Gravity, 1996
11. Crystalline Beams and Related Issues, 1996
12. The Spin Structure of the Nucleon, 1997
13. Hadron Colliders at the Highest Energy and Luminosity, 1998
14. Universality Features in Multihadron Production and the Leading Effect, 1998
15. Exotic Nuclei, 1998
16. Spin in Gravity: Is It Possible to Give an Experimental Basis to Torsion?, 1998
17. New Detectors, 1999
18. Classical and Quantum Nonlocality, 2000
19. Silicides: Fundamentals and Applications, 2000
20. Superconducting Materials for High Energy Colliders, 2001
21. Deep Inelastic Scattering, 2001
22. Electromagnetic Probes of Fundamental Physics, 2003
23. Epioptics-7, 2004
24. Symmetries in Nuclear Structure, 2004
25. Innovative Detectors for Supercolliders, 2003
26. Complexity, Metastability and Nonextensivity, 2004
27. Epioptics-8, 2004
28. The Physics and Applications of High Brightness Electron Beams, 2005

THE SCIENCE AND CULTURE SERIES PHYSICS

THE PHYSICS AND APPLICATIONS OF HIGH BRIGHTNESS ELECTRON BEAMS

Proceedings of the 46th Workshop of the
INFN ELOISATRON Project

Erice, Italy 9 — 14 October 2005

Editors
L. Palumbo
J. Rosenzweig
L. Serafini

Series Editor
A. Zichichi

World Scientific

NEW JERSEY · LONDON · SINGAPORE · BEIJING · SHANGHAI · HONG KONG · TAIPEI · CHENNAI

Published by

World Scientific Publishing Co. Pte. Ltd.
5 Toh Tuck Link, Singapore 596224
USA office: 27 Warren Street, Suite 401-402, Hackensack, NJ 07601
UK office: 57 Shelton Street, Covent Garden, London WC2H 9HE

British Library Cataloguing-in-Publication Data
A catalogue record for this book is available from the British Library.

THE PHYSICS AND APPLICATIONS OF HIGH BRIGHTNESS ELECTRON BEAMS

ISBN-13 978-981-277-216-9
ISBN-10 981-277-216-2

Printed in Singapore by World Scientific Printers (S) Pte Ltd

46th Workshop: The Physics and Applications of High Brightness Electron Beams
Erice (Trapani), Italy – 9-14 October, 2005

PREFACE
THE PHYSICS AND APPLICATIONS OF HIGH BRIGHTNESS ELECTRON BEAMS

LUIGI PALUMBO
Istituto Nazionale di Fisica Nucleare
Laboratori Nazionali di Frascati
Via E. Fermi 40, 00044 Frascati (Roma), Italy
Dip. Energetica
Universita' degli Studi di Roma "La Sapienza"
Roma, Italy

JAMES B. ROSENZWEIG
UCLA Dept. of Physics and Astronomy
405 Hilgard Ave., Los Angeles, CA 90095-1547
E-mail: rosen@physics.ucla.edu

LUCA SERAFINI
Istituto Nazionale di Fisica Nucleare
Sezione di Milano
Via Celoria 16, 20133 Milano, Italy

The ICFA-sponsored (ICFA Panel on Beam Dynamics, ICFA Panel on Advanced Accelerators) workshop on "The Physics and Applications of High Brightness Electron Beams" was held on October 9-14, 2005, in Erice, Sicily, Italy, at the Ettore Majorana Foundation and Center for Scientific Culture (EMFCSC). It was the 46th Workshop of the Series "INFN Eloisatron Project".

We had a fine international representation from the fields of intense electron sources, free-electron lasers, advanced accelerators, ultra-fast laser-plasma, beam-plasma and laser-beam physics, well according to the tradition of this workshop series, built up on the heritage coming from the merging of two ICFA-sponsored series, one on high brightness beams, and the other on free-electron laser physics. The most recent high brightness beam workshop was entitled "The Physics of High Brightness Beams", endorsed by the ICFA Panel or Novel and Advanced Accelerators, and was held in Los Angeles in November 1999. It concentrated on the emerging physics of intense beam systems, and was notable for the exciting synthesis of methods and concepts between electron beam and ion beams. On applications side, the final installment "Arcidosso Series" of workshops, held in Sept. 2001, was entitled "The Physics of, and Physics With, the X-ray Free-Electron Laser". This title foreshadows the merging of discussions between the physics of the enabling device (in this case the FEL), with the examination of the physics of the enabled applications. The first workshop merging the two series was held in July 2002 in Sardinia (Italy): the proceedings are available from World Scientific*.

As for the previous one, the mission of this workshop was stated by the increasingly critical role that high brightness electron beams are playing in two frontier fields: radiation generation methods and advanced acceleration schemes. Such state-of-the-art radiation production methods include various types of free-electron lasers, as well as inverse Compton scattering (ICS) of intense lasers, having diverse approaches to creating high peak and average power light sources. As they are capable of harder photon production, ICS sources are candidates not only for X-ray sources, but also high-energy physics applications. Likewise, high brightness beams are at the center of future accelerator schemes, e.g. based on high gradient wakefields, and electron cooling. Indeed, possibilities exist to create unique light sources based on advanced acceleration schemes, just as intense light sources enable advanced accelerator research. The goal of this workshop is to provide a comparative study of the generation, manipulating, modeling and measuring of high brightness electron beams, and the underlying methods linking the physics of these beam systems to the physics of advanced applications.

The workshop would not have been a success without its working group leaders (M. Ferrario, G. Travish, L. Giannessi, M. Uesaka) and their scientific secretaries (D. Filippetto, G. Gatti, A. Schiavi, A. Rossi).

Further debts are owed to the workshop local organization (Rossana Centioni, Silvia Giromini and Mary Jo Robertson), the EMFCSC local team (Fiorella Ruggiu and Pino Aceto), the program committee (C. Pellegrini, W. Barletta, M. Ferrario, P. Emma, D. Dowell, L. Giannessi, J. Murphy, P. O'Shea, J. Rossbach, T. Garvey, M. Eriksson, G. D'Auria, D. Giulietti), and the organizing committee (L. Palumbo, J. Rosenzweig, L. Serafini, G. Krafft, H. Braun, K-J. Kim, S. Bertolucci, I. Ben Zvi, A. Renieri, S. Milton, S. Chattopadhyay, J. Galayda, M. Poole, C. Bocchetta, K. Floettmann, P. Krejcik).

But it's certainly with Prof. A. Zichichi, President of the Ettore Majorana Foundation and Center for Scientific Culture, that we are mostly indebted: thanks to his kind invitation to join the "INFN Eloisatron Project" workshop series, all our participants could enjoy a unique combination of a fascinating land to visit, Sicily, an enchanting town to live, Erice, and a world-wide famous Center of Science and Culture, EMFCSC, to host the workshop activities.

Sicily, the largest island in the Mediterranean sea, displays spots of extreme nature, made of fire and volcanos, as well as marvellous beaches and sea landscapes. Its history is full of culture and merging civilizations, from

* "The Physics and Applications of of High Brightness Electron Beams", Ed. J.B. Rosenzweig, L. Serafini and Gil Travish (World Scientific, Singapore, 2003).

Phoenicians to Greeks, from Romans to Normans: a fusion whose the village of Erice is one of the best example. It is in the context of such a poetic and inspirating environment that EMFCSC developed through the last 43 years an impressive amount of scientific culture, making a deep impact within the world scientific community.

We hope that these proceedings give some flavor of the workshop plenary presentations and the discussions within the working groups, and also that they provide a unique reference to the status of the relevant research fields as well as their development into the future.

Luigi Palumbo
Rome

James B. Rosenzweig
Los Angeles

Luca Serafini
Milan

September 2006

CONTENTS

Plenary Sessions

RF DEFLECTOR BASED SUB-PS BEAM DIAGNOSTICS: APPLICATION TO FEL AND ADVANCED ACCELERATORS†

DAVID ALESINI

Laboratori Nazionali di Frascati-INFN, Via E. Fermi 40, 00044 Frascati, Italy

RF deflectors are very promising devices for longitudinal and transverse phase space characterization and beam manipulation in advanced accelerators. Measurement setups for longitudinal pulse shape as well as phase space and transverse beam slice emittance characterizations are described as well the main design criteria of traveling wave and standing wave RF deflectors used for beam diagnostics putting in evidence the advantages or disadvantages of the two devices in terms of performances, required power and power limitations.

1. Introduction

The characterization of the longitudinal and transverse phase space of the beam at the end of an injector is a crucial point in order to verify and tune all photoinjector performances. With the use of an RF deflector (RFD) it is possible to measure the bunch length [1,2] and, adding a dispersive system, the longitudinal beam phase space. Similarly, since the longitudinal beam distribution can be projected along a transverse coordinate, the transverse emittance of each longitudinal bunch slice can be measured using the quadrupole scan technique [3-5]. In the first section of the paper I shortly review the basics principles of the longitudinal and transverse beam phase space characterization using an RFD. In particular I illustrate the bunch length measurement and the sources of errors (par. 2.1), the longitudinal phase space characterization (par. 2.2) and the horizontal beam slice emittance measurement (par. 2.3). The possibility to obtain ultra short bunches with the use of two RFDs is described in par. 2.4 while the diagnostics with a circular polarized RFD is illustrated in par. 2.5. The second section of the paper is centered on the RFD structures. In particular I refer to the traveling wave devices (par. 3.1), to the standing wave devices (par. 3.2) and I compare the two structures in term of performances (3.3). In the par. 3.4 I address the RF measurements that can be done with an

† This work has been partially supported by the EU Commission in the sixth framework programme, contract no. 011935 EUROFEL-DS1.

RFD and (on par. 3.5) I discuss the power systems. The already obtained experimental results are finally summarized in par. 4.

2. Beam diagnostics and manipulation using RFD

The different types of measurements that can be done with RFDs are based on the property of the transverse voltage[1] (V_T) to introduce a correlation between the longitudinal coordinate of bunch and the transverse one (Fig. 1). The phase of the deflecting voltage is tuned in order to have zero crossing[2] of V_T in the center of the bunch giving a linear transverse deflection from the head and to the tail of the bunch. After the deflector the transverse kick (vertical, for example) results into a transverse displacement of the centroid of each bunch slice, proportional to the slice longitudinal position with respect to the bunch center.

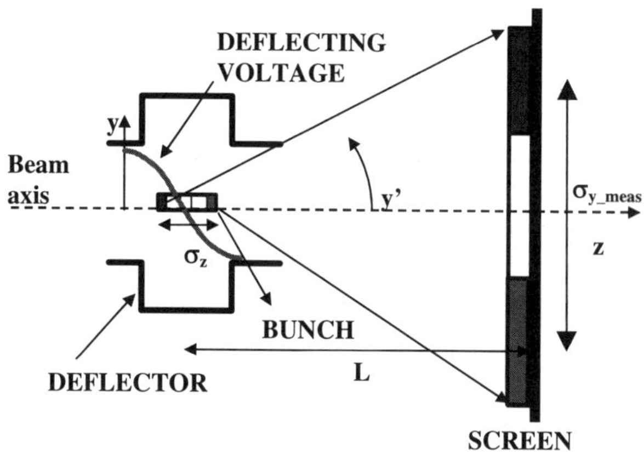

Figure 1. Longitudinal to transverse correlation induced by the RFD transverse voltage.

2.1. *Bunch length measurements*

The relation between the z longitudinal component within the bunch (z=0 corresponds to the bunch center) and the vertical position y on the screen is given by:

$$y = R_{34}\frac{V_T}{E/e}\sin\left(\frac{\omega_{RF}}{c}z+\varphi_{RF}\right)\underset{z<<\lambda_{RF}}{\cong} R_{34}\frac{V_T}{E/e}\sin\varphi_{RF} + \underbrace{R_{34}\frac{V_T}{E/e}\frac{\omega_{RF}}{c}\cos\varphi_{RF}}_{K}\, z \qquad (1)$$

[1] The integrated transverse Lorenz force per unit charge.
[2] As illustrated in the following paragraphs this condition is not strictly necessary.

where ω_{RF} is the angular frequency of the deflecting voltage, $V_\perp$ is the peak transverse voltage, E is the beam energy in eV units, φ_{RF} is the RF phase with respect to the bunch passage (at perfect zero crossing $\varphi_{RF}=0,\pi$) and R_{34} is the optical transfer function given by:

$$\begin{cases} R_{34} = \sqrt{\beta_{y\_defl}\beta_{y\_screen}}\sin \varDelta\Phi & \text{(general expression)} \\ R_{34} = L & \text{(simple drift space)} \end{cases}$$

(2)

where β_{y_defl} and β_{y_screen} are the vertical β-functions at the deflector and screen position, respectively, $\Delta\Phi$ is the phase advance between the deflector and the screen and L is the drift space of length L.

The position of the bunch centroid on the screen ($\bar{y}$) and its rms value (σ_y) are given by[3]:

$$\bar{y} \cong R_{34}\frac{V_T}{E/e}\sin\varphi_{RF}$$

(3)

$$\sigma_y \cong \underbrace{|R_{34}|\frac{V_T}{E/e}\frac{\omega_{RF}}{c}|\cos\varphi_{RF}|}_{|K|}\sigma_z$$

(4)

Equations (1)-(4) show that the longitudinal bunch distribution can be measured by measuring the transverse bunch distribution after the deflector.

In particular, looking at the previous equations, it is possible to directly calibrate the screen vertical coordinate measuring the bunch centroid position $\bar{y}$ on the screen for different values of the RFD phase φ_{RF} by assuming $|\sin\varphi_{RF}|<<1$. Looking, for example, at the case $\varphi_{RF}<<\pi$, from the previous expressions one has, in fact, that:

$$\bar{y} \cong R_{34}\frac{V_T}{E/e}\varphi_{RF} = \underbrace{R_{34}\frac{V_T}{E/e}\frac{\omega_{RF}}{c}}_{K'}z$$

(5)

The calibration coefficient K' multiplied by $\cos\varphi_{RF}$ is equal to the coefficient K that I have pointed out in equation (1) and (4) that gives the scaling factor between the screen coordinate y and the longitudinal position z.

---

[3] From now I suppose that is always satisfied the relation: $z<<\lambda_{RF}$.

Plotting the square of the measured σ_y on the screen as a function of *K*, one simply obtains the parabola of Fig. 2 with the coefficient of the quadratic term proportional to the square of rms bunch length:

$$\sigma_y^2 = \sigma_{y\_RFD}^2 = \sigma_z^2 K^2$$

(6)

The use of the quadratic form will be useful in the following considerations on errors effects.

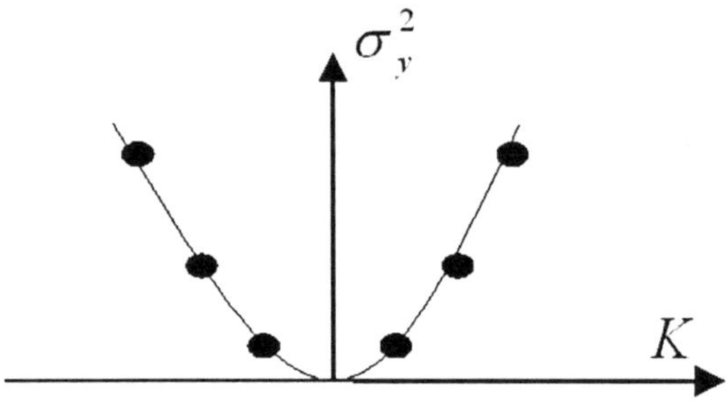

Figure 2. Example of the measured rms vertical dimension of the bunch at the screen as a function of *K*

2.1.1. *Effect of the vertical beam emittance*

Let us now investigate the different sources of errors in the previous measurement. First of all in (1) and (4) I have completely neglected the vertical emittance of the beam. In the real case, the transverse distribution of the bunch at the screen position is the superposition between the deflected beam profile and the vertical size of the bunch slices at the screen position (σ_{y_BEAM}), as illustrated in Fig. 3. It is possible to demonstrate that, at the screen position, the measured vertical rms distribution is given by [2]:

$$\sigma_y = \sqrt{\sigma_{y\_RFD}^2 + \sigma_{y\_BEAM}^2}$$

(7)

with σ_{y_RFD} given by (1) and $\sigma_{y\_BEAM} = \sqrt{\beta_{y\_screen}\varepsilon_y}$ where ε_y is the transverse vertical emittance of the beam. One can define the resolution length σ_{z_RES} as the bunch length that gives, on the screen, a distribution with rms vertical size exactly equal to $\sqrt{2}\sigma_{y\_BEAM}$. From the previous expressions it is easy to verify that:

$$\sigma_{z\_RES} = \frac{c(E/e)\sqrt{\varepsilon_y}}{V_T \omega_{RF} |\cos\varphi_{RF}| \sqrt{\beta_{y\_defl}} |\sin \Delta\Phi|}$$

(8)

In the case of a simple drift space this expression becomes:

$$\sigma_{z\_RES} = \frac{c(E/e)\sqrt{\beta_{y\_screen}\varepsilon_y}}{V_\perp \omega_{RF} |\cos\varphi_{RF}| L}$$

(9)

The plot of σ_y^2 on the screen as a function of *K* is, in this case, of the type sketched in Fig. 4. In the plot one can distinguish between two regions: low and high resolution, according to the previous considerations.

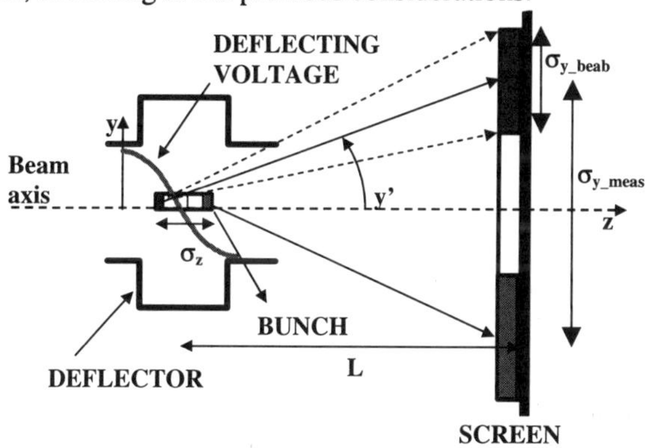

Figure 3. Longitudinal to transverse correlation induced by the RFD transverse voltage putting in evidence that the transverse profile after the deflector is given by the superposition of the deflected beam profile and the vertical proper dimensions of the bunch slices at the screen position.

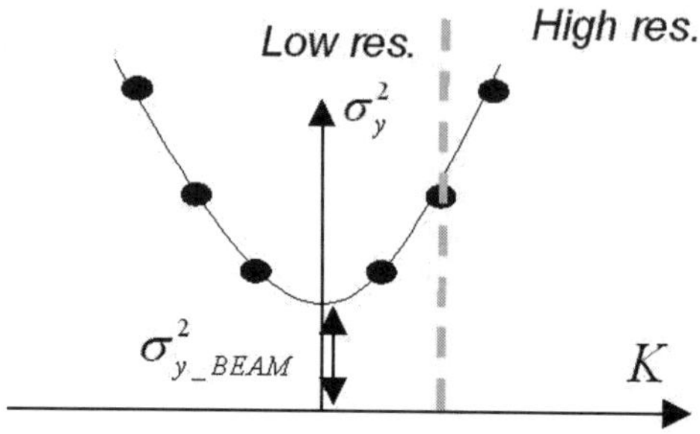

Figure 4. Example of the measured rms vertical dimension of the bunch at the screen as a function of *K* considering the effect of the vertical emittance of the beam.

2.1.2. *Bunch energy spread*

The case of an uncorrelated energy spread within the bunch can be treated as a perturbation similar to the previous one. If the energy spread is σ_{E1}/E and the beam has zero vertical emittance, it is easy to verify that the measured vertical dimension of the beam at the screen position is given by:

$$\sigma_y = \sqrt{\sigma^2_{y\_RFD} + \sigma^2_{y\_ES1}}$$

(10)

with:

$$\sigma_{y\_ES1} \cong R_{34} \frac{V_T}{E/e} \sin \varphi_{RF} \left(\sigma_{E1}/E \right)$$

(11)

This effects is in general negligible.

The case of a correlated energy spread within the bunch of the type:

$$E_c(z) = E\left(1 + \frac{(\sigma_{E2}/E)}{\sigma_z} z\right)$$

(12)

give the following perturbation in the measurement [2]:

$$\sigma_y = \left| \sigma_{y\_RFD} - \sigma_{y\_ES2} \right|$$

(13)

where:

$$\sigma_{y\_ES2} = R_{34}\frac{V_T}{E/e}\sin\varphi_{RF}\left(\sigma_{E2}/E\right)$$

(14)

Both effects can be cancelled, to the first order, by choosing $\varphi_{RF} = 0$ that requires a screen on axis screen.

The behaviour of σ_y^2 on the screen as a function of *K*, in the case of a correlated energy spread, clearly shows two different coefficients of the quadratic term for two different values of the RF phase ($\varphi_{RF} = \varphi_0$ and $\varphi_{RF} = \varphi_0 + \pi$) according to the formula [2]:

$$\sigma_y^2 = \sigma_z^2\left(1 - \frac{c}{\omega_{RF}}\frac{\left(\sigma_{E2}/E\right)}{\sigma_z}\tan\varphi_{RF}\right)^2 K^2$$

(15)

one obtains, in this case, something of the type sketched in Fig. 5. In the plot one supposes that the values of the positive values of the K factor are obtained for a given $\varphi_{RF} = \varphi_0$ while the negative values are obtained for $\varphi_{RF} = \varphi_0 + \pi$. By fitting the parabola in the two regions it is possible in principle to find the bunch length and the correlated energy spread.

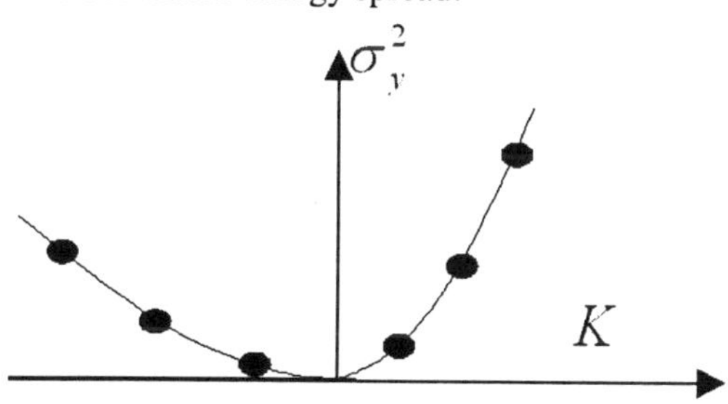

Figure 5. Example of the measured rms vertical dimension of the bunch at the screen as a function of *K* considering the effect of a correlated energy spread within the bunch.

2.1.3. *Effect of the wakefield*

Wakefield induced upstream the deflector by the accelerating structures or (especially with an off-axis kick) between the deflector and the screen can cause both a linearly correlated energy spread within the bunch (that can be corrected or taken into account, as shown in the previous case) or a linear correlation between y and z on the screen position ($y = \alpha z$). Neglecting the correlated energy spread the measured rms vertical dimension of the bunch is given, in this case, by:

$$\sigma_y^2 = \sigma_z^2 (K + \alpha)^2 + \sigma_{y\_BEAM}^2$$

(16)

Looking at the plot of σ_y^2 on the screen as a function of *K*, one obtains, in this case, something of the type sketched in Fig. 6 where the vertex of the parabola is shifted by quantities related to the linear correlation between the y and z plane and to the size of the beam at the screen.

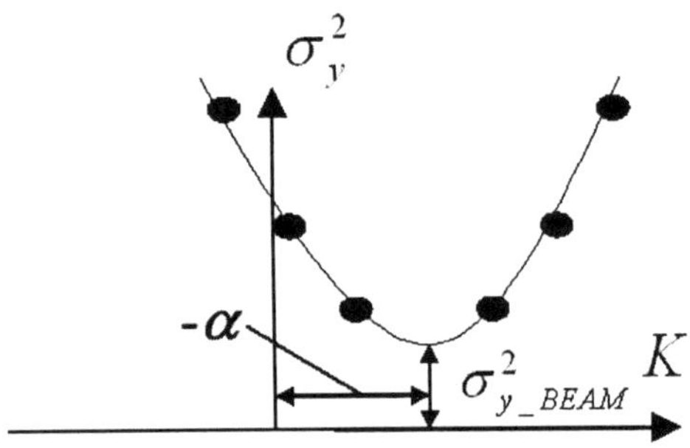

Figure 6. Example of the measured rms vertical dimension of the bunch at the screen as a function of *K* considering the effect of a correlation between the y and z planes and the size of the beam at the screen position.

2.2. *Longitudinal phase space measurements*

The conceptual scheme to measure the longitudinal phase space is sketched in Fig. 7. In this scenario, the bunch is deflected vertically by the RFD and horizontally by a magnetic dipole. The dispersion properties of the dipole allow to completely characterize the energy distribution of each bunch slice allowing the reconstruction of the bunch longitudinal phase space. In this case the energy spread resolution is simply given by:

$$\left.\frac{\Delta E}{E}\right|_{res} = \frac{\sqrt{\varepsilon_x \beta_{x\_screen}}}{D_{screen}}$$

(17)

where ε_x, β_{x_screen} and D_{screen} are the horizontal emittance, the horizontal β-function and dispersion at the screen position.

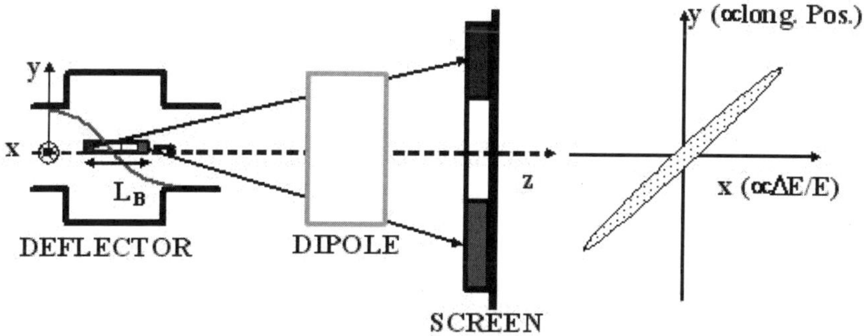

Figure 7. Conceptual scheme of the longitudinal phase space measurement setup using an RFD.

2.3. *Horizonthal slice emittance measurements*

With the deflector it is also possible to measure the horizontal beam slice emittance [3-5]. In this case the deflecting voltage gives the correlation between the longitudinal slice position and the vertical coordinate at the screen while the horizontal emittance can be measured with the quadrupole scan technique. The conceptual scheme is shown in Fig. 8.

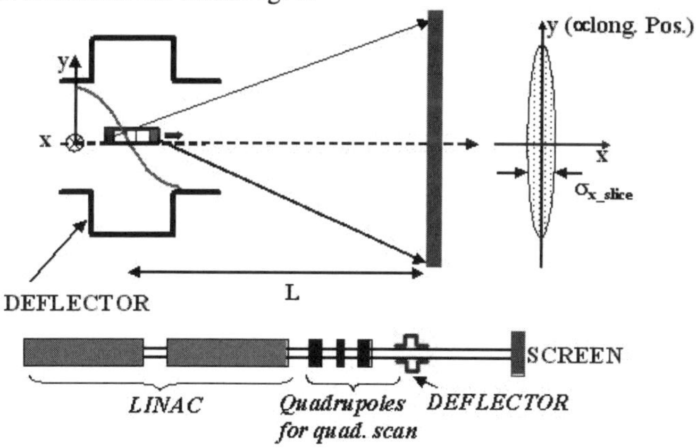

Figure 8. Conceptual scheme of the horizonthal beam slice emittance using an RFD.

2.4. *Beam manipulation with an RFD: ultra-short bunch generation*

RFDs allow to manipulate the beam in order to obtain, for example, ultra-short bunches. The conceptual scheme is given in Fig. 9. With the plot of the beam shape in the planes zy and zy'. The bunch deflected by the first RFD is cut by the collimator. The correction of the zy' correlation is finally realized by the second RFD.

With a simple calculation it is possible to demonstrate that the bunch length after the process is given by:

$$\sigma_z = \frac{wcE/e}{2V_{T1}\omega_{RF}R_{34\_1}}$$

(18)

where V_{T1} is the transverse deflecting voltage of the first RFD, w is the full width of the collimator aperture and $R_{34\text{-}1}$ is the optical matrix element between the RFD1 and the collimator. Moreover the deflector voltages and the optical functions must satisfy the following relations:

$$V_{T2} = -R_{44\_2}V_{T1}$$
$$R_{34\_2} = 0$$

(19)

where $R_{44\text{-}2}$ and $R_{34\text{-}2}$ are the optical matrix elements between RFD1 and RFD2.

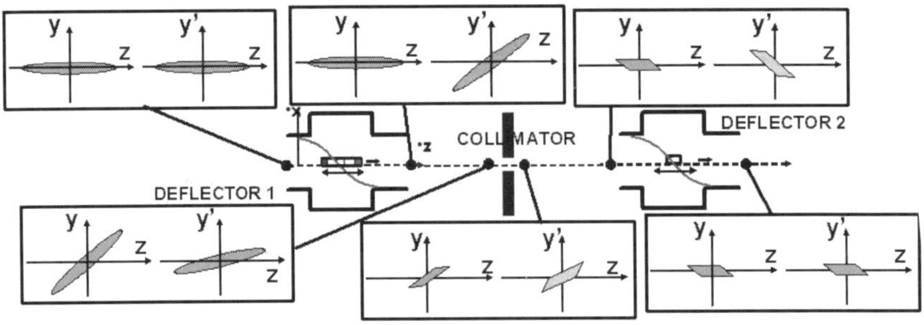

Figure 9. Conceptual scheme of the ultra-short bunch generation using an RFDs.

2.5. *Diagnostics with a circular polarized RFD*

The conceptual scheme of a circular polarized RFD used for beam diagnostics is reported in Fig. 10. Two deflecting cavities generate two different polarization of the deflecting field and, if the two deflecting voltages have the same amplitude and the phase between the two cavities is properly chosen, the total deflecting field seen by the bunch has a circular polarization. Each longitudinal bunch slice is therefore deflected in the xy plane of the screen by an angle ϑ proportional to the longitudinal slice position (as sketched in Fig. 11). The distance of each bunch slice from the origin (x=0, y=0) is inversely proportional to the slice energy while the radial thickness of each slice is proportional to the energy spread of the slice itself (assuming vanishing transverse size of the beam). In this way the longitudinal phase space is completely reconstructed in the polar coordinate system at the screen (r, ϑ). If L is the distance between the deflectors and the screen it is easy to verify that the longitudinal resolution and the energy spread resolutions are given by:

$$\sigma_{z\_RES} = \frac{c(E/e)\sqrt{\varepsilon\beta_{screen}}}{V_T \omega_{RF} L}$$

(20)

$$\left.\frac{\Delta E}{E}\right|_{RES} = \frac{(E/e)\sqrt{\varepsilon\beta_{screen}}}{L V_T}$$

(21)

where β_{screen} is the value of the β function at the flag position ([4]) and ε is the emittance of the beam.

The values of the longitudinal and energy spread resolutions that is possible to achieve are comparable with those that one can obtain from the measurement techniques discussed in the par. 2.1 and 2.2. Obviously, to measure the bunch length with circular polarized RFD, the used wavelength should be comparable with the bunch length. More details on diagnostics with circular polarized RFD and recent interesting experimental results obtained with a 17 GHz RFD can be found in [10,11].

---

[4] The largest between the horizontal and vertical one to be conservative.

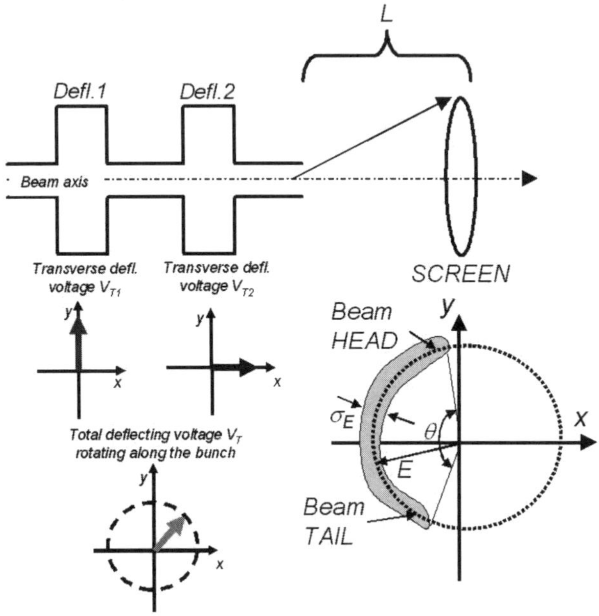

Figure 10. Conceptual scheme of the longitudinal phase characterization using a circular polarized RFD.

3. RF Deflecting structures

Different types of deflecting structures can be used for beam diagnostics. I will refer, in the present paper, to normal conducting structures.

As in the case of accelerating sections, the main differentiation is between the travelling wave (TW) and standing wave (SW) cavities. In the next paragraphs I will illustrate in detail the differences between the two type of structures in terms of electromagnetic (e.m.) properties and performances.

3.1. *TW RF deflectors*

The TW RFDs are disc loaded structures working on the TM_{11}-like mode. The single cell is sketched in Fig 11. In the TW structures both the electric (*E*) and the magnetic (*B*) field contribute to the total deflection, Fig. 11 shows the two field vectors (z is the propagation direction). The ratio between the *E* and *B* deflection depends on the geometry and in particular on the iris aperture *a* as shown in par. 3.1.1. Each cell has two planes of symmetry: the yz plane that is Perfect E and the xz plane that is Perfect H (see Fig. 11).

As in the case of conventional TW accelerating structures, it is possible to define some structure parameters that I report in the following for completeness and that one can find (referred to the accelerating structures) for example in [12] and in [13-15] referred to the deflecting ones:

a) *transverse series impedance* per unit length Z_T defined as:

$$Z_T = \frac{E_T^2}{P} = \frac{\left| \int_0^D \left(\tilde{E}_x - c\tilde{B}_y \right) e^{j\omega_{RF} z/c} dz \Big/ D \right|^2}{P} \quad \left[\frac{\Omega}{m^2} \right]$$

(22)

where P is the power flux along the structure, E_T is the total equivalent deflecting field, D is the cell period, ω_{RF} is the angular RF frequency, z is the longitudinal coordinate of the traveling wave and $\tilde{E}_x, \tilde{B}_y$ are the transverse complex components of the electric and magnetic field on axis (referred to the coordinate system of Fig. 11);

b) *transverse shunt impedance* per unit length r_T defined as:

$$r_T = \frac{E_T^2}{p_d} \quad \left[\frac{\Omega}{\mathrm{m}} \right]$$

(23)

where p_d is the average dissipated power per unit length given by:

$$p_d = \frac{P(z) - P(z+D)}{D} = \frac{1}{D} \int_S \frac{1}{2} R_s H_{\tan}^2 dS$$

(24)

where R_s is the surface resistance, $H_{\tan}$ is the tangential component of the H field on the cell surface and the integral is performed on the surface of a single cell;

c) the *attenuation* α per unit length defined as:

$$\alpha = p_d / 2P \quad \left[\mathrm{m}^{-1}\right]$$

(25)

d) the *quality factor Q* per unit length defined as:

$$Q = \omega_{RF} \frac{w}{p_d}$$

(26)

where w is the average stored energy per unit length given by:

$$w = \frac{1}{4D}\left(\int_{V\_cell} \varepsilon_0 E^2 dV + \int_{V\_cell} \mu_0 H^2 dV \right)$$

(27)

e) the *group velocity* v_g of the structure defined as the velocity of propagation of the energy in the structure and given by:

$$v_g = \frac{P}{w} \quad [\mathrm{m/s}]$$

(28)

f) the *filling time* τ_f of the structure is directly related to v_g with the usual formula:

$$\tau_f = \frac{L}{v_g} \quad [\mathrm{s}]$$

(30)

where L is the structure length.

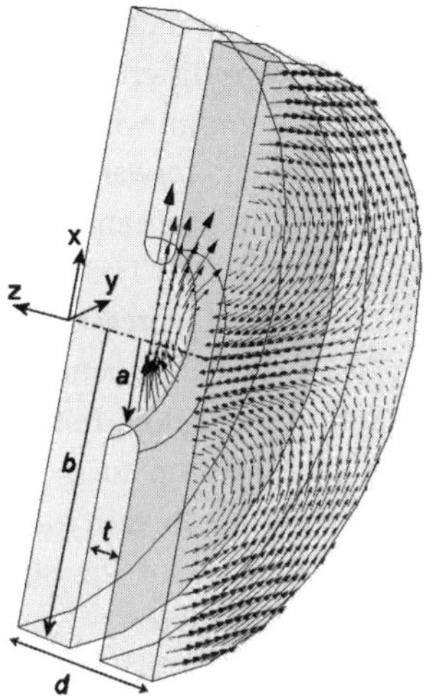

Figure 11. Single cell of a TW RFD with the *E* and *B* field lines.

3.1.1. *Single cell design criteria*

The length of each cell (D) is chosen in order to synchronize the phase velocity with the beam according to the formula:

$$D = \frac{v_b \Delta\phi}{\omega_{RF}}$$

(31)

where $\Delta\phi$ is the phase advance per cell of the TW deflecting mode and v_b is the particle beam velocity (equal to c in the case of ultra-relativistic particles). The external radius (*b*) of the cell is chosen in order to tune the working frequency of the cell at the correct generator frequency f_{RF}; the thickness (*t*) of the iris is not a critical parameter and can be chosen, for example, in order to minimize the peak surface electric field or the attenuation of the structure. Concerning the iris aperture (*a*), this is a critical dimension because it is the most important parameter to set the group velocity of the structure (and therefore the filling time), the deflection efficiency and the attenuation per unit length. Using the analytical expression valid in the case of small picth approximation (t<<D) [13-15] it is possible to calculate, for example, the behaviour of the previous quantities as a function of the dimension *a*. In Fig. 12 I have plotted, for example, the behaviour of the group velocity and the transverse deflecting voltage (normalized to the square root of the input power P) for a

copper structure of length L=1m working on the $2\pi/3$ mode at 2.856 GHz. In the plot it is possible to distinguish two different regions: the first one for small value of the aperture *a* (up to about 3 cm) in which the structure is backward and the other region in which the structure is forward[5]. By the analytical treatment it is possible to verify that the transition between the backward structure and the forward one occurs when $a/\lambda \cong 0.26$. The efficiency of the structure is strongly dependent on the iris aperture: it increases for small iris and it decreases if one increases the iris aperture of the structure except near the transition between the backward and the forward region. In the same plot are reported the previous quantities calculated by HFSS up to a=25 mm (for some particular values of the iris aperture). It is possible to verify that the analytical calculation gives a good estimate of the efficiency of the structure and group velocity especially for small values of the iris aperture.

Deflectors used as RF separators [13,14] require large iris apertures while those used for beam diagnostics (especially for low emittance beams) can have smaller iris apertures (and therefore higher transverse impedance) compatibly with the intensity of the induced wakefield, filling time and beam transverse dimensions.

A typical dispersion curve calculated by HFSS [16] is shown in Fig 13. It is also possible to calculate analytically the dispersion curve through an analytical expression reported in [13-15].

The behaviour of the deflecting field on axis can be calculated by e.m. numerical codes or, in the approximation discussed before, with analytical formulae. In case of phase velocity equal to c they exhibit a very simple expression:

$$\begin{cases} E_x = E_0 \dfrac{k^2 a^2}{4} e^{-j\frac{\omega_{RF}}{c} z} \\ cB_y = E_0 \left(\dfrac{k^2 a^2}{4} - 1 \right) e^{-j\frac{\omega_{RF}}{c} z} \end{cases}$$

(32)

From these formulae it is easy to verify that the ratio between the magnetic and the electric deflection depends on the iris aperture and that the total

[5] The reason why one has two different regions is because, for small values of the irises aperture a, the coupling between cells is essentially magnetic, as the irises aperture increases the coupling between the cells becomes also electric and the structure becomes progressively forward.

deflecting force is uniform over the iris aperture and constant along z. In the real case this is not exactly true because edge effects occur near the irises which perturb the field distribution giving a distortion of the total deflecting field. Moreover, because of the periodicity of the structure, the deflecting field has a modulation along the z axis given by the presence of all space harmonics. The integrated deflecting force calculated by HFSS as a function of the transverse coordinate is shown in Fig. 14 in the case of a 2.856 GHz RFD working on the $2\pi/3$ mode with a=20mm and t=10mm. The edge effects near the iris perturb the deflecting field by few percent with respect to the field on the center of the structure. In the choice of the iris aperture a this effect should be also taken into account. Concerning the longitudinal modulation of the deflecting field, its behaviour in a single cell for a 2.856GHz RFD is given in Fig. 15a as a function of the longitudinal coordinate and calculated in the center of the structure (with HFSS). It is possible to remark that the ratio between the magnetic and the electric deflection is about 1 and that even if the E and B field have a longitudinal modulation given by the irises periodicity, the total deflecting field has a smaller modulation due to the partial compensation between the two E and B contributions. The behaviour of the phase is shown in Fig 15b. Also in this case, even if the phase advance in the cell for the E and B field is not perfectly linear, the total deflecting field has an almost linear phase advance.

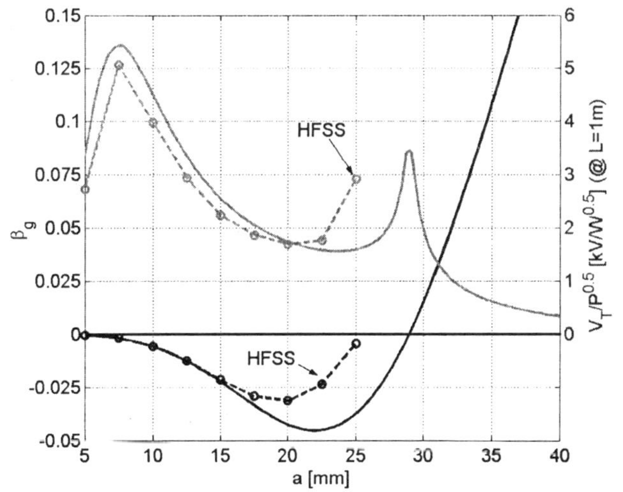

Figure 12. Behaviour of the group velocity and total as a function of the iris aperture a for a copper structure of length L=1m working on the $2\pi/3$ mode at 2.856 GHz (analytical and HFSS results).

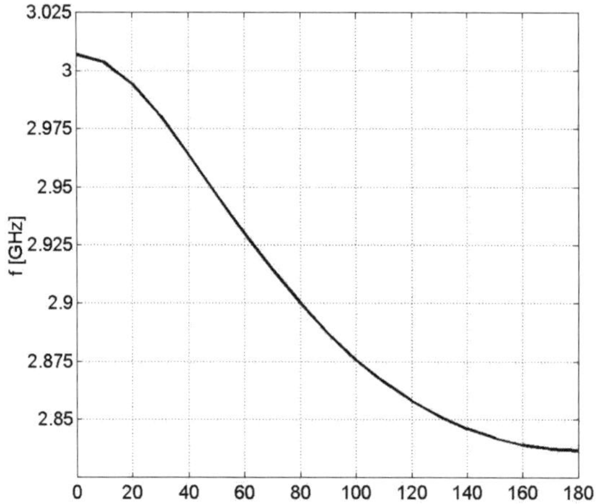

Figure 13. Dispersion curve of a TW RFD working on the $2\pi/3$ mode at 2.856 GHz.

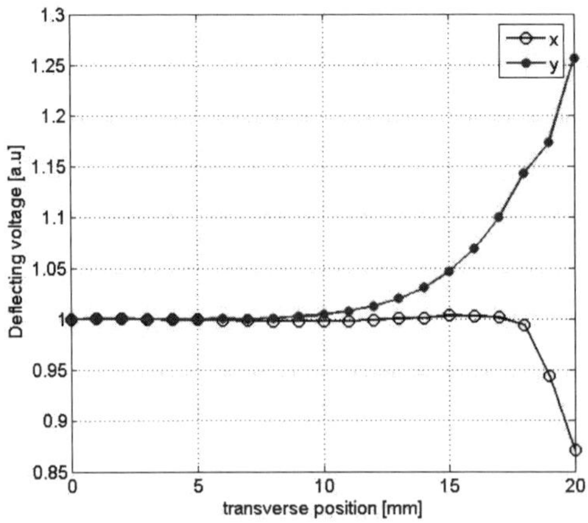

Figure 14. Deflecting voltage as a function of the transverse coordinate for a TW RFD at 2.856 GHz.

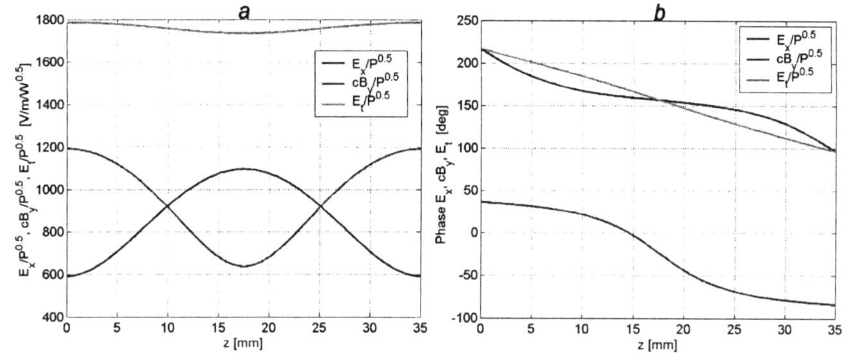

Figure 15. a) Amplitude of the B, E and total deflecting field as function of the longitudinal coordinate z; b) Phase of the B, E and total deflecting field E_T as a function of the longitudinal coordinate z.

Another important quantity to be considered in the single cell design is the maximum E field on the surface of the cell. The plot of the amplitude of E field on the perfect H plane (where it is maximum) is reported in Fig. 16 for two different cases. For small iris apertures (case (a)) the maximum occurs between the iris aperture and the outer radius while, for large iris apertures (case (b)), the maximum occurs on the edge of the iris aperture itself. In general, in the second case, the maximum electric field normalized to the deflecting one is much bigger because of the iris edge effect itself. There are some analytical formulae to estimate the maximum E on the surface as a function of the geometrical parameters [13]. As an example, the maximum of the surface electric field normalized to the deflecting one as a function of the iris aperture for a 2.856 GHz RFD, as calculated by the analytical formulae, is given in Fig 17. In the same plot I have reported the maximum surface electric field calculated by HFSS. Even if the analytical calculations underestimates the maximum surface E field (especially for big values of the iris apertures), it puts in evidence the large presence of two different regions for small and big irises apertures respectively. In the second region there is a strong increase of the maximum surface E field.

The last point that I would like to remark is the possibility of mode overlap. There are several possibilities in this sense:

a) overlap between the deflecting mode and other TW modes (as example the TE_{11}-like);

b) overlap between the deflecting $2\pi/3$ TM_{11}-like mode, for example, and the same mode with a different phase advance per cell. This case can occur if the dispersion curve has a double value for a given f_{RF};

c) the presence of the 90 deg tilted polarity of the TM_{11}-like mode;

Concerning the first two cases they can be, in general, avoided if one chooses a sufficiently small iris aperture ($a < \lambda/4$) [15]. Concerning the third case it should be pointed out that there are two possible polarities of the deflecting TM_{11}-like mode: the working one and the 90 deg tilted one. Since without azimuthal asymmetries the frequency of the two modes is the same, one can have a mode tilting along the deflector. To avoid this possibility one has to introduce an azimuthal asymmetry in the structure in order to split the two frequencies. The simplest ways are reported in Figs. 18: the first solution (Fig. 18a) consists on the introduction of some rods while the second one (Fig 18b) consists in the use of holes. In general the frequency separation that one can obtain with metallic rods is larger than with holes. For example with two rods of 2 mm of radius the frequency split[6] is $\approx +20MHz$ @ 2.856 GHz while with two holes of 6 mm radius it is $\approx -1MHz$. Fig. 19b shows the complete dispersion curves of the working mode and tilted polarity in the two previous cases. In order not to excite the tilted polarity the two dispersion curves should be well separated in frequency. This is the reason why the rods are, in general, preferable even if for high frequency structures, it is sometimes difficult to mechanically realize small diameter rods, and the solution of holes is necessary.

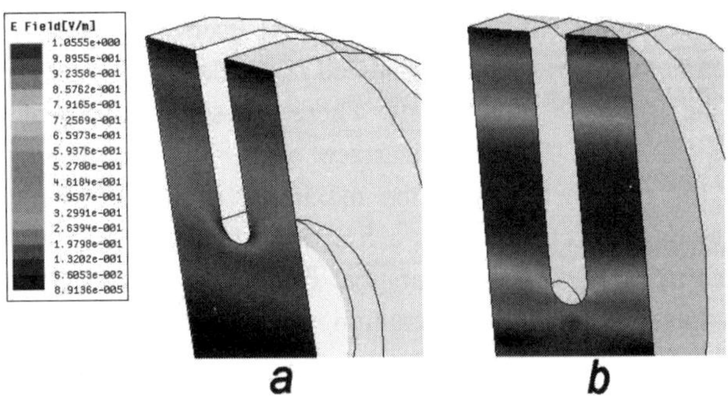

Figure 16. Amplitude of the electric field on the perfect H plane for two different cases: (a) small iris aperture; (b) large iris aperture.

[6] The frequency split is defined as the difference between the frequency of the 90 deg tilted polarity and those of the working mode.

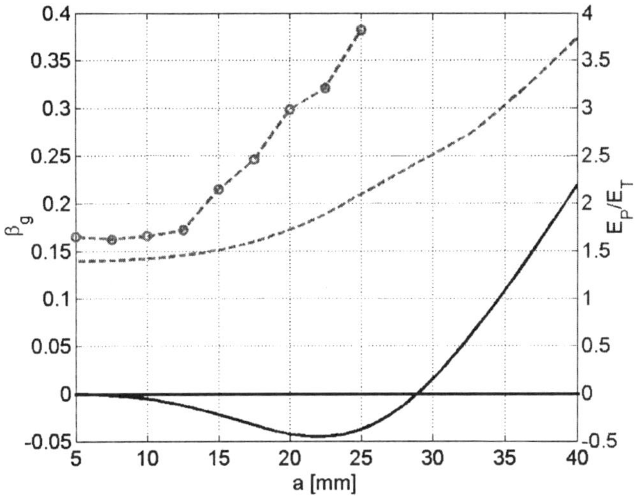

Figure 17. maximum of the surface electric field normalized to the deflecting one as a function of the iris aperture for a $2\pi/3$ mode at 2.856 GHz (analytical and HFSS results).

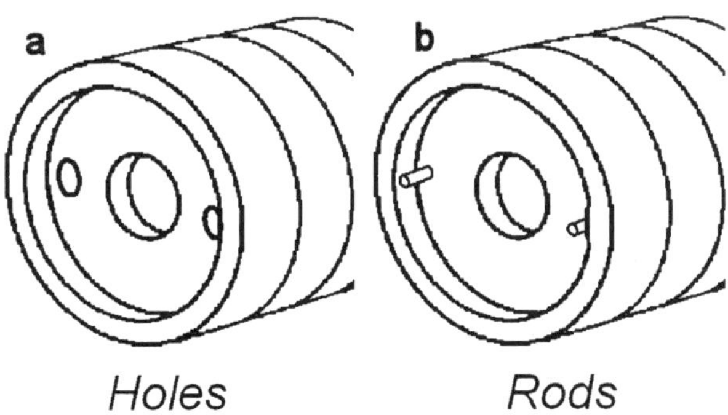

Figure 18: holes and rods to split the frequency of the 90 deg tilted polarity.

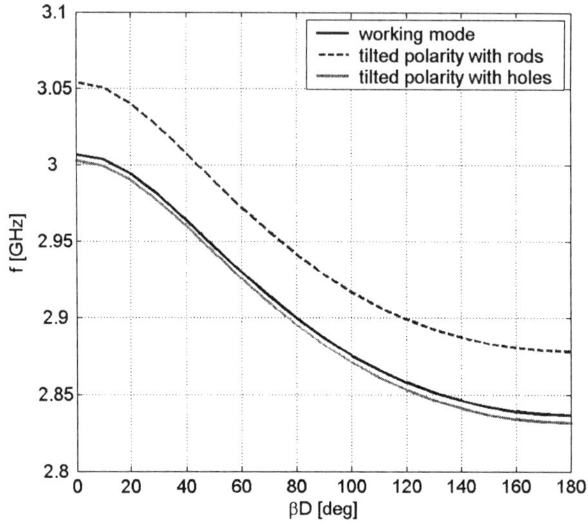

Figure 19. Dispersion curves of the working mode and 90 deg tilted polarity in the case of rods of 2 mm diameter and holes of 6 mm diameter.

3.1.2. *Input coupler design considerations*

The procedures to design the input coupler are equal to those used in the design of the accelerating structure. The only two things that I would like to remark are:

1 the more efficient structures are backward and, therefore, the input coupler is on the opposite direction with respect to the beam direction;

2 an asymmetric coupler can give, in the coupler cell, a distortion of the e.m. field giving a longitudinal component of the electric field on axis. In fact, in a perfectly symmetric cell, the longitudinal electric field is zero on the axis of the structure and increases linearly with the distance from the axis itself. An analytical expression of this components is:

$$E_z = jE_0 k\ x\ e^{-j\frac{\omega z}{c}} \tag{33}$$

This longitudinal component on axis can be partially compensated by putting the two input and output couplers on two opposite directions with respect to the transverse plane. Nevertheless, it is therefore easy to verify, also from an analytical calculation, that the longitudinal voltage given by the asymmetric coupler is in general negligible.

3.1.3. *Performances of the TW RFDs*

I have plotted in Fig. 20 the resolution length achievable with a TW RFD as a function of the input power for two different cases[7] (f_{RF}=2.856 GHz and f_{RF}=11.424 GHz) and for two different lengths of the structure (1 and 2 meters). The used parameters were: energy of the beam E_B=2 GeV, normalized emittance ε=2 mmmrad, iris aperture a=20 mm @ 2.856 GHz and a=5 mm @ 11.424 GHz; beta function at the deflector $\beta_{defl} = 20m$ and $\Delta\phi = 90°$.

For a given input power, in the case of an RFD at 11.424 GHz there is a gain in the achievable resolution given by the frequency scaling and also because the r_T of the deflector scales with $\sqrt{f}$. It is also important to point out that the attenuation constant α scales as $f^{3/2}$, and this is the reason why in the plot of Fig 20 the ratio between the resolution length with f_{RF}=2.856 GHz and f_{RF}=11.424 GHz in the case of L=1 and 2 m is different.

In Figs. 21 and 22 I have plotted the deflecting field and the maximum surface electric field as a function of the input power for the two previous cases. As observed before the ratio between the deflecting fields in the case of f_{RF}=2.856 GHz and f_{RF}=11.424 GHz decreases if one increases the total length of the deflector because of the different attenuation constant. Concerning the maximum electric field (for a given input power), this scales as f and, in the high frequency case, can became very high.

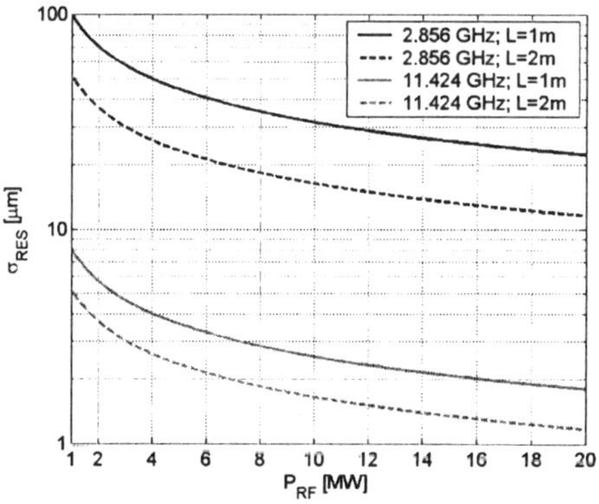

Figure 20. Resolution length achievable with a TW RFD as a function of the input power for two different cases.

[7] The transverse impedance and attenution were calculated by HFSS.

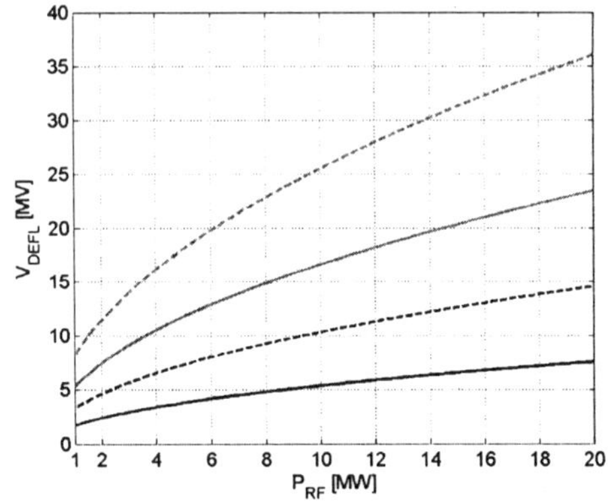

Figure 21. Deflecting field as a function of the input power with TW RFD for two different cases.

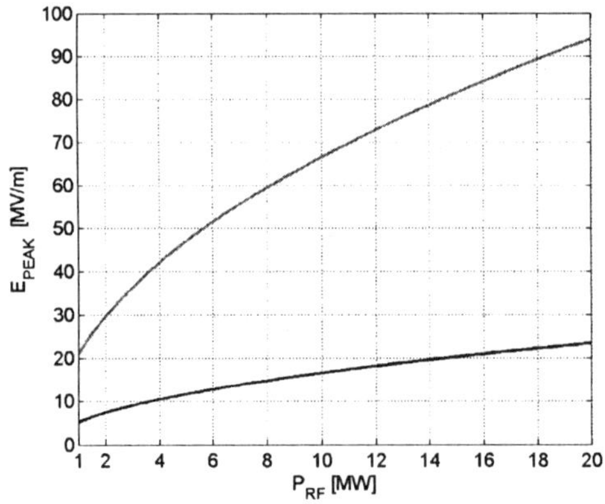

Figure 22. Maximum surface electric field as a function of the input power with TW RFD for two different cases.

3.2. *SW RF deflectors*

A single SW cell with the *E* and *B* field profiles is sketched in Fig. 23. As in the case of a TW RFD also in this case both *E* and *B* field components contribute to the total deflection. The ratio between the deflection given by the electric and the magnetic one can vary significantly as a function of the iris aperture and can go from about zero, in the case of very small beam pipe apertures, to 1, in the case of larger ones. The main parameters of a SW deflecting cavity are similar to those of an accelerating section and are reported below for completeness:

a) *transverse shunt impedance* R_T defined as:

$$R_T = \frac{\left|\int_0^L (E_x - cB_x) e^{j\omega_{RF} z/c} dz\right|^2}{2P_d}$$

(34)

where P_d is the average dissipated power in the cavity.

b) *quality factor Q* and R_T/Q

$$Q = \omega_{RF} \frac{W}{P_d}; \quad \frac{R_T}{Q} = \frac{\left|\int_0^L (E_x - cB_x) e^{j\omega_{RF} z/c} dz\right|^2}{2\omega_{RF} W}$$

(35)

where W is the average stored energy in the cavity.

Related to the quality factor is the *filling time* of the structure with the usual expression:

$$\tau_f = \frac{2Q}{\omega_{RF}}$$

(36)

To increase the transverse shunt impedance it is possible to consider multi-cell structures as in the case of multi-cell accelerating sections. The choice of the working mode (π or $\pi/2$ for example) follows the same criteria of an accelerating section.

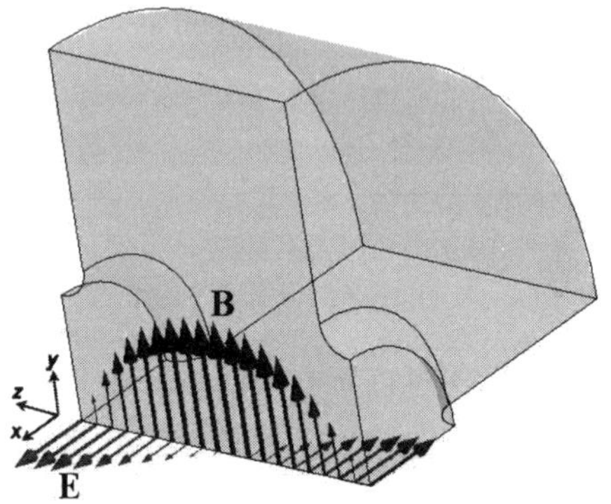

Figure 23. Single cell of a standing wave RFD with the *E* and *B* field lines.

3.2.1. *Single cell design of a multi cell structure and coupler design considerations*

The single cell design criteria are similar to those of a multi cell SW accelerating section. Choosing, for example, the π multi-cell mode, the cell length D should be equal to $\lambda_{RF}/2$, the iris aperture as small as possible in order to increase the transverse shunt impedance, while the external radius b should be chosen in order to tune the resonant frequency of the cavity at the correct value.

R_T/Q has been calculated as a function of the iris aperture a/λ_{RF} for a single cell of length $\lambda_{RF}/2$ using the e.m. code HFSS (Fig. 24). Similar considerations to those of the TW deflectors concerning the behaviour of the maximum surface electric field as a function of the iris aperture can be done. I have plotted in Fig. 25 the value of the Q factor and maximum surface electric field normalized to the dissipated power in the cavity as a function of the iris aperture for f_{RF}=2.856 GHz.

Concerning the mode overlap it is possible to make considerations similar to those of the TW case. Moreover in the case of a SW multi-cell section there is the problem related to the multi-cell modes with frequencies near the π mode one.

If *n* is the number of cells the nearest multi-cell mode is the $(n-1)\pi/n$. The frequency separation between this mode and the working one can be calculated by e.m. codes. A simple scaling law as a function of the number of cells n for a 2.856 GHz RFD with a=20mm is for example

$\Delta f \cong 116/n^2\ [MHz]$. With a symmetric central coupler (as that shown in Fig. 25 for the SPARC RFD [17]) it is possible to avoid exciting this unwanted $(n-1)\pi/n$ mode. In this case the nearest mode with the same polarity that can be excited by the coupler is the $(n-2)\pi/n$ mode.

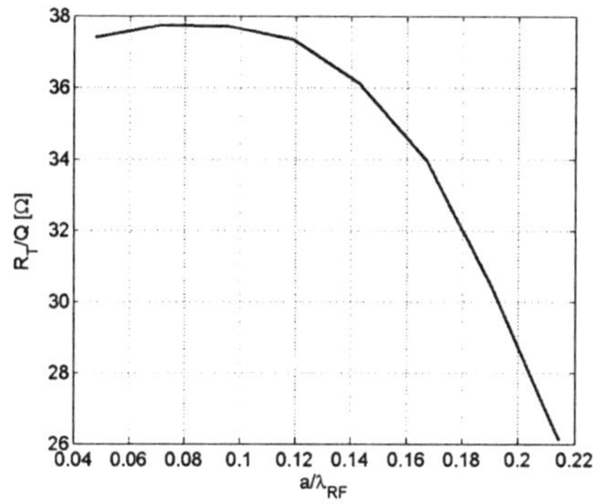

Figure 24. R_T/Q as a function of the iris aperture for a single cell of length $\lambda_{RF}/2$.

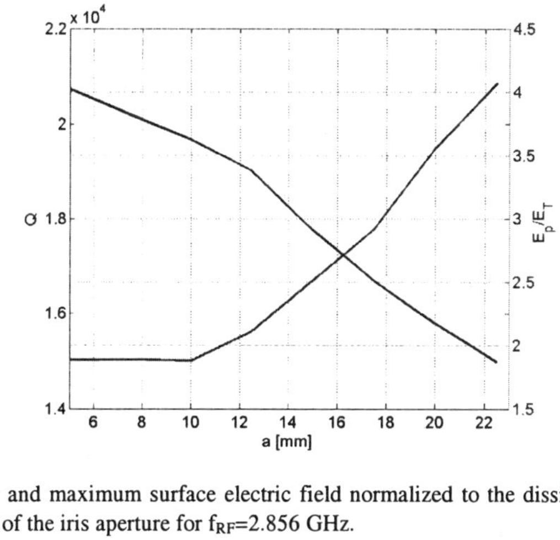

Figure 25. Q factor and maximum surface electric field normalized to the dissipated power in the cavity as a function of the iris aperture for f_{RF}=2.856 GHz.

3.2.2. *Performances of SW RFD*

I have plotted in Fig. 26 the resolution length achievable with a SW RFD as a function of the dissipated power in the cavity for two different cases (f_{RF}=2.856 GHz and f_{RF}=11.424 GHz) and for two different numbers of cells (5 and 9) while the beam parameters are the same as in the TW case. Comments similar to those in the TW case can be done concerning the behaviour of the resolution length that can be achieved with a SW RFD as a function of the number of cells or working frequency. I have also plotted in Figs. 27 and 28 the deflecting field and the maximum surface electric field as a function of the input power for the two previous cases. It is important to remark that high values of the maximum surface electric field can be reached in the high working frequency case (because it scales as $f^{3/4}$) and that it can limit the maximum input power in the structure.

It is also interesting to compare the performances of SW and TW devices comparing the Figs. 20-22, 26-28. More details can be found in the next paragraph.

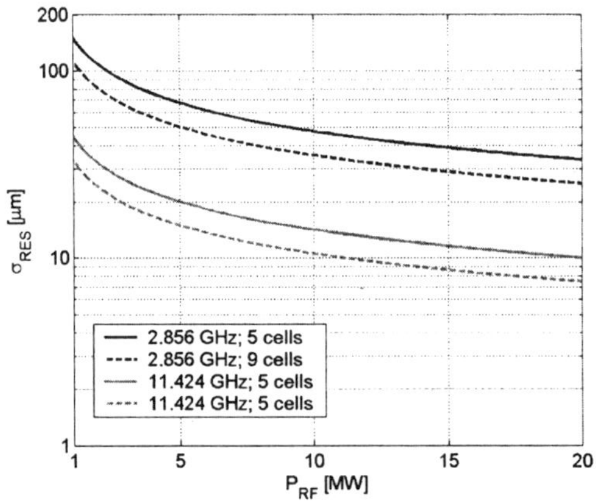

Figure 26. Resolution length achievable with a SW RFD as a function of the input power for two different frequencies (f_{RF}=2.856 GHz and f_{RF}=11.424 GHz) and two different number of cells.

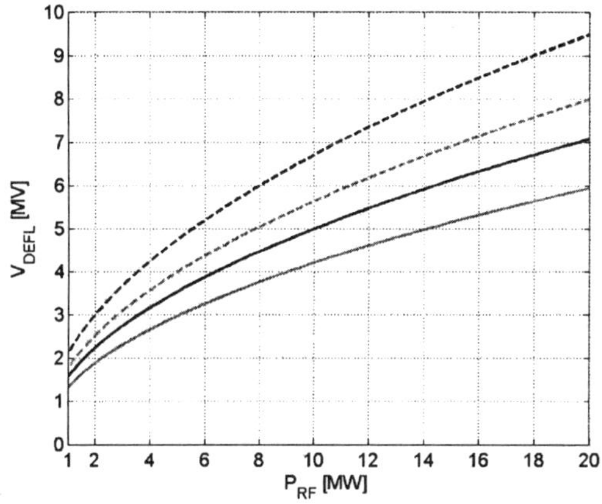

Figure 27. Deflecting field as a function of the input power for f_{RF}=2.856 GHz and f_{RF}=11.424 GHz and for two different numbers of cells.

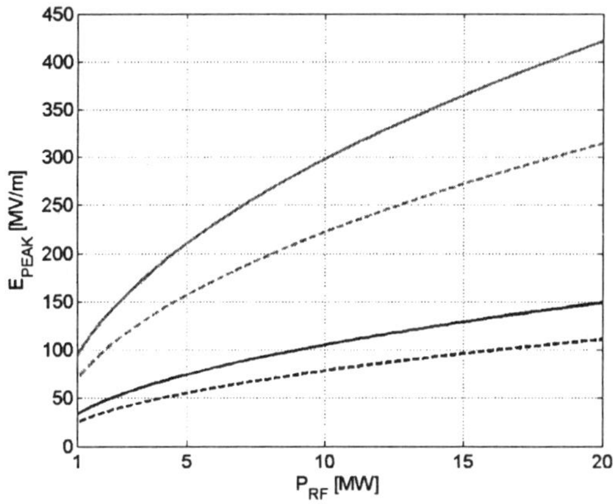

Figure 28. Maximum surface electric field as a function of the input power for f_{RF}=2.856 GHz and f_{RF}=11.424 GHz and for two different numbers of cells.

3.3. *Comparison between TW and SW devices*

The comparison between TW and the SW RFDs is summarized in Table 1 and is similar to that of a SW/TW accelerating section. The only thing that weremark is that, in general, a SW section could be more useful with respect to a TW section if one has a low input power, a long RF pulse length and one needs a compact device.

Table 1. Comparison between SW and TW RFD.

	SW	TW
Deflecting field per unit length	High	Low
Filling time	Proportional to the quality factor: generally longer than TW ($\approx \mu s$ @ 2.856 GHz)	Proportional to the group velocity and length: shorter than SW ($\approx 0.1\mu s$ @ L=1m)
Maximum number of cells and maximum lenght	The number of cells (for a π -mode) is limited to about 10-15 because of mode overlapping. Thus $L_{MAX@2.856GHz} \cong 50$-80cm	The maximum length should be 1÷2/ α . At 2.856 GHz this gives $L_{MAX} \cong 12$-25 m.
Total deflecting voltage vs. total length L	$\propto L^{1/2}$	$\propto \left(1-e^{-\alpha L}\right)/\alpha$
Power system	Circulator needed to protect the klystron	Circulator not necessary.
Maximum surface electric field.	It can limit the maximum input power especially at high working frequencies	Not a big constraint
Temperature sensitivity	High	Lower than SW
Wakefild intensity	Low	High
Resolution vs. frequency	$1/f^{5/4}$ (for a given number of cells)	$1/f^2$ (for a given total length and assuming small dissipations)

3.4. *Deflectors RF measurements*

The measurements that can be done on a TW o SW RFD are similar to those of an accelerating section (input port reflection coefficient, coupling coefficient,…). Concerning the bead pull measurements, the points that I remark is that, since the deflection is given by both the magnetic and the electric field, both components have to be measured for a correct characterization of the device. Two types of perturbing objects should be therefore used to measure the *B-E* field components. Results on RF measurements on a SW deflecting cavity can be found in [17].

3.5. *Feeding systems*

Examples of power feeding systems are shown in Fig. 29. The first two are the standard ones for SW and TW structures. In case c) the power is split out for an existing power line (see for example the SPARC case [17]) and therefore the attenuator and phase shifter are high power devices. The fourth system (d) refers to the circular polarized case. In this scheme one attenuator and phase shifter are low power while two other ones are high power devices.

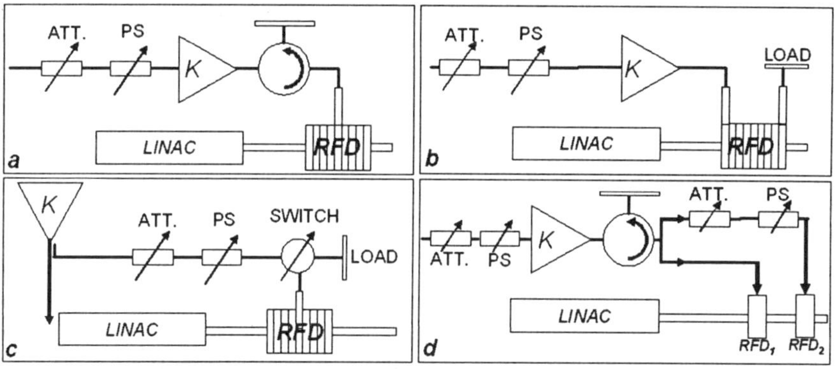

Figure 29. Examples of power feeding systems: a) TW RFD; b) SW RFD; c) power split; d) circular polarized RFD.

4. Experimental results

Several experimental results have been obtained in already existing LINACs using RFDs for beam diagnostics. As examples I would like to mention those obtained at SLAC [6], at CERN in the CTF3 project [7], at TTF [8,9] and at the Haimson Research Corporation (HRC) with circular polarized RFDs [10,11]. I have reported the main RFDs and beam parameters for these cases in Table 2.

Table 2: main RFDs and beam parameters of the obtained experimental results.

	RFD properties			Beam parameters	
	Type	Freq. [GHz]	Length [m]	Energy [GeV]	Bunch length [mm]
SLAC	TW	2.856	2.44	28	0.3-0.5
CTF3	TW	2.998	0.4	0.150(*)	0.5-2.5
TTF	TW	2.856	2.44	0.450	0.015
HRC	SW	17	2 cav.	0.015	0.05-5

(*)multi bunch

5. Conclusions

RFDs are very promising devices for longitudinal and transverse phase space characterization or beam manipulation in advanced accelerators. Measurement setups for longitudinal pulse shape as well as phase space and transverse beam slice emittance characterizations have been illustrated. I have also shown the main design criteria of traveling wave and standing wave RFDs used for beam diagnostics pointing out the advantages and drawbacks of the two devices in term of performances, required power and breakdown limitations. The choice between SW or TW devices depends on several parameters and is similar to that of the accelerating cavities. Measurements results already obtained at SLAC, CTF3 (CERN) or TTF (DESY) have been also presented.

Acknowledgments

I would like to thank L. Ficcadenti, A. Mostacci, L. Palumbo and J.B. Rosenzweig for helpful discussions on diagnostics with RFD and on both SW and TW deflecting structures properties and parameters. I would like also to thank C. Vaccarezza for the help on measurements simulations with tracking codes, G. Di Pirro who is responsible of the SPARC diagnostics and M. A. Preger who is the scientific responsible the accelerator division at LNF-INFN for the corrections and suggestions on the paper. Finally, I would like also thank J. Haimson for the material and references on circular polarized RFDs.

References

1. G.A. Loew and O. H Altenmueller, "Design and applications of RF deflecting structures at SLAC", PUB-135, Aug. 1965.
2. P. Emma, et al., "A Transverse RF deflecting structure for bunch length and phase space diagnostics", LCLS-TN-00-12, 2000.
3. D.H. Dowell et al, "Slice Emittance Measurements at the SLAC Gun Test Facility", SLAC-PUB-9540, 2002.
4. J.F. Schmerge et al, "Transverse-emittance measurements on a S-band photocathode RF electron gun", Nucl. Instr. & Meth. A 483, 2002, pp. 301-304.
5. X. Qiu et al ,"Demonstration of Emittance Compensation through the Measurement of the Slice Emittance of a 10- ps Electron bunch", Phys. Rev. Lett. 76, 20, 1996, pp. 3723-3726.

6. P. Emma, et al, "Bunch length measurements using a transverse RF deflecting structure in the SLAC LINAC", Proc. of EPAC 2002, Paris (France), 2002.
7. A. Ghigo, et al., "Commissioning and first measurements on the CTF3 chicane", Pro. of PAC 2005, Knoxville (Tennessee), 2005.
8. M. Huning, et al., "Observation of femtosecond bunch length using a transverse deflecting structure", Proc. of FEL Conference 2005, Stanford (California), 2005.
9. M. Rohrs, et al., "Measurement of slice-emittance using transverse deflecting structure", Proc. of FEL Conference 2005, Stanford (California), 2005.
10. J. Haimson, et al., "A Circularly Polarized Beam Deflector for Direct Measurement of Ultra Short Electron", AIP Conference Proceedings, 647, (New York: AIP Press, 2002), pp. 810-820.
11. J. Haimson, "Longitudinal Phase Space Measurements of Short Electron Bunches Using a 17 GHz Circularly Polarized Beam Deflector", AIP Conference Proceedings, 737, (New York: AIP Press, 2004), pp. 95-108.
12. G.A. Loew and R.B. Neal, "Linear Accelerators" edited by P. Lapostolle and A. Septier, Nort-Holland, Amsterdam, 1970.
13. P. Bernard and H. Lengeler, "On the design of disc-loaded waveguides for RF separators", CERN 68-30, Geneva, 1968.
14. G. A. Loew, O. H. Altenmueller and R.R. Larsen, "Investigations of Travelling-wave separators for the Stanford two-mile Linear Accelerator", SLAC Report 17, SLAC, Stanford, California, 1963.
15. Y. Garault, "Etude d'une classe d'ondes electromagnetiques guidees: les ondes EH. Application aux structures deflectrices pour les separateurs a onde progressive de particules relativistes", CERN 64-43, Geneva, 1964.
16. www.ansoft.com.
17. D. Alesini, et al., "RF Deflector design and measurements for the Longitudinal and transverse phase space characterization at SPARC", to be submitted to NIM.

PRODUCTION OF FEMTOSECOND PULSES AND MICRON BEAM SPOTS FOR HIGH BRIGHTNESS ELECTRON BEAM APPLICATIONS*

S. G. ANDERSON, D. J. GIBSON, F. V. HARTEMANN, J. S. JACOB, AND A. M. TREMAINE

Lawrence Livermore National Laboratory
7000 East Avenue
Livermore, CA 94551, USA

J. K. LIM, P. FRIGOLA, J. B. ROSENZWEIG, AND G. TRAVISH

University of California, Los Angeles
405 Hilgard Avenue
Los Angeles, CA 90095

Current and future applications of high brightness electron beams, which include advanced accelerators and beam-radiation interactions require both transverse and longitudinal beam sizes on the order of tens of microns. Ultra-high density beams may be produced at moderate energy (50 MeV) by compression and subsequent strong focusing of low emittance, photoinjector sources. We describe the implementation of this method used at the PLEIADES inverse-Compton scattering (ICS) x-ray source at LLNL in which the photoinjector-generated beam has been compressed to 300 fsec rms duration using the velocity bunching technique and focused to 20 μm rms size using an extremely high gradient, permanent magnet quadrupole focusing system.

1. INTRODUCTION

The performance of many high brightness electron beam applications depends not only on brightness, but also on achieving very high beam density. Examples of these applications include the driving of a plasma wake-field accelerator (PWFA) [1] and providing an electron beam for an ICS x-ray source

*This work was performed under the auspices of the u.s. department of energy by the university of california, lawrence livermore national laboratory under contract no. w-7405-eng-48, and by the university of california, los angeles under contract numbers de-fg-98er45693 and de-fg03-92er40693

[2]. The need for shorter bunch lengths, σ_z, and lower transverse emittances, $\varepsilon_{x,y}$, has driven work on photoinjector optimization, bunch compression systems, and the development of future electron beam sources. In addition, proportional scaling of the dimensions of the beam focal system must be considered to produce ultra-dense beams.

Examine in turn the two examples of PWFAs and ICS sources. In a PWFA, the maximum accelerating field, the "wave-breaking" field, achievable in a relativistic plasma wave is $E_{WB} = m_e c^2 k_p / e$, where k_p is the plasma wave number. If the driving beam is short ($k_p \sigma_z < 1$), then the wake is efficiently excited and the electric field of the plasma wave scales as $E \propto k_p^2 N_b$. Therefore, larger accelerating fields are obtained by operating a larger k_p, and shorter σ_z. The use of large plasma wavenumbers effects the transverse forces as well, as can be seen from the matched β-function associated with the plasma's ion-derived focusing,

$$\beta_{eq} = \sqrt{2\gamma} k_p^{-1}. \tag{1}$$

For low energy experiments, β_{eq} can be extremely short, and has already been in the range of several mm [3].

In the case of ICS sources, the rate of x-ray photons production, dN_x/dt, depends on the intensities of the overlapping electron and laser beams, and the duration of the interaction. Assuming a 180° interaction geometry, Gaussian laser and electron beam profiles, and that the beam dimensions do not change significantly during the interaction, the number of x-rays produces is

$$N_x = \frac{N_e N_L \sigma_T}{2\pi \left(\sigma_e^2 + \sigma_L^2\right)}, \tag{2}$$

where N_e and N_L are the number of input electrons and photons, respectively, σ_e and σ_L are the beam rms sizes, and σ_T is the total Thomson cross section. The x-ray source brightness, often a more important figure of merit for applications, is given by

$$\mathcal{B}_x = \frac{N_x}{(2\pi)^{5/2} \sigma_x^2 \sigma_{x'}^2 \sigma_\tau \,(0.1\%\mathrm{BW})}. \tag{3}$$

Since the source size, σ_x, divergence, $\sigma_{x'}$ and duration, σ_τ are all strongly coupled to the electron bunch phase space, it is clear that the electron beam brightness, in addition to density, should be maximized for x-ray production.

Magnetic compression schemes have proven successful in producing sub-ps beams generated by photoinjectors. The effect of the compression process on the transverse phase space, however has been determined to be damaging, both at low energies due to space-charge forces [4], and at higher energies due to coherent synchrotron radiation [5]. This observed emittance growth runs counter to the goal of increasing the beam brightness that motivates the use of compression to begin with. The "velocity bunching" scheme, proposed by Serafini and Ferrario [6], may preserve the transverse phase space quality, while compressing the beam to sub-ps bunch length [7]. This scheme, as opposed to the path-length dependence of magnetic compression systems, is an extension of the commonly employed technique of RF rectilinear compression. As design trajectory bending is not needed in this system, one may avoid the phase space degrading effects observed in magnetic compression experiments on photoinjector-derived beams.

In order to create minimum β-functions small enough for modern high-brightness electron beam applications, one must look into the use of much shorter focal length lenses. This is motivated by the need to avoid chromatic aberrations due to energy spread, and emittance growth due to residual space-charge effects in the beam if it is expanded for final focusing. Permanent magnet-based quadrupoles (PMQs) can access high-field gradients, and have been under intensive study recently [8]. They offer the advantages of simplicity and being relatively inexpensive when compared to alternatives such as superconducting systems, or plasma lenses.

In this paper we describe the implementation of both the velocity bunching technique [9] and a PMQ based final focus system [10] at the PLEIADES ICS X-ray source [11], a collaboration between LLNL and UCLA. Both of these measures have improved the performance of the X-ray source.

2. VELOCITY BUNCHING

In the velocity bunching method, an energy/phase correlation is imparted to the electron beam and removed smoothly, through phase slippage and acceleration, inside of an rf linac section. In this method the beam is rotated through one quarter of a phase space oscillation. The electron trajectories in the sinusoidal longitudinal electric field component of the RF wave in a traveling wave structure can be tracked and plotted in (ϕ, γ) phase space using the electron equations of motion; $\frac{d\gamma}{dz} = -\alpha k \sin\phi$, and $\frac{d\phi}{dz} = k\left[1 - \frac{\gamma}{\sqrt{\gamma^2-1}}\right]$.

Particles injected at or near $\phi_0 = 0$ will begin to slip into an accelerating phase. As the particles accelerate to ultra-relativistic velocity, the phase slippage slows, and is eventually arrested. In terms of the particle's initial phase space coordinates (ϕ_0, γ_0), the asymptotic value of the slipping phase is $\phi_\infty \cong \cos^{-1}\left[\cos\phi_0 - \frac{1}{2\alpha\gamma_0}\right]$.

The phase space rotation occurs as a result of the varying orientation of the phase contour lines. If two particles are injected at $\phi_0 = 0$ with a difference in phase but no spread in energy, such that they align parallel to a phase contour, then they will remain parallel to that contour as they slip in phase and accelerate. As the particles approach ϕ_∞, their orientation becomes nearly parallel to the γ axis, having rotated by nearly 90 degrees.

The PLEIADES photoinjector and linac consist of a 1.6 cell photocathode RF gun followed by four SLAC style 2.5 meter, S-band traveling wave sections. While the linac is capable of producing 100 MeV electrons, it is typically run using only the first two or three sections for X-ray production, resulting in beam energy between 20 and 70 MeV. Velocity bunching has been preformed with this system using the travelling wave section immediately following the gun to compress the beam, while the remaining sections were used to accelerate the beam on-crest, thereby reducing the relative energy spread.

The longitudinal dynamics in the experiment agreed well with those found in simulations. A 250 pC bunch was compressed from an initial duration of 3 ps (rms) to a final length of 280 fs. The bunch length was measured using a coherent transition radiation (CTR) based interferometer [12,13]. The measured CTR autocorrelation signal is shown in Fig. 1. The energy spread of the compressed beam, accelerated to 50 MeV, was 0.5%. Again, in agreement with simulations, and higher than the nominal uncompressed value of 0.2%.

Transversely, the increasing space-charge forces resulting from the increasing beam current in the compressor must be matched with external focusing to minimize space-charge induced emittance growth. This external focusing was provided in the experiment by a solenoid field along the length of the compressing rf section. The emittance of the compressed electron beam was measured using the quadrupole scanning method for various settings of the solenoid field strength. The results of those measurements are shown in Fig. 2. As expected, the emittance shows a very strong dependence on the compressor focusing, growing by a factor of 3 in $\varepsilon_{n,x}$ with a 20% change in solenoid current. A minimum of 11 mm mrad in both emittances was found to occur at the 10 A solenoid setting in this scan.

The velocity compressed beam was used in conjunction with a Ti:Sapphire-based, 500 mJ, 50 fs (FWHM), 800 nm laser pulse for ICS

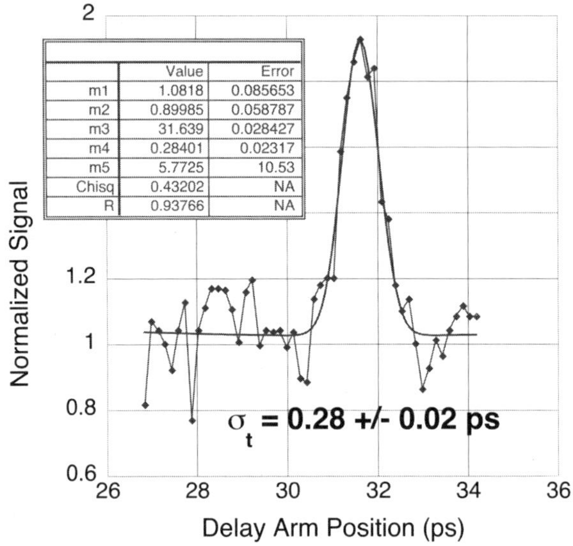

Figure 1. Autocorrelation of the optimally compressed electron bunch using the CTR based interferometer. The measured rms bunch length is 0.28 ± 0.02 psec.

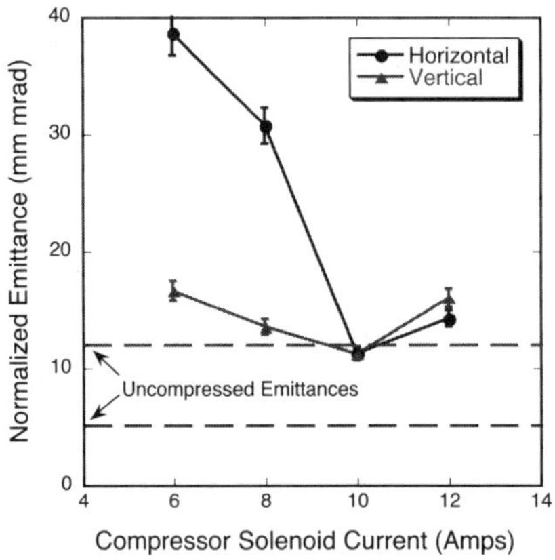

Figure 2. Measured horizontal and vertical normalized emittances of the 280 fsec beam versus the compressor solenoid strength.

x-ray generation. The compressed beam was measured to be roughly 50% larger in x and y than the uncompressed beam at final focus. This increase is due to both the increased energy spread of the compressed bunch, and emittance growth. A comparison of the resulting x-ray pulses is shown in

Fig. 3. While the increased final focus spot size reduces the total number

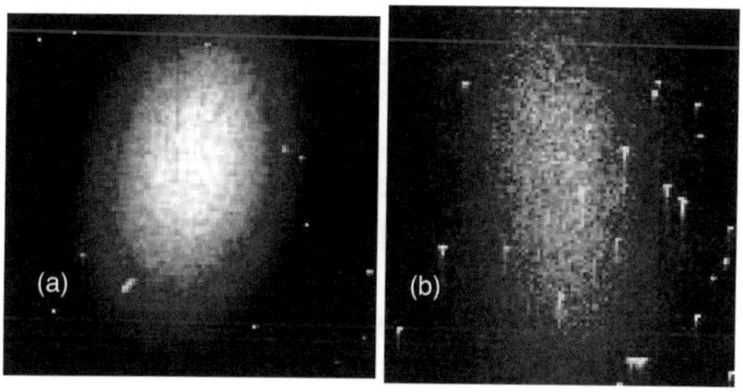

Figure 3. Single shot, false-color X-ray beam images measured by a CCD for the (a) uncompressed and (b) compressed electron beams.

of scattered photons, the *brightness* of the 300 fsec bunch is increased by 70% from that of the x-rays produced without velocity bunching.

3. PMQ FOCUSING

PMQs were chosen to deliver the extremely high magnetic field gradients desirable in the PLEIADES final focus system. A 16-segment Halbach configuration [14] was simulated with the 3D magnetostatic code, RADIA, using PM material (NdFeB) with a 1.22 T remanent field, 5 mm inner diameter, and a 15 mm outer diameter. The field gradient produced in the simulation was $B' = 573$ T/m, slightly lower than the theoretical prediction for this design due mainly to three-dimensional (edge) effects. The quadrupole length was chosen based on the theoretical and simulation predictions so that the focusing phase advance in a single lens is less than one-half for a 70 MeV beam ($k_q l_q \leq 0.5$).

A series of RADIA simulations were performed in order to evaluate the performance of the magnet design in the presence of fabrication and assembly errors. The following results were obtained: (1) a magnetization easy-axis angular orientation error is allowable up to 2% before a notable change develops between the magnetic center and the mechanical center; (2) the wedge shape error (the wedge angle) is limited and allowed up to again 2%; and finally (3) bore radius variations of ± 0.05 mm from the nominal 2.5 mm radius produced the field gradient variation of 3% from the ideal B'. These results were then used to apply magnetic field errors in ELEGANT simulations of the final focus using input beam parameters typical of the

PLEIADES experiment. The most significant source of emittance growth was found to be rotation (skew) error between successive magnets. A 10 mrad rotation error was found to result in a doubling of the emittance of the beam at focus, with a corresponding increase in spot size.

The fabrication of PMQs with the required precision proved to a technically challenging task. Nevertheless, a simplified procedure may be given as follows: a long trapezoidal wedge which is pre-magnetized to a Br =1.22 T is machined to a specified design shape from a larger magnet block through wire electrical discharge machining. The 16 magnetized wedges are then coupled together into an equally long aluminum keeper tube. A thin layer of cohesion material is applied, strongly joining neighboring magnets. The tube keeper and a bonding glue together ensure structural stability of the PMQ under extreme magnetic repulsions. The long assembly is then cut into six identical PMQs, with mechanical length of 10 cm, to ensure consistent performance between all PMQs. In particular, construction of a single, long magnet assembly assured that the relative azimuthal (roll) errors of individual units may be essentially eliminated. Finally, the inner diameter of bore is slowly ground to the specified value of 5 mm with a diamond cutter. An illustration which schematically summarizes the PMQ fabrication procedure is shown in Fig. 4. One of the produced magnets is shown

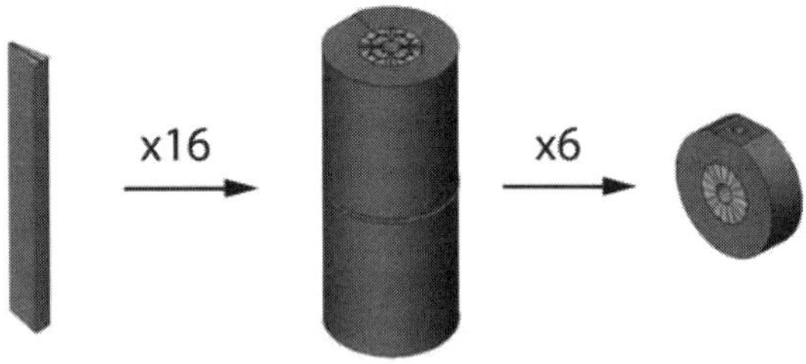

Figure 4. An illustration of the PMQ manufacturing process showing an individual wedge, the assembled 16 wedge rod, and a single PMQ cut from the rod.

in Fig. 5.

Measurements of the individual magnets were performed using both a Hall probe and the pulsed wire technique [15]. These measurements were found to agree well with RADIA simulations, as Fig. 6 demonstrates. The Hall probe measured field gradient was $B' = 560$ T/m in the center of

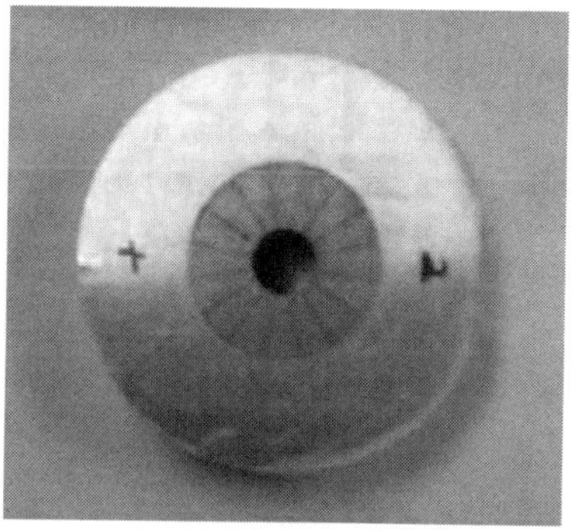

Figure 5. An assembled PMQ.

the PMQ bore region, while the pulsed wire measurements verified the co-location of the mechanical and magnetic centers to within 25 μm.

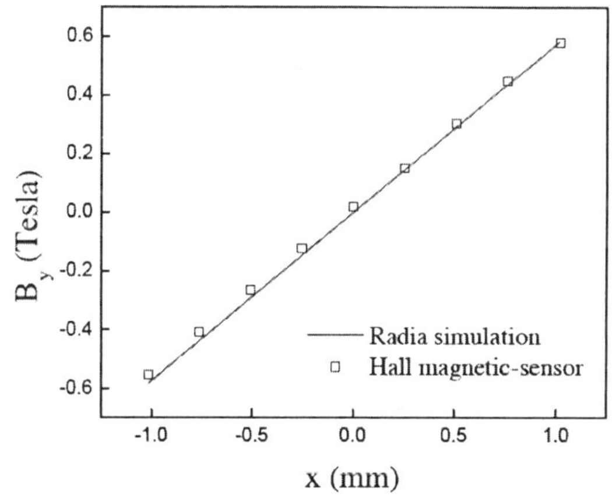

Figure 6. A comparison of PMQ field profile measurements using a Hall probe to RADIA simulation results.

For the final focus system, a simplified thin lens triplet design was analyzed. Assuming a symmetric beam waist at both the focus and the input to the triplet, and given a minimum achievable lens focal length f_{min}, it can be shown that optimal focusing is found in an asymmetric design in which the lens focal lengths are $2f_{min}$, $-f_{min}$, f_{min}. Based on this analysis, and verification of the (F-DD-FF) configuration using TRACE3D simulations, the PMQ "mover system" was designed to independently position (thereby providing tunability) 3 separate lenses, the first being a single PMQ, and the following two consisting of two concatenated magnets.

The PMQ mover assembly, illustrated in Fig. 7, was built and later

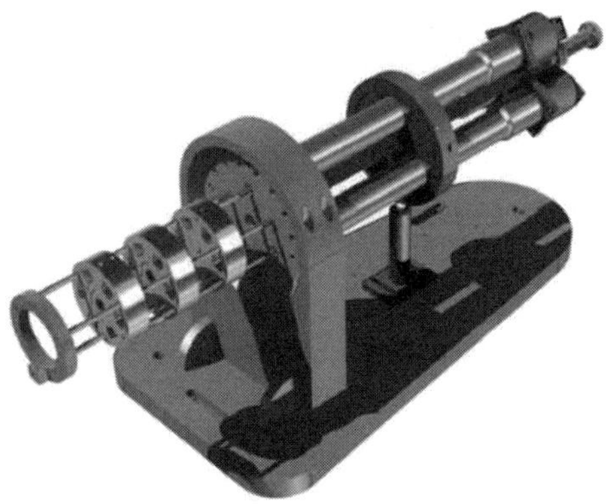

Figure 7. A rendered CAD drawing of the final-focus assembly.

measured to position the magnets with a level of precision as described above, and installed in the PLEIADES interaction beamline. The assembly consists of a rail system which maintains the alignment of each magnet, and a set of push-rods connected to stepper motor driven vacuum linear motion feedthroughs.

The PMQ system has successfully focused beams from 50 to 90 MeV, demonstrating the tunability of this system. The beam spot size at the focus was measured using optical transition radiation (OTR) generated as the electrons impact a polished metal surface. Figure 8 show a CCD image of the final focus produced using a 200 pC, 70 MeV beam. The spot size, 20 μm, is a significant improvement over that obtained with the previously employed electromagnet based final focus system. The final focus tunability has been used to produced ICS x-rays over the range from 40 to 140 keV.

References

1. P. Chen, J.M. Dawson, R.W. Huff, and T. Katsouleas, Phys. Rev. Lett. **54**, 693 (1985); P. Chen, J.M. Dawson, R.W. Huff, and T. Katsouleas, Phys.

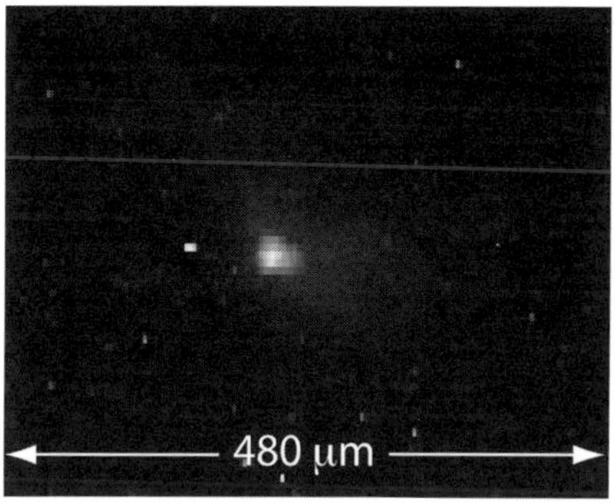

Figure 8. False color CCD image of the PMQ focused electron beam. The RMS beam size is 20 μm in both x and y.

Rev. Lett. **55**, 1537 (1985); J.B. Rosenzweig, B. Breizman, T. Katsouleas, and J.J. Su, Phys. Rev. A **44**, R6189 (1991).

2. R.W. Schoenlein *et al.*, Science **274**, 236 (1996); I.V. Pogorelsky *et al.*, Phys. Rev. ST Accel. Beams **3**, 090702 (2000).
3. N. Barov *et al.*, Phys. Rev. ST Accel. Beams **3**, 011301 (2000).
4. H. Braun *et al.*, Phys. Rev. Lett. **84**, 658 (2000); S.G. Anderson, J.B. Rosenzweig, P. Musumeci, and M.C. Thompson, Phys. Rev. Lett. **91**, 074803 (2003).
5. J.S. Nodvick and D.S. Saxon, Phys. Rev. **96**, 180 (1954); K. Ishi *et al.*, Phys. Rev. A **43**, 5597 (1991); S. Heifets, G. Stupakov, and S. Krinsky, Phys. Rev. ST Accel. Beams **5**, 064401 (2002).
6. L. Serafini and M. Ferrario, in Physics of, and Science with, the X-ray Free-Electron Laser, AIP Conf. Proc. No. 581 (2001).
7. X.J. Wang, X. Qiu, and I. Ben-Zvi, Phys. Rev. E **54**, R3121 (1996); P. Piot, L. Carr, W.S. Graves, and H. Loos, Phys. Rev. ST Accel. Beams **6**, 033503, (2003).
8. W. Lou *et al.*, in Proceedings of the 1997 Particle Accelerator Conference, 3236, (1998); J.T. Volk *et al.*, in Proceedings of the 2001 Particle Accelerator Conference, 217, (2001).
9. S.G. Anderson *et al.*, Phys. Rev. ST Accel. Beams **8**, 014401 (2005).
10. J.K. Lim *et al.*, Phys. Rev. ST Accel. Beams **8**, 072401 (2005).
11. D.J. Gibson *et al.*, Physics of Plasmas **11**, 2857, (2004).
12. A. Murokh *et al.*, Nucl. Instrum. Methods Phys. Res., Sect. A **410**, 452 (1998).
13. RadiaBeam Technologies, `http://www.radiabeam.com/`.
14. K. Halbach, Nuclear Instruments and Methods **169**, 1 (1980); K. Halbach, Nuclear Instruments and Methods **198**, 213 (1982).
15. R. Warren, Nucl. Instrum. Methods Phys. Res., Sect. A **272**, 257 (1988).

WAKEFIELDS OF SUB-PICOSECOND ELECTRON BUNCHES*

K. L. F. BANE

Stanford Linear Accelerator Center,
2575 Sand Hill Road,
Menlo Park, CA 94025, USA
E-mail: kbane@slac.stanford.edu

We discuss wakefields excited by short bunches in accelerators. In particular, we review some of what has been learned in recent years concerning diffraction wakes, roughness impedance, coherent synchrotron radiation wakes, and the resistive wall wake, focusing on analytical solutions where possible. As examples, we apply formulas for these wakes to various parts of the Linac Coherent Light Source (LCLS) project. The longitudinal accelerator structure wake of the SLAC linac is an important ingredient in the LCLS bunch compression process. Of the wakes in the undulator region, the dominant one is the resistive wall wake of the beam pipe.

1. Introduction

In recent accelerator projects one often finds high energy bunches of electrons or positrons that are simultaneously short, intense, and have small emittances. In the International Linear Collider (ILC) trains of 3 nC, 300 μm long bunches are accelerated through 10 km of linac, on their way the final focus and the collision point. In the Linac Coherent Light Source (LCLS), a SASE FEL, a 1 nC bunch is compressed to a length of 20 μm, accelerated in 500 m of linac, before entering the undulator for lasing [1]. In both cases a major challenge is keeping in check wakefields that are induced in various parts of the accelerator and that tend to increase emittances, thereby degrading luminosity (in the former case) or lasing (in the latter).

In August 1987 a workshop entitled "Impedance Beyond Cut-Off" was held at the Lawrence Berkeley Laboratory. In the foreword to the proceedings to this workshop, the editor noted that "latest designs of many ad-

*This work is supported by the Department of Energy, contract DE-AC02-76SF00515.

vanced accelerator systems ... involve charged particle beams of extremely short duration in time (a few tens of picoseconds and shorter)" [2]. Nowadays, a similar statement can still be made, but with the words "tens of picoseconds" probably changed to "hundreds of femtoseconds." Over the intervening years, as projects have been calling for ever shorter bunches the understanding of short-bunch wake/high frequency impedance has also been increasing. This report is a review of some of what has been learned about this subject in recent years.

One way of measuring "short" is to take the ratio σ_z/a, where σ_z is the rms bunch length and a is beam pipe radius. According to this simple criterion both the ILC bunch in the main linac and the LCLS bunch in the final linac are short ($\sigma_z/a = 0.009$ in both cases). In general, we can say that as bunches become shorter the longitudinal wakefields become stronger and the transverse wakes weaker. As bunches become shorter/their frequency content becomes higher, new sources of wakefields become important—*e.g.* the roughness wake—and familiar wakefields display unfamiliar behavior—*e.g.* the resistive wall wake. Numerically calculating a short-range wake/high frequency impedance can be a difficult task. Often, however, analytical approximations exist—especially for cylindrically symmetric structures—that can still be accurate. One factor that helps is that the broad-band (averaged) impedance suffices, since the (broad) bunch spectrum smooths out resonances.

In this report we present in some detail wakes that become important for short bunches, including diffraction wakes, roughness impedance, coherent synchrotron radiation (CSR) wake, and the resistive wall wake. We focus on analytical solutions, and on longitudinal over transverse wakes. Formulas presented here can be used to set up an impedance budget, to order in importance the various wake sources in a project. However, for some objects, especially 3D objects, numerical calculations will still be necessary.

1.1. *The LCLS*

As concrete examples for the wake formulas, throughout the text we will consider parts of the LCLS project. Figure 1 gives a schematic of the LCLS, displaying important machine parameters and the beam properties of energy, rms bunch length, rms relative energy spread (σ_δ), at various locations. The LCLS comprises an rf gun, four S-band accelerator regions (Linacs-0 to -3), an X-band structure, two chicane bunch compressors (named BC1 and BC2), and an undulator. The bunch charge is

$eN = 1$ nC. At the end of BC1, at energy $E = 250$ MeV, the bunch shape is roughly Gaussian, with rms length $\sigma_z = 0.19$ mm. At the end of BC2, $E = 4.54$ GeV. Beginning at this location the rms length $\sigma_z = 22$ μm; the bunch shape, however, is "double horned"–it consists of a 3 kA flat-top, with leading and trailing spikes 12–16 kA high, each of rms length ~ 2 μm. The final energy is 14.1 GeV.

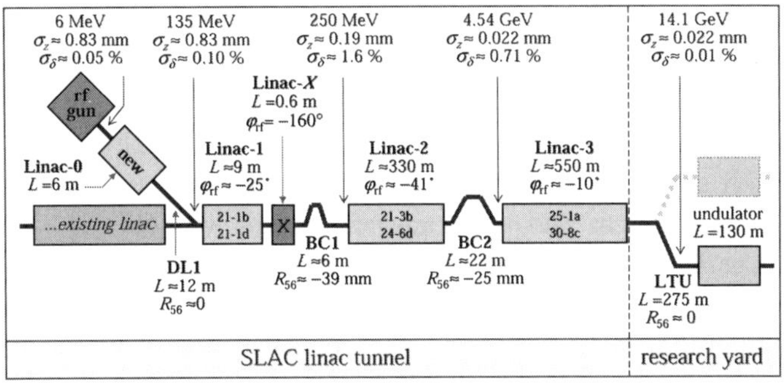

Figure 1. Schematic of the LCLS.

In the LCLS the longitudinal wakes tend to be more important than the transverse wakes. Also, in a SASE FEL, like the LCLS, we are not sensitive to projected emittance; amplification is sensitive to slice properties (~ 1 μm scale *vs.* a bunch rms of 20 μm). Wakefield effects, however, are head-tail effects, and only weakly affect slice properties directly. Nevertheless, in order not to interfere with lasing, there is a limit in allowable induced energy change in the undulator, to be less than a few times the Pierce parameter ($\lesssim 0.3\%$). In addition, there are bunch specifications on bandwidth ($< 0.1\%$), and on slice alignment ($< \sigma_x$), with σ_x transverse rms beam size.

In this report, for examples, we consider the effects of the accelerator structure wake in the linacs, the CSR wake in the BC2 chicane, and the wakes of transitions, roughness, and resistive wall in the beam pipe of the undulator. Note that in the linacs the wakes are an integral part of the compression process that determines the final phase space of the beam. In the chicane, CSR wake induced energy change results, through dispersion, in emittance growth. In the undulator region, the resistive wall wake of the 130 m beam pipe dominates; originally the beam pipe was meant to be round with $a = 2.5$ mm, made of stainless steel coated with copper.

2. WAKES AND IMPEDANCES

We begin by reviewing basic properties of wakes and impedances. The reader is referred to A. Chao's book for more details [3].

Consider a point particle of unit charge moving at the speed of light c through a structure, that is followed, at distance s, by a test particle, that is also moving at c. The longitudinal wake $W(s)$ is the voltage *loss* experienced by the test particle, typically given in units [V/C] for a single structure, in [V/C/m] for a periodic one. The wake is zero if the test particle is in front ($s < 0$). For a bunch of longitudinal charge distribution λ_z, the *bunch wake* $\mathcal{W}(s)$—the voltage *gain* for a test particle at position s—is given by

$$\mathcal{W}(s) = -\int_0^\infty W(s')\lambda_z(s-s')\,ds' \; . \tag{1}$$

The average of minus the bunch wake, $-\langle\mathcal{W}\rangle$, gives the loss factor; the rms $\mathcal{W}_{rms}$ gives the energy spread increase: $\Delta E_{rms} = eNL\mathcal{W}_{rms}$, with L the length of structure (in the periodic case).

The impedance is the Fourier transform of the wake:

$$Z(k) = \frac{1}{c}\int_0^\infty W(s)e^{iks}ds \; , \tag{2}$$

with k the wave number. In a cylindrically symmetric structure the impedance/wakes excited by a particle off-axis vary as $\cos m\theta$, with m an integer and θ the azimuthal angle; the dominant transverse wakefield is the dipole ($m = 1$) wake. Here the transverse wake W_x is defined as the dipole force experienced by the test particle per unit offset of the driving particle, given in [V/C/m^2] in the periodic case. The transverse impedance Z_x is, by convention, taken to be i times the Fourier transform of W_x. For the special case of the resistive wall wake the longitudinal ($m = 0$) and transverse ($m = 1$) wakes and impedances are related by

$$W_x(s) = \frac{2}{a^2}\int_0^s W(s')\,ds' \; , \qquad Z_x(k) = \frac{2}{a^2k}Z(k) \; . \tag{3}$$

These relations, as we will see, hold also for many of the short-range wakes/high-frequency impedances considered in this report, though they are not, in general, valid.

We sometimes denote an impedance/wake as being *resistive*, *inductive*, or *capacitive*, from analogy to simple circuit element properties. By these labels we mean objects for which $\mathcal{W}(s)$ is proportional to $\lambda_z(s)$, $\lambda_z'(s)$, and $\int_{-\infty}^{s} \lambda_z(s')\,ds'$, respectively. Note that these imply that $W(s)$ is proportional to, respectively, $\delta(s)$, $\delta'(s)$, and $H(s)$, with $H(s)$ the unit step function ($= 0$ for $s < 0$; 1 for $s > 0$).

2.0.1. *Asymptotic Behavior of $Z(k)$, $W(s)$, and $\mathcal{W}(s)$ (see e.g. [3])*

As will be seen in the following section, in some cases the asymptotic behavior of the wake $W(s)$ or the impedance $Z(k)$ is not integrable. We find *e.g.* asymptotic wakes of the form $W \sim -As^{-\alpha}$, with $1 < \alpha < 2$ and A a constant (an example is the longitudinal, long-range resistive-wall wake). In such cases it appears that one cannot simply convert between asymptotic forms of W and Z, or from the asymptotic form of W to that of $\mathcal{W}$. However, we know that such asymptotic behavior cannot continue forever, and both W and Z are always integrable: $\int_{-\infty}^{\infty} W(s)\,ds = Z(0) = 0$ and $\int_{-\infty}^{\infty} Z(k)\,dk = 2\pi W(0)$ is finite (at least cut-off by finite beam energy). This knowledge allows us to make such conversions, either through differentiation under the integral sign or through integration by parts.

For example, consider the asymptotic wake form discussed above. The asymptotic behavior of the bunch wake can be obtained using integration by parts (we need to assume the bunch shape is smooth):

$$\mathcal{W}(s) = \frac{A}{\alpha - 1} \int_0^{\infty} \frac{d\lambda_z(s-z)}{dz} \frac{dz}{z^{\alpha-1}} \,. \tag{4}$$

We see that knowledge of the very short-range wake, how it behaves as $s \to 0$, is not necessary.

2.1. *Considerations for Short Bunches*

Considerations especially relevant for short bunches are:

Catch-up distance: In calculating the effect on a beam, the wake is typically taken to act instantaneously. For very short bunches there can be a significant lag between the generation of radiation by the head of a bunch and its effect on tail particles. When a head particle passes a vacuum chamber object, such as the beginning of a cavity, that information cannot arrive at a tail particle until a downstream distance $z = a^2/2s$, where a is the beam pipe radius and s is the distance between the two particles. For example, if $a = 1$ cm and $s = 20$ μm, then the catch-up distance is 2.5 m.

Transients: Similarly, for periodic structures, the interaction with a short bunch will entail an initial transient region before the steady-state wake is reached (see the example shown in Fig. 2). For a Gaussian bunch with length σ_z the transient regime will last $z \sim a^2/2\sigma_z$. As a bunch becomes shorter, the transients become ever more important.

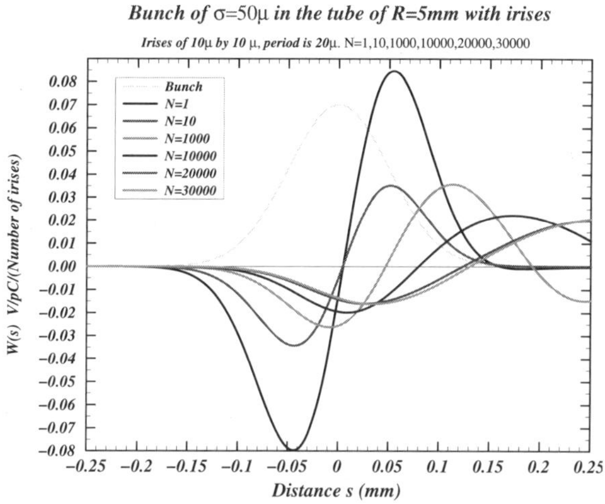

Figure 2. Simulation of wake per period generated by a bunch in a tube with N small corrugations, showing the passage from transient to steady-state behavior; the bunch shape is given in yellow (from Ref. [5]).

Limiting value of wake: For periodic, cylindrically symmetric structures whose closest approach to the axis is a, the steady state wakes have the property

$$W(0^+) = \frac{Z_0 c}{\pi a^2} \quad \text{and} \quad W_x'(0^+) = \frac{2 Z_0 c}{\pi a^4}\,, \tag{5}$$

with $W_x(0^+) = 0$, where $Z_0 = 377\ \Omega$. This is true for a resistive pipe [3], a disk-loaded accelerator structure [4], a pipe with small periodic corrugations [5,6], and a dielectric tube within a pipe [7]; it seems safe to assume that it is generally true. For a non-round structure the result will be a different constant [*e.g.* for 2 parallel, resistive plates separated by $2a$, $W(0^+) = \pi Z_0 c/16a^2$ [8]], but again a constant dependent on transverse dimensions and independent of material properties. Note that these formulas give an upper limit to the rate of energy loss and to the dipole wake that can be induced by a short bunch.

Finite energy: The impedance will drop sharply to zero for frequencies $k > \gamma/a$, with γ the Lorentz energy factor. In the time domain when

$\sigma_z < a/\gamma$, σ_z should be replaced by a/γ in wakefield formulas. For example, if $a = 1$ cm and energy $E = 14$ GeV this occurs when $\sigma_z = 0.4$ μm.

3. TYPES OF SHORT BUNCH WAKES

We discuss four types of wakes that tend to be important for short bunches: diffraction wakes, roughness impedance, coherent synchrotron radiation wakes, and resistive wall wakes. In all cases the structures considered are cylindrically symmetric with a beam pipe radius a. Impedances presented here are valid at high frequencies, wakes at short distance. Note that for all expressions of (point charge) wakes given, it is implied that they are multiplied by the unit step function $H(s)$.

3.1. *Diffraction Wakes*

When a short bunch ($\sigma_z/a \ll 1$) passes by an abrupt change in the beam pipe, such as at the beginning of a cavity, waves begin to radiate from the corner in a manner that can be described by the diffraction theory of light. If the corner is part of a shallow step or shallow cavity Fraunhoffer diffraction applies, if it is part of a deep cavity Fresnel diffraction applies. A third important example, a periodic array of deep cavities, is a model for a multi-cell (disk-loaded) accelerator structure.

3.1.1. *Steps and Cavities*

Consider beam pipes of radius a and b with $b > a$. For an abrupt step connecting the pipes, Heifets and Kheifets have shown that the high frequency impedance $Z \approx Z_0 \ln(b/a)/\pi$ ($Z_0 = 377$ Ω) when moving from the pipe of radius a to that that of radius b, and $Z \approx 0$ when moving from b to a [9]. Combining the two cases (in either order) to make a pair of shallow transitions or a shallow cavity (see Fig. 3a), they obtain for impedance, $Z = Z_0 \ln(b/a)/\pi$, implying a resistive short-range wake:

$$W(s) = \frac{Z_0 c}{\pi} \ln(b/a)\delta(s) \; . \tag{6}$$

For the transverse case the impedance, $Z_x = Z_0(a^{-2} - b^{-2})/(\pi k)$, and the wake is capacitive [10]:

$$W_x(s) = \frac{Z_0 c}{\pi} \left(\frac{1}{a^2} - \frac{1}{b^2} \right) \; . \tag{7}$$

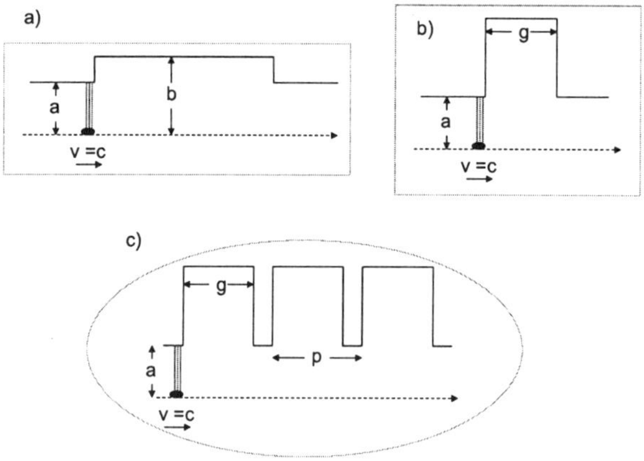

Figure 3. Structure geometry for diffraction wakes: (a) shallow transition pair, (b) deep cavity, (c) periodic array of deep cavities. The on-axis bunch and its electric field are indicated in red.

For a deep cavity with beam pipes (see Fig. 3b) the high frequency impedance, $Z \propto \sqrt{g/k}/a$, and the wake [3,11]

$$W(s) = \frac{Z_0 c}{\sqrt{2}\pi^2 a}\sqrt{\frac{g}{s}} \; . \tag{8}$$

For the transverse case $Z_x \propto g^{1/2}(ka)^{-3/2}$, and

$$W_x(s) = \frac{2^{3/2} Z_0 c}{\pi^2 a^3}\sqrt{gs} \; ; \tag{9}$$

the longitudinal and dipole cases are related according to Eqs. 3. Note that a cavity is considered shallow (deep) to a bunch of total length ℓ if $g \gg (\ll)[2(b-a)^2/\ell]$. This is a statement that radiation, generated by the head of the bunch as it enters the cavity, can (cannot) reflect off the outer wall and return in time to meet the bunch tail.

In accelerators with longer bunches, tapering of transitions is often used as a way of reducing their wakefield effect. For short bunches, however, radiation is diffracted forward at a shallow angle, $\theta \sim \sigma_z/a$, and until the taper angle becomes comparable to this quantity the tapering has little effect. In the undulator region of the LCLS there are transitions, flanges, etc, objects that generate diffraction wakes; however, with $\sigma_z = 20$ μm and $a = 2.5$ mm, tapering would have no effect unless $\theta \lesssim 0.5°$.

In a first step in generating an impedance budget for a project it is sometimes useful to calculate the first and second moments of the bunch wake for the various impedance objects; these are then compared to estimate the relative importance of the different objects for–energy loss, energy spread, transverse wake, etc. In Table 1, the first and second moments of wake forms for Gaussian bunches, including those for diffraction wakes and other wakes discussed later in this report, are given.

Table 1. First and second moments of (Gaussian) bunch wakes, $\langle \mathcal{W}_g \rangle$ and $(\mathcal{W}_g)_{rms}$, corresponding to common short-bunch wake forms $W(s)$. Here the s dependence alone is considered. $H(s)$ is the unit step function ($= 0$ for $s < 0$; 1 for $s > 0$).

Wake form, $W(s)$	Wake example	$\langle \mathcal{W}_g \rangle$	$(\mathcal{W}_g)_{rms}$
Circuit Models:			
Resistive: $\delta(s)$	Shallow transition pair, W	$\frac{1}{2\sqrt{\pi}\sigma_z}$	$\frac{0.111}{\sigma_z}$
Capacitive: $H(s)$	Shallow transition pair, W_x	$\frac{1}{2}$	$\frac{1}{\sqrt{3}}$
Inductive: $\delta'(s)$	Roughness, inductive model, W	0	$\frac{1}{\sqrt{6\pi}\,3^{1/4}\sigma_z^2}$
Power Law, s^α:			
$\alpha = -\frac{1}{2}$	Deep cavity, W	$\frac{0.723}{\sqrt{\sigma_z}}$	$\frac{0.292}{\sqrt{\sigma_z}}$
$\alpha = \frac{1}{2}$	Deep cavity, W_x	$0.489\sqrt{\sigma_z}$	$0.374\sqrt{\sigma_z}$
$\alpha = -\frac{3}{2}$	Resistive wall, low freq., W	$\frac{-0.489}{\sigma_z^{3/2}}$	$\frac{0.516}{\sigma_z^{3/2}}$
$\alpha = -\frac{4}{3}$	CSR, W	$\frac{-0.758}{\sigma_z^{4/3}}$	$\frac{0.532}{\sigma_z^{4/3}}$

In the undulator region of the LCLS there are 30 rectangular-to-round transition pairs. A 10 by 5 mm rectangular cross-section becomes a round pipe with $a = 4$ mm in the BPM's. A pessimistic calculation considers a cylindrically symmetric transition pair with $a = 2.5$ mm and $b = 4$ mm and uses Eq. 6 and the "Resistive" entry in Table 1. The rms induced energy spread, for all transition pairs, is then estimated to be ($E = 14$ GeV, $\sigma_z = 20\ \mu$m): $\Delta E_{rms}/E = 0.02\%$. A full 3D simulation of the real geometry also finds a resistive wake, but with $\Delta E_{rms}/E = 0.013\%$ [12]. In the undulator region of the LCLS, the wake effect of the transitions is small compared to that of the resistive wall wake (discussed later), and can be ignored.

3.1.2. *Periodic Structure*

For an array of cavities with period p (see Fig. 3c), the asymptotic, high frequency impedance is given by [4,13],

$$Z(k) \approx \frac{iZ_0}{\pi k a^2}\left[1+(1+i)\frac{\alpha(g/p)\,p}{a}\left(\frac{\pi}{kg}\right)^{1/2}\right]^{-1} , \tag{10}$$

with $\alpha(x) \approx 1 - 0.465\sqrt{x} - 0.070x$; the real part of the impedance $\mathcal{R}e(Z) \sim k^{-3/2}$. Inverse Fourier transforming, one obtains an analytical expression for the very short-range wake:

$$W(s) \approx \frac{Z_0 c}{\pi a^2}\exp\left(\frac{2\pi\alpha^2 p^2 s}{a^2 g}\right)\operatorname{erfc}\left(\frac{\alpha p}{a}\sqrt{\frac{2\pi s}{g}}\right) , \tag{11}$$

with erfc the compliment error function.

Obtaining the short-range wake numerically and fitting to a simple function, we obtain a result that is valid over a larger s range and over a useful range of structure parameters [14]:

$$W(s) = \frac{Z_0 c}{\pi a^2}\exp\left(-\sqrt{s/s_1}\right) \quad , \tag{12}$$

with

$$s_1 = 0.41\frac{a^{1.8}g^{1.6}}{p^{2.4}} . \tag{13}$$

The result is valid for $s/p \leq 0.15$, $0.34 \leq a/p \leq 0.69$, and $0.54 \leq g/p \leq 0.89$. For the SLAC linac (a = 11.6 mm, g = 29.2 mm, p = 35.0 mm) s_1 = 1.5 mm. It has been numerically verified that, for a Gaussian bunch in a periodic accelerator structure, the steady state result becomes valid after a distance of $z \approx a^2/2\sigma_z$ [15]. For Linac-3 of the LCLS where σ_z = 20 μm, the distance to steady-state is 3.4 m if the bunch were Gaussian, is $\sim$ 35 m for the actual double horned distribution; but even this distance is small compared to the 550 m length of linac, and the transient region can be ignored.

The high frequency transverse impedance can be shown to be related to the high frequency longitudinal impedance (Eq. 10) according to $Z_x = 2Z/ka^2$ [16]. As with the longitudinal case, the short-range dipole wake was obtained numerically for a set of parameters and then fit to a simple function [17]:

$$W_x(s) = \frac{4Z_0 c s_2}{\pi a^4}\left[1-\left(1+\sqrt{\frac{s}{s_2}}\right)\exp\left(-\sqrt{\frac{s}{s_2}}\right)\right] , \tag{14}$$

with $s_2 = 0.17a^{1.79}g^{0.38}/p^{1.17}$. This equation is valid over the same parameter regime as in the longitudinal case.

Real linear accelerator structures are often not perfectly periodic; so-called "constant gradient" (*e.g.* the SLAC linac) or "detuned" structures have gradual variations in cell geometry along the structure. However, for the short-range wake/high frequency (broad-band) impedance of such structures the equations given here still apply (with possible slight adjustments of parameters, where *e.g.* a is replaced by effective iris radius, etc). The point charge wakes and (Gaussian) bunch wakes in the SLAC linac are plotted in Fig. 4.

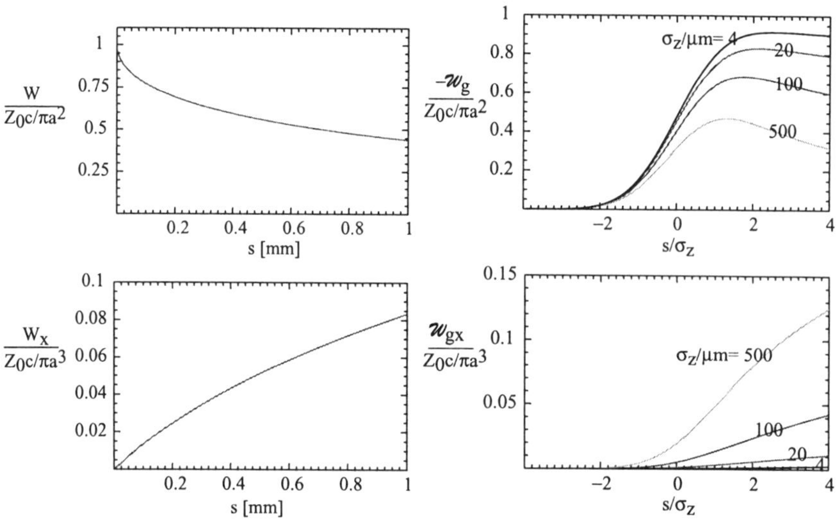

Figure 4. For the SLAC linac: the point charge wakes, W and W_x (left plots), and the bunch wakes for Gaussian bunches, $\mathcal{W}_g$ and $\mathcal{W}_{gx}$, for various bunch lengths (right plots). The scaling factors $Z_0c/\pi a^2 = 0.27$ GV/(nC-km) and $Z_0c/\pi a^3 = 23$ MV/(nC-mm-km).

In the LCLS the longitudinal wakes induced in the linacs are integral parts of the longitudinal phase space manipulation process and thus need to be known accurately. The LCLS beam line has been designed with the aid of a macro-particle, longitudinal phase space tracking program, *LiTrack* [18], using the function $W(s)$ for the SLAC linac as an input. Measurements at SLAC confirmed such calculations down to $\sigma_z \sim 0.5$ mm many years ago [19], and down to $\sigma_z \sim 50$ μm in more recent times [20,21].

In Linac-1 and 2 of the LCLS the sum of the wake and the applied

RF wave generates the energy/longitudinal position correlation needed for bunch compression. For example, in Linac-2 the bunch phase offset is $\sim -40°$ (ahead of the rf crest), the rms bunch shape is roughly Gaussian with $\sigma_z = 190\ \mu$m, and the total structure length is 330 m. The energy/position correlation induced by the RF is ~ 250 MeV/mm, by the wake ~ -100 MeV/mm. In Linac-3, which is 550 m long, where the bunch shape is double horned with rms 20 μm, the wake cancels a residual correlation of 850 MeV/mm, left at the end of BC-2. Finally, note that it is due the non-linear, cubic behavior in $\mathcal{W}(s)$ in Linac-2 that the final LCLS bunch shape contains the unavoidable horns.

3.2. *Roughness Impedance*

A metallic beam pipe with a rough surface has an impedance that is enhanced at high frequencies. Two approaches to modeling the impedance of a rough surface are with: (i) a random collection of bumps on a surface and (ii) a wall with small periodic corrugations.

An early model of roughness impedance assumes a random, non-interacting collection of bumps of various shapes (see sketch Fig. 5), with the total impedance given by the sum of the individual impedances. Consider a beam pipe of radius a on which there is a small hemispherical bump of radius h. At low frequencies $k \ll 1/h$ the impedance is inductive and given by [22]

$$Z(k) = ikc\mathcal{L}_1 = ik\frac{Z_0 h^3}{4\pi a^2}\ , \tag{15}$$

with $\mathcal{L}_1$ the inductance of the bump. If the hemispherical bump is replaced by one of a different shape but of about the same size, the above equation is multiplied by a form factor of order 1. For many randomly distributed bumps the inductance per unit length $\mathcal{L}/L$ can be written as [23]

$$\mathcal{L}/L = \frac{2\alpha f a\mathcal{L}_1}{h^2} = \frac{\alpha f Z_0 h}{2\pi ac}\ , \tag{16}$$

where α is the filling factor of bumps and f is the effective form factor. The rms wake of a Gaussian bunch is given by $\mathcal{W}_{rms} \approx 0.06c^2\mathcal{L}/L\sigma_z^2$.

The simple idea of this model has been systematized so that one can, from measurements of the contour of a surface, obtain the inductance per length of the surface [24]

$$\mathcal{L}/L = \frac{Z_0}{2\pi ca}\int_{-\infty}^{\infty}\frac{k_z^2}{\sqrt{k_\theta^2+k_z^2}}S(k_z,k_\theta)\,dk_z dk_\theta\ , \tag{17}$$

Figure 5. Random collection of bumps model of roughness impedance.

where $S(k_z, k_\theta)$ is the spectrum of the surface profile—the square of the absolute value of the Fourier transform of the surface variation—and k_z, k_θ, are the wave numbers in the longitudinal and azimuthal directions. Finally, note that since, for an inductive model, the wake of a bunch is given by the derivative of the bunch distribution, such a model cannot be applied to a rectangular or other non-smooth bunch shape.

The second modeling approach is to consider a beam pipe with small periodic corrugations (see Fig. 6). The motivation for such a model came from numerical simulations of many randomly placed, small cavities on a beam pipe; it was found that, in steady state, the short range wake is very similar to the truly periodic case [5]. Consider a beam pipe with small, rectangular, periodic corrugations, with h the half-depth, g the gap ($= p/2$), and p the period. In the case $h/p \gtrsim 1$ the wake is dominated by one mode of relatively low frequency ($k_0 \ll 1/h$) [5,6]:

$$W(s) \approx \frac{Z_0 c}{\pi a^2} \cos k_0 s \quad \text{with} \quad k_0 = \frac{2}{\sqrt{ah}} \,. \tag{18}$$

For smooth bunches, with $k_0 \sigma_z$ large, the wake for this model becomes inductive with $\mathcal{L}/L = Z_0 h/4ac$, very similar to the first model. However, even for non-smooth distributions this model can be applied. Note that the dipole wake for this model is also given by a single mode at the same frequency.

This single resonator model is valid when the depth-to-period of the surface roughness is not small compared to 1. However, measurements of copper surfaces with good finish show that the opposite tends to be true: that the depth-to-period $\lesssim 0.01$ (see Fig. 7) [25]. As the depth-to-period ratio becomes small the dominant, low frequency mode is replaced by many weak, closely spaced modes beginning just above $k = \pi/p$ [26]. For a sinusoidally oscillating wall with amplitude h and period p, where $h \ll p$, the wake was

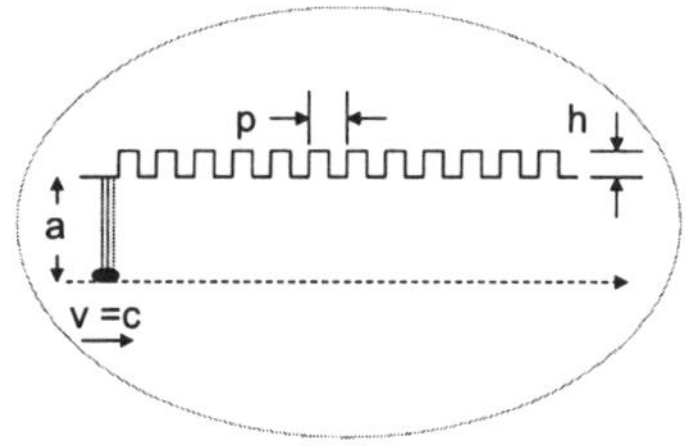

Figure 6. Geometry for small corrugations model of roughness impedance.

found by Stupakov to be [27]:

$$W(s) = \frac{Z_0 c h^2 k_1^3}{4\pi a} f(k_1 s) , \tag{19}$$

$$f(\zeta) = \frac{1}{2\sqrt{\pi}} \frac{\partial}{\partial \zeta} \frac{\cos(\zeta/2) + \sin(\zeta/2)}{\sqrt{\zeta}} , \tag{20}$$

with $k_1 = 2\pi/p$ (the function $f(\zeta)$ is shown in Fig. 8a). The formula is valid for $s \gtrsim k_1 h^{4/3} a^{2/3}$. For bunch length $\sigma_z \lesssim k_1^{-1}$ all we need is the (small s) leading order term:

$$W(s) = -\frac{Z_0 c h^2}{16\pi^{3/2} a} \left(\frac{k_1}{s}\right)^{3/2} . \tag{21}$$

In this case the wake of a Gaussian bunch

$$\mathcal{W}_g(s) = W(\sigma_z) w_g(s/\sigma_z) , \tag{22}$$

where the form factor w_g is shown in Fig. 8b. Note that for given h the bunch wake here is weaker, by $\sim h/p$, than for the single mode model described earlier. Note also that the wake has the form of the long-range resistive wall wake; it can be described as the wake of a metal with conductivity $\sigma = 16/(Z_0 h^4 k_1^3)$.

In the undulator region of the LCLS, $\sigma_z = 20$ μm and $a = 2.5$ mm. If we believe Fig. 7 is representative of the undulator beam pipe surface roughness ($h \sim 0.5$ μm, $p \sim 100$ μm) then our parameters are in the regime of Stupakov's model. In this case the roughness wake is small, only ~ 0.15 the strength of the resistive wall wake (discussed below).

3.3. *CSR Wake*

The effect on a bunch of coherent synchrotron radiation can be described in terms of a wakefield, although there are some differences from normal

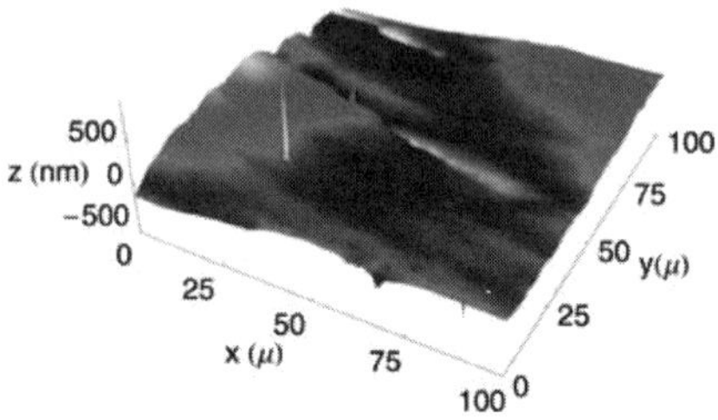

Figure 7. A sample surface profile measured with an atomic force microscope (from Ref. [25]).

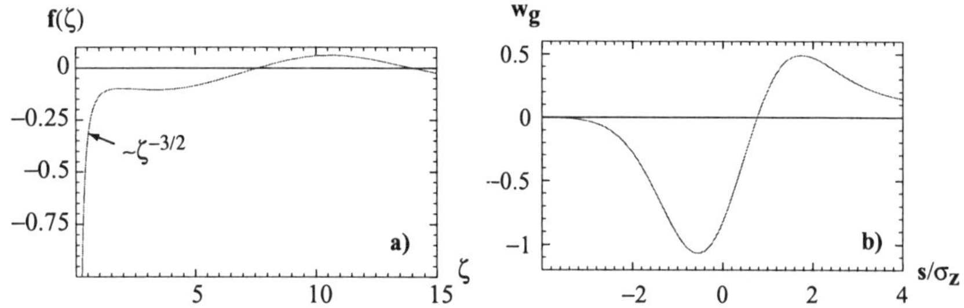

Figure 8. The wake of shallow, periodic corrugations ($h \ll p \ll a$): form factor for wake a), for Gaussian bunch wake [$k_1\sigma_z \lesssim 1$] b).

wakes. For example, for an ultra-relativistic particle moving in a circle of radius R in free space (see Fig. 9), the wake (for a test particle on the same path) is non-zero *ahead* of the radiating particle ($s < 0$), because the radiation moves on a shorter, straight path. The wake experienced by the test particle is sketched in Fig. 10a. For $(-s) \gg R/\gamma^3$ it is given by [28,29]

$$W(s) = -\frac{Z_0 c}{2 \cdot 3^{4/3} \pi R^{2/3} (-s)^{4/3}} \qquad s < 0 \;, \tag{23}$$

while $W(0^-) = Z_0 c \gamma^4 / 3\pi R^2$. The wake of a Gaussian bunch[28,31]

$$\mathcal{W}_g(s) = W(-\sigma_z) w_g(s/\sigma_z) \;, \tag{24}$$

where the form factor $w_g(s/\sigma_z)$ is plotted in Fig. 10b. The impedance, the Fourier transform of the wake, is [28,29,30]

$$Z(k) = \frac{Z_0}{2 \cdot 3^{1/3} \pi} \Gamma\left(\frac{2}{3}\right) e^{i\pi/6} \frac{k^{1/3}}{R^{2/3}} \;, \tag{25}$$

with $\Gamma(2/3) = 1.35$, valid to very high frequencies ($k \sim \gamma^3/R$).

For particles moving in a circle through a beam pipe the wake will be modified, an effect that can be calculated using image charges in the time

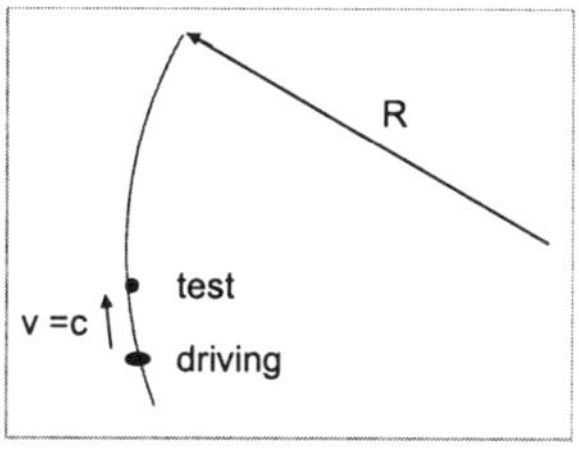

Figure 9. Geometry for CSR wake calculation: the driving and test particles follow a circular path of radius R in free space.

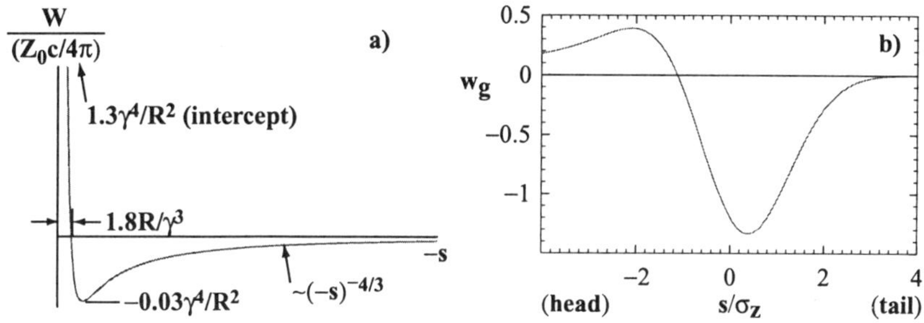

Figure 10. a) Sketch of the CSR wake of a particle, in free space, moving on a circle of radius R; b) bunch wake form factor for a Gaussian bunch.

domain [28], or be dealt with in the frequency domain [31]. The pipe suppresses the wake of a bunch, provided that $\sigma_z/a \gtrsim (a/R)^{1/2}$ [31]. If we consider the beam in the last bend of BC2 of the LCLS, and take $\sigma_z = 20\ \mu$m, $a = 1$ cm (here the half the vertical beam pipe aperture), and $R = 15$ m, we see that the bunch length is 13 times too short to feel the effect of shielding.

Chicanes are composed of 3 or 4 bends with drifts in between. Instead of one continuous circle, the beam moves in circular arcs (in the bends) and straight lines (outside the bends). Analytical solutions for the 1D wake of an ultra-relativistic particle entering, traversing, and leaving a bend without shielding have been derived [32]. There are transients when the beam enters a bend and also after it leaves. The catch-up distance is the distance a test particle, ahead of the radiating particle, needs to travel to catch up to the radiated wave. The distance of the incoming transient region is $z \approx (24R^2\sigma_z)^{1/3}$ (assuming z/R is small) [29].

One can consider the potential energy change (the "compression work") that the beam undergoes in being compressed. The average kinetic energy change to balance this is approximately (assuming the compression factor is large, for a Gaussian bunch) [33,34]

$$\langle \Delta E \rangle = -\frac{eNZ_0c}{4\pi^{3/2}\sigma_z} \ln\left(\frac{\gamma\sigma_z}{\sigma_x + \sigma_y}\right) , \tag{26}$$

where beam sizes are final quantities; the rms spread $\Delta E_{rms} \approx -0.4\langle \Delta E \rangle$. Interestingly, for example chicanes, Eq. 26 was found to roughly agree with the *total* energy change obtained by detailed CSR simulations [34].

To simulate the CSR force in a chicane compressor, computer programs that slice the bunch into macro-particles and solve the Liénard-Wiechert equations have been written [35,36,37], as has one that employs a frequency domain method [38]. In these programs the bunch can have transverse as well as longitudinal dimensions, shielding can be added, and the orbit/forces can be computed in a self-consistent manner. These programs, however, can be time consuming to run. The 1D analytical wake formulas for a particle entering, traversing, and leaving a bend, when used in a tracking program such as *ELEGANT* [39], give results that are quick to obtain and agree reasonably well with more detailed simulations for typical beam/chicane parameters [37]. Reference [37] gives a comparison of some simulation programs.

CSR induced energy spread in a chicane, through the dispersion, leads to emittance growth. For the BC-2 chicane of the LCLS, we can estimate the rms energy spread increase by taking 0.4 times the Eq. 26 result (a Gaussian approximation, with $E = 4.5$ GeV, $\sigma_z = 20$ μm, $\sigma_x = \sigma_y = 25$ μm), yielding $\Delta E_{rms}/E = 0.018\%$. *ELEGANT* simulation gives 0.016% (good agreement), and a resulting x emittance growth of 38% [40]. Finally, note that CSR has been a much studied topic in recent years; this has only been a cursory view of the subject.

3.4. *Resistive Wall Wake*

The impedance of a round metallic beam pipe of radius a has long been known to be [3]

$$Z(k) = \left(\frac{Z_0}{2\pi a}\right)\left[\frac{\lambda}{k} - \frac{ika}{2}\right]^{-1} , \tag{27}$$

where parameter λ is given by

$$\lambda = \sqrt{\frac{Z_0\sigma|k|}{2}}[i + \mathrm{sign(k)}] , \tag{28}$$

with σ the (dc) conductivity of the metal. Inverse Fourier transforming the impedance, one obtains the wake [3,41]

$$W(s) = \frac{4Z_0c}{\pi a^2}\left(\frac{e^{-s/s_0}}{3}\cos\frac{\sqrt{3}s}{s_0} - \frac{\sqrt{2}}{\pi}\int_0^\infty \frac{dx\, x^2 e^{-x^2 s/s_0}}{x^6+8}\right) , \tag{29}$$

where the characteristic distance s_0 is given by

$$s_0 = \left(\frac{2a^2}{Z_0\sigma}\right)^{\frac{1}{3}} \tag{30}$$

(see Fig. 11). The long-range asymptote, corresponding to the low frequency part of the impedance, is given by [3]

$$W(s) = -\frac{c}{4\pi^{3/2}a}\sqrt{\frac{Z_0}{\sigma}}\frac{1}{s^{3/2}} , \tag{31}$$

a formula that is valid provided that $s \gg s_0$. Note that, unlike the long range wake, the general wake formula does not scale as $\sigma^{-1/2}$. The transverse wake can be obtained from the longitudinal wake using Eq. 3.

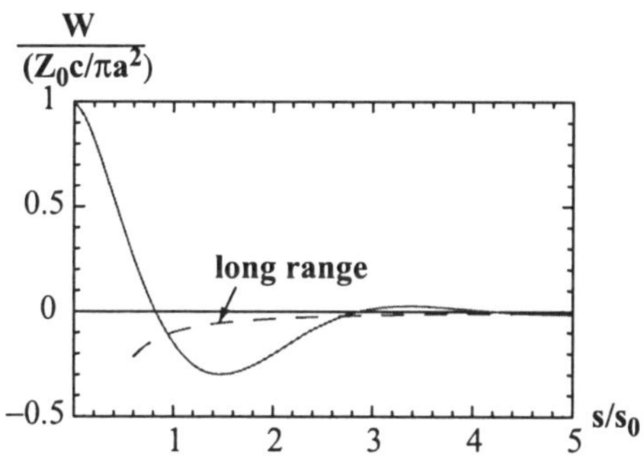

Figure 11. Plot of Eq. 29, the resistive wall wake. The long-range asymptote (Eq. 31) is shown in dashes.

For the LCLS in the undulator region, the rms bunch length $\sigma_z = 20\ \mu$m, and $a = 2.5$ mm; if the beam pipe is made of copper ($\sigma = 6.5\times10^7\ \Omega^{-1}\,\mathrm{m}^{-1}$) then $s_0 = 8.0\ \mu$m; if it is made of aluminum ($\sigma = 4.2\times10^7\ \Omega^{-1}\,\mathrm{m}^{-1}$), $s_0 = 9.2\ \mu$m. Thus we see that $\sigma_z \not\gg s_0$. In addition, the bunch in the undulator region is not Gaussian; it is a double-horned distribution with very high frequency content ($k \sim [0.05, 0.5]\ \mu\mathrm{m}^{-1}$, whose upper limit $\gg 1/s_0$). Thus the long-range wake will not suffice, and the general solution is needed.

The problem with using Eq. 29 for the short bunch wake is that it assumes that the conductivity of the metal wall is constant, *i.e.* independent of frequency. It is believed, however, that electrical conductivity in normal metals, up to the highest frequencies of interest to us ($k \sim 0.5\ \mu\text{m}^{-1}$), is described by the Drude-Sommerfeld free-electron model of conductivity [42,43]. According to this model the dc conductivity of a metal is given by $\sigma = ne^2\tau/m$, with n the density of conduction electrons; with e the charge, τ the relaxation time, and m the mass of the electron. In this model the ac conductivity, a response to applied oscillating fields, is given by

$$\tilde{\sigma} = \frac{\sigma}{1 - ikc\tau} \ . \tag{32}$$

The dc wake equation, Eq. 29, is valid provided $\Gamma = c\tau/s_0 \ll 1$. For copper at room temperature $c\tau = 8.1\ \mu$m, for aluminum $c\tau = 2.4\ \mu$m; if $a = 2.5$ mm: $\Gamma = 1.0$ (Cu), $\Gamma = 0.26$ (Al). We see that the dc wake equation is not valid for LCLS undulator beam pipe properties.

The calculation of the resistive wall wake including ac conductivity again involves (numerically) inverse Fourier transforming the impedance (Eq. 27), but with σ in parameter λ (Eq. 28) replaced by $\tilde{\sigma}$ [41]. For the special case $\Gamma \gtrsim 1$ (and s not too large) the wake can be approximated as a damped resonator

$$W(s) \approx \frac{Z_0 c}{\pi a^2} e^{-s/c\tau} \cos\left[\sqrt{2k_p/as}\right] \qquad [\Gamma \gtrsim 1] \ , \tag{33}$$

with $k_p = \sqrt{Z_0\sigma/c\tau}$, the plasma frequency of the metal. This approximation can *e.g.* be used for the LCLS undulator beam pipe if it is copper. In Fig. 12 we plot the ac wakes for the LCLS undulator beam pipe, for both copper and aluminum. Note that the ac wake of aluminum damps more rapidly than that of copper, due to its smaller value of τ. The dc wake for copper is also shown, for comparison (remember, this result is not in accord with the free electron model and thus non-physical).

The energy change, $\Delta E(s)$, induced by the resistive wall wake in the undulator region of the LCLS, for cases of copper and aluminum beam pipes, is shown in Fig. 13 ($\Delta E > 0$ indicates energy gain). Here energy $E = 14$ GeV, pipe length $L = 130$ m, and the beam shape is double horned with rms length 20 μm (the bunch shape is superimposed by the dashed curve in the figure). The sharp rise of the leading horn in the bunch shape results in wakes that ring; the ringing amplitude is about twice as large for copper than aluminum, due to the difference in relaxation times in the two metals. Note that other wakefield contributors in the undulator region,

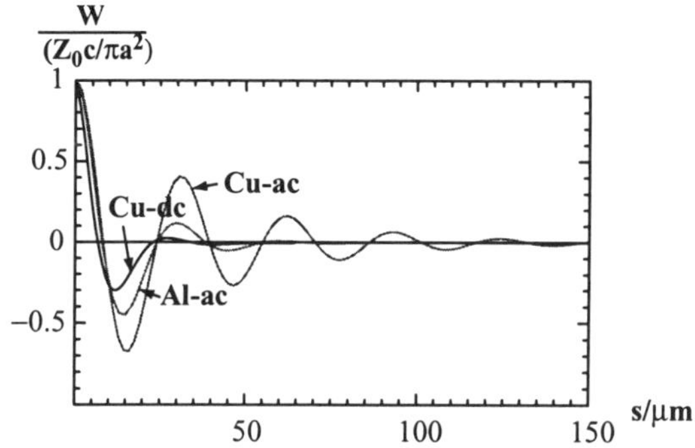

Figure 12. The dc and ac wakes for copper, and the ac wake for aluminum, for the LCLS undulator beam pipe ($a = 2.5$ mm).

such as wall roughness, transitions, and BPM's, are weak in comparison to the resistive wall, and can be ignored. From FEL simulations for LCLS parameters, it appears that particles within a window $\Delta E/E \lesssim 0.3\%$ will lase [44]. Because of uncertainties in the emittance and energy spread within the horns of the beam, it is not clear what parts of these will lase. As for the region of beam between the horns, for copper the energy deviation is large compared to 0.3%, for aluminum it is smaller. It appears, according to FEL simulations, that more of the beam between the horns will lase in the case of aluminum than in the case of copper [44].

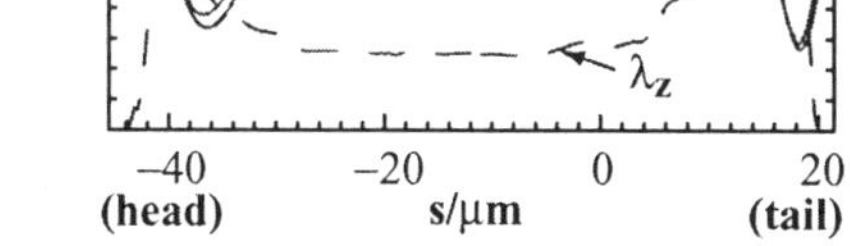

Figure 13. For LCLS parameters: the total energy variation induced within the 1 nC bunch in the 130 m long LCLS undulator beam pipe. Results are given for round beam pipes of copper and aluminum. Superimposed is the bunch shape, with the head to the left (the dashes).

The resistive wall wake of a flat beam pipe (rectangular with width much greater than height) has also been computed [45]. Keeping the vertical aperture of the LCLS undulator beam pipe fixed, the wake effect can be reduced by about 30% by going to a flat geometry (this is the chosen LCLS beam pipe shape, aluminum is the chosen surface material). Another effect that can be noticeable at high frequencies is the (room temperature) anomalous skin effect; here it appears to be a $\sim 15\%$ effect [45]. Finally, one may ask the question, What confidence do we have that the free-electron model is valid in the frequency range of interest to us, namely k in the range [0.05,0.5] μm^{-1}? Reflectivity measurements can be used to study this question. However, over our frequency range of interest, such measurements are difficult to do with sufficient accuracy, since the reflectivity of metals becomes very nearly 1; little data over this range is found in the literature. Recently, however, reflectivity measurements were performed at the synchrotron light source at Brookhaven on Cu and Al samples; the results are in reasonably good agreement with our assumed free-electron model [46].

4. Conclusion

We have discussed wakefields excited by short bunches in accelerators. In particular, we have reviewed some of what has been learned in the last 10–15 years concerning diffraction wakes, roughness impedance, coherent synchrotron radiation wakes, and the resistive wall wake. We have studied the importance of these wakes in the Linac Coherent Light Source (LCLS) project. The longitudinal accelerator structure wake in the SLAC linac is an important ingredient in the bunch compression process. Of the wakes in the undulator region of the LCLS, the dominant one is the resistive wall wake. To weaken its effect, the undulator beam pipe was chosen to be aluminum instead of copper, and of flat geometry instead of round.

Acknowledgements

The author thanks G. Stupakov for discussions about short-range wakes, P. Emma for discussions on the LCLS and its phase space manipulation. In addition, he thanks P. Emma for Fig. 1, A. Novokhatski for Fig. 2, and G. Stupakov for Figs. 5 and 7.

References

1. "LCLS Design Report," SLAC-R 593, 2002.

2. S. Chattopadhyay, editor, *Part. Accel.* **25**, 1990.
3. A. Chao, *Physics of Collective Beam Instabilities in High Energy Accelerators* (Wiley & Sons, New York, 1993).
4. R. Gluckstern, *Phys. Rev.* **D39**, 2780 (1989).
5. A. Novokhatski and A. Mosnier, in *Proceedings of the Particle Accelerator Conference, Vancouver, Canada, 1997* (IEEE, Piscataway, NJ, 1997), p. 1661.
6. K. Bane and A. Novokhatski, SLAC-AP 117, March 1999.
7. K.-Y. Ng, *Phys. Rev.* **D42**, 1819 (1990).
8. H. Henke and O. Napoli, in *Proceedings of the European Particle Accelerator Conference, Nice, France, 1990* (Editions Frontieres, Gif-sur-Yvette, France, 1991), p. 1046.
9. S. Heifets and S. Kheifets, *Rev. Mod. Phys.* **63**, 631 (1991).
10. E. Gianfelice and L. Palumbo, *IEEE Trans. Nucl. Sci.* **32**, 2, 1084 (1990).
11. K. Bane and M. Sands, *Part. Accel.* **25**, 73 (1990).
12. K. Bane and I. Zagorodnov, SLAC-PUB-11388, August 2005.
13. K. Yokoya and K. Bane, in *Proceedings of the Particle Accelerator Conference, New York, NY, 1999* (IEEE, Piscataway, NJ, 1999), p. 1725.
14. K. Bane, A. Mosnier, A. Novokhatski, K. Yokoya, in *Proceedings of the International Computational Accelerator Physics Conference, Monterey, CA, 1998* (SLAC, 1998) SLAC-R-580, p. 137.
15. K. Bane, M. Timm, T. Weiland, in *Proceedings of the Particle Accelerator Conference, Vancouver, Canada, 1997* (IEEE, Piscataway, NJ, 1997), p. 515.
16. A. Fedetov, R. Gluckstern, M. Venturini, *Phys. Rev. ST-AB* **2**, 064401 (1999).
17. K. Bane, SLAC-PUB-9663, March 2003.
18. K. Bane and P. Emma, in *Proceedings of the Particle Accelerator Conference, Knoxville, TN, 2005* (IEEE, Piscataway, NJ, 2005), p. 4266.
19. K. Bane, F.-J. Decker, J. Seeman, F. Zimmermann, in *Proceedings of the Particle Accelerator Conference, Vancouver, Canada, 1997* (IEEE, Piscataway, NJ, 1997), p. 1876.
20. K. Bane *et al*, in *Proceedings of the Particle Accelerator Conference, Portland, OR, 2003* (IEEE, Piscataway, NJ, 2003), p. 3126.
21. C. Barnes, Stanford University, PhD Thesis, SLAC-R-799, March 2006.
22. S. Kurennoy, *Phys. Rev.* **E55**, 3529 (1997).
23. K. Bane, C.K. Ng, A.W. Chao, in *Proceedings of the Particle Accelerator Conference, Vancouver, Canada, 1997* (IEEE, Piscataway, NJ, 1997), p. 1738.
24. G. Stupakov, *Phys. Rev. ST-AB* **1**, 064401 (1998).
25. G. Stupakov, R. Thomson, D. Waltz, R. Carr, *Phys. Rev. ST-AB* **2**, 060701 (1999).
26. M. Chatard-Moulin and A. Papiernik, in *Proceedings of the Particle Accelerator Conference, San Francisco, CA, 1979*; published as *IEEE Trans. Nucl. Sci.* **NS-26**, 3, p. 3523.
27. G. Stupakov, in *Proceedings of the Physics of, and Science with, the X-Ray Free-Electron Laser, Arcidosso, Italy, 2000* (Amer. Inst. Phys., Melville, NY, 2000), p. 141.

28. J.B. Murphy, S. Krinsky, R. Gluckstern, *Part. Accel.* **57**, 9 (1997), and in *Proceedings of the Particle Accelerator Conference, Dallas, TX, 1995* (IEEE, Piscataway, NJ, 1996), p. 2980.
29. Ya. S. Derbenev, J. Rossbach, E. L. Saldin, V. D. Shiltsev, DESY-TESLA-FEL 95-05, DESY, 1995.
30. R. L. Warnock and P. Morton, SLAC-PUB-4562, March 1998, and *Part. Accel.* **25**, 113 (1990).
31. R. Warnock, in *Proceedings of 4th Advanced ICFA Beam Dynamics Workshop, Tsukuba, Japan, 1990* (KEK, Tsukuba, Japan, 1990), p. 151, and in *Proceedings of the Particle Accelerator Conference, San Francisco, CA, 1991* (IEEE, Piscataway, NJ, 1991), p. 1824.
32. E. Saldin, *et al*, *Nucl. Inst. Meth.* **A398**, 373 (1997).
33. M. Dohlus, talk given at CSR-Workshop-2002 at Zeuten, Germany; see http://www.desy.de/csr/csr_workshop_2002.
34. K. Bane and A. Chao, *Phys. Rev. ST-AB* **5**, 104401 (2002).
35. A. Kabel, *et al*, *Nucl. Inst. Meth.* **A455**, 180 (2000).
36. R. Li, *Nucl. Inst. Meth.* **A429**, 310 (1998).
37. L. Giannessi, *Phys. Rev. ST-AB* **6**, 114802 (2003).
38. T. Agoh and K. Yokoya, *Phys. Rev. ST-AB* **7**, 054403 (2004).
39. M. Borland, *Phys. Rev. ST-AB* **4**, 070701 (2001).
40. P. Emma, private communication.
41. K. Bane and M. Sands, in *Proceedings of Micro Bunches Workshop Upton, NY, 1995* (Amer. Inst. Phys., Woodbury, NY, 1995), p. 131.
42. N. Ashcroft and N. Mermin, *Solid State Physics*, (Harcourt Brace Publishers, Orlando, FL, 1976).
43. E. Shiles *et al*, *Phys. Rev.* **B4**, 1612 (1980).
44. Z. Huang and G. Stupakov, in *Proceedings of the Particle Accelerator Conference, Knoxville, TN, 2005* (IEEE, Piscataway, NJ, 2005), p. 2059.
45. K. Bane and G. Stupakov, in *Proceedings of the Particle Accelerator Conference, Knoxville, TN, 2005* (IEEE, Piscataway, NJ, 2005), p. 3390.
46. K. Bane, G. Stupakov, J. Tu, SLAC/LCLS-TN-05-6, February 2005.

DIAMOND SECONDARY EMITTER

I. BEN-ZVI, T. RAO, A. BURRILL, X. CHANG, J. GRIMES, J. RANK, Z. SEGALOV, AND J. SMEDLEY

Brookhaven National Laboratory, Upton, New York

We present the design and experimental progress on the diamond secondary emitter as an electron source for high average power injectors. The design criteria for average currents up to 1 A and charge up to 20 nC are established. Secondary Electron yield (SEY) exceeding 200 in transmission mode and 50 in emission mode have been measured. Preliminary results on the design and fabrication of the self contained capsule with primary electron source and secondary electron emitter will also be presented.

1. Introduction

In recent years, there has been considerable interest in developing high brightness photo injectors with high average current, even up to 1 A, for variety of applications [1-3]. One of the key issues to be addressed for such injectors is the generation of the required high current reliably and reproducibly. The injector should be capable of delivering charge up to tens of nC, with a bunch length of a few to tens of ps at a repetition rate in the range of tens to hundreds of MHz. Furthermore, the transverse and longitudinal emittance of the electron beam should be very small. A typical approach has been to irradiate a suitable photocathode material with a laser in the presence of an electric field so that the emitted electrons can be quickly accelerated to relativistic velocities. The choices of the field, cathode, and laser are dictated by the application. The high field achievable in a superconducting RF (SRF) cavity makes it an attractive candidate for in CW operation. Metal photo cathodes have been used very successfully in low average current injectors. Although these cathodes have low quantum efficiency (QE) and require UV photons, they are rugged and inert and hence are the preferred cathode material in a number of injectors [4-6]. However, to deliver high average currents with the present state-of-the-art laser systems, the cathode should have higher QE than metals while preserving the ruggedness of the metal cathode. In addition, the cathode material should be relatively insensitive to the contaminants and should not contaminate its environment either, especially that of the SRF cavity. It should be capable of emitting high charge and current, have a fast response to laser pulse, a low

thermal emittance, long life and be simple to install. Since the requirements on the laser energy and power are directly linked to the QE of the cathode, higher the QE, more relaxed are the performance requirements on the laser.
Semiconductors such as alkali antimonide and Cs:GaAs have been investigated successfully and shown to yield a QE of >10% at visible and near IR wavelengths [7,8] . However, these cathodes require an ultra high vacuum, and a complicated load-lock system for installing into the RF injector. In addition, Cs, an element typically present in these cathodes, may contaminate the cavity over time and degrades its performance. In this paper we present the possibility of isolating these cathodes from the accelerating structure and simultaneously enhancing the current/charge more than one hundred fold by using a diamond layer between the cathode and the accelerating structure. Although the concept has been developed for photo cathodes and RF injectors, it can be easily adapted to other primary electron sources and accelerating structures.

2. Concept:

The schematic of the Secondary Emission Enhanced Photoemitter (SEEP) is shown in Figure 1.

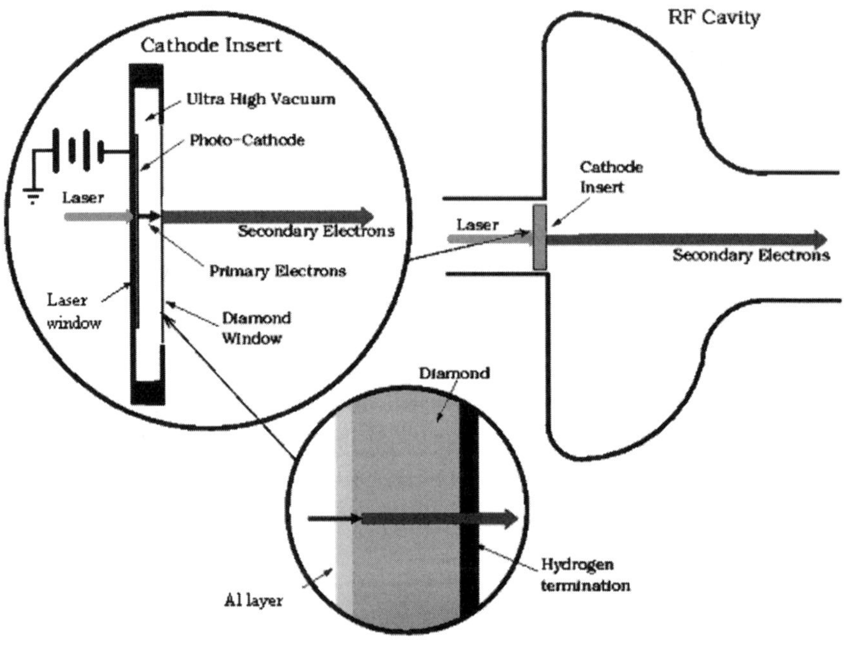

Figure 1. Schematic of the SEEP. See text for details

Figure 1a shows a cathode insert inside a RF injector. The cathode insert consists of a high QE primary photocathode, a diamond window, with an insulating spacer between the two, as shown in Figure 1 b. As shown in Figure 1c, the cathode side of the diamond window is coated with a thin metal layer while the cavity side is hydrogenated to result in a Negative Electron Affinity (NEA) surface. A voltage bias in the range of 5 to 10 kV is applied between the primary photocathode and the metal layer on diamond. The laser light passes through the substrate of the primary photocathode and irradiates it, releasing electrons. These electrons are accelerated by the applied field between the primary cathode and diamond, pass through the metal layer into the diamond, to create a shower of secondary electrons. These secondary electrons drift through the diamond in the presence of the RF field in the cavity to emerge from the NEA surface in to the RF cavity to be accelerated further. This scheme has a number of advantages, as listed below:

- Significant reduction in the number of primary electrons required to meet the current /charge requirement, resulting in a corresponding reduction in the laser power/energy requirement
- Increase in the life time of the primary photocathode due to reduction in current/charge delivered
- Isolation of the cathode from the cavity vacuum, increasing the life time of the primary cathode
- Isolation of the cavity from the cathode, preserving the performance of the cavity
- Low energy spread of the secondary electrons due to thermalization in diamond and emission from NEA surface
- Simplified transport of the capsule that can be exposed to air

3. Design criteria:

The design of the capsule is determined by the operating parameters of the injector and the properties of the diamond. Table 1 lists the properties of diamond relevant to this application.

Table 1 Diamond properties relevant to this application

Crystal structure	Diamond Best orientation for NEA is (111)
Number of atoms in 1 cm^3	$1.764*10^{23}$
Density	3.515 g/cm^3
Dielectric constant (10^2-10^4 Hz)	5.7
Lattice constant	3.567A
Breakdown field	10^8-10^9 V/m
Mobility electrons	≤ 2200 $cm^2V^{-1}s^{-1}$
Mobility holes	≤ 1800 $cm^2V^{-1}s^{-1}$
Electron thermal velocity	~10^5 m/ s
Melting point (@ p=125kbar)	4373 °C
Specific heat	0.52 J g^{-1} ° C^{-1}
Thermal conductivity	6-20 W cm^{-1} °C^{-1}
Thermal expansion, linear	$0.8 \cdot 10^{-6}$ °C^{-1}
Saturated electron drift velocity	$2.7 \cdot 10^7$ cm/s

Diamond has ideal properties for this application. SEY higher than 100 has been measured [9] in the reflection mode. The ability to form NEA surface helps both in the high SEY and the narrow energy distribution of the secondary electrons [10]. The high breakdown field, mobility and saturation velocity of diamond are very important in high field RF environment. The high thermal conductivity helps to dissipate the heat generated by the RF, primary and secondary electron current. With the strength of diamond, it is possible to form free standing structures capable of withstanding a one atmosphere pressure differential across a diamond as thin as 10 micron. For our application, in order to minimize RF heating and reduce charge trapping, we have chosen undoped preferably large grain size (> 10 micron or single crystal diamond with impurity concentration << 1 ppm. (111) surface of diamond results in the best NEA surface and high electron emission, however, as we will show later, other orientations work as well. In the following section, calculations used to determine the dimensions of diamond to deliver 0.200 -1 A current and dissipate the heat generated in the process will be discussed.

Dimensions of the diamond window:

The thickness is determined by the transport of the electrons through the diamond and phase requirement in RF cavity, the strength of the window for holding pressure during installation as well as the ability to dissipate the heat generated in the process. The calculations below are directed towards two applications: electron cooling of ion beams at RHIC (RHIC II) and the ERL project at BNL.

For the ERL, the electron beam parameters are, maximum charge of 1.4 nC at repetition rate of 352 MHz, into a 703 MHz cavity (high current case). For magnetized electron cooling of RHIC [2], the electron beam parameters are, maximum charge of 20 nC (high charge case) at a repetition rate of 9.4 MHz emitted in to the 703 MHz cavity at 35 deg. Since the secondary electrons generated at the cathode side need to drift through the diamond in the presence of the RF field and be emitted in to the cavity at the appropriate phase, they should be generated at ~ 5 degrees, with drift time of ~ 30 degrees, corresponding to 120 ps for 703 MHz. The electron drift velocity in diamond under high external electric field has been experimentally measured [11]. The data shows that when the external field is larger than 2MV/m the electron drift velocity is saturated at about v_D=2.7x10^{7}cm/s and is independent of temperature. This means that the secondary electron bunch drift velocity in diamond is almost constant except when the charge density is very high and the space charge force is comparable to RF field. The optimum diamond thickness for our example is estimated to be:

$$t_{Dmd} < 120 ps \times v_D = 32 \mu m \quad (1)$$

The diameter of the diamond is determined by the required current density, emittance and capability to dissipate the heat generated. Optimization of the current density and emittance dictates the transverse dimension of the electron beam to be ~ 5 mm for 1.4 nC and ~ 7 mm for the 5 nC charge. The heating of the diamond is caused by a number of sources: the power dissipated by the primary electrons, power dissipated by the secondary electrons passing through the diamond in the RF field, resistive heating by the electron replenishment current flowing through the metal film, power dissipation by the RF currents through the aluminum film and motion of charge carriers induced by impurities. Table I and II list the contribution from each of the effects.

Table I. Thermal contribution for high charge case

R	10mm	12.5mm	15mm
Primary	6.3 (W)	6.3(W)	6.3(W)
Secondary	7.6(W)	7.6(W)	7.6(W)
RF	7.5 (W)	20.0(W)	48.6(W)
Replenishment	0.042(W)	0.046(W)	0.054(W)
Total	21.4 (W)	33.9(W)	62.5(W)

Table II. Thermal contribution for high current case

R	2.5mm	5mm	7.5mm
Primary	33.3(W)	33.3(W)	33.3(W)
Secondary	40.2(W)	40.2(W)	40.2(W)
RF	0.046(W)	0.67(W)	3.4(W)
Replenishment	0.025(W)	0.023(W)	0.023(W)
Total	73.6(W)	74.2(W)	76.9 (W)

A code was written to simulate the temperature distribution across the diamond due to these four contributions, taking into account, the correct temperature dependent material properties, such as thermal conductivity and resistivity. The intermittent nature of the current is accounted for by considering peak and average current effects. In each of the simulations, the edge of the diamond was assumed to be clamped at a given temperature and the transverse temperature distribution from the center (R=0) to the edge was simulated. The results of the simulation for different radii and thickness of diamond, thickness of metallic film and edge temperature are shown in figures 2-7. The electron beam parameters for the high charge case are: charge 20 nC, PRF 9.4 MHz, RF peak field 15 MV/m, electron bunch length 40 ps, primary electron energy 10 keV, and SEY 300. For the high current case, the charge is 1.42 nC and PRF 703 MHz while other parameters remain the same. For subsequent measurements and capsule design, the physical dimensions of diamond were chosen to be 30 μm thickness and 10 mm diameter with ~ 7 mm diameter active area for electron emission.

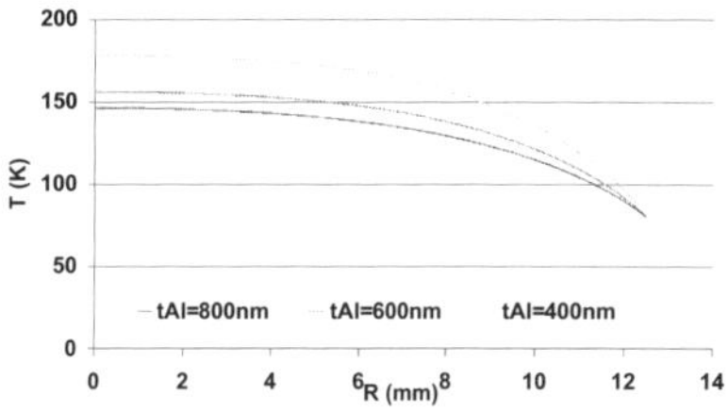

Figure 2 Temperature distribution for different Al thickness. R=12.5 mm, High Charge regime

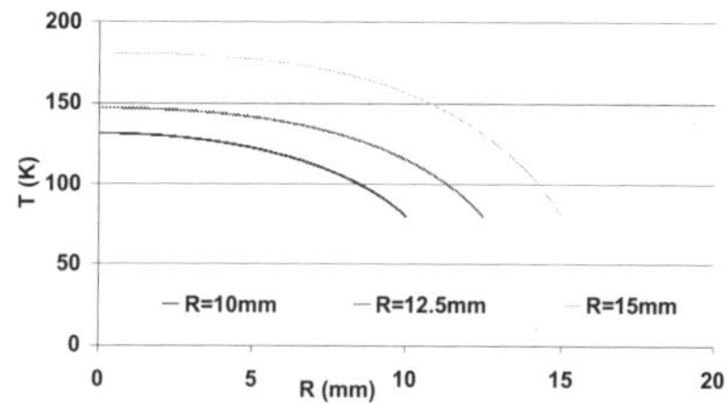

Figure 3. Temperature distribution for different radii of diamond. Al thickness is 800 nm, high charge regime

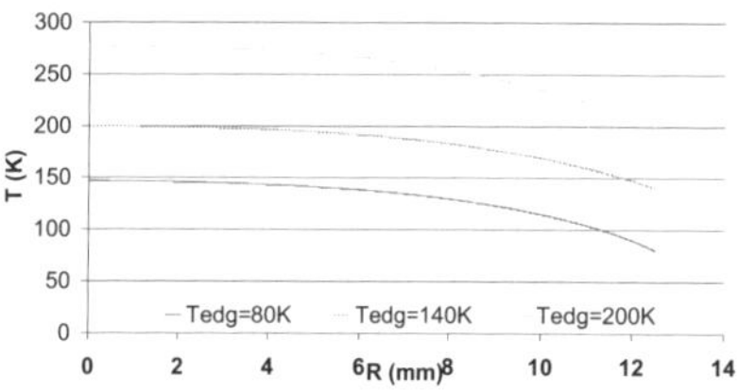

Figure 4 Temperature distribution for different edge temperatures. Al thickness 800 nm, R=12.5 mm, high charge regime

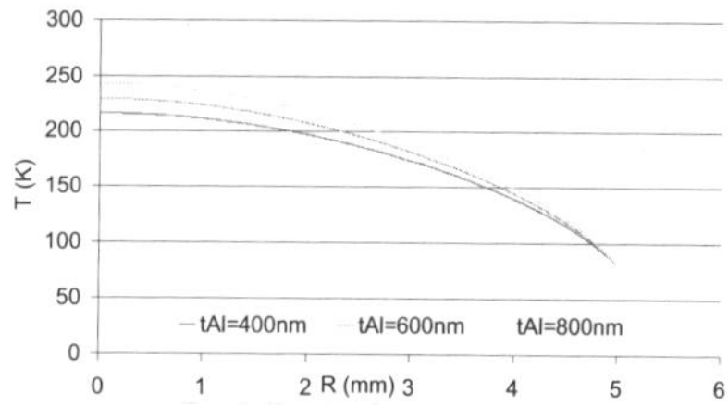

Figure 5 Temperature distribution for different Al thickness. R= 5mm, High current regime

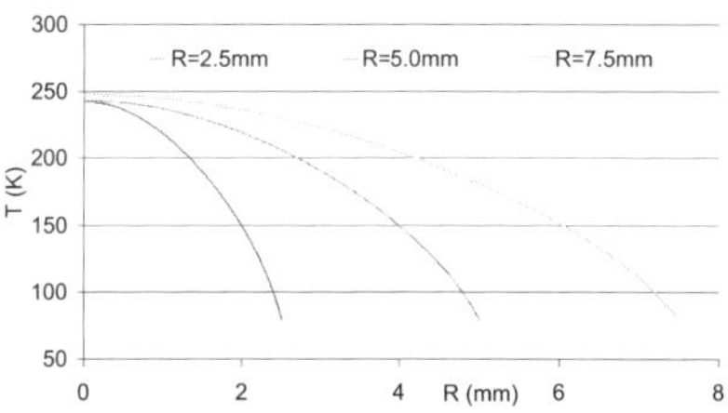

Figure 6. Temperature distribution for different radii. Al thickness 800 nm, high current regime

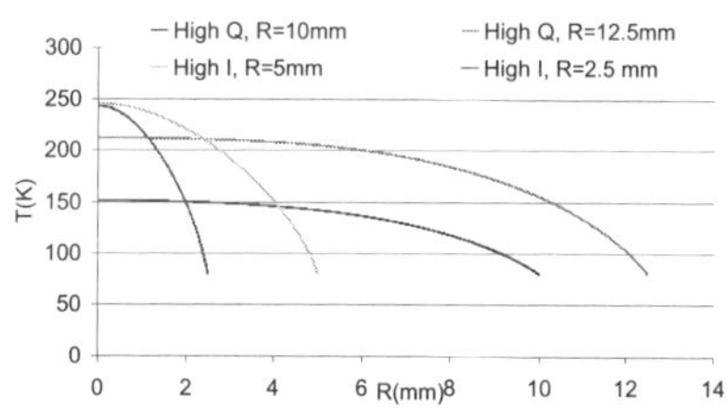

Figure 7. Temperature distribution for 30 nm Au for different radii, high charge(Q) and high current(I) regime

Metal layer:

Since the diamond is chosen to be undoped, a conducting layer on the cathode side of the diamond is needed to provide the replenishment current. Careful analysis and experiments of thin metal films on dielectric material at the cathode center of the RF cavity [12] indicate that even with a metal layer as thin as 10 nm, the RF penetration through the film is small and the metal acts rather as the return path for the RF current. Furthermore, since the primary electrons need to pass through this layer before impacting the diamond, energy loss of the primary electrons in this layer not only adds to the heating effect, but also reduces the SEY. The material and its thickness should be chosen to minimize the energy reduction of the primary electrons as well as the resistance. Other considerations are the ability to make good electrical and thermal the contact with diamond, ruggedness and compatibility with fabrication process and the primary electron source. Presently aluminum, gold and Ti/Pt/Au are being considered with the thickness ranging from 800 Å to 300 Å.

Hydrgenation:
For efficient electron emission, the cavity side of the diamond needs to be an NEA surface. It has been shown that cesiated or hydrogenated diamond [13] result in NEA surface with < 1 eV and ~ 0.3 eV negative affinity respectively. Since the presence of cesium is not desirable in a SRF cavity, for our application, the diamond will be hydrogenated to provide the NEA surface.

Primary Cathode:
Although the concept is valid for any primary electron source, our application requires short electron bunches with tens of pC and up to a few mA average current. Semiconductor photo cathodes such as CsK_2Sb [7, 14] and GaAs:Cs [8] have shown to meet these requirements. In subsequent design considerations, CsK_2Sb will be considered as cathode of choice, mostly because of its prompt emission.

4. Secondary Electron Beam Quality:

Unlike in a conventional RF injector, in SEEP, the beam quality is modified by passage of the electrons through the diamond. The energy distribution and bunch length are modified by collisions with other electrons and phonons. Monte Carlo simulations [15] of electron transport in the presence of high electric field in the diamond indicate that for the fields of ~10 MV/m, fields anticipated in our application, the average energy of the electrons is in the range of 100 meV and the drift velocity saturates to ~ 10^5 m/s. Hence if the space charge field does not reduce the RF field significantly, a condition easily met in our application, the electron bunch length would not change during the passage through the diamond. However, since the secondary emission process is statistical, a fraction of the primary electron may straggle [16] over a distance of 0.4 micron, giving rise to a bunch lengthening of ~ 2.6 ps. This value can be reduced by lowering the primary electron energy, at a price of reduced amplification. Hence the electron beam emitted from the diamond is expected to have a very low emittance.

5. Experiments:

The primary goals of the experiments were to investigate a number of diamond samples to determine the critical parameters of the diamond, establish sample preparation and hydrogenation processes, measure the SEY in transmission and emission mode, and investigate the critical steps towards the fabrication of the capsule.

Sample preparation:

Chemical treatment:
All the samples were chemically treated prior to measurement to remove metallic and organic impurities from the surface. The process is as follows:
15 minutes in fuming solution of saturated CrO_3 in H_2SO_4
2 minutes in NH_3OH in ultrasonic bath
2 minutes in HCl: H_2O (1:10) in ultrasonic bath
5 minutes in boiling H_2SO_4: H_2O_2 (5:1)
2 minutes in boiling NH_3OH: H_2O_2 (1:1)
2 minutes in boiling HCl:H_2O_2: H_2O (1:1:4)
5 minutes in DI water in ultrasonic bath
Each step is followed by DI water rinse, and the final step is drying by nitrogen gas jet.
Immediately after chemical treatment, 300 Å -500 Å Al was evaporated on either side of the sample for transmission measurements. For emission measurements, the samples were coated with 300 Å of gold on one side and hydrogenated on the opposite side. Care was taken to minimize the time the sample was exposed to air.

Hydrogenation:
The photograph of the chamber used for hydrogenation is shown in Figure 8. The chamber consists of a stainless steel sample holder, a tungsten filament, and a leak valve to introduce molecular hydrogen into the chamber. The sample holder and tungsten filament can be heated electrically to 900 C and 1800 C respectively. The temperature was measured using a calibrated pyro viewer and the pressure with the cold cathode gauge. The chamber is also equipped with a fused silica window for viewing and illuminating the sample. An anode suitably biased can also be positioned in front of the sample to enable photoemission measurements from the sample.
After metallization, the sample was inserted in the hydrogenation chamber and pumped till the base pressure is in the range of 1-3 nTorr. The sample holder and the sample are then heated to 800 C for 30 minutes in order to clean the sample as well as break the C-O bonds on the surface in readiness for hydrogenation. Once the sample is cooled and the vacuum is returned to nTorr level, 2 μTorr of

H_2 is introduced in to the chamber and the tungsten filament is heated to 1800 C for 15 minutes. The interaction between the hot filament and the molecular hydrogen gas releases atomic hydrogen that attaches to the free dangling carbon bonds on the surface of the diamond, thereby hydrogenating it and generating a NEA surface. The sample is then removed from the hydrogenation chamber and installed in the Secondary Electron Yield (SEY) test chamber.

Figure 8. Photograph of the hydrogenation cell

Experimental arrangement:

Two sets of measurements, SEY in transmission mode and SEY in emission mode, were made on the samples. For the transmission measurements, the sample with metallic coating on both the polished faces was sandwiched between two stainless steel holders (electrodes) that are electrically isolated from each other by a kapton layer. Each of these holders makes contact with the metallic coating on its side of the diamond and can be biased up to 5 kV. The holder on the primary side has a small hole to let the primary electron through. The signal of the primary current arriving on one side of the diamond and the secondary current leaving the opposite side of the diamond are fed to an oscilloscope. Current due to either electron or hole as the charge carrier can be measured by using bias of appropriate polarity on the electrodes. The schematic of the arrangement is shown in Figure 9. For emission measurements, only the primary side of the diamond is metalized. The opposite face is hydrogenated. The secondary electrode is not in contact with the diamond, but is positioned away from it. The width of the vacuum gap is determined by the thickness of the diamond and the voltage applied such that the field across the diamond is sufficient to transport the electrons through it.

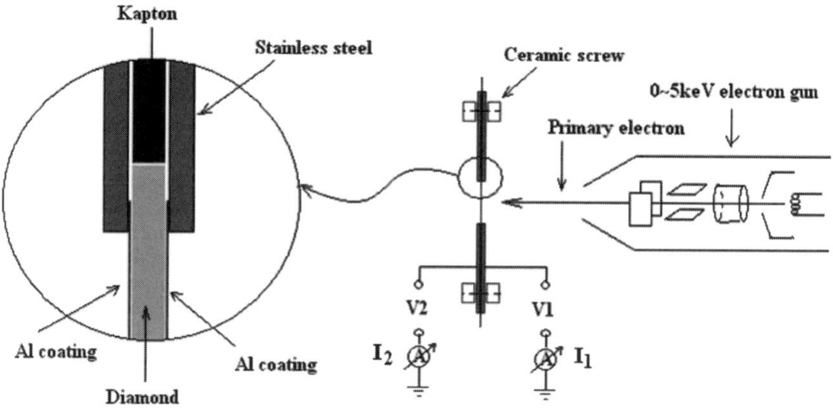

Figure 9. Schematic of the experimental arrangement. For the transmission measurements, both the surfaces of the diamond are coated with Al. The metal electrodes contact the Al coating on both sides. For emission measurements, the surface facing the electron gun is coated with Au and the opposite surface is hydrogenated. The electrode on primary side is in contact with Au and the electrode on secondary side is isolated from diamond by the vacuum gap.

The sample in the holder is mounted on a linear manipulator and installed in a vacuum chamber equipped with an electron gun and energy analyzing system. For the measurements described below, the energy analyzing system was not used. The electron beam from the gun is used as the primary electron and has the following parameter ranges: electron energy-1-5 keV, current 10 nA-10 μA, radius of the focused beam is estimated to be ~ 150 μm. The energy of the electrons entering the diamond is reduced by the loss in the metallic layer and increased by the bias on the primary electrode.

Experimental Results:

A number of samples from different vendors and different crystalline properties were tested. Since the crystal dimensions were different, for each of the samples, even though the general lay out of the sample holder remained the same, the size of the hole in the holder on the primary side was varied to accommodate the varied sample sizes. Table III lists the sample parameters, experimental arrangement and results.

Table III. Specifications and the performance of different diamond samples tested

Vendor	Sample parameters	Ranges of Experimental parameters	Results
Element-Six Diafilm OP	CVD Poly crystalline, 6 mm diameter 0.2 mm thick, central 4 mm coated with Al, 20 nm on primary side and 50 nm on secondary side. 2.5 mm hole in the primary electrode	Primary current: 10 nA-5 μA, Energy 1 – 6 keV. Bias across diamond to 5 kV	No electron or hole current
Apollo Diamonds	CVD single crystal, Type II a, 6 mm diameter, 0.2 mm thick central 4 mm coated with Al, 30 nm on primary side and 50 nm on secondary side. 2.5 mm hole in the primary electrode,	Primary current: 100 nA to 8.5 μA Electron energy 1-8 keV Field across diamond : 10-20 MV/m	SEY Transmission mode up to 200 SEY decreases with increasing primary current indicating saturation of secondary current. Neither the rise and fall times were prompt. Response time was in the order of minutes. Gain vanished when sample was cooled to 80 K
Element-Six detector grade	CVD, single crystal, low impurity concentration, 4 mm square with 0.25 mm thick	Primary current 50 nA Energy: 4 keV Field across diamond up to 6 MV/m	SEY transmission mode electron current very high, strong dependence on the field. SEY transmission mode hole current ~30, nearly linear dependence to field
UralAlmazInvest	Natural Type II a single crystal (110)	Data shown below	

	3x2.64x0.16 mm^3, 60 ppm nitrogen, central 2 mm coated with 30 nm thick Al, 1.5 mm hole in the primary electrode for transmission measurements	
Element-Six Diafilm EI (electronic grade)	CVD Polycrystalline, low impurity concentration, 10 mm diameter, 0.45 mm thick, one side coated with 30 nm Au, other side hydrogenated	Data shown below

Measurements with natural diamond:
The natural diamond sample (UralAlmazInvest) is single crystal (110) with physical dimensions of 3x2.64x0.16 mm^3 both sides polished to an optical quality finish. The vendor specifications indicate the nitrogen concentration to be 60 ppm and other impurity levels to be below measurable level. Optical and scanning electron microscope studies of the sample indicate very good surface quality. X ray spectroscopy confirms this sample to be single crystal (110).
For transmission measurements, the central 2 mm of chemically treated sample was coated with 30 nm Al. The secondary electron yield was measured as a function of primary electron energy, current and the field in the diamond at 300 K and at 80 K. The results are shown in Figures 10 and 11. Maximum gain of 220 has been measured with primary electrons of 5 keV energy. The difference in the maximum gain between the 300 K and 80 K measurements is due not to the temperature, but to the spatial variation typically found in the natural diamond. The gain has a very small dependence on the primary current, the cause of it still under investigation. The observed threshold on the primary electron energy is due to the energy loss in the Al layer. At both the temperatures, the gain shows a very steep increase at low fields and a lower rate of increase at higher fields. If we define the gain value at this knee point as the maximum SEY for this primary electron energy, the SEY is linearly dependent on the primary electron energy as expected, with an intercept of ~ 2 keV and a slope of 14. The intercept is a measure of the energy lost in the Al layer and agrees well with the calculated energy loss in 30 nm of Al. The slope is the

energy required to form an electron hole pair in diamond and is consistent with other reports [17].

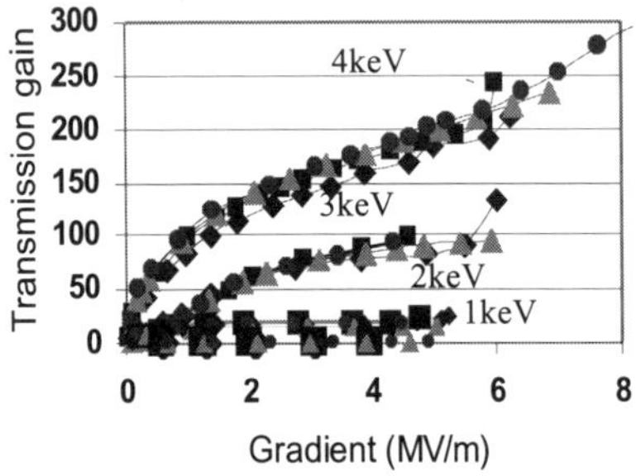

Figure 10. Gain measurements for different primary currents and energies in transmission mode at 300 K

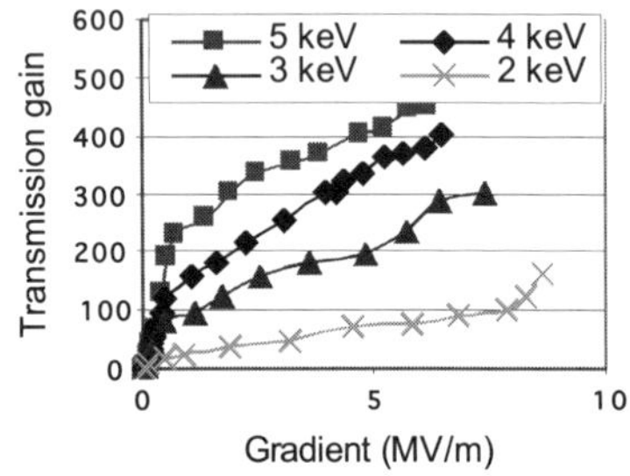

Figure 11. Gain measurements at different energies in transmission mode at 80 K

For emission measurements, the center of one of the surfaces was coated with 5 nm of Cr followed by 15 nm of Au and the opposite face was hydrogenated. The metalized surface was in contact with the primary electrode. The secondary electrode was electrically isolated from the sample through a 0.025 mm vacuum gap. Care was taken to shield the metal with kapton to prevent arcing. The gain was measured as a function of primary electron energy, current and the field on diamond. The gain curve for 50 nA current for different electron energies is shown in Figure 12. A maximum gain of 38 was measured, with an electron energy of 5 keV and a field of 1.7 MV/m. The gain is limited by the maximum field that can be applied to the sample. Although this demonstrated gain is lower than that measured in the transmission mode, it is already sufficient to lower the required laser power significantly. All the gain curves have a threshold field of ~ 1 MV/m, the origin of which is still under investigation.

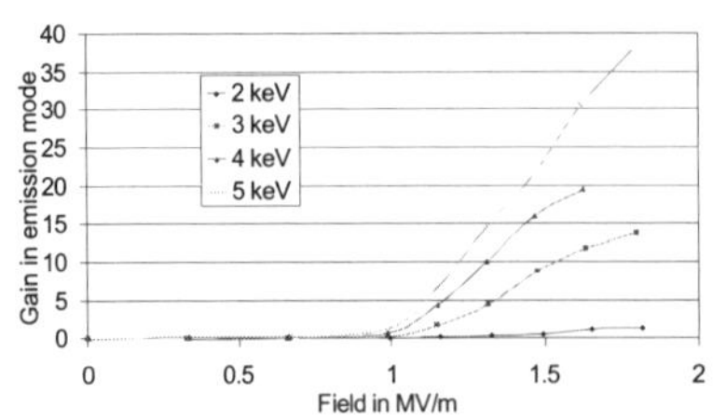

Figure 12. Gain measurement from natural diamond, in emission mode for different primary energies, with 50 nA primary current at 300 K

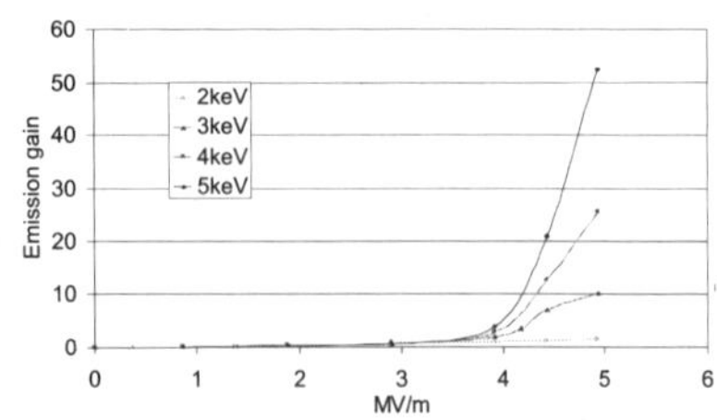

Figure 13. gain measurements on CVD polycrystalline sample for primary current of 100 nA at 300 K different primary energies

Measurements with CVD polycrystalline diamond:
The diamond sample used for these measurements was a polycrystalline (Diafilm EL from Element-Six) sample, 10 mm diameter and 0.45 mm thick, both sides polished to an optical quality finish. The vendor specification indicates that all the impurity levels are below measurable limit. X ray crystallography of similarly grown, unpolished sample shows presence of (110) and (111) orientation, although the ratio is expected to change during polishing, owing to the higher hardness of the (111). After chemical treatment, one of the polished surfaces was coated with 30 nm of Au and the other surface was hydrogenated. Experimental arrangement was similar to that of the natural diamond, except for a larger vacuum gap of 0.088 mm. Gain in the emission mode as a function of primary current and energy are shown in Figures 13. Gain as large as 50 has been measured with primary electron energy of 5 keV. As in the natural diamond, the gain curves have a threshold field, but of a larger magnitude (~ 4 MV/m) in this sample.

6. Capsule Fabrication:

In SEEP, a highly sensitive primary cathode and a specially prepared diamond need to be integrated to form a capsule that can be seamlessly installed in the (SRF in our application) injector. The first step towards the design and fabrication of such a capsule is investigating the techniques to bond diamond to Nb which can then be installed in the SRF cavity. Brazing was chosen to be bonding process and the first diamond was brazed to Nb successfully by a commercial vendor (Delaware Diamond Knives). Subsequently, a number of diamond samples were brazed to Nb in house, in a vacuum oven and reproducible brazing technique has been established. Presently we are in the process of polishing and thinning the brazed diamond to the required dimension and finish. In parallel, a capsule with metal photocathode is being designed and built to test the entire fabrication process.

7. Conclusion:

Theoretical calculations indicate that SEEP using diamond as a secondary emitter is a very promising technique to achieve very high average current, high quality electron beams. Electron beams with charges up to 5 nC and current up to 0.5 A can be delivered from ~ 30 micron thick diamond coated with 800 nm Al from a radius up to one cm. Heat generated during the process can be easily dissipated by cooling the edges to liquid nitrogen temperatures. Our experimental results show SEY as high as 37 and 52 in the emission mode from natural single (110) crystal and high purity CVD poly crystal respectively. The SEY measurements on different samples indicate that the magnitude of the SEY depends strongly on both the quality of the crystal and on the impurity level. The mechanisms leading to high threshold fields for the gain and the role of impurities, crystal grain boundaries and the cause for lower SEY in emission mode in comparison to the transmission mode are under investigation. Future plans include investigating the temporal response of SEY, and demonstration of higher currents, measuring secondary electron energy distribution, and establishing good metal/diamond interface and testing in RF cavity. Diamond has been brazed successfully to Nb numerous times and efforts are underway to fabricate the capsule and test the emission from the capsule.

Acknowledgments

The authors would like to thank J. Walsh and D. Pate for their expert technical assistance. This manuscript has been authored by Brookhaven Science Associates, LLC under Contract No. DE-AC02-98CH10886 with the U.S. Department of Energy. The United States Government retains, and the publisher, by accepting the article for publication, acknowledges, a world-wide license to publish or reproduce the published form of this manuscript, or allow others to do so, for the United States Government purposes.

References

1. Litvinenko V. N. , Beavis D., et al. "High Current Energy Recovery Linac at BNL", Proc. of PAC 05, Knoxville TN, p 2242
2. Ben-Zvi I. , Brennan J. , et al. "R&D towards cooling of the RHIC Collider" NIM A 532 (2004) 177–183
3. Workshop on Photo-injector for Energy Recovery Linac, January 22 & 23, 2001 Brookhaven National Laboratory , Upton, NY, http://www.nsls.bnl.gov/organization/Accelerator/highlights/workshop/PERL%20Injector%20Workshop%20Proceeding.pdf
4. Srinivasan-Rao T. , Schill J., Ben-Zvi I. , and Woodle M. , "Sputtered Magnesium as a photocathode Material for RF Injectors" Review of Scientific Instruments, 69, (1998), pp 2292-2296
5. http://www-ssrl.slac.stanford.edu/lcls/parameters.html
6. Suberlucq G., Proc. Of EPAC 2004, Lucerne, Switzerland, (2004) p 64
7. Dowell D., Bethel S.Z., Friddell K. D., " Results from the average power laser experiment photocathode injector test" NIM A 356, (1995),167
8. Rao T. et al. "Photocathodes for the energy recovery Linacs" NIM A 557 (2006), Pp 124-130
9. Yater J. E. and Shih A. ,"Secondary electron emission characteristics of single-crystal and polycrystalline diamond", J. Appl. Phys., Vol. 87, No. 11, 8103, 1 June 2000
10. Yater J. E. , Shih A., Butler J. E. Pehrsson P. E. " Transmission of low energy electrons in boron doped nanocrystalline diamond films" J. Appl. Phys. Vol. 93, (2003) Pp 3082-3089
11. Nebel C E and Stutzmann M *Properties, Growth and Applications of Diamond* (Emis Data Reviews Ser. 26) ed M H Nazare and A J Neves (London: INSPEC)(2001) p 40
12. Zhao Y. , Ben-Zvi I. , et al. "The Penetrability of a Thin Metallic Film Inside the RF Field", Proc. Of PAC 05, Knoxville TN p 3073

13. Minoru Niigaki, Toru Hirohata, et al., "Extremely High Quantum Photoyield from Cesiated Polycrystalline Diamond Films", Jpn. J. Appl. Phys. Vol. 37 (1998) pp. L1531-L1533
14. Burrill A., Ben-Zvi I., et al. "Multi-Alkali Photocathode Development at Brookhaven National Lab for Application in Superconducting Photoinjectors", Proc. Of PAC 05, Knoxville TN, p 2672
15. Watanabi T., Teraji T., Ito T., " Monte Carlo simulations of electron transport properties of diamond in high electric fields using full band structure", J. Appl. Phys. Vol. 95 (2004) p. 4866
16. Chuan-Jong Tung, "CSDA Ranges of Electrons in Metals", Chiinese Journal Of Physics, Vol. 17, no. 1, Spring 1979
17. Pernegger H. Roe S. et al. "Charge carrier properties in synthetic single crystal diamond measured with the transient-current technique" J. Appl. Phys. 97 (2005) p 073704

PARAMETRIC OPTIMIZATION FOR AN X-RAY FREE ELECTRON LASER WITH A LASER WIGGLER

R. Bonifacio[1,3,*], N. Piovella[1,2], M.M. Cola[1,2]

[1] *INFN-Sezione di Milano, via Celoria 16, 20133 Milano, Italy*

[2] *Dipartimento di Fisica, Università degli Studi di Milano,via Celoria 16, 20133 Milano, Italy*

[3] *Centro Brasileiro de Pesquisas Fìsicas, Rio de Janeiro -Brasil*

ABSTRACT

In this paper we optimize the experimental parameters to operate a Free Electron Laser with a laser wiggler in the Angstrom region. We show that the quantum regime of the Self Amplified Spontaneous Emission (Quantum SASE) may be reached with realistic parameters. The classical SASE regime is also discussed and compared with the quantum regime.

PACS: 41.60.Cr; 42.50.Fx

* *e-mai address:* rodolfob@cbpf.br.

INTRODUCTION

It has been shown that the quantum effects in a Free Electron Laser (FEL) are ruled by the quantum FEL parameter $\overline{\rho} = \rho(mc\gamma/\hbar k)$ [1], where ρ is the classical FEL parameter [2]. The classical analysis is valid only for $\overline{\rho} >> 1$, whereas for $\overline{\rho} < 1$ the quantum effects dominate [3,4]. In particular, in the quantum Self Amplified Spontaneous Emission (SASE) mode operation, the quantum purification of the radiation spectrum has been predicted [3,4], i.e. the broad and chaotic spectrum predicted in the classical SASE [5] and observed experimentally [6] shrinks to a very narrow spectrum when $\overline{\rho} << 1$.

It has been suggested that a quantum SASE FEL could be constructed using a laser wiggler [7,8] in a Compton backscattered configuration, instead of the static wiggler used in the current classical SASE experiments [6]. In a laser wiggler configuration, a low-energy electron beam back scatters the photons of a counter-propagating high power laser, with a frequency up-shifted by a factor $4\gamma^2$. The use of a laser wiggler has been discussed in the past by Gallardo et al. [9] in a classical theory .

In the following, we propose a way to optimize the experimental parameters for an X-ray FEL with a laser wiggler, showing the main differences between the parameters necessary to operate in the classical and in the quantum regimes. The analysis shows that the quantum regime appears, in general, more feasible than the classical regime for the state-of-art of electron beams and lasers technology. We stress that only in the quantum SASE regime a temporally coherent X-ray source could be realized, contrarily than in the classical SASE regime whose chaotic spectrum is temporally incoherent. Furthermore it is clear that a quantum FEL with a laser wiggler not only would be coherent but two or three orders of magnitude smaller in size (possibly table-top) and cost. The results of the present parametric study are rather encouraging with respect to a future realization of a quantum SASE X-ray FEL source.

PARAMETRIC OPTIMIZATION

For an FEL lasing at the wavelength λ_r with a laser wiggler with a wavelength λ_L and wiggler parameter a_0, the resonance condition reads:

(1) $$\gamma = \sqrt{\frac{\lambda_L(1+a_0^2)}{4\lambda_r}}.$$

which differs from the usual FEL resonance by a factor two, since in a laser wiggler the static wiggler period λ_w is replaced by $\lambda_L/2$. In fact, as it is well known, the exact resonance conditions in the static and laser wiggler are

(2) $$\lambda_r = \lambda_w \frac{1-\beta_\parallel}{\beta_\parallel} = \lambda_L \frac{1-\beta_\parallel}{1+\beta_\parallel},$$

which implies, for $\beta_\parallel \sim 1$,

(3) $$\lambda_w = \lambda_L \frac{\beta_\parallel}{1+\beta_\parallel} \approx \frac{\lambda_L}{2}.$$

The quantum FEL parameter $\bar{\rho}$ [1] is related to the classical FEL parameter ρ [2] by

(4) $$\rho = \bar{\rho}\frac{\lambda_C}{\gamma\lambda_r} = \bar{\rho}\frac{2\lambda_C}{\sqrt{\lambda_r\lambda_L(1+a_0^2)}},$$

where $\lambda_C = h/mc = 0.024$ Å is the Compton wavelength and we used Eq.(1) to eliminate γ. In Eq.(4) ρ is given by

(5) $$\rho = \frac{1}{2\gamma}\left[\frac{I}{I_A}\left(\frac{k\lambda_L a_0}{4\pi\sigma}\right)^2\right]^{1/3}.$$

where $I_A \approx 17\,\text{kA}$ is the Alfven current. Assuming as current density the peak current I divided by $2\pi\sigma^2$ or $\pi\sigma^2$, the parameter k is 1 or $\sqrt{2}$ for a transversally gaussian or for a flat top shape of the current, respectively, where σ is the beam radius. Eq. (5) is a generalization of the usual expression to a laser wiggler (see eq. (3)).

From (5) and (6) with some algebra we obtain

(6) $$I = 3.10^2 \frac{\bar{\rho}^3\sigma^2}{k^2\lambda_r^3\lambda_L^2 a_0^2}$$

The units, from now on, will be λ_r(Å), λ_L(μm) and σ(μm). We note that the electron current is proportional to $\bar{\rho}^3$, so that, going from the quantum to the classical regime, if $\bar{\rho}$ increases for instance by a factor 10, the current increases by a factor 10^3. This is the general reason why the use of a laser wiggler may be much more convenient in the quantum regime ($\bar{\rho} < 1$) than in the classical regime $\bar{\rho} >> 1$.

The relation between a_0 and the laser power P, in agreement with [9], is

(7) $$P(TW) = \left(\frac{Ra_0}{2.4k\lambda_L}\right)^2$$

where R is the minimum radius of the laser and P is in TW and k is 1 or $\sqrt{2}$ for a transversally gaussian (with beam section $2\pi R^2$) or for a flat top (with beam section πR^2) profile of the laser, respectively. From eqs. (6) and (7), we obtain the important relation between the electron current and the laser power:

(8) $$I(A) \approx 50\bar{\rho}^3 \frac{R^2\sigma^2}{k^4\lambda_L^4\lambda_r^3}\frac{1}{P(TW)}$$

As shown in ref.[7], the gain length and the cooperation length can be written in the form

(9) $$L_g = \frac{\lambda_L}{8\pi\rho}\sqrt{\frac{1+\bar{\rho}}{\bar{\rho}}}, \qquad L_c = \frac{\lambda_r}{8\pi\rho}\sqrt{\frac{1+\bar{\rho}}{\bar{\rho}}}$$

where the factor $\sqrt{1+\bar{\rho}}$ in the numerator has been added by hand to obtain the classical expression when $\bar{\rho} >> 1$ and the quantum expression when $\bar{\rho} << 1$. Note that eq.(9) has a factor 8π instead of 4π in the denominator, since a laser wiggler is assumed (see eq.(3)).

Since the interaction length L_{int}, given by the time duration τ of the laser pulse, is approximately twice the Rayleigh range Z_L of the laser, and requiring that it is larger than the gain length L_g, we can write the following relations:

$$L_{int} = c\tau \approx 2Z_L = a_1 L_g \ , \ a_1 \geq 1, \tag{10}$$

where

$$Z_L = \frac{4\pi R^2}{\lambda_L}. \tag{11}$$

Hence the total energy of the laser pulse is given by

$$U = P\tau = a_1 P \frac{L_g}{c}. \tag{12}$$

From Eqs. (10) and (11), we obtain the following self-consistent value of the laser rms radius at the focus:

$$R = \sqrt{\frac{a_1 L_g \lambda_L}{8\pi}}. \tag{13}$$

Note that the power gain length is half of the value given by (9). So, if for instance a_1=5, then the interaction length is ten times the power gain length.

Concerning the requirement on the emittance, a very important geometrical matching condition is the following

$$\beta^* \equiv \frac{\gamma\sigma^2}{\varepsilon_n} \geq Z_L, \tag{14}$$

where ε_n is the normalized beam emittance. Eq. (14) imposes that the electron beam is contained in the laser beam, provided $\sigma \leq R$, and that the electron beam does not diverge appreciably in a Rayleigh range Z_L. From Eqs. (11) and (14), it follows:

$$\varepsilon_n \leq \varepsilon_n^{(hom)} = \frac{\gamma\lambda_L}{4\pi}\left(\frac{\sigma}{R}\right)^2. \tag{15}$$

This is the correct condition on the emittance to be satisfied in a laser wiggler, which becomes quite restrictive when $\sigma << R$.

As discussed in ref. [8], the Pellegrini-Kim emittance criterium for the FEL radiation, $\varepsilon_n < \gamma\lambda_r/4\pi$, does not apply in a laser wiggler, since it would imply $\beta^* > Z_r$ where Z_r is the Rayleigh range of the FEL radiation, so that the emitted radiation would get outside of the electron beam, making impossible the amplification process. To forbid this, we should reverse the criterion, i.e. $\beta^* < Z_r$ and

$\varepsilon_n > \gamma\lambda_r/(4\pi)$. For simplicity we are assuming an equal radius for the radiation and the electron beams.

Furthermore, we should impose

(16) $$\sigma = \frac{R}{a_2}$$

with $a_2 > 1$ as we explained above. From eq. (15) it follows

(17) $$\varepsilon_n^{(\mathrm{hom})} = \frac{\gamma\lambda_L}{(4\pi)a_2^2}.$$

Up to now, all we have written is valid both in the classical and in the quantum regime. In both the cases, the condition on the energy spread is

(18) $$\frac{\Delta\gamma}{\gamma} < \Gamma,$$

where Γ is the FEL line width. In ref.[3] we have estimated that the line width in the quantum regime is

(19) $$\Gamma = 4\rho\sqrt{\bar{\rho}} \quad \text{if} \quad \bar{\rho} < 1 \quad ,$$

whereas in the classical regime it is the well known expression

(20) $$\Gamma = \rho \quad \text{if} \quad \bar{\rho} >> 1 \; .$$

Emittance is one of the causes of the energy spread increasing. In fact, since the resonance wavelength depends on the divergence angle θ, according to

(21) $$\lambda_r = \frac{\lambda_L(1 + a_0^2 + \gamma^2\theta^2)}{4\gamma^2}, \qquad \text{with } 0 \le \theta \le \frac{\varepsilon_r}{\sigma},$$

we have

(22) $$\frac{\Delta\lambda}{\lambda} \approx \frac{2\Delta\gamma}{\gamma} \approx \frac{\varepsilon_n^2}{\sigma^2(1 + a_0^2)} \le 2\Gamma$$

Hence, we obtain the following 'inhomogeneous' condition for emittance:

(23) $$\varepsilon_n \le \varepsilon_n^{(\mathrm{in\,hom})} = \sigma\sqrt{2\Gamma(1 + a_0^2)}$$

which, using eq. (9), is equivalent to

(24) $$\varepsilon_n \le \alpha\frac{\gamma\lambda_r}{4\pi}\sqrt{\frac{Z_r}{L_g}}$$

where $\alpha = 2$ for the classical case and $\alpha = 4$ for the quantum case, in agreement with refs.[8,10].

We remark that the inequality (15) must be strictly satisfied, otherwise the FEL action is destroyed. The inequality (23) or (24) arises from an inhomogeneous broadening of the resonance which may reduce the emission deteriorating the gain, since only the electrons whose θ is small enough will participate to the radiation process [11].

Another cause which may contribute to the broadening of the resonance is the intensity fluctuations in the laser wiggler, i.e. the fluctuations in the wiggler parameter a_0. Using Eq. (1) and imposing (18) we obtain, very simply,

(25) $$\frac{\Delta a_0}{a_0} \le \frac{1+a_0^2}{a_0^2}\Gamma .$$

Finally, the peak power in an FEL is given by [2]

(26) $$P_r = P_{beam}(\rho |A|^2)$$

where A is the dimensionless field amplitude in the 'universal scaling' and $P_{beam} = (I/e)mc^2\gamma$ is the beam power. Classically, at saturation, $|A|^2 \approx 1$ [2], so that eq. (26) can be written

(27) $$P_{sat} \approx \rho P_{beam} \approx (I/e)\hbar\omega\bar{\rho} .$$

Eq.(27) shows that $\bar{\rho}$ is the average number of the emitted photons per electron at saturation. In the quantum regime $\bar{\rho} < 1$ and at saturation [3], $|A|^2 \approx 1/\bar{\rho}$ and eq.(26) yields

(28) $$\bar{P}_{sat} \approx \frac{\rho}{\bar{\rho}} P_{beam} \approx \frac{\hbar\omega}{mc^2\gamma} P_{beam} \approx (I/e)\hbar\omega$$

Note that the power in the quantum regime is larger than the one predicted in the classical theory. The meaning of Eq. (28) is transparent: in the quantum regime each electron emits a single photon. In conclusion, the number of emitted photons in the classical and in the quantum regimes is respectively

(29) $$N_{ph} = \frac{Q}{e}\hbar\omega$$

and

(30) $$\bar{N}_{ph} = \frac{Q}{e}\hbar\omega\bar{\rho} .$$

CLASSICAL VERSUS QUANTUM REGIME

In order to discuss some specific example for the classical and the quantum regimes, we take as independent the following system of six parameters: $\bar{\rho}$, λ_r(Å), λ_L(μm), a_0 (the wiggler parameter), a_1 (the number of amplitude gain lengths in the interaction region $2Z_L$) and a_2 (the ratio between the laser and the radiation beam radius at the focal point). The other parameters are deduced self-consistently as follows. Using eq. (1) we deduce γ. With eq. (9) we calculate L_g, with eq. (13) we deduce R and than σ from (16). Introducing these values in eqs. (6), (7) and (12) we calculate the current I, the power P and the total energy U. Finally we calculate the limit values on emittance from (17) and (23) and the number of photons from eqs. (29) and (30), for a given beam charge Q. In the table 1 we report the results of the optimization with λ_r=2Å, λ_L=0.8 μm, a_1=5 and a_2=2, both for a quantum case, with $\bar{\rho}=0.2$ and a_0=0.1, and for a classical case, with $\bar{\rho}=2$ and a_0=0.8. Furthermore, we assume Q=1 nC.

$\bar{\rho}$	0.2	2
a_0	0.1	0.8
ρ	$7.55\cdot10^{-5}$	$5.93\cdot10^{-4}$
γ	32	40
Γ	$1.35\cdot10^{-4}$	$6\cdot10^{-4}$
$\Delta a_0 / a_0$	$1.36\cdot10^{-2}$	$1.5\cdot10^{-3}$
L_g (mm)	1	0.06
Z_L (mm)	2.6	0.17
R (μm)	12.9	3.25
σ (μm)	6.4	1.6
P (TW)	0.23	0.92
Energy (J)	3.9	1.0
τ (ps)	17.1	1.1
I (A)	990	985
Photons' number	$6.2\cdot10^{9}$	$12.4\cdot10^{9}$
$\varepsilon_n^{(hom)}$	0.5	0.6
$\varepsilon_n^{(in\ hom)}$	0.11	0.07

Table 1

From our results it appears that in the quantum regime the gain length and the laser Rayleigh range are appreciably longer than in the classical regime, so that in the quantum regime a longer duration time and a larger energy of the laser are required. The current in both cases is of the order of 1kA, however the requirement on the current density in the classical case is an order of magnitude larger since an electron beam radius of 1.6 micron is very small. We note also a more stringent condition for the inhomogeneous emittance and the laser fluctuations in the classical case. Therefore, on the basis of this example, we would conclude that it is easier to operate in the quantum regime. We stress that in the quantum regime the emitted radiation has the important property of high temporal coherence with no spiking, whereas for the classical regime, with $\overline{\rho}$=2, one would have nearly 3000 random spikes. This is the fundamental difference between the two regimes. A 3D quantum model for a FEL with a laser wiggler will be discussed elsewhere.

REFERENCES

[1] R. Bonifacio and F. Casagrande, Nuclear Instruments and Methods in Physics A 237, 168 (1985).

[2] R. Bonifacio, C. Pellegrini and L.M. Narducci, Opt. Commun. 50, 373 (1984)

[3] R. Bonifacio, N. Piovella, G.R.M. Robb, Nuclear Instruments and Methods in Physics A 543, 645 (2005).

[4] R. Bonifacio, N. Piovella, G.R.M. Robb, A. Schiavi, Phys. Rev. ST Accel. Beams 9, 090701 (2006).

[5] R. Bonifacio, L. De Salvo, P. Pierini, N. Piovella, and C. Pellegrini, Phys. Rev. Lett. 73, 70 (1994).

[6] S.V. Milton, et al., Science 292, 2037 (2001); V. Ayvaziyan, et al., Phys. Rev. Lett. 88, 104802 (2002).

[7] R. Bonifacio, Nuclear Instruments and Methods A 546, 634 (2005).

[8] R. Bonifacio, M. Ferrario, G.R.M. Robb, N. Piovella, A. Schiavi, L. Serafini, Proceedings of the 27th International Free Electron Laser Conference, 71 (2005).

[9] J. C. Gallardo, R. C. Fernow, R. Palmer, C. Pellegrini, IEEE J. Quantum Electron. 24, 1557 (1988), see eq. (4) with k=1 for a gaussian shape.

[10] R. Bonifacio, L. de Salvo Souza and B. W. J. McNeil, Optics Comm. 93, 179 (1992).

[11] N. Piovella, R. Bonifacio, Nuclear Instruments and Methods in Physics A 560, 240 (2006).

Needle cathodes for high-brightness beams

C. H. BOULWARE, J. D. JARVIS, H. L. ANDREWS, AND C. A. BRAU
Vanderbilt University, Nashville, TN 37235 USA

At the tips of sharp needles, the surface electric field is enhanced by many orders of magnitude. This intensifies thermionic emission and photoemission of electrons through the Schottky effect, and reduces the effect of space charge. The increased current density improves the brightness of electron sources by orders of magnitude. In addition, at very high fields (>10^9 V/m), field emission and photo-field emission produce very high current density. Arrays of needles can be used to achieve high total current.

1. Background

Owing to the high fields that surround their tips, needle cathodes make interesting sources of high-brightness electron beams. For example, at the tip of an elliptical needle protruding from a flat cathode, the field is

$$E \approx E_0 \frac{2a/R}{\ln(4a/R)} \tag{1}$$

where E_0 is the field at large distances from the needle, a the length of the needle, and R the radius of the tip [1]. For a needle with tip radius 10 μm and length 10 mm in an electric field 1 MV/m, the surface field is 200 MV/m. Although needle cathodes are most important for applications where the total current is low, their high brightness compensates for their low current in a variety of applications where current would usually be thought to be most important. In addition, arrays of needle cathodes can produce high-current beams with brightness comparable to that of individual needles.

For the purposes of the following discussion, we review briefly the meaning of the term "brightness" as it applies to electron beams. Specifically, brightness refers to the current density in the six-dimensional transverse phase space of the electron beam. Thus, it is a local property of the beam in the transverse phase space. More commonly, we compute and measure the total current and the emittance of an electron beam. These are global measures, or moments, of the transverse phase space, and are not uniquely related to the brightness, or even to the peak brightness. However, for a known distribution of electrons in phase space the relationship between the brightness on the one hand and the current and emittance on the other can be computed and we find that for a Gaussian distribution the normalized peak brightness B_N is related to the normalized rms emittance ε_{rms} by the formula [2]

$$B_N = \frac{1}{4\pi^2} \frac{I}{\varepsilon_{rms}^2} \quad (2)$$

where I is the total current.

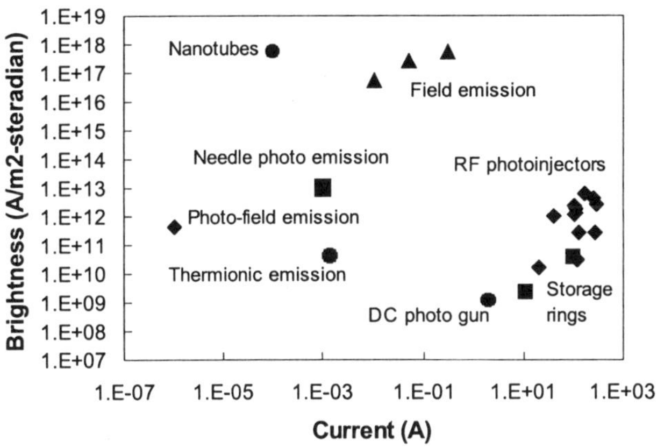

Figure 1. Normalized brightness of various sources of electrons [3]

The normalized brightness of various electron-beam sources is summarized in Figure 1 [3]. As shown there, the peak current of various sources spans eight orders of magnitude, while the peak brightness spans ten orders of magnitude! At the bottom of the graph we find the dc photocathode gun used in the high-power free-electron laser experiments at Jefferson Laboratory [4]. Storage rings do better, but the brightness of storage rings is limited by quantum effects in the synchrotron emission that damps the ring and makes it stable. At higher electron energy, the damping is faster but the brightness is smaller. Still higher brightness and current are provided by rf photoelectric injectors. These are commonly the injectors of choice for high-power, short-wavelength FELs. But the highest brightness shown in Figure 1, by far, is produced by field emission from needle cathodes. Perhaps most remarkable is the current from individual tips: up to 1 A total current from tungsten needles (up to 10 A in nanosecond pulses), and up to 1 mA (cw!) from individual carbon nanotubes only 30 nm in diameter!

To illustrate the importance of brightness, as opposed to total current, we consider two examples. The first is a Compton-backscatter x-ray source formed by directing a high-power pulsed laser against a pulsed electron beam. The figure of merit for most applications is the peak spectral brilliance of the x-rays, which is the number of photons per unit area, per unit solid angle, per unit time, per unit relative energy (or bandwidth). It is easily shown that if the betatron motions are small, so the angle the electrons make to the axis is less than $1/\gamma$, where γ is the Lorentz factor, then the peak spectral brilliance B_ν is

$$B_\nu = \frac{2\pi\sigma_T}{hq_e c^2}\frac{\gamma^2 U_L}{\tau_e + \tau_L} B_N \tag{3}$$

where σ_T is the rms transverse size of the electron beam, h Planck's constant, q_e the electron charge, c the velocity of light, U_L the energy in the laser pulse, τ_e the duration of the electron pulse, and τ_L the duration of the laser pulse. In this limit, the spectral brilliance of a Compton x-ray source is proportional to the brightness of the electron beam and is independent of the total current.

Another example that illustrates the importance of brightness is a high-power, short-wavelength, free-electron laser, such as the LCLS now under construction at SLAC [5]. Provided that the current exceeds some minimum value, so the diameter of the optical beam is controlled by gain guiding, then it is easy to show that the gain is

$$\mu = 3^{3/4} a_W \left[JJ\right]\left(2\pi \frac{1+a_W^2}{\gamma^3}\frac{B_N}{I_A}\right)^{1/2} \tag{4}$$

where a_W is the normalized rms wiggler magnetic field, $\left[JJ\right]$ the usual Bessel function factor (of order unity) that appears in the gain equation to account for the figure-eight motion of the electrons in a strong wiggler field [6], and I_A the Alfven current [1]. For gain guiding, it is necessary that the total current satisfy the requirement $I > \gamma^2 \lambda_L^2 B_N$, where λ_L is the laser wavelength, which is equivalent to the requirement $2\pi\varepsilon_{rms} / \gamma\lambda_L > 1$. These simple results agree with more extensive computations [7] and are satisfied by the LCLS. In this case, the gain is independent of the current and depends only on the brightness. Thus, we see that in important cases, the brightness of the electron beam is more important than the current.

2. Electron emission

The mechanisms by which electrons are emitted from a cathode are summarized in Figure 2. These include thermionic emission, photoemission, and field emission. In addition, combinations of these mechanisms are possible, such as thermal field emission and photo-field emission.

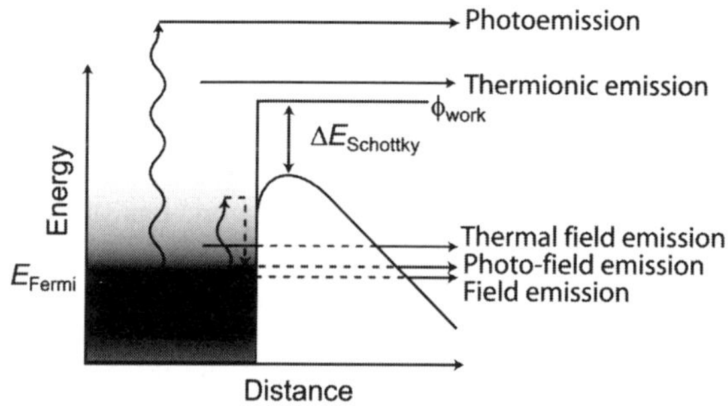

Figure 2. Mechanisms for the emission of electrons.

For thermionic emission, the cathode is raised to some high temperature. The Fermi distribution of electrons in the cathode develops a high-energy tail that extends above the work function of the cathode, and electrons with sufficient energy escape from the surface. The current density J is given by the Richardson-Dushmann equation [8]

$$J = AT^2 e^{-\phi_w / k_B T} \tag{5}$$

where T is the temperature, k_B Boltzmann's constant, and ϕ_w the work function (the energy required to remove an electron at the Fermi level from the cathode). Richardson's constant is

$$A = \frac{4\pi m k_B^2 q_e}{h^3} = 1.20 \text{ MA/m}^2\text{-K}^2 \tag{6}$$

in which m is the electron mass and h Planck's constant. When the electric field at the surface is large, the electrical potential falls off rapidly outside the surface. Due to the attraction of the electron for its image charge, the potential barrier at the surface becomes rounded off and is depressed as shown in the figure. This is called the Schottky effect, and it lowers the escape energy by the amount [9]

$$\Delta W = \sqrt{\frac{q_e E}{4\pi\varepsilon_0}} \tag{7}$$

where ε_0 is the permittivity of free space. For the needle considered earlier, the Schottky effect reduces the barrier by 0.8 eV. At a temperature of 1000 K, this increases the thermionic emission by the factor 10^4. At higher fields, tunneling (thermal field emission) increases the thermionic emission further, as discussed below [10]. Thus, the enhanced field at the tip of a needle has a substantial effect on the current density. The normalized brightness of thermionic emission is given by the formula [11]

$$B_N = \frac{mc^2 J}{2\pi k_B T} \tag{8}$$

which increases in proportion to the current density.

Since the development of lasers, photoemission has become an important source of high-brightness electron beams. As we were taught by Einstein, an electron may be liberated from the surface of a cathode when it is struck by a photon with energy greater than the work function. However, when the electric field at the surface is sufficiently strong, the Schottky effect lowers the barrier, as described above. This reduces the threshold photon energy required to liberate an electron, and above the threshold energy it raises the probability that an electron will be emitted. Just above threshold, the emission probability, called the quantum efficiency, follows the approximate relation [12]

$$\eta_{QE} \sim K\Delta E^2 \tag{9}$$

where $\Delta E = h\nu - \phi_w + \Delta W$ is the energy of the photon above the threshold value, K is a constant that depends on the material, and $h\nu$ is the photon energy. When we include the Schottky effect, we see that the quantum efficiency can be measurably enhanced by the intense fields at the tip of a needle. This threshold law seems to remain valid at least several eV above threshold [13].

The constant that appears in (9) can be understood in terms of the "three-step model." According to Spicer [14], the emission probability is the product of three factors. The first is the probability that the electron that absorbs the photon will be excited to an energy level above the barrier. The second is the probability that the electron will migrate to the surface without losing too much energy. The third is the probability that the electron reaching the surface will have enough momentum in the direction normal to the surface to pass over the barrier. The first and third factors are each proportional to the excess energy above the barrier, and lead to the threshold law (9). The second factor in the three-step model depends mostly on the material, since the electron is presumed to diffuse to the surface with a mean free path determined by electron-lattice collisions and lose energy with a mean free path determined by electron-electron collisions. The high quantum efficiency of semiconductor cathodes, compared with metals, is due to the low probability of energy-loss collisions. In the simplest three-step model, the electrons are assumed to lose no energy in electron-lattice collisions, but to lose enough energy to fall below the barrier in electron-electron collisions. The energy distribution of the emerging electrons is therefore the same as that of the photoexcited electrons, which is essentially uniform from zero up to the photon excess energy ΔE. When the photon energy is well above threshold, the brightness of the beam is reduced by the broad electron energy distribution.

Two effects limit the current density that can be obtained from a photocathode. These are laser damage to the cathode surface and space charge. For metals, laser damage occurs when the laser heats the surface temperature above the melting point. However, for repetitively pulsed lasers, damage accumulates over a period of time due to heating and cooling of the surface. This thermal cycling is observed to cause grain growth that produces a rough surface. For 5-ns pulses, which are characteristic of many Q-switched lasers, the observed damage threshold for tungsten and yttrium is on the order of 10^{12} W/m^2. This corresponds to a surface temperature rise on the order of 1000 K. Considerations based on heat conduction suggest that the damage threshold scales inversely as the square root of the laser pulse length.

Space charge is a well-known effect described by the Child-Langmuir law. For a spherical geometry, which represents roughly the geometry near the tip of a needle, the space-charge-limited current density is [11]

$$J = \frac{4\varepsilon_0}{9}\sqrt{\frac{2e}{m}}\frac{V^{3/2}}{d^2} \tag{10}$$

where V is the voltage across the diode and d the anode-cathode spacing. For laser spot sizes small compared with the radius of the needle tip, three-dimensional effects increase the current density, as observed in planar geometries. However, when space charge is important, the transverse electric field caused by the space charge causes the beam to expand to fill a much larger solid angle than that of the laser spot on the needle [15]. This effect is already apparent at a total current of 1 mA from a 10-μm spot in a surface field on the order of 10^7 V/m.

Field emission of electrons occurs when the surface electric field is very high, on the order of 10^9 V/m or more. Under these circumstances, the potential barrier at the surface of the metal is thin enough for the electrons to escape by tunneling. At low temperatures (room temperature and below), the current is dominated by electrons near the Fermi level, and the current density is given by the expression [16]

$$J = \frac{q_e^2}{8\pi h}\frac{E^2}{\phi_w}\exp\left[-\frac{8\pi\sqrt{2mq_e}}{3h}\frac{\phi_w^{3/2}}{E}f(y)\right] \tag{11}$$

where $f(y)$ is a dimensionless function introduced to account for image forces near the surface and $y = \sqrt{q_e E / 4\pi\varepsilon_0\phi_w^2}$. As a function of the field, the current is nearly a straight line on a so-called Fowler-Nordheim plot, as illustrated in Figure 3 [17].

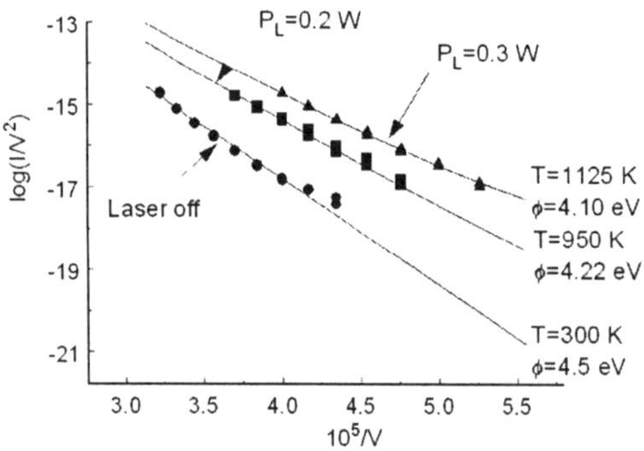

Figure 3. Fowler-Nordheim plot of the field emission current I as a function of the voltage V for a tungsten needle.

At higher temperatures, electrons at higher energies, where the barrier is thinner and tunneling is easier, contribute to the emission. When this process, called thermal field emission, is important, the current shown on a Fowler-Nordheim plot becomes curved. At sufficiently high current density, the current is reduced by space-charge effects. Since the emission is very sensitive to the surface field, the effects of space charge appear in field emission at much lower current density than in thermionic or photoemission. Nevertheless, a peak current as high as 6 A (in a microsecond pulse) has been obtained from a needle with a 1-μm tip radius [3].

The brightness that has been achieved using field emission is shown in Figure 1. The highest brightness shown there was achieved in nanosecond pulses from tungsten needles, but remarkably, comparable brightness has been achieved cw using carbon nanotubes [18]. Lower values of the brightness and current shown in Figure 1 indicate what has been achieved from tungsten needles in microsecond pulses and cw [19].

Another mechanism for emission of electrons is photo-field emission. As shown in Figure 3, the emission observed when the laser illuminates the tip follows a nearly straight line on a Fowler-Nordheim plot. This indicates that the electrons are emitted from low-lying levels, near the Fermi level but sufficiently above it (as indicated by the reduced work function) to tunnel out where the barrier is thinner. The current observed with the laser on cannot be explained by thermal field emission alone at the (relatively low) temperatures measured for the tip of the laser-heated needles [20].

3. Arrays of needles

Larger currents should be possible by using arrays of needles. An array of individually gated diamond pyramids fabricated at Vanderbilt University is shown in Figure 4 [21]. By adding another electrode above the gate electrode, it should be possible to focus the beam emerging from each tip to maintain the brightness. If this is not done, the electron trajectories from neighboring emitters cross, and the emittance is seriously degraded. In the same way, arrays of carbon nanotubes have been formed as shown in Figure 5 [22].

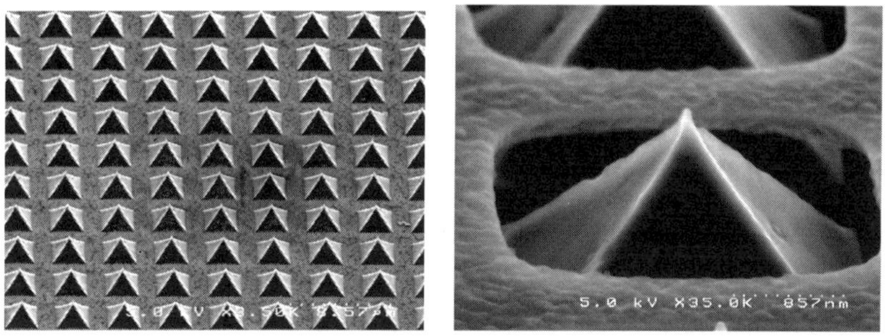

Figure 4. Array of gated diamond pyramid field emitters.

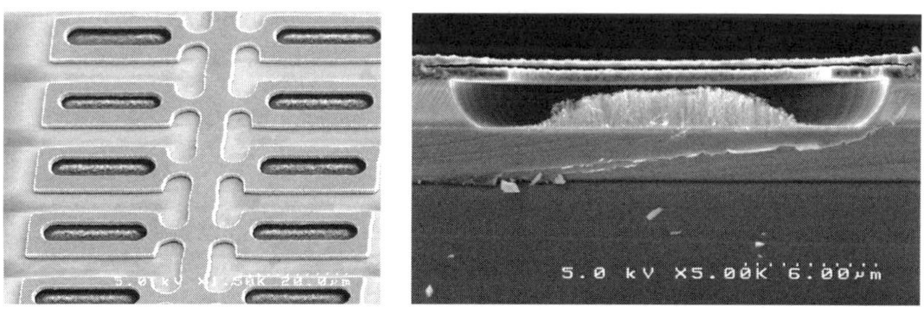

Figure 5. Array of gated nanotube field emitters.

4. Conclusion

In conclusion, we see that needle cathodes can be used as interesting, practical sources of electron beams of very high brightness. Typically, the current from needle cathodes is low, but it is also true that in many applications, such as high-power FELs and Compton-backscatter light sources, brightness is more important than total current. Moreover, arrays of needle cathodes can be used to increase the total current. Finally, we see that several mechanisms can be used to produce electron emission. These include thermionic emission, photoemission, and field emission. An interesting process that combines photoexcitation with field emission is photo-field emission. The physics of this process deserves further study.

Acknowledgement

This work was supported by Office of Naval Research under Grant N000149411023 and by the Medical Free Electron Laser Program of the Department of Defense under grant number F49620-01-1-0429.

References

1 C. A. Brau, *Modern Problems in Classical Electrodynamics* (Oxford, New York, 2004).
2 C. A. Brau, "What Brightness Means," in *The Physics and Application of High Brightness Electron Beams*, J. Rosenzweig, G. Travish, and L. Serafini, editors, World Scientific, Singapore, 2003.
3 Data have been taken from a variety of sources, including
J. F. Schmerge, "Summary for Working Group A on High Brightness Beam Production," in *The Physics and Application of High Brightness Electron Beams*, J. Rosenzweig, G. Travish, and L. Serafini, editors, World Scientific, Singapore, 2003.
P. H. Piot, "Review of Experimental Results on High-Brightness Photo-Emission Electron Sources," in *The Physics and Application of High Brightness Electron Beams*, J. Rosenzweig, G. Travish, and L. Serafini, editors, World Scientific, Singapore, 2003.
V. N. Litvenenko, *et al.*, *Nucl. Instr. Meth. A* **375**, 46 (1996).
M. E. Coupre, *et al.*, *Nucl. Instr. Meth.*, *A* **296**, 13 (1990).
W. P. Dyke *et al.*, *J. Appl.Phys.* **31**, 790 (1960).
G. N. Fursey, *Field emission in Vacuum Microelectronics* (Kluwer/Plenum, New York, 2005).
J. M. Bonard, *et al., Appl. Phys. A* **69**, 245 (1999).
4 G. R. Neal, *Phys. Rev. Lett.* **84**, 662 (2000).
5 C. Pellegrini, *Nucl. Instr. Meth. A***475**, 1 (2001).
6 W. B. Colson, "Classical free-electron laser theory," in *Laser Handbook*, Volume 6, W. B. Colson, C. Pellegrini, and A. Renieri, Eds., (North-Holland, Amsterdam, 1990).
7 C. Pellegrini, "X-Band Microwave Undulators for Short Wavelength Free-Electron Lasers," presented at *FEL 2005*, Stanford University, August 21-26, 2005.

8 O. W. Richardson, *The Emission of Electrons from Hot Bodies* (Longmans Green, London, 1921).
9 W. Schottky, *Physik. Z.* **15**, 872 (1914).
10 E. L. Murphy and R. H. Good, Jr., *Phys. Rev.* **102**, 1464 (1956).
11 J. D. Lawson, *The Physics of Charged-Particle Beams* (Oxford, New York, 1988).
12 L. A. Dubridge, *Phys. Rev.* **43**, 727 (1933).
13 T. Srinivasen-Rao, J. Fischer, and T. Tsang, *J. Appl. Phys.* **69**, 3291 (1991).
14 W. E. Spicer and A. Herrera-Gomez, "Modern Theory and Applications of Photocathodes," *SLAC-PUB-6306*, August, 1993.
15 John Lewellen, private communication.
16 R. H. Fowler and L. W. Nordheim, *Proc. Roy. Soc. A* **119**, 173 (1928).
17 C. Hernandez Garcia and C. A. Brau, *Nucl. Instr. Meth. A* **429**, 257 (1999).
18 J. M. Bonard, *et al., Appl. Phys. A* **69**, 245 (1999).
19 G. N. Fursey, *Field emission in Vacuum Microelectronics* (Kluwer/Plenum, New York, 2005).
20 C. Hernandez Garcia and C. A. Brau, *Nucl. Instr. Meth. A***429** 257 (1999).
21 A. Wisitsora-at, W. P. Kang, J. L. Davidson, D. V. Kerns, and T. Fisher, J. Vac. Sci. B, Vol. 21 (1), pp. 614-617, 2003.
22 Q. Li, J. F. Xu, K. Yu, T. Feng, X. Wang, X. H. Liu, W. P. Kang, J. L. Davidson, *J. Vac. Sci. B* **21**, 1684 (2003).

NON LINEAR EVOLUTION OF SHORT PULSES IN FEL CASCADED UNDULATORS AND THE FEL HARMONIC CASCADE

LUCA GIANNESSI†

ENEA Centro Ricerche Frascati, Via E. Fermi 45, 00044 Frascati (Rome) Italy tr

PIETRO MUSUMECI

INFN, ROMA 1 Piazzale Aldo Moro 2, 00185 Roma, Italy

Free Electron Laser (FEL) devices based on sequence of amplifiers with an harmonic relation between the resonant frequencies of each section, have been proposed to extend to shorter wavelengths the FEL operating range. When this "cascade" of undulators is operated at intensities above saturation, a short optical pulse propagates in a solitary wave-like regime. The pulse is up-converted in frequency at the transitions between the cascade sections, according the resonance central frequency of each undulator. In this paper we briefly review the main dynamical aspects of the FEL cascade operating in this regime and we study the case where the undulators of the cascade are not resonant at the fundamental frequency but have one of the higher order harmonics at a common frequency. In these conditions high harmonic multiplication factors in the final stage may be obtained with a relatively low energy electron beam.

1. Introduction

1.1. *Cascaded free electron lasers*

A cascaded Free Electron Laser (FEL) amplifier is a sequence of FEL amplifiers where the beam and the radiation produced in one section are used in a successive section which has its central resonant frequency tuned at one of the higher harmonics of the first. These schemes have been proposed in order to combine the FEL amplification with the non-linear behavior of the FEL dynamics and to generate higher order harmonics of the resonant field. The simplest cascade is composed by two undulator sections, where the first is resonant at ω_1 and the second is resonant at $\omega_2 = n\,\omega_1$ with n integer. In this layout the first section modulates the current density while in the second section the emission process at the selected higher order harmonic takes place. The

† Work partially supported by EU contract EUROFEL.

underlining idea, introduced in [1] consists in the fact that the current density modulation, which occurs during the FEL optical amplification, has higher order harmonic components which are responsible of the enhancement of the higher harmonics emission and which generates radiation when the beam is injected in an undulator with the central frequency tuned to match the density modulation periodicity. The concept of modulator/radiator FEL was studied by several authors in the last two decades [2, 3, 4, 5]. This idea, based on a single stage modulator/radiator, was successfully tested at BNL/ATF with n=2 [6]. In this experiment a CO2 laser was used to modulate the beam, and radiation was observed at 5 μm. In another experiment at BNL, a modulator was seeded by a Ti:Sa laser with ω_l=800nm and the spectrum after the radiator at 266 nm (n=3) was recorded[7]. One of the interesting aspects pointed out by this experiment is that the amplified FEL signal at $3\omega_l$ preserves the coherence/spectral properties of the input seed. The possibility of using the FEL as a frequency up-conversion device appears a valid solution in order to reach the soft x-ray spectral region, without the limitations in terms of spectral line width and stability of Self Amplified Spontaneous Emission (SASE) sources. The idea of a cascade initiated at visible/UV wavelength from solid state lasers sources has stimulated several proposals based on multiple stages cascades FEL [8, 9, 10, 11]. At the same time important progresses in the field of strong laser-matter interaction have lead to the generation of high-order harmonics by intense laser pulses propagating in low pressure gases. This technique is being well controlled and the generation of radiation down to 12 nm [12] has been demonstrated. An intense tunable source as the High-order Harmonics of a Ti:Sa laser generated in a Gas (HHG), represents a good candidate to feed a FEL amplifier cascade with the final operating wavelength down to the water window range [13, 14, 15].

One of the limitation of multiple stage cascaded devices is related to the evolution of the signal to noise ratio through the stages of the cascade. This aspect was pointed out in [16] and consists in the fact that any phase fluctuation of the field at ω is magnified by n at $\omega' = n\ \omega$. In terms of intensities, the effect scales unfavorably as n^2 thus limiting to few stages the possibility of preserving the signal to noise ratio of the seed through the cascade. The situation appears less critical when the limited bandwidth associated to the FEL process is taken into account [11], but the issues on the stability and preservation of longitudinal coherence in a multiple stage cascade are still consistent.

1.2. *Superradiance in FELs*

An alternative to these schemes has been proposed in ref.[17] where the cascade of undulators is designed to induce the transition of the propagating pulse in the superradiant regime [18]. In the superradiant regime a short and intense pulse of radiation with a self-similar shape propagates through the undulator. The pulse leading edge travels at the velocity of light and slips over the beam current (which may be assumed uniform in this picture) by N resonant optical periods every N undulator periods. The pulse length is related to its intensity by the relation

$$\tau \approx \frac{c\tau_s}{2}\frac{\lambda}{\lambda_u} \tag{1}$$

where τ is the pulse length, λ is the optical wavelength, λ_u is the undulator period and τ_s is the synchrotron oscillation period. The latter can be expressed as a function of the system parameters as

$$\tau_s = \frac{2\pi}{\omega_s} = \frac{\lambda_u}{c}\frac{\sqrt{1+K^2/2}}{\sqrt{2 f_B(\xi) K K_R}} \tag{2}$$

where K is the normalized undulator strength and where $f_B(\xi) = J_1(\xi) - J_0(\xi)$ is the planar undulator Bessel function factor with $\xi = K^2/4(1+K^2/2)$. The parameter K_R represents the dimensionless normalized vector potential associated with the radiation wave. The pulse dynamics obey a number of scaling laws derived in the limit of a one dimensional model and assuming that the superradiant pulse travels at the velocity of light in vacuum. The pulse energy grows as $z^{3/2}$, the peak power grows as z^2 and the pulse length decreases as $z^{-1/2}$, being z the longitudinal position of the pulse along the undulator.

In this regime the pulse leading edge modulates the e-beam energy, emission takes place in the front part of the pulse where the strong optical field and the undulator dispersion has converted the energy modulation into a density modulation. In the rear side the particles following synchrotron orbits in the longitudinal phase space, shift out of phase with the radiation and part of the radiation energy is re-absorbed by the electrons. This mechanism explains the reason why the pulse length is strictly related to the synchrotron period depending on the laser intensity (Eq.(1)). For a more comprehensive view of particles and field dynamics in FEL in superradiant regime the reader is addressed to ref.[17, 18, 19, 20].

1.3. *The superradiant cascade*

In a cascaded FEL operating in the superradiant regime the distinction between the modulator and the radiator is not any more appropriate: the steps of energy modulation, density modulation and emission take place simultaneously in different parts of the optical pulse. As in steady state FEL dynamics, the density modulation induces intense emission of radiation at higher order harmonics, but this occurs during the pulse propagation along all the undulator. A radiation pulse propagating in superradiant regime in a cascaded FEL, is able to propagate with a self similar shape, and up-converting itself in frequency while it passes from a cascade section to another. Furthermore a fluctuation in intensity of the seed initiating the cascade, may induce a shift in the longitudinal coordinate of the position where a given power level is reached, which scales as z^2. At the same time the pulse length at that position may change by a factor $z^{-1/2}$ and with the same relation varies the longitudinal extension of the distribution of density modulation of the electron current, which is responsible for transferring the phase and amplitude information from an undulator to another at a cascade transition. The peak amplitudes of the bunching coefficients are completely independent from the seed intensity, once superradiant state is established. According to the simple scaling relations quoted in the previous section, this process is then expected to show a lower sensitivity to external parameter variations with respect to a conventional cascade operating below saturation, where the scaling relation with the longitudinal coordinate is exponential. This concept was the base of a multiple stages cascade designed with a frequency multiplication factor exceeding 50, with the final wavelength at 5 nm [17].

1.4. *The "harmonic cascade"*

One of the factors limiting the extension of the cascade to higher frequency multiplication, is related to the necessity of operating all the stages of the cascade with the same electron beam energy. The change in the resonant wavelength at the fundamental harmonic in the different sections of the cascade must be compensated by variations of the undulator period and *K* parameter. An alternative is the "harmonic cascade" concept. In order to explain the concept, let us at first analyze the case of a conventional cascade composed by two undulators (as the one of ref.[1]) with resonances ω_1 and $\omega_2 = n\omega_1$. In the first undulator, where the modulation at ω_1 occurs, a coupling at the higher frequency $\omega_2 = n\omega_1$ is already present for n odd. The coupling coefficient is indeed given by the expression [21]

$$j_n = 2\pi^2 \left(\frac{N}{\gamma}\right)^3 [\lambda_u K f_b(n,\xi)]^2 r_0 \rho \tag{4}$$

Where N is the number of undulator periods, γ is the beam energy in m_0c^2 units, λ_u is the undulator period, K is the undulator parameter, $f_b(n,\xi) = J_{\frac{n-1}{2}}(\xi) - J_{\frac{n+1}{2}}(\xi)$ with $\xi = K^2/4(1+K^2/2)$, ρ is the electron density and r_0 is the classical electron radius, is the Alfven current. In Fig.1 it is shown the behavior of the function $f_b(n,\xi)$ for the first 7 harmonics at K=2, 3 and 5.

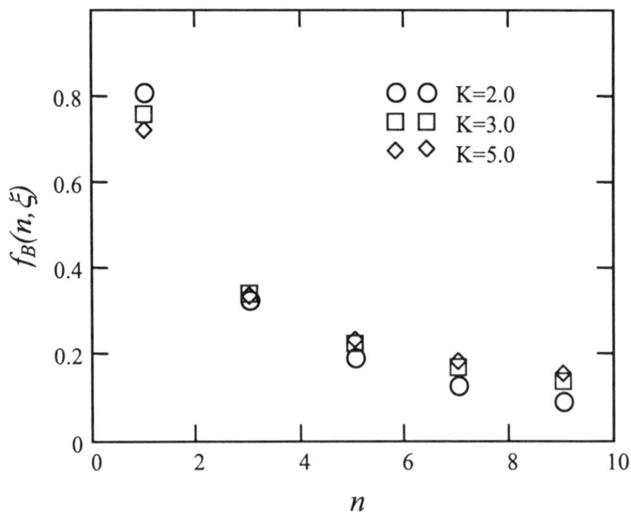

Fig. 1 Plot of the function $f_b(n,\xi)$ for the first 9 odd harmonics at different values of K

When K is sufficiently large, coupling on the higher order harmonics is already present in the first undulator with resonance ω_1. One of the reasons why the cascade design based on two undulators with the second resonant at $\omega_2 = n\omega_1$ provides higher power emission at the harmonic wavelength is that the field at ω_1 is out of resonance in the second undulator section, and the modulated e-beam interacts only with the field at the frequency ω_2. The synchrotron oscillations induced by saturation of the field at the first harmonic, which would inhibit the amplification of all the higher order harmonics, are suppressed and the harmonic field may grow up to full saturation at the harmonic wavelength.

A similar effect can be obtained with two undulators which have their main resonances (n=1) which differ of a relatively small amount. This can be obtained by tuning the second undulator with one of its higher order harmonics equal to $n\omega_1$, i.e. , $m\omega_2 = n\omega_1$ with $m \neq n$. A schematic diagram of the frequency spectrum of the two undulators is sketched in Fig. 2.

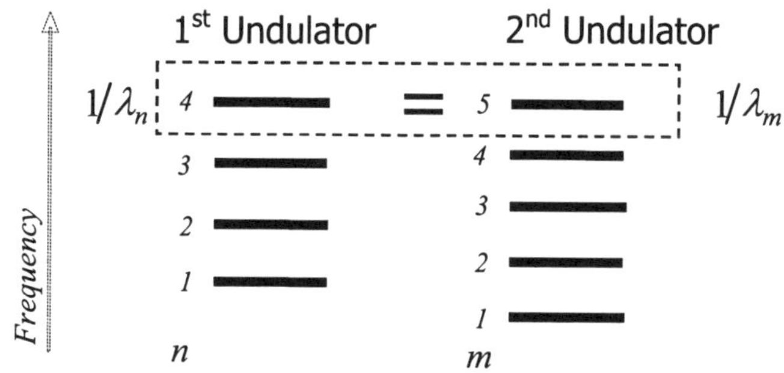

Fig. 2 Frequency spectrum of the harmonic cascade FEL

The two undulators have their main resonant frequencies differing by the ratio m/n. The two undulator parameters are close each other and the output frequency $\omega = n\omega_1 = m\omega_2$ is a higher harmonic in both the undulator sections. This scheme is expected to be particularly interesting when operated in superradiant regime. The reason is that in this case the energy exchange between the current and the laser field is proportional to the slippage length, i.e. the difference of the distances covered by the light and by the electrons in the undulator. The superradiant pulse propagates at the velocity of light while the current travels at a velocity given by

$$\beta_z \cong 1 - \frac{1}{2\gamma^2}\left(1 + \frac{K^2}{2}\right) \tag{5}$$

The slippage length, equal to $N\lambda_0$ being $\lambda_0 = \lambda_u/2\gamma^2\left(1 + K^2/2\right)$ the fundamental resonant wavelength, is independent from the harmonic order. In the second undulator the radiation is slipping over the current at a velocity λ_0 per undulator period even if the radiation is amplified at the harmonic wavelength, which is *m* times smaller and a higher efficiency is expected. In order to study the behavior of the harmonic cascade FEL, we have solved numerically the equations governing the harmonic cascaded FEL dynamics and we have analyzed some example cases. The next section is devoted to a description of the equations and of the problems related to their numerical solution. In the last section we discuss specific configurations of harmonic cascaded FELs.

2. Simulation of the harmonic cascade.

The particles dynamics in the harmonic cascade configuration, is governed by the FEL pendulum equation where the coupling with higher order harmonics is included. In order to implement the higher harmonics in a FEL dynamics simulation, the slowly varying envelope approximation has to be applied to the field components at the different harmonics. The field may be written as

$$E(z,t)=\sum_n \widetilde{E}_n(z,t)\,e^{in[kz-\omega t]} \tag{6}$$

We have implicitly assumed that the field is non-zero only in narrow spectral regions centred around the harmonics $n\omega$ and the terms $\widetilde{E}_n(z,t)$ represent the "slow" components at the different harmonics. After a change of variables from the laboratory frame to the beam frame moving with velocity β_z i.e. $(z,t)\rightarrow(\zeta=z-\beta_z ct, \widetilde{z}=\beta_z ct)$, the equation for the evolution of the field slowly varying component at the harmonic n reads (in SI units)

$$\left(\frac{\partial}{\partial\widetilde{z}}+(1-\beta_z)\frac{\partial}{\partial\zeta}\right)\widetilde{E}_n(\widetilde{z},\zeta)=-\frac{e_0\rho(\zeta)}{2\varepsilon_0}\frac{K}{\gamma}f_B(n,\xi)\left\langle\exp(in\vartheta_i(\widetilde{z},\zeta))\right\rangle_\zeta \tag{7}$$

where $\gamma=\langle\gamma_i\rangle_i$ is the mean beam energy in m_0c^2 units, $(\vartheta_i=(k+k_u)z-\omega t+\varphi,\gamma_i)$ are the coordinates describing the particles longitudinal phase space position, $k_u=2\pi/\lambda_u$ is the wave-vector corresponding to the undulator period, $e_0\rho(\zeta)$ is the longitudinal current density. The average $\langle\ \rangle_\zeta$ is intended over a region of size λ at the position with coordinate ζ. The problem of FEL coupling on higher order harmonics was analysed in Ref.[21]. The multi-frequencies pendulum equation for the for the *i-th* electron dynamics may be written as

$$\begin{aligned}\frac{d\gamma_i}{d\widetilde{z}}&=\frac{K}{2\gamma}\frac{e_0}{m_0c^2}\sum_n\widetilde{E}_n f_B(n,\xi)\cos(n\vartheta_i+\varphi)\\ \frac{d\vartheta_i}{d\widetilde{z}}&=k_u-k\frac{1+K^2/2}{2\gamma^2}\end{aligned} \tag{8}$$

In the case $n=1$ the solution of the coupled equations (6) and (7) is treated extensively in literature [22, 23, 24]. The extension to the multiple harmonic case is numerically implemented in *Perseo*, a library of functions that can be simply programmed to set up FEL simulations in a wide variety of practical configurations (ref. http://www.perseo.enea.it). One of the issues in the solution of the coupled system of equations (5) and (6) is related to the correct numerical representation of the electron beam distribution. The typical number of simulated macroparticles per optical period in the evaluation of the average of eq. (6), is a few thousands. In a real beam this number is $10^5 - 10^7$ electrons, depending on the wavelength and on the current density. In order to correctly reproduce the emission properties of a real beam, the frequency components of numerical representation of beam must resemble those of the real distribution in the simulation bandwidth [25]. The numerical "shot" noise is generally reduced by representing the beam as an ensamble of beamlets, each one with macroparticles uniformly distributed as

$$\vartheta_i = \vartheta_0 + \frac{2\pi}{n_p} i + \delta_i , \qquad i = 0, 1, ..., n_p - 1 \tag{8}$$

where the number n_p is the number of simulated macroparticles in one optical period, and δ_i are coefficients corresponding to a shift of the macroparticles from the regular distributed positions applied in order to induce an artificial shot noise equivalent to that of a "real" beam [26]. In the simulation of an harmonic cascaded FEL, the beam propagates in two (at least) undulators, where the ratio between the resonant wavelengths are a rational number. In order to avoid the unwanted effects associated to a non uniform beam distribution, the length of the beamlet must be equivalent to an integer number of optical cycles in both the undulators. In order to satisfy this condition the simulated particles have been distributed in the interval $(0, 2\pi\, m)$ in the first undulator and $(0, 2\pi\, n)$ in the second undulator, whose respective resonant wavelengths are in the ratio $\lambda_1/\lambda_2 = n/m$. In these conditions it is not straightforward to set the coefficients δ_i introducing the shot noise, consistently in both the undulators. The shot noise is not expected to give relevant effects in a simulation starting from a high intensity seed. For this reason the δ_i have been set to zero and the beam shot noise is suppressed.

2.1. *Examples of "harmonics cascade" FELs*

The simulation is set up in *Perseo* configured for modeling an undulator composed by two segments with different parameters. The first section has the main resonance at the wavelength $\lambda_{1,1} = 266nm$ while the second section is tuned at $\lambda_{2,1} = 200nm$ (we used the notation $\lambda_{um\#,harmonic\#}$). The beam is modulated and bunched up to reproducing the phase space pattern typical of a superradiant state as described in ref.[17]. The 4th harmonic of the main modulation periodicity matches the 3rd harmonic of the second undulator. Despite the high order of harmonic multiplication, the two undulators differ only by a small K parameter variation which shifts the resonance of about 30%, as it can be observed in Tab. 2 where the main parameters of the two undulators are listed. We have considered the parameters of the SPARC FEL and we have selected the seeding wavelength equal to the third harmonic of a Ti:Sa solid state laser. This source is the cathode driving laser and is synchronized with the electron bunch. It allows the generation of the intense and short pulses required to set up a superradiant state in the short magnetic distance corresponding to a single undulator segment of the SPARC FEL.

Table 2. Parameters used for the simulation of the harmonic cascade with *n=4, m=3*

1st Undulator Period	2.8 cm
1st Undulator K parameter	1.955
1st Undulator number of periods	77
1st Undulator resonant wavelength	266 nm (4th @66nm)
2nd Undulator Period	2.8 cm
2nd Undulator K parameter	1.538
2nd Undulator number of periods	385
2nd Undulator resonant wavelength	200 nm (3rd @66nm)
Input seed power	10 MW
Input seed wavelength	266 nm
Input pulse length (FWHM)	100 fs
Beam energy	200 MeV
Beam Energy Spread	20 KeV
Beam current	110 A
Slice emittance (ε_x=ε_y)	1 mm-mrad

In Fig. 3 the pulse shape and the bunching coefficients after the modulation section are shown. The seed pulse is modified by the FEL interaction. The 1st bunching coefficient exceeding 0.75 is an indication of saturation at the fundamental wavelength. As expected in superradiant conditions a strong bunching at the 4th harmonic in the front side of the pulse is present. This 4th harmonic modulation is resonant with the 3rd harmonic in the second undulator.

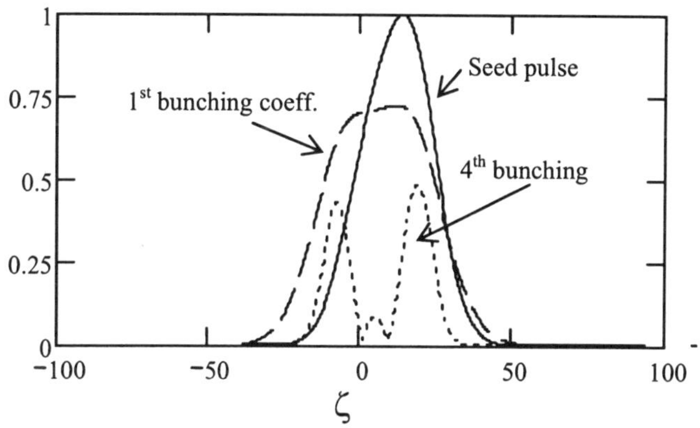

Figure 3. Pulse shape after the modulation section and bunching coefficients at the 1st and the 4th harmonics

In Fig. 4 the pulse energy growth in the second undulator with the resonance at 200 nm is shown. In the first 4 m the existing modulation with an harmonic component at 66 nm induces the growth of the superradiant pulse. The superradiant pulse slips over the e-bunch and is fed by the high quality beam electrons that had no FEL interaction except for the SASE induced in the first segment. After approximately five meters the pulse energy scaling vs.z becomes that typical of the superradiant regime $E \propto z^{3/2}$ and at the end of the undulator the pulse energy is larger than 4 μJ.

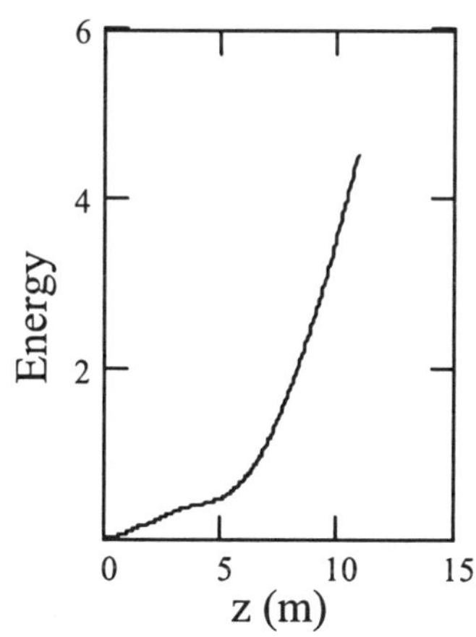

Figure 4. Laser pulse energy at 66 nm (3rd harmonic) in the second undulator segment

The pulse shape is shown in Fig. 5. The pulse length at the end of the harmonic cascade is 30 fs (fwhm) and the peak power exceeds the 130 MW level. The spectrum at the end of the undulator is shown in Fig. 6. The spectral width is 0.3nm (fwhm) corresponding to ≈ 0.5% of the central wavelength, approximately a factor two above the Fourier limit.

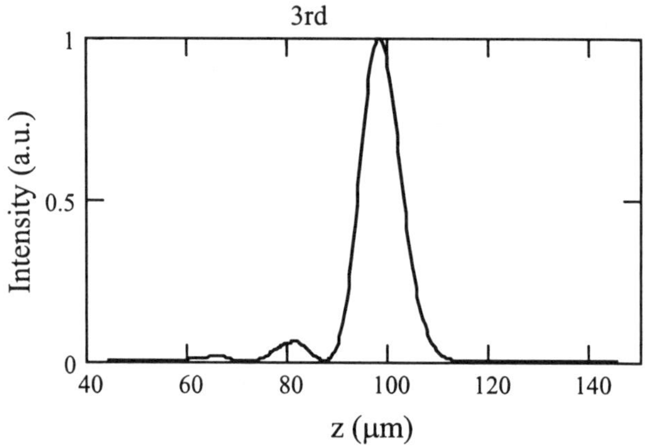

Fig. 5 Pulse shape at the end of the SPARC undulator. The total puse energy is 4.5 μJ and the pulse length is about 30 fs (fwhm).

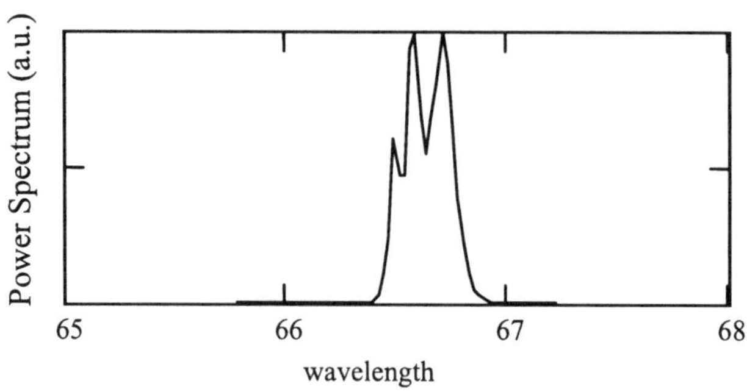

Fig. 6 Power spectrum of the pulse shown in fig. 4, corresponding to the position at the end of the cascade.

The same final wavelength may be obtained by setting the K parameter in the second section to K= 1.95. In this case the fourth harmonic of the first section is resonant with the fifth harmonic in the second one (i.e. $n=4$ and $m=5$). With all the other parameters set as in the previous case, the output energy is 4.6 μJ and the pulse duration at the end of the undulator is about 50 fs. Despite of the lower gain factor associated to the higher harmonic order, the output energy in this case is higher than in the previous example, and the reason is the slightly longer slippage length, proportional to the fundamental wavelength which is 332nm in place of 200 nm as in the previous case.

As a final example we consider the case of the reference case with the parameters of Table 2 (*n=4* and *m=3)*, where the input intensity is reduced to 2 MW, and the number of undulator periods of the second stage is increased to 693. In this condition we have studied the effect of detuning the second stage from perfect resonance. In Fig. 7.a and 7.b it is shown a contour plot of the laser intensity as a function of the coordinate ζ along the bunch (horizontal axis) and of the coordinate $\tilde{z}$ along the undulator. In Fig. 7.a the two undulators are matched with $4\omega_1=3\omega_2$. In Fig. 7.b, we have introduced a shift of $2*10^{-3}$ in the resonant frequency of the second stage. The intensity is normalized at each $\tilde{z}$ in order to observe the pulse evolution without the effect of the intensity growth.

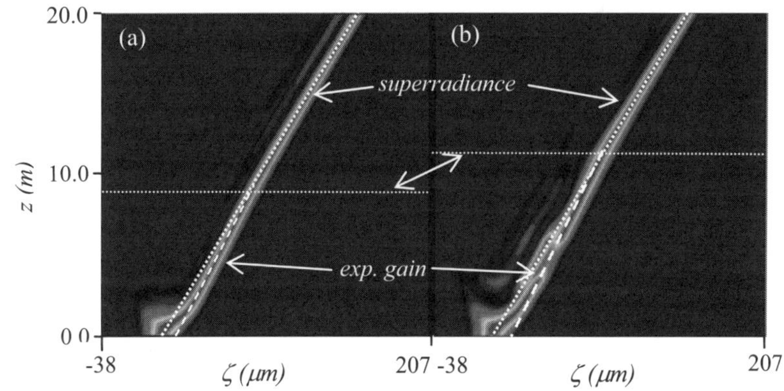

Fig. 7 Contour plot of the pulse intensity evolution in the second stage of the cascade. In (a) and (b) the simulation parameters are the same as in Table 2 except for the input power (2MW) and the number of periods of the second section (N=693). In (b) the second stage is detuned by reducing the undulator parameter by 0.2%.

As pointed out in ref. [17] the exponential gain regime may be recognized by observing the slope of the intensity trace in Figs. 7.a,b. The slope is indeed inversely proportional to pulse velocity and while in the exponential growth regime the pulse is slowed down by the gain process, in superradiant regime the pulse propagates approximately at the velocity of light. This slope change corresponding to the transition may be observed both in Fig. 7.a and 7.b. In Fig. 7.b the effect of the frequency mismatch is compensated by an increased distance of propagation along the undulator exponential gain regime. The evolution of the pulse energy and of the pulse length are shown in Fig. 8 and Fig.9 respectively.

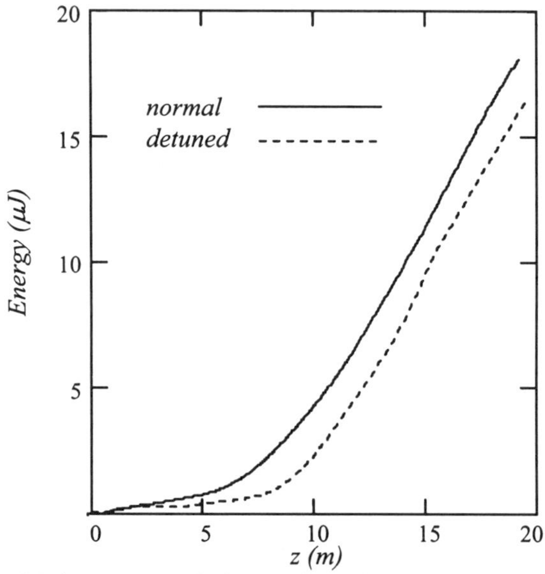

Fig. 8 Pulse energy vs. z in the normal case (as in Fig. 7.a, solid line) and detuned case (as in Fig. 7.b, dashed line).

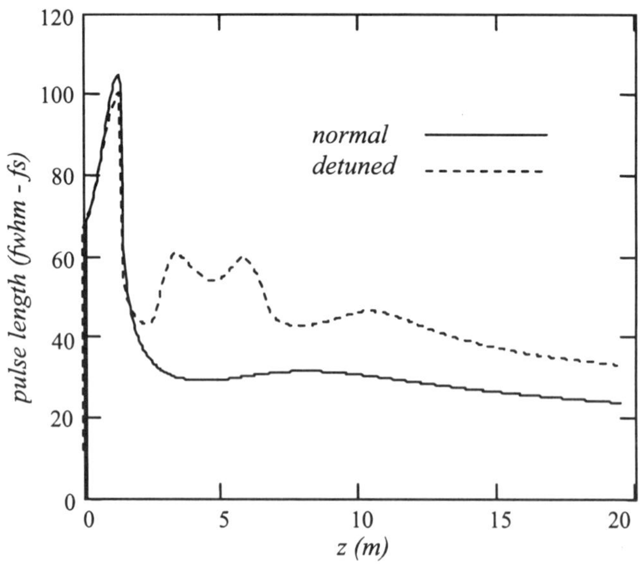

Fig. 9 Pulse length (fwhm) vs. z in the normal case (as in Fig. 7.a, solid line) and detuned case (as in Fig. 7.b, dashed line). Final pulse length is 23 fs, 18 mJ

The frequency detuning has caused a reduction of the final pulse energy from 18μJ to 16μJ and an increase of the pulse length from 23fs to 26fs. This is a small effect if we consider that the detuning of 0.2% is comparable to the bandwidth of the exponential gain spectrum. The reason is the self-compensation of the mismatch between the two sections operated by the longer propagation in the exponential gain regime.

3. Conclusions

The harmonic cascaded FEL is a promising scheme to extend to the short wavelength the operation range of a Free Electron Laser. The wavelength of 66nm has been obtained in simulations with a beam energy of 200 MeV, compatible with the SPARC linac energy. Schemes implementing higher order harmonics in the second section may provide even shorter wavelength without the limitations related to the period length and the FEL gain. One of the aspects that characterize superradiance is indeed the minor role played by the gain process. The FEL may not even enter in exponential gain regime and the energy growth of the superradiant pulse is proportional to the slippage length and to the beam current, i.e. to the number of electrons that are trapped by the moving bucket of the saturated FEL. Despite the emission is occurring at the nth harmonic resonant wavelength in the second section, the slippage length is that of the fundamental harmonic, which is n times larger. The 1D time dependent simulations exploited with Perseo have shown that the basic principle is correct and looks promising. The code Perseo has been cross checked with GENESIS 1.3 in different occasions, and the two codes have provided compatible results when compared in compatible conditions. Diffraction effects in superradiance have been studied in ref.[27], but the layout studied in this context has to be simulated including the effect of the radiation on the higher order harmonics, and this is possible with Perseo, but it is not possible at this moment with GENESIS. A modification of GENESIS to treat higher harmonic lasing is required in order to test the behavior of this system in 3D and to get additional confidence in this scheme by cross checking the predictions of two codes. This will be the subject of forthcoming work where the effects of diffraction may be analyzed together with the sensitivity of the harmonic cascade to beam parameters as energy spread, emittances and beam current.

Acknowledgments

This work has been partially supported by the EU contract no.011935 EUROFEL.

References

1. I. Boscolo and V. Stagno, Il Nuovo Cimento B 58, 267 (1980)
2. R. Bonifacio, R. De Salvo Souza, P. Pierini, E. T. Scharlemann NIM A 296, 787 (1990)
3. L. H. Yu, Phys. Rev. A 44, 5178 (1991)
4. F. Ciocci et al. IEEE J.Q.E. 31, 1242 (1995)
5. H. Freund, NIM A 475, 373 (2001)
6. L. H. Yu et al. Science 289, 932 (2000)
7. L. H. Yu et al. Phys. Rev. Lett. 91, 74801 (2003)
8. S. Werin et al. Design consideration for the coherent radiator FEL in the MAX IV proposal, in proceedings of the 26th International Free Electron Laser Conference and 11th User Workshop, http://www.JACoW.org, 190 (2004)
9. D. Kramer, A Multistage HGHG scheme for the BESSY Soft X-Ray FEL Multiuser Facility, in proceedings of the 26th International Free Electron Laser Conference and 11th User Workshop, http://www.JACoW.org, 108 (2004)
10. C. Bocchetta et al., FERMI@ELETTRA: A seeded harmonic cascade FEL for EUV and Soft X-Rays, in proceedings of the 27th International Free Electron Laser Conference, http://www.JACoW.org, 682 (2005)
11. W. M. Fawley, W. A. Barletta, J. N. Corlett, A. Zholents, Simulation studies of a XUV-Soft X-Ray harmonic cascade FEL for the proposed LBL recirculated linac, in proceedings of the 2003 Particles Accelerator Conference, http://www.JACoW.org, 923 (2003)
12. E. J. Takahashi et al. IEEE J.Q.E. 10, 1315 (2004)
13. M. E. Couprie et al. The Arc en Ciel FEL proposal, in proceedings of the 27th International Free Electron Laser Conference, http://www.JACoW.org, 55 (2005)
14. G. Lambert et al., Seeding the FEL of the SCSS phase 1 facility with the 13th laser harmonic of a Ti:Sa laser produced in Xe gas, in proceedings of the 27th International Free Electron Laser Conference, http://www..JACoW.org, 224 (2005)
15. L. Giannessi et al., Status of the seeding experiment at SPARC, in proceedings of the 27th International Free Electron Laser Conference, http://www..JACoW.org, 63 (2005)
16. E. L. Saldin, E. A. Schneidmille, M.V. Yurkov, Optics Communication 202, 169 (2002)

17. L. Giannessi, P. Musumeci, and S. Spampinati, Journal of Applied Physics 98, 43110 (2005)
18. R. Bonifacio, B. W. J. McNeil and P. Pierini, Phys. Rev. A 40, 4467 (1989)
19. R. Bonifacio, L. De Salvo Souza, P. Pierini, and N. Piovella, NIM A 296, 358 (1990)
20. R. Bonifacio, N. Piovella, B. W. J. McNeil, Phys. Rev. A 44, 3441 (1991)
21. W. B. Colson, G. Dattoli, F. Ciocci, Phys. Rev. A 31, 828 (1985)
22. K. J. Kim, Nucl. Instrum. Methods Phys. Res. A 250, 396 (1986)
23. R. Bonifacio, C. Pellegrini, and L. M. Narducci, Opt. Commun. 50, 373 1984
24. W. B. Colson, Laser Handbook vol 6, *Classical Free Electron Laser Theory*, 301 North-Holland, Amsterdam, 1990
25. L. Giannessi, Phys. Rev. ST-Accel. Beams 6, 114802 (2003)
26. W. M. Fawley, Phys. Rev. ST Accel. Beams 5, 70701 (2002)
27. L. Giannessi, S. Spampinati, P. Musumeci, Non Linear Pulse Evolution In Seeded And Cascaded Fels, in proceedings of the 27th International Free Electron Laser Conference, http://www.JACoW.org, 211 (2005)

HIGH BRIGHTNESS LASER INDUCED MULTI-MEV ELECTRON/PROTON SOURCES*

D. GIULIETTI†, E. BRESCHI, M. GALIMBERTI, A. GIULIETTI, L.A. GIZZI, P. KOESTER, L. LABATE, P. TOMASSINI
Intense Laser Irradiation Laboratory,IPCF-CNR, Pisa and INFN-Pisa, Italy

PH. MARTIN, T. CECCOTTI, P. DE OLIVERA, P. MONOT
Physique à Haute Intensité, CEA-DSM/DRECAM/SPAM, Saclay, FRANCE

M. BORGHESI, L. ROMAGNANI, S. KAR
Queen's University of Belfast

S. BERTOLUCCI, M. CALVETTI
Laboratori Nazionali di Frascati, INFN, Italy

A. SCHIAVI
Dipartimento di Energetica, Università "la Sapienza", Roma, Italy

O. WILLI
Heinrich-Heine Universitat Dusseldorf, Germany

The chirped pulse amplification (CPA) technique has opened new perspectives in the radiation-matter interaction studies using ultra-short laser pulses at ultra-relativistic intensities. In particular the original idea, proposed by Tajima and Dawson, of accelerating electrons by the huge electric fields of plasma waves which develop in the wake of a laser pulse propagating in a plasma, become feasible. Some laboratories all over the world have produced by such a technique collimated electron busts of hundreds of MeV along acceleration lengths of a few hundreds of microns. In other experiments, using thin solid targets, intense bursts of energetic protons have been at the same time detected. The proton acceleration mechanism is essentially based on the Coulomb force appearing at the thin solid target surface as a consequence of the previous escape of the energetic electrons from the target. In the paper some experimental results will be presented as well as the opportunities the INFN PLASMONX project will offer in this research field at LNF.

* This work is partially supported by INFN project PLASMONX.
† Also at Physics Department of Pisa University, Largo B. Pontecorvo n.3, 56127, Pisa, Italy. E-mail : danilo.giulietti@df.unipi.it

1. Introduction.

The current slowing down of further development of the experimental research in high energy particle physics is mainly due to the gigantism of conventional particle accelerators and their cost hardly sustainable, even by international consortia as the CERN.

The size of conventional accelerators producing high energies particles have to be very large in both cases of linear and circular acceleration. In fact, the maximum allowed accelerating electric field must be below the material breakdown threshold which is typically of the order of several tens of MV/m. In the case of a 1TeV accelerator, this condition sets an acceleration length not less than 20Km for an accelerating field of 50MV/m. In the case of circular accelerators, relatively large dimensions are still required, because of the synchrotron radiation losses.

This breakdown limitation does not apply to plasmas which are therefore regarded as the ideal medium in which charged particles can be accelerated by means of electric fields that are order of magnitude higher than those of conventional accelerators [1]. After the first pioneering experiments on plasma accelerators carried out in the early eighties, impressive progress in this field is being achieved today, following the advent of the ultra-short pulse lasers and the Chirped Pulse Amplification (CPA) technique [2]. In fact, thanks to CPA, pulses of a few joules in tens of femtoseconds can now be produced and used to induce the judge electric fields that can accelerate charged particles up to 1GeV in a few mm length. In the last years theoretical and experimental activity has been developed achieving more and more encouraging results concerning the energy and the quality of the electron/proton beam produced by laser-driven acceleration mechanisms [3, 4, 5, 6, 7, 8, 9, 10,11,12,13,14].

2. Laser driven acceleration.

Electron plasma waves (EPW) provide the basic conditions for the generation of accelerating electric fields in plasmas. Several schemes can be employed to excite EPW in plasmas. A simple approach is based upon intense laser pulses to induce strong perturbations in the electron density distribution. Initially, the so-called laser beat-wave technique was adopted (see [15] and references therein) to produce sufficiently strong electron density perturbations and the subsequent formation of relativistic electron plasma waves.

More recently, with the advent of CPA technique, a single, very intense laser beam can be focused inside a plasma to generate an EPW along the laser propagation path, leading to a substantial reduction of the complexity of the experimental set up. If the longitudinal size of the laser pulse is about half (or less) of the wavelength of the natural oscillation mode (λ_p) of the plasma, a high amplitude plasma wave develops quasi-resonantly on the wake of the pulse, excited by the action of the longitudinal ponderomotive force [16]. Since λ_p depends on the plasma density (n_e) through the plasma frequency (ω_p):

$$\lambda_p = \frac{2\pi c}{\omega_p} \propto \frac{1}{\sqrt{n_e}}$$

the condition for the quasi-resonant growth can be expressed as follows:

$$\tau_{laser} c \approx \frac{\lambda_p}{2} \Leftrightarrow \tau_{laser} \approx \frac{T_p}{2} \Leftrightarrow n(cm^{-3}) \approx \frac{3 \cdot 10^{-9}}{\tau_{laser}^2(s)}$$

Where τ_{laser} and T_p are the laser pulse duration and the period of the electromagnetic field respectively. The detailed description of the EPW excitation and evolution, along with the dynamics of the laser pulse inside the plasma, can be achieved by means of fully relativistic particle codes.
The Laser Wake-Field Acceleration (LWFA) is considered a quasi-resonant process, in the sense that the resonant condition is not so severe. The crucial issue here is the identification of key experimental parameters that allow a significant control of the interaction conditions required to ensure reproducibility of acceleration performance. In the following we identify two schemes in which the aspects of quality and control are faced with different approaches.

2.1. *Electron Acceleration in E P W by Self-Trapping.*

In this case the energy gain of the electrons depends on the conditions of the injection in the EPW (i.e. electron velocity and phase of the wave) and those of their expulsion from the wave. This scheme has been widely studied experimentally and electron energies above 100 MeV level have been achieved and reproduced in a number of laboratories world-wide.
In a recent experiment [17] performed at the Laboratoire d'Optique Appliquée, very collimated bunches of high energy electrons have been produced by focusing a super-intense femtosecond laser pulse in sub-millimetre under-dense plasmas. The density of the plasma, pre-formed with the laser exploding-foil technique, was mapped using Nomarski plasma interferometry. The electron

beam was fully characterised: up to 10^9 electrons per shot were accelerated, most of which in a beam of aperture below 10^{-3} sterad, with energies up to 40 MeV. These measurements, which are well modelled by 3-D numerical simulations, validate a reliable method to generate ultra-short and ultra-collimated electron bunches. Fig.1 shows a typical radiochromic film result [18] after a single exposure to the electron beam generated by focusing the CPA pulse at an intensity of $\approx 8\text{x}10^{19}\text{W/cm}^2$ in the plasma produced by the explosion of a $1\mu\text{m}$ thick target. These images show a small central spot very closely aligned with the laser propagation axis. This feature is visible on all the seven layers while other features surrounding this bright spot are only visible on the first layer. These results suggest that besides an intense electron flux of lower energy in a ring of $\theta_{\text{ring}} \approx 18$ deg aperture, there is a bunch of very collimated, high energy electrons accelerated along the laser propagation axis, $\theta_{\text{bunch}} < 1$ deg.

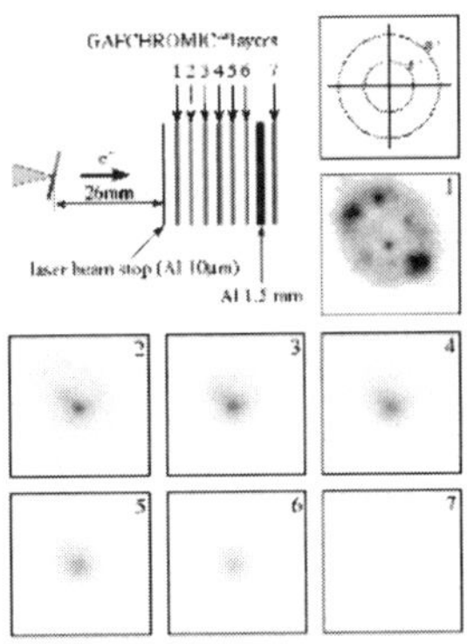

Fig. 1 - Densitometer scans of the radiochromic films (see insert) after exposure to the energetic electron beam produced by interaction of an ultraintense laser pulse with a preformed plasma located at 26 mm from the first film. The CPA pulse intensity in the plasma is $8\text{x}10^{19}\text{W/cm}^2$. The angular scale is also shown for reference.

The experimental results have been compared with the predictions of numerical simulations carried out using a three-dimensional particle-in-cell code with a plasma density profile and a laser pulse intensity distribution very close to the experimental ones. The angular distribution obtained from the simulations clearly shows the main feature of that observed experimentally, namely the central, collimated electron beam is well reproduced. A typical spectrum of the accelerated electrons is shown in Fig.2 (black dots). According to this plot, the electron beam consists of a sizeable number of electrons with energies from a few MeVs up to 40 MeV. Also plotted in Fig.2 (solid line) is the calculated spectrum of the very collimated, high energy electrons accelerated along the laser propagation axis, obtained from the PIC simulation. There is a very good agreement between the experimental spectrum of the high energy electron beam and the simulated one.

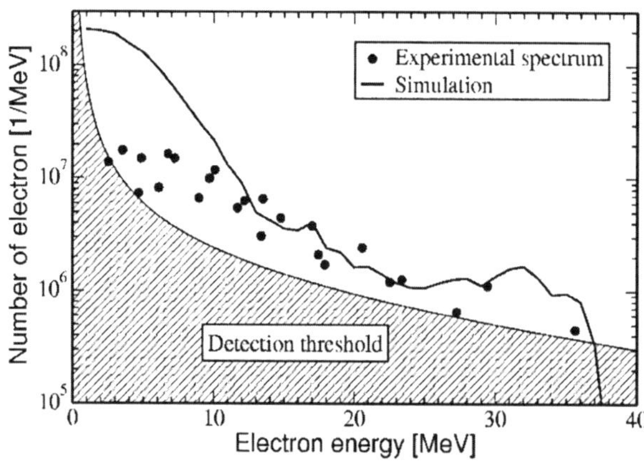

Fig. 2 - Spectrum of the electrons in the collimated beam. Filled circles represent the experimental electron spectrum. Also shown is the detection threshold of the electron spectrometer. The solid line is the spectrum of the electrons in the collimated beam as obtained from the PIC simulation.

The kind of target used to produce the plasma in which the electric field of the Langmuir waves can accelerate the electrons is one of the main point to be addressed in the experiments. In fact the conditions for an effective and controlled acceleration concern the laser parameters (intensity, pulse duration, focal spot diameter, pulse/pre-pulse contrast ratio,...) as well as the plasma parameters (dimensions, density distribution, temperature,...). Gas jets are now regarded as the most promising targets because they allow high repetition rate for the laser-target interactions and satisfactory control of the electron density distribution of the produced plasma [19, 20].

In a recent experiment at SLIC facility (CEA-Saclay) the interaction of 60fs laser pulses with gas jet targets has been studied by using optical diagnostics and detectors for energetic particles and X/γ radiation. The maximum laser intensity was, in the focal spot, 10^{19}Wcm^{-2}. In this regime of fairly relativistic

laser intensities, different experimental parameters have been varied to match the best conditions for the production of energetic electron bunches.
The experiment was performed with a 10 TW Ti:Sa laser system operating in the CPA mode and delivering up to 0.6 J in 60-fs pulses at 800-nm. The contrast of the main laser pulse on both the nanosecond and picosecond scale were measured by a third order auto-correlator. The nanosecond contrast, due to Amplified Spontaneous Emission (ASE), was found to start about 1ns before the main pulse with a contrast ratio of $\approx 10^{-6}$, while the *pedestal* of the femtosecond pulse, due to the pulse compression, had a contrast ratio of $\approx 10^{-4}$ for a duration of about 1ps before the main pulse. These values are typical of most high power femtosecond lasers.
The pulse was focused in the gas jet using a 200 mm, silver coated off-axis parabolic mirror. The numerical aperture of the focusing optics was f/2.5 and the maximum focused nominal intensity was 10^{19} Wcm^{-2} in a slightly elliptical focal spot whose horizontal and vertical diameters were 8μm and 10μm FWHM respectively. The corresponding normalised field parameter in the focal spot was $a_o = 2.2$. A fraction of the main femtosecond pulse was frequency-doubled using a 2 mm thick, type I KDP (Potassium Dihydrogen Phosphate) crystal and used as an optical probe propagating perpendicular to the main pulse, in a Mach-Zehnder interferometer configuration. The probe pulse is expected to have a pulse duration of approximately 120 fs, due to group velocity dispersion in the 2mm frequency doubling crystal. The gas-jet was delivered by a 1 mm diameter, cylindrical nozzle. The maximum gas pressure in the valve was about 8 bar corresponding to a neutral gas density in the jet greater than 10^{19} atoms/cm^{-3}. The value of neutral gas density as well as the plasma density was obtained from a Fourier analysis [21] of the fringe pattern taken with the Mach-Zehnder interferometer.
A systematic study of the energetic electron production has been performed for different plasma densities (changing gas type (He, N_2) and pressures), plasma shapes (changing the nozzle) and for different focusing points with respect to the gas jet (see Fig.3).
The major production of energetic electrons has been observed when the main laser pulse was focused at the border of the gas plume (right side in the Fig.3) and for the highest plasma densities attained with the employed apparatus (N_2 pressure in the valve at about 8 bar). Moreover the production of energetic electrons was rather unpredictable, even if clearly correlated with the enhanced propagation of the fs laser pulse, most likely due to the on set of the beam self-focusing.

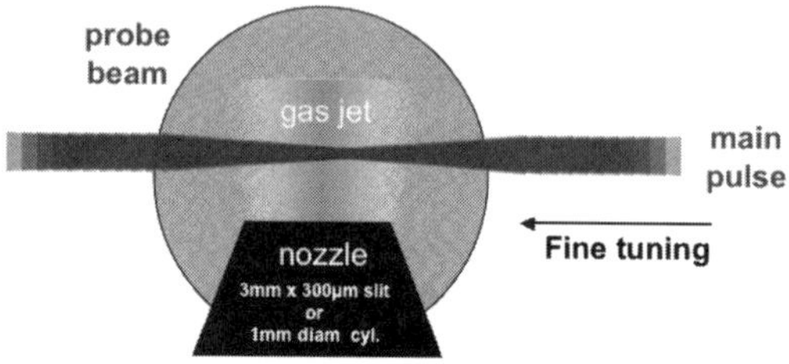

Figure 3. Detail of the gas jet and the interaction region.

To get a quantitative information about the energy spectrum of the produced electrons a plot of the released energy as a function of the incident electron energy is calculated for each layer of the detector by means of MonteCarlo simulations.

The measured dose (obtained by scanning the radiochromic films) is then converted into electron energy distribution. In Figure 4 the number of electrons produced per MeV as a function of their energy in MeV are plotted for four shots to evidence the variability of the acceleration process. As one can see in some shots energetic electrons up to 3 MeV have been produced.

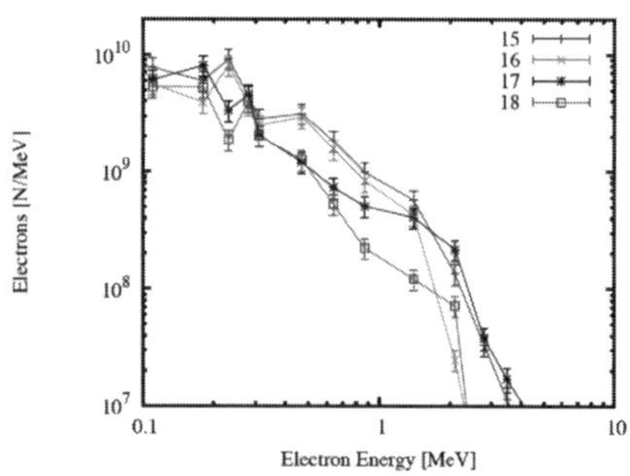

Figure 4. Electron energy distribution at 7.6 bar N_2 pressure in the gas valve.

From the space distribution of the dose on the first radiochromic film, that was 7mm apart from the laser produced plasma, come out the electrons were emitted forward in a cone of about 60^0 full aperture.

2.2. *Suitably shaped pre-formed plasma.*

Several teams are investigating, by theoretical and experimental approaches, different possibilities of reducing the energy spread of the electrons accelerated by laser plasma techniques.

In 1998 S.Bulanov et al [22] proposed a novel scheme for the production of high-quality electron beams in LWFA in which a controlled longitudinal non-linear wave-breaking is induced by an electron density depletion. This proposal was supported by both analytical and numerical results in a one dimension configuration. In fact, if the plasma density decreases in the direction of the pulse propagation (z coordinate), the wave number k increases with time and the very simple relation holds:

$$\frac{\partial k_p}{\partial t} = -\frac{\partial \omega_p}{\partial z}$$

The resulting decrease of the phase velocity, at the interface between two uniform density regions, makes the wave to break, even when its initial amplitude is below the wave-breaking threshold. As a result of the break at the interface between the two regions, fast electrons from the wave crest are trapped by the wave and are pre-accelerated into the lower density region where the wake field remains regular. An energy balance argument shows that in the weakly inhomogeneous case the relative density of fast electrons is approximately given by the ratio

$$\frac{v_{ph}}{\omega_{pe} L}$$

where ω_{pe} is the plasma frequency at the interface and L its inhomogeneity length. In the following the results of PIC simulations concerning such an acceleration scheme will be shortly presented [23]. In Fig.5 the electron density map of a pre-formed plasma, consisting of two quasi uniform regions with a sharp transition, is reported. After a smooth vacuum-plasma transition, the longitudinal profile of the electron density reaches a first plateau (n_e=2.1 10^{19}cm^{-3}) and then it decreases abruptly with a scale length L=2μm to a second plateau (n_e=1.1 10^{19}cm^{-3}). In the same figure the electron plasma wave excited by a laser pulse propagating in the plasma is apparent, as well as the first bunch of accelerated electrons (see the arrow).

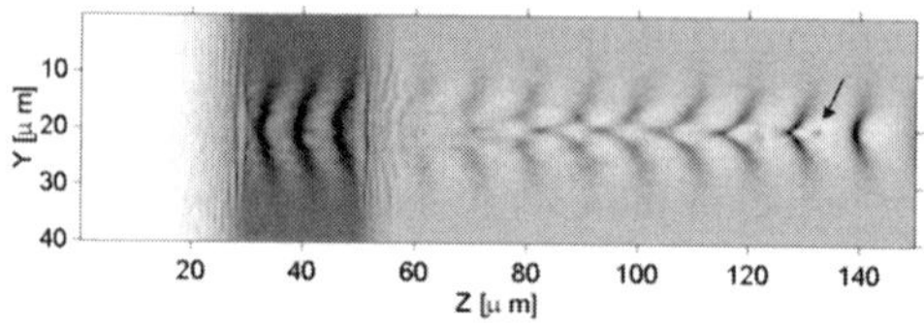

Fig.5 Electron density map of the pre-formed plasma, with a sharp density depletion, crossed by an intense laser pulse: 17fs, λ=1μm, I=2.5 10^{18}W/cm^2. The excited electron plasma wave as well as the accelerated electron bunch (arrow) is apparent.

The energy spectrum of the accelerated electrons is reported in Fig.6. We can see that, selecting the electrons with energy exceeding 7MeV, the resulting bunch of $N_e \approx 10^8$ electrons with energy E≈10MeV is characterized by a remarkable good quality and has an energy spread ΔE/E≈5%, transverse and longitudinal size δy≈1μm, δz≈5μm respectively and emittances $\varepsilon_{rms}^{tr} = 9 \cdot 10^{-2}\, mm \cdot mrad \quad \varepsilon_{rms}^{lon} = 2 mm \cdot KeV$.

The possible applications of such high quality beams include injection of electrons in a LINAC accelerator for the production of high energy beams and the production of ultra-short pulses of X rays with linear and non linear Thomson backscattering.

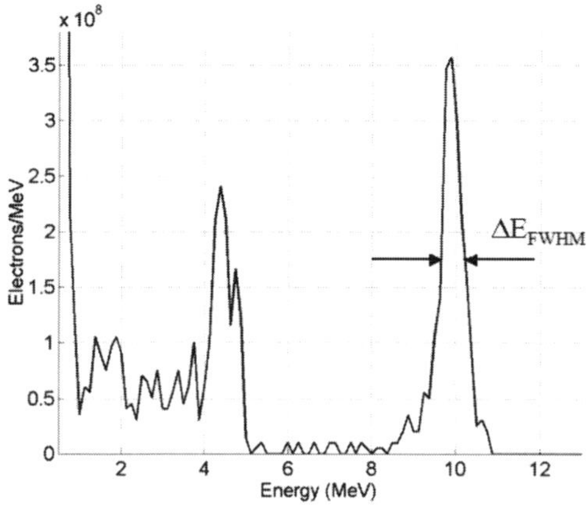

Fig.6 Energy spectrum of the accelerated electrons in pre-formed plasma with a sharp density depletion.

2.3. *Acceleration of Externally Injected Electrons.*

The idea here is to inject electron bunches produced by an external source in the electron plasma wave in such a way that the electrons experience a huge accelerating field and their energy is thus increased in a very short distance. In this configuration, the external source of electron bunches can be either a conventional photo-injector or a plasma photo-injector as described in previous sections.

As in the case of self-injection, also external injection was initially pursued using the so-called laser beat-wave driven relativistic plasma wave. However, experiments carried out so far using intense, ultrashort CPA pulses have demonstrated acceleration gradients which are up to two orders of magnitude larger than those achieved in laser beat-wave experiments and suggest a perspective of much more compact accelerators [24], provided that schemes for relatively long interaction length are demonstrated.

The injection of monochromatic electron bunches from a LINAC into plasma waves, previously excited by a super-intense and ultra-short laser pulse propagating in a pre-formed plasma, is one of most ambitious goals of three sub-disciplines: accelerator physics, plasma physics and laser physics. This acceleration scheme can allow, in principle, high energy and monochromatic electron bunches to be produced in a very short distance. However severe constraints have to be fulfilled to get the required low values of emittance and energy spread. In fact the electron bunch length have to be much shorter than the plasma wave wavelength and the injection phase has to be controlled within a few degrees of the plasma oscillation. For instance in the case of a plasma wavelength of 200-300 μm, the electron bunches have to be shorter than 10μm and syncronized with the laser pulse within less than 100fs.

Here we describe the result of a numerical simulation in which an electron bunch produced by a LINAC, at a given energy, is injected into a plasma for further acceleration. The simulation technique is based upon a test-particle code that assumes a quasi-static field approximation. Transverse fields are neglected, while non-linear effects are treated in a 2-dimensional representation. The injected electron bunch has a charge of 21 pC at an energy of 130 MeV, in an 8μm transverse size and a longitudinal size of 12 μm. The energy spread is assumed to be 0.1%.

From the point of view of the laser-plasma accelerator, the simulation uses a 4J, 100 fs duration pulse at 800 nm wavelength, focused in a 50 μm focal spot. This laser pulse is focused in a preformed plasma at a density of $1.5x10^{17}$ cm^{-3}. The maps of Figure 7 show the electron density distribution of the electron plasma wave excited by the laser and the associated longitudinal electric field.

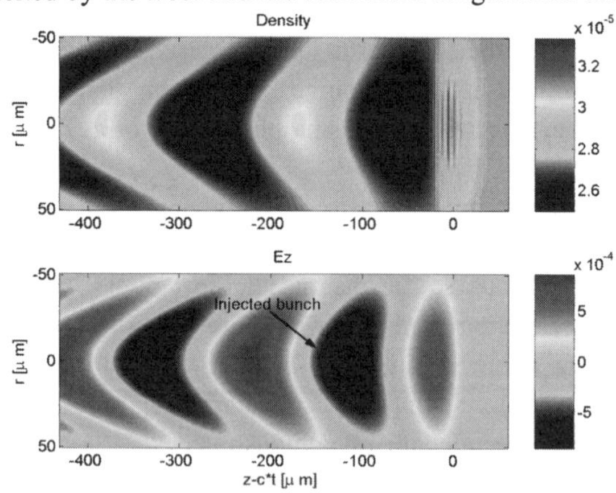

Figure 7 - (top) Map of the electron density, normalized to the critical density at 800 nm, showing the charge separation of the electron plasma wave induced by the high intensity laser pulse. (bottom) Map of the longitudinal electric field associated to the EPW. Also visible in the frame is the bunch of externally injected electrons.

The bunch, also visible in the map, has a longitudinal extent which is small compared to the wavelength of the EPW. The energy distribution of the electrons at the exit of the plasma is shown in Figure 8 for two different plasma acceleration lengths of 20 mm and 100 mm.

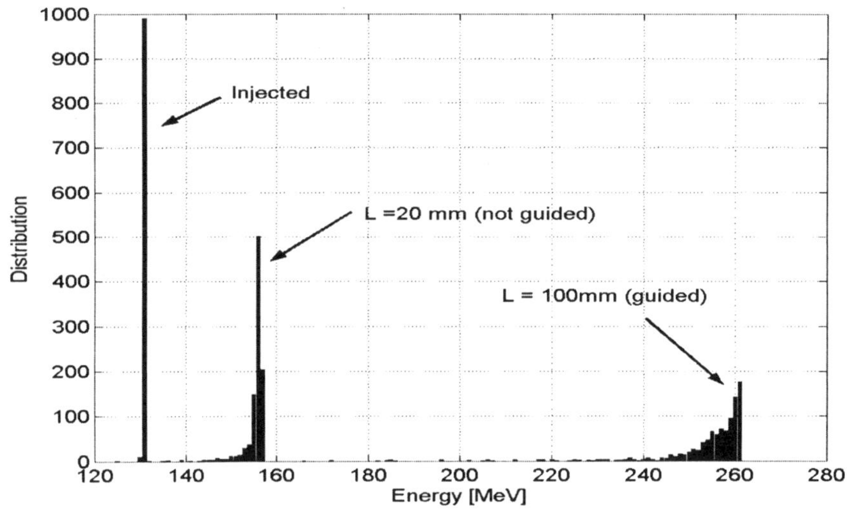

Figure 8 - Electron energy distribution at the exit of a plasma of 1.5×10^{17} cm^{-3}, for an acceleration length of 20 mm, corresponding to twice the Rayleigh length Z_R of the focusing optics. Also shown in the plot is the result of an accelerator length of 100 mm, assuming a laser pulse guiding over 5 Z_R. The energy spread of the output electrons is 3% (rms).

These results provide us with a well defined range of parameters that are the ground on which the performance of the laser facility proposed in the PLASMONX project will operate (see section **4.**). Together with the calculation tools benchmarked with test experiments carried out so far, they supply a powerful and robust framework for the R&D activity on laser-driven acceleration at LNF.

3. Laser produced proton beams.

When an ultra-intense laser pulse is focused on a solid target a sizeable fraction of the laser energy is transferred to charged particles [25]. In particular, the generation of multi-MeV proton beams has been observed. Such proton beams are now under consideration for application in radiotherapy as well as medical diagnostic. In spite of the growing interest on laser produced proton beams, the physical mechanisms of generation and acceleration of such protons is only

partially understood, even if it undoubtedly concerns the electric field produced by the fast electron escaping the irradiated target.
During an experimental campaign at CLF, the Vulcan Nd:glass laser operating in the Chirped Pulse Amplification mode was used to investigate this mechanism. The targets used for the production of the proton beams were 1 mm wide and about 1 cm long Al foils. The target thickness was 25, 100 or 250 μm. The energy of the laser pulse was about 100 J. The diameter of the focal spot on target was about 10 μm (FWHM) containing almost 40% of the energy. The duration of pulse was 1 ps allowing intensities up to 10^{19} Wcm^{-2}. The protons were detected using several layers of Radio-Chromic Films (RCF). A schematic experimental setup is shown in Figure 9.

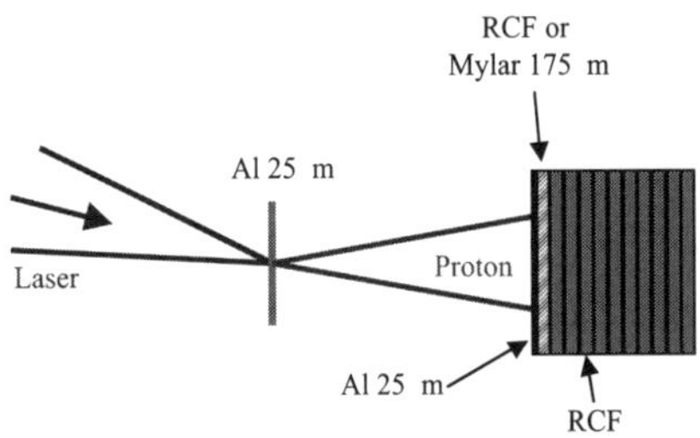

Figure 9. Experimental setup for the detection and characterization of proton beams.

A fundamental step towards the experimental characterization of laser produced proton beams is the use of a detection technique capable of providing good spatial and spectral resolution in single shot measurements and in the presence of the large electromagnetic noise typical of high power, ultrashort laser interaction experiments.
The detection technique based on RCF exploits the proportionality between the optical density (OD) variation of active layers and the energy deposited locally by the impinging particle beam (absorbed dose). The main disadvantage in the use of RCF is that from the signal recorded it is not possible to discriminate between different types of ionizing particles (electrons, protons or other kind of ions). Therefore additional measurements, possibly based upon track particle detectors (e.g. CR39) are used to identify the nature of particles.
A straightforward interpretation of data from RCF does not provide immediately a complete spectral information. A numerical analysis is required in order to reconstruct the proton beam spectrum. An algorithm was developed [26,27] to reconstruct the energy-dependent proton density. The main idea behind the

algorithm is to reconstruct the proton angular distribution and spectrum at a particular sampling energy. The choice of this energy value is obtained by comparison with the response of the detector to monochromatic and mono-directional protons beams. The numerical expression of the detector response is calculated by means of a Montecarlo simulation.
Let us consider data from a typycal shot taken in the case the target was a 25 μm thick Al foils. The generated proton beam was detected using a stack of RCF type MD-55 by Gafchromic. The numerical reconstruction of the proton beam profile is reported in Figure 10 where the angular distribution of the protons at the sampling energies for both shots are shown. The space resolved spectrum shows a ring where the proton density is enhanced, with a decreasing radius for increasing proton energy, as previously observed with CR39 detectors [28]. This is a clear evidence that protons are emitted with a quasi-monochromatic component localized at the center of the beam.

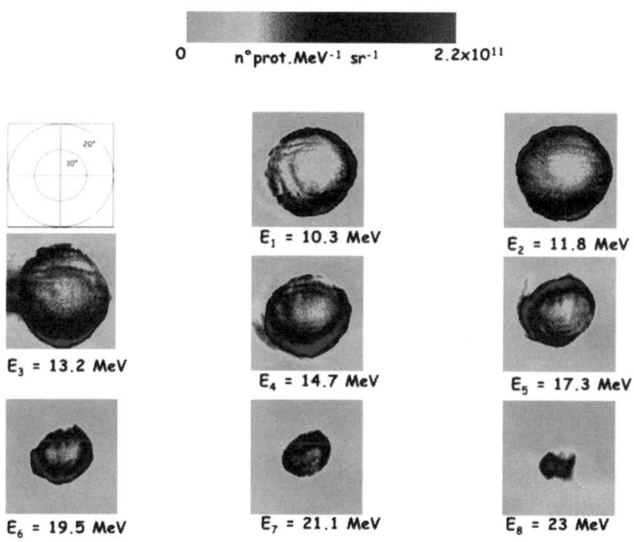

Figure 10. Angular distribution at different proton energies.

4. PLASMONX

The aim of the PLASMONX (**PLAS**ma acceleration and **MON**ochromatic **X**-ray production) project is to provide LNF with a world-class, high-power laser facility suitable for the development of an innovative, high-gradient acceleration technique based upon super-intense and ultra-short laser pulses, and X/γ-ray tunable sources, using Thomson scattering of optical photons by energetic electrons. The facility will be built in close interaction with the SPARC project [29] presently in progress at LNF. The main purpose of the facility will consist in R&D activity aimed at the following objectives:
demonstration of high-gradient acceleration of electrons injected into electron plasma waves excited by ultra-short, high power laser pulses;
development of a monochromatic and tunable X-ray source in the 20-1000 keV range, based upon Thomson Scattering of laser pulses by relativistic electrons.
Both lines require very bright and short electron bursts together with very high power laser pulses of sub-100fs duration. Studies at the forefront of research in this field consistently indicate that the power level and the focused intensity for a laser system appropriate for these studies are >10 TW and $>10^{19}$W/cm^2 respectively.
An important point to face in this kind of experiments concerns the design of the target on which the super-intense laser pulse is focused. In fact while plasmas produced from gas jet targets are very suitable for electron acceleration, thin foils are been fruitfully experimented in high energy proton/ion production.
The R&D activity carried out at this facility will be open to some extent to external users, in particular for what concerns bio-medical oriented uses [30] of the available sources for diagnosis and therapy.

5. Conclusions.

Laser-driven acceleration is now emerging worldwide as a mature approach to particle acceleration. The paper reports on recent results obtained in the major European facilities involved in this field. The idea of embedding a highly advanced laser laboratory in a LINAC dedicated laboratory is a necessary step in order to establish this technology in Italy. With the ongoing SPARC activity dedicated to FEL development at LNF, the PLASMONX initiative is most appropriate and timely. Chances of success of the proposed program and the possibility of achieving high impact scientific results in the field of compact accelerators and ultra-short X-ray sources are now very high and call for a prompt transition from planning to execution.

6. Acknowledgments.

We would like to thank the staff of the LOA, CLF-RAL and SLIC facilities for their invaluable and friendly support. We also acknowledge the LASERLAB EUROPE Trans-national Access Programme and the CINECA High Performance Computing.

References

11 Tajima T. and Dawson J.M., Phys.Rev. Lett, **43**, 267. (1979).
2 Strickland D. and Mourou G., Opt. Comm., **56**, 219.(1985).
3 Katsouleas T., Nature, **431**, 515 (2004).
4 S.P.D. Mangles et al., Nature, **431**, 535 (2004).
5 C.G.R. Geddes et al., Nature, **431**, 538 (2004).
6 J. Faure et al., Nature, **431**, 541 (2004).
7 V. Malka et al., Laser and Particle Beams, **20**, 217-221 (2004).
8 H. Hora et al., Laser and Particle Beams, 1**8**, 135-144 (2000).
9 K. Nakajima, Laser and Particle Beams, **18**, 519-528 (2000).
10 F. Dorchies et al., Laser and Particle Beams, **17**, 299-305 (1999).
11 K. Nakajima, Laser and Particle Beams, **18**, 519-528 (2000).
12 F. Dorchies et al., Laser and Particle Beams, **17**, 299-305 (1999).
13 S.V. Bulanov, V.S. Khoroshkov, Plasma Phys. Rep. **28**, 453, (2002).
14 M. Borghesi et al, Phys. Rev. Lett., **88**, 135002, (2002).
15 C. Joshi et al., Nature, **311**, 525 (1984).
16 E. Esarey et al., IEEE Trans. On Plasma Sci. **24** (2) 252 (1996).
17 D. Giulietti et al., Phys. Plasmas, **9**, 3655 (2002).
18 M. Galimberti et a., Rev. Sci. Instrum. **76**, 053303 (2005)
19 J. Faure, V. Malka, J.-R. Marques, et al., Phys. Plasmas **10**, 2039, (2003)
20 C.B. Schroeder, E. Esarey, C.G.R. Geddes et al., Phys. Plasmas **7**, 3009, (2000)
21 M. Takeda, H. Ia, and S. Kobayashi, J. Opt. Soc. Am. **72**, 156, (1982); K.A. Nugent, Appl. Opt. **18**, 3101 (1985).
22 Bulanov S.et al., Phys. Rev.E, 1998
23 Tomassini P. et al, Phy. Rev. Special Topics-A&B, **6**, 121301, 2003
24 F. Dorchies et al., Phys. Plasmas **6**, 2903 (1999).
25 S.J. Gitomer et al., Phys. Fluids, **29**, p. 2679 (1986)
26 E. Breschi et al., Nuclear Instr. and Methods A, **522**, p. 190 (2004)
27 E. Breschi et al., Laser and Part. Beams, **22**, p. 393 (2004)
28 E.Clark *et al*, Phys. Rev. Lett., **84**, p. 670 (2000)

29 D. Alesini et al., NIM-A **507** (2003) 345-349.
30 D. Giulietti, The Congress of Società Italiana di Fisica, Brescia, 2004.

EMITTANCE LIMITATION OF A CONDITIONED BEAM IN A STRONG FOCUSING FEL UNDULATOR

Z. HUANG, G. STUPAKOV

Stanford Linear Accelerator Center, Stanford, CA 94309

S. REICHE

University of California at Los Angeles, Los Angeles, CA 90095

Various methods have been proposed to condition an electron beam in order to reduce its emittance effect and to improve the short-wavelength free electron laser (FEL) performance. In this paper, we show that beam conditioning does not result in a complete elimination of the emittance effect in an alternating-gradient focusing FEL undulator. Using a one-dimensional model and a three-dimensional simulation code, we derive a criteria for the emittance limitation of a perfectly conditioned beam that depends on the focusing structure.

1. INTRODUCTION

A primary factor limiting the performance of short-wavelength, high-gain free electron lasers (FELs) is the electron beam transverse emittance. In order to generate transversely coherent radiation with the diffraction limited emittance $\lambda_r/4\pi$ (λ_r being the radiation wavelength), the ideal electron transverse emittance ε should be less than $\lambda_r/4\pi$ for the efficient beam-radiation interaction. This requirement is usually not met in the x-ray wavelength region. The spread in the transverse betatron motion degrades the resonant FEL interaction and limits the short-wavelength reaches of an x-ray FEL.

To mitigate this problem, it has been proposed to "condition" an electron beam prior to the undulator by increasing each particle's energy in proportion to the square of its betatron amplitude [1]. This conditioning enhances the FEL gain by reducing the axial velocity spread within the electron beam generated over the undulator, due to both energy spread and finite transverse emittances.

The original proposal [1] to condition the beam utilizes a set of cavities in TM_{210} mode immersed in a focusing lattice. Later approaches to the problem invoke a slow TM waveguide mode internal to the undulator [2], and an energy chirp in the beam in combination with a chromatic delay line [3,4,5,6]. It has also been shown that an attempt to condition the beam on a short distance is accompanied by a head-tail focusing variation which can result in the large effective transverse emittance growth [4]. Although this emittance growth can be avoided [6], the pace of conditioning becomes much slower, and the required length of the conditioner considerably increases. Other proposals are based on using lasers and include Thomson backscattering [7], and interaction of the laser with the electron beam in two dedicated undulators [8]. Recently, beam conditioning using nonlinear RF acceleration is also discussed [9].

In the FEL gain analysis of the original proposal [1], the undulator natural focusing is assumed to confine the electron beam in both transverse directions. This results in the complete elimination of the emittance effect for a perfectly conditioned beam. In this paper, we study a more practical situation when the transverse focusing is provided by alternating-gradient quadrupoles in the undulator, as found in typical x-ray FEL designs. We show that although conditioning eliminates the effect of the average slippage of electrons relative to the radiation phase, there remain phase oscillations with the period given by the focusing lattice. These phase oscillations can affect the FEL performance and impose a limitation on the beam emittance even for a perfectly conditioned beam, albeit this emittance limitation is much relaxed as compared to a unconditioned beam.

2. emittance effect in a strong focusing undulator

Although a magnetic undulator can provide natural focusing in both transverse planes [10], the focusing strength is typically too weak for the high-energy electron beam that drives an x-ray FEL. Thus, alternating-gradient quadrupole magnets are inserted in undulator breaks to provide the necessary strong focusing, usually in the form of a FODO lattice (consisting of repetitive focusing-undulator-defocusing-undulator cells). The horizontal betatron motion is given by

$$\begin{aligned} x_\beta(z) &= \sqrt{2J_x\beta_x}\cos\Phi_x(z)\,, \\ p_x(z) &\equiv \frac{dx_\beta}{dz} = -\sqrt{\frac{2J_x}{\beta_x}}\left[\sin\Phi_x(z)+\alpha_x\cos\Phi_x(z)\right]\,, \end{aligned} \tag{1}$$

where J_x is the horizontal action of the electron, α_x and β_x are the Twiss parameters, and p_x denotes the angle of the orbit with the z axis. The second term in Eq. (1) is ignored in Ref. [11] under the smooth approximation. Its importance in FEL dynamics and emittance compensation (i.e., conditioning) is pointed out in Ref. [12].

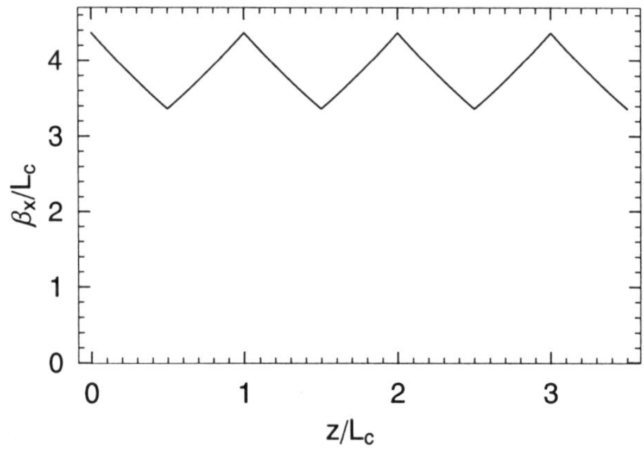

Figure 1. Variation of the beta function along the distance of the lattice for the phase advance per cell of 15 degrees. The derivative $d\beta_x/dz$ is close to the values ± 2, but the deviation of β_x from the average value $\bar{\beta}$ is relatively small.

To avoid a large beam size variation in the undulator, the FODO lattice is usually designed to have a small phase advance per cell. In this case, the average betatron wavelength $2\pi\bar{\beta}$ is much larger than the FODO cell length L_c. In a first approximation the beta function is constant over the length of the undulator, and the accumulated betatron phase is

$$\Phi_x(z) = \phi_x + \int_0^z \frac{ds}{\beta_x(s)} \approx \phi_x + \frac{z}{\bar{\beta}}, \tag{2}$$

where ϕ_x is the initial electron phase. More importantly for the emittance compensation, such a FODO lattice has [12]

$$\alpha_x = -\frac{1}{2}\frac{d\beta_x}{dz} \approx \pm 1 + O\left(\frac{L_c^2}{\bar{\beta}^2}\right) \tag{3}$$

with the sign alternating per half cell. Fig. 1 illustrates the variation of the beta function in a FODO lattice with 15 degrees phase advance per cell.

Under this short-cell-length approximation, the square of the horizontal divergence is

$$p_x^2 \approx \frac{2J_x}{\bar{\beta}}\left[1 \pm \sin\left(\frac{2z}{\bar{\beta}} + 2\phi_x\right)\right]. \tag{4}$$

Similarly in the vertical y direction, we have

$$p_y^2 \approx \frac{2J_y}{\bar{\beta}}\left[1 \mp \sin\left(\frac{2z}{\bar{\beta}} + 2\phi_y\right)\right]. \tag{5}$$

Here the signs $\pm$ becomes $\mp$ due to the focusing-defocusing asymmetry. We have also assumed that the average beta function is the same in both directions.

The FEL interaction is a resonant phenomenon that depends critically on the evolution of the electron phase relative to the co-propagating radiation field. The so-called ponderomotive phase is defined as $\theta = (k_r + k_u)z - ck_r\bar{t}$, where $k_r = 2\pi/\lambda_r$ is the radiation wavenumber, $k_u = 2\pi/\lambda_u$, λ_u is the undulator period, and $\bar{t}$ is the electron arrival time at the location z averaged over the undulator period. The rate of the phase change is

$$\frac{d\theta}{dz} = (k_r + k_u) - \frac{ck_r}{v_z}, \tag{6}$$

where v_z is the undulator-period-averaged longitudinal velocity and is given by

$$\frac{v_z}{c} \approx 1 - \frac{1}{2\gamma^2} - \frac{v_\perp^2}{2} = 1 - \frac{1 + K^2/2}{2\gamma^2} - \frac{p_x^2 + p_y^2}{2}. \tag{7}$$

Here K is the undulator parameter and is approximately constant. Its weak dependence on x and y can be neglected as long as the external focusing of the FODO lattice is much stronger than the natural focusing of the undulator. Making use of the resonant condition

$$k_r = \frac{2\gamma_r^2 k_u}{1 + K^2/2}, \tag{8}$$

we can write Eq. (6) as

$$\begin{aligned} \frac{d\theta}{dz} =& 2k_u\frac{\Delta\gamma}{\gamma_r} - \frac{k_r}{2}(p_x^2 + p_y^2) \\ =& 2k_u\frac{\Delta\gamma}{\gamma_r} - k_r\frac{(J_x + J_y)}{\bar{\beta}} \\ & -k_r\left[\pm\frac{J_x}{\beta}\sin\left(\frac{2z}{\bar{\beta}} + 2\phi_x\right) \mp \frac{J_y}{\beta}\sin\left(\frac{2z}{\bar{\beta}} + 2\phi_y\right)\right]. \end{aligned} \tag{9}$$

Since the last term of Eq. (9) is oscillatory with the FODO lattice period (cell length), the main accumulating effects on the ponderomotive phase are the first two terms. Neglecting this oscillatory term, the phase equation in such a FODO cell is identical to that in a natural focusing undulator [13], as noted in Refs. [12,14]. For a beam with finite energy spread and emittance, the first two terms introduce phase slippage of the electron relative to the radiation and result in the phase spread of the beam. Denoting $\langle\rangle$ as the average over the beam, we have $\langle J_x \rangle = \varepsilon_x$ and $\langle J_y \rangle = \varepsilon_y$, where $\varepsilon_{x,y} = \varepsilon_n/\gamma_r$ are the beam transverse emittances. In order to not significantly degrade the FEL performance, the emittance-induced phase spread over one FEL power gain length L_G should be less than unity, i.e.,

$$\frac{2k_r\varepsilon_n}{\gamma_r\bar{\beta}}L_G < 1\,, \quad \text{or} \quad \varepsilon_n < \gamma_r\frac{\lambda_r\bar{\beta}}{4\pi L_G}\,. \tag{10}$$

This situation changes if the initial electron energy can be conditioned to its transverse betatron amplitudes such that

$$2k_u\frac{\Delta\gamma_0}{\gamma_r} = k_r\frac{(J_x + J_y)}{\bar{\beta}}\,, \tag{11}$$

the dominant emittance effect (i.e., the second term in Eq. (9)) is then removed from Eq. (9). Averaging over J_x and J_y, the (transversely) correlated energy spread required to condition a beam with the normalized emittance $\gamma_r\varepsilon_x = \gamma_r\varepsilon_y = \varepsilon_n$ is

$$\langle\Delta\gamma_0\rangle_c = \frac{\lambda_u}{\lambda_r}\frac{\varepsilon_n}{\bar{\beta}}\,. \tag{12}$$

However, when the beam is perfectly conditioned to satisfied Eq. (12), the last oscillatory term in Eq. (9) is no longer negligible and can play a limiting role with a large enough emittance. We discuss quantitatively this remaining emittance effect on the FEL performance for a conditioned beam in the following two sections.

3. 1D model of the ponderomotive phase oscillation

The ponderomotive phase oscillations in Eq. (9) depend on the transverse variables through J_x, J_y, ϕ_x and ϕ_y and is a three-dimensional (3D) problem. To isolate the oscillation effect and to simplify the problem, we study a heuristic one-dimensional (1D) model with the following phase equation

$$\frac{d\theta}{dz} = 2k_u\delta + \frac{k_r\varepsilon_n}{\gamma_r\bar{\beta}}f(z)\,, \tag{13}$$

where $\delta = (\Delta\gamma - \Delta\gamma_0)/\gamma_r$ is the FEL-induced energy change, and $f(z)$ represents the oscillatory behavior introduced by a conditioned beam in the FODO lattice (with a cell length much smaller than the betatron wavelength), i.e.,

$$f(z) = \begin{cases} +1 & \text{when } (n-1)L_c \le z < (2n-1)\frac{L_c}{2}\,, \\ -1 & \text{when } (2n-1)\frac{L_c}{2} \le z < nL_c\,, \end{cases} \tag{14}$$

and $n = 1, 2, 3....$ In the absence of the FEL interaction (i.e., when $\delta = 0$), the phase oscillates between 0 and θ_0, with the maximum phase deviation

$$\theta_0 = \frac{k_r \varepsilon_n L_c}{2\gamma_r \bar{\beta}} \,. \tag{15}$$

Using the FEL Pierce parameter ρ [15], we introduce a scaled distance $\tau = 2k_u \rho z$ and a scaled energy $\eta = \delta/\rho$. Equation (13) becomes

$$\frac{d\theta}{d\tau} = \eta + \xi \,, \quad \text{with} \quad \xi = \frac{k_r \varepsilon_n}{2\gamma_r k_u \rho \bar{\beta}} f(\tau) \,. \tag{16}$$

The FEL-induced energy change is

$$\frac{d\eta}{d\tau} = a e^{i\theta} + \text{complex conjugate} \,, \tag{17}$$

where a is the slowly-varying radiation field amplitude (properly scaled by ρ). Neglecting any transverse dependence and considering the radiation field at the resonant frequency ck_r, the 1D field equation is

$$\frac{da}{d\tau} = -\langle e^{-i\theta} \rangle \,. \tag{18}$$

This set of coupled equation can be solved by averaging over the fast oscillation if its period is much smaller than the field gain length, similar to the undulator-period averaging procedure for a planar undulator. To illustrate this process, we define

$$\tilde{\theta} = \theta - \int^{\tau} \xi(\tau') d\tau' \,, \tag{19}$$

Equation (16) can be written as

$$\frac{d\tilde{\theta}}{d\tau} = \eta \,. \tag{20}$$

Equations (17) and (18) become

$$\frac{d\eta}{d\tau} = a e^{i\tilde{\theta}} \exp\left(i \int_0^{\tau} \xi(\tau') d\tau' \right) + \text{complex conjugate} \,, \tag{21}$$

$$\frac{da}{d\tau} = -\langle e^{-i\tilde{\theta}} \rangle \exp\left(-i \int_0^{\tau} \xi(\tau') d\tau' \right) \,. \tag{22}$$

Treating η, $\tilde{\theta}$, and a as slowly-varying variables, we can average Eqs. (21) and (22) over the oscillation period L_c to obtain

$$\frac{d\eta}{d\tau} = a e^{i\tilde{\theta}} A(\theta_0) + \text{complex conjugate} \,, \tag{23}$$

$$\frac{d\bar{a}}{d\tau} = -\langle e^{-i\tilde{\theta}} \rangle A(-\theta_0) \,, \tag{24}$$

where

$$A(\theta_0) = \int_0^{L_c} \frac{dz}{L_c} \exp\left(i \int_0^{\tau} \xi(\tau')d\tau'\right)$$
$$= \int_0^{L_c} \frac{dz}{L_c} \exp\left(i\theta_0 \int_0^{z} \frac{2ds}{L_c} f(s)\right) . \quad (25)$$

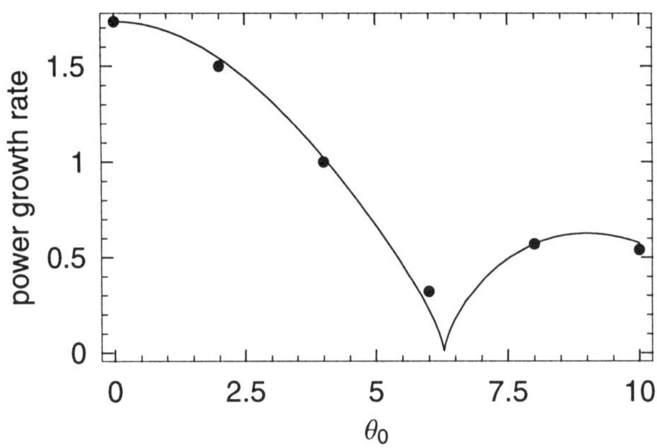

Figure 2. 1D FEL power growth rate under a rapid phase oscillation with the maximum phase deviation θ_0. Solid curve is the solution of Eq. 26, and the symbols represent 1D simulation results.

Equations (20), (23) and (24) are the FODO-cell averaged FEL equations and can be solved with the usual techniques (see, e.g., Ref. [15]). Assuming that $a(\tau) \propto e^{-i\mu\tau}$, we obtain a cubic equation for the complex growth rate μ:

$$\mu^3 = |A(\theta_0)|^2 . \quad (26)$$

The FEL power growth rate $2\text{Im}\mu = \sqrt{3}|A(\theta_0)|^{2/3}$ versus the maximum phase deviation θ_0 is shown in Fig. 2. It agrees well with the 1D simulation results by solving the original set of Eqs. (16), (17) and (18) for an electron beam without any initial energy spread. The FEL growth rate is degraded when

$$\theta_0 = \frac{k_r \varepsilon_n L}{2\gamma_r \beta} > 1 . \quad (27)$$

4. 3D genesis simulations

The above 1D model shows that when the maximum phase oscillation amplitude exceeds unity, the FEL gain will be degraded as compared to the ideal case. Therefore, for a perfectly conditioned beam that satisfies Eq. (12) in a FODO lattice, the FEL performance will still be affected by the emittance-induced phase oscillations when the normalized emittance ε_n exceeds a critical emittance given by

$$\varepsilon_n^c \equiv \gamma_r \frac{\lambda_r \bar{\beta}}{\pi L_c}, \tag{28}$$

Since this emittance criteria is derived with a heuristic 1D model, we examine it using the 3D FEL code GENESIS [16]. The electron beam and undulator parameters are given in Table 1. A typical FEL power evolution (P versus z) is shown in Fig. 3 (blue dashed curve). The local power growth rate $1/P(dP/dz)$ is also plotted in the same figure (green solid curve) and is oscillatory with a period equaling to $L_c/2$ as a result of emittance-induced phase oscillations.

Table 1. GENESIS FEL simulation parameters of conditioned beams in a strong focusing undulator.

Parameter	Symbol	Value
electron energy	$\gamma_r mc^2$	20 GeV
flattop bunch current	I_{pk}	3 kA
transverse norm. emittance	ε_n	0.1 to 19 μm
conditioned energy spread	$\langle\Delta\gamma_0\rangle_c$	given by Eq. (12)
uncorrelated energy spread	σ_{γ_0}	0
undulator period	λ_u	3 cm
undulator parameter	K	4.67
average beta function	$\bar{\beta}$	4.8/9.4/19 m
FODO cell length	L_c	2.64/5.04/9.83 m
FEL wavelength	λ_r	1.0 Å

We extract the FEL power gain length by averaging a relatively constant $1/P(dP/dz)$ over many oscillation periods in the GENESIS simulations. The resulting gain length versus the normalized emittance is plotted in Fig. 4 for $\bar{\beta} = 19$ m and for three different FODO cell lengths. We also compare the simulation results with the theoretical gain length for a conditioned beam in the absence of any phase oscillation. The latter is obtained by solving the usual 3D eigenmode equation (see, e.g., Ref [17]) for the fundamental Gaussian mode without any energy and angular spreads. As shown in Fig. 4, the simulation results agree with the 3D theory that does not take into account phase oscillations up to the critical emittances predicted by Eq. (28). For emittances exceed these critical values, the gain lengths extracted from simulations start to increase faster than the ideal case.

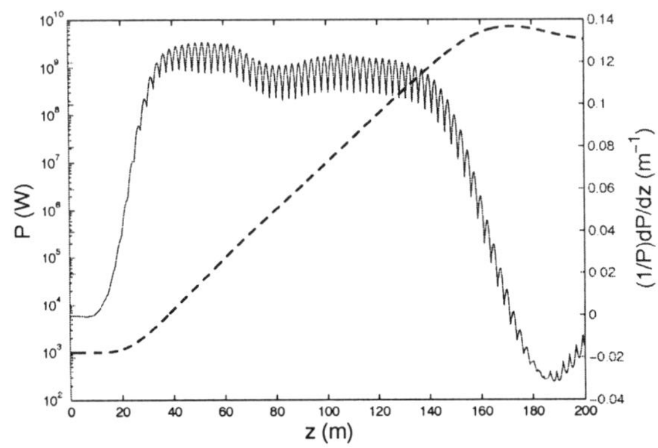

Figure 3. (Color) GENESIS amplifier run of a conditioned beam for $L_c = 5.04$ m, $\bar{\beta} = 19$ m and $\varepsilon_n = 10$ μm. The blue dashed curve is the FEL power P at $\lambda_r = 1$ Å, while the green solid curve is the local growth rate $1/P(dP/dz)$.

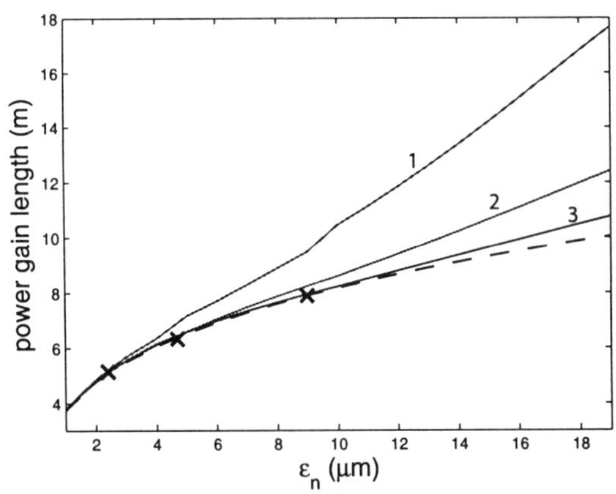

Figure 4. (Color) FEL power gain length from GENESIS simulations at $\bar{\beta} = 19$ m for $L_c = 2.64$ m (blue solid curve 1), $L_c = 5.04$ m (green solid curve 2), $L_c = 9.83$ m (red solid curve 3), and from 3D theory without any phase oscillation (black dashed curve). The colored symbols represent the corresponding critical emittance for each lattice (determined by Eq. (28)) when the phase oscillation is expected to increase the gain length.

We have also performed GENESIS simulations by varying the average beta function $\bar{\beta}$ while keeping the same focusing structure (with $L_c = 2.64$ m). Figures 5 and 6 show the gain length comparisons for $L_c = 2.64$ m at $\bar{\beta} = 9.4$ m and 4.8 m, respectively. Combining with Fig. 4, we see that Eq. (28) is reasonably accurate in predicting the critical emittances for different beta functions and FODO lattices.

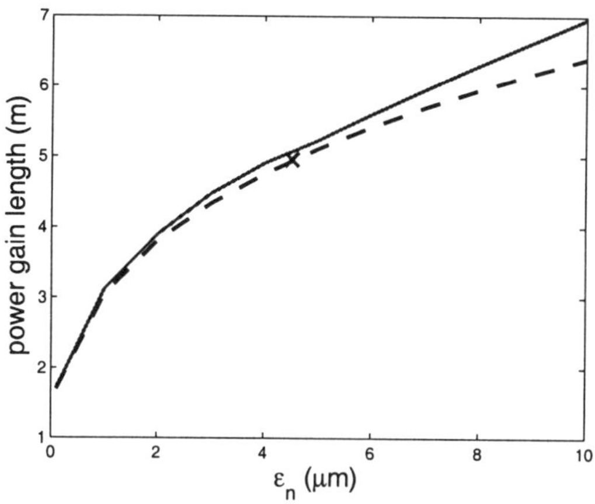

Figure 5. (Color) FEL power gain length from GENESIS simulations at $\bar{\beta} = 9.4$ m for $L_c = 2.64$ m (red solid curve), and from 3D theory without any phase oscillation (black dashed curve). The symbol represents the corresponding critical emittance (determined by Eq. (28)) when the phase oscillation is expected to increase the gain length.

As the ratio of the average beta function to the cell length becomes too large (e.g., when $\bar{\beta} = 39$ m and $L_c = 2.64$ m or when $\bar{\beta} = 19$ m and $L_c = 1.44$ m), the gain length extracted from the GENESIS simulation starts to deviate from the 3D theory (without any phase oscillation) at a smaller emittance than that predicted by Eq. (28). A plausible explanation is given as follows. In x-ray FELs the transverse coherence builds up slowly as many higher-order transverse modes are excited by an electron beam with a relatively large emittance. To some degrees a periodic modulation in the electron beam size along the undulator is beneficial to the build-up of the transverse coherence (namely higher order modes are suppressed by a varying beam envelope [18]). When the ratio of the beta function to the cell length becomes too large, this higher-order suppression mechanism is not effective since the beam size is almost constant, and the growth of the radiation is contributed from both the fundamental mode as well as higher-order modes. In this case, the comparison between simulations (including higher-order modes) and the theory (for the fundamental mode) is no longer valid.

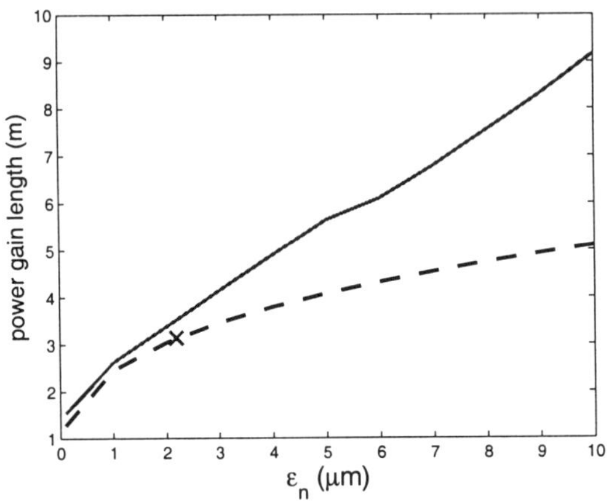

Figure 6. (Color) FEL power gain length from GENESIS simulations at $\bar{\beta} = 4.8$ m for $L_c = 2.64$ m (red solid curve), and from 3D theory without any phase oscillation (black dashed curve). The symbol represents the corresponding critical emittance (determined by Eq. (28)) when the phase oscillation is expected to increase the gain length.

5. Summary

In this paper, we study the FEL performance for a perfectly conditioned beam in a strong focusing undulator consisting of FODO cells. We develop a heuristic 1D model for the emittance-induced phase oscillation and show that the maximum phase deviation should be less than unity in order not to degrade the FEL gain. This criteria limits the maximum emittance that may be conditioned in such a focusing channel and is confirmed by the 3D GENESIS simulations over a reasonably wide parameter range. Therefore, instead of the usual emittance criteria for the unconditioned beam as given by Eq. (10), the emittance of the conditioned beam in a FODO lattice with an average beta function $\bar{\beta}$ and a cell length L_c must satisfy

$$\varepsilon_n < \gamma_r \frac{\lambda_r \bar{\beta}}{\pi L_c} . \tag{29}$$

That is, the emittance requirement for a conditioned beam is much relaxed but not without any limitation.

6. Acknowledgments

We thank A. Zholents for useful discussions. This work was supported by Department of Energy contract DE–AC02–76SF00515.

References

1. A. Sessler *et al.*, Phys. Rev. Lett., **68**, 309 (1992).
2. P. Sprangle *et al.*, Nucl. Instrum. Methods A **331** (1993) 6.
3. N. A. Vinokurov, Nucl. Instrum. Methods A **375** (1996) 264.
4. P. Emma and G. Stupakov, Phys. Rev. ST Accel. Beams, **6**, 030701 (2003).
5. P. Emma and G. Stupakov G. Proceedings of 9th European Particle Accelerator Conference, p. 503 (2004)
6. A. Wolski, G. Penn, A. Sessler, and J. Wurtele Phys. Rev. ST Accel. Beams **7**, 080701 (2004)
7. C. B. Schroeder, E. Esarey, and W. P. Leemans Phys. Rev. Lett. **93**, 194801 (2004)
8. A. Zholents, Phys. Rev. ST Accel. Beams **8**, 050701 (2005)
9. G. Stupakov and Z. Huang, in *Proc. of the FEL'05*, Stanford, CA, USA, p. 176 (2005).
10. E. Scharlemann, J. Appl. Phys. **58**, 2154 (1985).
11. L.-H. Yu, C. Huang, D. Li, and S. Krinsky, Phys. Rev. A **51**, 813 (1995).
12. S. Reiche, Nucl. Instrum. Methods A **445**, 90 (2000).
13. In a natural focusing undulator, $\alpha_{x,y} = 0$ in Eq. (1), but the transverse variations of the undulator parameter K together with the first term in the bracket of Eq. (1) result in the phase equation $d\theta/dz = 2k_u\Delta\gamma/\gamma_r - k_r(J_x + J_y)/\beta_n$, where $\beta_n = \gamma_r\lambda_u/\pi K$ is the natural focusing beta function. See also Ref. [10].
14. E. Saldin, E. Schneidmiller, and M. Yurkov, Nucl. Instrum. Methods A **475**, 86 (2001).
15. R. Bonifacio, C. Pellegrini, and L. Narducci, Opt. Commun. **50**, 373 (1984).
16. S. Reiche, Nucl. Instrum. Methods A **429**, 243 (1999).
17. M. Xie, Nucl. Instrum. Methods A **445**, 59 (2000).
18. S. Reiche, Ph. D. dissertation, DESY (1999).

SCALED MODELS: SPACE-CHARGE DOMINATED ELECTRON STORAGE RINGS*

R.A. KISHEK, G. BAI, S. BERNAL, D. FELDMAN, T.F. GODLOVE, I. HABER, P.G. O'SHEA, B. QUINN, C. PAPADOPOULOS, M. REISER, D. STRATAKIS, K. TIAN, C.J. TOBIN, AND M. WALTER

Institute for Research in Electronics and Applied Physics, University of Maryland, College Park, MD 20742, USA

New coherent radiation sources in the hard X-ray and Terahertz regimes promise exciting new developments in science, as previously dark areas of the spectrum are brightly illuminated. Ultra-short, ultra-bright radiation packets can probe the structure of matter, and image chemical and biological processes well beyond the present state of the art. Production of this coherent radiation, however, places an unprecedented challenge on the production and acceleration of high-quality electron beams. To deliver a nano-Coulomb of charge with an emittance of less than one micron, while transporting the beam through long sections of acceleration and compression, is the prerequisite for unlocking the gates of this promising new science. Using a low-energy electron storage ring, we deliberately enhance the space charge force while slowing down the time-scale to easily measurable levels so as to maximize our understanding of the particle dynamics necessary for producing bright beams.

1. Introduction

Our ability to engineer structures at the nanometer scale is premised on our ability to accurately image objects as small as individual atoms and molecules. Diffraction effects limit the utility of light at optical wavelengths at sub-micron scales. Light in the hard X-ray wavelengths, however, does not suffer this resolution problem and hence is much more suited to imaging the nano-world. Indeed X-ray crystallography has been successfully used to for decades to analyze the crystal structure of a vast number of materials, including biological materials such as some proteins. However not all materials of interest can be crystallized. This means that a brighter source of X-rays is required to acquire enough scattering data to deduce their 3-D structures. The demand on X-ray brightness is furthermore increased if dynamic imaging of the fast nano-scale processes is desired. To take a snapshot of a chemical reaction taking place at

* This project is funded by the US Dept. of Energy grant numbers DEFG02-94ER40855 and DEFG02-92ER54178, and by the US Dept. of Defense Office of Naval Research.

sub pico-second timescales, for instance, a burst of X-rays in the femto-second range is needed. Thus the radiation needs be concentrated in time as well as space and wavelength. The demands on spectral brightness, defined as number of photons per unit time, per unit area, per unit solid angle, per unit bandwidth, become extreme.

Increases in X-ray source brightness have been achieved in the past by the discovery of synchrotron radiation, which is emitted by relativistic electron beams undergoing bending, and subsequently by the insertion of undulators or wigglers in the synchrotron sources. A major advance in the brightness of X-ray sources can be obtained if the electron beam entering an undulator is bunched such that the radiation emitted is coherent [1]. Free Electron Lasers (FELs) based on this principle of coherent undulator radiation promise a sufficiently bright, and tunable, X-ray source for probing the structure of matter and for imaging nuclear and chemical processes. Many such machines are under construction or being planned worldwide [2-5]. The brightness of the photon beam in an FEL is ultimately determined by the brightness of the driving electron beam [6-7]. The applications motivating the new X-ray FELs such as the Linac Coherent Light Source (LCLS) at Stanford [2] demand the delivery of a nano-Coulomb of charge to the undulator within a few femto-seconds, while at the same time maintaining transverse rms emittance (normalized) within one micron, and longitudinal energy spread within 10^{-4} of the beam energy.

The generation and transport of electron beams with such brightness is severely constrained by the mutual repulsion among the beam electrons, known as space charge forces. While these forces become negligible at ultra-relativistic energies, this is not the case in and near the source, where they are typically comparable in magnitude to the externally applied fields. Hence the beam dynamics near the source is usually dominated by space-charge forces, a regime in which single-particle behavior gives way to collective interactions, with the beam essentially resembling a nonneutral plasma capable of carrying waves [8-9]. This implies that perturbations in the beam density profile that inevitably are created at the cathode can, under certain conditions, be amplified and then become frozen-in as the beam is accelerated. The fluid nature of a space-charge-dominated beam implies that, even if somehow the density perturbations are smoothed out in configuration space, they will likely remain in phase space, appearing for instance as energy modulations. During subsequent beam manipulation stages, such as doglegs for injection or the compression chicanes needed to shorten the bunch length and increase peak brightness [2], these perturbations may result in unwanted coherent synchrotron radiation [10] which would lead to an undesirable beam distribution going into the FEL undulator.

Recent simulation studies have shown that this effect can be mitigated by artificially enhancing the beam's longitudinal energy spread, but an experimental test of that concept remains to be seen [11].

In order to maintain the coherence of FEL emission, transverse oscillations of the electrons need to be severely constrained, requiring that the rms emittance of the electron beam be inversely proportional to the wavelength of the lased light [12]. For X-ray wavelengths, this translates into a beam emittance at the undulator of about one micron, which is at present the emittance that can be obtained from the best and brightest electron sources. Controlling emittance is therefore a top priority that is nevertheless made considerably difficult by the intense space charge near the source, which is likely to contribute to emittance growth. It has been conjectured recently that, to mitigate space charge forces, it is best to generate an ellipsoidal electron bunch in which space charge forces are linear, and various mechanisms for creating such a bunch have been proposed [13-14]. Again, it would be of great interest to experimentally test these concepts and to experimental assess the effects of bunch shapes on emittance growth.

Understanding space charge effects, therefore, is of primary relevance to the development of the bright electron sources needed to build probes for nanotechnology and biological research. Bright electron sources are also required for other applications, such as future linear colliders. Space charge effects become even more important for heavier particles such as muons, hadrons, and especially heavy ions, where the beams stay non-relativistic for a greater part of the accelerator. It is the understanding of space charge forces in ion and hadron storage rings that originally motivated the development of the University of Maryland Electron Ring (UMER) [15] as a low-cost scaled model and a platform for code benchmarking. The design and construction of UMER and its relevance to ion accelerators are discussed in detail in Ref. [15]. This article is rather concerned with the relevance of UMER to the development of bright electron sources.

The approach taken at the University of Maryland is to implement scaled experiments using modest-cost nonrelativistic electron beams whereby meaningful experiments on space charge dynamics can be conducted over a relatively long path length in a university setting. The construction of UMER has been recently completed and the machine is currently in the process of multi-turn commissioning. The 10 keV electron beam energy ensures that space charge forces can be easily accessed, while minimizing radiation effects, keeping the overall costs down, and enabling the use of inexpensive printed-circuit magnets [16]. An adjustable current (0.1-100 mA) provides a range of intensities spanning the entire range of space charge intensities. At the same time, the low beam energy and relatively long bunch length (5-100 ns) implies that beam can be easily observed with inexpensive diagnostics, as opposed to the complex measurement techniques needed to characterize femto-second beams.

UMER is augmented with a separate setup, the Long Solenoid Experiment (LSE) [17] for investigating longitudinal beam dynamics and the evolution of energy spread due to Coulomb collisions in a straight geometry. It consists of a 1.5-m long-solenoid with associated matching lenses, and a pair of diagnostic chambers at each end. In addition to testing diagnostics, it is used to perform experiments on transverse-longitudinal energy transfer, and on the evolution of longitudinal perturbations. The straight geometry removes the complications to the physics that would arise from dispersion.

This publication will briefly review the UMER design, methods of particle generation, and diagnostics. We provide an update on the commissioning, and conclude with a discussion of possible UMER experiments of relevance to bright electron sources.

2. UMER Design

Fig. 1: Photograph of UMER.

The design of UMER, photographed in Fig. 1, was discussed in detail in ref. [15] and is summarized in Table 1. The 10 keV beam energy implies UMER is free from radiation problems and the voltages involved are modest, making it an ideal machine for students. The ring circumference is divided into a 36-cell alternating-gradient (FODO) lattice, each cell containing a single 10° bend. Three induction modules (under construction) will provide longitudinal focusing. At the injection β (average beam velocity / speed of light) of 0.2, the beam is nonrelativistic, implying it is highly sensitive to space charge forces. As discussed in the next section, a variable beam current in the range 0.1 to 100 mA provides a control over the strength of the space charge force, allowing us to access a wide portion of parameter space.

TABLE I. UMER Design Specifications at 10 keV

Beam Energy	10 keV
$\beta\ (= v/c)$	0.2
Beam Current	≤ 100 mA
Generalized perveance	≤ 0.0015
rms Emittance, normalized	≤ 3.0 μm
Pulse Length	5-100 ns
Ring Circumference	11.52 m
Lap time	197 ns
Pulse repetition rate	60 Hz
Mean beam radius	≤ 1 cm
FODO period	0.32 m
Zero-current phase advance, σ_o	76 °
Zero-current Betatron tune, ν_o	7.6
Tune Depression	≥ 0.2

The beam intensity can be quantified in a number of ways, for example using the dimensionless intensity parameter, χ, defined in ref. [15] as the ratio of the space-charge force at the beam edge to the external focusing force. The tune depression is expressed in terms of χ as $\frac{k}{k_0} = \sqrt{1-\chi}$, where k is the betatron wave number with space charge, and k_o is the same in the absence of space charge. Since a beam can be thought of as a nonneutral plasma capable of carrying waves, one can also define a plasma frequency or wavenumber, k_p, expressed in terms of the intensity parameter as $\frac{k_p}{k_0} = \sqrt{2\chi}$. With a zero-current phase advance per period of $\sigma_0 = 76°$, a measured rms normalized emittance of ε_n = 2-3 μm and beam radius a = 0.3 - 1 cm, χ in UMER can be made to vary from 0.2 to the extremely space charge dominated 0.96.

Longitudinally, the UMER beam consists of a single pulse whose length can be varied from 5-100 ns. Since the circulation time for a 10 keV electron round the 11.52-m UMER circumference is 197 ns, typically only a single pulse is injected at any one time, at a repetition rate of 10-60 Hz. The pulsed injection system relies on a single pulsed dipole surrounding a wide glass gap, specially designed with a low number of conductors to minimize its inductance. Due to space limitations and the space charge forces in the most intense UMER beam, this configuration forces us to use a single quadrupole for both arms of the Y-shape, sharing it between injection and circulation [18-20].

3. The UMER Source And Bunch Shaping

All beams are born space-charge-dominated at the source. Whereas subsequent acceleration and emittance growth will reduce the beam intensity, previous experience has shown that any irregularities at the source can be 'frozen-in' and preserved for a relatively long time downstream [21]. Imperfections in the particle distribution at the source, a hot spot for instance, may subsequently resurface in the form of unwanted halo, or seed instabilities. In order for the UMER beam to serve as a flexible model of other accelerators, the UMER source is provided with many adjustments to allow the creation of a tailored particle distribution right from the start.

An adjustable Anode-Cathode (A/K) gap enables us to vary the perveance, and hence the strength of the space charge forces. Similar adjustments in beam intensity can be made using an aperture wheel immediately at the exit of the anode. The wheel can be rotated in vacuum to select from among a number of apertures. Since beam current scales as the area of the aperture while emittance scales as the radius, changing the aperture size is a quick and simple way of varying the beam intensity. In addition, introduction of asymmetric or non-round apertures is a simple way of generating exotic distributions for particular experiments. For example, we use an aperture with five outlets in a quincunx pattern (that of the number five on the face of a die) to generate a highly nonlinear pattern of five beamlets in the transverse x-y plane.

A DC 10 kV potential is applied across the 25 mm A/K gap, while a fine grid close to the cathode is biased negatively to suppress emission. The beam is extracted at 60 Hz by applying a 60 V positive pulse to the grid with a width corresponding to the desired beam pulse length. The grid-cathode distance of 0.15 mm, roughly of the order of the grid wire thickness, presents a formidable challenge to the simulation of the gun. As we discovered, the grid significantly affects the beam dynamics near the cathode, where the particles are moving at low energies. A significant outcome of the grid is the formation of oscillating virtual cathodes [22] accompanied by a hollowed velocity distribution, first measured experimentally with a pepper-pot [23]. This behavior has been demonstrated using high-resolution particle simulations using the WARP code [24], which draw heavily upon advanced techniques such as adaptive mesh refinement [22]. Introducing a hollowed velocity distribution into WARP simulations of the remainder of the injector was observed to significantly improve agreement with the measured data [25].

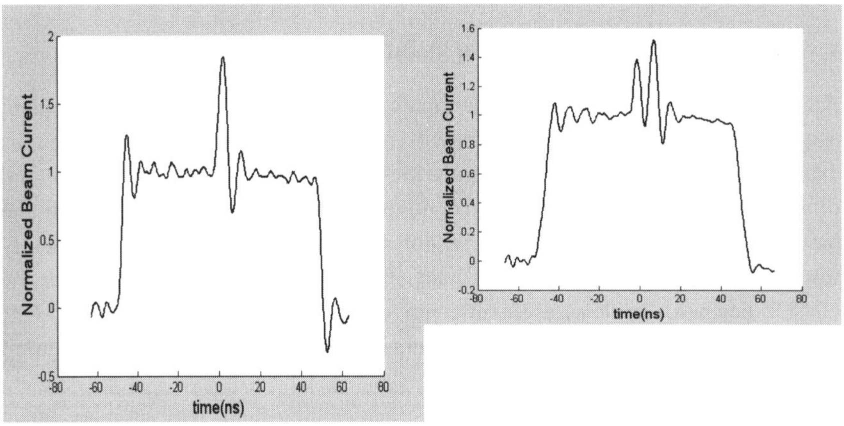

Fig. 2: Oscilloscope traces of beam current vs. time from UMER showing a typical 5 ns laser-induced perturbation (in this case highly nonlinear, with amplitude 20 mA) superimposed over a 100 ns, 20 mA rectangular thermionic main beam; (a) near the source; (b) 2/3 of the way around the first turn of UMER, showing the splitting of the perturbation into slow and fast waves [27].

While presenting a complication, the cathode grid affords us additional means for manipulating the particle distribution. For one thing, small changes in the bias voltage on the grid can be used to fine tune the beam current and emittance, hence the intensity parameter and operating point of the UMER beam. Larger changes in the bias voltage can drastically alter the transverse velocity distribution of the beam, as well as introduce fast modulations in the beam current due to the virtual cathode oscillations. Finally, by applying a perturbation to the grid pulse voltage, we can induce localized longitudinal perturbations in the beam density and/or velocity [26]. These perturbations are an extremely useful tool for modeling imperfections in the beam bunch shape and testing its stability.

Another innovative method we have developed to introduce perturbations is the application of photo-emission, in combination with thermionic emission, to place a density bump on the beam [Fig. 2]. A mirror in the first diagnostic chamber in the injector is used to direct a 5-ns uv laser pulse onto the cathode. This method is extremely flexible, as it allows us, by using a combination of masks, filters, and lenses in the path of the laser beam, to arbitrarily vary the timing of the perturbation relative to the main beam, their relative intensities, and also the transverse shape and size of the perturbation [27]. Monitoring the evolution of a typical perturbation such as the one in Fig. 2 over the first turn of UMER, we observed the current pulse splitting into a forward-traveling "fast" wave, and a backward-traveling "slow" wave. By measuring the rate of separation of those pulses, we get an accurate measurement of the wave speed [27], which agrees nicely with the theoretical predictions [8] for sufficiently small initial perturbations. This experiment has been simulated using the WARP code and the actual experimental measurement of the initial pulse. The simulation agrees well with the experiment [27]. By splitting the laser pulse and letting its parts travel through different path lengths, we can generate up to four perturbations on the beam [28].

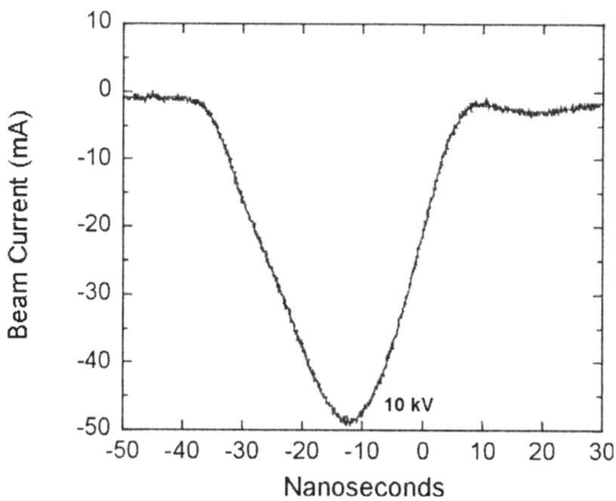

Fig. 3: Beam current from a UMER bunch with parabolic line charge density.

Significant for the purpose of modeling high-brightness electron sources is our ability to generate long bunches in which the line charge density is parabolic, implying a linear longitudinal space charge force. Figure 3 illustrates an early parabolic beam we obtained by simply applying a parabolic pulse shape to the grid. We are able to obtain more sophisticated parabolic bunches using the laser perturbation technique, with which we may be able to also vary the transverse density profile. Using such techniques we can come close to generating the ideal ellipsoidal bunches suggested in Ref. [13].

4. Diagnostics

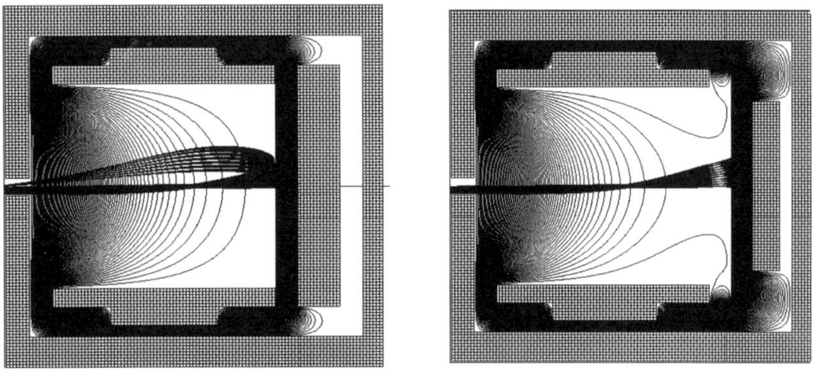

Fig. 4: SimIon simulation of the UMD Energy Analyzers, showing the structure of the device and the equipotentials, as well as some representative particle trajectories: (left) 2nd generation EA; (right) 3rd generation EA [30].

In order to fulfill UMER's mission of serving as an experimental testbed for a wide variety of problems, UMER possesses a formidable array of diagnostics which allow us to frequently measure the beam in 6-D, time-dependent detail. The diagnostics available are summarized in Table 2 below. (Diagnostics listed in italics have already been beam-tested on UMER during phased installation and will be re-installed on UMER along with the extraction line and End Diagnostic Chamber). Most diagnostics were developed in-house specifically for the parameters of the UMER beam. Collectively, they enable us to measure the 6-D phase space. Combinations of Phosphor Screens (P-Screens) and Capacitive Beam Position Monitors (BPMs) are distributed all around the ring at intervals of 64 cm, and are mounted on feedthroughs that allow them to be interchanged under vacuum. The BPMs provide time-resolved information on the beam position, current, and eccentricity, while the P-screens provide a time-integrated detailed image of the beam from which beam size, centroid, and rotation angle information can be extracted, among other things. In two locations the P-screens are replaced with Optical Transition Radiation (OTR) screens, which, combined with a sophisticated gated camera, provide the time-resolved equivalent of P-screen pictures. The P-screens are proving to be most versatile diagnostics, their only drawback being that they intercept the beam and hence can be read only during the first turn. For example, computer-aided tomographic reconstruction enables us to translate a large number of P-screen photographs into a detailed transverse phase space map of the beam at the location of the screen [29].

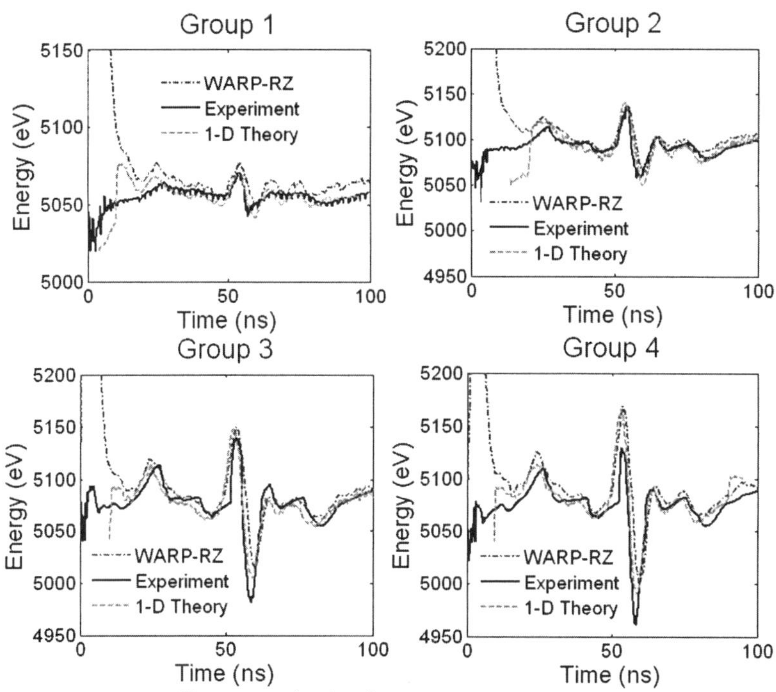

Fig. 5: Beam energy profiles measured using the energy analyzer on the LSE for different beam parameters, each starting with a perturbation at the source, showing the splitting of the perturbation into slow and fast waves. The red line corresponds to an r-z WARP simulation starting with the same perturbation and the same transverse focusing lattice, while the green line corresponds to a 1-D cold fluid model [26].

Another important diagnostic is the compact, high spatial and temporal resolution energy analyzer. This device is an advanced in-house design based on a parallel-plate retarding-voltage concept, except it introduces a collimating cylinder which improves the energy resolution by two orders of magnitude [30]. A typical particle tracking simulation using the code SIMION, shown in Fig. 4, indicates a factor of 100 enhancement of energy resolution over the traditional parallel plate design. This device has been used to measure the energy spread close to the LSE source and further downstream, after transport through a 1.5-meter long solenoid [17]. The results indicate a general agreement between experimental measurements of energy spread and predictions from the collisional Boersch effect and longitudinal-longitudinal effect. A remarkable anomaly is the faster increase in energy spread for higher beam densities than theoretical predictions [31].

When applied to analyzing the energy distribution of a beam on which an initial perturbation was applied, the energy analyzer clearly demonstrated the evolution of this perturbation into slow and fast waves, with mean energy respectively smaller and greater than the mean beam energy [26]. The energy analyzer results furthermore demonstrated very good agreement with WARP simulation results [26]. Figure 5 shows examples of this splitting for different beam currents on the LSE.

5. Commissioning Update

During the construction of UMER, we have adopted the philosophy of phased commissioning, in which we have installed a few sections at a time and used the end diagnostic chamber to carefully characterize the beam around the perimeter. After closing the ring, multi-turn operation is gradually becoming a reality as we deploy and debug the pulsed injection system. At the extreme space charge limit of UMER, our goal is to circulate for 10 turns. Noting that we can vary the beam current and emittance at the source using various means, we are starting commissioning with lower currents and working our way up. Our ultimate goal for lower intensities is to circulate 100 turns. Note however that our least intense beam still contains significant space charge, its tune depression k/ko of 0.82 implying a space charge tune shift of 1.4.

As of the time of writing (Dec. 2005), the UMER beam registers five pulses on the first beam position monitor, with the second pulse displaying nearly no particle loss. This indicates that we are starting to lose beam somewhere in the second turn, yet having a fraction of particles make it into the fifth turn. We expect this result to continue to improve as we fine-tune the injection pulser electronics, and as we improve the beam control. Injection is perhaps the most complex task, given the shared quadrupole mentioned above and given the Earth magnetic field (~ 0.5 Gauss pointing straight down) which breaks the symmetry of the injection. At present we are working through an advanced singular value decomposition (SVD) matrix method to achieve a closed-orbit correction for the beam centroid [32]. In order to improve the correction matrix, we employ an automated quadrupole-scan technique to multiply the number of beam position monitors at our disposal. In addition, we are working through advanced rms matching and rotation correction schemes.

In order for the beam to survive beyond a few turns at full beam current, we intend to deploy three induction modules around the ring, which would provide the necessary kicks to the beam ends to counter the space charge force and keep the beam confined. The induction gaps can be used as further means to manipulate and perturb the beam.

6. Conclusion

Two factors make UMER a unique tool that provides a valuable testbed for understanding the space charge beam dynamics in bright beam sources. First is our ability to generate and manipulate a wide variety of beam distributions in a controlled fashion. In section 3 we have reviewed our ability to generate beer-can-type rectangular bunches as well as ellipsoidal-like parabolic bunches, our ability to vary the beam parameters, and our ability to manipulate the beam distribution transversely and longitudinally, e.g., by adding perturbations. The second factor is our ability to carefully measure the 6-D phase space at different points along UMER, enabling us to measure the detailed evolution of that initial distribution which we have carefully created. At present our diagnostics operate individually, e.g., the energy analyzer provides information about the longitudinal phase space, tomography maps the transverse phase space integrated over the entire pulse, while OTR is still at an experimental stage. A unique and exciting opportunity exists when we combine all three diagnostics (OTR, energy analyzer, and tomography) to produce time-resolved 6-D phase space maps.

At this juncture it is clear that UMER provides a valuable test bench for carrying out some of the experiments needed to design brighter electron sources whose beams will maintain their brightness despite the downstream manipulations in the FEL. For instance we can perform detailed bunch-shaping studies to verify the predictions of the studies in Refs. [13-14]. Another idea is to experimentally test the role of energy spread in damping space charge waves and smoothing the beam distribution before acceleration [11]. A third possibility is to use UMER as a testbed for benchmarking space-charge codes that will be used to design and simulate future beam sources.

Having emphasized the flexibility of UMER, its density of diagnostics, and the unique opportunity it presents to advancing high-brightness electron source research, we would like to make it clear that UMER is a research tool available to scientific community. We invite collaborations from other scientists interested in modeling space charge dynamics in high-brightness beams and injectors using specially-designed experiments on UMER.

Acknowledgements

We wish to thank D. Dowell, P. Emma, C. Limborg-Duprey, O.J. Luiten, K. Floetmann, C. Brau, J. Rosenzweig for beneficial discussions. We are indebted to Dave Grote, Alex Friedman, and Jean-Luc Vay for their excellent support of the WARP code.

References

1. G.L. Carr, Michael C. Martin, Wayne R. McKinney, K. Jordan, George R. Neil and G.P. Williams, Nature, **420**, 153-156 (2002).
2. John N. Galayda, Proc. 8th International Conference on X-ray Lasers, May 2002, Aspen, CO, SLAC-PUB-9847 (2002).
3. V. Ayvazyan, *et al.*, *"First operation of a Free-Electron Laser generating GW power radiation at 32 nm wavelength,"* to be published in European Physics Journal D (2005).
4. R. Brinkmann, Proceedings of LINAC 2004, Luebeck, Germany, p. 2 (2004).
5. L. Serafini, *et al.*, *"Space Charge Issues in High Brightness Electron Beams for X-ray FEL's,"* Proc. Coulomb 2005, Senigallia, Italy, Sep. 2005.
6. James B. Rosenzweig, Fundamentals of Beam Physics (Oxford: Oxford University Press, 2003).
7. C. Brau, Proc. ICFA workshop on High Brightness Beams, Erice, Italy, (2005).
8. Martin Reiser, Theory and Design of Charged Particle Beams, (New York: John Wiley & Sons, Inc., 1994).
9. Ronald C. Davidson and Hong Qin, Physics of Intense Charged Particle Beams in High Energy Accelerators (Singapore: World Scientific, 2001).
10. H.H. Braun, R. Corsini, L. Groening, F. Zhou, A. Kabel, T.O. Raubenheimer, R. Li and T. Limberg, Phys. Rev. ST Accel. Beams **3**, 124402 (2000).
11. Z. Huang, *et al.*, Phys. Rev. ST Accel. Beams **7**, 074401 (2004).
12. Charles A. Brau, Modern Problems in Classical Electrodynamics (Oxford: Oxford University Press, 2003).
13. O.J. Luiten, S.B. van der Geer, M.J. de Loos, F.B. Kiewiet, and M.J. van der Wiel, Phys. Rev. Lett. **93**, 094802 (2004).
14. C. Limborg-Duprey, *et al.*, Proc. FEL 2005, Monterrey, CA, (2005).
15. M. Reiser, P.G. O'Shea, R.A. Kishek, *et al.*, Proc. 1999 IEEE Particle Accelerator Conference, New York City, NY, 234 (1999); P.G. O'Shea, M. Reiser, R.A. Kishek, *et al.*, Nuclear Instruments and Methods **A464**, 646-652 (2001); P.G. O'Shea, R.A. Kishek, M. Reiser, *et al.*, Laser and Particle Beams **20**, 599 (2002).
16. W.W. Zhang, S. Bernal, H. Li, T. Godlove, R.A. Kishek, P.G. O'Shea, M. Reiser, V. Yun, and M. Venturini, PRST-AB **3**, 122401 (2000).

17. Y. Cui, Y. Zou, M. Reiser, R.A. Kishek, I. Haber, S. Bernal, and P.G. O'Shea, Physical Review ST - Accelerators & Beams, **7**, 072801 (2004).
18. H. Li, R.A. Kishek, S. Bernal, T. Godlove, M. Walter, P.G. O'Shea, and M. Reiser, Proceedings of the 2003 IEEE Particle Accelerator Conference, Portland, OR, edited by J. Chew, P. Lucas and S. Webber, IEEE Cat. No. **03CH37423C**, 1676 (2003).
19. G. Bai, S. Bernal, I. Haber, R.A. Kishek, B. Quinn, M. Reiser, T.C.J. Tobin, M. Walter, and P.G. O'Shea, Proceedings of the 2005 IEEE Particle Accelerator Conference, Knoxville, TN, ed. C. Horak, IEEE Cat. No. **05CH37623C**, 3733 (2005).
20. M. Walter, D. Lamb, S. Bernal, T. Godlove, I. Haber, R. Kishek, H. Li, P. O'Shea, B. Quinn, and M. Reiser, Nuclear Instruments and Methods A **544**, 374-377 (2005).
21. I. Haber, S. Bernal, C.M. Celata, A. Friedman, D.P. Grote, R.A. Kishek, B. Quinn, P.G. O'Shea, M. Reiser, and J.-L. Vay, Nuclear Instruments and Methods A **519**, 396-404 (2004).
22. I. Haber, S. Bernal, R.A. Kishek, P.G. O'Shea, M. Reiser, Y. Zou, A. Friedman, D.P. Grote, and J.-L. Vay, Proc. 2005 IEEE Particle Accelerator Conference, Knoxville, TN, ed. C. Horak, IEEE Cat. No. **05CH37623C**, 2908 (2005).
23. S. Bernal, H. Li, T. Godlove, I. Haber, R.A. Kishek, B. Quinn, M. Reiser, M. Walter, Y. Zou, and P.G. O'Shea, Physics of Plasmas **11(5)**, 2907 (2004).
24. D.P. Grote, A. Friedman, I. Haber, S. Yu, Fus. Eng. & Des. **32-33**, 193-200 (1996); Friedman, *et al.*, Nuclear Instruments and Methods **A 544** (2005).
25. R.A. Kishek, S. Bernal, C.L. Bohn, D. Grote, I. Haber, H. Li, P.G. O'Shea, M. Reiser, and M. Walter, Physics of Plasmas **10 (5)**, 2016 (2003).
26. K. Tian, Y. Zou, Y. Cui, I. Haber, R.A. Kishek, M. Reiser, and P.G. O'Shea, Physical Review Special Topics - Accelerators & Beams, **9**, 014201 (2006).
27. Yijie Huo, Masters Thesis, University of Maryland, College Park (2004).
28. J.G. Neumann, J.R. Harris, B. Quinn, and P.G. O'Shea, Review of Scientific Instruments, **76**, 033303 (2005).
29. H. Li, Ph.D. Thesis, University of Maryland, College Park (2004).
30. Y. Cui, Y. Zou, A. Valfells, M. Walter, I. Haber, R.A. Kishek, S. Bernal, M. Reiser, and P.G. O'Shea, Review of Scientific Instruments, **75(8)**, 2736 (2004).
31. Y. Zou, Y. Cui, M. Reiser, and P. G. O'Shea, Physical Review Letters **94**, 134801 (2005).
32. M. Walter, *et al.*, Phys. Plasmas, submitted for publication (2006).

HIGH BRIGHTNESS BEAM APPLICATIONS: ENERGY RECOVERED LINACS*

GEOFFREY A. KRAFFT

Center for Advanced Studies of Accelerators, Jefferson Lab, 12000 Jefferson Ave. Newport News, VA 23606, USA

In the first part of the paper some general statements are made regarding applications suitable for utilizing energy recovered linacs (ERLs) by contrasting their potential performance to that of single pass linacs and storage rings. As a result of their potential for extremely good beam quality in combination with high average beam current, ERLs have been used and considered as drivers of both free electron laser and partially coherent photon sources, from THz through X-rays; as a suitable technology for high energy electron cooling; and as a continuous or semi-continuous electron beam source for high energy colliders. At present, beam requirements tend to be highly matched to end use requirements. By reviewing some of the many examples which have either been reduced to practice, or are being explored presently, one can develop an appreciation for the wide range of parameters being considered in ERL applications.

1. Introduction

Recently, energy recovered linacs have been considered for a wide variety of purposes, well beyond their initial development as drivers for high average power free electron lasers. These developments have led to new and interesting problems in the field of high brightness beams as new high average beam power electron sources must be developed in order to drive the ERL, and some of the applications require very low emittance in addition. In this paper the recent progress in energy recovered linacs will be reviewed, with particular emphasis on the electron sources that have been built and that are being contemplated for these applications.

In brief, the paper begins with a description of the beam energy recovery process. Next, energy recovered linacs are compared and contrasted to the usual workhorses in electron accelerator physics: the single pass linacs and the storage rings. This discussion leads naturally to a review of the presently existing ERLs, which up to now have been developed as free electron laser (FEL) drivers. Next we discuss the types of electron sources being proposed for ERLs. In

* This work supported by the United States Department of Energy under Contract DE-AC05-84-ER40150

contemplating the requirements on the electron injectors for ERLs, it becomes obvious that the brightness requirements depend on the application, the most stringent being for anticipated recirculated linac X-ray sources. Therefore, the primary end use applications will be discussed in turn, referring to the most recent ideas concerning the electron sources for the energy recovered linac. Their implications on high brightness beam sources will be addressed.

2. Beam Energy Recovery

Beam energy recovery can be most simply explained by reference to Figures 1 and 2, which present the idea in the simplest case of a two-pass recirculated linac with a single accelerating pass and a single decelerating pass. Figure 1 provides a schematic diagram of the more common "front-to-back" beam recirculation geometry, where "front-to-back" refers to the fact that the second pass beam proceeds from the front of the linac to the back of the linac. Suppose a two-pass linac exists, with the recirculation path length, as defined in Figure 1, chosen to be an integer plus ½ accelerating mode RF wavelengths long. Suppose further that there is no beam loss between passes. Because the current load phasors are equal and opposite, there is no RF beam load on the cavities within the recirculation loop. Physically, the energy for acceleration of the first pass beam no longer comes from the incident RF power on the accelerating cavity, but is transferred directly out of the decelerating second pass beam via the RF field of the cavity. That this transfer can be done very efficiently, and almost all of the decelerating beam energy passes to the accelerated beam is demonstrated by recent experiments involving superconducting recirculated RF linacs.

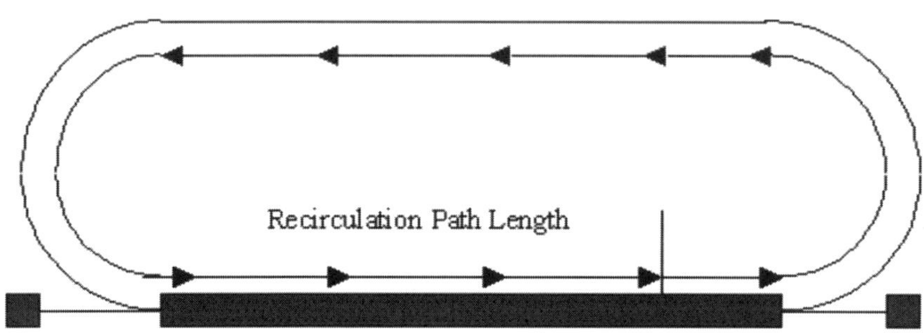

Figure 1. Two pass recirculated linac in "front-to-back" beam recirculation arrangement. Beam energy changes depending on the pass number.

There is another way to achieve perfect energy transfer between the accelerating beam pass and the decelerating beam pass, as shown schematically in Figure 2 [1]. Rather than having the beam load phasors cancel by a 180 degree difference in the RF phase between accelerating and decelerating beams, one could just as well have the beam load cancel by having the beam velocities oppositely directed with the arrival timed to be on the same phase on the two beam passes. In order for this situation to be achieved the recirculation path length as defined in the figure must be an integer number of RF wavelengths to the center of each accelerating cavity. Such "back-to-front" or reflex beam recirculation was actually used on the earliest recirculated linac where energy recovery was performed [2].

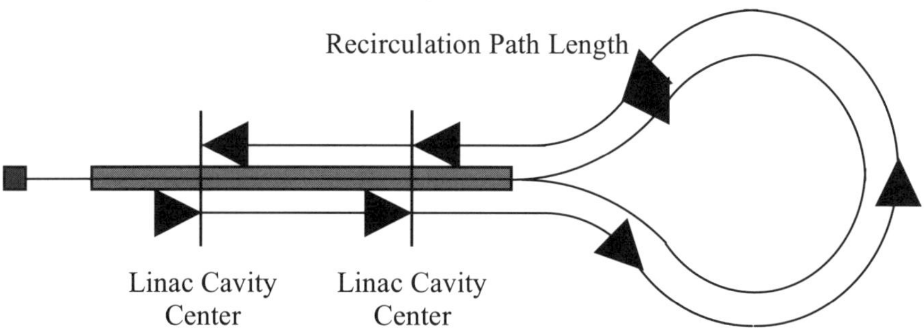

Figure 2. Two pass recirculated linac in "back-to-front" or reflex beam recirculation geometry. Here, the beam energy is the same for both passes between the accelerating cavities.

A convenient way to characterize the benefit of the beam energy recovery is in terms of the RF-beam multiplication factor, discussed more thoroughly in [3] and [4]. The multiplication factor

$$k = P_{ave,beam} / P_{RF}$$

gives the ratio between the average beam power at its end use and the RF power required to accelerate the beam. By the first law of thermodynamics, any single pass linac has $k < 1$. If one considers normal conducting recirculators, the multiplication factor is less than 1 because of the large RF power needed to maintain the accelerating field in the RF cavities. For superconducting accelerating structures, because of the possibility to operate at high loaded Q values, one has the possibility of having the multiplication factor close to 1 even in non recovered linacs such as the CEBAF accelerator at Jefferson Lab [5], where the multiplication factor is about .995 for a perfectly matched beam current, and it is usually about 0.8 at more common operating beam loads. The first recirculated linac with a multiplication factor that exceeded 1 on a continuous basis was the superconducting Jefferson Lab IR DEMO FEL [6], with a multiplication factor around 16. This device has recently been upgraded and operates with a multiplication factor around 30. A storage ring with superconducting accelerating cavities has a multiplication factor around 1000.

Table 1 presents a comparison between the beam properties possible in single pass linacs and the beam properties possible in storage rings. One notices an advantage to single pass linacs in that the ultimate electron emittance may be better and the achievable beam pulse lengths may be much shorter than in storage rings, whereas the electron storage rings have a large advantage in their ability to handle high average current beams. Recirculated linacs that are energy recovered, because of the possibility to achieve high multiplication factors, are quite interesting because they contain the possibility of combining the good emittance and short pulses of linacs with the high average current capability of storage rings, all in a single technology. The arrows in this table extend from present experience to numbers that have been proposed in various ERL projects.

Table 1. Comparison of Accelerator Types.

	High Energy Electron Linac	High *k* Super-conducting Linac	Storage Ring	Unit
Accelerating Gradient	>50	10-20	NA	MV/m
Duty Factor	<1%	1	1	
Average Current	<1	10→100	1000	mA
Average Beam Power	0.5	1→1000	3000	MW
Multiplication Factor	<1	33→200	1000	
Normalized *rms* Emittance	1	1	4	mm mrad
Pulse Length	0.1	0.1	20	psec

3. Existing and Planned ERL Free Electron Lasers

Early work on beam energy recovery was performed at Stanford University on the Superconducting Accelerator [7]. A primary motivation of this work was to provide a path to higher average beam current and higher average power free electron lasers utilizing the higher efficiency possible at a higher multiplication factor. Their early work did not include operating the FEL oscillator inside the beam energy recovery loop, but the work did anticipate many of the problems later encountered when the FEL oscillator was included in the loop.

It took more than a decade before an oscillator was enclosed inside the recirculation loop of an energy recovered recirculated linac, accomplished at the Jefferson Lab IR DEMO FEL, followed sometime later by the Japan Atomic Energy Research Institute (JAERI) FEL at Tokai [8]. These devices achieved several kW average photon beam power from devices handling of order 200 kW average electron beam power. More recently, a similar average beam power has been achieved at much higher average beam current in a normal conducting energy recovered linac at Budker Institute of Nuclear Physics called MARS [9]. Higher beam powers of order 1 MW have recently been used to drive a 10 kW scale FEL laser oscillator utilizing superconducting accelerating cavities [10].

These successes have led to a veritable plethora of more advanced FEL projects. The furthest along is the 4GLS project at Daresbury [11], which proposes, in addition to a non-energy recovered IR FEL, to construct a non-energy recovered DUV-FEL of a conventional SASE type and an energy recovered DUV light source. A small-scale energy recovered linac, including an enclosed oscillator, is being built at Daresbury as a prototype. The Korean Atomic Energy Reseach Institute (KAERI) has plans to build an energy recovered IR FEL similar to the device at JAERI, and the National High Magnetic Field Laboratory at Florida State University plans an ERL based on the IR DEMO FEL parameters, but featuring more advanced beam optics designs. Also, there is a French proposal for an ERL based roughly on the 4GLS model, in a less complete state of development. Of course, plans are being developed to extend the average power in the FELs to the 100 kW level with 10 MW average electron beam power drivers.

Because of the long wavelength of the operating oscillator FELs, the emittance requirements for the devices are relatively modest. For example, the Jefferson Lab FELs operate at of order 10 mm mrad normalized *rms* emittance, JAERI has an emittance of 40 mm mrad, and MARS has an emittance of 20 mm mrad. Because of the relatively modest emittance requirements of present devices, because of the prior existence of DC photocathode guns that could without much further development provide the requisite beams for the FEL, and because DC guns may be easily run in a "CW" mode, the Jefferson Lab FEL and its likely offshoots are based on DC photocathode gun technology. JAERI and MARS have utilized the usual thermionic emission sources coupled with high extraction voltage in the formation of their beams. 4GLS, at least for the present, has chosen to base their sources for the IRFEL and ERL ring on the DC photocathode sources, copying development work being done at Cornell University to extend the performance of DC guns into new brightness regimes [12].

4. Electron Source Technology for ERLs

Because beam for the energy recovered linac must be continuously provided in high duty factor applications, there is a great need for developing advanced high average current electron sources to drive the ERL. Many technologies have been explored to solve this problem, and it is fair to say that no clearly obvious "winning" technology has emerged to the point that future discussion on this topic is fruitless. Actually, this situation accounts for some of the interest and excitement in the field generally. In this section the possibilities will be discussed, with their advantages and disadvantages. In the following sections of this paper, where some of the potential applications of ERLs are discussed, the technology choices being made will be related to the application.

The types of guns being studied presently fall into 4 broad categories: DC photocathode guns, RF photocathode guns, SRF photocathode guns, and hybrid DC-SRF photocathode guns. As mentioned in the previous section, all of the present ERLs have been driven by DC electron sources, and the IR DEMO FEL was driven by a DC photocathode source. Advantages of this technology choice include: natural integration into CW ERLs because of the DC accelerating field; existing sources performing at the 10 mA, 100 pC bunch charge level; good thermal emittance of the photocathode; and a natural upgrade path in average current by increasing the photocathode drive laser power by increasing the pulse repetition rate. A primary disadvantage of this choice at present is that it has never achieved transverse emittance, and hence brightness, comparable to the best RF photocathode sources. This situation may soon change, as a 5-15 MeV ERL electron source is being constructed at Cornell University that is expected to have markedly better emittance than previous DC photocathode guns. Interestingly, an essential element of this design is the application of emittance compensation techniques similar to those that have been applied in RF photocathode guns for a number of years [12].

Normal conducting RF photocathode guns have been developed, built, and operated over a number of years. The best RF guns have the best beam brightness from photocathode guns, accounting for their overwhelming predominance throughout this conference. However, these best results have been obtained only in low duty factor pulsed RF guns. Pulsing the RF allows the accelerating field in the cavity to be greater than may be possible in ERL applications requiring CW operations. The highest duty factor normal conducting RF gun was operated for the Boeing FEL [13], at 25%. Here, the average accelerating gradient was 5 MV/m and the normalized *rms* emittance was several 10s of mm mrad, only comparable to the present DC data. However, an advanced normal conducting RF gun is being developed by Los Alamos and AES [14] as an FEL driver. It is anticipated that the average accelerating gradient will be 7 MV/m and the emerging normalized *rms* emittance will be 6.5 mm mrad with 100 mA of average beam current.

Given the inherent limitations in the accelerating gradient of CW RF guns tied to ones inability to cool them beyond a certain point, it is natural to contemplate removing such limitations by using superconducting RF (SRF) acceleration cavities instead normal conducting RF cavities. This should allow one to operate at higher accelerating fields, with better extracted emittance. The main difficulty of such an approach is that one must integrate, in a single device, both the photocathode, and any preparation activities needed to keep the photocathode emitting, and the relatively touchy (e.g., the surface must be free of contaminants) superconducting material forming the accelerating cavity. On the other hand it should be stated there is broad recognition among practitioners that the ideal source for many ERL applications would be an RF photocathode gun based on superconducting accelerating cavities. Work in this direction was covered in talks by Stefan and Sekutowicz at this conference, where many of the difficulties in and progress towards achieving an SRF gun were reviewed. Unfortunately for applications to projects being planned now, SRF guns are the least developed path, even though substantial progress is being made.

5. Future Applications

The future applications of ERLs that are presently being contemplated fall into four broad categories: higher average power FELs, photon sources in the DUV and X-ray wavelength bands, sans the FEL, electron coolers, and electron-ion colliding beam accelerators for high energy and nuclear physics experiments where the electron beam originates in an ERL.

In addition to the projects mentioned in Section 3, work will continue on producing higher average power FELs by increasing the electron beam power energy recovered to the level of 10 MW, corresponding to photon beam powers of order 100 kW. Compared to the present experience, this jump in performance should be relatively straightforward to achieve by increasing the beam current to the level of 100 mA by increasing the bunch repetition rate in the FEL. In particular, the bunch charge does not have to increase much beyond 100 pC to provide the requisite average current. Such a low bunch charge is well within the present operating experience of CW electron sources. In the further future, to achieve 100 MW beam power, it will be necessary to increase the charge-per-bunch substantially, to levels commonly investigated within the X-FEL community. As seen in more detail below, designs at higher levels in charge-per-bunch have tended to take advantage of the good impedance characteristics of superconducting cavities with lower operating frequency and larger physical size.

Because of the relative ease in scaling up the beam current in an energy recovered recirculated linac to the 100 mA level with existing DC photocathode sources, and because the beam current in typical storage ring radiation sources is also around 100 mA, it is natural to ask whether energy recovered linacs might be used in order to produce a recirculated linac light source with unique properties [15, 16]. This idea is being vigorously pursued at Cornell University [17] and 4GLS [11], and with varying degrees of seriousness at other laboratories around the world.

For high average X-ray brightness and flux in this application, there is a requirement for CW operation of the electron source. To maximize the X-ray brightness, as in X-FELs, one would like to minimize the electron beam emittance in both transverse planes in the ERL. Thus one is led to photocathode electron sources and large initial accelerating gradients. The fact that every accelerating phase may be populated with a beam bunch means that at 100 mA beam current only relatively modest charge-per-bunch, less than 100 pC, is needed, along with beam emittances around 1 mm mrad normalized *rms*. It is presently perceived that both DC photocathode guns and normal conducting RF photocathode guns may be able to produce beams with the required properties in the near term, and the hope, as a result of the development efforts mentioned previously, SRF photocathode guns will become excellent electron sources.

Compared to normal conducting RF or SRF photocathode guns, more modest developments are required to produce DC photocathode guns with emittance characteristics superior to those in the present CW FELs. By capitalizing on the lower thermal emittance inherent in GaAs photocathodes, by applying emittance compensation and drive laser pulse shaping techniques similar to those used in X-FEL guns, and by optimizing the placement of individual beam-line elements by application of an "evolution" optimization algorithm, the calculated normalized *rms* beam emittance from an injector designed for the Cornell ERL has 0.1 mm mrad at 80 pC beam charge [12]. The ERL gun at Cornell is being assembled presently and should be producing its first beam soon [18], and should produce high average brightness electron beams well beyond the requirements of the large recirculated linac light source.

At this point one should mention that there is a long-term proposal to energy recover the spent beam from an X-FEL [19, 20]. The main benefit of this approach is that it is now possible to run the repetition rate of the beam bunches up to 1 MHz, allowing the average brilliance to be increased by a factor of 10^5 in a machine that retains the huge peak brilliances of an X-FEL. Here, a major development task is found in building the high average electron beam power source to retain the exceptional beam quality required for the X-FEL. The beam line for recovering the energy of the DESY FEL is presently laid out in a reflex geometry, implying that rather elaborate collision avoidance beam optics be installed throughout the linac. To the best of the author's knowledge, no one has explored the perhaps more natural design, which features front-to-back recirculation and the X-FEL undulator in the return straight of the recirculation loop, although 4GLS has a similar geometry for their non-energy recovered DUV-FEL.

Till now, electron beam coolers have been based on DC electron beam technology, including spent electron beam energy recovery by suppressed collector voltage operating modes [21]. It has not been necessary to operate the coolers at beam energies beyond several MeV. With the advent of the RHIC, whose luminosity might be increased by the application of electron cooling, it becomes necessary to consider arrangements at higher beam energy. Because the application requires beam energies higher than a few MeV and because the electron cooling rate is higher the higher the (average) electron current, it becomes natural to consider a CW RF linac as a beam energy source and an energy recovered linac as a means to achieve the high average electron current, no longer limited by the CW RF power required in a single pass arrangement. Plans of this sort are being pursued at Brookhaven National Lab, where electron coolers based on ERLs at around 50 MeV and up to 0.1 A average current are being developed.

There are several special features required of the electron sources of this application. In contrast to the FEL and light source applications, where high average current is obtained by filling every accelerating phase with a relatively low charge bunch, because the ion bunch repetition rate in RHIC is only about 30 MHz, it makes no sense to accelerate with an electron bunch repetition rate of the cooler exceeding this same frequency. To obtain high average current, a charge-per-bunch of 4 nC has been chosen for this application. Therefore, by comparison to the previous applications, much greater care must be taken in dealing with the usual single bunch collective phenomena, for example wakes, coherent synchrotron radiation, and space charge effects, that afflict high charge-per-bunch beams. An example, dealing with injecting such beams onto the linac axis is provided in a separate contribution to this conference [22].

For this reason, and separately because, SRF cavities of lower frequency and larger physical size have less transverse and longitudinal impedance, and higher thresholds against multipass beam breakup instability, the cooler linacs are anticipated to operate at around 800 MHz, a frequency where previous SRF development work has occurred yielding SRF cavities with suitable performance characteristics. For a RHIC cooler the source should have a normalized *rms* emittance of 10s of mm mrad and an average beam current of 100 mA. Cooler development efforts have centered on building an SRF electron source for the ERL.

In the further future, electron-ion colliders based on an energy recovered linac for the electrons are being considered for nuclear physics facilities. There are two basic ideas being explored presently. In the first idea called eRHIC, an ERL would be added at the Brookhaven site, colliding off the ion beam stored in one of the RHIC ion storage rings [23]. In the second idea called ELIC, the present Jefferson Lab nuclear physics accelerator would be upgraded to make an energy recovered linac and two figure eight storage rings added [24]. The first ring would be added as an electron circulater ring, the second would act as an ion storage ring. It is felt that by utilizing electron cooling to maintain high ion densities in the ring, and by allowing the electrons to be more highly disrupted by their beam-beam collisions with the ions than is possible in storage ring colliders, it may be possible to increase the luminosity by one or two orders of magnitude beyond that possible from a more conventional storage ring collider.

Another advantage of the ERL-ring collider is the relative ease with which the beam polarization can be manipulated, at the ERL electron source, by comparison to a storage ring. The present Jefferson Lab accelerator performs experiments with 80 percent electron polarization, the beam polarization being produced by the circular polarization of the incident drive laser on a strained GaAs photocathode electron source. But, this source operates at much lower charge-per-bunch and average current than would be required for a high luminosity collider. As beam polarization manipulation and control seems a highly desirable feature of these advanced colliders, there is a requirement to develop highly polarized high average current electron sources with beam properties comparable to those emerging from the best current photocathode sources.

A little thought shows that in collider applications it may be possible to reduce the average current required from a source, polarized or otherwise, at the cost of an additional "circulater ring" with its fast beam injection and ejection systems. The circulater ring is designed not to store an electron beam for a long time, say exceeding many radiation damping times as in a storage ring, but to store the electron beam for a time that is short compared to the radiation damping time, say 100 turns of a typical-sized storage ring. If one could store and collide the electron beam for 100 turns in the circulater ring without so much degradation of the electron beam properties that the beam energy in the circulating beam can not be energy recovered, then one has the means of reducing the average current needed from the source by a factor of 100. This reduction may be accomplished because the electron source would need to deliver beam for one turn to completely fill the ring, but then could be off during the period of the 99 = 100 – 1 turns when the subsequent collisions during the fill occurred. The spent beam is extracted and energy recovered in the conventional way, making sure that the decelerating and accelerating beam for the next filling pulse are co-incident on the ERL simultaneously.

A concise summary of the source requirements for the eRHIC collider is of order 10 nC of charge at 30 MHz repetition rate, with emittances at around 10 mm mrad normalized *rms*. It is anticipated that some form of advanced polarized SRF gun will be utilized as the electron source. The ELIC beam parameters are of order 1 nC at 1 GHz repetition rate with similar *rms* emittances to eRHIC. The high average current of order 1 A needed from a CW electron source for ELIC may be reduced to the level of around 10 mA by the application of a 100 turn circulater ring. In this case, similar charge-per-bunch and macropulse currents are needed during a macropulse length equal to the ring filling time. By applying the circulater ring, the average current requirements are within reach of DC photocathode guns.

6. Summary

In this paper, the idea of an energy recovered linac has been introduced and its potential properties compared and contrasted to single pass linacs and storage rings. Energy recovery linacs have provided an elegant and powerful means to achieve high average power in free electron lasers. The pioneering energy recovered linac driven FELs have established many of the fundamental principles of ERLs. The multitude of ERL projects and proposals worldwide promises an exciting next decade as:

1. The three currently operating ERL-FELs will reach higher performance.
2. At least five more ERLs are in serious planning stages and will likely be constructed.
3. New advanced concepts are being explored; most of the applications need new high average brightness electron beam sources.

Acknowledgments

Ivan Bazarov, Ilan Ben-Zvi, Lia Merminga, Charles Sinclair, Todd Smith, and Susan Smith contributed slides and other material for this summary talk.

References

1. Tigner, M., *Nuovo Cim.* **37**, 1228 (1965)
2. Schriber, S. O., *et al., IEEE Trans. Nucl. Sci.* **NS-24**, 1061 (1977)
3. Merminga, L., D. R. Douglas, and G. A. Krafft, *Annu. Rev. Nucl. Part. Sci.* **53**, 387 (2003)
4. Krafft, G. A., *Proc. of the 2002 Joint Accelerator School,* Long Beach, CA, pg. 301 (2002)
5. Leemann, C. W., D. R. Douglas, and G. A. Krafft, *Annu. Rev. Nucl. Part. Sci.* **51**, 413 (2001)
6. Neil, G. R., *et al., Phys. Rev. Lett.* **84,** 662 (2000)
7. Smith, T. I., *et al., Nucl. Instrum. Methods A* **259**, 1 (1987)
8. Hajima, R., *et al., Nucl. Instrum. Methods A* **507**, 115 (2003)
9. Kulipanov, G., A. Skrinsky, and N. Vinokurov, *Nucl. Instrum. Methods A* **467**, 16 (2001)
10. Behre, C., *et al., Nucl. Instrum. Methods A* **528**, 19 (2004)
11. Poole, M. W., and E. A. Seddon, *Proc. of the 2004 European Part. Acc. Conf.*, 455 (2004)
12. Bazarov, I. and C. K. Sinclair, *Phys. Rev. ST Accel. Beams*, **8**, 034202 (2005)
13. Dowell, D. H., *et al., Applied Physics Letters* **63**, 2035 (1993)
14. Nguyen, D. C., *et al., Nucl. Instrum. Methods A* **528**, 71 (2004)
15. Kulypanov, G. N., A. N. Skrinsky, N. A. Vinokurov, *J. Synchrotron Rad.* **5**, 176 (2002)
16. ICFA Beam Dynamics Newsletter, Vol. 26, Edition on Advances in Recirculated Linac Light Sources, G. Krafft and Y. Zhang, Issue Editors, (2001)
17. Gruner, S., *et al., Reviews of Sci. Instrum.* **73**, 1402 (2002)
18. Sinclair, C. K., recent communication
19. Krafft, G. A., and J. J. Bisognano, *Proc. of the 1989 Part. Acc. Conf.*, 1256 (1989)
20. Sekutowicz, J., *et al., Phys. Rev. ST Accel. Beams*, **8**, 010701 (2005)
21. Nagaitsev, S., *et al., Proc. of the 2002 European Part. Acc. Conf.*, 1094 (2002)
22. Litvinenko, V., these proceedings
23. Ptitsyn, V., *et al., Proc. of the 2004 European Part. Acc. Conf.*, 923 (2004)
24. Derbenev, Y., *et al., Proc. of the 2004 European Part. Acc. Conf.*, 893 (2004)

MAXIMIZING BRIGHTNESS IN PHOTOINJECTORS*

C.LIMBORG-DEPREY†

SLAC, 2275 Sand Hill Road, MS18,
Menlo Park,CA,94025, USA

H.TOMIZAWA

SPring-8, 1-1-1 Kouto, Mikazuki-cho,Sayo-gun,
Hyogo 679-5198, Japan

If the laser pulse driving photoinjectors could be arbitrarily shaped, the emittance growth induced by space charge effects could be totally compensated for. In particular, for RF guns, the photo-electron distribution leaving the cathode should be close to a uniform distribution contained in a 3D-ellipsoid contour. For photo-cathodes which have very fast emission times, and assuming a perfectly uniform emitting surface, this could be achieved by shaping the laser in a pulse of constant fluence and limited in space by a 3D-ellipsoid contour. Simulations show that in such conditions, with the standard linear emittance compensation, the emittance at the end of the photo-injector beamline approaches the minimum value imposed by the cathode emittance. Brightness, which is expressed as the ratio of peak current over the product of the two transverse emittance, seems to be maximized for small charges. Numerical simulations also show that for very high charge per bunch (10nC), emittances as small as 2 mm-mrad could be reached by using 3D-ellipsoidal laser pulses in an S-Band gun. The production of 3D-ellipsoidal pulses is very challenging, but seems worthwhile the effort. We briefly discuss some of the present ideas and difficulties of achieving such pulses.

1. Introduction

The high brightness quality of beam produced in RF photoinjectors comes from the possibility of achieving very good emittance compensation [1].
After emittance compensation, the total emittance has four contributions, the cathode emittance, the non-linear space charge emittance, the RF emittance and the chromatic emittance as shown in equation (1).

$$\varepsilon_{tot} = \sqrt{\varepsilon^2_{cathode} + \varepsilon^2_{RF} + \varepsilon^2_{chrom.} + \varepsilon^2_{non-lin.\,space\,charge}} \qquad (1)$$

For the two types of photoinjectors discussed in this paper [2,3], which as of now use flat top laser distributions, the cathode emittance and the non-linear

* This work is supported by US Department of Energy contract DE-AC02-76SF00515
† contact email: limborg@slac.stanford.edu

space charge emittance each share half of the emittance budget, when the beamline is optimized, while the other contributions are nearly negligible. Small RF and chromatic emittances are the consequence of the short pulse lengths (about 10-15 degrees for the S-Band and L-Band systems discussed in this paper) required to meet high peak currents (~50-100A) at the end of the injector. The non-linear space charge emittance could nearly be totally suppressed by using 3D-ellipsoidal laser pulses. Indeed, for 3D-uniform ellipsoidal electron bunch distributions, the space charge forces are linear. So emittance compensation, which is based on linear focusing, can work perfectly. The total final emittance can then only be limited by the cathode emittance. In this paper, beam dynamics in photoinjector driven by 3D-ellipsoidal laser pulses are reviewed. Numerical simulations run for the optimization of an S-Band RF photoinjector with the goal of emittance minimization are presented. They include results for charges varying from 0.2nC to 10nC. The very promising results obtained suggest that the effort required to produce 3D-ellipsoidal laser pulses is worthwhile.

2. EMITTANCE

2.1.1. *Cathode Emittance*

For metallic cathodes, the "thermal emittance" can be computed based on the laser energy and the surface barrier potential of the metal. For the semi-conductor cathode, this calculation, in addititon, requires the knowledge of the electron affinity [4]. For copper cathode, the most common material for S-Band guns, the "thermal emittance" per unit spot radius can be computed theoretically to be 0.3 mm.mrad per mm. For Cs_2Te, the most common material for L-Band guns, the "thermal emittance" is calculated to be 0.43 mm.mrad. Several electron beam based measurements have shown that the "cathode emittance" is larger than the "thermal emittance" for both types of guns. For both copper and Cs_2Te the measured value is close to 0.6 mm.mrad per mm radius [5,6]. Many reasons could explain the difference between the "cathode emittance", which is a measured quantity, and the "thermal emittance" which is a theoretical value, such as surface roughness, scattering of the electrons, oxidation of the surface, two photon absorption ... No comprehensive explanation of these large discrepancies exists to our knowledge yet.

The "cathode emittance" is proportional to the radius of the laser spot. A radius as small as possible is desirable. However, the limit on the radius is imposed by the "image space charge" at the cathode. A too small radius is also undesirable when non-linear space charge forces are present as the magnitude of those latter increases with charge density. Those two mechanisms usually define the optimum radius for cylindrical laser pulses. As non-linear space charge forces are inexistent in the case of 3D-ellipsoidal pulse, the "image charge" limit alone imposes the minimum radius, which in turn allows to use smaller radii than those which are usually chosen for cylindrical pulses. The space charge limit is described by the two equations of formulae (2). They express the fact that the accelerating field on the cathode, E_{RF}, needs to overcome the image charge field E_s given by Gauss's law.

$$|E_{RF}| > |E_s| \quad \text{with } E_s = \frac{\sigma}{\varepsilon_o} = \frac{Q}{\pi r^2 \varepsilon_o} \tag{2}$$

where Q is the bunch charge, r the laser radius spot size ,ε_o the vacuum permittivity

Typically for a 1nC charge and a 1mm radius spot size, the cathode field needs to exceed 36 MV/m. For a 30 degree operating phase, the field should be of at least of 72MV/m. The operating phases need to be in the 20-40 degrees range to have a small energy spread to achieve good emittance compensation.

2.1.2. *RF Emittance*

The RF emittance has been described analytically in [7]. Its scaling is reproduced in (3), which shows that it is proportional to the product of the beam volume by the rms bunch length σ_z. This emittance is as small as 0.05 mm-mrad for the LCLS nominal parameters, i.e. an S-Band gun operated at 120MV/m and a cylindrical laser pulse 10ps long with a radius of 1mm.

$$\varepsilon_{RF} \sim \frac{E_{peak}}{f_{rf}} \sigma_r^2 \sigma_z^2 \tag{3}$$

with f_{rf} the gun RF frequency, E_{peak} the peak accelerating voltage on the cathode, σ_r the rms radius, σ_z the rms bunch length.

2.1.3. *Chromatic Emittance*

The chromatic emittance is a consequence of the energy dependence of the solenoid focusing. Equation (4) shows that it is proportional to the energy spread ($\Delta p/p$), the square of the rms beam size σ_r, and the inverse of the solenoid focal length f.

$$\varepsilon_{chrom.} \propto \frac{\Delta p}{p} \frac{\sigma_r^2}{f} \tag{4}$$

Since the space charge induced emittance growth can totally be suppressed with ellipsoidal pulses, the combination of RF and chromatic emittances can be evaluated numerically by setting to zero the "cathode emittance". For our LCLS case, it was determined to be of 0.15mm.mrad for the 1nC case with 1.2 mm radius and 10ps of cylindrical beam. Obviously, this method slightly underestimates the value when space charge is present, for that reason we assumed in the computation leading to the orange curve of figure 6 that this value corresponded to a 1mm radius.

2.1.4. *Non-linear Space Charge Emittance*

The non-linear space charge emittance scales like the charge density and depends on a form factor F. For a 3D-ellipsoidal uniform pulse, the form factor F can be forced to zero after appropriate choice of the emittance compensation tuning parameters. For all other pulses, the form factor F is always non-zero and will depend on the shape and uniformity of the charge distribution at emission.
For the cylindrical shape laser pulse, the optimum radius results from a compromise between tolerable space charge force (non-linear space charge), and "cathode emittance". This compromise is summarized in equation (5).

$$\varepsilon_{tot} = \sqrt{\left(\varepsilon_{cathode}\,\sigma_r\right)^2 + \left(F\frac{Q}{\sigma_r^2\sigma_z}\right)^2} \quad (5)$$

$\varepsilon_{cathode}$ the cathode emittance normalized to laser spot size, and F a form factor which depends on the charge density distribution at emission, so directly from the laser pulse shape and local fluence.

2.1.5. *An example: the LCLS nominal tuning using cylindrical pulse*

For the LCLS photoinjector, using its nominal 1.2mm radius, 10ps long cylindrical laser pulse, the optimum tuning gives equal contributions from the "cathode emittance" (for 0.7 mm-mrad) and from the non-linear space charge emittance (for 0.7 mm-mrad) to a total emittance which is 1 mm-mrad. A smaller total emittance of 0.85 mm-mrad can be obtained if the "cathode emittance" is reduced to 0.5 mm-mrad by using a radius of 0.85 mm instead of 1.2 mm. But the laser pulse length has to be stretched to a 20ps pulse duration [8]. The peak current is then 60A instead of the required 100A peak current. The 0.85 mm.mrad is the ultimate minimum for the LCLS injector operated with 1nC because a further reduction of the radius requires longer bunches and which increases the RF emittance to a significant level, following the scaling law already presented with equation (3).

3. 3D-ellipsoidal pulses

3.1. *Beam dynamics*

Figures 1-2-3 and 4 illustrate the benefits of the 3D-ellipsoidal laser shape case compared to that of the cylindrical laser pulse. In figure 1 and 2, we plotted the evolution of the transverse phase space, the longitudinal phase space and the focusing parameter r'/r between. Figure 1 shows the case of the cylindrical pulse and figure 2 that of the 3D-ellipsoidal pulse. The first row gives the beam distribution at the exit of the gun, i.e., at the location where the solenoid starts refocusing the beam. The second row gives the distribution when the beam is entering the linac, i.e., when the emittance compensation is nearly completed since the normalized phase space starts getting frozen. The seven temporal slices have been color coded. The suppression of the non-linear space charge force for the 3D-ellipsoial distribution is visible in the r'/r graph. The r'/r becomes very non-linear at the head and tail of the cylindrical beam since the longitudinal space charge force distorts the square longitudinal profile into a parabolic one, varying the charge density with respect to that of the longitudinal core of the bunch. For the 3D-ellipsoidal bunch, the charge density remains uniform across the whole volume of the bunch as the beam is transported.

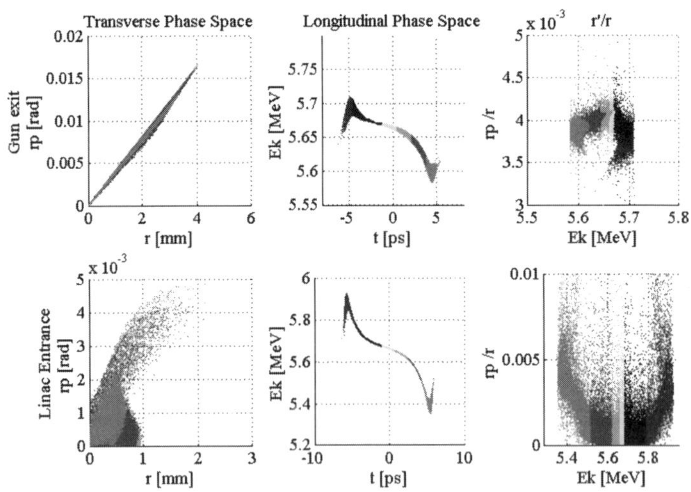

Figure 1- Emittance compensation for a cylindrical laser pulse

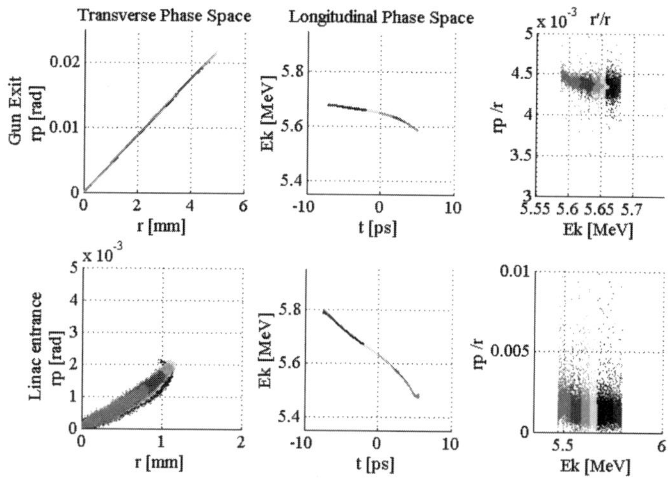

Figure 2: Emittance compensation for a 3D ellipsoidal laser distribution

The impressive reduction of slice emittance and increase in peak brightness for the 3D-ellipsoial shape bunch compared to the cylindrical one are given in figure 3 (a) and (c) for the S-Band gun and in figure 4 (a) and (c) for the L-Band gun. Graph-b of figure 3 shows that the peak current profile of the 3D-ellipsoidal reaches nearly the same level as that of the cylindrical shape. Finally in graph (d) of figure 3, the very linear longitudinal phase space of the 3D-ellipsoidal beam is compared to that of the cylindrical beam, the 2nd order term coming from the RF curvature has been removed. The linearity of the longitudinal phase space will allow reducing the strength of the high harmonic linearization which precedes the bunch compressors.

3.1.1. *S-Band*

A comparison of the best optimization of the LCLS photoinjector beamline for the 1nC charge for both the cylindrical and 3D-ellipsoidal cases are shown in figure 3. This optimization was obtained for a 120MV/m gun peak field and a cathode emittance of 0.6 mm-mrad per mm.

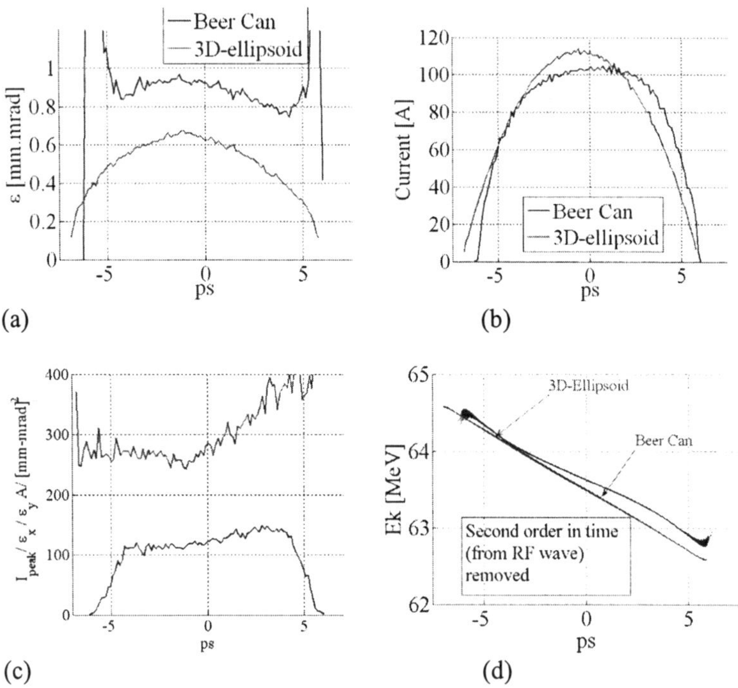

Figure 3: Comparison of beam parameters for 3D ellipsoidal distribution and cylindrical beam (also called "Beer can")-

3.1.2. *L-Band*

The optimization results presented in figure 4 compare the cylindrical shape and the 3D-ellipsoidal case for the TTF2 layout. The RF gun peak field is 40MV/m and the thermal emittance is 0.43 mm-mrad per mm. The author apologizes for not updating the results after the measured cathode emittance of 0.6 mm-mrad per mm was known.

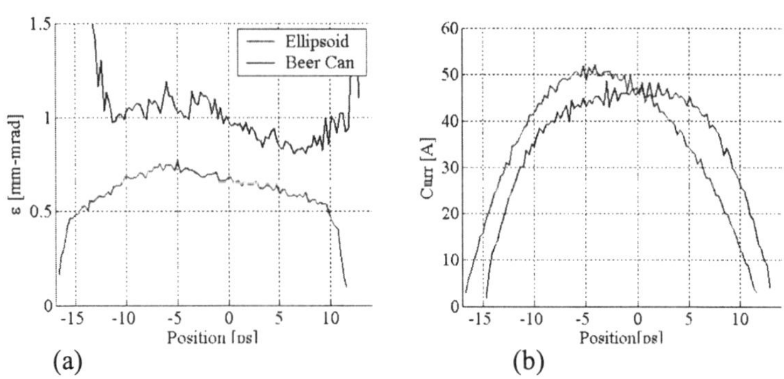

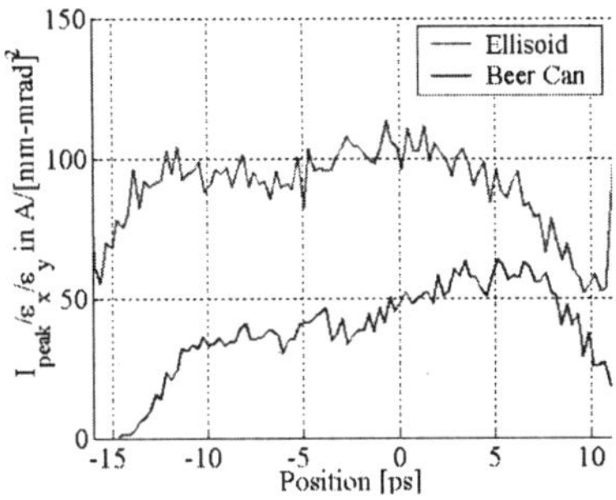

(c)

Figure 4: Comparison slice emittance (a) peak current (b) and peak brightness(c) for 3D ellipsoidal distribution and cylindrical beam for the L-Band gun –

The improvement of the stability of the emittance with variation of the parameters has been described in [9] and is not repeated here.

3.2. *Emission in simulations*

3.2.1. *Approximations in multiparticle tracking codes*

The results presented in this paper are based on numerical simulations. In the next few paragraphs, we discuss the validity of the representation of physics in the standard multiparticle tracking codes [10,11,12] at emission and in the low energy regime corresponding to the few mm of travel after the cathode. We justify why those computations should give results close to reality. It was already mentioned that the measured "cathode" emittance is larger than the theoretical "thermal" emittance. In the simulations presented here, we used the value of the measured "cathode emittance". This value is likely to be reduced once better understanding of the source of degradation is known.

3.2.2. *Image Charge*

In multiparticle tracking codes, the electromagnetic boundary condition derived from Maxwell's equations associated with the cathode wall is modeled by the image charge of individual emitted particles. It has been verified using PIC (Particle-In-Cell) codes [13], which base their computation on the direct resolution of Maxwell equations, that the image charge model gives accurate results for a 100MV/m accelerating field on the cathode. Even if this study should be repeated at lower gradient (~30 MV/m or less) for completeness, we could conclude that the "image charge" model gives a valid approximation for the beam dynamics calculations.

3.2.3. *Space Charge at low energy*

For the standard multiparticle tracking codes [11,12,13], space charge forces are computed in the frame of the centroid of the bunch. A Lorentz transformation is performed on all the particles using the particular Lorentz factor of the centroid. However, the energy spread is larger than 50% in the first few mm of acceleration. The studies described in [13] also show that the error in computation coming from that approximation with respect to results from PIC codes is in fact very small. In the studies described [13], which assumed an accelerating field of 100MV/m on the cathode, tracking codes even gave emittance results larger than those obtained with PIC codes. The IMPACT code [14], which is a multiparticle tracking code offering the possibility of slicing the bunch in energy to perform multiple Lorentz transformations for different energy slices, shows very small emittance differences for the standard LCLS configuration (with RF field on cathode of 60MV/m at emission time, i.e. a 120MV/m peak field) .

In the low energy regime, when the energy spread is larger than 50%, the space charge force is not exactly linear across the bunch emitted from a 3D-shaped uniform ellipsoidal laser pulse. Accordingly, the bunch will not be maintained exactly as a uniform 3D-ellipsoid. This can be observed by using very small meshes in space and time and very large numbers of particles. This should be further studied with a code like IMPACT or directly with PIC codes. However, the suspected small asymmetry in longitudinal profile observed could be compensated for by small modifications on the shaping of our 3D-ellipsoidal laser pulse.

3.2.4. *Limits imposed by Gauss's law*

Gauss's law is a one-dimensional formula, which assumes the whole charge concentrated in a single narrow slice and a single radius. The limit on the radius given by equation (2) is well defined. However, for the 3D-ellipsoidal pulse the radius varies following a *circ* function from 0 to a maximum r value and back to zero along the pulse as described in equation (8). For a 60 MV/m cathode field (120MV/m at 30 degrees launch), and a 1nC, the hard edge on the radius limit is 0.77 mm. In our simulations, the beam will still be emitted for slightly smaller radii than 0.77 mm. But more importantly, for radii between 0.75 mm to 0.85 mm the bunch profile is going to be distorted and the electron bunch will lose the symmetry of the driving laser pulse, as shown in figure 5. Consequently, the limit radius is slightly larger than that predicted by applying Gauss's law. This possibly explains the offset of the blue curve with respect to the orange curve in figure 6.

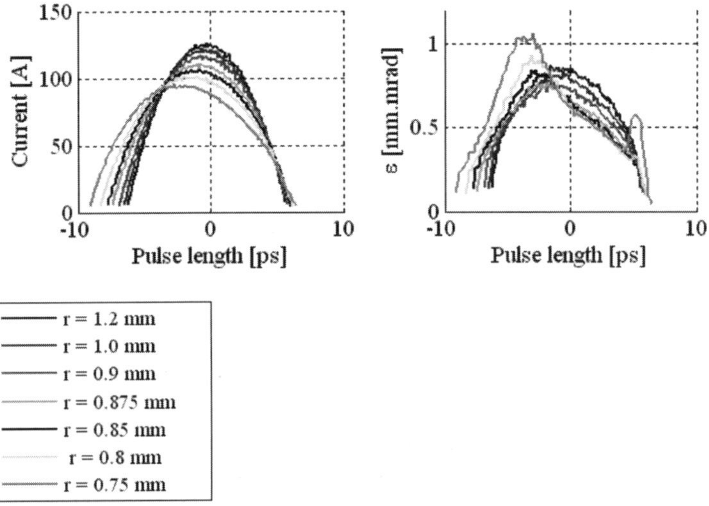

Figure 5- Two graphs showing respectively the final peak current along the bunch at the end of the LCLS beamline, and slice emittance along the bunch for different maximum radii r of the 3D-ellipsoid; the beamline parameters are 1nC, and 120 MV/m peak accelerating field in the gun and 32 degrees launch phase.

4. OPTIMIZATION

The optimization of a beamline requires finding appropriate laser beam dimensions and beamline parameters to minimize the projected emittance at a location where the energy is high enough so the space charge is not a problem. One usually searches a tuning for which both normalized emittance and beam sizes are slightly converging at that high energy location.

4.1.1. *Optimum brightness*

The brightness scales like the ratio of the peak current over product of the transverse normalized emittances. Optimizing directly on the brightness is difficult as it requires finding the combination of small emittances and high peak current. Those two start acting against one another when very high peak current gets required. The study presented here primarily focused on minimizing the emittance using ideal ellipsoidal laser shapes. For a few of the points of figure 6, an attempt to increase the peak current resulted in larger emittances, and above all lower brightness values.

The minimization of emittance was performed for the S-Band gun version to be run at the LCLS. As recent modifications implemented on the LCLS gun [15] should allow 140MV/m peak field, this value was used in the simulations presented here. Tuning parameters such as the maximum radius, the solenoid strength, the injection phase, the linac gradient were varied. The linac gradient value was chosen to match the invariant envelope criteria [16]. Simulations were run down the entire LCLS beamline after the end of the second S-Band accelerating section to reach energies above 120MeV. In figure 6, the emittance value given by the best optimization for each of the charges has been plotted in blue. The theoretical curve showing the minimum cathode emittance based on the minimum radius imposed by Gauss's law is shown in red following equation (6). The orange curve is computed by adding the chromatic and RF emittance terms as shown in equation (7) to the cathode emittance shown in the red curve. The chromatic and RF emittance were combined in a single term, scaled from the numerical value of 0.15 mm-mrad obtained for the 1nC, r = 1mm case after switching the cathode emittance to zero. A better estimate would have been made by running individual charge cases and switching the cathode emittance off for each case.

$$\varepsilon_{cathode} = 0.6\sqrt{\frac{Q}{\varepsilon_o \, \pi \, V_{RF}}} \text{ mm-mrad} \quad (6)$$

$$\sqrt{\varepsilon_{chr.}^2 + \varepsilon_{RF}^2} = 0.15\frac{r^2}{r_o^2} \text{ mm-mrad with } r_o = 1\text{mm} \quad (7)$$

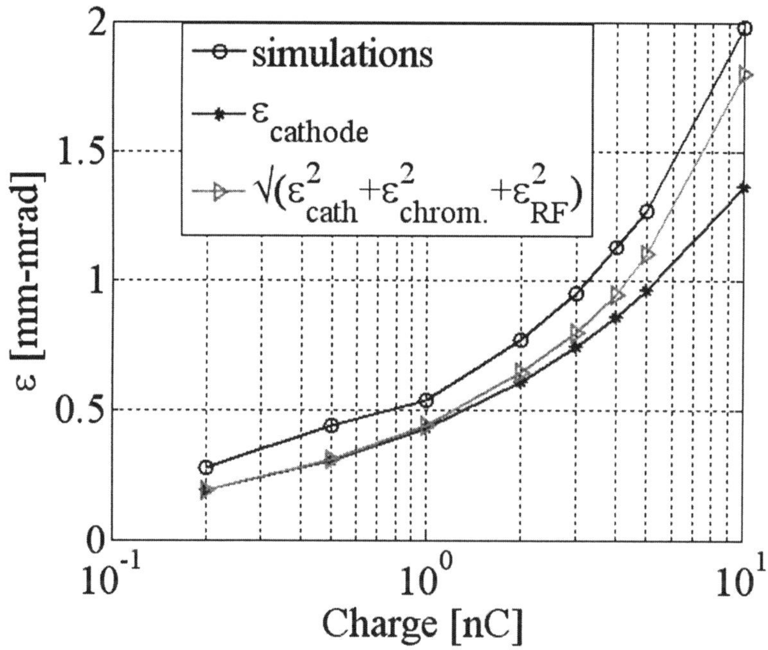

Figure 6 – Projected emittance optimized from simulations for charges varying from 0.2nC to 10nC using 3D-ellipsoidal laser pulses in the blue curve– Limit imposed by the cathode emittance is represented in red- Limit imposed by the combination of cathode emittance, RF emittance and chromatic emittance is given in orange- A cathode emittance of 0.6 mm-mrad per mm radius spot size has been assumed

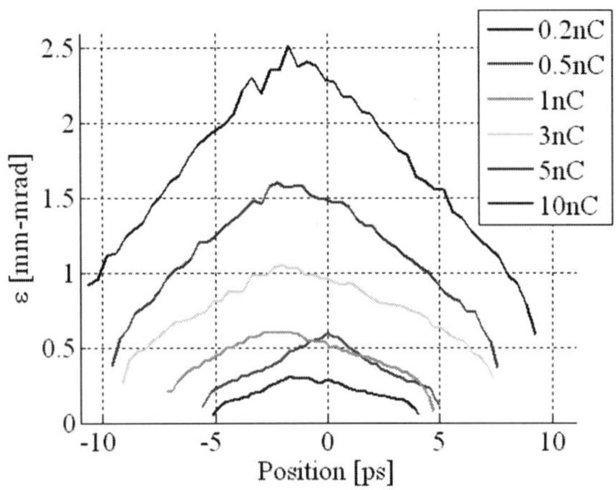

Figure 7 (a) - Slice emittance for optimum tuning –

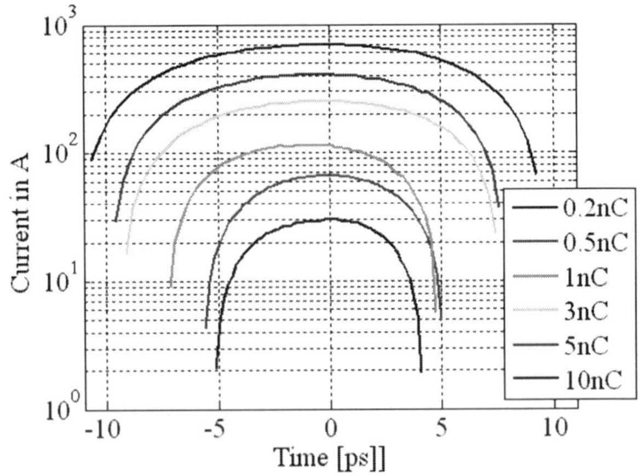

Figure 7 (b) - Peak current at end beamline -

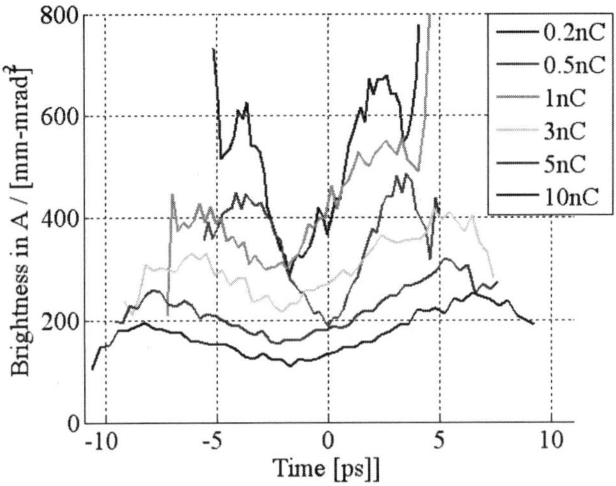

Figure 7 (c) - Brightness expressed as the ratio of peak current over product of transverse slice emittances

In figure 7 (a-b-c), the slice emittance, peak current and brightness along the bunch are shown at the end of the beamline for the different charges.
Those results show that the maximum brightness is obtained for 0.2nC and 1nC and not for the high charges. We don't claim this result to be definitive as more efforts should probably be put in the optimization of the 0.5 nC case.

In figure 8 (a-b), the evolution along the beamline of the rms emittance and rms beam sizes have been plotted for the two extreme charges studied of 0.2nC and 10nC. They show a nice smoothly decreasing beam size and emittance with no emittance oscillation for the 10nC case. For the 0.2nC case, we claim that an emittance of 0.28mm-mrad is achievable even if the present tuning does not show a nice converging beam in the graph presented here.

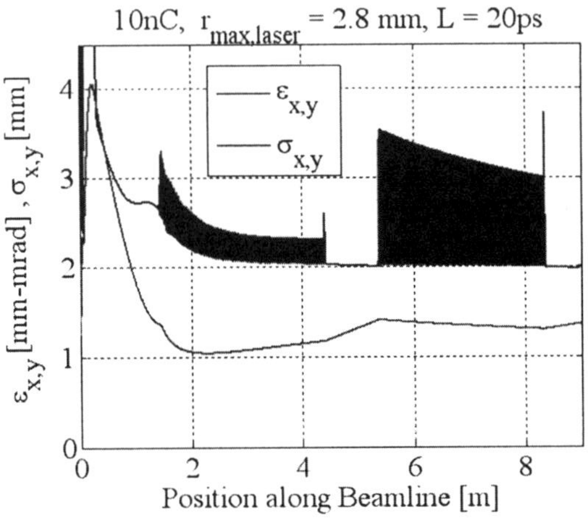

Figure 8 (a): Transverse beam for 10nC tuning

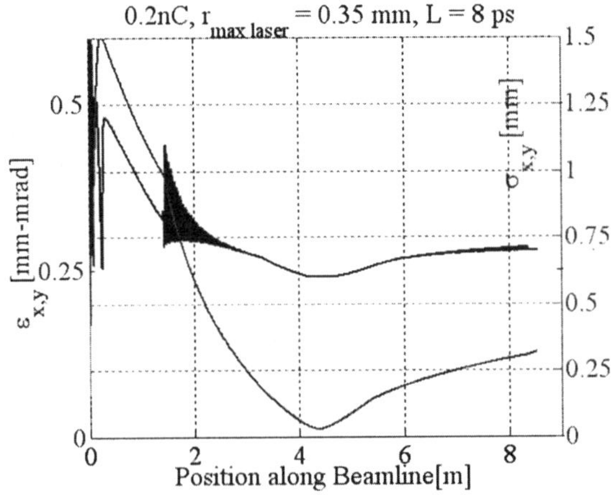

Figure 8 (b): Transverse beam for 0.2 nC tuning

4.1.2. *Photoinjector as an X-Ray FEL driver*

An X-Ray FEL injector needs to provide electron bunches with at the same time very small slice emittances and large peak current for the SASE mechanism to develop and reach saturation in the shortest undulator length as possible. Ideally, one wants to have the combination of small emittance and large peak current for as many slices as possible. Flat top distributions had been the preferred option until now as the peak current and emittance are about constant for the core 70% of the beam. A high current spike is present at the head and tails of the bunch but with large slice emittance. The use of 3D-ellipsoidal laser pulses would not suppress the spike as this one is generated from the combination of parabolic longitudinal profile and bunch compression. However, the very small emittances produced at the head and tail for those types of pulses would help making the spike more attractive for the lasing. Start-to-End simulations are required to demonstrate this statement as CSR needs also to be included.

5. 3D-ellipsoidal Laser Pulse Production

Producing a laser pulse with a radius varying from zero to the maximum radius and back to zero in less than 10ps, with constant fluence is very challenging. Difficulties of producing uniform 3D-ellipsoidal laser pulses have been described in [9]. We briefly report here on three of the possible solutions and illustrate the expected results with electron beam simulations.

5.1.1. *Pulse stacker*

The pulse temporal stacking approach requires the summation or stacking of multiple beamlets, each of which has a specified time delay, transverse radius and pulse energy. Assume that each beamlet is temporally Gaussian but transversally uniform (flattop). Fluence would then be held constant in time only in a discrete way (beamlet peak-to-beamlet peak). The summation in the overlap region partially smoothes the discrete fluence levels. Each beamlet must have its own transverse profile shaper with imaged transport downstream of the shaper. Each beamlet must also have independent optical delay and independent pulse energy control. Fluence preservation over all beamlets requires that the energy of a given beamlet scales linearly with its transverse area. The uniform transverse profile guarantees constant fluence within any beamlet Interference effects will introduce additional intensity and power variations (ripples) which can be minimized by alternating beamlet polarization between 'S' and 'P' cases. Complexity, size and cost can escalate rapidly with the number of beamlets used (for example, 10 or more). The efficiency of such a system was estimated to be at the 10^{-2} level for UV wavelengths.

In figure 9, we compare simulations results obtained for 8 and 12 beamlets. Twelve trials were computed for each case using random phases between the beamlets to take into account the interference effects.

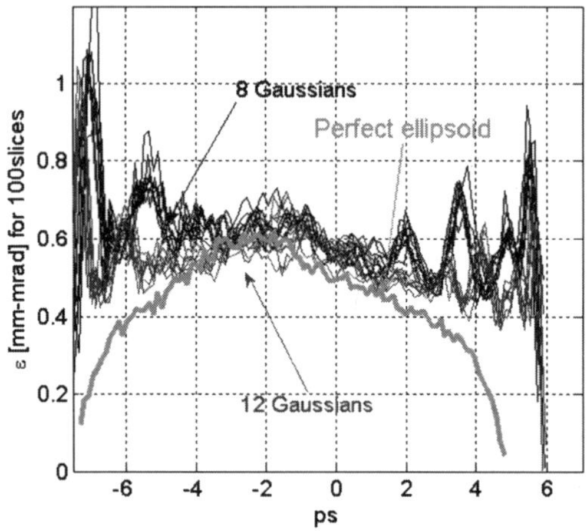

Figure 9: Slice emittance from simulations for a pulse stacker with 8 Gaussians (blue) and 12 Gaussians (red) compared with an ideal 3D-ellipsoidal shape

5.1.2. *Spectral masking*

An energy-chirped laser beam reflecting on a grating will project its enery-time dimension into a spatial dimension. The efficacy of spectral masking to control the transverse beam shape and diameter can be considered in terms the uniqueness of the time-space (one dimensional) correlation downstream of a diffraction grating in the plane of dispersion. Uniqueness limits are attributed to the nonzero beamsize projected on the grating and can be evaluated with a uniqueness function, U. Discussion of the uniqueness function has been given in [9] and proves that such an optical layout allows to generate a elliptical cylinder. The intersection of 4 ellipical cylinders rotated by 45 degrees gives a good approximation of an ellipsoid, with an octagonal cross-section instead of a disk. If the number of such elliptical cylinders is increased to 6, with 30 degrees between them, the approximation is perfect. Results of the beam dynamics for these two configurations are given in figure 10.

The efficiency of a 2 stage (8 gratings) configuration is unfortunately estimated to be as low as 10^{-10}.

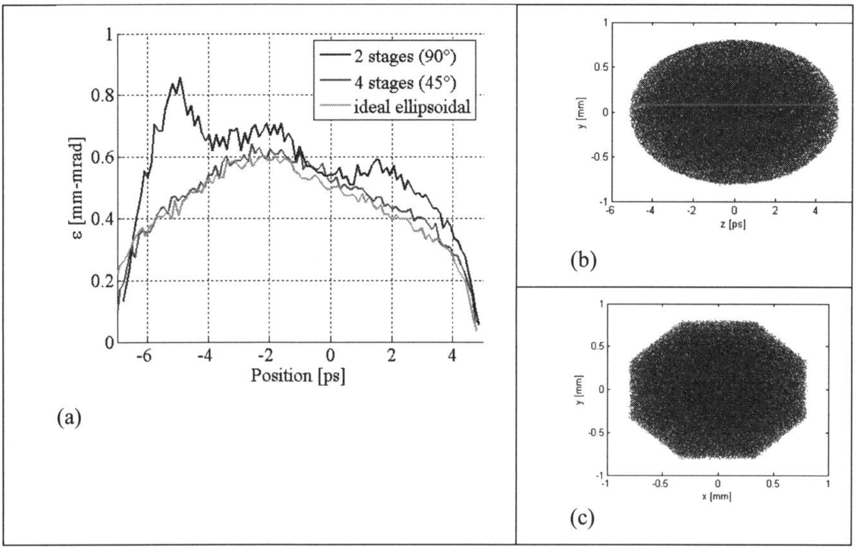

Figure 10: (a)Slice emittance from simulations from pulse built with spectral masking. The cross section of the 4 stage configuration are given in (b) for the z-y plane and (c) for the x-y plane.

5.1.3. *Fiber Optics bundle*

It has been suggested by H.Tomizawa that for transmissive cathode a straight forward solution would exist for generating 3D-ellipsoidal pulses. Here we summarize those concepts from H.Tomizawa in a very simplified version. More details will be given in a coming paper [18].

H.Tomizawa described in [19] how the combination of deformable mirrors with fiber bundle allows to produce any temporal and spatial shape. A schematic is given in figure 11, which exhibits the principle. Beamlets are sent from the deformable mirrors to the appropriate fiber in r location. H.Tomizawa has demonstrated experimentally the technique for generating homogenized transverse profile while maintaining the input longitudinal profile. The next step consists in proving that the algorithm would work if one wants to limit the radius r(z) following equation (8)

$$r(z) = r_{\max}\sqrt{1-\frac{z^2}{L^2}} \quad (8)$$

Obviously, because of its short working distance, this technique only works for transmissive cathodes, and the fiber bundles need to terminate at the cathode.

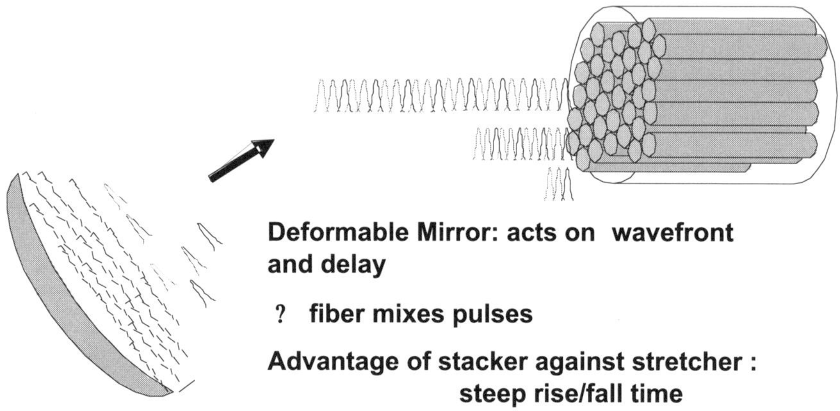

Figure 11- Schematic illustrating the system combining deformable mirror and fiber bundle for generating 3D-ellipsoidal pulses

6. Conclusions

The great improvement of emittance and brightness performances when using 3D-ellipsoidal laser pulses instead of cylindrical shapes has been demonstrated. For 3D-ellipsoidal laser pulses, we demonstrated that the maximum peak brightness follows is in fact obtained in for 0.2nC and 1 nC for the S-Band gun. However, more simulations are required which would aim at maximizing peak current, by for instance using velocity bunching.

The great benefits of such pulses make them worthwhile investigating thoroughly the possibility of producing them. Another possibility to produce ellipsoidal emitted electron bunches is described in [17].

References

1. B.Carlsen, "New Photoelectric Injector Design for the Los Alamos Natioanl Laboratory XUV FEL Accelerator", NIM A285 (1989) 313-319

2. LCLS Photoinjector http://www-ssrl.slac.stanford.edu/lcls/cdr/lcls_cdr-ch06.pdf
3. TESLAPhotoinjector, http://tesla.desy.de/new_pages/ TDR_CD/PartII/accel.html
4. K.Floettmann, "Note on the Thermal Emittance of Electrons Emitted by Cesium Telluride Photo Cathodes", Feb 1997, TESLA-FEL, 97-01
5. W.Graves et al. "Measurement of Thermal Emittance for a copper PhotoCathode", PAC01 Proceedings
6. V.Mitchell "Thermal Emittance Measurments", ICFA X-Ray FELs Commissioning workshop, Zeuthen April 2005
7. K.J.Kim "RF and space charge effects in laser driven RF electron guns". NIM in Physics Research A275 (1989) 201-218
8. C.Limborg-Deprey et al. "Modifications of the LCLS PhotoInjector Beamline" EPAC 04 Proceedings, Lucerne, July 2004
9. C.Limborg-Deprey "Optimum Electron Distributions For Space Charge Dominated Beams", ERL05 Proceedings, Jlab, published in NIM A 557 (2006) 106-116
10. L.Young, J.Billen, PARMELA, LANL Codes, laacg1.lanl.gov/laacg/services/parmela.html
11. K.Floetmann,ASTRA, http://www.desy.de/~mpyflo/Astra_dokumentation/
12. S.B Van der Geer, M.J de Loos, GPT, http://www.pulsar..nl/gpt
13. E.Colby, V.Ivanov, C.Limborg "Simulation Issues for RF PhotoInjectors", ICAPS 2002
14. Qiang J. et al. "A three-dimensional Quasi-Static Model for High Brightness Beam Dynamics Simulations", to be published in PRST-AB in summer 2006
15. L.Xiao, "Dual Feed RF Gun Design for the LCLS", PAC 05, Knoxville, USA
16. M.Ferrario, "Homdyn study for the LCLS RF Photoinjector", "Physics of high Brightness beams", World Scientific (2000), P534
17. O.J.Luiten "these proceedings"
18. LimborgC. Tomizawa. H , "Interest in generating 3D-ellipsoidal laser pulses " to be submitted to PRST_AB
19. H.Tomizawa, "Adaptive Shaping System for both spatial and temporal profiles of a highly stabilized UV pulse for RF PhotoInjectors", ERL05 Proceedings

ULTRACOLD ELECTRON SOURCES*

O.J. LUITEN, B. J. CLAESSENS, S. B. VAN DER GEER, M. P. REIJNDERS, G. TABAN, AND E. J. D. VREDENBREGT

Eindhoven University of Technology
Den Dolech 2,
P.O. Box 513, 5600 MB Eindhoven, Netherlands
E-mail: o.j.luiten@tue.nl

Ultra-cold plasmas with electron temperatures of ~ 10 K can be created by photoionization just above threshold of a cloud of laser-cooled atoms. Recently it was shown [7] by GPT particle tracking simulations that an ultra-cold plasma has an enormous potential as a pulsed bright electron source. Here we discuss these results in the framework of normalized 6D brightness, which allows us to make a proper comparison both with the performance of pulsed, radio-frequency photo-emission sources and with the performance of continuous, needle-like field-emission sources. In addition we speculate on the possibility of using ultra-cold plasmas to realize quantum degenerate electron beams, constituting the ultimate limit in electron beam brightness.

1. Introduction

The present state-of-the-art in brightness of pulsed, high-current electron beams is supplied by radio-frequency (rf) photoguns [1]. In these devices charges of, typically, 0.1–1 nC are produced by (sub)picosecond photoemission at the cathode surface, in a spot size of the order of 1 mm. This gives rise to space-charge fields of 1–10 MV/m, a significant fraction of the typical acceleration fields of 100 MV/m. Space-charge forces generally depend in an irregular, nonlinear way on the particle position within the bunch. This in turn gives rise to distortion of the bunch, which results in unrecoverable brightness degradation.

The obvious solution is reducing space charge forces by increasing the spot size on the photocathode. Unfortunately, however, the photoemission

*This work is part of the research programme of the Stichting voor Fundamenteel Onderzoek der Materie (FOM), financially supported by the Nederlandse Organisatie voor Wetenschappelijk Onderzoek (NWO)

process itself gives rise to initial velocities which are randomly distributed in all three directions, so that the photocathode effectively behaves as a thermionic source with an effective electron temperature $kT \approx 0.5$ eV, i.e. $T \approx 5 \times 10^3$ K [1]. This translates into an initial, "thermal", emittance ε_{th} proportional to the spot size:

$$\varepsilon_{th} \approx \frac{R}{2}\sqrt{\frac{kT}{mc^2}}, \tag{1}$$

with m the elctron mass, c the speed of light, and R the initial radius. For a typical initial radius of $R = 1$ mm this results in $\varepsilon_{th} \approx 0.5$ mm mrad. The best performing rf photoguns produce beams with an emittance of 1 mm mrad, containing approximately equal contributions due to nonlinear space-charge forces and the thermal emittance [1].

Following the pioneering work of Serafini [2] on production of electron bunches in the blow-out regime, it was recently shown by Luiten *et al.* [3] that there is a practical way to substantially reduce the space-charge problem *without* increasing the initial radius: by careful shaping of the photoexcitation laser pulse it is possible to generate electron bunches with a uniform, ellipsoidal charge distribution, so-called waterbag bunches, which are characterized by perfectly linear space-charge fields. Immediately after initiation the bunch will still expand rapidly due to Coulomb forces, but the expansion is linear and will therefore not contribute to irreversible emittance growth. Several groups are investigating this idea further by numerical simulations [4,5,6] and are working on its experimental realization.

The implementation of waterbag bunches in rf photoguns may lead to an improvement of the emittance by about a factor of two, i.e. the value of the thermal emittance of a 1 mm initial radius beam. Although this amounts to a substantial and important improvement, the thermal emittance cannot be reduced further by reducing the initial radius. This is because rf photoguns are operating in a regime where the image-charge and self fields cannot be neglected. If the initial radius is reduced further, the image-charge and self field strengths become comparable to the acceleration field strength, leading to 100% energy spread and severe distortion and lengthening of the bunch. In addition there is the obvious practical limitation of laser damage to the cathode.

The only way to reduce the thermal emittance without reducing the initial radius, is by lowering the effective electron temperature of the source. Recently, we proposed to extract electron bunches from an ultracold plasma (UCP) [7], which can be fashioned from a cloud of laser-cooled atoms by photo-ionization just above threshold [8]. Such plasmas are characterized by electron temperatures down to 10 K [9], i.e. 2–3 orders of magnitude lower than photoemission sources. This implies a potentially enormous gain in brightness.

In a recent publication [7] we showed by means of particle tracking simulations with the General Particle Tracer (GPT) code [15] that an ultra-cold plasma has an enormous potential as a bright electron source. In this paper we will discuss those results in the framework of normalized 6D brightness. To this purpose a Lorentz-invariant definition of normalized 6D brightness is presented in Sec. 2. The connection of the normalized 6D brightness to the more commonly used transverse or reduced brightness is discussed. This leads to the introduction of a brightness chart, which allows a proper comparison between the performances of completely different electron sources and a clear way to visualize thermal limitations. In Sec. 3 the recently published [7] particle tracking simulation results of extraction of electron bunches from a UCP are summarized. The calculated performance of UCP-based sources is compared to the state-of-the-art performance both of continuous sources developed for electron microscopy, and of rf photoguns. Finally, in Sec. 4 we speculate on the possibility of reaching the ultimate fermi-degenerate quantum limits of 6D brightness with ultracold electron sources.

2. Brightness: definitions and state-of-the-art

The ultracold electron source concept represents a radical break with tradition in several ways. To enable a proper comparison with existing sources, we therefore first carefully examine how particle beam quality can be defined correctly. Here we follow, and slightly expand on, the treatment by Brau [10].

The average quality of a particle beam can be expressed in terms of a single, comprehensive figure of merit: the *average* 6D brightness $\bar{B}$, also referred to as spectral brightness or brilliance, which is given by the current density per unit solid angle and per unit energy spread:

$$\bar{B} \equiv \frac{I}{\Delta A \, \Delta\Omega \, \Delta U}. \tag{2}$$

Here I is the beam current, ΔA the surface area of the beam cross section, $\Delta\Omega$ the solid angle subtended by the beam, and ΔU the beam energy spread. For a given beam energy the average brightness $\bar{B}$ can be considered as a conserved quantity, independent of beam size and divergence. The quantity $\bar{B}$ represents an average over the *local* 6D brightness distribution function B:

$$B(x, x', y, y', U, t) = \frac{\partial^6 Q}{\partial x\, \partial y\, \partial x'\, \partial y'\, \partial t\, \partial U}. \tag{3}$$

where Q is charge, t is time, and the beam angles $x' \equiv v_x/v_z$ and $y' \equiv v_y/v_z$ are defined in terms of particle velocities v_x, v_y, and v_z, in the usual paraxial approximation for a beam pointing in the z-direction. The 6D brightness B thus defined, however, is not an invariant quantity: if the average energy of the particles is increased, the beam divergence decreases, leading to an increase of B. To enable a proper comparison between beams of different energy it is convenient to switch to phase space coordinates $(\vec{r}, \vec{p})$, with momentum $\vec{p} \equiv \gamma m \vec{v}$, Lorentz factor $\gamma \equiv 1/\sqrt{1-\beta^2}$, and $\beta \equiv v/c$. For a beam pointing in the z-direction we have $v \approx v_z \gg v_x, v_y$, so that $\partial/\partial x' = \beta\gamma\, mc\, \partial/\partial p_x$, $\partial/\partial t = \beta c\, \partial/\partial z$, and $\partial/\partial U = (\beta c)^{-1}\, \partial/\partial p_z$. It then follows that

$$B(\vec{r}, \vec{p}) = \beta^2\gamma^2 m^2 c^2 \frac{\partial^6 Q}{\partial x\, \partial y\, \partial z\, \partial p_x\, \partial p_y\, \partial p_z}. \tag{4}$$

Now it is easy to see that the 6D brightness B is proportional to the Lorentz-invariant local phase space density $f(\vec{r}, \vec{p})$, normalized to the number of particles in the bunch $N = \int f(\vec{r}, \vec{p})\, d^3r\, d^3p$:

$$B(\vec{r}, \vec{p}) = e\beta^2\gamma^2 m^2 c^2\, f(\vec{r}, \vec{p}), \tag{5}$$

with e the elementary charge. We thus arrive at a natural, Lorentz-invariant definition of the *normalized* local 6D brightness B_n:

$$B_n(\vec{r}, \vec{p}) \equiv e\, m^2 c^2 f(\vec{r}, \vec{p}) = \frac{1}{\beta^2\gamma^2} B(\vec{r}, \vec{p}). \tag{6}$$

Generally, the distribution function $f(\vec{r}, \vec{p})$ is not known. The simplest model for a particle bunch is a 6D Gaussian phase space distribution aligned along the x, y, and z-axes:

$$f(\vec{r}, \vec{p}) = f(0,0)\, \exp\left(-\frac{x^2}{2\sigma_x^2} - \frac{p_x^2}{2\sigma_{p_x}^2} - \frac{y^2}{2\sigma_y^2} - \frac{p_y^2}{2\sigma_{p_y}^2} - \frac{z^2}{2\sigma_z^2} - \frac{p_z^2}{2\sigma_{p_z}^2} \right). \tag{7}$$

For a Gaussian distribution the peak normalized 6D brightness $\hat{B}_n \equiv B_n(0,0)$ can be expressed in terms of the total charge Ne and the rms quantities:

$$\hat{B}_n = \frac{m^2 c^2}{(2\pi)^3} \frac{N\, e}{\sigma_x\, \sigma_{p_x}\, \sigma_y\, \sigma_{p_y}\, \sigma_z\, \sigma_{p_z}}. \tag{8}$$

For a thermal distribution we have $\sigma_{p_x}^2 = \sigma_{p_y}^2 = \sigma_{p_z}^2 = mkT$, so the peak normalized 6D brightness of a Gaussian distributed thermionic source is given by:

$$\hat{B}_n = \frac{N\,e\,m^2c^2}{(4\pi^2 mkT)^{3/2}\,\sigma_x\,\sigma_y\,\sigma_z} = \frac{n_0 e}{mc}\left(\frac{mc^2}{2\pi kT}\right)^{3/2}, \tag{9}$$

where n_0 is the maximum density. Applying Eq. (8) to a beam that is either expanding or being focussed in any of the three directions, would lead to a significant underestimation of the actual peak brightness, due to correlations between position and momentum coordinates. For this reason the rms normalized emittances ε_x, ε_y, and ε_z are introduced:

$$\epsilon_x \equiv \frac{1}{mc}\sqrt{< x^2 >< p_x^2 > - < x\,p_x >^2}, \tag{10}$$

where the brackets <> denote averaging over the entire bunch. The emittances ϵ_y and ϵ_z are defined analogously. We thus arrive at a practical estimate for the peak normalized 6D brightness, based on measurements of the total charge and the three rms emittances:

$$\hat{B}_n \equiv \frac{1}{mc}\frac{N\,e}{(2\pi)^3\epsilon_x\,\epsilon_y\,\epsilon_z}. \tag{11}$$

In the remainder of this paper the peak normalized 6D brightness, operationally defined by Eq. (11), will be used for quantifying beam quality achieved in experiments or particle tracking simulations. For a source characterized by an effective temperature T and spatial dimensions σ_x, σ_y, and σ_z, Eq. (9) supplies us with the maximally achievable peak normalized 6D brightness.

For many applications, however, the peak current and the transverse normalized emittances are the most important parameters, while the energy spread is less critical. For this reason the most often used figure of merit for overall bunch quality is the *transverse* brightness $B_\perp$, which is defined by [10]:

$$B_\perp \equiv \frac{\hat{I}}{4\pi^2\epsilon_x\epsilon_y}, \tag{12}$$

where $\hat{I}$ is the peak current. In the absence of position-momentum correlations in the z-direction, the transverse brightness $B_\perp$ is simply proportional to the product of the peak normalized 6D brightness $\hat{B}_n$ and the rms energy spread:

$$B_\perp = \sigma_U\,\sqrt{2\pi}\;\hat{B}_n, \tag{13}$$

where $\sigma_U = mc^2\sigma_\gamma = mc^2\sqrt{<(\gamma - <\gamma>)^2>}$. Here we have used the fact that in the absence of position-momentum correlations $\epsilon_z = \sigma_z\,\sigma_{p_z}/mc = \sigma_t\,\sigma_U/mc$. Note that the transverse brightness $B_\perp$ is not a Lorentz-invariant quantity and is therefore not suited for comparing beams of different energy. For comparison purposes it is useful to note that at non-relativistic energies, such as in the field of electron microscopy, the beam quality is usually expressed in terms of the so-called *reduced* brightness B_r [14], which is related to the transverse brightness by a simple proportionality factor: $B_r = 2eB_\perp/mc^2$.

In order to do justice to both the normalized 6D brightness B_n as a fundamental figure of merit, and to the customary practice of expressing beam quality in terms of the transverse brightness $B_\perp$, we plot in Fig. 1 a brightness chart showing the relationship between the transverse brightness $B_\perp$ and the rms energy spread σ_U. In the brightness chart the diagonal lines are lines of constant normalized 6D brightness B_n, as follows from Eq. (13). This way of presenting beam quality was also used by De Jonge et al. [12] for the electron optical performance of carbon nanotube (CNT) electron sources. In Fig. 1 the relation between $B_\perp$ and σ_U is plotted for various high-brightness electron sources. The highest brightness measured at present, both $B_\perp$ and B_n, is produced by CNT field-emitters, recently developed for electron microscopy [12]. CNTs can produce up to 1 μA of continuous current from a few nm^2 source area. For completeness also the performance of tungsten Cold Field Emission Guns (CFEG) is indicated, representing the state-of-the-art in conventional sources for electron microscopy.

For many relativistic applications, such as the X-ray Free Electron Laser (XFEL), high currents are required [11], but not at the brightness level of continuous needle-like sources (see Fig. 1). For this purpose pulsed, (sub)picoseond photo-emission-based guns have been developed [1]. In particular, state-of-the-art rf photoguns can produce peak currents up to a few 100 A from a few mm^2 source area, at the expense of transverse brightness.

The filled solid circles on the brightness chart in Fig. 1 show what can be experimentally achieved today. Assuming an effective source temperature of $kT \approx 0.5$ eV, the continuous-current CFEG en CNT sources are already close to the thermal brightness limit. This is however not the case for rf photoguns: assuming $kT \approx 0.5$ eV the achieved normalized 6D brightness is still more than four orders of magnitude removed from the thermal limit. In Fig. 1 a diagonal solid line indicates the thermal limit of the normalized 6D brightness for the case of photoemission of a 1 mm radius, 1 nC bunch

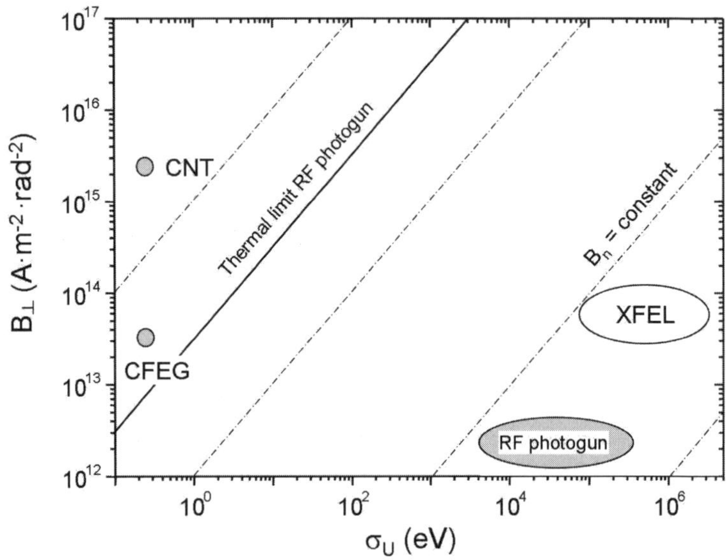

Figure 1. Transverse brightness $B_\perp$ versus rms energy spread σ_U for various electron sources. The requirements for XFEL operation are indicated by a solid, open circle. The diagonal lines are lines of constant normalized 6D brightness B_n. The thermal limit of a $Q = 1$ nC, $R = 1$ mm, $\tau = 10$ ps rf photogun is indicated by the diagonal solid line.

with a 10 ps laser pulse. The huge gap between the thermal limit and state-of-the-art performance is almost exclusively due to degradation of the longitudinal emittance $\epsilon_z = \sigma_t \, \sigma_U/mc$ whose thermal limit can be shown to be equal to $kT\,\tau/mc$, with τ the rms duration of the photoemission process. Bunch lengths of the order of τ can be achieved by magnetic compression techniques, but at present only with an energy spread of $10 - 100$ keV, i.e. more than four orders of magnitude larger than at the source. In principle there is more than sufficient room for improvement, but in practice increasing the transverse brightness by less than a single order of magnitude, roughly required for XFEL operation, constitutes a major challenge. An interesting recent development is a hybrid approach, in which the needle-like sources developed for electron microscopy are operated in a pulsed-current mode. In this way 10 ns electron pulses of 0.1 A current are extracted by pulsed photo-emission from micron-sized needle cathodes, suggesting a brightness comparable to continuous CNT emitters [13]. Such currents are however not yet sufficient for applications such as XFELs.

3. Extraction of electrons from an ultracold plasma: simulations

The strategy for improving the brightness that both photo-emission and field-emission based sources have in common up to now, is to increase the current density at the source. We propose a completely different approach [7], which aims instead at reducing the electron temperature T. This new source concept is based on pulsed extraction of electrons from an Ultra Cold Plasma (UCP).

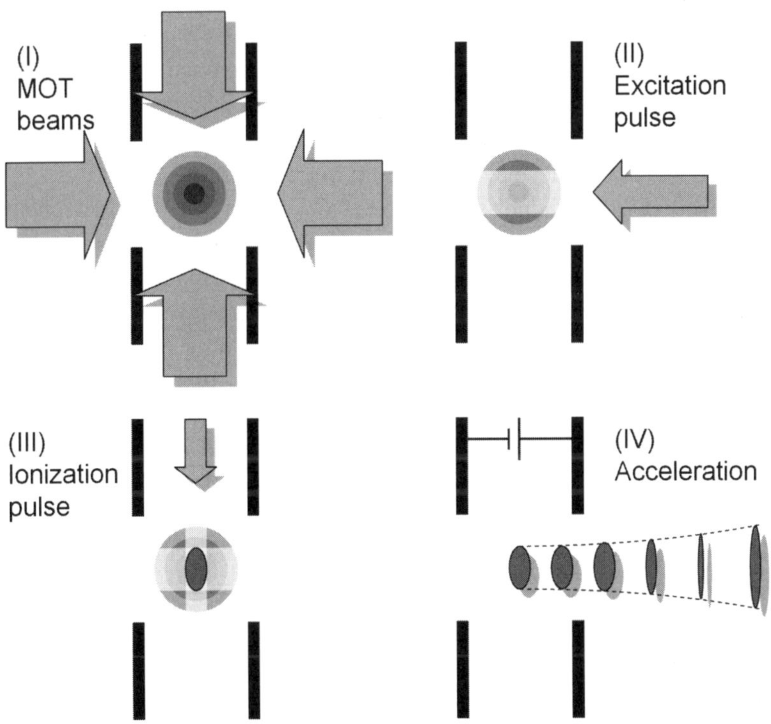

Figure 2. Schematic of the four-step procedure to realize a pulsed UCP electron source.

A UCP source may be realized in practice using a four-step procedure, illustrated schematically in Fig. 2:
(I) A cold ($T < 10^{-3}$ K) cloud of atoms is trapped in a Magneto-Optical Trap (MOT) in a volume of a few mm^3 with densities up to 10^{18} m^{-3} [16].
(II) Part of the cold atom cloud is excited to an intermediate state with a quasi-continuous, μs laser pulse.

(III) Then, a pulsed laser beam propagating at right angles to the excitation laser, ionizes the excited atoms only within the volume irradiated by both lasers. Here a UCP is formed [8]. In this way charges up to 100 pC with an electron temperature of 10 K [9] can be created in a volume of ~ 1 mm^3. (IV) The bunches are extracted by an electric field at least an order of magnitude stronger than what is minimally required for pulling the electrons and ions apart. For a 1-mm-sized 100 pC bunch this typically means applying a voltage of 1 MV across a 1 cm gap, which should be switched on extremely rapidly ($<$ 1ns) to prevent space-charge-induced emittance growth during acceleration. The extracted bunches are automatically compressed in the drift space after acceleration because the electrons in the rear of the bunch experience a larger potential difference than those in the front. This ballistic bunching leads to sub-ps bunch lengths.

It is interesting to note that the electrons are extracted much faster than the time it takes for thermal equilibrium between ions and electrons to occur. For this reason it is not strictly necessary to extract the electrons from a UCP to achieve ultra-low electron temperatures. The initial electron velocities are determined by the excess energies of the ionizing photons, which also depend on the thermal motion of the atoms through the Doppler shift. In a room temperature atomic gas Doppler broadening results in a spread of (ionizing) photon energies of the order of 10 K. In other words: if the electron extraction is sufficiently fast, a 10 K electron beam can in principle also be produced by narrow-band photo-ionization of a room temperature atomic gas. There are however two reasons to stick to UCPs: first, UCPs possibly allow electron temperatures substantially lower than 10 K, which may have important, fundamental consequences, as will be discussed in Sec. 4; second, a room temperature gas is hard to confine, resulting in an extended region with a relatively high density of neutral atoms, which may lead to deterioration of the electron beam through electron-atom collisions. Ultra-cold trapped atomic gases are by their nature compatible with, and actually require, ultra-high vacuum conditions.

The fact that the initial electron density is proportional to the product of the intensities of the excitation and the ionization laser beams in the region of overlap, allows shaping of the initial charge distribution in three dimensions. As was shown recently [3], space-charge induced emittance growth may be virtually eliminated by the combination of lowering the dimensionality of the bunch and proper shaping. Without minimizing the detrimental effects of space charge forces, the benefits of low temperatures would be lost. There are two initial geometries which are favorable:

first, the *pancake* bunch (bunch length L much smaller than bunch radius R) with a *half-circle* radial charge density distribution [3]

$$\rho(r) \propto \sqrt{1-(r/R)^2}. \tag{14}$$

Such a distribution automatically evolves by longitudinal expansion into a uniform, ellipsoidal bunch, which is characterized by perfectly linear space charge fields and thus zero space-charge-induced emittance growth. This is the initial distribution also proposed for rf photoguns. The second favorable initial distribution is the *cigar* bunch ($R \ll L$) with a *parabolic* longitudinal charge density distribution, which will evolve by transverse expansion into a uniform, ellipsoidal bunch:

$$\rho(z) \propto 1-(2z/L)^2. \tag{15}$$

In spite of the ideal initial distributions, rapid acceleration is still necessary, because initially the electron bunch is still subjected to nonlinear forces due to the ion cloud left behind in the trap.

The starting point of the simulations is a MOT containing rubidium atoms in a spherically symmetric gaussian density distribution with an rms radius of 2 mm and a density in the center of 1×10^{18} m^{-3}. To create a pancake bunch, a fraction of the atoms is excited with a radial distribution given by Eq. (14), with R=2 mm. The ionization laser beam then cuts out a longitudinal slice of 15 μm thickness. Assuming an overall ionization efficiency of 50%, this results in 10 pC charge. To create a cigar bunch, the atoms are excited within a radius of $R = 80\,\mu$m from the axis. Subsequently, the ionization laser cuts out a parabolic longitudinal density profile with a total length of $L = 1$ mm, resulting in 1 pC of charge. The initial electron temperature of both bunches is set at 10 K.

For the accelerating stage, a cylindrically symmetric field geometry is assumed in which both the cathode and the anode are thin conducting plates with a circular hole of 5 mm radius, separated by a distance $d =$ 20 mm, as is shown in Fig. 3(a). A voltage of 1 MV is applied across the diode, with a risetime of 150 ps.

In Fig. 3(a) the acceleration electric field geometry is shown, indicated by equipotential lines, as well as the radial bunch envelope as a function of z. In Fig. 3(b) and (c), respectively, the rms normalized emittance ϵ and the rms bunch length σ_t (in fs) are plotted as a function of z. The pancake bunch is started upstream at $z = -5$ mm. As a result the bunch is initially focused by the non-uniform electric field, as can be seen in Fig. 3(a), thus partially compensating for the radial space charge forces and the defocusing

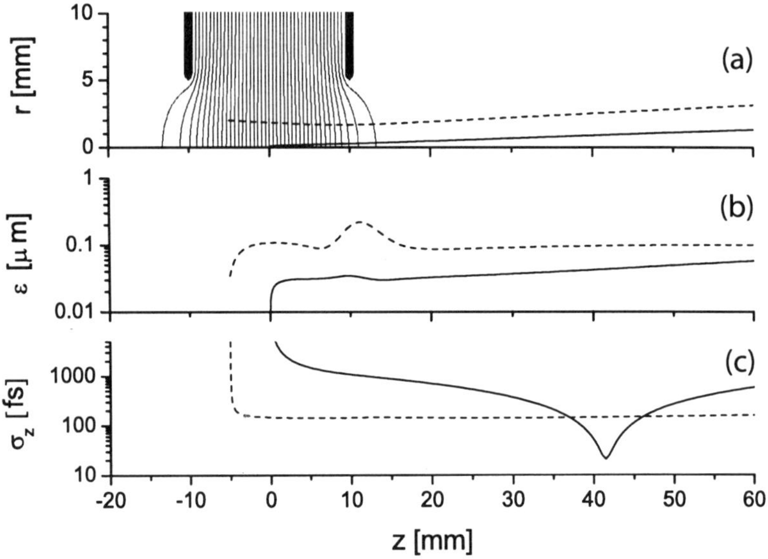

Figure 3. (a) Field geometry and radial bunch envelope as a function of z; (b) rms normalized emittance ε as a function of z; (c) rms bunch length σ_t as a function of z. Solid lines: cigar bunch; dashed lines: pancake bunch.

exit kick of the diode. The final energy of the pancake bunch is 470 keV. The cigar bunch is started at $z = 0$, as its small initial radius does not require any additional focusing. The final energy of the cigar bunch is 270 keV.

As is shown in Fig. 3(b), ultra-low normalized emittances are achieved of the order of $\varepsilon \approx 0.1\,\mu$m for the pancake bunch and even lower for the cigar bunch. The behavior of ϵ as a function of z is similar for both configurations. Initially $\varepsilon \approx 0.04\,\mu$m for the pancake bunch and $\varepsilon \approx 0.0015\,\mu$m for the cigar bunch, in agreement with Eq. (1). After initiation, ε first rises sharply due to space charge forces and then levels off to slow monotonous growth, only briefly interrupted by a temporary rise while passing through the non-uniform field in the hole of the anode.

Figure 3(c) shows that sub-ps bunch lengths can be realized. The pancake bunch almost immediately reaches a bunch length value of $\sigma_t = 150$ fs, corresponding to $I = 25$ A, which is maintained over many cm of its trajectory. This steady behavior reflects a balance between space charge force repulsion and ballistic bunching, which is relatively weak due to the small acceleration potential difference experienced by pancake bunches. The rms energy spread of the pancake bunch is $\sigma_U = 0.85$ keV.

Compression due to ballistic bunching is particularly efficient for the cigar bunch: at $z = 42$ mm an rms bunch length $\sigma_t = 20$ fs is achieved, resulting in a peak current $I > 25$ A. The rms energy spread of the cigar bunch at the position of the bunch length minimum is $\sigma_U = 7.5$ keV.

The combination of ultra-low emittances and ultra-short bunch lengths results in extremely high transverse brightness values: after leaving the diode, the pancake bunch attains a constant value $B_\perp \geq 5 \times 10^{13}$ A/(rad^{2}m^2). The cigar bunch performs even better: at the bunch length minimum, $z = 42$ mm, $B_\perp \approx 5 \times 10^{14}$ A/(rad^{2}m^2), within an order of magnitude from the best CNT performance [12].

In Fig. 4 the brightness chart $B_\perp$ versus σ_U is plotted, including the results of the UCP simulations for both the cigar and the pancake initial geometries. The results of both geometries are characterized by the same value of the normalized peak 6D brightness $\hat{B}_n$. As the initial electron density n_0 and temperature T are identical, the thermal normalized 6D brightness limit, as given by Eq. (9), is also the same for both geometries. The UCP thermal limit is indicated in the brightness chart by a diagonal solid line. The achievable 6D normalized brightness is mainly limited by the remaining space-charge induced emittance growth, associated with the initial distributions, which are not fully perfect uniformly charged ellipsoids. These results may therefore be improved even further.

It is interesting to note that the thermal normalized 6D brightness limit of the UCP sources is comparable to that of rf photoguns, as can be seen by comparing Fig. 1 and 4: the much higher initial density of rf photoguns is nicely compensated by the reduced temperature of the UCP sources. The high initial 6D brightness of the rf photogun is largely due to its low longitudinal emittance ε_z, many orders of magnitued lower than its initial transverse emittance ε_x. This low value of ε_z is directly connected to the extreme aspect ratio $L/R \ll 1$, which, unfortunately, is very hard to maintain in practice. This results in an enormous, unrecoverable increase of ε_z during the first stages of acceleration. In the proposed UCP sources the emittances are more evenly distributed over the three degrees of freedom, with less extreme aspect ratios. This makes it possible to approach the thermal limit more closely.

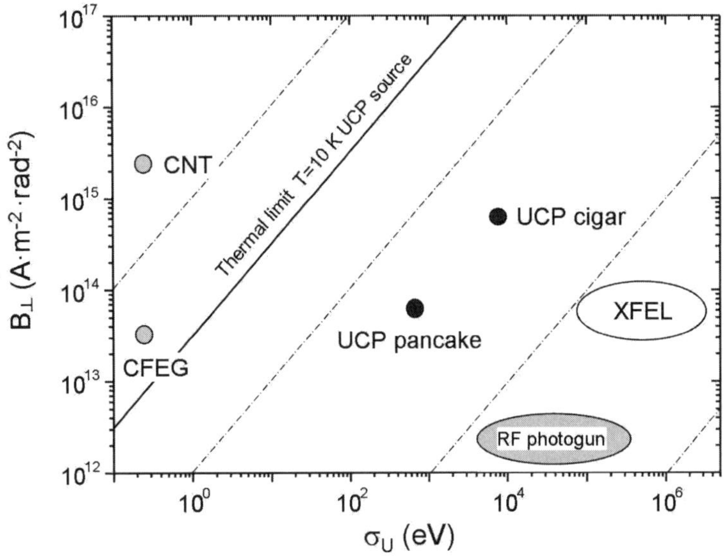

Figure 4. Transverse brightness $B_\perp$ versus rms energy spread σ_U for various electron sources, including UCP sources with different initial geometries. The thermal limit of the UCP sources is indicated by the diagonal solid line.

4. Towards the ultimate limits of electron beam brightness

On the basis of the particle tracking simulations discussed in the previous section we conclude that UCP sources have enormous potential for advancing the state-of-the-art in ultra-short electron bunch brightness. These simulations were based on conservative numbers for the initial electron temperature, $T = 10$ K, and density, $n_0 = 5 \times 10^{17}\,\mathrm{m}^{-3}$. The electron temperature immediately after ionization can be as low as 10^{-3} K, after which rapid correlation heating occurs to about 10 K. If the extremely low initial temperature could be maintained, for example by rapid extraction, this would imply a further increase of the 6D source brightness (9) by *six* orders of magnitude. In the face of such mind-boggling numbers it is appropriate to first reflect on what is the theoretical ultimate brightness that can be achieved.

As was shown in Sec. 2, the normalized 6D brightness B_n is simply another word for the phase space density f of the accelerated bunch. For electrons the maximum phase space density is fundamentally limited by the Pauli exclusion principle, which says that no two fermions may occupy the same quantum state. The maximum phase space density f_{max} of a fermi system is reached at $T = 0$. For a fermi system in a single spin state it is given by $f_{max} = h^{-3} = (2\pi\hbar)^{-3}$, where $\hbar$ is Planck's constant [17]. Using Eq. (6) we thus find that the ultimate quantum limit of the 6D normalized brightness of a fermi-degenerate polarized electron beam is given by

$$B_{n,max} = \frac{e\,m^2c^2}{h^3} = 6.4 \times 10^{18}\,\mathrm{A \cdot m^{-2} \cdot rad^{-2} \cdot eV^{-1}}. \qquad (16)$$

This can be expressed more conveniently in terms of the ratio of the transverse brightness $B_\perp$ and the rms energy spread σ_U, which can never exceed a maximum value, given by

$$\frac{B_\perp}{\sigma_U} \leq 1.6 \times 10^{19}\,\mathrm{A \cdot m^{-2} \cdot rad^{-2} \cdot eV^{-1}}. \qquad (17)$$

For an unpolarized electron beam, with equal occupation of both spin states, $B_{n,max}$ is twice as high. The degree of quantum degeneracy at finite temperatures can be expressed in terms of the product $n_0\lambda_{th}^3$, with the thermal De Broglie wavelength given by $\lambda_{th} = h/\sqrt{2\pi mkT}$. As follows from Eq. (7), $n_0\lambda_{th}^3 = 1$ corresponds to full degeneracy. In Fig. 5 the brightness chart $B_\perp$ versus σ_U is plotted for all electron sources discussed in the previous sections, but now on a scale which includes the fermi-degenerate brightness limit.

For a temperature $T = 10^{-3}$ K, the electron thermal De Broglie wavelength amounts to $\lambda_{th} = 2.4\,\mu$m. Already for densities less than $n_0 = 1 \times 10^{17}\,\mathrm{m^{-3}}$ this would imply a degeneracy parameter equal to $n_0\lambda_{th}^3 = 1$. In other words, the initial conditions which are easily possible in present-day UCPs could already be well into the fermi-degenerate regime! Then the question is of course whether temperatures $\sim 10^{-3}$ K can be maintained in practice. The typical timescale for correlation heating in a UCP is the inverse plasma frequency $\omega_p^{-1} = \sqrt{m\epsilon_0/n_0e^2}$ [9]. For an electron density $n_0 = 5 \times 10^{17}\,\mathrm{m^{-3}}$ the extraction of the bunch should therefore occur considerably faster than $\omega_p^{-1} = 25$ ps after ionization. A possible way to achieve this is by not photo-ionizing the cold atoms, which takes several ns because of the required narrow bandwidth, but by exciting them to a highly excited state instead. From there they can be field-ionized by a high-voltage pulse, which simultaneously extracts the electrons from the plasma. If the excited level is chosen such that in a field of $\approx 10^7$ V/m only its highest Stark-shifted sublevel is ionized, then a pancake slice of electrons is extracted from the plasma within a few ps. Since this is much faster than

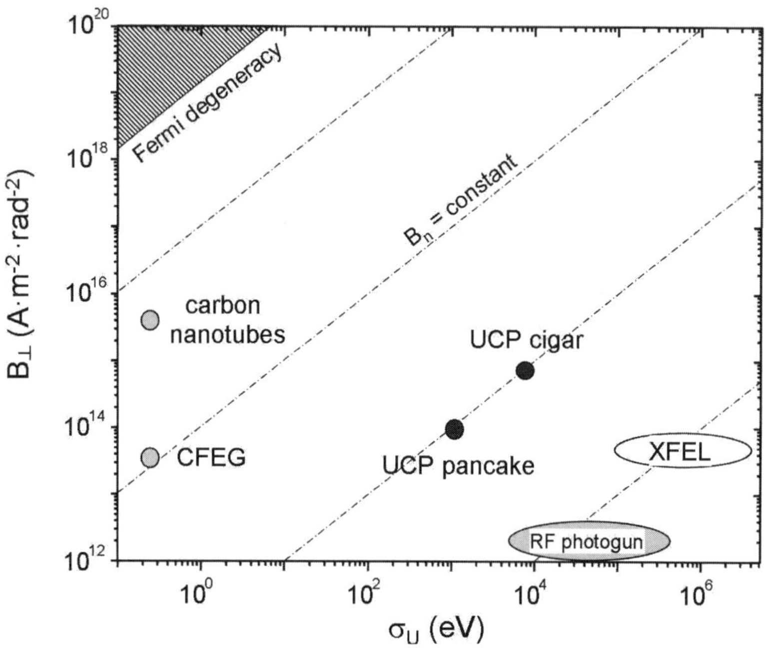

Figure 5. Transverse brightness $B_\perp$ versus rms energy spread σ_U for various electron sources. The hatched area indicates the region forbidden by the Pauli exclusion principle for a polarized electron beam.

the time scale at which plasma heating occurs, a fermi-degenerate electron beam may thus be realized.

Acknowledgments

We thank Marnix van der Wiel, Herman Beijerinck, and Richard van de Sanden for originally suggesting the idea of using a UCP as a bright electron source. Stimulating discussions with Marnix van der Wiel on cold electron sources and on all other matters concerning beam brightness are gratefully acknowledged. This work is part of the research programme of the Stichting voor Fundamenteel Onderzoek der Materie (FOM), financially supported by the Nederlandse Organisatie voor Wetenschappelijk Onderzoek (NWO).

References

1. For a recent overview, see P. Piot, in *The physics and Applications of High Brightness Electron Beams*, edited by J. Rosenzweig, G. Travish, and L. Serafini (World Scientific, Singapore, 2003), p. 127.

2. L. Serafini, *AIP Conf. Proc.* **413**(1), 321 (1997); L. Serafini, *AIP Conf. Proc.* **279**, 645 (1993).
3. O.J. Luiten, S.B. van der Geer, M.J. de Loos,F.B. Kiewiet, M.J. van der Wiel, Phys. Rev. Lett. **93**, 094802 (2004).
4. O. J. Luiten, S. B. van der Geer, M. J. de Loos, F. B. Kiewiet, M. J. van der Wiel, Proc. 2004 Eur. Part. Acc. Conf., Luzern, Switzerland, (2004) p. 725.
5. C. Limborg-Deprey, P. R. Bolton, Nucl. Instr. and Meth. A **557**, 106 (2006).
6. J. B. Rosenzweig, A. M. Cook, R. J. England, M. Dunning, S. G. Anderson, and M. Ferrario, *Emittance compensation with dynamically optimized photoelectron beam profiles*, Nucl. Instr. and Meth. A **557**, 87 (2006).
7. B.J. Claessens, S.B. van der Geer, G. Taban, E.J.D. Vredenbregt, and O.J. Luiten, Phys. Rev. Lett. **95**, 164801 (2005).
8. T.C. Killian, S. Kulin, S. D. Bergeson, L. A. Orozco, C. Orzel, and S. L. Rolston, Phys. Rev. Lett. **83**, 4776 (1999); for a recent review, see T.F. Gallagher *et al.*, J. Opt. Soc. Am. B **20**, 1091 (2003).
9. S.G. Kuzmin and T.M. O'Neil, Phys. Rev. Lett. **88**, 065003 (2002).
10. C. A. Brau, *The physics and applications of high brightness electron beams*, edited by J. Rosenzweig, G. Travish, and L. Serafini, World Scientific (Singapore), (2003), p. 20.
11. *TESLA, the superconducting electron-positron linear collider with an integrated X-ray laser laboratory*, DESY technical design report (2001), available from `http://www-hasylab.desy.de/facility/fel/`.
12. N. de Jonge, M. Allioux, J.T. Oostveen, K.B.K. Teo, W.I. Milne, Phys. Rev. Lett. **94**, 186807 (2005).
13. C.H. Garcia and C.A. Brau, Nucl. Instr. Meth. Phys. Res. **A483**, 273 (2002).
14. P.W. Hawkes and E. Kasper, *Principles of Electron Optics II: Applied Geometrical Optics* (Academic Press, London, 1996).
15. http://www.pulsar.nl/gpt.
16. W. Ketterle, K.B. Davis, M.A. Joffe, A. Martin, and D.E. Pritchard , Phys. Rev. Lett. **70**, 2253 (1993).
17. See, for example, K. Huang, *Statistical Mechanics*, 2nd Ed., Ch. 11 (John Wiley & Sons, 1987).

SCALING LAWS OF STRUCTURE-BASED OPTICAL ACCELERATORS

AMIT MIZRAHI, VADIM KARAGODSKY AND LEVI SCHÄCHTER

Department of Electrical Engineering,
Technion - Israel Institute of Technology,
Haifa 32000, ISRAEL
E-mail: levi@ee.technion.ac.il

A structure-based laser accelerator harnesses technological progress developed by the laser and optical fiber industries, potentially facilitating a compact and efficient system. In the optical regime, dielectrics sustain higher electric fields and gradients of the order of a few GV/m may become available, but the acceleration structures are different than those used in the microwave regime. Various dielectric structures have been analyzed, and from the pure accelerator parameters perspective (gradient, interaction impedance, group velocity, wake-fields), their performance is of great promise. Operation similar to current linear accelerators may lead to a prohibitively low efficiency. Therefore, including a feedback attached to each module may improve the efficiency from a few percents to higher than 90% – in fact, the efficiency is limited only by the constraints on the stability of the optical system. Single mode operation in the optical regime imposes that at least one of the dimensions of each micro-bunch ought to be sub-micronic leading to a stringent constraint on the emittance and thus on the luminosity. Attempting to increase the latter, imposes high energy density in the vacuum tunnel as well as in its adjacent dielectric layer(s). This, in turn, is bounded by the maximum stress, temperature increase and heat dissipation, dielectrics can sustain at these scales.

1. Introduction

Motivated by the availability of solid-state lasers with increasing wall-plug to light efficiencies, optical acceleration of charged particles is a subject of growing interest. Acceleration is facilitated by laser light rather than by microwave radiation, and accordingly, the acceleration structure must be made of dielectric materials as these have lower loss and are less susceptible to breakdown comparing to their metallic counterparts. An example of an *open* optical structure is the LEAP[1] crossed laser beam experiment where the interaction between the crossed laser beams and the particles is limited by slits to satisfy the Lawson-Woodward theorem[2,3]. Another example is the *travelling* wave acceleration structure, where a laser pulse is guided in a dielectric structure with a vacuum tunnel bored in its center. This concept can be implemented by a two-dimensional photonic band-gap structure[4], and recently, it was suggested[5] to use Bragg reflection waveguides[6,7,8], designed specifically for the speed-of-light mode.

It was demonstrated[5] that optical Bragg acceleration structures, either planar or cylindrical, having typical transverse dimensions of a few microns, exhibit high performance as acceleration structures and, therefore, seem to be promising candidates for future optical accelerators; Fig. 1 illustrates the planar version of such a structure. The laser light is confined in a vacuum core of width $2D_{\text{int}}$, so that the wave propagates along the z axis, and no variations are assumed along the y axis $(\partial/\partial y \equiv 0)$. The core is surrounded by dielectric layers with alternating permittivity, having a width equal to the transverse quarter-wavelength $[\lambda/(4\sqrt{\varepsilon_r - 1})]$ with the exception of the innermost layer. This first layer is a *matching layer* and it plays a crucial role facilitating the structure to support the speed-of-light TM mode required for the acceleration process.

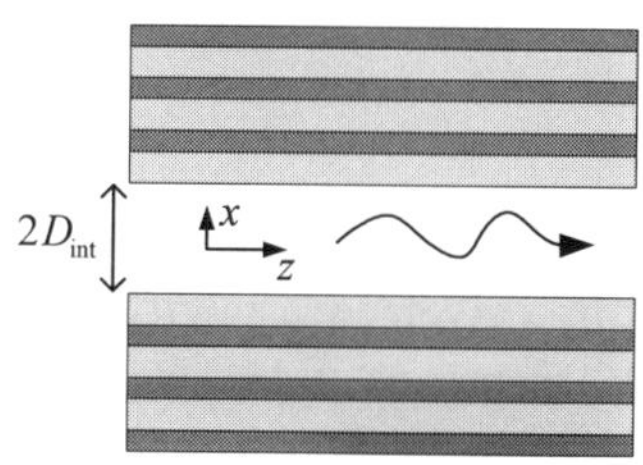

Figure 1. Schematic of a planar Bragg acceleration structure.

Transition from operation at radiation wavelengths of a few centimeters to a few microns requires examining a wide spectrum of phenomena which are insignificant in the former regime. For example, structures that operate at a wavelength of a few centimeters are machined today with an accuracy of microns. In the future, it will not be possible to maintain a ratio of 4-5 orders of magnitude between the operating wavelength and the achievable tolerance, since this would entail engineering of a surface at the atomic level. As a result, the size of irregularities may be of the same order of magnitude as the micro-bunches, and they may generate wake fields[9,10] that, in turn, may alter the dynamics of electrons. Fortunately, the electromagnetic properties of materials at wavelengths that are significantly smaller than 0.1 μm do not differ dramatically from these of the vacuum.

Consequently, a reduction on the sensitivity to manufacturing tolerances may be expected.

As a central component of a future optical accelerator, the acceleration structure, ought to withstand the manifestations of the two most important constraints imposed by the machine specifications, namely, the luminosity and the operating gradient. Being a measure of the number of colliding particles per second at the interaction point, the luminosity sets the lower limit to the energy level the particles are exposed to. Together with the operating gradient and the constraint of single mode operation, they determine the minimal electromagnetic energy density in the acceleration tunnel and its close vicinity. While obviously the average number of particles per second is at least as in a machine designed to operate at microwave wavelengths, in an optical acceleration structure, the volume where most of the electromagnetic energy is confined, is reduced by several orders of magnitude. Consequently, the potential impact of conversion efficiency, electromagnetic stress on the structure, heat flow or temperature increase may become a significant obstacle, and it is our goal in this study to determine the main scaling laws of these processes.

2. Efficiency Considerations

In an acceleration structure driven by a microwave source, a typical bunch consists of an order of 10^{10} electrons and its size is measured on 1 mm scale. In the optical range, the transverse dimensions are reduced by 2-3 orders of magnitude implying that if the particles' density is maintained the same, the number of accelerated electrons is reduced by 6-7 orders of magnitude. Consequently, following the same approach as in the microwave regime, may lead to prohibitively low efficiencies since each laser pulse accelerates only a few thousand electrons. As a result, for maintaining the same luminosity, the repetition rate ought to be increased correspondingly. Alternatively, it is possible to accelerate a *train* of micro-bunches rather than a single micro-bunch and release the constraint on the repetition rate.

Before further discussing this concept, let us briefly review[11] the efficiency estimate of an acceleration module of length d driven by an average laser power P_L generating a gradient G_0 at the location of the electrons. By virtue of the linearity of Maxwell's equations, these two quantities are related, and the so-called interaction impedance, $Z_{\rm int} = |G_0\lambda|^2/P_L$, characterizes any acceleration structure operating at a frequency corresponding to a vacuum wavelength λ. This laser pulse accelerates a point-charge (q)

that as it moves in an arbitrary acceleration structure, it generates an electromagnetic wake. Associated with this wake there is a decelerating electric field ($E_{\rm dec}$) which again, by virtue of the linearity of Maxwell's equations, must be proportional to the charge, namely, $E_{\rm dec} = \kappa q$, where the value of κ depends on the details of the structure, and it will be referred to as the *wake-coefficient*. By virtue of the fundamental loading theorem, the amplitude of the trailing wake is twice this value, namely, $E_{\rm W} = 2E_{\rm dec} = 2\kappa q$.

In case of a uniform dielectric medium filling the entire space except a vacuum tunnel of radius R, along which the point-charge propagates, the wake-coefficient is $\kappa = 1/(2\pi\varepsilon_0 R^2)$; for more realistic structures the reader is referred to detailed analysis of Bragg acceleration structure in Ref 5. In a closed structure, the wake generated by the bunch has projections on each one of the eigen-modes of the structure. Bane and Stupakov[12] have discussed the connection between the various parameters (group velocity $c\beta_{\rm gr}$ and interaction impedance $Z_{\rm int}$) and the projection of the wake on the fundamental (superscript F) mode

$$\kappa^{(F)} = \frac{\beta_{\rm gr}}{1-\beta_{\rm gr}} \frac{Z_{\rm int}}{\sqrt{\mu_0/\varepsilon_0}} \frac{\pi}{4\pi\varepsilon_0\lambda^2} . \tag{1}$$

In other words, this is the coefficient that given the charge of the bunch, determines the amplitude of the fundamental mode generated by the bunch $\left(E^{(F)} = 2\kappa^{(F)}q\right)$. In the absence of an accelerating gradient G_0, the electron bunch is decelerated along a distance d in the structure, and the loss of kinetic energy is $\Delta U_{\rm KIN} = -q^2\kappa d$. Consequently, when the gradient is not zero, the net change in the kinetic energy of the bunch traversing the same structure is given by $\Delta U_{\rm KIN} = q(G_0 - q\kappa)$. As a reference, the total electromagnetic energy stored in the structure is $U_{\rm EM} = P_L\tau_{\rm EM}$ wherein $\tau_{\rm EM} \equiv (d/c)(1/\beta_{\rm gr} - 1)$. This last expression takes into consideration the requirement that the point-charge and the electromagnetic pulse ought to overlap during the time the former spends in the structure. With these two energy definitions, the efficiency of the acceleration process may be determined by

$$\eta \equiv \frac{\Delta U_{\rm KIN}}{U_{\rm EM}} = \eta_{\rm max}\frac{4q\,(q_0 - q)}{q_0^2} , \tag{2}$$

where $q_0 = G_0/\kappa$ is the charge for which the effective gradient vanishes, and the maximum value of the efficiency is $\eta_{\rm max} \equiv \kappa^{(F)}/\kappa$ occurring for $q = q_{\rm opt} \equiv q_0/2$.

As a typical example, consider the parameters of a photonic band-gap structure as calculated in Ref. 4: $Z_{\rm int} \simeq 20\ \Omega$, $\beta_{\rm gr} \simeq 0.6$ and $R \simeq 0.7\lambda$;

according to these values the *maximum efficiency* is 6%. Assuming that the threshold for breakdown limits the power to 7 kW, the optimal number of electrons to be accelerated for reaching this efficiency is 6×10^4. Evidently, considerations of energy spread will lead to charges and efficiencies below these optimum values.

Although this result relies on an idealized model, it hints what the possible solution may be: first, we observe that more than 90% of the electromagnetic energy is wasted, and therefore, by *"recycling"* part of this energy we may improve significantly the efficiency of an optical system. Moreover, the analytic result from the above reflects the fact that the efficiency is determined by the relative projection of the wake on the fundamental mode. This quantity may be enhanced by *splitting* the bunch into a train of micro-bunches each one separated by the wavelength of the fundamental – see Fig. 2.

What makes the implementation of these two concepts feasible, is the fact that rather than accelerating a single bunch, the wake being broad-band and thus useless from the perspective of energy recycling, a macro-bunch consisting of many hundreds of micro-bunches is accelerated – each micro-bunch being one laser wavelength (λ) apart from its neighbor. This makes the wake *virtually coherent* with the laser field. Consequently, at the output of a given travelling wave acceleration module, the two field components (laser and wake) are practically inseparable and the total electromagnetic field may be extracted out of the acceleration structure, amplified by an active medium and re-injected into the travelling wave acceleration module.

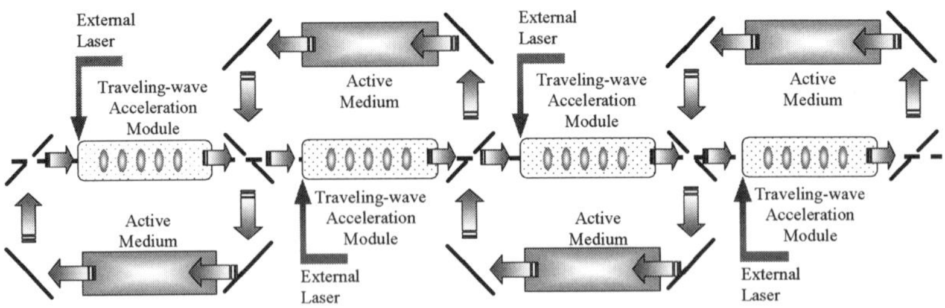

Figure 2. Recovery scheme for achieving high efficiency operation of an optical accelerator.

Obviously, the accelerating field ought to be tapered in order to compensate for the varying beam-loading. In practice, for a self-sustained solution of the field in a given acceleration module, it is necessary to provide from the outside a tapered field whereas all the energy loss has to be compensated by the feedback loop. However, these details and many more were considered in Ref. 11 and will be omitted here. The outcome of this analysis had demonstrated that indeed by combining multiple micro-bunches and a feedback loop it is possible to enhance both the efficiency as well as the number of electrons to be accelerated. Two important constraints are imposed: first the length of the accelerating module needs to be equal to the bunch length ($d \simeq M\lambda$). Second, the group velocity which makes the external laser pulse to have the simplest shape is $\beta_{\rm gr} \simeq 0.5$. According to the (loaded) quality factor of each module the efficiency may exceed the 90% accelerating in the process about 1000 micro-bunches each one consisting of a number of electrons limited by the (technological) ability to taper the external laser pulse.

3. Electromagnetically Induced Stress in a Planar Bragg Structure

The high energy electromagnetic field guided within an optical acceleration structure gives rise to forces on the dielectric materials sometimes referred to as "radiation pressure". In this section, we shall examine these forces on the dielectric layers of the planar optical Bragg acceleration structure. Our treatment here relies on macroscopic electromagnetic concepts, where the material is represented by a polarization density, with associated polarization volume or surface currents and charges. The local Lorentz force may then be calculated directly, as resulting from the interaction of the electromagnetic field and these effective charges and currents. Particularly, at the frequency of laser light, the *time-averaged* force is the quantity of interest.

Let us consider a planar acceleration structure guiding laser light and examine the time-averaged forces on this structure. In the vacuum core, the speed-of-light TM mode guided by the acceleration structure, assuming a time-dependence of $\exp(j\omega t)$, is of the form[5]

$$\begin{aligned} E_z &= E_0 \exp\left(-j\frac{\omega}{c}z\right), \\ E_x &= E_0\left(j\frac{\omega}{c}x\right)\exp\left(-j\frac{\omega}{c}z\right), \\ H_y &= \frac{E_0}{\eta_0}\left(j\frac{\omega}{c}x\right)\exp\left(-j\frac{\omega}{c}z\right). \end{aligned} \tag{3}$$

The *total* transverse force exerted on the mirrors may be found by integrating the time-averaged Maxwell stress-tensor[13] over a closed surface. Within the vacuum core, the time-averaged Maxwell stress-tensor component $\langle T_{xx}\rangle$ for a TM mode reads

$$\langle T_{xx}\rangle = \frac{1}{4}\varepsilon_0|E_x|^2 - \frac{1}{4}\varepsilon_0|E_z|^2 - \frac{1}{4}\mu_0|H_y|^2\,, \tag{4}$$

and given the expressions in Eq. (3), we are left with

$$\langle T_{xx}\rangle = -\frac{1}{4}\varepsilon_0|E_0|^2. \tag{5}$$

Enclosing one mirror by a rectangular surface, only the T_{xx} component contributes to the integral, as the T_{xz} component has zero contribution due to symmetry, and the T_{xy} component is identically zero. Assuming that the laser field decays to zero at $x = \pm\infty$, the time-averaged transverse pressure exerted by the guided mode on the Bragg mirror located at $x = D_{\rm int}$ is

$$\langle F_x\rangle = \frac{1}{4}\varepsilon_0|E_0|^2\,. \tag{6}$$

Hence, for a given accelerating gradient E_0, the total pressure is repelling and is independent of the details of the structure, that could be, for example, a dielectric loaded metallic transmission line. Assuming that the gradient of interest is $E_0 = 1$ GV/m, the total pressure is $\langle F_x\rangle \simeq 2.2\times10^{-6}$ N/μm$^2 \simeq 2.2\times10^{6}$ N/m^2 – we shall discuss the implications of this result in the last section.

Beyond the total pressure exerted by the radiation field on the confining walls, it is important evaluate the force(s) on each dielectric layer as well as the stress occurring at the interface between two layers. Evaluation of the time-averaged volume force densities of the accelerating TM mode is performed next based on the expression for the Lorentz force-density. Its *transverse* component may be shown to be given by

$$\langle f_x\rangle = \frac{1}{2}\Re\left[j\omega\varepsilon_0(\varepsilon_r-1)E_z^*\mu_0H_y\right]\,. \tag{7}$$

Bearing in mind that H_y/E_x inside the dielectric layers is a real quantity, the *longitudinal* component of the volume force density vanishes since

$$\langle f_z\rangle = \frac{1}{2}\Re\left[-j\omega\varepsilon_0(\varepsilon_r-1)E_x^*\mu_0H_y\right] = 0\,. \tag{8}$$

In the configuration of interest, the third component of the volume force density is identically zero $\langle f_y\rangle \equiv 0$ for the mode of interest (TM).

The surface forces are due to polarization surface charges at the interfaces between the layers. Defining $E_{x,\nu}^{(+)}$ as the x-component of the electric field in layer ν at the boundary with layer $\nu+1$, we obtain for the surface pressure at that boundary

$$\langle F_{x,\nu}\rangle = \frac{1}{2}\varepsilon_0 |E_{x,\nu}^{(+)}|^2 \left(\frac{\varepsilon_\nu^2}{\varepsilon_{\nu+1}^2} - 1\right) . \tag{9}$$

This total force may be conveniently divided into two contributions from the two polarization densities of each layer at the interface. The force associated with the polarization of layer ν is given by

$$\langle F_{x,\nu}^{(-)}\rangle = \frac{1}{2}\varepsilon_0 |E_{x,\nu}^{(+)}|^2 \left(\varepsilon_\nu^2 - 1\right) > 0 , \tag{10}$$

and the force on the polarization surface charge of layer $\nu+1$ is

$$\langle F_{x,\nu}^{(+)}\rangle = -\frac{1}{2}\varepsilon_0 |E_{x,\nu+1}^{(+)}|^2 \left(\varepsilon_{\nu+1}^2 - 1\right) < 0 , \tag{11}$$

and $\langle F_{x,\nu}\rangle = \langle F_{x,\nu}^{(-)}\rangle + \langle F_{x,\nu}^{(+)}\rangle$. In the two above inequalities, we have assumed that $\varepsilon_\nu > 1$, leading to the conclusion that the effect of these forces is to pull each of the two layers at the interface towards the other.

The total pressure is the sum of all transverse forces, and explicitly

$$\langle F_x\rangle = \sum_{\nu=0}^{\infty}\left[\langle F_{x,\nu}\rangle + \int_{x_\nu}^{x_{\nu+1}} dx\, \langle f_x(x)\rangle\right] . \tag{12}$$

where x_ν denotes the boundary between layer ν and layer $\nu+1$, and $\nu = 0$ denotes the core. Practically, the above equation may be verified numerically by carrying out the summation up to some layer where the fields have sufficiently decayed to be considered negligible.

The transverse volume force density and the surface forces for a structure made of SiO_2 ($\varepsilon_r = 1.45^2$) and Si ($\varepsilon_r = 3.45^2$) and core half-width $D_{\rm int} = 0.3\lambda_0$ are shown in Fig. 3. The frames in the left column correspond to a structure having the SiO_2 as the matching layer adjacent to the core, and the right column corresponds to a structure with Si as the layer adjacent to the core. When the lower refractive index is used for the matching layer, the maximum of the volume force density is obtained inside the second layer [Fig. 3 (c)], whereas in the second case, the maximum is obtained inside the matching layer [Fig. 3 (d)]. The surface force density $\langle F_{x,\nu}\rangle$, which peaks at the vacuum-dielectric interface, is shown in Fig. 3 (e)-(f) together with the one-sided force densities $\langle F_x^{(+)}\rangle$ and $\langle F_x^{(-)}\rangle$. Inside the periodic part of the Bragg mirror, E_x vanishes every other interface[5], and, accordingly, all the 3 surface force densities vanish as well.

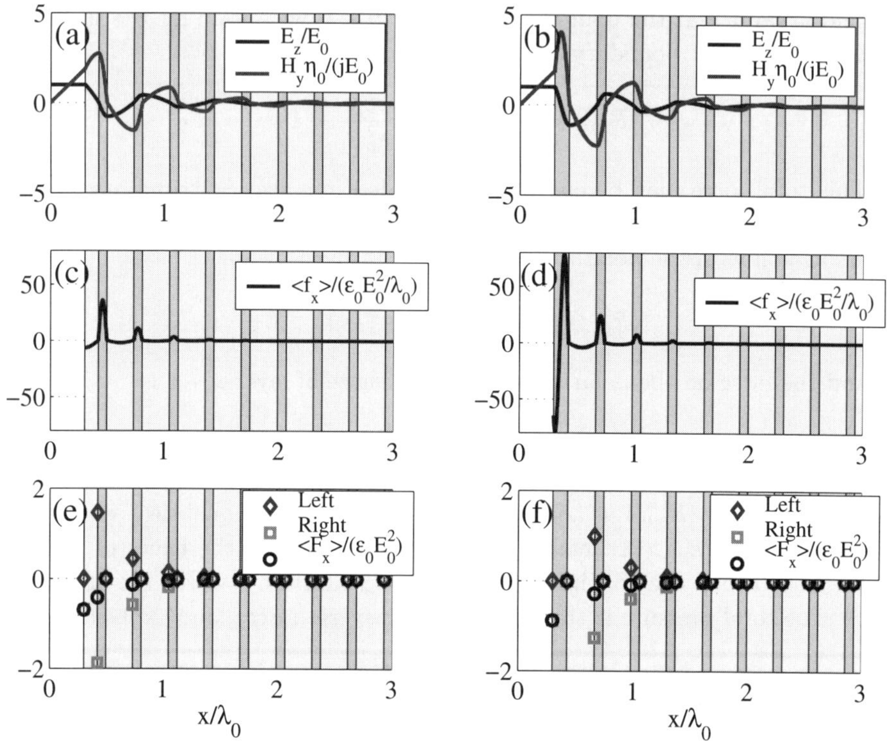

Figure 3. Force profiles in Bragg structures with SiO_2 and Si layers. The electromagnetic field (a-b), volume force density (c-d), and surface force densities (e-f), are shown.

Let us briefly examine the implications of the above results on the acceleration structure. Assuming that one Bragg mirror in the structure is of dimensions 1 mm $\times$ 1 mm $\times$ 50 μm and $E_0 = 1$ GV/m, the total transverse force on the mirror is 2.2 N. If we consider a material density of about 2 gr/cm^3, we obtain that this force is by no means negligible since it is 6 orders of magnitude larger than the gravitational force on the mirror. On the other hand, from the perspective of material strength and the possibility of crack formation, it is reasonable to assume that a pressure below E/π, where E is Young's modulus, may be sustained without damage to the structure[14]. Young's modulus for SiO_2 is 72.6 GN/m^2, whereas for Si it is 162 GN/m^2 (see Ref. 15). Since the total pressure in our case is of the order of 10^6 N/m^2, the electromagnetic forces are about 4 orders of magnitude below the theoretical threshold E/π.

4. Thermal Considerations

As indicated in the Introduction, reduction of the optical wavelength entails a corresponding increase in the energy density in the vacuum channel as well as in its close vicinity. Naturally, the question we are about to briefly address in this section is the impact of the optical field on the thermal processes in an acceleration structure[16]. As in the second section, we consider a planar optical Bragg acceleration structure as analyzed in Ref. 5. This structure will form the foundation of a model that allows to establish a set of approximate *analytic* expressions describing the temperature and the heat flux distributions in typical optical acceleration structures.

Starting from basic principles, it is well known that the temperature distribution is determined by the diffusion equation which reads

$$\left(\frac{\partial}{\partial t} - \frac{1}{D}\nabla^2\right)\Delta T = \frac{1}{D\sigma_{\mathrm{T}}} P_{\mathrm{loss}}, \tag{13}$$

wherein ΔT is the temperature change, $\sigma_{\mathrm{T}}[\mathrm{W}/(°\mathrm{K\,m})]$ is the thermal conductivity, $D[\mathrm{sec}/\mathrm{m}^2]$ is the thermal diffusion coefficient and P_{loss} is the dissipated power density. The following set of approximations allows us to significantly simplify the solution of Eq. (13) without notably affecting its accuracy:

- The temperature and the dissipated power density variations are assumed to be one-dimensional in the cross-sectional direction.
- The periodic dielectric layers, which have alternating thermal conductivities $\sigma_{\mathrm{T},1}$ and $\sigma_{\mathrm{T},2}$, may be represented by a uniform and effective thermal conductivity ($\sigma_{\mathrm{T,eff}}$), defined by $\dfrac{\Delta_1 + \Delta_2}{\sigma_{\mathrm{T,eff}}} \equiv \dfrac{\Delta_1}{\sigma_{\mathrm{T},1}} + \dfrac{\Delta_2}{\sigma_{\mathrm{T},2}}$, where Δ_1 and Δ_2 are the widths of the various layers in each period.
- The big difference in the typical repetition rate period (about 1 nsec) of the electromagnetic pulses, in comparison to the typical diffusion time which is of the order of 10^{-4} sec, allows us to assume a quasi-stationary regime ($\partial_t \simeq 0$).
- According to Ref. 5, the profile of the dissipated power density may be approximated by

$$P_0 \exp\left(-2\,\frac{x - D_{\mathrm{int}}}{x_c}\right), \tag{14}$$

where D_{int} is the cross-sectional coordinate of the interface between the first layer and the vacuum core and x_c is given by

$$x_c = \frac{\lambda_0}{4}\left[\frac{1}{\sqrt{\varepsilon_1 - 1}} + \frac{1}{\sqrt{\varepsilon_2 - 1}}\right]\left|\ln\left[\frac{\varepsilon_1}{\varepsilon_2}\sqrt{\frac{\varepsilon_2 - 1}{\varepsilon_1 - 1}}\right]\right|^{-1}, \tag{15}$$

ε_1 being the permittivity of the layer adjacent to the core, while the remainder of the layers are alternating with permittivities ε_2 and ε_1.

It has already been mentioned in the previous section that the field in the vacuum tunnel, is of the form $E_z = E_0 \exp\left(-j\frac{\omega}{c}z\right)$, and accordingly, the peak value of the dissipated power may be approximated by

$$P_0 \cong \frac{1}{2}\omega\varepsilon_0\varepsilon_1 \tan\delta E_0^2 \left[1 + \left(\frac{1}{\varepsilon_1}\frac{\omega}{c}D_{\text{int}}\right)^2\right] \tag{16}$$

The approximate solution may be further improved by using $\kappa_T P_0$ instead of P_0, where κ_T is a form-factor given by

$$\kappa_T = \frac{\int_{D_{\text{int}}}^{D_{\text{ext}}} dx P_{\text{exact}}(x)}{\int_{D_{\text{int}}}^{D_{\text{ext}}} dx P_0 \exp\left(-2\,\frac{x - D_{\text{int}}}{x_c}\right)} \tag{17}$$

$P_{\text{exact}}(x)$ being the exact dissipated power profile.

Relying on these assumptions, we may now find the temperature profile and the heat flux distribution, defined as $Q = -\sigma_{\text{T,eff}}\frac{\partial \Delta T}{\partial x}$, by solving Eq. (13) subject to the following boundary conditions: (i) no heat can escape from the first dielectric layer into the vacuum core, namely $\left.\frac{\partial}{\partial x}\Delta T\right|_{x=D_{\text{int}}} = 0$. (ii) The so-called Newtonian boundary condition, entailing that the amount of heat extracted from the outermost layer by a convective coolant is proportional to the temperature difference between this layer and the coolant, namely

$$\Delta T(x = D_{\text{ext}}) = -\frac{\sigma_{\text{T,eff}}}{h}\left.\frac{\partial}{\partial x}\Delta T\right|_{x=D_{\text{ext}}}, \tag{18}$$

where D_{ext} is the cross-sectional coordinate of the interface between the outermost layer and the coolant, and $h[\text{W}/^\circ(\text{Kcm}^2)]$ is the heat transfer coefficient. The limit $h \to \infty$ in the last condition represents a perfect cooling, which reduces the temperature of the boundary to the temperature of the coolant, i.e.,

$$\Delta T(x = D_{\rm ext}) = 0. \tag{19}$$

The expressions for the temperature and the heat flux distributions dictated by the solution of Eq. (13) under the approximations and the boundary conditions above read

$$\begin{aligned}\Delta T(x) = \frac{\kappa_T P_0}{\sigma_{\rm T,eff}} \frac{T_{\rm p}}{T_{\rm rr}} \frac{x_{\rm c}}{2} \Bigg\{ (D_{\rm ext} - x) - \frac{x_{\rm c}}{2} \left[\exp\left(-2\frac{x - D_{\rm int}}{x_{\rm c}}\right) - \exp\left(-2\frac{D_{\rm ext}}{\right.\right. \\ + \frac{\sigma_{\rm T,eff}}{h} \left[1 - \exp\left(-2\frac{D_{\rm ext} - D_{\rm int}}{x_{\rm c}}\right)\right]\Bigg\} \\ Q(x) = \kappa_T P_0 \frac{T_{\rm p}}{T_{\rm rr}} \frac{x_{\rm c}}{2} \left[1 - \exp\left(-2\frac{x - D_{\rm int}}{x_{\rm c}}\right)\right]\end{aligned} \tag{20}$$

wherein $T_{\rm p}$ denotes the electromagnetic pulse duration, whereas $T_{\rm rr}$ represents the repetition rate period. Clearly, the temperature is highest at $x = D_{\rm int}$, where there is no cooling, and the heat flux is maximal at the external wall $x = D_{\rm ext}$, which is the closest point to the coolant. Explicitly, the maximal temperature and heat-flux values are:

$$\begin{aligned}\Delta T_{\rm max} \equiv \Delta T(x = D_{\rm int}) = \frac{\kappa_T P_0}{\sigma_{\rm T,eff}} \frac{T_{\rm p}}{T_{\rm rr}} \frac{x_{\rm c}}{2} (D_{\rm ext} - D_{\rm int}) \left(1 - \frac{x_{\rm c}/2 - \sigma_{\rm T,eff}/h}{D_{\rm ext} - D_{\rm int}}\right) \\ Q_{\rm max} \equiv Q(x = D_{\rm ext}) = \kappa_T P_0 \frac{T_{\rm p}}{T_{\rm rr}} \frac{x_{\rm c}}{2}\end{aligned} \tag{21}$$

Figure 4 is a representative result which compares the exact solution of Eq. (13) and the approximations presented above. It is quite evident that the exact and the approximate solutions are virtually indistinguishable. Further analysis shows that for $h = 3$ W/(°Kcm^2) the maximal temperature in a Silica-Zirconia structure is $356 \times 10^7 \tan\delta_1$[°K] and the maximal heat-flux is $9.22 \times 10^9 \tan\delta_1$[W/cm^2], while in the Silica-Silicon structure these quantities are $66.4 \times 10^7 \tan\delta_1$[°K] and $1.74 \times 10^9 \tan\delta_1$[W/cm^2], wherein $\tan\delta_1$ is the loss tangent of the first layer (Silica). If we assume a loss factor similar to that existing in communication optical fibers: 1 dB/km ($\tan\delta \sim 10^{-11}$), the temperature increase will be less than 1°K and the heat flow will be a fraction of 1 W/cm^2 , which is more than three orders of magnitude lower than the known technological limit for heat dissipation. Obviously, using materials with a significantly higher loss-tangent may lead to unacceptable temperatures and temperature gradients.

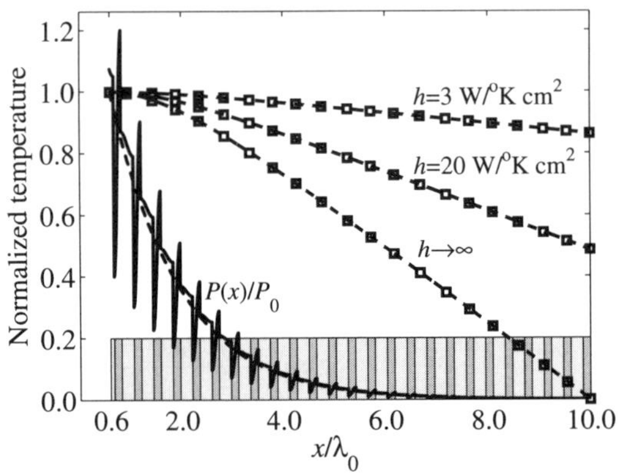

Figure 4. Temperature profile for different values of h in the silica-zirconia structure. For each value of h, the dashed line is the approximate curve, and the exact solution is indicated by square markers. The dissipated power profiles, approximate (dashed line) and exact (solid line), are also plotted. All quantities are normalized by their maximal approximate value.

5. Discussion

While examining the main constraints on Bragg acceleration structures we concluded that regarding the efficiency, for an acceptable value it is necessary "recycle" energy and accelerate trains of micro-bunches, each micro-bunch being located one wavelength apart from its neighbor. Obviously, the accelerating field ought to be tapered in order to compensate for the varying beam-loading. For a self-sustained solution of the field in a given acceleration module, it is necessary to provide the tapering by an external source (laser), whereas all the energy loss has to be compensated by the feedback loop. In order to satisfy the self-sustainability condition two important constraints must be imposed: first, the length of the accelerating module needs to be equal to the bunch length. Second, the group velocity which leads to the simplest shape of the external laser pulse is $\beta_{\rm gr} \simeq 0.5$.

The total electromagnetic force exerted on one mirror by the flowing laser light is of the order of 10^6 N/m^2. This significant force tends to push the mirrors apart, and may be 6 orders of magnitude larger than the gravitational force acting upon them. Nevertheless, the internal pressure is expected to be about 4 orders of magnitude lower than the acceptable values for which the structure remains undamaged.

As of thermal effects, if we assume a loss factor similar to that existing in communication optical fibers than the present estimates indicate that we are on safe ground: 1 dB/km ($\tan\delta \sim 10^{-11}$), the temperature increase will be less than 1°K and the heat flow will be a fraction of 1 W/cm^2 , which is more than three orders of magnitude lower than the known technological limit for heat dissipation. Obviously, using materials with a significantly higher loss-tangent may lead to unacceptable temperatures and temperature gradients.

Acknowledgements

This study was supported by the Israel Science Foundation and the United States Department of Energy.

References

1. Y. C. Huang, D. Zheng, W. M. Tulloch and R. L. Byer, *Appl. Phys. Lett.* **68**, 753 (1996).
2. J. D. Lawson, *IEEE Trans. Nucl. Sci.* **26**, 4217 (1979).
3. P. M. Woodward, *J. IEE* **93**, 1554 (1946).
4. X. E. Lin, *Phys. Rev. ST Accel. Beams* **4**, p. 051301 (2001).
5. A. Mizrahi and L. Schächter, *Phys. Rev. E* **70**, p. 016505 (2004).
6. R. P. Larsen and A. A. Oliner, A new class of low loss reactive wall waveguides, in *IEEE-MTT Int. Microwave Symp. Dig.*, (1) (Boston, MA, 1967).
7. P. Yeh and A. Yariv, *Opt. Commun.* **19**, 427 (1976).
8. P. Yeh, A. Yariv and E. Marom, *J. Opt. Soc. Am.* **68**, 1196 (1978).
9. S. Banna, D. Schieber and L. Schächter, *Appl. Phys. Lett.* **84**, 723 (2004).
10. S. Banna, D. Schieber and L. Schächter, *J. Appl. Phys.* **95**, 4415 (2004).
11. L. Schächter, *Phys. Rev. E* **70**, p. 016504 (2004).
12. K. L. F. Bane and G. Stupakov, *Phys. Rev. ST Accel. Beams* **6**, p. 024401 (2003).
13. J. A. Stratton, *Electromagnetic Theory* (Mcgraw-Hill, New York, 1941).
14. B. Lawn, *Fracture of brittle solids*, second edn. (Cambridge University Press, UK, 1995).
15. M. Bass, E. W. Van Stryland, D. R. Williams and W. L. Wolfe, *Handbook of Optics Volume II* (McGraw-Hill, New York, 1995).
16. V. Karagodsky, A. Mizrahi and L. Schächter, *Submitted for publication* .

HIGH BRIGHTNESS BEAMS – APPLICATIONS TO FREE-ELECTRON LASERS

SVEN REICHE

University of California, Los Angeles
Deptartment of Physics & Astronomy
Los Angeles, CA 90095-1547

Free-Electron Lasers as high-brilliance radiation sources, rely on a high quality of the electron beam driving the FEL process. The amount of energy, transferred from the electrons to the radiation field, and thus the efficiency of the FEL depends on the provided beam parameters. The presentation discusses the impact of various beam parameter and how current designs of FEL injector try to accomplish the demands on the beam quality for reaching saturation.

1. Introduction

A characteristic parameter for any kind of light source is the brilliance, which is defined as the number of photons per second, frequency interval, area and divergence angle and typically classifies the generation of the light source. The third generation is well established and there exist several facilities around the world, based on storage ring synchrotron radiation. The 4th generation light source was proclaimed[1] as a significant improvement in the brilliance, with radiation pulses in the femtosecond regime instead of the 3rd generation picosecond pulses. In addition, the light source should be transversely and longitudinally coherent.

Linac based Free-Electron Lasers are a possible candidate for fulfilling the criteria of a 4th generation light source. The key is that Free-Electron Lasers can saturation within a single pass of an electron bunch, starting from the spontaneous emission of that bunch. This defines the operation principle of an Self-Amplifying Spontaneous Emission (SASE) FEL. Because the SASE FEL does not require seeding nor an oscillator set-up to accumulate gain, the radiation wavelength can be extended to the X-ray regime.

Ongoing research since the nineties reduced the radiation wavelength of the SASE FEL down to less than 30 nm (e.g. VISA[2], LEUTL[3], DUV

FEL[4] and TTF[5]) reaching saturation. The trend is towards even shorter wavelengths with two projects currently under construction – LCLS[6] and the European X-ray FEL Project[7] –, with radiation in an Ångstrom regime.

All single-pass FELs rely on a high-brightness beam to enhance the efficiency of the FEL, transferring energy from the electron bunch to the radiation field. The electron beam parameters of several running or planned FELs are listed in Tab. 1. This paper discusses the importance of the beam brightness to the Free-Electron Laser. The next section gives an overview of the FEL principle and how beam parameters such as current, energy spread or emittance effect the FEL performance. Certain conditions are derived, which have to be fulfilled in order to achieve lasing in the FEL. The remaining sections deal with how the electron bunches are produced and transport from the electron source to the FEL.

Table 1. Beam parameters of recent and planned FELs

Name	Type	λ [nm]	E [MeV]	I [kA]	ϵ_n [mm·mrad]	Status
UCLA	SASE	12000	18	0.17	5.3	lased
VISA	SASE	840	70	0.25	2.3	lased
LEUTL	SASE	530/385	220/260	0.63	8.5/7.1	lased
TTF-FEL	SASE	6-120	1000(450)	2.0	3.0	lased(30nm)
DUV FEL	HGHG	400/266	<200	0.5	4.0	lased
LCLS	SASE	0.15-1.5	<14000	3.4	1.2	under const.
Euro. XFEL	SASE	6-0.085	<20000	5.0	1.4	under const.
SCSS	SASE	3.6	1000	2.0	2.0	planned
BESSY	HGHG	1.2-50	1000-2300	1.75	1.5	planned
FERMI	HGHG	10-100	500-1000	2.5	1.5	planned
SPARXINO	SASE	1-10	300-1000	2.0	1.0	planned
PAL	SASE	0.3	3700	3.5	1.2	planned

2. FEL Principle

This section gives an overview of the FEL physics with a focus on the dependance of electron beam parameters. An electron, injected into an alternating magnetic field of an undulator or wiggler, emits radiation. The slippage of the radiation field over one undulator period λ_u with respect to the electron determines the resonant radiation wavelength

$$\lambda = \frac{\lambda_u}{2\gamma^2}\left(1 + a_u^2\right) \quad , \tag{1}$$

where γ is the Lorentz factor of the electron energy and $a_u = (e/2\pi mc)B_u\lambda_u$ is the rms undulator parameter, with B_u the on-axis field, c the speed of light and e and m the electron charge and mass.

The emission on a spontaneous, incoherent level is called undulator radiation. FEL radiation is an enhancement of the undulator radiation by increasing the coherence level due to the formation of microbunches. The spacing of the microbunches is the resonant radiation wavelength. Bunching is caused by the collective instability of the FEL, where the emitted radiation acts back on the electrons and modulates their energy. Because the electron motion in the undulator is dispersive, the energy modulation is transformed into a bunching in the current. The emission level of the undulator radiation is enhanced. The physics is similar to the motion in an rf bucket, but with the difference that the size of the FEL bucket grows with the accumulated radiation and its phase adjusts automatically to the positions of the microbunches.

The 'time'-scale of the collective instability of the FEL is given by the gain length

$$L_g = \frac{\lambda}{4\pi\rho} \quad , \tag{2}$$

where the FEL parameter ρ[8] is mainly a measure for the brightness of the driving electron beam. In the 1D limit, where diffraction effects of the radiation field are negligible, the FEL parameter is given by

$$\rho = \frac{1}{\gamma}\left[\frac{I}{I_A \Sigma_b}\frac{(\lambda_u f_c a_u)^2}{16}\right]^{\frac{1}{3}} \quad , \tag{3}$$

with I and Σ_b the current and transverse beam size of the electron beam, respectively, and $I_A \approx 17$ kA is the Alvén current. The correction factor f_c arises due to the reduced coupling of the electron beam to the radiation field in a planar undulator. Typical values for ρ are well below unity.

The FEL amplification stops at saturation, when maximum current modulation is achieved. In the longitudinal phase space, the initially continuous distribution has been rotated to an upright position in each bucket. From this picture it can easily be seen that a larger initial energy spread degrades the FEL performance, because the bunching is more spread out, reducing the coherence level of the FEL radiation. Because the maximum relative width in energy of the bucket is ρ[8], the energy spread σ_γ has to fulfill the constraint

$$\frac{\sigma_\gamma}{\gamma} \ll \rho \quad . \tag{4}$$

Otherwise the bucket is smoothly filled and rotation provides nothing.

Emittance effects are similar to those of the energy spread. The magnetic field of the undulator couples velocity spread in the transverse direction to a spread in the longitudinal direction[9]. The constraint for the normalized emittance ϵ_n is

$$\epsilon_n \ll \frac{4\lambda\beta}{\lambda_u}\rho \quad , \tag{5}$$

where the average beta-function β is a measure for the focusing strength within the undulator. Although a stronger focusing reduces the beam size Σ_b and, thus, increases the ρ parameter, it enhances the emittance effect. For any given emittance value there is an optimum value for the β-function, which balances out both effects. As an example, Fig. 1 shows the dependence of the saturation length and power on the value of the β-function for the case of the LCLS FEL. The optimum value for an emittance of 1.2 mm·mrad is 18 m.

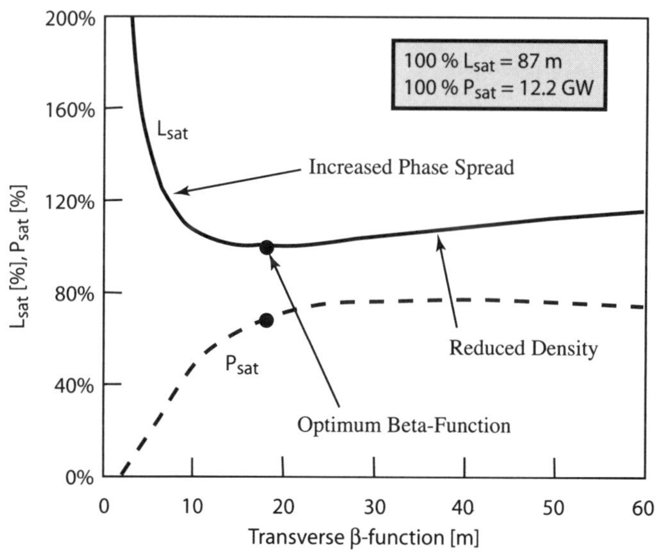

Figure 1. Saturation power and length as a function of the focusing. Optimum value is 18 m.

Because the radiation field is emitted within the finite area of the electron beam, the emitted radiation diffracts. The field amplitude at the location of the electron beam is reduced and, thus, the FEL amplification is inhibited. On the other hand diffraction spreads out the phase and amplitude information in the transverse direction, which is essential to achieving transverse coherence. For best coupling, the electron beam envelope should match that of the radiation beam, resulting in a second condition –

$$\frac{\epsilon_N}{\gamma} \approx \frac{\lambda}{4\pi} \tag{6}$$

– for the normalized emittance. However this condition is not as crucial as the first and is indeed violated for the planned X-ray FELs LCLS and the European X-ray FEL.

For an FEL with strong diffraction effects, where the Rayleigh length $z_r = \Sigma_b^2 k/2$ is shorter than the gain length, the definition of the 1D ρ parameter looses its validity. The FEL performance depends rather on the beam current than on the electron density ($\propto I/\Sigma_b$). The FEL theory can be extended to include diffraction effects, which allows one to define an effective ρ-parameter[10]. All scaling laws (gain length, constraints for beam parameters) remain the same.

The cooperation length

$$l_c = \frac{\lambda}{4\pi\rho} \tag{7}$$

is a measure for the longitudinal coherence of the FEL pulse[11]. In particular, for VUV and X-ray FELs l_c is much smaller than the bunch length and various parts of the electron bunch amplify independently from each other. To guarantee amplification over the entire electron bunch, each of these independent slices has to provide sufficient beam quality. The value of the projected beam parameters such as energy spread and emittance are less useful for determining the beam quality effecting the FEL performance. Indeed, these secondary parameters can violate the constraints, given above, but the pulse lases, because they are fulfilled for each slice.

3. Beam Quality Issues

A high electron beam quality is essential for a successful operation of an Free-Electron Laser. All requirements, given in the section above, can be combined in the demand for a high density of the 6D phase space distribution. Because a linear accelerator does not allow for beam cooling as is the case for a storage ring, a sufficient phase space density has to be generated at the electron beam source. The phase space density is correlated to the beam brightness and, thus, Free Electron Lasers are commonly based on high-brightness electron beam sources, which typically excludes the use of thermionic guns. A common choice is an rf photo-electron gun. Typical parameters are a charge of around 1 nQ, rms bunch length of around 3 ps and transverse, normalized emittances of 1 mm·mrad or above.

Three effects contribute to the transverse emittance[12]: space-charge, rf curvature and initial thermal velocity spread. The first two contributions scale monotonically with the emitted charge. Lowering the charge reduces the transverse emittance and initial bunch length. Because the beam brightness is actually increase by this, the FEL performance improves[13]. The choice of operation at 1 nQ was set, when Free-Electron Lasers were discusses as the 4th generation light sources[1]. Indeed, recent experiments[2] uses a much lower charge around a few hundreds of pC to shorten the FEL saturation length. The lower limit is given by the resolution of the beam diagnostics along the beam line.

Emittance compensation is essential to achieve small values for the transverse emittances. Using this scheme a focussing solenoid compensates the space charged field so that the electrons have minimal transverse momenta, when the acceleration diminishes the space charge effects. For best performance the charge distribution should be homogeneously, so that the transverse space charge field is predominantly linear in the transverse position. Otherwise the fluctuation in the applied kicks by the field increases the emittance, which are not compensated by the solenoid field. This effect puts a stringent requirement on the transverse uniformity of the laser profile[14] as well as for the cathode itself.

The transverse emittance is the most crucial parameter, when designing and building an FEL. Only for FELs with a wavelength of 10 microns or longer, a larger emittance would actually enhanced the FEL performance, because diffraction effects are reduced and the beam couples better to the fundamental mode, while higher modes are suppressed. Phase spread effects due to the transverse emittance dominate for UV and X-ray FELs. The uncorrelated betatron oscillation of all electron yields a longitudinal spread, as discussed above. The change in the longitudinal velocity is constant for natural focusing undulators[15] as well as for FODO lattices[16], where the cell length is shorter than the gain length, and does not depend on the betatron phase of the electrons. Theoretically, it can be compensated by a correlation between energy and the amplitude of the betatron oscillation[17]. However possible conditioning schemes (e.g. by higher order mode rf cavities) have the flaw that with the introduced energy correlation other beam parameters are effected as well (e.g. correlated transverse kick along the bunch due to the time dependence of the magnetic field in the cavities), which negates the benefits from the conditioned beam[18].

The FEL gain is less sensitive to the energy spread than to the emittance. The required energy spread can easily be achieved by the current generation of photo injectors. Even the growth of the energy spread within the undulator due to the quantum fluctuation of the spontaneous radiations[20] does not effect the FEL performance significantly. It also allows to operated at a larger energy spread than provided by the rf photo electron guns to suppress the microbunch instability in the bunch compressor (see below)[19].

Longitudinal homogeneity of the beam parameters are not essential for the FEL process if the cooperation length is smaller than the bunch length, but can affect the output properties of the FEL. If the variation in the slice beam energy is larger than the FEL parameter ρ the spectrum becomes broader. Experiments, based on the spectral purity of the radiation, place a monochromator prior to the experimental station. Due to the wider spectrum less photons will pass the acceptance of the monochromator, thus reducing the available flux for the experiment. Another issue is that most diagnostics are based on projected values, in particular the matching to the focusing structure. With variation in the slice centroid position and angle as well as in the divergence/convergence of the beam envelope, the individual slices are not matched to the undulator focusing lattice. Mismatch typically results in a shift in the wavelength, reduced coupling between the slice and the radiation field and thus reduces the gain for the slice. This is most critical if only a small part of the bunch is lasing, such as in the TTF experiment, where less than 20% contributes to the FEL amplification[5]. Fig. 2 shows the measure current profile using a deflecting cavity.

4. Beam Generation and Transport

For all current single-pass high gain Free-Electron Lasers the driving beam is created with an rf photo-electron gun, where a laser pulse illuminates the cathode within an rf cavity. The electrons are generated by the photo-electric effect and accelerated by the electric field in the cavity. Typical design of the rf guns uses standing wave structures with a one and a half cell layout[21] and an rf frequency in the S- or L-band regime. A solenoid focuses the electron beam, compensating the defocusing effects of the space charge field. Emittance compensation[22] balances the solenoid field with

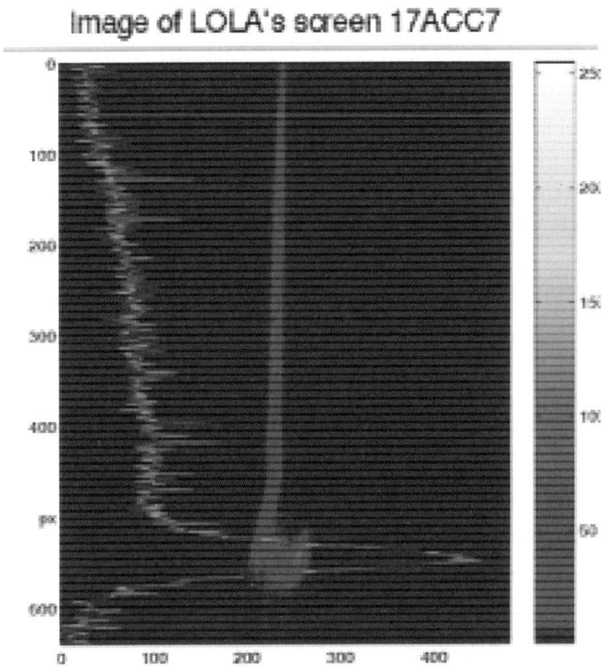

Figure 2. Longitudinal charge distribution and current profile for the electron beam at the entrance of the TTF-FEL Phase II (Courtesy of Deutsches Elektronen Synchrotron).

the space charge field, so that the transverse emittance is minimized when further acceleration reduces the sensitivity to the space charge forces. Because the head and the tail has less charge than the core, the solenoid overcompensates the space charge. As a result the local, longitudinal sliced phase space distributions in the transverse plane are not aligned to each other. The projected emittance is larger than the individual slice emittance. As already mentioned above, projected beam properties are not useful for describing FEL performance.

The main acceleration is accomplished by a series of rf cavities. Due to the rf curvature as well as wakefields within the rf structure the longitudinal energy profile becomes modulated. For a given charge and a certain bunch charge both effects can cancel each other but only in the linear term of the effective energy change along the bunch. Higher order modulation remains. However it has been shown that a cavity at a high frequency can compensate the second order effects as well[23] and the bunch compression is more homogeneous along the bunch. Because the operation of the machines requires the cancellation of rf-curvature and wakefields, it implies tolerances on the fluctuation of charge and rf phases. It also limits the charge range in order to control saturation power and length without changing the machine settings.

The bunch compressor is the most crucial part in the beam manipulation, because there are two effects, which can spoil the beam quality. Both are based on the coherent spontaneous radiation (CSR)[24] within the magnetic chicane. The CSR causes that the emitted radiation of the tail of the bunch acts back on the head of the bunch, because the radiation travels on a straight path and can catch up with the bunch, propagating a long a curved trajectory. The first effect is caused by the fact that a particle, which changes its energy within the chicane, has a residual dispersive effect at the end of the chicane. Because the CSR effect has different amplitude along the bunch, the main effect is a mismatch of the slice phase space distribution and, thus, a growth in the projected emittance. In addition there is also a slight increase in the slice emittance itself due to the non-uniformity of the effective CSR field in the transverse direction. The second effect is a collective instability[25] very similar to that driving the FEL. Any modulation in the beam current modulates the beam energy. Due to the dispersive effects the energy modulation enhances the current modulation, and thus the emission of the coherent synchrotron radiation. For the planned X-ray FELs LCLS and the European X-Ray FEL the overall gain of this microbunch instability is small[26] and can be further reduced if the energy spread is slightly increased in the injector part of the accelerator [27]. This also suppresses the disruption in the phase space distribution within the bunch compressor due to the energy modulation by the longitudinal space charge in the low energy part of the machine.

Within the undulator itself wakefields can further effect the FEL amplification. This effect is distinct from the effect of the wakefields of the linear accelerator. The latter result in a variation in the initial energy of each slice and, thus, the resonant wavelength. This effect can be regarded as static while the undulator wakefields acts dynamically during the amplification. Once the electron enters the undulator, the slice energy defines the radiation wavelength. Once defined, the slice has to stay within the FEL bandwidth, thus requiring that the energy change by the undulator wakefields are fulfilling the constraint[28]

$$\frac{d\gamma}{dz} < \frac{\rho^2\gamma}{\lambda_u} \quad . \tag{8}$$

If it is not fulfilled the slice stops amplifying before reaching saturation and the emission is shifted to different wavelengths.

All together – the undulator wakefields, the variation in the slice emittance and energy spread as well as the offset and mismatch of each slice – can yield a strong modulation of the FEL radiation pulse[29]. Fig. 3 shows the expected profile of the LCLS FEL. The main features are the two gaps at the head and tail of the bunch. They are caused by wakefields within the undulator as well as strong degradation of slice emittance and energy spread in the preceeding bunch compressors. On the other hand, the enhanced power level at the head of the pulse arises from non-linear terms in the compression scheme, which locally increases the current up to 15 kA.

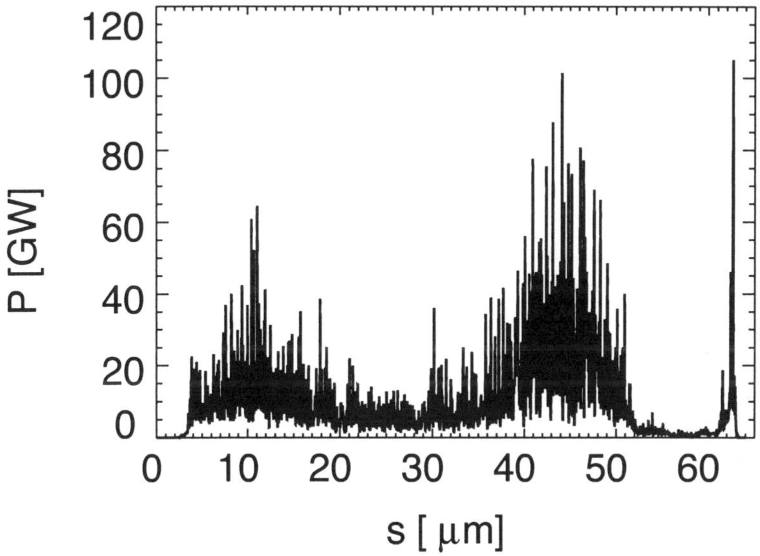

Figure 3. Simulated power profile of the LCLS FEL pulse at the undulator exit. The initial pulse (right) exceeds the expected power level due to a high local current of about 14 kA. The gaps arise due to wakefields in the undulator and insufficient slice emittances.

A more technical aspect is the control on the FEL pulse length. It is determined by the bunch charge and degree of compression. For X-ray FELs the bunch length will be much longer than the cooperation length. Experiments, relying on longitudinal coherence, have to place a monochromator before their experimental station. However there are classes of experiments, which requires FEL pulses no longer than 50 fs [30]. Common solutions use a monochromator to select a subsection of a chirped FEL pulse, which is produced by a chirped electron beam. The energy chirp of the bunch must be larger than the FEL bandwidth and is of the order 0.5-2%. This set-up is more feasible than compressing the bunch down to 50 fs while preserving the beam quality.

The duty cycle of the FEL depends on the acceleration technology. Normal conducting cavities have large thermal losses if a high repetition rate is desired. Also wakefields will affect succeeding bunches in a pulse train with short separation between each pulse. Superconducting cavities are much less effected by losses and beam loading, but comes with the cost of the cryogenic system. However super conducting rf guns[31] are still under development, because their usage prohibits solenoids for emittance compensations.

5. New Concepts

The primary demand on the electron beam for a Free-Electron Laser is a high current and a low emittance. In addition a high repetition rate is desirable if the FEL operates as a user facility. Currently this is limited by the transverse emittance and bunch compression using a magnet chicane.

The minimal emittance is defined by the rf gun and the scaling shows that going to a higher rf frequency decrease the minimal value. However, reliable cavities at shorter wavelength are more difficult to manufacture. In addition the smaller aperture of these cavities would strongly enhance the wakefield effects. Concepts to replace the rf gun totally, such as plasma guns or injectors[32], are still in their initial development phase.

Bunch compressors are the strongest source of emittance degradation, because they deflect the electron beam from the primary axis of the linear accelerator. High current rf guns are not feasible because the strong space charge field of non-relativistic electrons would yield emittances not suitable for FELs. A promising concept is velocity bunching[33], where the accelerating cavities, succeeding the gun, are slightly detuned. The initial phase space distribution in the longitudinal plane is rotated till maximum current is achieved. The pay-off is a reduced energy gain over these initial cavities, which can always be compensated by additional accelerating section. Also the bunch is short in an early stage of acceleration and all wakefields adds up more coherently. On the other hand the transverse emittance is unaffected and preserved.

An alternative concept to FELs based on linear accelerator are energy recovery linacs (ERL)[34], based on super-conduction rf-technology. An electron beam is injected into a circular ring, where it is first accelerated and then injected into an FEL. Finally the bunch reenters the rf cavities but with a 180 degree phase difference. The electron energy is transferred back to the rf field, before the low-energy beam is dumped. This set-up combines the high repetition rate and energy stability of a storage ring FEL with the high beam brightness of a linac based FEL to achieve saturation within a single pass.

References

1. J. Als-Nielsen, "Scientific Opportunities for the 4th generation Light Sources, Hard X-ray", Proc. of Workshop on 4th Generation Light Sources, ESRF Report, 1996, Grenoble
2. A. Tremaine *et al.*, Phys. Rev. Lett. **88** (2002) 204801
3. S.V. Milton *et al.*, Science **292** (2001) 2037
4. L.H. Yu *et al.* Phys. Rev. Lett. **91** (2003) 074801
5. V. Ayvazyan *et al.*, Phys. Rev. Lett. **88** (2002) 104802
6. *Linac Coherent Light Source (LCLS) Design Study Report*, SLAC-R-521, UC-414 (1998)
7. *TESLA - Technical Design Report* DESY 2001-011, ECFA 2001-209, TESLA Report 2001-23, TESLA-FEL 2001-05, Deutsches Elektronen-Synchrotron, Hamburg, Germany (2001)
8. R. Bonifacio, C. Pellegrini and L.M. Narducci, Opt. Comm. **50** (1984) 373
9. J. Rossbach *et al.*, TESLA-FEL 95-06, Deutsches Elektronen Synchrotron, Hamburg, Germany
10. E.L. Saldin, E.A. Schneidmiller and M.V. Yurkov, Opt. Comm. **97** (1993) 272
11. R. Bonifacio and F. Casagrande, Nucl. Inst. & Meth. **A 239** (1985) 29
12. J.B. Rosenzweig, E. Colby, Proc. Conf. Adv. Accel. Conf. **AIP 335** (1995) 724
13. C. Pellegrini *et al.*, Nucl. Inst. & Meth. **A 475** (2001) 328
14. F. Zhou *et al.*, Proc. of the European Part. Accel. Conf., 2002, Paris
15. E.T. Scharlemann, J. appl. Phys. **58** (1985) 2154
16. S. Reiche, Nucl. Inst. & Meth. **A 445** (2000) 90
17. A.N. Sessler, D.H. Whittum, L.H. Yu, Phys. Rev. Lett. **68** (1992) 309
18. P. Emma and G. Stupakov, Phys. Rev. STAB **6** (2003) 030701
19. Z. Huang *et al.*, Phys. Rev. STAB **7** (2004) 074401
20. M. Sands, Phys. Rev. **97** (195) 470
21. D.T. Palmer *et al.*, Proc. of the Part. Accel. Conf., Vancouver (1997) 2843
22. B.E. Carlsten, IEEE Catalog no. 89CH2669-0
23. M. Borland *et al.*, Nucl. Inst. & Meth. **A 483** (2002) 268
24. Y.S. Derbenev, J. Rossbach, E.L. Saldin and V.D. Shiltsev, TESLA-FEL 95-05, Deutsches Elektronen Synchrotron, Hamburg, Germany

25. S. Heifits,S. Krinsky and G. Stupakov, SLAC-PUB-9165, SLAC, Stanford, USA (2002)
26. Z. Huang, M. Borland, P. Emma, K.-J. Kim, Proc. of the 24th FEL Conf., 2002, Argonne
27. Z. Huang, K.-J. Kim, Phys. Rev. ST Accel. Beams 5, 074401 (2002)
28. S. Reiche and P. Emma, Proc. of the 24th FEL Conf., 2002, Argonne
29. S. Reiche *et al.*, Nucl. Inst. & Meth. **A 483** (2002) 70
30. G.K. Shenoy, J. Stoehr (eds.),*LCLS - The First Experiments*, Stanford (2000)
31. D. Janssen *et al.*, Proc. of the Part. Accel. Conf., 1999, New York
32. D. Umstadter, J.K. Kim, E. Dodd, Phys. Rev. Lett. **76** (1996) 2073
33. L. Serafini *et al.*, Proc. of the Part. Accel. Conf., 2001, Chicago
34. Maury Tigner, Nuovo Cimento **37** (1965) 1228

CONCEPTION OF PHOTO-INJECTORS FOR THE CTF3 EXPERIMENT

R. ROUX

Laboratoire de l'Accélérateur Linéaire, IN2P3-CNRS, Université de Paris-Sud,

B.P 34, 91898 Orsay, France

In the framework of the CLIC Test Facility (CTF3) under development at CERN, LAL-Orsay is responsible for the construction of two photo-injectors for two different linacs. One, dedicated for the so called "drive beam linac" must fulfil very demanding specifications. The RF gun has to provide a high quality beam composed of more than 2,000 bunches containing 2.33 nC of charge each in one RF pulse. The model adopted is inspired from the CERN 3 GHz 2-1/2 cell type IV RF gun. We will summarize all the studies performed on the RF design and on the beam dynamics. The vacuum issue has been also carefully investigated. The constraints on the second photo-injector are less severe since it must be operated with one or 64 bunches of 0.5 nC each for the so called "probe beam linac". It will also be a 2.5 cell gun at 3 GHz but its design will be substantially different with respect to the former. This last project has only recently begun and therefore we will show only the preliminary results of the RF and of the beam dynamics simulations.

1. Introduction

For 5 years, the LAL is involved in the CLIC-Test-Facility 3 (CTF3) under construction at CERN. The goal of this accelerator is to test and valid the concepts of the multi-TeV linear collider CLIC, proposed by the CERN [1]. Schematically this accelerator is based on two linacs. One, the drive beam linac at 3 GHz, must provide a 150 MeV electron beam with high charge which passing through resonant decelerating structures should produce high RF power at 30 Ghz. Then, this power is sent to the 30 GHz travelling sections of the main linac where the electron beam is accelerated to the TeV energy for particle physics.

The drive beam linac of CTF3 was successfully operated with a thermionic gun built by the LAL [2,3] but it was decided to put a photo-injector in place of it because of the many advantages they have [4]. This project is a part of a Joint Research Activity, called PHIN, itself part of the Coordinated Accelerator

Research in Europe*. Other work packages are defined in the context of an R&D program on photo-cathodes and the construction of a custom laser to drive the photo-injector. In addition, the LAL is in charge of the construction of a second photo-injector for the probe beam linac which is used as the main linac in CLIC. The rest of the linac should be built by a French institute, the CEA-Saclay in collaboration with the CERN. The first part of this paper is dedicated to the RF design and the beam dynamic simulation of the PHIN photo-injector. In a second part, the photo-injector of the probe beam will be described mostly on the aspects which differ to the previous.

2. The PHIN photo-injector

LAL owns a good knowledge in the fabrication of photo-injectors. The first one home made, CANDELA with 1,5 cell, was dedicated to accelerator physics [5]. A second photo-injector for a small linac, ELYSE, according to a CERN type model, was built for radiolysis experiments in chemistry [6]. Now, for the CTF3 experiment, we decided to benefit from the CERN experience with CTF2 [7]. So, this gun is based on a previous proto-type CERN RF gun (2.5 cells, so called "type IV") which was built and tested with beam at CERN [8]. However, it needed further study in order to meet the specifications of the CTF3 drive beam linac (see table 1).

Table 1: Beam and RF gun parameters desired for CTF3.

RF frequency (GHZ)	2.99855
RF power (MW)	30
Beam energy (MeV)	5 – 6
Beam current (A)	3.51
Pulse train duration (μs)	1.548
Bunch spacing (ns)	0.6666
Charge/bunch (nC)	2.33
Repetition rate (Hz)	5
Bunch length, FWHM (ps)	< 10
rms energy spread (%)	< 2
Normalized emittance (πmmmrad)	< 25
Vacuum pressure (mbar)	$< 2.10^{10}$

In the following, we show the results of the RF simulations performed with a 2-D electromagnetic code, SUPERFISH, and a 3-D code, HFSS, which

* We acknowledge the support of the European Community-Research Infrastructure Activity under the FP6 "Structuring the European Research Area" programme (CARE, contract number RII3-CT-2003-506395)

allowed us to obtain the main characteristics such as the cavity dimensions and the axial field distribution. Beam loading is also evaluated. Then simulations of the beam dynamics in the gun with the PARMELA code are summarised. In addition the vacuum aspects have been thoroughly investigated, as in previous experiments with RF photo-injectors at CERN the pressure has been found to increase dramatically at high levels of extracted charge (≈1 μC).

2.1 *2D RF Simulations*

The SUPERFISH simulations are used essentially to determine the physical dimensions of the photo-injector cavity. First, the main objectives are to check if the original design is compatible with the CTF3 parameters. We have studied, for example, the influence of the angle of the cathode wall as well as the shape of the iris. Of course, any change to the cavity must be compensated in order to keep the resonant frequency at the required value. The CERN design of the RF gun was optimised for higher charge by choosing the angle of the half-cell wall around the photo-cathode, to provide additional transverse focusing. According to electron beam dynamic simulations (see 1.3), this angle should be reduced to 3.4 ° rather than the previous value 8° in order to keep the energy spread within the desired limits.

Moreover, we have changed the shape of the iris from circular to elliptical to decrease the surface electric field and therefore it minimizes electrical breakdown and dark current levels [9]. Furthermore, we introduced a slight asymmetry in the walls of the cavities for mechanical reasons and to try to reduce multipactor hazards. The new design is shown schematically in Fig. 1.

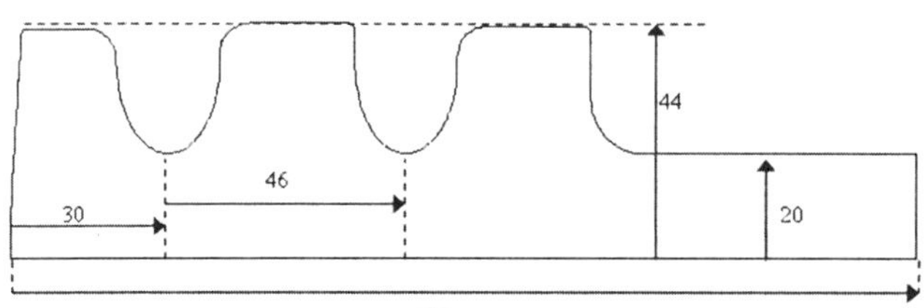

Figure 1: SUPERFISH design of the PHIN gun, lengths are in mm.

The evidence of the efficiency of the elliptical iris for the reduction of the surface electric field with respect to the circular one is illustrated in Fig. 2. At first sight, the gain is not obvious since the average field is roughly the same in the two cases. But the peak electric field, one the irises, in the elliptical design is 15 % lower with respect to the circular case. Now, the yield of the field emitted electrons by the cavity walls is strongly non linear with the electric field.

Therefore the use of the elliptical iris should be helpful for the reduction of discharge problems due to field emission.

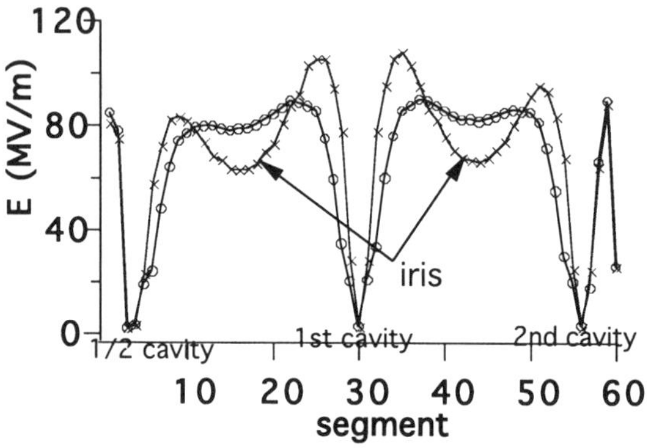

Figure 2: Surface electric field as a function of the SUPERFISH segment on the circumference of the model in figure 1. The surface electric field is zero in the mid-cells and maximum on the iris. The circles show the elliptical iris case and the crosses correspond to the circular iris case. The axial electric field is 85 MV/m.

The parameters of this RF gun are shown in table 2. Finally, the main difficulty in designing the gun is to obtain good axial electric field "flatness" in every cell of the structure. This requirement is rather difficult to meet as the cells are strongly coupled and any change in the radius of one cell induces a variation of the electric field in every cell. The best adjustment we have obtained is shown in Fig. 3.

Table 2: RF photo-gun parameters

Resonance frequency (GHz)	3.00305
Shunt impedance R_s (MΩ)	6
Quality factor Q	14530

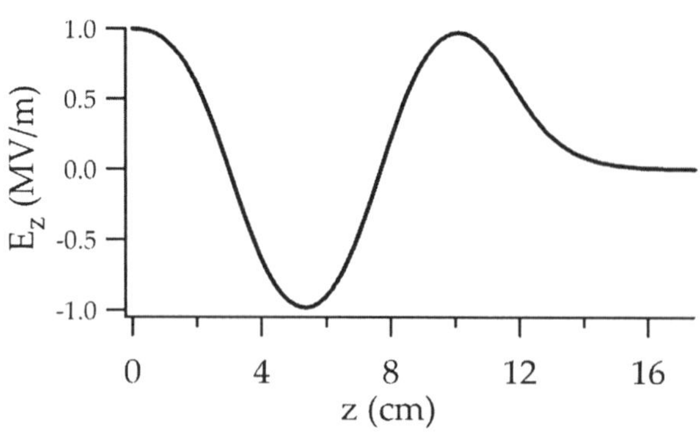

Figure 3: normalized axial electric field of the RF photo-gun.

2.2 *Calculation of the Beam-Loading*

It is well known that the electron beam interacting with the impedance of the RF structures induces a beam loading voltage which can be detrimental for gun operation [10]. An important issue to consider is the HF matching of the photo-injector. The reflection factor depends on the coupling, β. In presence of the electron beam, the coupling is given by:

$$\beta = \frac{\beta_0}{1 + \frac{P_{beam}}{P_{RF}}} \tag{1}$$

where β_0 is the unloaded cavity coupling, P_{RF} is the RF power dissipated into the cavity walls and $P_{beam} = V*I$ is the power of the electron beam at the output of the RF gun.

For the nominal current and electric field, the beam energy is 5.5 MeV hence P_{beam} is 19.3 MW. The necessary RF power is 9.6 MW; therefore, to get $\beta = 1$ and no reflection in presence of the beam, the RF gun must be over-coupled with $\beta_0 = 3$. It also means we need roughly 30 MW of RF power to compensate for the electron beam consumption. Besides the matching, the beam-loading induces a supplementary energy spread over the train of bunches. The beam loading voltage evolves as cavity voltage induced by the RF input power. Both grow exponentially with the time constant $\tau = 2Q_L/\omega_r$ where Q_L and ω_r are respectively, the cavity loaded quality factor and the resonant frequency. To avoid a decrease of the accelerating gradient over the train of bunches, one needs to inject the beam during the rise time of the electric field. In this way, the induced voltage is compensated by the increase of the RF input power resulting in a constant accelerating voltage. The time t_0 of injection is obtained from the evolution of the electric field due to the HF input power:

$$\begin{aligned} E(t) &= E_{max}(1 - \exp(-\frac{t}{\tau})) \\ t_0 &= -\tau \ln(1 - \frac{E(t_0)}{E_{max}}) \end{aligned} \tag{2}$$

with $E(t_0)$, the accelerating peak electric field, e.g. 85 MV/m and E_{max} the electric field which would be established corresponding to 30 MW if no beam was injected. But without beam, the RF gun is not matched and the reflection factor is 25 %. So, only 22.5 MW goes into the gun which gives $E_{max} = 130$ MV/m. Using (2), one finds $\tau = 0.385$ μs and $t_0 = 0.408$ μs.

2.3 *3D RF Simulations*

The RF power is transmitted to the photo-gun via wave-guides and enters the gun through coupling holes whose aperture must be adjusted to, in principle, minimise the reflected power. In our case, as explained in the previous paragraph, we must design the gun to get a coupling factor, β, to be 2.9 which will allow the gun to be at critical coupling in the presence of a beam at the nominal current of 3.51 A. Moreover, for beam emittance preservation, the electric field must be kept symmetric around the axis. Consequently we decided to connect two couplers symmetrically with respect to the horizontal plane. The 3-D design is shown in Fig. 4.

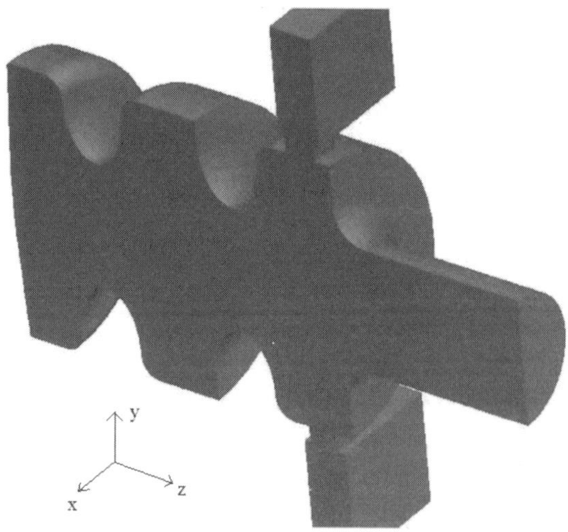

Figure 4: Half view of 3-D model of the RF photo-gun.

To leave enough space to install a solenoid around the gun it was decided to connect the couplers to the output cell and to use wave-guides whose inner height is 14 mm rather than the 3 GHz standard 34 mm. The shape of the coupling aperture is in the form of a racetrack and its dimensions are approximately 25mm x 10mm. This model was obtained after many simulations since three conditions need to be satisfied in simultaneously: the resonant frequency, the coupling and the field-flatness. For instance, according to the simulations it is necessary to have the radius of the last cell 1 mm smaller than in the second cell.

One last issue to check is the transverse symmetry of the electric field in the coupling cell. Thanks to the dual coupling, the field is symmetric vertically with respect to the (x,z) plane. But, inevitably, there is a difference of the field amplitude between the x and y direction. It adds a quadrupolar component to the acceleration in the transverse plane which leads to a degradation of the emittance. One solution to counteract this problem is to have a completely symmetrical coupling, namely a coaxial coupling with a hollow cylindrical antenna going into the RF gun along the axis of the cut-off tube. This scheme is now becoming standard equipment on the photo-injectors dedicated for free electron laser experiments striving to reach ultra-low emittance of ~1 μmrad [11]. But, in our case, we need a strong over-coupling which pushes the antenna, according to the simulations, deeply into the RF gun. The antenna would be practically at only 4 mm away from the iris between the middle cell and the coupling cell in an area of high electric field. In addition, the coaxial coupler has to support 30 MW of HF power. We considered all these features would enhance the breakdown hazards, hence we gave up this option. Nevertheless, it is possible to improve the electric field symmetry in the case of the dual waveguide coupling with a special design of the cavity. It was proposed by J. Haimson [12] to use a racetrack shape for the coupling cell in order to damp the asymmetry between both transverse directions (see drawing in figure 5a). Results of the HF simulations with this design are showed in figure 5b. First, it appeared that the field asymmetry is below 1 % in the case of the usual cylindrical shape. But, between 0 and 6 mm, where most of the particles are accelerated, the numerical noise, due to the finite number of tetrahedron, is dominating and makes difficult to analyse the results.

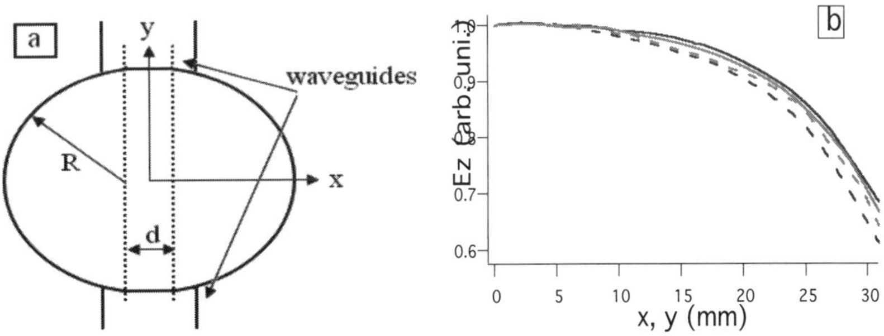

Figure 5: left, drawing of the section of the coupling cell; right, magnitude of the longitudinal electric field as a function of the distance in both transverse directions, red lines stand for the pure cylindrical cavity, green lines are for the racetrack cavity, plain lines represent the field along the y axis while the dash lines is along the x axis.

The goal is to reduce as much as possible the relative difference of the electric magnitude in the area of interest. With the racetrack shape of the cavity, if one looks far away from the axis at 25 mm, where the asymmetry is dominant with respect to the noise, the asymmetry is reduced by a factor 3. Then, it is very likely that the field asymmetry is reduced in the same proportions in the space volume where the electron beam is present. The best reduction of the asymmetry, shown in figure 5, is obtained with a very small value of the straight part of the racetrack, only 0.6 mm is necessary.

2.4 *Dynamics of the Electron Beam*

The simulations with PARMELA* presented in this section are based on the real longitudinal electric field calculated with SUPERFISH for the 2D model shown in figure 1. Unfortunately, it is not possible to extract the electric field of the HFSS model and to use it in PARMELA. Most of the simulations are done with 600 macro-particles for the study of the beam performances as a function of all parameters as the accelerating phase and gradient. Once the operation point is well defined, we will give beam performances with the maximum precision reachable with ten thousand particles. The results are often shown in the x direction because of the cylindrical symmetry and generally there are zero losses in the gun.

Several aspects of the beam dynamics have been investigated. First, calculations on the gun alone will be presented. Then, a study of the compensation of the beam emittance increase due to the space charge forces will be presented. The last simulations show the influence of the laser profile shape on the performances of the gun.

2.4.1 The natural behavior

The CERN design of the RF gun was optimised for higher charge, e.g. the choice of the angle of the wall around the photo-cathode. Therefore, we decided to study the influence of this parameter on the beam dynamics. The results are summarised in table 3.

Table 3: results of the simulations with PARMELA with 600 particles and the nominal current of CTF3 (3.51 A) as a function of the wall-angle around the photo-cathode. The widths are rms values and the rms emittance ε_T is normalised.

Wall angle	0°	3.4°	8.3°
σ_x (mm)	3.4	3.25	3.04
ε_T (π mm-mrad)	20.6	20.7	22.3
σ_z (mm)	1.11	1.15	1.2
rms $\Delta E/E$ (%)	0.6	0.75	2.6

* PARMELA from Los Alamos National Laboratory, modified by B. Mouton at LAL.

The case of an angle of 8.3° is incompatible with the gun specifications and the case of 3.4° seems less interesting with respect to a vertical wall. However, with a higher current, a small wall-angle is advantageous as it adds a focusing force which compensates the increase of the space charge forces.

So, as we want to maintain the possibility of operating at a higher current, we decided to adopt this small angle on the photo-cathode wall. Then we investigated the optimum operation point for the photo-injector as a function of the two HF parameters: the phase, ϕ, and the magnitude of the electric field, E. As far as the phase is concerned, it was found that at 55° we have the maximum energy gain and the minimum emittance. At 35° the energy spread is minimized. We chose ϕ = 35° because it is more important to improve the energy spread than the little increase of energy gain at 55° and the slight degradation of the emittance can be largely compensated as we will see in next sections. As for the choice of the electric field magnitude, simulations showed that the emittance could be reduced by 10 % at 120 MV/m with respect to the nominal gradient, 85 MV/m. In addition the bunch length and the beam radius are smaller and the beam energy 40 % bigger. However, because of the beam-loading issue, 120 MV/m would require a power of 50 MW which is beyond the limit of the available HF generator and breakdowns hazards would be enhanced. So, the operating gradient will be 85 MV/m.

Results of the simulations for the gun model shown in Fig. 1 are illustrated in Fig. 6. The beam parameters at the output of the gun are summarised in table 4.

Table 4 : Beam parameters for a beam charge of 2.3 nC and a Gaussian laser spot σ_r = 1.4 mm (rms), σ_t = 4 ps and 1000 particles.

E (MeV)	5.452
ε_x (πmmmrad)	19.6
σ_x (mm)	3.2
σ_z (mm, ps)	1.07, 3.56
rms ΔE/E (%)	0.36

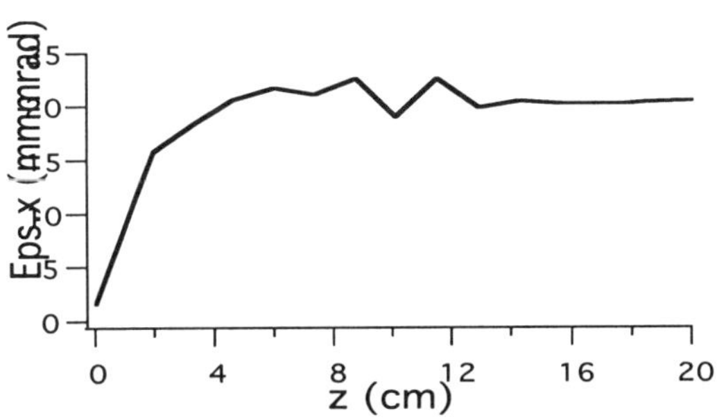

Figure 6: horizontal normalized emittance as a function of the longitudinal axis in the gun.

The comparison of Table 4 and Table 1 shows the performances of the photo-injector fulfil the specifications of the drive beam of CTF3. The electron beam dynamic is dominated by space charge effect. In Fig. 6, most of the emittance growth occurs in the first half-cell when the beam energy is below 1 MeV. Thanks to the magnetic focusing, it is possible somehow to reduce the emittance degradation.

2.4.2 Compensation of the space charge forces

Due to the space charge forces, the emittance grows linearly with the distance until the beam enters the accelerating section. At the output, the emittance is "frozen" because the space charge force is strongly damped at high energy. Therefore, to keep the emittance at the lowest possible value, it is necessary to use a transverse focusing between the output of the gun and the input of the section to compensate the transverse defocusing effect of the space charge. One possible technique, proposed by E. Carlsten [13], is the use of a magnetic lens. Simulations have been performed with a SLAC type section at the nominal current and 1.4 mm of laser spot size and with two coils for which the positions are variable (see figure 7). In all cases, thanks to a bucking coil, the magnetic field is less than one gauss on the photo-cathode. Besides, the fact that we used a SLAC section in the simulations has no influence on the conclusions which can be drawn from this study. For each couple of coil positions, we found the optimum value of the magnetic field for a good compensation of the emittance. Figure 7 shows that the biggest reduction of the emittance is obtained when the coils are placed near the photo-cathode. The only drawback is that it implies a stronger magnetic field, 0.25 T for the case where the coils are at z = 2 cm and 10 cm.

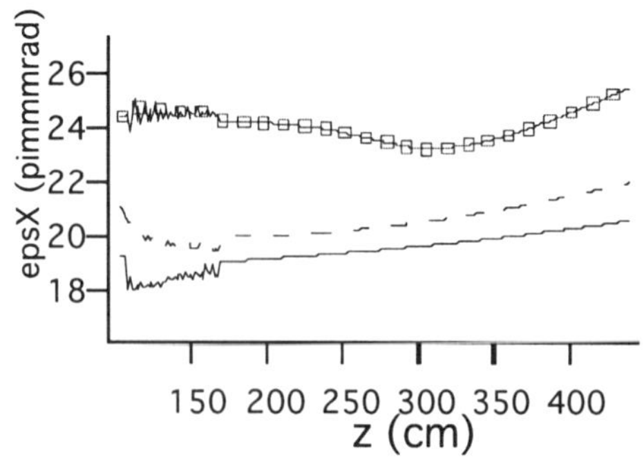

Figure 7: Emittance as a function of distance for several kinds of magnetic field. The SLAC section is located at z = 107 cm and has a length of 63 cm and a gradient of 30 MV/m. The position of the coils is variable; square points, coil at z = 15 cm and B_z = 0.21 T; dashed line, 2 coils at z = 10 cm and 20 cm and B_z = 0.23 T; plain line, 2 coils at z = 2 cm and 10 cm and B_z = 0.25 T.

Space constraints around the gun are very stringent because of a heavy pumping system which will be detailed later. So, it was decided to install the bucking coil close to the photo-cathode plane and the focusing coil just downstream the waveguide coupled to the last cell. On this base, simulations of the coils have been performed with POISSON, the magnetic arrangement is showed in figure 8a and the corresponding beam dynamic in figure 8b.

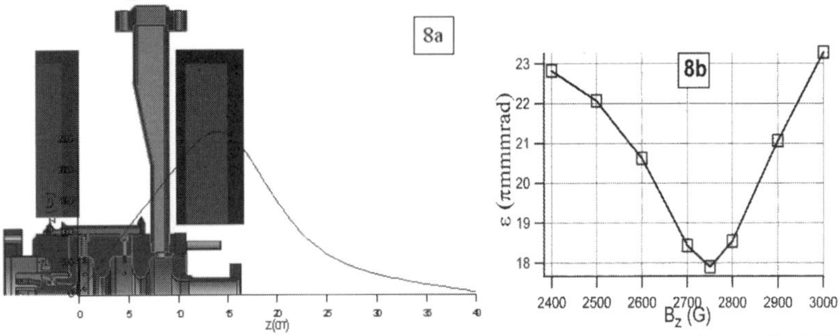

Figure 8: a, half section of the coils and magnetic field as a function of the longitudinal distance, grey blocks stands for iron and red is for copper; b, normalized emittance as a function of the peak magnetic field, PARMELA simulations with a Gaussian laser.

The dimensions of the coils which produce the required field lead to have 2 pancakes with 100 spires for the bucking coil and 4 pancakes with 208 spires for the focusing solenoid (for a square conductor of 6mm side). The coils need a current of roughly 250 A from the power supply. Unfortunately, it was impossible to get the maximum of the magnetic field closer than the output of the gun. Nevertheless as it is illustrated in figure 8b, the compensation of the emittance growth due to space charge forces is quite effective and is obtained for $B_z = 0.275$ T. The minimum rms normalized emittance is 18 πmmmrad at a distance of roughly one meter from the photo-cathode and can be kept at this value if a traveling wave section is located at this coordinate.

2.4.3 Influence of the laser profile shape

The presence of the coils allowed one to cancel the emittance growth which would have occurred otherwise in the drift tube between the output of the gun and the entrance of the first section. To reduce more significantly the emittance, it is proposed to decrease the space charge forces thanks to a square profile of the laser pulse instead of a gaussian [14]. Indeed, the space charge effect is proportional to the electronic density and a more homogenous profile as a square profile should be less harmful to the emittance. In the PARMELA at our disposal, the only one available profile is the Gaussian. But it is still possible to simulate a square pulse thanks to a very large rms width and cutting the profile well inside the width. The simulation will not represent exactly the laser pulse because it is not possible to choose a rise time and a fall time. The principle of the simulation is illustrated in figure 9.

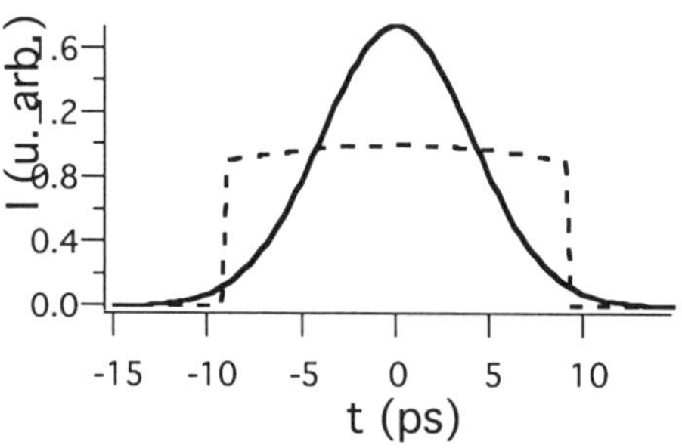

Figure 9: PARMELA simulations with different profiles of the laser temporal pulse; plain line, Gaussian shape with σ_t = 4.2 ps; dashed line, Gaussian shape, with σ_t = 20 ps over the interval [-9, 9] ps, which is equivalent to a square laser profile.

During the simulations, it appeared that a very flat profile is not the best suited for the emittance. It induces beam losses in the gun. So, we looked for the optimum temporal and transverse widths to get the minimum emittance which is 11 πmmmrad, almost a factor 2 lower with respect to the case with a true Gaussian shape of the laser pulse. Then, we again investigated the minimization of the emittance with the coils with these square profiles and we found 9.6 πmmmrad. The performances of the photo-injector are now optimised as a function of all parameters, a simulation with the maximum number of particles possible in our home PARMELA, 10000 particles, which lasted 10 hours, gave the final results summarised in table 5.

E (MeV)	5.452
$\varepsilon_{x,y}$ (πmmmrad)	5.6
$\sigma_{x,y}$ (mm)	1
σ_z (mm, ps)	1.43, 4.76
rms ΔE/E (%)	1.3

Table 5 : Beam parameters for a beam charge of 2.3 nC at the position of the minimum emittance (z = 150 cm) with the coils compensating the space charge effect, a square laser spot and 10000 particles.

Obviously, the energy of the beam did not change but the emittance, with a square 3D distribution of the laser pulse taking into account the coils and ten thousands of particles, is almost reduced by a factor 4 with respect to the initial results. This good performance is obtained to the cost of a small increase of the energy spread and bunch length because of the wider square distribution, but it is quite acceptable in regard to the specifications.

2.5 *Vacuum Calculations*

Vacuum considerations are a critical aspect of the project as the experiments performed at CERN showed it occurred an exponential growth of the pressure inside the gun as a function of the total extracted charge, whatever the bunch-train distribution is. Therefore, a thorough study of the residual pressure has been carried out to understand how we can attain a good vacuum in the RF photo-gun. The most relevant results are presented here but the overall analysis is given elsewhere [15]. A simplified model of the gun was used in the MONTE-CARLO based calculations. A pumping system is placed at 230 mm downstream of the gun and the out-gassing rate for the OFHC copper is assumed to be 5.10^{-11} $mbar.l.s^{-1}.cm^{-2}$ (N_2 equivalent) for a standard cleaning and after roughly one hundred hours of pumping. Simulations showed that the pressure can not be lower than 2.10^{-9} mbar in the first half-cell because of the poor conductance of the photo-injector. In order to improve the quality of vacuum, it was decided to use NEG (Non Evaporative Getter) coating on the vacuum chamber at the output of the gun. In addition, we plan to install around the gun a cylindrical chamber which the inner surface is coated with NEG. It is connected to the vacuum of the gun thanks to 42 ducts drilled in the walls of the half-cell and the middle cell of the gun. A scheme of the design is showed in figure 10.

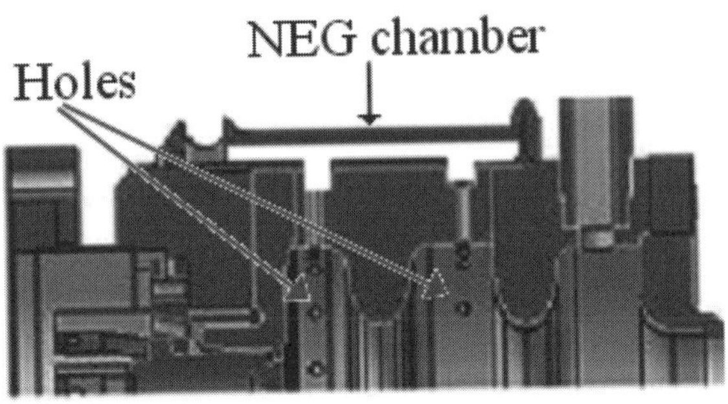

Figure 10: scheme of the NEG vacuum chamber around the photo-injector with the holes into the walls of the gun connecting both volumes.

Obviously a pre-pumping system is needed for this NEG chamber. In order to not to perturb the magnetic field, the ionic pumps must be exported outside the external radius of the coils. As a consequence, all the mechanical arrangement is very tight. Results of the simulations with this NEG pumping are illustrated in figure 11.

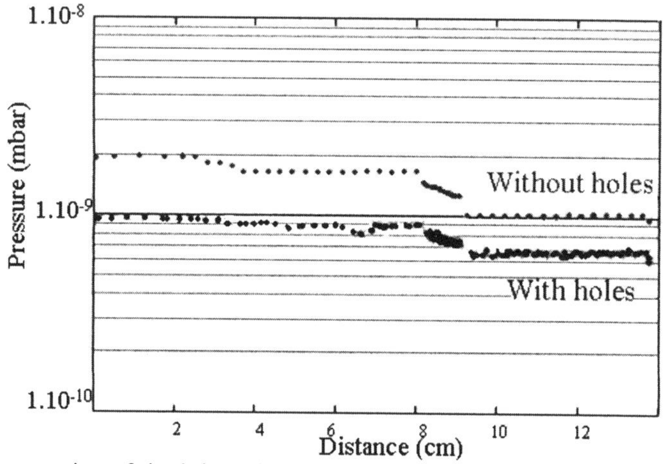

Figure 11: comparison of simulations of the residual pressure between the cases with and without NEG pumping.

The extra pumping coming from the NEG allows one to reduce the residual pressure by a factor 2, achieving 10^{-9} mbar which is not satisfying yet. It is difficult to do better by lack of available space in the cell walls which hold also the cooling channels. In addition to the improvement of the static pressure, the NEG pumping can be helpful during the commissioning of the gun. As the pumping velocity is higher, time recovery of normal pressure after a sudden out-gassing is significantly shorter (one minute less). But the only way to reach static pressure below 10^{-10} mbar is to lower the out-gassing rate of the copper. One solution is a high temperature bake (800 °C) of the photo-injector. In this case, calculations predict a residual pressure of 10^{-11} mbar in the 1st cell. Simulations with HFSS showed the impact of the presence of the 42 holes is negligible on the HF parameters (5 % less on Q).

3. The photo-injector of the probe beam

The goal of the CTF3 accelerator is to test the technical challenges which arise in the CLIC project. Among them, the 30 GHz accelerating sections must do the proof a beam can be propagated with a very high gradient, e.g. 150 MV/m, and with low losses. So, one linac in CTF3, the so called "probe beam linac", must provide such a beam. One agreement was found between CERN and two French institutes, LAL/IN2p3 from Orsay and DAPNIA/CEA-Saclay to build this linac. CERN provides advices and material (as the LIL accelerating sections), LAL is in charge of the construction of a photo-injector and DAPNIA should build the rest of the linac. In order to save time and money it was decided to take the PHIN gun as a start reference for the design of the probe beam photo-injector. Technical specifications required by CERN for the photo-injector are given in table 6.

Table 6: technical specifications for the probe beam photo-injector.

RF frequency (GHZ)	2.99855
Beam energy (MeV)	5 – 6
Charge/bunch (nC)	0.5
Bunch length, FWHM (ps)	<10
Rms Energy spread (%)	< 2
Normalized emittance (πmmmrad)	< 20
Number of bunches	Variable 1-64
Bunch spacing (ps)	333.3
Vacuum pressure (mbar)	$<2.10^{-10}$

The biggest differences with respect to the requirements for the drive beam photo-injector are the charge per bunch, which is 4 times lower, and the number of bunches which can be several order of magnitude smaller. As a consequence, the tolerances on the design of photo-injector can be considerably relaxed. First, the charge per bunch and the duration of the bunch train (21 ns at maximum) are so small that the beamloading can be neglected. So there is no need to have overcoupling between the waveguides and the gun. The coupling apertures will be surely smaller therefore the dissymmetry of the electric field between the two transverse directions will be also reduced. It was decided not to use a racetrack shape of the coupling cell making the design simpler and cheaper for the machining. However this issue will be carefully checked. We also reduced the aperture of the irises down to 15 mm instead of 20 m because the electron beam radius should be significantly smaller with respect of that in the PHIN photo-injector. Finally the pumping system can be probably simplified. In the previous photo-injector a cylindrical NEG chamber is installed around the gun leading to a very complex mechanical assembling, since it was vital to improve the quality of vacuum (see 1.5). But in the case of the probe beam the total extracted charge, even in the worst case of 64 bunches, is two orders of magnitude lower than in the PHIN photo-injector. Therefore, it is probably safe to give up the idea to provide a NEG chamber in the probe beam photo-injector. We keep the option of a NEG coating on the downstream vacuum chamber of the gun because it requires no additional machining. HF design studies begun later than in the PHIN gun and are still under way. Preliminary results are given in table 7 for a design which is very similar to that of the drive beam photo-injector except the modifications above discussed.

Table 7: results of the 2D HF simulations with SUPERFISH and beam dynamic simulations with PARMELA of the probe beam photo-injector.

Q	14400
R_s (MΩ)	7.8
E_{acc} (MV/m)	80
HF Power (MW)	5.2
Beam energy (MeV)	5.35
Rms Energy spread (%)	0.3
Normalized emittance (πmmmrad)	6.5
Bunch length (ps), FWHM	8
Ream radius (mm), rms	2.3

It is worth to notice that the required HF power is reduced by a factor 2 with respect of that in the PHIN photo-injector for a gradient which is slightly lower. So, we expect to have fewer troubles during the commissioning of the gun and the normal operation. According to the simulations all the electron beam parameters fulfil the specifications. The obtained emittance is for a Gaussian laser pulse, we could probably get a lower value with a square shape of the laser (see 1.4.3), close to the thermal emittance limit. Finally, it is foreseen to install coils on the photo-injector for the compensation of the emittance growth induced by the space charge forces.

4. Conclusion

The HF design and the electron beam dynamic of the drive beam photo-injector have been thoroughly investigated. The beam-loading induced by the beam is quite manageable and performances of the electron beam fulfill the CTF3 requirements. Since there is a tremendous total extracted charge which has to be produced, lot of efforts, as the envelop around the gun with NEG coating, have been carried out to improve the quality of the vacuum. Detailed technical drawings are finished and a prototype is being tested at LAL. Once HF measurements will have been confronted to the simulations the gun will be ordered to the manufacturer. So we hope the full equipped photo-injector will be installed and commissioned in the middle of 20006.

The photo-injector for the probe beam linac is less challenging. The HF design and the technical drawings are almost finished as we took benefit from the work achieved on the previous gun. Predictions of the electron beam performances are in agreement with the CTF3 requirements. The construction should take place during 2006 and the delivery to CERN one year after.

Acknowledgments

The fabrication of two photo-injectors fully equipped does not only rely on the shoulders of the author but are the result of a team work. So, I would like to thank the drive beam photo-injector project leader, G. Bienvenu; B. Mouton for his considerable help in the simulations and computers; the engineers in Vacuum, B. Mercier and C. Prevost; the mechanical designers, J. Prevost and M. Desmons and a newcomer but enthusiastic in HF simulations, J. Brossard. I also thank T. Garvey who gave me the opportunity to present this work and A. Variola for his suggestions of the paper.

References

[1] H. H. Braun *et al*, CERN yellow report, CERN-99-06 (1999).
[2] R. Corsini *et al*, proc. of the 9th Eur. Part. Acc. Conf. EPAC 2004, Lucerne, Switzerland, 05-09 Jul. 2004, p. 39.
[3] G. Bienvenu *et al*, proc. of the 8th Eur. Part. Acc. Conf. EPAC 2002, Paris, France, 03-07 Jun. 2002, p. 1753.
[4] H. H. Braun *et al*, proc. of the Part. Acc. Conf., PAC 2001, Chicago, U. S. A., June 18-22 2001, p. 39.
[5] C. Travier *et al*, Nucl. Inst. and Methods, Vol A 393, pp451-454, 1997.
[6] T. Garvey *et al*, proc. of the 8th Eur. Part. Acc. Conf. EPAC2002, Paris, pp 254 – 256, 2002.
[7] H. H. Braun, 8th CLIC/CTF3 meeting, Sept.2003, CERN, http://ctf3.home.cern.ch/ctf3/New_collab_meet.htm
[8] R. Bossart, H. Braun, M. Dehler, J.-C. Godot, Nucl. Inst. and Methods, Vol A 375, ABS7-ABS8, 1996.
[9] J. Luiten *et al*, proc. of the 9th Eur. Part. Acc. Conf. EPAC2004, Lucerne, Switzerland, p. 725, 2004.
[10] D. Boussard, Cern Accelerator Courses, 5th Advanced Physics Course, Rhodes, Greece 1993, p. 415 (1995).
[11] B. Dwersteg, K. Flötmann, J. Sekutowicz, Ch. Stolzenburg, Nucl. Instr. Meth. in Phys. Research A 393 (1997), 93.
[12] J. Haimson, B. Mecklenburg and E. L. Wright, proceedings of the Advanced Accelerator Concepts conference 1998, AIP conference proceedings 398, pp. 898-911.
[13] B. E. Carlsten, Nucl. Instr. Meth. in Phys. Research A 285 (1989), 313.
[14] F. Zhou *et al*, proc. of the 8th Eur. Part. Acc. Conf. EPAC2002, Paris, p. 275, 2002.
[15] B. Mercier, C. Prevost, R. Roux, "Presentation of the vacuum simulations on the photo-injector for the CLIC Test Facility", to be submitted to Vacuum.

SUPERCONDUCTING RF PHOTOINJECTORS; AN OVERVIEW

J. SEKUTOWICZ
Deutsches Elektronen-Synchrotron, Notkestrasse 85, 22603 Hamburg, Germany

R&D programs based on various approaches, at BNL, FZ-Rossendorf, Beijing University and AES generate optimism that low emittance continuous wave operating SRF-guns and injectors will be available in the near future. This contribution covers present status of these programs and it presents recently proposed alternative approach for the SRF-gun design with thin lead spot as the photo-cathode for generation of electrons in a bulk niobium cavity operating at 2K.

1. Introduction

Remarkable improvement in RF performance of superconducting cavities[1, 2] over the past decade has made feasible continuous wave (cw) and near-cw operations of superconducting electron linacs at high accelerating gradients. Several future projects assume these modes of operation, for example superconducting accelerators proposed as drivers for FEL facilities[3, 4] and the electron cooling facility proposed for Relativistic Heavy Ion Collider at Brookhaven National Laboratory[5]. Both operation modes require injectors operating at cw or near-cw and providing low emittance electron beams. An example of such injector (the so called split injector) is discussed in Ref. 6. The most demanding component of a cw injector is a cw operating RF-gun delivering highly populated (~1 nC) low emittance bunches. RF-guns, both working at room temperature and superconducting, in order to generate highly populated low emittance bunches have to be operated at high accelerating gradients to suppress emittance diluting space charge effects. Normal conducting RF-guns face difficulties in meeting this requirement in the cw or near-cw mode. Their copper walls dissipate many kilowatts of power to fulfill high gradient condition even when they operate at low pulse repetition rate (low duty factor). It is a technical challenge to increase significantly the duty factor of normal conducting RF-guns while having sufficient cooling and keeping them thermally stable during operation.

Superconducting RF-guns (SRF-guns) dissipate orders of magnitude less power than the normal conducting devices. They can be easily operated at high duty factor but the challenge here is the integration of a non super-conducting photo-cathode material with a super-conducting cavity in a way preserving its original high intrinsic quality factor Q_o (small cryogenic losses). One possible solution to this problem, which will be shown later (based on choke filter) has been investigated at Forschungszentrum Rossendorf (FZR)[7]. Other approach, very attractive and technically feasible for miliampere-class SRF-guns, is to use a superconducting metal as the photo-cathode. In this case difficulty arises from low quantum efficiency (QE) of the super-conducting materials. Low QE must be compensated with shorter wavelength and higher energy pulses of the illuminating laser. This has been studied first for niobium cathode at BNL and will be reported in next sections. A complementary approach, with lead[8], is discussed in the last section of this overview.

2. Projects

2.1. *A. Overview of Designs*

2.1.1. *Rossendorf 1.3 GHz SRF- gun*

The FZR R&D program, which began in 1998, was the first worldwide effort following the original work at University of Wuppertal by A. Michalke[9]. Figure 1 shows Rossendorf SRF-gun cavity and its assembly in a cryostat[10]. The gun consists of 3.5-cell 1.3 GHz bulk niobium cavity, a mechanism for exchanging and precise positioning of the Cs_2Te cathode and a choke filter preventing leakage of the RF. The specs for the gun are listed in Table 1. The gun in its final version will operate in three various modes: the mode for the BESSY FEL facility, the FZR laser mode and the FZR nuclear physics mode. A simplified half-cell prototype of the gun (see Fig. 2) was built for performance tests of sc cavity with:

Figure 1. FZR superconducting RF-gun and its assembly in the cryostat. *Courtesy of FZR.*

Table 1. Spec and measured parameters of the FZR gun

Parameter	BESSY	FZR-L	FZR-NP	Measured
E [MeV]	5	9.5	9.5	0.85
ΔE [keV]	-	5	-	8
q [nC]	2.5	0.077	1	0.022
Bunches/s [10^6]	0.025	13	1	26
I_b [mA]	0.063	1	1	0.52
ε [μrad]	1.5	1	1.5	1

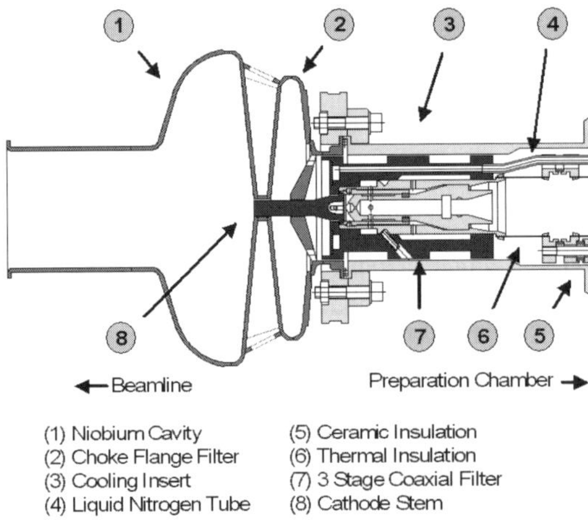

Figure 2. Tested half-cell prototype of the FZR gun. *Courtesy of FZR.*

the choke filter, the cathode exchanger and the Cs_2Te cathode exposed to high e-m fields in a clean super-conducting cavity. Two RF-tests have been conducted at 4.2 K and 2 K. In the 4.2 K test the intrinsic quality factor of 2.5 10^8 was measured when the RF electric field on the cathode E_{cath} reached 22 MV/m. In the superfluid test at 2 K, Q_o was 5 10^9 when E_{cath} was 46 MV/m.

The prototype was also used in the experiment to generate the first electron beam by that type of RF-gun and to test the cathode performance. The last column of Table 1 displays measured data of the beam generated with the prototype. The specifications for: the cathode operation temperature T_{cath}, its QE at wavelength λ_{Ph}, spot size Ø, electric field on the cathode E_{cath} and all these data measured in the experiments are listed in Table 2. The beam experiment showed that the life time of the cathodes in a clean superconducting

environment is about 50 days. Table 2 contains also specs and parameters for the illuminating laser: the single pulse energy E_{pulse} and the average laser power P_{laser}.
In 2005 two 3.5-cell cavities were built. The first one was made of low 40 RRR niobium. This cavity was built to test applicability of the surface preparation procedures at DESY and to verify coupling to the input line and the HOM damping scheme. The second prototype was made of high 250 RRR niobium and will be used to verify a whole concept of the gun.

2.1.2. *RF-gun at Institute of Heavy Ion Physics*

The SRF injector R&D program at IHIP of Peking University began in 2001. The RF-gun at IHIP consists

Table 2. Specs and measured data for the Cs_2Te cathode and illuminating laser of the FZR RF-gun.

	T_{cath} [K]	QE [%]	λ_{Ph} [nm]	E_{pulse} [μJ]	P_{laser} [W]	Ø [mm]	E_{cath} [MV/m]
BESSY	78	1	262	1.19	0.03	3.0	25
FZR	78	1	262	0.5	0.5	3.0	25
Measured	78	0.3	260	0.06	1.5	2.0	22

of DC Pierce gun and 1.3 GHz 1.5-cell niobium cavity. The cavity and the RF-gun assembly in a cryostat are shown in Figure 3 along with its cross-section in Figure 4. The Cs_2Te cathode in this design is exposed to a DC electric field rather than to RF field as it is done in the FZR gun. The emitted electrons are first accelerated in the DC field before they enter the accelerating superconducting cavity via 8 mm tube. The tube is very deep in cutoff frequency for 1.3 GHz and thus prevents leakage of the RF power towards the DC section.

Figure 3. IHIP superconducting RF-gun cavity and its assembly with Pierce DC gun in the cryostat. *Courtesy of IHIP PU.*

The gun was used in 2003 for the first time to generate electron beam. The specs and experimentally measured data[11] are listed in Table 3. Table 4 shows specs and test result of the Cs_2Te cathodes. The electric field E_{cath}, resulting from the 70 kV voltage applied to the Pierce gun, was 2.5 MV/m. Observed cathode life times were more than 100 hours before they began to lose their QE.

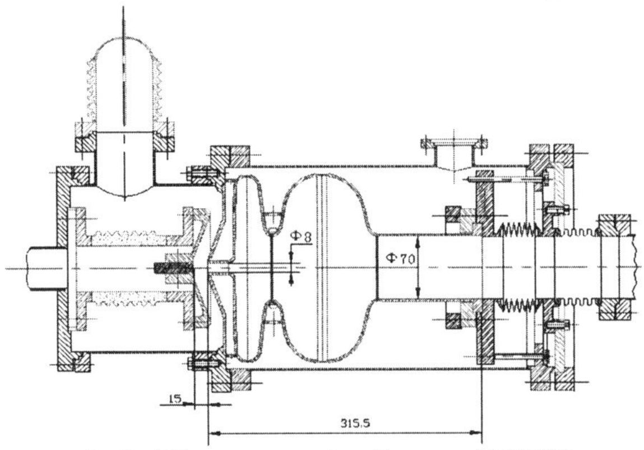

Figure 4. IHIP superconducting RF-gun cross-section. *Courtesy of IHIP PU.*

Table 3. Specs and measured data of the IHIP RF-gun

Parameter	Specs	Measured
E [MeV]	2.61	0.58
ΔE [keV]	30	35
q [nC]	0.06	0.001
Bunches/s [10^6]	17	81
I_b [mA]	1.0	0.08
ε [μrad]	3.3	2.7

Table 4. Specs and measured data for the Cs_2Te cathode and laser of the IHIP RF-gun.

	T_{cath} [K]	QE [%]	λ_{Ph} [nm]	E_{pulse} [μJ]	P_{laser} [W]	Ø [mm]	E_{cath} [MV/m]
Spec	273	1	266	0.015	1.2	5.6	2.7
Measured	273	1	266	0.010	0.8	6.0	2.7

The 1.5-cell cavity was RF-tested in 4.2 K helium bath only. In the test the cavity reached accelerating gradient E_{acc} of 5 MV/m with the intrinsic quality f actor of 10^8.

As a continuation of the R&D program, IHIP group has proposed to replace 1.5-cell cavity with 3.5-cell and to make technical modifications to the DC part enabling 100 kV operation. The proposed modified gun and its parameters are shown in Fig. 5 and Table 5, respectively.

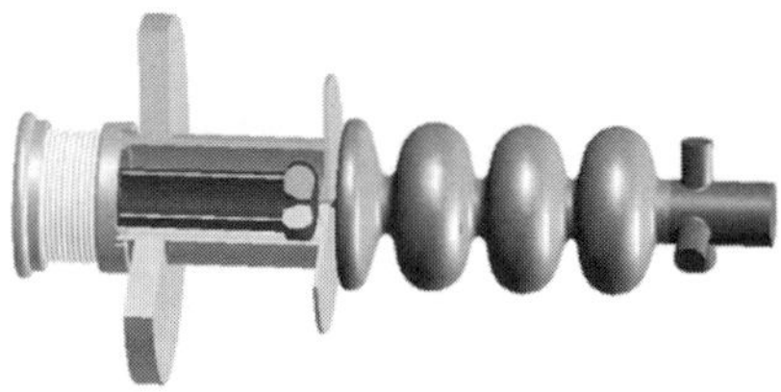

Figure 5. IHIP 3.5-cell RF-gun. *Courtesy of IHIP.*

Table 5. Parameters for the 3.5-cell IHIP RF-gun

E_{acc}	[MV/m]	15
V-DC	[kV]	100
I_b	[mA]	1
E	[MeV]	4.9
ΔE	[keV]	110
ε	[μrad]	3.4

2.1.3. *1.3 GHz RF-gun at Brookhaven National Laboratory*

In 2002 BNL proposed an approach alternative to the ones presented above. The RF-gun concept was based on the photoemission directly from the superconducting niobium. The electrons were emitted from the back wall of 1.3 GHz bulk niobium half-cell cavities (Fig. 6) which were illuminated in the center with a laser light. Two cavities were built by Advance Energy Systems (AES) to investigate QE for niobium and its enhancement caused by various surface cleaning methods[12]. The laser cleaning procedure at room temperature improved significantly QE of niobium samples. The measurement data at 300K presented in Ref. 13 are partially listed in Table 6 showing QE at three wavelengths of the illuminating laser. Such good results have not been observed at 2 K and 4 K. The highest QE for cold niobium was 10^{-5}. An explanation for such low QE could be the contamination of the niobium surface with residual gases. Even extensive laser cleaning did not improved QE significantly.

Both cavities have been RF-tested at 2 K at JLab. The best measured performance was $Q_o = 3\ 10^9$ at 46 MV/m of the peak electric field at the emitting spot location[14].

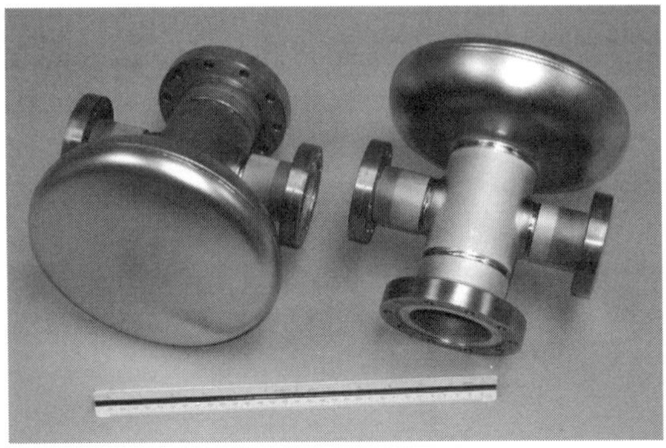

Figure 6. BNL half-cell RF-guns. *Courtesy of BNL.*

Table 6. QE of niobium cathode at 300 K

Energy Density	QE $[10^{-4}]$		
$[mJ/mm^2]$	193 nm	248 nm	266 nm
0 (no cleaning)	0.7	0.020	0.005
0.76	3.5	0.149	0.075
1.45	4.0	0.215	0.120
2.52	3.8	0.178	0.107

Recently, one of the cavities was mechanically re-worked. The quarter-wave choke filter was attached to make it useful for QE investigation (Fig. 7). The measurement program for various cathode materials deposited on a plug inserted in the choke with and without the diamond multiplier will begin this year.

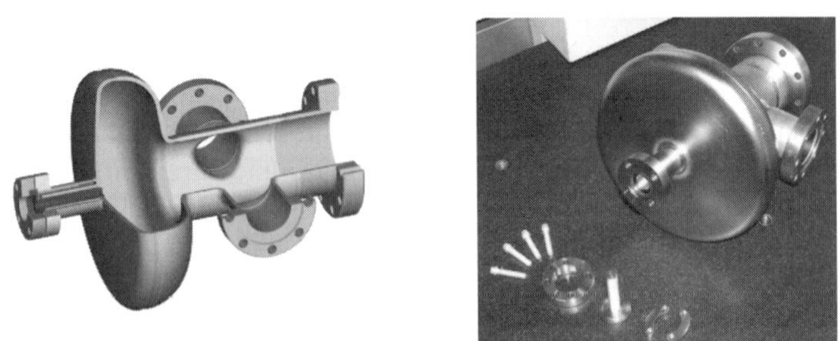

Figure 7. 1.3 GHz half-cell cavity with choke filter for QE investigations. *Courtesy of BNL.*

2.1.4. *703.75 MHz RF-gun; Collaboration between BNL and AES*

Since 2004 collaboration between BNL and AES develops superconducting RF-gun in the several hundreds miliampere class. Two guns of this kind will be needed to generate electron beams for the electron coolers for two rings of RHIC. Other potential future applications for the high current RF-gun are electron injectors for linacs driving the MW-class FELs. The version for the RHIC cooler (see Fig. 8) consists of the half-cell cavity operating at 703.75 MHz, two input couplers, quarter wave choke filter and the mechanism for the cathode assembly. Tables 7 and 8 show specs for the beam parameters and the cathode performance respectively. The cathode performance was specified with and without diamond multiplier. The studies on cavity shape optimization continue[15]. The recently presented results show that with a proper cathode positioning into the opening in the center of slightly conical back wall emittance can be as low as 2 μrad (much better than the specified value). The prototype cavity will be built at AES in 2006.

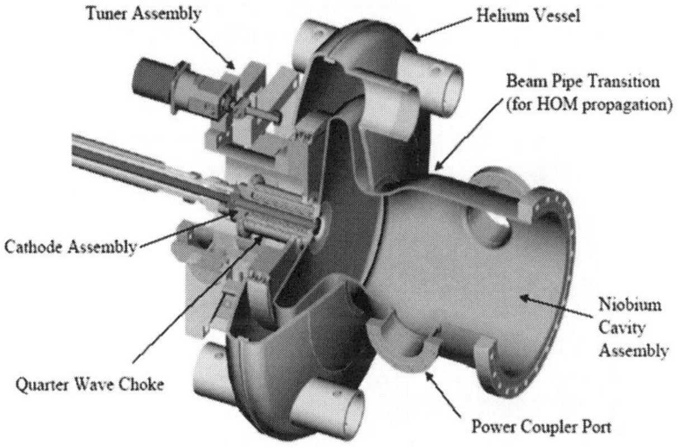

Figure 8. 703.75 MHz superconducting RF-gun. *Courtesy of AES.*

Table 7. Specs 703.75 MHz RF-gun.

Parameter		Without Diamond	With Diamond
E	[MeV]	2	2
ΔE	[keV]	62	62
q	[nC]	1.33	1.33
Bunches/s	[10^6]	351.8	703.75
I_b	[A]	0.5	1.0
ε	[μrad]	5	5

Table 8. Spec for an alkali cathode of 703.75 MHZ gun.

	QE [%]	λ_{Ph} [nm]	E_{pulse} [μJ]	P_{laser} [W]	Ø [mm]	E_{cath} [MV/m]
No diamond	5	527	0.071	25	2.0	40
With diamond	500	527	0.0006	0.2	2.0	40

3. Emittance growth compensation

3.1. *Compensation of the ε growth with E field*

The electric field pattern in the vicinity of the cathode is very sensitive to the positioning and misplacing of the plug in the opening of the back wall. The computed example for Ø 8 mm plug in the Ø 12 mm opening is shown in Fig. 9. Two cathode positions differ by 3 mm. At the same stored energy in the cavity, accelerating component E_z in the front of the cathode is 3 times stronger for the cathode with 1 mm recess. The radial component E_r is order of magnitude stronger for the 1 mm recess and it has opposite sign to E_r for 4 mm recess. The electric field components vary fast vs. plug position. It suggests that positioning of the emitting spot, plug shape and its location, its parallelism to axis must be

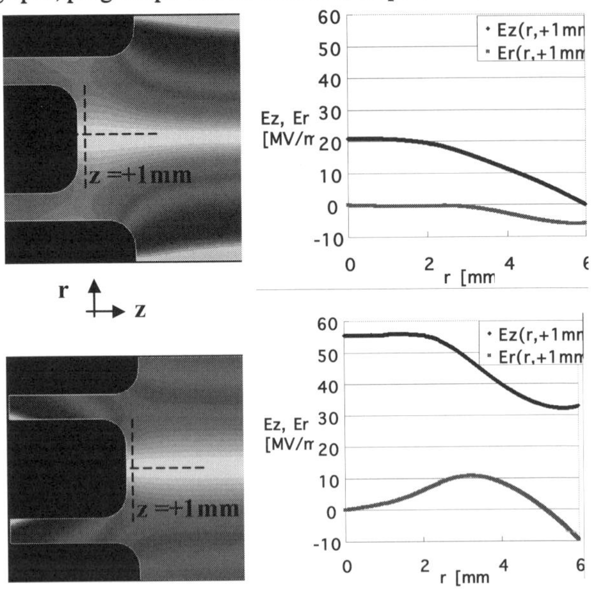

Figure 9. Electric field pattern for two cathodes recessed 4 mm in the back wall (upper diagrams) and 1 mm (lower diagrams). Plots show electric field components E_z and E_r as function of the radial position at the plane 1 mm in the front of the cathodes.

done very precisely to compensate for the emittance growth with E_r in the region where the beam has still very low energy. FZR and BNL groups based emittance growth compensation on recessed cathodes. In addition both groups proposed conical shape of thc back wall; 1° FZR[16] and 7° BNL[15] to enhance E_r. Table 9 shows minimal emittance and the optimal recess for both RF-guns.

Table 9. ε for recessed cathode and conical back wall

	BNL 703.75 MHz gun	FZR 3.5-cell 1.3 GHz gun
ε [μrad]	1.99	1.49
Recess [mm]	3	2

3.2. *Compensation of the ε growth with B field*

3.2.1. *Static B Field*

Solenoidal B field is commonly and very effectively used for compensation of the emittance growth in normal conducting RF-guns. In contrary, in the superconducting guns a static B field cannot act on the beam near the cathode. There are two problems associated with B field in the case of SRF-guns. First, B field does not penetrate cavity interior through its superconducting wall (due to the Meissner effect). Secondly, the presence of magnetic flux causes degradation in the cavity performance or if B is bigger than the critical field for Nb (~180 mT) it leads to a quench. In the split injector, mentioned before, a spatial separation of a sc cavity and a solenoid solves both problems. Fig. 10 shows the SRF-gun region and the solenoid position in that concept. The solenoid,

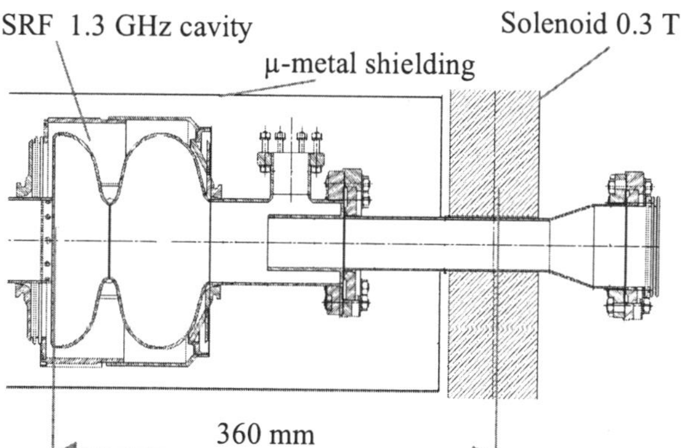

Figure 10. SRF-gun region and the solenoid position in the split injector.

positioned downstream after the gun, acts on the beam after it is accelerated in the superconducting cavity. Shielding around the cavity reduces B well below its residual natural value of 50 μT.

3.2.2. *Solenoidal Resonant Mode*

A new concept of the emittance growth compensation with solenoidal resonant mode has been proposed at FZR[17]. In a superconducting RF-gun cavity the solenoidal monopole (TE011) is excited additionally to the accelerating mode. The magnetic flux of the solenoidal mode acts on beam in the same way as the static magnetic flux of the solenoid in the split injector. One should note that in both concepts strength of B field (~300 mT) and its distance to the cathode (~360 mm) are very similar. Table 10 shows simulation results for 3.5-cell FZR gun cavity and 1 nC bunch. As it is reported in Ref. 17 normalized transverse emittance below 1 μrad could be obtained for a rather broad range of phase difference between the accelerating and the solenoidal mode.

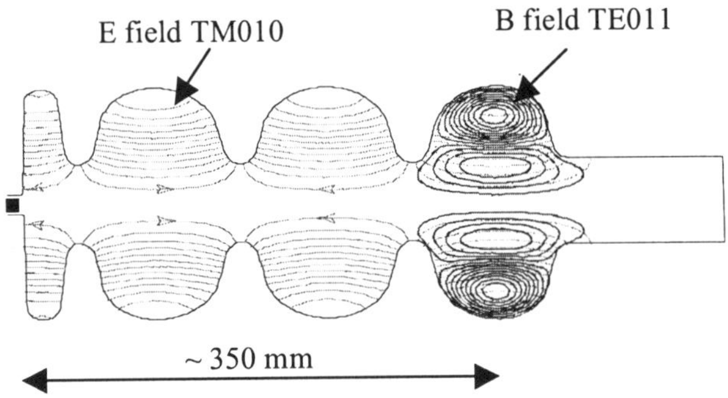

Figure 11. Electric field of the accelerating TM010 mode (red lines) and magnetic flux of the solenoidal TE011 mode in 3.5-cell FZR cavity. *Courtesy of FZR.*

Table 10. Emittance compensation with TM011 mode.

TE011 Frequency	[GHz]	3.8
ε	[μrad]	0.98
B_{TE011} on axis of TE011 mode	[mT]	324
Max B on wall, $[B^2_{TE011} + B^2_{TM010}]^2$	[mT]	144

4. Lead-Niobium RF-gun

Low quantum efficiency of niobium was the motivation for testing alternative superconductors as photo-emitters to continue the BNL concept of all superconducting RF-gun. In this section we will report progress on mutual efforts of DESY, BNL, Stony Brook University, JLab, Institute for Nuclear Studies (INS) in Poland and SLAC to build a hybrid Pb-Nb SRF-gun. Lead, which critical temperature 7.2 K is not very different from niobium, is a commonly used superconductor in accelerator technology. It demonstrates much higher QE than niobium. Tests results of lead's QE at room temperature were reported by the collaboration in Ref. 13. The recent summary of QE measurements vs. photon energy E_p for niobium, bulk lead and lead samples deposited with various techniques is shown in Fig. 11. The highest QE of 0.55% was measured for the arc-deposited sample illuminated with 6.5 eV photons. For 5.8 eV photons (fifth harmonics of 1064 nm infrared laser) the QE of that sample was still 0.25%. The QE of the electroplated and magnetron deposited samples at this photon energy is 0.17%. Emitting 1 nC bunch at this QE will require 3.4 μJ energy per pulse only. For cw operation of XFEL, as it was proposed in Ref. 18, the average laser power of 3.4 W at 214 nm will be required to generate 1 nC bunches at 1 MHz repetition rate. More details on lead QE measurements are presented in Ref. 19.

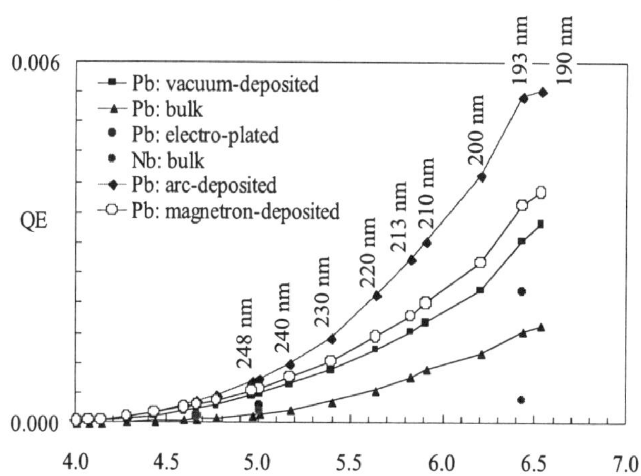

Figure 11. Measured QE of lead deposited with various coating methods. Bulk Pb and Nb data are displayed for comparison.

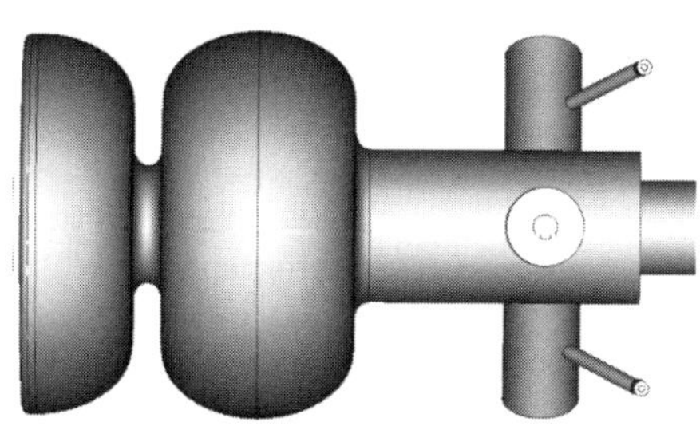

Figure 12. 1.5-cell SRF-gun cavity with two HOM couplers and port for the input coupler.

In this hybrid Pb-Nb gun small emitting spot of lead (Ø<3 mm) will be located in the center of back wall of the half-cell of 1.5-cell cavity (Fig.12) which will be made of high purity niobium. The cavity equipped with two HOM couplers, input coupler and pickup probe will be assembled in a cryostat as shown in Figure 10. The gun is designed to be implemented in the split injector, so the solenoidal static B field will be used for the emittance growth compensation Two half-cell cavities have been built to measure lead QE at 2 K and to test performance of superconducting cavities illuminated by laser. Both cavities are shown in Fig.13. The left one was built at JLab. This cavity has an opening in the center of the back wall which is vacuum tight closed with a niobium plug which can be coated with an emitting material. This cavity is very helpful to test various coating methods and various superconductors as photo-emitters. It can be an alternative solution to the second cavity, built at DESY (shown in Fig. 13 right), in which technically difficult coating is done directly on the back wall. An additional difficulty in this version is that the emitting spot must withstand cleaning procedures. Two of them could degrade the QE: chemical treatment and high pressure water rinsing. Unfortunately, both procedures are essential for good performance of the superconducting cavity which in that concept should operate at 60 MV/m peak field. Two features make this version very attractive. Smooth back wall does not enhance locally electric field in the vicinity of the cathode and there is no electric contact in high field region which may reduce

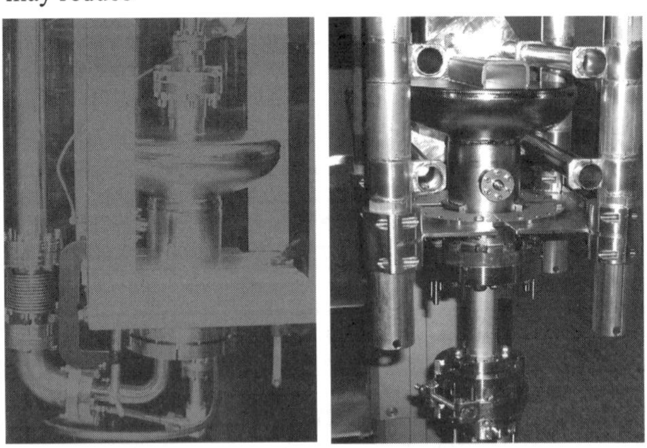

Figure 13. Test half-cell cavities built at JLab (left) and at DESY (right).

intrinsic Q of the cavity. Both cavities were very preliminarily tested last year but did not demonstrate the expected performance. The JLab cavity reached the peak field of 20 MV/m without coating and 30 MV/m with lead coated plug, but intrinsic Q was 8 10^8 only due to insufficient cooling of the plug. The DESY cavity without coating reached the peak field of 60 MV/m with Qo of 4 10^9. Cavity performance with coating was very poor. The cavity quenched at low field due to low purity of the coating. Improved versions of both cavities will be tested this year.

Acknowledgments

The author wants to express gratitude to T. Rao, J. Smedley and R. Calaga (BNL), P. Kneisel and A. Bogacz (JLab), J. Hao and K. Zhao (IHIP), D. Janssen and J. Teichert (FZR), J. Langner and P. Strzyzewski (INS) for all information and data used in this overview.

References

1. H. Weise, "Superconducting RF Structures- Test Facilities and Results", PAC03, Portland 2003, USA.
2. P. Kneisel et al., "First Cryogenic Tests with JLab New Upgrade Cavities", SRF2003, Lübeck 2003, Germany.
3. A. Büchner et al., "The ELBE-Project at Dresden-Rossendorf", Proceedings EPAC2000, 26-30 June 2000, Vienna, Austria.
4. J. A. Clarke et al., "Prospects for a 4th Generation Light Source for the UK", Proceedings PAC2001, 18-21 June 2001, Chicago, USA.
5, I. Ben-Zvi, "Electron Cooling for RHIC", Proceedings PAC2001, 18-21 June 2001, Chicago, USA.
6. M. Ferrario, "Low emittance photo-injectors", NIM A472, (2001), 303-308.
7. D. Janssen, P. vom Stein, "The Development of a Superconducting RF Gun: Status of the Drossel Collaboration", Proceedings PAC97, 12-16 May 1997, Vancouver, Canada.
8. J. Sekutowicz et al., "Nb-Pb Superconducting RF-Gun", TESLA-FEL Report 2005-09, DESY, Hamburg, Germany.
9. A. Michalke et al., "First Operation of the High Quantum Efficiency Photocathodes Inside Super-conducting Cavities", in Proceedings of the EPAC92, March 24-28, 1992, pp. 1014-1016, Berlin, Germany.
10. D. Janssen et al., "First Operation of a Superconducting RF-gun", NIM A507 (2003) 314-317.
11. R. Xiang et al., "Experimental Investigations Of DC-SC Photoinjector at Peking University", NIM A528 (2004) 321–325.
12. T. Srinivasan-Rao et al., "Design, Construction and Status of All Niobium Superconducting Photo-injector at BNL", PAC03, Portland 2003, USA.
13. J. Smedley et al., "Photoemission Properties of Lead", Proceedings EPAC2004, 5-9 July 2004, Lucerne, Switzerland.
14. P. Kneisel, private communication.
15. R. Calaga et al., "High Current Superconducting Gun at 703.75 MHz", Proceedings SRF2005, 10-15 July 2005, Ithaca, USA.
16. D. Janssen, V. Volkov, "RF Focusing–an Instrument for Beam Quality Improvement in Superconducting RF Guns", NIM A452 (2000), 34-43.

17. D. Janssen et al., "Status of the 3 1/2 Cell Rossendorf Superconducting RF Gun", Proceedings FEL2004, August 29th - September 3rd 2004, Trieste, Italy.
18. J. Sekutowicz et al., "Proposed continuous wave energy recovery operation of an x-ray free electron laser", PRST-AB, Volume 8, 2005.
19. J. Smedley et al., "Superconducting Photocathodes", this Proceedings.

STATUS AND PERSPECTIVES OF PHOTO INJECTOR DEVELOPMENTS FOR HIGH BRIGHTNESS BEAMS

FRANK STEPHAN
DESY, Platanenallee 6, 15738 Zeuthen, Germany
E-mail: frank.stephan@desy.de

The development and reliable operation of high brightness electron sources are key issues for next generation short wavelength light sources and a future linear collider. This paper will concentrate on photo emission based injector developments for various projects which require individual solutions and face specific problems and challenges. A selective overview of developments and experimental results will be presented demonstrating that for a large parameter range dedicated photo injectors are able to fulfill the specific stringent requirements on beam quality and other beam parameters, e.g. the time structure of the beam.

1. Introduction

There is no unique way to build electron sources for high brightness beam applications. On the contrary, for each application, e.g. Self-Amplified Spontaneous Emission – Free Electron Lasers (SASE-FELs) or Energy Recovery Linacs (ERLs), the technical realization of the electron source depends on a set of key parameters, which in most cases are defined by the users of the application and the technology choice for the subsequent linac (e.g. normal conducting or super conducting). These key parameters include

- operation mode: pulsed or continuous wave (CW) operation
- single bunch charge
- time structure of the beam: single pulses or pulse trains, single bunch repetition rate, length of the pulse trains and pulse train repetition rate
- normalized transverse emittance required for the application
- whether further bunch compression is needed after the source, what are the required longitudinal phase space parameters at the application ?

These parameters do not only define the injector part but require a sophisticated design from the source to the application of the electron beam, e.g. the undulator of an FEL. This paper concentrates on source developments. However, the parameters of the longitudinal phase space might be changed significantly by elements further downstream. For example, the bunch length, energy spread and longitudinal cmittancc will change when bunch compressors and higher harmonic RF accelerating systems for linearizing the longitudinal phase space are applied at a higher energy level, as will be shown later. Therefore, brightness of the electron beam will not be defined in this paper but the beam quality will be characterized by the measured or designed normalized transverse emittance at a certain beam energy level for a given charge.

In the next three sub-sections, the three basic photo electron sources will be briefly defined and their basic advantages and disadvantages will be discussed. Then a generic injector layout will be described, as it is currently used for most of the new photo injector proposals. After this introduction 14 different photo injector projects are presented, their parameter sets are listed and their challenges are described briefly. The parameter range for bunch charges is between 1 pC and 10 nC, the time structure ranges from 10 Hz single bunches to 1.3 GHz CW operation, and the measured or designed emittances range from 0.1 mm mrad to 11 mm mrad. The average current ‹I› for the projects ranges over approximately nine orders of magnitude, from nano-Amperes almost up to Ampere currents. The different projects are grouped into three categories, high, medium and low average current for $\langle I \rangle \geq 1$mA, 1mA $> \langle I \rangle >$ 1μA, and $\langle I \rangle \leq 1$ μA, respectively.

The selected photo injector projects are by no means exhaustive and simply represent a personal selection of interesting projects on the basis of which specific problems and solutions relating to photo injector developments can be discussed.

Throughout this paper emittance is always discussed as normalized transverse emittance if not explicitly stated otherwise.

1.1. *DC Photo Electron Guns*

In DC photo electron guns the photo cathode is placed at a static negative potential of several 100 kV. Because of the specific geometry of cathode and anode and their separation, the accelerating field at the cathode is of the order of a few MV/m. Two important advantages of DC photo electron guns are a) good vacuum conditions with pressures down to 10^{-12} mbar and b) the rich operating experience in many different applications. Because of the good vacuum environment, negative electron affinity (NEA) cathodes like e.g. GaAs can be used in DC photo electron guns. If illuminated with the proper laser wavelength, this material has been shown to result in a low kinetic energy of the photo emitted electrons at the cathode providing a low thermal emittance for a given laser spot size. Early measurements of the transverse thermal energy for GaAs cathodes reported numbers down to 34 meV [1] while recent measurements at 90 K showed a most probable transverse kinetic energy of the emitted electrons of about 10 meV [2]. A clear disadvantage of DC photo electron guns is the relatively low accelerating gradient at the cathode. In order to reduce the space charge force and overcome the space charge limit for charge extraction from the cathode, the emitting laser pulse is either large in diameter or long in time. Since an increased transverse laser spot size results in an increased thermal emittance, in general, the laser pulse length is increased for extracting significant bunch charges. To accommodate later RF acceleration the long electron bunches are usually compressed with lower frequency buncher cavities, which may result in a deterioration of the transverse emittance. New developments in this respect will be mentioned in Sections 2.1 and 2.4.3. After the buncher cavity the electron beam still has low momentum and therefore requires further booster acceleration with proper matching of the beam.

1.2. *Normal Conducting RF Guns*

In this kind of electron source the photo cathode is located at the backplane of a normal conducting (NC) RF cavity. The first cell has a reduced length (half cell) so that a high accelerating surface field is obtained at the cathode. The cavity is surrounded by a main solenoid for space charge compensation [3, 4, 5]. In some cases a bucking solenoid in the rear of the cathode plane is needed to zero the magnetic field of the main solenoid. The high accelerating gradient at the cathode, the subsequent accelerating cells and the solenoid arrangement for the space charge compensation allow operation of this electron source with a high bunch charge as well as higher beam energy from the source compared to DC guns. A wealth of experience has been gathered for operating NC RF guns at many different locations and this type of electron source has experimentally proven to provide very good emittances (see below). Unfortunately, only medium vacuum conditions of the order of 10^{-9} mbar have been realized during operation at high average RF power up to now. A second disadvantage is that the water cooling of the cavity limits the average RF power used for accelerating the bunch charge. This RF power can either be used for obtaining high accelerating gradient, for operating at high duty-cycle or for a reasonable compromise between both options. Nevertheless, with proper choice of resonant frequency, cavity and photo cathode material as well as sophisticated cooling channel design a broad range of average beam currents with high beam quality can be produced as will be demonstrated later.

1.3. *Super Conducting RF Guns*

This photo injector type is schematically very similar to the version previously discussed. The major difference is that the accelerating cavity is super conducting (SC). The low RF losses imply that the cavity can be operated with a high RF duty cycle or in most cases even in CW mode so that a high average beam power can be achieved and the beam energy from the source can be significantly higher than for the DC photo cathode gun case. The vacuum condition in the SC gun is expected to be reasonably good so that NEA cathodes such as GaAs which offer low thermal emittance could be considered an option for normal conducting cathodes in an SC gun. This approach has however not been put into practice yet. On the other hand, the SC cavity also implies that the space charge compensating solenoid can not be placed around the cavity itself. This often results in a split injector design, where the solenoid is placed downstream of the cold-warm transition from the SC gun cavity. To improve the focusing within the cavity, electrical RF focusing is often applied in the cathode region, e.g. [6]. Nevertheless, most simulations up to now have shown that this concept limits the charge per bunch if very good beam quality is required. A new approach to applying a magnetically focusing RF mode which can overcome this limitation will be presented in Section 2.3. For SC RF guns the photo cathode is a more complicated issue than in NC RF guns. Should the cathode be normal conducting or super conducting? If it is NC, how can the NC cathode be isolated from the SC cavity and how can a high accelerating field be maintained at the cathode? If it is SC, what cathode material with which quantum efficiency is used and what average laser power can be deposited on the cathode before it switches to normal conductivity? Some SC RF gun projects with their specific solutions will be discussed later in this paper. A more detailed discussion of SC RF gun developments is presented elsewhere [7, 8]. Compared to the other two basic photo injector types the SC RF guns are a new development so that the operating experience for this kind of electron source is very limited.

1.4. *Generic Photo Injector Layout*

In order to construct a high brightness electron beam injector the following general design considerations are applied, e.g. [9, 10]: The individual laser pulses illuminating the photo cathode are appropriately shaped in transverse space as well as temporally. To counteract the space charge induced emittance growth a solenoid field is applied in most cases, either around the gun cavity itself or later downstream. To exit the space charge dominated regime of a low energy electron beam (few MeV) a second accelerating section is applied to boost the beam to a few 10 or even a few 100 MeV. In this case special care has to be taken to place the booster in the appropriate position and apply the appropriate accelerating gradient to the low energy beam in order to conserve and even further improve the beam quality originating from the source [5].

Simulation studies, see e.g. [10], have shown that fairly long laser pulses should be extracted from the photo cathode if a low normalized emittance is required. The long electron pulses acquire some RF curvature during the acceleration in the booster which increases the longitudinal emittance and can produce a spiking current distribution with strong CSR effects during compression of the beam in a bunch compressor. In order to prevent this, a higher harmonic cavity can be applied to linearize the longitudinal phase space upstream of the bunch compressor section which then results in a more flat-top current distribution. Within the injector part different diagnostics components are essential to measure the drive laser and electron beam properties. After the injector, the electron beam will in many cases obtain further acceleration and will pass through further bunch compression stages before it arrives at its application, e.g. a SASE-FEL.
The optimization of the many injector parameters, e.g. cathode laser pulse transverse and longitudinal profile, RF gradient and emission phase in the gun, solenoid strength and location, booster position, gradient, and phase, higher harmonic acceleration section and bunch compressor parameters, is a many dimensional task with many correlations. As one example for such a multivariate injector optimization, an 'upgraded VUV-FEL injector' case is studied using ASTRA simulations [11]. The basic difference to the actual VUV-FEL setup [12] is the increase of the accelerating gradient at the cathode from about 40 MV/m to 60 MV/m. A schematic of the VUV-FEL injector layout is shown in Figure 1. For the numerical optimization [13] the gun, solenoid, booster, and third harmonic sections are kept at their original location. The longitudinal photo cathode laser profile is assumed to have a rise and fall time of 2 ps and a flat-top length of 20 ps. An accelerating gradient of 60 MV/m and an initial kinetic energy of the photo emitted electrons of 0.55 eV are assumed at the Cs_2Te cathode. The simulations were carried out with a charge of 1 nC and the second half of the cryomodule (cavities 5-8) is assumed to run at high peak accelerating field of 40 MV/m. The multivariate optimization of the parameters rms laser spot size on the cathode, peak accelerating field in the first half of the cryomodule (cavities 1-4), phases of the gun and the super conducting cavities, and strength of the gun solenoid yield 0.53 mm, 25.8 MV/m, -2.0 degrees, -15.3 degrees, and 225.3 mT, respectively. The phases are given with reference to the maximum accelerating phases. The resulting evolution of the beam parameters as a function of the distance to the photo cathode is shown in Figures 2 and 3.

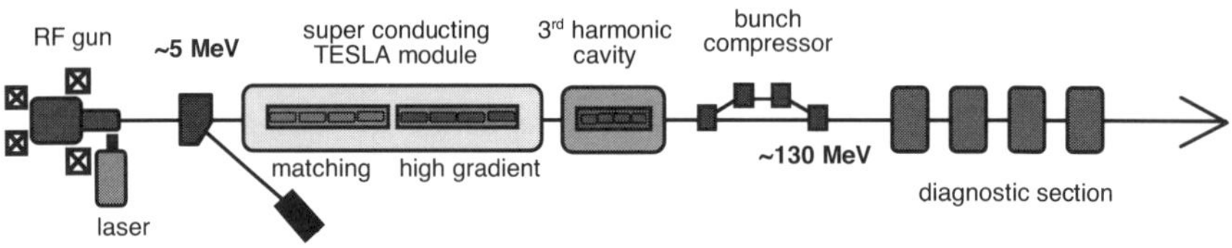

Figure 1. Schematic layout of the VUV-FEL injector, see text.

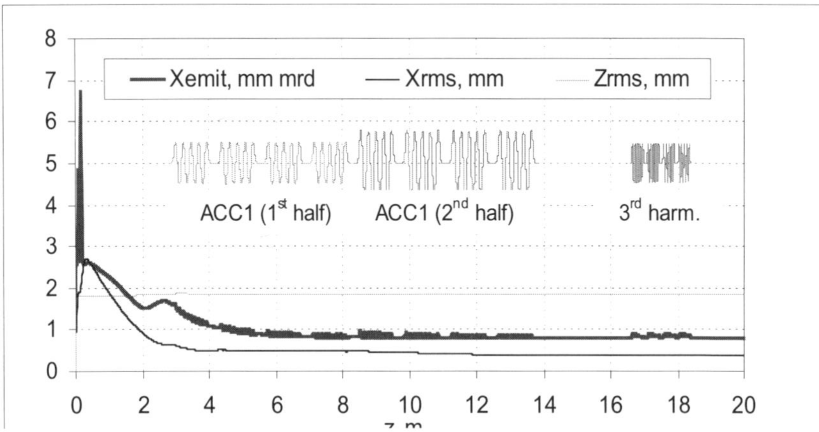

Figure 2. Normalized projected transverse emittance (Xemit), rms beam size (Xrms) and rms bunch length (Zrms) as a function of the distance from the photo cathode for the 'upgraded VUV-FEL injector' case, see text.

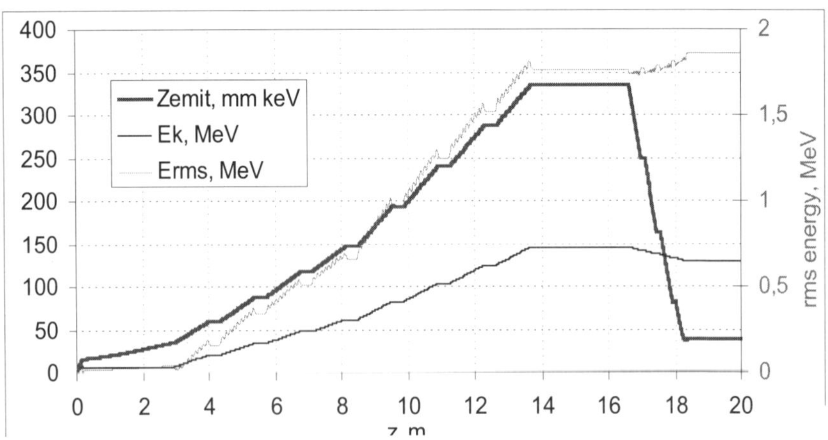

Figure 3. Longitudinal emittance (Zemit), mean beam energy (Ek), and rms energy spread (Erms, right axis) as a function of the distance from the photo cathode for the 'upgraded VUV-FEL injector' case, see text.

Figure 2 illustrates the usual projected emittance oscillation in the gun as well as the proper matching of the booster and the incoming electron beam. After the third harmonic section a normalized projected emittance of 0.81 mm mrad including thermal emittance is predicted. In Figure 3 the strong reduction of the longitudinal emittance by using the third harmonic cavity is visible.

2. High Average Current Electron Sources

In this section several photo injector developments providing average beam currents of at least 1 mA are listed. Three experimentally measured and eight design goal parameter sets are summarized in Table 1. The projects are described briefly in the following subsections.

2.1. *DC Photo Cathode Gun Development at Cornell*

The DC photo electron source with the maximum average current achieved for DC guns so far is running at JLab. Some of their operation parameters are listed in Table 1 a) [14].

For the operation of an energy recovery linac based X-ray source at Cornell, a DC photo cathode injector design was published in 2003 [15]. The injector schematic is shown in Figure 4 and the design parameters are listed in Table 1 b). Compared to the operation parameters at JLab the charge is reduced by about 40 % but the estimated transverse emittance is improved by about one order of magnitude. In addition, every bucket of the 1.3 GHz RF accelerating section will be filled so that the average current increases to 100 mA.

Figure 4. Schematic of the photo electron injector for the Cornell ERL-based X-ray source.

In 2005, an improved design of the Cornell injector was published [16], improving the transverse emittance of the above mentioned operation mode by about one order of magnitude and also allowing for a high charge operation mode of 1nC with a transverse emittance of about 1 mm mrad at 13 MeV beam energy, see Table 1 b) entry marked with "new" and Table 1 c). This impressive improvement in the design was obtained by a multivariate optimization of almost all injector parameters. Twenty-two decision variables had been defined, incorporating all the physical constraints of the problem, e.g.

Table 1: Photo injector developments for a high average current of at least 1 mA. All emittance numbers quoted in this table are projected normalized transverse emittances. For details on the projects refer to the text.

	a)	b)	c)	d)
Location	Jlab, USA	Cornell, USA		Boeing, USA
gun type	DC gun	DC Gun		4 cell NC RF Gun
Data or simulation	exp. results	design goals		exp. results
operation mode		low charge	new: high charge	-
pulsed / CW	CW	CW	CW	-
single bunch charge	122 pC	77 pC	1 nC	1 – 7 nC
single bunch rep rate	75 MHz	1300 MHz	1 – 10 MHz	27 MHz
length of bunch train	-	-	-	8.3 ms
bunch train rep rate	-	-	-	30 Hz
DC voltage / gap	350 kV / 10.57 cm	~ 600 kV / 5 cm	~ 800 kV / 5 cm	-
average current	9.1 mA	100 mA	1 mA	6.7 – 47 mA
Norm. trans. emittance (rms)	~ 8 – 10 mm mrad @ 10 MeV	old:< 1 mm mrad, new: 0.1 mm mrad @ 13 MeV	~ 1 mm mrad @ 13 MeV	5 – 10 mm mrad @ 5 MeV
RF frequency	-	-	-	433 MHz

	e)	f)	g)
location	FZ Rossendorf, Germany		
gun type	½ cell SC RF gun	3.4 cell SC RF gun	
data or simulation	exp. results	design goals	
operation mode	-	ELBE	high charge
pulsed / CW	CW	CW	CW
single bunch charge	1-20 pC	77 pC	1 nC
single bunch rep rate	26 MHz	13 MHz	1 MHz
average current	≤ 130 μA	1 mA	1 mA
norm. trans. emittance (rms)	2.5 mm mrad @ 4 pC, 900 keV	1.5 mm mrad @ 9.5 MeV	2.5 mm mrad @ 9.5 MeV
RF frequency	1.3 GHz	1.3 GHz	1.3 GHz

	h)		i)	j)
location	BNL, USA		JLab, USA	LANL, USA
gun type	½ cell SC RF gun		DC gun + SC RF booster	2 ½ cell NC RF gun
data or simulation	design goals		design goals	design goals
operation mode	high current	high charge	-	-
pulsed / CW	CW	CW	CW	CW
single bunch charge	1.42 nC	10 nC	0.133 nC	1.0 nC
single bunch rep rate	351.87 MHz	10 MHz	748.5 MHz	100 MHz
average current	500 mA	100 mA	100 mA	100 mA
norm. trans. emittance (rms)	2.4 mm mrad @ 2 MeV	11.1 mm mrad @ 3.2 MeV	1.2 mm mrad @ 7.7 MeV	4 mm mrad @ 2 MeV
RF frequency	703.75 MHz		booster: 748.5 MHz	700 MHz

the field strengths and locations of the elements, the RF phases, the transverse and longitudinal laser profiles on the cathode, the bunch charge, and the thermal energy of the electrons leaving the photo cathode. The ASTRA code was used for tracking the electrons and the optimization was run for the two cases 80 pC and 800 pC to obtain maximum average brilliance and maximum photon flux per pulse respectively. For details of the optimization procedure and the simulation results see reference [16]. No single parameter is responsible for the impressive emittance improvement compared to the old design of 2003 but the multivariate optimization changed a large proportion of the design parameters significantly. This shows the importance and the benefit of a multivariate optimization resulting from the strong correlation in the high dimensional design parameter space. From the simulation results it is also interesting to note that the optimized transverse laser profiles for both charge cases are truncated elliptical distributions and that the temporal laser profile for the 800 pC case is a flat-top profile with short rise/fall time and in the case of 80 pC is a flat-top profile with short rise/fall time but with significant depletion in the center of the flat-top. These numerically optimized laser shapes are significantly different from the 3D ellipsoidal distributions currently being discussed [17, 18, 19].

The experimental schedule for the realization of the design foresees that the first photo electrons from the new DC gun will be produced in spring 2006, that the laser system for producing the electron bunches with 1.3 GHz repetition rate will not be available before mid-2006 and that the cryomodule to accelerate the beam to about 13 MeV should be available in 2008 [20].

Finally, it is interesting to mention that this electron source would be capable of producing polarized electrons with high beam quality if the appropriate photo cathode laser and cathodes are used.

2.2. *NC RF Gun Results from Boeing*

Although the measurements with a 433 MHz normal conducting RF gun at Boeing were performed as long ago as 1992 [21], they still represent the highest average current produced to date, see Table 1 d). The gun was operated with long pulse trains at a duty cycle of 25 % with an average RF power of 600 kW. In spite of the high duty cycle and the low RF frequency, the re-entrant design of the gun cavity allowed a fairly high peak accelerating field of about 25 MV/m at the cathode.

2.3. *SC RF Gun Results and Developments at FZ Rossendorf*

The first beam from a super conducting RF gun was demonstrated in 2002 at the Forschungszentrum Rossendorf, near Dresden [22]. Some operating parameters are listed in Table 1 e). A particularity of this RF gun is that the cathode is normal conducting and coated with Cs_2Te. This enables high quantum efficiency and relaxes the requirements on the laser system. The super conducting ½ cell cavity was operated over a period of 7 weeks and showed no degradation of its quality factor.

After this proof-of-principle experiment, the design of a 3.4 cell SC RF gun began [23]. It consists of a specially designed cathode cell and three modified TESLA shape cells. The design requirements for this cavity are listed in Table 1 f) and Table 1 g) for the operation mode of the ELBE FEL at Rossendorf and a high charge mode for test purposes respectively [24]. An ASTRA simulation of the 1 nC operation mode with a solenoid outside of the gun cryostat and a subsequent TESLA module with eight SC 9-cell cavities [23] predicts a transverse projected emittance of about 1.7 mm mrad, well below the design goal.

For the 3.4 cell cavity a new idea to improve the space charge compensation process within the SC gun cavity was developed, too [25, 26]. This idea utilizes an additional magnetically focusing RF mode which is localized in one of the cavity cells. Care was taken that the sum of the surface fields is below the quench limit. Simulations predict that for a charge of 1 nC a transverse emittance between 0.8 and 1.0 mm mrad can be obtained depending on the RF phase of the magnetically focusing RF mode [25].

Two 3.4 cell SC gun cavities have been produced and tuned. The surface treatment is ongoing at DESY. The first beam from this gun is expected towards the end of 2006.

2.4. *Results and Developments of various US Laboratories together with Advanced Energy Systems, Inc.*

Advanced Energy Systems, Inc. (AES) is very active in the field of electron source developments. In collaboration with different research laboratories they produce different kinds of electron sources. Here, four examples are presented.

2.4.1. *Full Niobium ½ Cell SC Gun*

In collaboration between BNL, JLab and AES two half-cell full niobium SC cavities at an RF frequency of 1.3 GHz were designed and put into operation. For this design a special cathode plug was omitted and the photo electrons were extracted from the Niobium backplane of the cavity. At 2 K a quality factor of up to 2.5×10^9 at a peak field of about 55 MV/m has been obtained. The quantum efficiency at 4 K was measured to be 2×10^{-6} at 266 nm laser wavelengths and 1×10^{-5} at 248 nm, both lower than anticipated from room temperature measurements [27]. One cavity has now been modified to allow for a demountable cathode and corresponding cathode tests as explained in the next sub-section.

2.4.2. *½ Cell SC Gun with NC Cathode*

In collaboration with BNL a half cell SC cavity at an RF frequency of 703.75 MHz is also under development [28, 29, 30]. It aims for a high average current and a reasonably small transverse emittance for ERL applications. Some goal parameters of the design are listed in Table 1 h). Two different operation modes for high current and high charge are listed. Because the electron beam goes through four merger dipoles right after the gun, the projected horizontal and vertical emittances at the entrance of the linac are somewhat different and the table gives the geometrical average of both planes. For this project, the cathode is normal conducting and utilizes a quarter wave choke joint concept to prevent RF leakage from the accelerating cell. This design allows testing of the new development of a diamond amplifier cathode that is ongoing at BNL and promises high quantum efficiency [31]. The cavity development is currently in its final design stage and the first beam from the 703.75 MHz gun is expected in 2007.

2.4.3. *DC Photo Cathode Gun plus SC RF Booster*

A DC photo cathode gun combined with a super conducting RF booster at an RF frequency of 748.5 MHz is under development in collaboration with JLab [29]. The emittance compensation solenoid is placed between both components and a separate buncher cavity is omitted. Some design parameters of this work are given in Table 1 i). The gun is operated to fill every RF bucket of the booster. A particularity of the booster is that it incorporates a third harmonic cavity to linearize the longitudinal phase space after the booster. The assembly of the cryomodule is in progress now and first tests are planned for 2006.

2.4.4. *2 ½ Cell NC RF Gun for CW Operation*

A collaboration led by LANL has developed a 2½ cell normal conducting RF gun. The design parameters for a 1 nC CW operation are listed in Table 1 j) [29, 32]. The special challenge of this electron source is the extremely high thermal load of about 720 kW in the NC RF gun needed to produce a beam with reasonable emittance in continuous wave mode operation. In order to cope with this requirement the cavity is not fabricated from standard copper but from

Glipcop® although this material is more difficult to braze. In addition, the larger surface compared to higher frequency guns helps to provide sufficient water cooling. To obtain a reasonable vacuum in the cavity allowing a multi-alkali cathode operation a third non-resonant cell of the cavity is used as a pumping manifold. The gun was produced in 2005 and high power tests will start in 2006 at LANL.

3. Medium Average Current Electron Sources

In Table 2 five different photo injector developments at three different locations are listed which provide average beam currents between 1 mA and 1 μA. The projects are described briefly in the following subsections.

Table 2. Photo injector developments for a medium average current between 1 μA and 1 mA. In the first column the measured projected emittance is cited while the other columns give the design slice emittances. For details on the projects refer to the text.

	a)	b)	c)	d)	e)
location	CEA, France	DESY (PITZ), Germany		BESSY, Germany	
gun type	1 cell NC RF gun	1.5 cell NC RF gun		1.5 cell NC RF gun	SC RF gun
data or simulation	exp. results	design goals		design goals	
operation mode	-	VUV-FEL	European XFEL	phase 1	phase 2
pulsed / CW	pulsed	pulsed	pulsed	pulsed	CW
single bunch charge	1 – 3 nC	1 nC	1 nC	2.5 nC	2.5 nC
single bunch rep rate	72 MHz	1 – 9 MHz	5 MHz	333 kHz	> 25 kHz
length of bunch train	≤ 140 μs	≤ 800 μs	≤ 650 μs	3 bunches	-
bunch train rep rate	1 – 10 Hz	1 – 10 Hz	10 Hz	1 kHz	-
average current	≤ 300 μA	≤ 72 μA	≤ 32.5 μA	7.5 μA	> 62.5 μA
norm. trans. emittance (rms)	~ 1–1.5 mm mrad @ 19 MeV	2 mm mrad @ 1 GeV	1.4 mm mrad @ 20 GeV	1.5 mm mrad @ 2.3 GeV	1.5 mm mrad @ 2.3 GeV
RF frequency	144 / 433 MHz	1.3 GHz	1.3 GHz	1.3 GHz	1.3 GHz

3.1. *NC RF Gun Results from ELSA*

The ELSA facility at CEA has demonstrated the production of very good emittances from a 144 MHz single cell normal conducting gun cavity [33]. Table 2 a) summarizes some of the parameters measured up to 3 nC, although measurements have been published up to approximately 15 nC. Downstream of the 144 MHz gun cavity the electron beam is accelerated in a 433 MHz linac yielding up to 19 MeV. The very low gun and linac RF frequencies allow the use of a 60 ps long drive laser pulse generating electron bunches down to about the same length if the charge is not too high (≤ 7 nC). While the long drive laser pulse has the advantage of reducing the charge density allowing the production of good emittances at reasonable charge, the long electron bunch needs compression before acceleration in a higher frequency linac (L-band or even S-band) with higher accelerating gradients. The projected emittance numbers were measured with a quad scan method at 19 MeV and linear space charge effects in the drift between quadrupole and screen have been taken into account (points marked with "corrected" in Figure 5). The measurements reported in [33] are given in Figure 5.

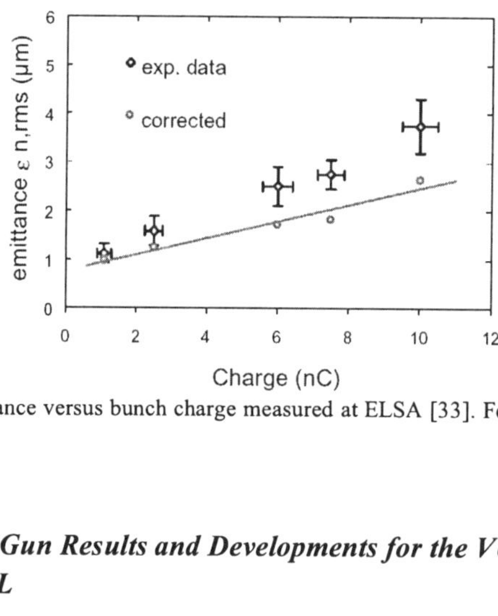

Figure 5. Projected emittance versus bunch charge measured at ELSA [33]. For more details see the text.

3.2. *PITZ: NC RF Gun Results and Developments for the VUV-FEL and European XFEL*

The electron source developments for the VUV-FEL at DESY and the European XFEL take place at the Photo Injector Test Facility at Zeuthen (PITZ). In Table 2 b) and Table 2 c) some of the design goals of both projects are listed. The specialties of the source developments for both free electron lasers are a) the generation of long electron bunch trains and b) the stringent requirements on the transverse emittance. In the table the required slice emittance at the entrance of the undulator system for both applications is given. In order to allow for some emittance deterioration from the injector to the undulator (e.g. in the bunch compressor sections) and to operate with a reasonable safety margin, the requirements at the injector are even more stringent. In the case of the XFEL a slice emittance of approximately 0.9 mm mrad is desired for 1 nC charge from the injector. At PITZ, the operation of the RF gun with RF pulses of 1 ms length, 10 Hz repetition rate and a maximum cathode gradient of approximately 42 MV/m corresponding to an average RF power of 30 kW was demonstrated in 2004 [34]. The available laser system is able to produce flat-top laser pulse trains of 800 μs length with a repetition rate of 10 Hz. The individual laser pulses in these trains can be shaped to flat-top pulses transversely and temporally and are repeated with a repetition rate of 1 MHz [35]. This laser system allowed the production of small projected emittances of about 1.3 mm mrad in one plane and a geometrical average of both planes of approximately 1.6 mm mrad for 1 nC charge [36]. While these measurements have been taken at an electron momentum of 5.2 MeV/c at PITZ, a copy of this gun was characterized at PITZ and then installed at Hamburg, now delivering a low emittance beam for the VUV-FEL. Measuring at a beam momentum of 127 MeV/c, projected emittances of 2.1 mm mrad for 1 nC charge are regularly obtained [37]. If 10 % of this bunch charge, which is located in the tails of the distributions and which does not in any case contribute to the SASE lasing, is neglected then a projected emittance of 1.4 mm mrad is regularly measured.

Since the required improvement of the electron source to reach the XFEL goal for the injector mentioned above is still significant, an upgrade and extension program was started at PITZ in 2004. This PITZ2 program aims at the preparation and commissioning of the necessary hardware to reach the XFEL requirements. This includes increasing the accelerating field on the photo cathode to 60 MV/m, improving the laser system to produce temporal flat-top pulses with rise/fall time of $\leq$ 2 ps, and confirming as well as experimentally optimizing the conservation of a small emittance beam from the electron source to higher beam energy. This requires an upgrade of the RF and laser systems as well as the design, construction and installation of a booster cavity with appropriate diagnostics beam line for higher beam energies. The PITZ2 development program is still ongoing but first results have been obtained. A preliminary version of the booster cavity and a diagnostics beam line has been put into operation, currently yielding a maximum electron beam momentum of 13.7 MeV/c. A new 10 MW klystron was used to condition the gun cavity to about 57 MV/m at the cathode for short RF pulses. Before photo electrons could be generated with this high gradient, the gun cavity re-developed a previous vacuum leak that was first sealed in 1998 and the operation had to be stopped. A new gun cavity (prototype number 3) was installed at PITZ in December 2005 and the final version of the normal conducting booster cavity allowing for maximum energy gain with limited RF power is planned to be installed towards the end of 2006. The laser system providing short rise/fall time of the individual pulses in the 800 μs long pulse train will be installed in spring 2007.

A common challenge to all photo injector projects is the generation of an electron beam starting with a low thermal emittance. At PITZ, a sequence of thermal emittance measurements for Cs_2Te cathodes was performed. Figure 6 shows the measured emittance for a bunch charge of 2-3 pC and a short Gaussian laser pulse with rms length of about 3 ps. Simulations confirmed that space charge and RF effects make only a small contribution to the measured emittance [38] so that the numbers in Figure 6 mainly represent the thermal emittance for an rms laser spot size at the photo cathode of 0.48 – 0.55 mm. Two main conclusions can be drawn from the measurements shown in Figure 6: Firstly, the thermal emittance increases with the accelerating gradient at the cathode at the moment of emission. This can be explained by the Schottky effect [38]. Secondly, the thermal emittance depends on the cathode properties and status. While cathode 43 was fairly fresh when the measurement was made, cathode 61 had already been in operation for several months. Apart from small differences between the photo cathodes themselves, the effect of poisoning of the active material, e.g. due to the interaction of Cs_2Te with the remaining gas in the cavity, could explain an increase of the electron affinity resulting in a reduction of the kinetic energy of the photo electrons at the cathode surface and therefore the lower thermal emittance [39, 40]. With the worldwide progress being made on the development of low emittance sources, the emittances measured are getting constantly closer to the thermal emittance limit so that more studies on different cathode types and the overall emission process are becoming increasingly important.

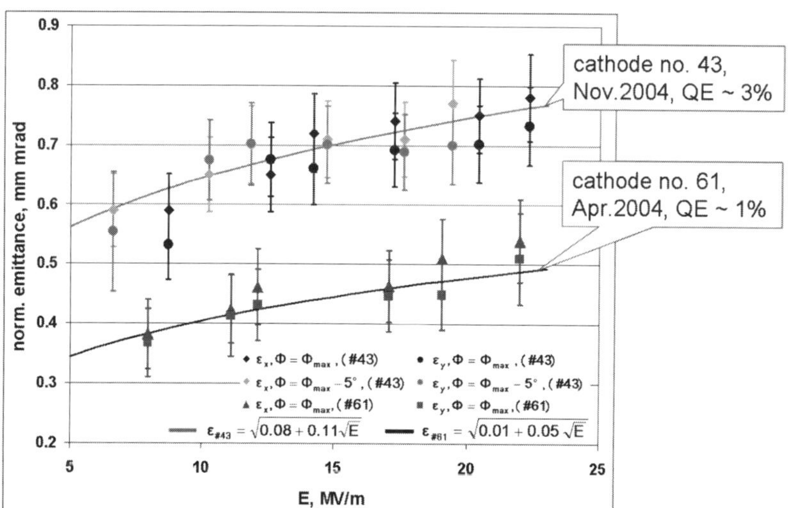

Figure 6. Measurements of the thermal emittance of Cs_2Te cathodes as a function of the accelerating field at the cathode at the moment of emission for two different photo cathodes. For more details see the text.

3.3. *NC and SC RF Gun Developments for BESSY-FEL*

For the injector of the BESSY-FEL two design options are considered. For the commissioning of the facility a normal conducting RF gun with short bunch trains at 1 kHz is foreseen. Later, it is planned that an SC RF gun will be operated in CW mode. Some design goals for both electron sources are given in Table 2 d) and e) [41]. The emittance figures quoted represent the central slice emittance at the entrance of the undulator system. Simulations show that these goals can be reached [42]. A prototype of the NC gun has been produced and it is foreseen that it will be tested with up to 60 kW average RF power at the PITZ facility in 2006. In the case of the SC gun, the physics design for a 1.625 cell cavity is ready. In addition, BESSY is strongly involved in the R&D on SC guns at Rossendorf where it is also foreseen that the BESSY-FEL bunch mode will be tested.

4. Low Average Current Electron Sources

In this section photo injector developments providing average beam currents of less than 1 μA are presented. Three examples have been selected, the measurements or design parameters of which are listed in Table 3. The projects are described briefly in the following subsections.

Table 3. Photo injector developments for a low average current below 1 μA. In columns a) and b) projected emittances are given while in column c) the design goals for slice emittance are listed. For details on the projects refer to the text.

	a)	b)	c)
location	SHI+FESTA, Japan	SPARC, Italy	LCLS, USA
gun type	1.6 cell NC RF gun	1.6 cell NC RF gun	1.6 cell NC RF gun
data or simulation	exp. results	design goals	design goals
pulsed / CW	pulsed	pulsed	pulsed
single bunch charge	0.25 – 1.2 nC	1.1 nC	1.0 nC / 0.2 nC
single bunch rep rate	10 Hz	1 – 10 Hz	120 Hz
average current	2.5 – 12 nA	1.1 - 11 nA	120 nA / 24 nA
norm. trans. emittance (rms)	~ 0.8 – 1.4 mm mrad @ 14 MeV	< 2 mm mrad @ 155 MeV	1.0 / 0.8 mm mrad @ 135 MeV
RF frequency	2856 MHz	2856 MHz	2856 MHz

4.1. *Emittance Results from the NC RF Gun at SHI + FESTA*

One of the best measured emittances for 1 nC charge was presented in 2002 by a Japanese collaboration of Sumitomo Heavy Industries, Ltd (SHI), Femtosecond Technology Research Association (FESTA) and Waseda University [43]. Some of the parameters obtained are listed in Table 3 a). A particularity of their facility is that they used a Ti:Saphire laser together with a temporal pulse shaping system using gratings and a liquid crystal spatial light modulator. With this they where able to produce different kinds of temporal laser pulse shapes: either Gaussian-like or square-like pulses. The rise / fall time of the square-like pulses was estimated to be less than 1.5 ps. In addition, the transverse laser distribution was shaped and a kind of flat-top distribution in the horizontal direction and a narrow Gaussian distribution in the vertical direction was obtained with 1.2 mm and 0.4 mm FWHM respectively. The electron beam is produced from an S-Band gun with space charge compensating solenoid and subsequently accelerated with a 70 cm long standing wave linac. The projected emittance measurements shown in Figure 7 were made by applying a quadrupole scan at 14 MeV beam energy. The results were obtained from a Gaussian fit to the background subtracted signal distributions. The clear advantage of the square-like pulses with respect to the Gaussian-like pulses is demonstrated. At a charge of 1 nC a horizontal normalized projected emittance of about 1.2 mm mrad was obtained.

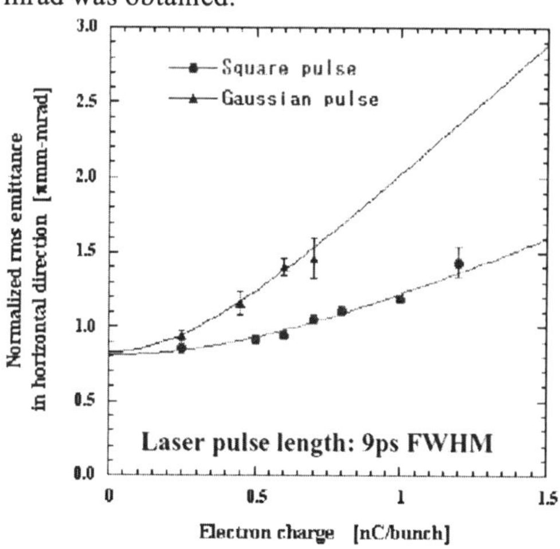

Figure 7. Experimental results from SHI+FESTA [43]. For details see the text.

More recently, a compact setup containing an RF gun, a phase optimized accelerating tube and a bending section were operated at Osaka University to produce very short electron bunches [44]. An rms bunch length of 98 fs was obtained with a bunch charge of 0.17 nC.

4.2. *NC RF Gun Injector Developments for SPARC*

An example of the operation of a copy of the BNL/SLAC/UCLA gun in Europe is the SPARC project in Frascati. Table 3 b) lists some of the design goals. The quoted emittance number is the normalized projected emittance at the end of the linac. The envisaged normalized slice emittance for a 300 μm slice is < 1 mm mrad [45]. One of the interesting new beam diagnostics developments from the SPARC project is the design and construction of the so-called emittance-meter [46]. This device consists of an emittance measurement system based on the pepper-pot technique which is located between two long bellows to allow a scan of the measured emittance over a distance of 1.2 m. This is especially interesting for the region right after the gun where a significant evolution of the transverse emittance takes place. After some laboratory tests at Frascati the emittance-meter was installed at PITZ for first commissioning and operation [47]. The functionality was demonstrated successfully and in early fall 2005 the device was transported back to Italy for installation into the SPARC photo injector setup. The full SPARC facility is now under installation and the first beam is expected in spring 2006.

4.3. *NC RF Gun Developments for LCLS*

The BNL/SLAC/UCLA 1.6 cell S-band cavity is the most copied and operated RF gun type around the world. The LCLS project at SLAC is also based on this kind of cavity. Table 3 c) lists some of the design goals for this project. The baseline design is with 1 nC charge and a slice emittance of 1.0 mm mrad after the injector. 1.2 mm mrad is the requirement at the undulator. An interesting option is to reduce the charge to 0.2 nC and fulfill the loose requirement of 0.8 mm mrad at the injector [48, 49]. According to simulations cited above the X-ray pulse energy for the 1 nC and 0.2 nC cases are quite similar, within about a factor of two, with at least 10^{12} photons per pulse. This is because the large wakefields in the 1 nC case only allow saturation for a fraction of the bunch. The low charge case also produces a shorter X-ray pulse of about 70 fs FWHM as compared to the 200 fs FWHM in the 1 nC case.

To further improve the performance of the 1.6 cell S-band gun several modifications to the cavity have been suggested for the LCLS project [50]. These include a larger separation between the 0 and the π mode of the cavity. By increasing the mode separation from 3.5 MHz to 15 MHz the extracted emittance is slightly improved and in particular a simplified tuning and a higher stability is expected. Another modification of the LCLS gun is a dual RF feed with coupling slits along the z-axis (beam direction) and increased radius on the inside surface of the coupler aperture. This ensures that the pulse heating should stay below 50 degrees. The quadrupole moment introduced by the couplers is

minimized by applying a racetrack cell shape [51]. To increase the quantum efficiency (QE) of the metal photo cathode an in-situ hydrogen beam cleaning of the cathode surface has been suggested [52]. This method prevents the usual laser cleaning with fluence close to the damage threshold and rastering of the small diameter laser beam which can potentially damage the cathode. On an unbaked copper sample it was demonstrated that an illumination with a total H-ion charge of about 2 mC is already sufficient to reach a QE of $2\cdot10^{-5}$ at a wavelength of 255 nm, which is the specification for the LCLS.
In 2006 the LCLS injector will be installed in its tunnel and first beam tests with the new setup are expected.

4.4. *Other Injector Developments*

There are many other interesting electron source developments worldwide which are not covered here due to the limited length of this contribution. Of special interest is e.g. the development at Eindhoven, the Netherlands, for 3D ellipsoidal electron distributions which might be extracted from a photo cathode or even from cold atoms as covered in other papers in these proceedings [19, 53]. Besides the source developments based on photo cathodes which are the subject of this paper, injector developments based on other kinds of electron extraction e.g. the special thermal emission injector for the SCSS in Japan [54] or the field emission based electron source for the LEG project in Switzerland [55] are of interest as well. The LEG project for example aims at a normalized emittance of less than 0.1 mm mrad at a charge of 0.2 nC and an average current of 2 nA [56].

5. Summary

There is no unique way to build an electron source for high brightness beams. On the contrary, for each application of the electron beam a specific set of key parameters characterizing the required beam properties is defined, mainly by the users of the application. These specific requirements in many cases require individual solutions for the technical realization of the electron source.
Following an introduction to DC photo guns, normal conducting and super conducting RF guns and a discussion of a general photo injector layout, 14 different photo injector projects were reviewed. They demonstrate that a large parameter space on beam quality, time structure of the beam and average current can be addressed by different photo injector developments. Simulations predict very good performance of all three basic photo injector types. Experimental progress is visible on the different subsystems (guns, lasers systems and beam diagnostics) as well as on the measured beam quality. Nevertheless, the desired

value of normalized transverse emittance for many projects of around 1 mm mrad at a charge of 1 nC has not yet been obtained anywhere experimentally. Although the different approaches towards reaching this goal have been well defined by simulations in the mean time, verifying these designs in experiments for the first time remains a challenge. Since the measured values of transverse emittance are getting smaller and smaller, research on the emission process from the cathode defining the thermal emittance as a lower limit of the transverse emittance is becoming increasingly more important.

Acknowledgments

The author would like to thank I. Ben-Zvi, H. Bluem, L. Catani, D. Dowell, R. Ganter, D. Janssen, M. Krasilnikov, F. Marhauser, D. Nguyen, J. Rathke, Ch. Sinclair, T. Srinivasan-Rao, J. Teichert, A. Todd, and the members of the PITZ collaboration for providing material and many useful discussions. Special thanks also to J. Bähr, J. Fitzgerald, K. Flöttmann and M. Krasilnikov for reading and commenting on the manuscript.

References

1. B.M. Dunham, L.S. Cardman, and C.K. Sinclair, proc. PAC1995, Dallas, 1030-1032 (1995).
2. D.A. Orlov et al., *Nucl. Instr. Meth.* **A532**, 418-421 (2004).
3. B.E. Carlsten, *Nucl. Instr. Meth.* **A285**, 313-319 (1989).
4. B.E. Carlsten, *Part. Acc.* **49**, 27-65 (1995).
5. L. Serafini and J.B. Rosenzweig, *Phys. Rev. E* **55**, 7565-7590 (1997).
6. D. Janssen, V. Volkov, *Nucl. Instr. Meth.* **A452**, 34-43 (2000).
7. J. Sekutowicz, "Superconducting RF Photoinjectors", these proceedings.
8. J. Teichert, "Status and Future Prospects of SRF Gun Developments", proc. Nanobeams2005, Kyoto, Japan, 2005.
9. M. Ferrario et al., "HOMDYN study for the LCLS RF photo-injector", proc. 2nd ICFA Advanced Accelerator Workshop on the Physics of High Brightness Beams, UCLA, USA (1999).
10. Ph. Piot et al., "Conceptual design of the XFEL photoinjector, TESLA FEL Report 2001-03.
11. ASTRA manual, http:\\www.desy.de\~mpyflo\Astra_documentation.
12. The TESLA Test Facility FEL team, "SASE FEL at the TESLA Facility, Phase 2", TESLA FEL Report 2002-01.
13. M. Krasilnikov, private communication.
14. Ch. Sinclair, private communication.
15. I.V. Bazarov and C.K. Sinclair, proc. PAC2003, Portland, 2062-2064 (2003).
16. I.V. Bazarov and C.K. Sinclair, *Phy. Rev. ST – AB* **8**, 034202 (2005).

17. O.J. Luiten et al., *Phys. Rev. Lett.* **93**, 094802-1 – 094802-4 (2004).
18. C. Limborg-Deprey, proc. FEL2005, Stanford, 418-421 (2005).
19. B.v.d. Geer et al., "Conceptual Design of a 1kA, 100fs, 1 micron split RF-photoinjector", these proceedings.
20. C.K. Sinclair, "High Brightness, High Average Current Injector Development at Cornell", talk at the FEL2005, Stanford, 2005.
21. D.H. Dowell et al., *Appl. Phys. Lett.* **63** (15), 2035-2037 (1993); D.H. Dowell, S.Z. Bethel and K.D. Friddell, *Nucl. Instr. Meth.* **A356**, 167-176 (1995).
22. D. Janssen et al., *Nucl. Instr. Meth.* **A507**, 314-317 (2003).
23. D. Janssen et. al., *Nucl. Instr. Meth.* **A528**, 305-311 (2004).
24. J. Teichert, private communication.
25. D. Janssen and V. Volkov, proc. EPAC2004, Lucern, 330-332 (2004).
26. K. Flöttmann, D. Janssen and V. Volkov, *Phy. Rev. ST – AB* **7**, 090702 (2004).
27. T. Rao et al., "Design, construction and performance of all niobium superconducting radio frequency electroninjector", submitted to *Nucl. Instr. Meth.*
28. T. Rao, private communication.
29. A. Todd, "State-of-the-art electron guns and injector designs for energy recovery linacs", proc. of ERL2005, Newport News, to be published in *Nucl. Instr. Meth.* **A**.
30. I. Ben-Zvi et al., proc. FEL2004, Trieste, 355-358 (2004).
31. I. Ben-Zvi, "Diamond secondary emitter status and plans", these proceedings.
32. D. Nguyen, private communication.
33. J.-G. Marmouget, A. Binet, Ph. Guimbal and J.-L. Coacolo, proc. EPAC2002, Paris, 1795-1797 (2002).
34. J. Bähr et al., proc. FEL2004, Trieste, 385-388 (2004).
35. F. Stephan et al., proc. FEL2004, Trieste, 347-350 (2004).
36. V. Miltchev at al., proc. FEL2004, Trieste, 399-402 (2004).
37. K. Honkavaara, proc. FEL2005, Stanford, 411-417 (2005).
38. V. Miltchv, "Measurements of thermal emittance for Cs2Te photocathodes at PITZ", proc. Nanobeams2005, Kyoto, Japan, 2005.
39. J.H. Han, "Dynamics of Electron Beam and Dark Current in the Photocathode RF Guns ", PhD thesis, Hamburg University, 2005.
40. V. Miltchev, "Investigations on the transverse phase space at a photo injector for minimized emittance", PhD thesis, Humboldt University Berlin, 2005.
41. The BESSY Soft X-ray Free Electron Laser, Technical Design Report, March 2004, eds.: D. Krämer, E. Jaeschke, W. Eberhardt, ISBN 3-9809534-0-8, BESSY, Berlin (2004).
42. F. Marhauser, proc. EPAC2004, Lucern, 315-317 (2004).

43. F. Sakai et al., "Production of low-emittance electron bunch from a photoinjector with a temporally shaped laser pulse", proc. 24th Advanced ICFA Beam Dynamics Workshop on Future Light Sources, Spring-8, Japan (2002); J. Yang et al., proc. EPAC2002, Paris, 1828-1830 (2002).
44. J. Yang, "Ultra-low emittance electron source & ultashort bunch electron source", proc. Nanobeams2005, Kyoto, Japan, 2005.
45. see http://www.lnf.infn.it/acceleratori/sparc/parameters.pdf
46. A. Cianchi et al., proc. EPAC2004, Lucern, 2622-2624 (2004).
47. L. Catani et al., proc. FEL2005, Stanford, 652-655 (2005).
48. C. Limborg-Deprey and P. Emma, proc. PAC2005, Knoxville, 2140-2142 (2005).
49. P. Emma et al., proc. PAC2005, Knoxville, 344-346 (2005).
50. C. Limborg, et al., "RF Design of the LCLS Gun", LCLS-TN05-3.
51. L. Xiao et al., proc. PAC2005, Knoxville, 3432-3434 (2005).
52. D.H. Dowell et al., proc. FEL2005, Stanford, 172-175 (2005).
53. J. Luiten et al., "Ultracold electron source", these proceedings.
54. T. Shintake, "Thermionic Gun for X-ray FEL", proc. Nanobeams2005, Kyoto, Japan, 2005.
55. R. Ganter, "Low Emittance Gun Project", these proceedings.
56. R. Ganter, private communication.

RESULTS FROM THE UCLA/FNPL UNDERDENSE PLASMA LENS EXPERIMENT

M. C. THOMPSON*, H. BADAKOV, J. B. ROSENZWEIG, G. TRAVISH†

Department of Physics and Astronomy,
University of California, Los Angeles
Los Angeles, CA 90095
E-mail: thompson93@llnl.gov

H. EDWARDS, R. FLILLER, G. M. KAZAKEVICH, P. PIOT, J. SANTUCCI

Fermi National Accelerator Laboratory,
Batavia, Illinois, 60510

J. LI AND R. TIKHOPLAV

University of Rochester,
Rochester, New York 14627

A gaussian underdense plasma lens with peak density 5 x 10^{12} cm^{-3} and a full width half maximum (FWHM) length of 2.2 cm has been used to focus a relativistic electron beam. This plasma lens is equivalent in strength to a quadrupole magnet with a 150 T/m field gradient. The lens focused a 15 MeV, 16 nC electron beam with initial dimensions $\sigma_{x,y} \approx 650$ μm and $\sigma_z \approx 6.5$ mm onto an optical transition radiation (OTR) screen ~2 cm downstream of the lens. The average transverse area of the plasma focused electron beam was typically demagnified by a factor of 23. The evolution of the beam envelope in the area near the beam waist was measured for both round beams and asymmetric beams with x:y aspect ratios as large as 1:5. The light from the OTR screen in the round beam case was also imaged into a streak camera in order to directly measure the correlation between z and σ_r within the beam.

1. INTRODUCTION

Plasma lenses are of great interest to the accelerator physics community because they can theoretically provide radially symmetric focusing gradi-

*Current Affiliation: Lawrence Livermore National Laboratory
†Spokesman for the collaboration at Erice.

ents on the order of 10 MT/m, which exceeds the strength of conventional quadrupole lens by several orders of magnitude [1–3]. Increasing focusing gradients to this level would greatly aid in achieving the small spot sizes and high luminosity necessary for future e^+e^- colliders such as the planned International Linear Collider (ILC). Additionally, it has been shown that adiabatic plasma lenses can overcome the synchrotron-radiation limit on final focus spot size [4]. The underdense regime of plasma lens operation, in which the density of beam electrons n_b is greater than half of the plasma density n_p, is considered the most advantageous for the focusing of electron beams in many scenarios [2,4].

The most commonly considered regimes of plasma lens operation are the overdense ($n_b << n_p$) and underdense ($n_b > n_p/2$) regimes [3]. In the overdense case the plasma essentially cancels the beam space charge and allows it to focus under its magnetic self forces. Since the beam magnetic self forces are generally not uniform over the length of the bunch or linear over its breadth, overdense lenses typically have significant abberations. In the underdense case the strong space charge of the electron beam ejects the plasma electrons entirely, leaving a uniform ion column. It can be shown that the radial focusing force of this ion column is given by

$$F_{radial} = -2\pi n_p e^2 r, \tag{1}$$

where e is the electron charge and r is the radial distance from the axis. An underdense plasma lens is hence free of spherical aberrations in the ideal limit, and is equivalent in strength to a quadrupole focusing magnet with a magnetic gradient of (3 x 10^{-11})n_p T/m. A plasma lens with the relatively modest density of 5 x 10^{12} cm^{-3} therefore has the strength of a 150 T/m quadrupole. State-of-the-art superconducting quadrupoles, by comparison, have maximum gradients on the order of 250 T/m. Plasma lenses also focus both transverse beam axes simultaneously. The drawback of plasma lenses in either regime of operation is that only about 80 % of the beam is focused. Due to the finite time it takes for the plasma to react to the beam space charge, the head of the beam is not subject to the plasma focusing forces.

The focal length f of a thin plasma lens is given by $f = 1/Kl$, where K is the lens focusing strength and l is the lens length. For an underdense plasma lens

$$\frac{1}{f} = Kl = \int \frac{2\pi r_e n_p(l)}{\gamma} dl, \tag{2}$$

where r_e is the classical electron radius, γ is the Lorentz factor, and $n_p(l)$ is the function that describes the variation in plasma density along the electron beam's path through the lens.

2. EXPERIMENTAL METHODOLOGY

Our underdense plasma lens experiment, which evolved from earlier proposals [5,6] and follows previous electron plasma lens work in other regimes [7-10], was conducted at the Fermilab NICADD Photoinjector Laboratory (FNPL) at FNAL. A schematic of the experiment is shown in Fig. 1. A bulk

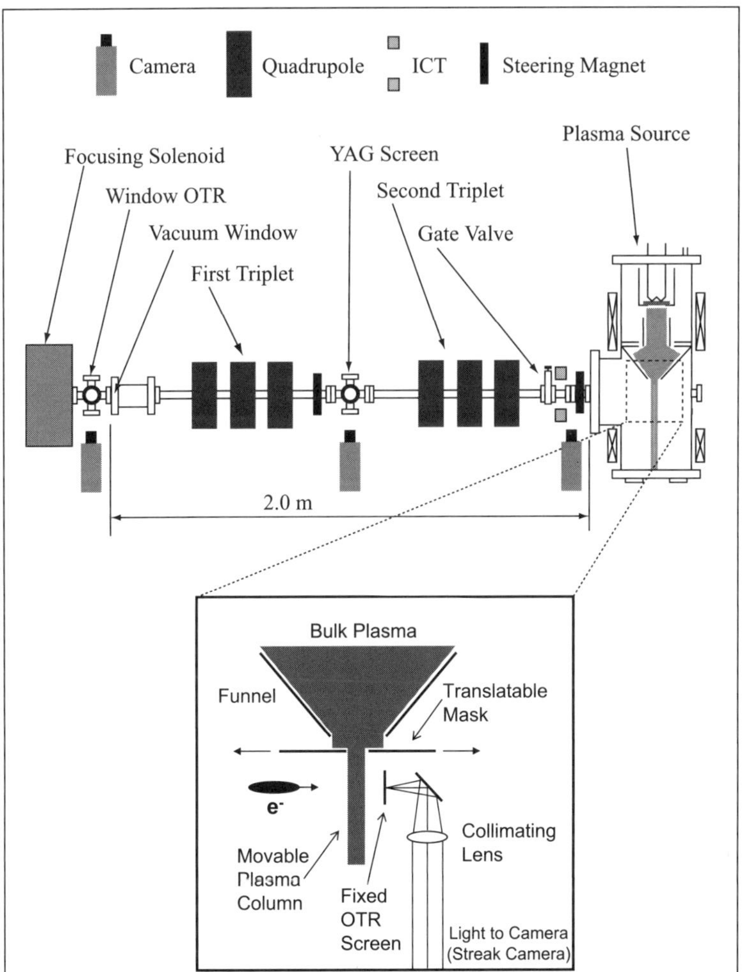

Figure 1. Schematic of the plasma lens experiment beamline and apparatus.

Table 1. Experimental Parameters

Peak Plasma Density	4.7 x 10^{12} cm^{-3}
Plasma Thickness (σ_p)	9.9 mm
Beam Charge	16 nC
Beam Duration (σ_t)	22 ps
Initial Beam Radius (σ_r)	950 μm
Beam Emittance (ϵ_n)	90 mm-mrad
Peak Beam Density	2.2 x 10^{12} cm^{-3}

column of plasma approximately 5 cm FWHM is produced using our pulsed argon discharge plasma source [11]. The gaussian plasma lens is formed by selecting a slice of this static plasma column with a translatable mask with a 1.25 cm wide slit. This arrangement allows the distance between the plasma lens column and the fixed OTR diagnostic screen to be varied without altering the OTR collection optics. Measurements have shown that the plasma lens column properties, listed in Table 1, vary only slightly as the mask is translated through the 1 cm region of interest. An in-vacuum, 160 mm focal length, 2 inch diameter lens provides point-to-parallel collimation of the OTR produced when the electron beam emerges from the back of a polished aluminum foil. The OTR is then transported either 30 cm to a conventional CCD camera with a 75 mm compound camera lens or 7 m to a streak camera where the light is imaged onto the streak slit using a single 280 mm focal length lens.

The electron beam for the experiment is provided by the FNPL electron accelerator. The FNPL accelerator is a 15 MeV electron linac [12] consisting of a normal conducting L-band RF gun with a cesium telluride photo-cathode and a 9-cell superconducting accelerating cavity. After acceleration, the electron beam is propagated at a tight focus through a 10μm thick aluminum vacuum foil, used to isolate the plasma experiment vacuum from the rest of the beam line, and refocused into the plasma lens experiment, Fig. 1.

The plasma lens experiment parameters are listed in Table 1. Note that $n_{b,peak} \approx n_{p,peak}/2$, which puts the experiment just on the boundary of the underdense regime at the entrance to the plasma lens. This is a lower bound, however, since the beam is substantially denser compared to the edge of the gaussian plasma column which it first encounters and, since the lens is relatively thick, the beam core compresses somewhat by the time it reaches the center of the plasma lens. It is therefore clear that the beam focusing is governed by underdense dynamics at the threshold of the regime. Given the values in Table 1, Eq. 2 predicts $f = 1.5$ cm for the gaussian plasma lens. The observed focal length was $f \cong 1.9$ cm.

The agreement between these two values is reasonable considering that the gaussian plasma lens does not satisfy the thin lens criteria ($f >> l$), which is the limit assumed in Eq. 2.

3. UNDERDENSE PLASMA FOCUSING

Three distinct series of measurements were made during this experiment: time integrated measurements of round focused spots, time resolved measurements of round focused spots, and time integrated measurements of focused beams with ratios of x dimension to y dimension of up to about 1:5. Each of these measurements provides information on a different aspect of the underdense plasma lens performance and behavior.

3.1. *Time Integrated Measurements of Round Beam Focusing and the Focused Beam Envelope*

The focusing effect of our underdense plasma lens was strong and highly repeatable. Figs. 2 and 3 show the effect of plasma focusing on the electron beam spot size as observed with the conventional CCD camera. Note that the focused beam is imaged at the lens focus, $f = 1.9$ cm, and shows only the core of the beam since our camera lacked sufficient dynamic range to simultaneously record the unfocused portion of the beam along with the focused core. The unfocused beam is nearly round with x FWHM = 1630 μm and y FWHM = 1540 μm. The underdense plasma lens focuses this beam down to x FWHM = 260 μm and y FWHM = 420 μm which is a

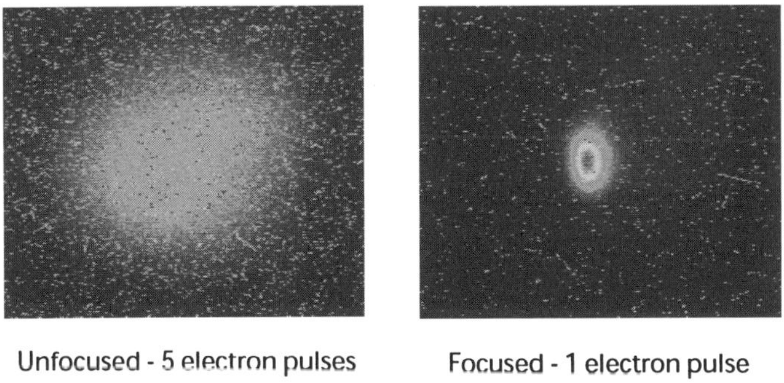

Figure 2. False color images of the unfocused (left) and focused (right) electron beam spots as observed on the OTR screen after the plasma lens. The unfocused image is shown at five times normal intensity in order to provide sufficient image contrast. Both images use the same scaled-to-intensity color map.

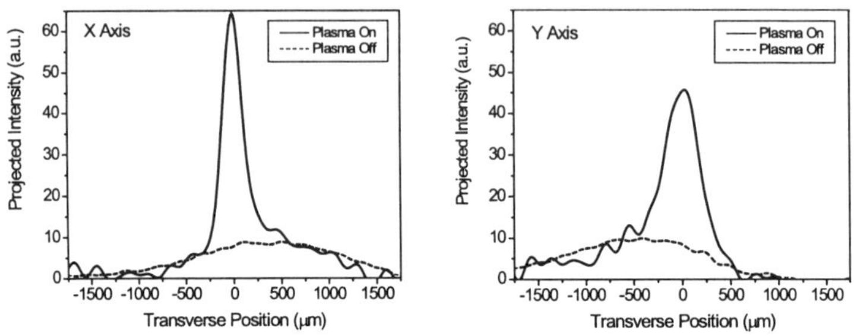

Figure 3. Plots of the image intensity of the beam images shown in Fig. 2 (normalized to 1 electron pulse) in the x axis (left) and y axis (right).

demagnification of 6.3 in x and 3.66 in y. The average transverse area of the beam is consequently reduced by a factor of 23. This area reduction equates to a boost in the core beam density from 2.2 x 10^{12} cm^{-3} in the unfocused case to 5 x 10^{13} cm^{-3} in the focused case.

Measurements of the focused beam envelope were made by taking spot size measurements at various lens / OTR screen spacings, Fig. 4. The variation of the beam spot size near the focus should be describable (neglecting space charge effects) by the beam envelope equation:

$$\frac{d^2\sigma_x}{dz^2} + K\sigma_x = \frac{\epsilon_x^2}{\sigma_x^3}, \tag{3}$$

where σ_x is the rms beam size in x, ϵ_x is the geometric beam emittance in x, and K is once again the focusing strength. The equivalent equation for the other transverse direction can be written by substituting y for x. We are

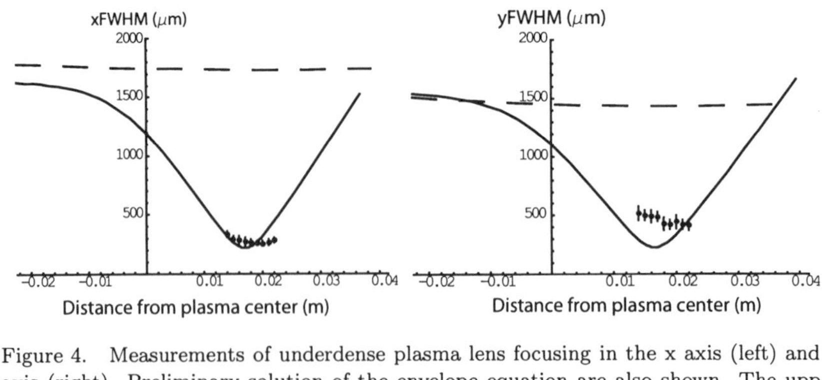

Figure 4. Measurements of underdense plasma lens focusing in the x axis (left) and y axis (right). Preliminary solution of the envelope equation are also shown. The upper dashed lines are solutions to the beam line transport model, while the lower solid lines are solutions to the lens model.

using Eq. 3 to model both the transport of the beam from the vacuum foil to the plasma lens, and the effect of the plasma lens itself. The results of our preliminary modeling efforts clearly illustrate the strong focusing effect of the plasma lens and are shown, along with the beam envelope data, in Fig. 4. We should be able to refine the rough agreement between the data and the solutions to Eq. 3 as the data analysis progresses. It should also be possible to extract quantitative information about the plasma lens' aberrations from the beam envelope analysis.

3.2. *Time Resolved Measurements of Focusing*

A series of time resolved measurements of the plasma focusing were made by imaging the beam OTR light onto the slit of a Hamamatsu series C5680 streak camera. An example of one the images recorded on the streak camera, along with an analysis of that shot, is shown in Fig. 5. As expected, the intensity profile of the focused beam in the time domain remains roughly gaussian while the beam is radially larger at the head than in the middle or tail. Width measurements were made by examining 5 ps wide swaths of the beam. The time resolution of the streak camera had previously been measured to be below 2 ps. While the time domain behavior of the plasma focusing is in general agreement with theoretical predictions, the raw data does not have as large a change between head and tail diameter as expected from either theory or the time integrated measurements. This apparent lack of transverse resolution seems to originate from limitations in the performance of our OTR light transport and may result from slight optics misalignments, the large angles inherent in such a strongly focused

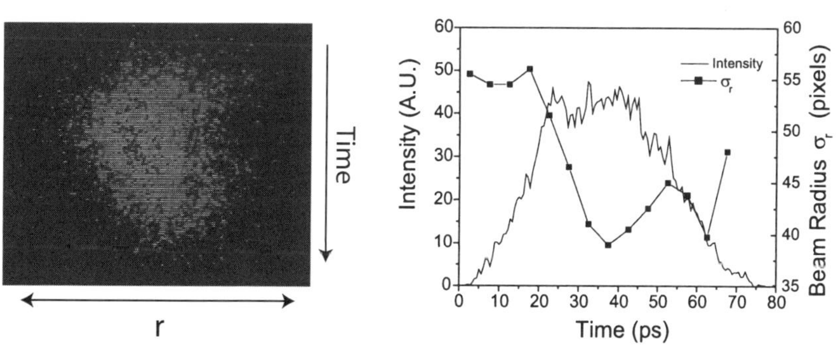

Figure 5. A false color streak camera image of the plasma focused beam (left) with initial analysis of the image (right).

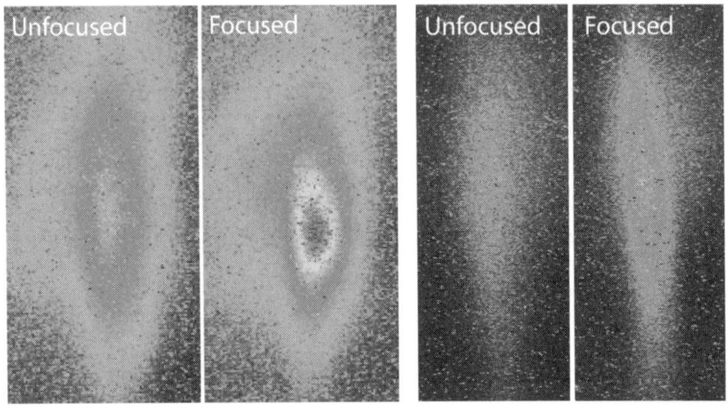

Figure 6. False color images of a beams with an initial aspect ratio of 1:3.4 (at left) and 1:4.5 (at right). In each set, both images show the same number of beam pulses.

beam, and other effects. Analysis of the streak camera data is ongoing.

3.3. *Elliptical Beam Focusing*

Our studies of underdense plasma lens focusing also included an examination of elliptical beam focusing. Fig. 6 shows examples of time integrated measurements of elliptical beams with 1:3.4 and 1:4.5 aspect ratios between the x and y dimensions. A clear plasma focusing effect is visible in both cases. The highest aspect ratio examined was 1:4.5, largely because it was impractical to transport flat beams with large asymmetric emittances to the experiment. We plan to make a quantitative analysis of the elliptical beam data in a manner similar to round beam analysis discussed above.

The motivation behind these measurements of modestly asymmetric beams is the desire to understand how the dynamics of underdense plasma focusing change as the beam aspect ratio increases. The development of such an understanding is vital if underdense lenses are to be applied in a ILC context where aspect beam ratios at the interaction point are 1:100. Our measurements are a start, but future experiments with flat, very high aspect ratio beams will be needed.

4. Conclusions

We have measured extreme demagnification of a high brightness electron beam by a strong underdense plasma lens. The collected data include both time integrated and time resolved measurements of the focused beam spot and its envelope. While data reduction and analysis are ongoing, the general behavior of the plasma lens matches our theoretical expectations very well. A full analysis of this underdense plasma lens experiment will be reported elsewhere [13].

5. ACKNOWLEDGEMENTS

This work was performed under the auspices of the US Department of Energy under Contract No. DE-FG03-92ER40693 and W-7405-ENG-48.

References

1. P. Chen. *Part. Accel.*, **20**, 171 (1987).
2. P. Chen, *et al.*, *Phys. Rev. D*, **40**, 923 (1989).
3. J. Su, *et al.*, *Phys. Rev. A*, **41**, 3321 (1990).
4. P. Chen, *et al.*, *Phys. Rev. Lett.*, **64**, 1231 (1990).
5. H. Suk, *et al.*, In *Proc. PAC 1999*, 3708 (1999).
6. M. C. Thompson, *et al.*, In *Proc. PAC 2001*, 4014 (2001).
7. H. Nakanishi, *et al.*, *Phys. Rev. Lett.*, **66**, 1870 (1991).
8. G. Hairapetian, *et al.*, *Phys. Rev. Lett.*, **72**, 2403 (1994).
9. R. Govil, *et al.*, *Phys. Rev. Lett.*, **83**, 3202 (1999).
10. P. R. Bolton, *et al.*, Int. J. Mod. Phys. A **18**, 2843 (2003).
11. M. C. Thompson, *et al.*, *Rev. Sci. Instr.*, **76**, 013303 (2005).
12. J.-P. Carneiro, *et al.*, In *Proc. PAC 1999*, 2027 (1999).
13. M. C. Thompson, *et al.*, *Phys. Rev. Lett.*, (to be submitted).

MEDICAL APPLICATION OF MULTI-BEAM COMPTON SCATTERING MONOCHROMATIC TUNABLE HARD X-RAY SOURCE

Mitsuru Uesaka[1], Katsuhiro Dobashi[2], Fumito Sakamoto[1], Atsushi Fukasawa[1], Haruyuki Ogino[1], Tomohiko Yamamoto[1], De Meng[1], Toshinobu Miyoshi[2], Junji Urakawa[3], Toshiyasu Higo[3], Mitsuo Akemoto[3]

[1]Nuclear Professional School, the University of Tokyo,
2-22 Shirakata-Shirane, Tokai, Naka, Ibakaki, 319-1188 Japan

[2]National Institute of Radiological Sciences,
4-9-1 Anagawa, Inage, Chiba-shi, Chiba, 263-8555 Japan

[3] High Energy Accelerator Research Organization,
1-1 Oho, Tsukuba, Ibaraki, 305-0801 Japan

E-mail: uesaka@nuclear.jp

Compton scattering hard X-ray source which consists of an X-band (11.424 GHz) electron linear accelerator and YAG laser is under construction at Nuclear Professional School, the University of Tokyo. Monochromatic hard X-rays are required for variety of medical and biological applications. Our scheme of the hard X-ray source is to produce a monochromatic hard X-ray via collision between 35 MeV electron beam and 2.5 J/10 nsec Nd:YAG laser. In order to increase the efficiency of the X-ray yield, we adopt a laser pulse circulation system. In our case, the laser pulse circulation system can increase the X-ray intensity of up to 10 times. Main features of our scheme are to produce monochromatic tunable hard (10-40 keV) X-rays with the intensities of 10^8-10^9 photons/sec. In addition, X-ray energy can be changed with rapidly by 40 ms by introducing two different wavelength lasers (YAG fundamental (1064 nm), 2nd harmonic (532 nm)) and optical switch. This quick energy change is indispensable to living specimens and very difficult by a large SR light source and others. Dual-energy X-ray CT and subtraction X-ray CT are available to determine 3D distribution of atomic number density and electron density, and specified atomic distribution, respectively. Here, the construction status of the X-band beam line and the application plan of the hard X-ray are described and discussed.

1. Introduction

Hard X-rays of 10-40 keV are now very useful in medical science, biology and material science. For example, Dynamic IVGAC [1,2] and monochromatic

X-ray imaging, CT. In addition, Dual energy X-ray CT [3] and Subtraction imaging with contrast agent and Dual energy X-ray are realized that require two monochromatic hard X-ray.

Intense hard X-rays are generated by a third generation light source. However, most SR sources are too large to be applied and used widely for public usage of the monochromatic hard X-ray. Therefore, we are developing a compact monochromatic hard X-ray (10-40 keV) source based on laser-electron collisions with the X-band (11.424 GHz) linac system [4-8]. One to ten percent narrow band X-rays are generated by collimating scattered photons that are related to the energy and scattering angle.

The final target of this study is an integrated system for medicine. This system is equipped with an X-band RF-source and a moving arm including an X-band linac, Q-switch laser system and X-ray detector.

A multi-bunch electron beam generated by a thermionic-cathode RF gun is collimated and compressed temporally by an alpha-magnet and accelerated by X-band accelerating structures. The electron beam is bent by the achromatic bends and focused at the collision point (CP). About a 10 ns hard X-ray is generated via Compton scattering upon laser-electron collision. After the collision, the electron beam is bent and decelerated by an X-band decelerating structure. The decelerated electron beam with the energy lower than 1 MeV is injected to a beam dump. In addition, this system can generate dual energy monochromatic hard X-ray using two Nd:YAG laser by turn.

The laser system for collision is composed of a Q-switch Nd:YAG laser and a laser pulse circulating system to increase X-ray yield.

To demonstrate that the proposed X-ray source can be realized and will be useful in medicine, an X-band linac beam line for the proof-of-principle experiment shown in Fig. 1 is under construction. The X-ray yield by the electron beam and Q-switch Nd:YAG laser of 2.5 J/10 ns is 10^7 photons/RF-pulse 10^8 photons/sec in 10 pps). X-band beam line for the demonstration is under commissioning. We also design to adopt a technique of laser circulation to increase the X-ray yield up to 10^8 photons/pulse (10^9 photons/s).

In this paper, we present the design and numerical analysis of the X-ray source system to demonstrate hard X-ray generation and its applications.

Fig. 1: Schematic illustration of Compact Hard X-ray source based on thermionic-cathode X-band RF-gun, X-band accelerating structure and Q-switch Nd:YAG laser.

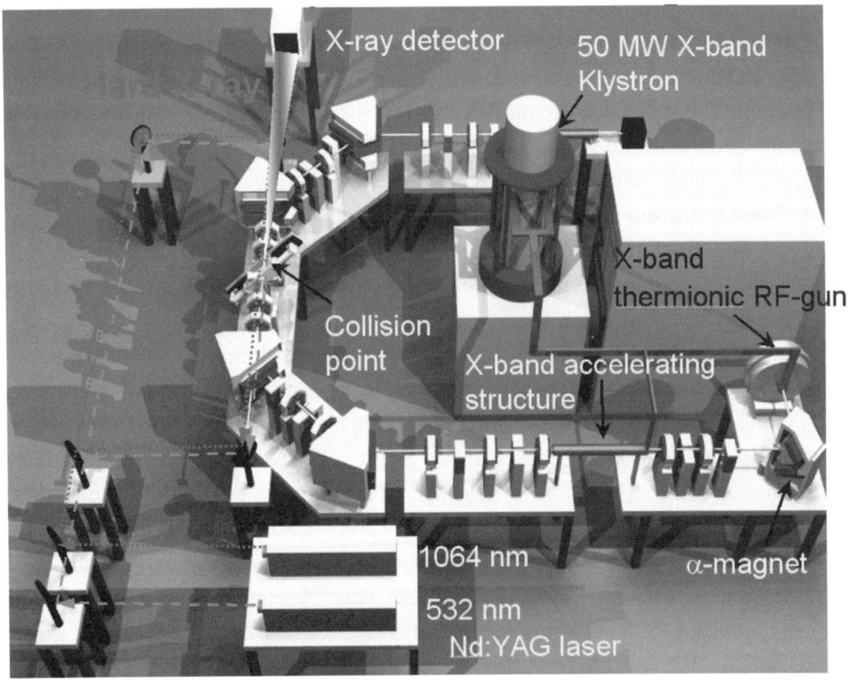

2. Brief Review of Linac-laser Compton scattering X-ray sources

Early experimental demonstrations of the proof-of-principle systems were done at LBL[9,10], BNL[11], KEK-JAERI-U.Tokyo[12] and so on in 1990s. Then, use and distribution of table-top Ti:Sappire lasers, CO_2 lasers and Nd:glass laser and RF photoinjectors promote their development in the world. Especially, not only the fundamental but also the second harmonics are produced recently[13]. Among them, several systems are specialized for medical- and bio-scientific uses such as monochromatic X-ray imaging, dual-energy-Xray-CT, substract-X-ray-CT, angiography, Auger cascade therapy, protein structural analysis, etc. Those systems are summarized in Table.1. Especially, the pioneering works on the medical monochromatic X-ray imaging was performed by F.Carrol et al [16]. However, most of them use the scattering between a single electron bunch and a single laser pulse so that they suffer a lack of X-ray intensity up to 10^8 photons/s. Dynamic imaging, angiography and protein structural analysis have been found to be difficult with such intensity. In order to overcome this problem, the solution would be multiple scattering between multi-electron-bunches and multi-laser-pulses which could generate the X-ray intensity of more than 10^9 photons/s. Let us define the first and second generations for the single and multi-scatterings as follows. The first generation established the following.

-First demonstration and application,

-Intensity up to 10^8 photons/s,

-Intensity fluctuation due to the time-jitter between electron and laser pulses.

On the other hands, the second generation is going to realize the following.

-Multi-scattering of electron- and laser-pulses

-Intensity of more than 10^9 photons/s

-A variety of applications for medicine, protein structural analysis, nondestructive evaluation and nuclear engineering

We would like to report the development of the typical second generation system which is under way at University of Tokyo in the following chapters. Moreover, KEK is developing a multi-bunch-S-band-phtoinjector for the global linear collider, where more than 100 nC per RF pulse is produced[19]. This system must be the second candidate for the second generation.

Table 1: Performance of Linac/Laser Inverse Compton X-ray Source for Medical Use

Laboratory	Electron Energy	Charge	Wavelength & Power of Laser	X-ray Energy & Intensity	Uncertainty of X-ray Intensity
SHI[15]	14 MeV (S-band)	400 pC Single Bunch	800nm, 300 mJ. (Ti:Sapphire, 300 ps)	3.5 keV < 1.0E+4 photons/shot	50%
MXI Systems Inc[16]	25 MeV (S-band)	500 pC Single Bunch	1 μm, 20 J (Nd:Glass)	12~50 keV 1.0E+8 photons/sec	50%
LLNL[17]	57 MeV (S-band)	250 pC Single Bunch	780 nm, 400 mJ (Ti:Sapphire, fs)	40~80 keV 1.0E+7 photon/ssec	50%
SLAC[18]	60 MeV (S-band)	500 pC Single Bunch	800 nm, 300 mJ (Ti:Sapphire, fs)	20~85 keV 1.0E+8 photons/s	80%
Univ. of Tokyo /NIRS/KEK[5]	35 MeV (X-band)	20 pC×10000 Multi Bunches	1 μm, 2 J, (Nd:YAG, 10 nsec)	33 keV 1.0E+8 photons/s	<10%

3. X-BAND –YAG-laser Compton scattering X-ray source of University of Tokyo

Compact hard X-ray source based on the X-band linac that we propose is shown in Fig. 1. Multi-bunch beam generated by thermionic-cathode RF-gun is accelerated by X-band accelerating structures. The beam is bent and focused at the collision point. About 10 ns hard X-ray is generated via Compton scattering on laser-electron collision.

3.1. *X-band linac*

X-band linac is applied to the compact hard X-ray source. RF-wavelength of the X-band is 1/4 of S-band (2.856 GHz). However, the maximum filed gradient as ~ 40 MV/m enable remarkable compactness.

An X-band accelerating structure with 0.7 m long is used for the X-ray source. The technologies for X-band accelerating structure developed for future linear colliders [20] at KEK and SLAC are fully adapted for this development. At first, the RDS type accelerating structure has been adopted, which is already under manufacturing.

We adopt PPM type X-band Klystron (Toshiba-E3768A) designed for linear colliders [20]. Klystron Modulator is under design to fit this X-ray source. RF power is above 50MW in 1 μs.

The beam optics of the X-band linac beam line was designed using SAD code [21]. Beam parameters at the collision point (CP) are shown in Table 2.

3.2. *Laser system*

To concentrate on R&D of the accelerator, we choose a commercial and reliable laser for laser-electron collision. To realize such a compact system, we adopt two Q-switch Nd:YAG laser with the intensity 2.5 J/pulse(1.4J/pulse for second harmonics), the repetition rate 10 pps, the pulse length 10 ns(FWHM) and wavelength of 1064 nm(fundamental).

In second step, to switch the X-ray energy immediately, we add the laser system. Each laser system for fundamental and second harmonics is shot by turn. Then we can generate dual energy monochromatic X-ray by turn with repetition rate of 12.5 Hz as shown in Fig. 2.

Advantage of this dual energy X-ray source is the switching of X-ray energy with high repetition rate that is very useful for dynamic subtraction imaging and dual energy X-ray CT. To increase X-ray yield, we have to design the technique of circulation of laser pulse, which enhance the luminosity by 10 times.

First experimental demonstration of the laser circulation was done by the second harmonics of 25 mJ YAG laser to realize the increase of the collisional laser energy by about 10 times. The circulation of the fundamental was also performed. The full laser energy system is under construction [14].

3.3. *X-ray yield and properties*

The X-ray yield per bunch is calculated with cross section of Compton scattering and Luminosity. Energy distributions and energy due to scattering angle of the X-ray for each wavelength of laser light with electron beam energy 35MeV are shown in Fig. 3. Property of the X-ray is summarized in Table 3.

Figure 4 shows the mean energy and rms energy spread of the available X-ray for each laser wavelength when scattered photons are collimated with each collimated angle.

Table 2: Beam parameters at the collision point.

Beam energy	35 MeV
Charge/bunch	20pC
bunches/RFpulse	10^4
Beam size(rms)(x,y)	100, 100 μm
Beam emittance(x,y)	10, 10 πmm.mrad

Fig. 2: Concept of dual energy X-ray generation system.

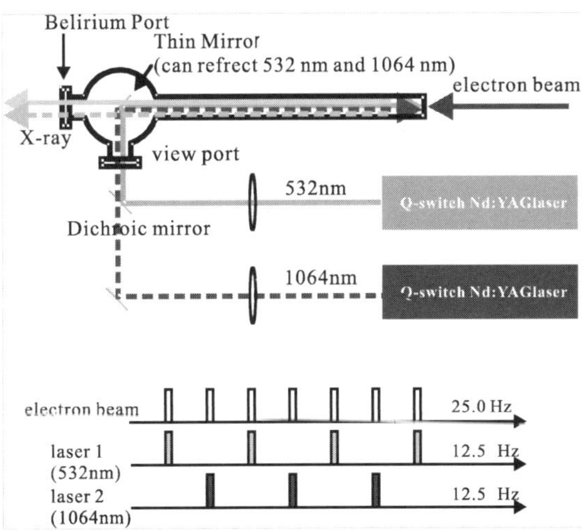

Fig. 3: Energy spectrum of X-ray for laser wavelength of 1064 and 532 nm with electron beam energy 35 MeV.

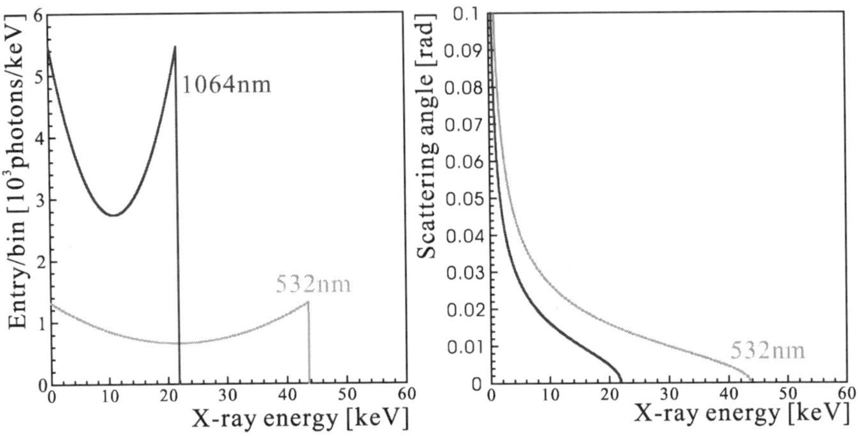

Fig. 4: Energy spread in rms and mean energy of available X-ray with each collimated angle.

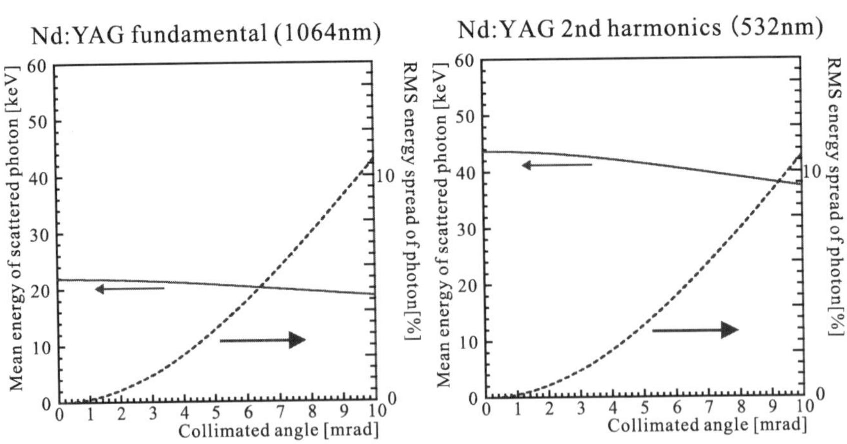

4. TEST of X-BAND RF SOURCE and X-BAND THERMIONIC CATHODE RF-GUN

Test of RF generation and RF aging of the X-band klystron was finished. Peak power of RF is estimated to 10 MW per RF output port of the klystron. Total output power is 20 MW and pulse length 600 ns, repetition rate 10 pps.

We have done RF-aging of the X-band thermionic cathode RF-gun and established the nominal energy of 2 MeV at 5.5 MW 100ns RF pulse as shown in Fig.5.

After the RF-gun experiment, the klystron-aging continues to reach the RF parameter up to 50 MW, 1 μs, 5 pps. Then, X-band accelerating structure is installed, acceleration test and X-ray generation will be performed.

Fig. 5: Input - and output - RF- singles and beam - signals before and after the alpha magnet for 5.5 MW 100 ns RF pulse and 2 MeV beam energy.

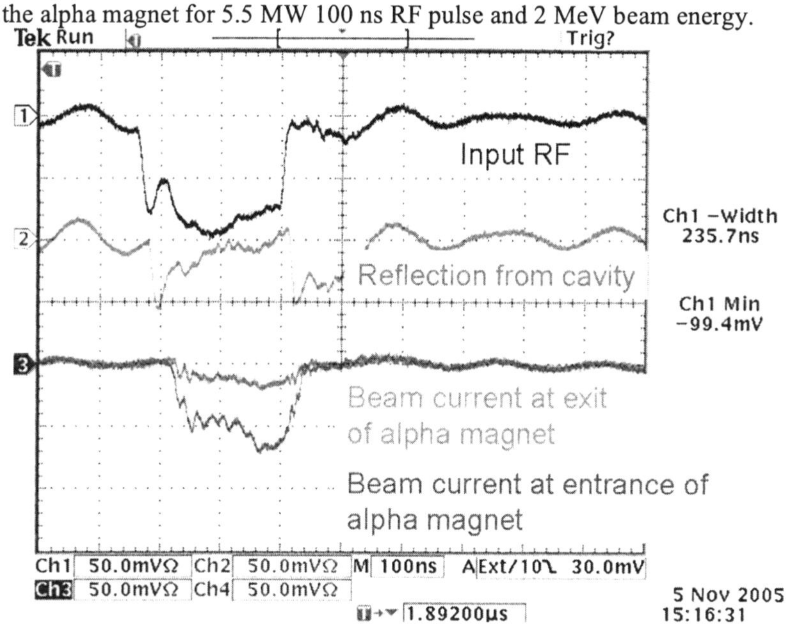

5. DUAL-ENERGY X-RAY CT

Theoretical Background

Effective atomic number Z_{eff} and electron density ρ_e are obtained by linear attenuation coefficients of a material using two monochromatic X-rays with different energies. A linear attenuation coefficient μ of a material is approximately written as a function of atomic number Z and X-ray energy E using a formula proposed by Jackson and Hawkes [22], as follows;

$$\mu(Z,E) = \rho \frac{N_A}{A} Z\left\{4\sqrt{2}Z^4\alpha^4\left(\frac{mc^2}{E}\right)\phi_0 \sum_{nu} f_n u' + \sigma_{KN} + \frac{Z\left(1-z^{b-1}\right)}{Z^{r2}}\sigma_{SC}^{coh}(Z',E')\right\} \quad (1)$$
$$= \rho_e\left(Z^4 F(Z,E) + G(Z,E)\right)$$

where, ρ is mass density, N_A is Avogadro's number, A is atomic mass, $f_n u'$ is the collection terms for photoelectric absorption cross section, σ_{KN} is the Klein-Nishina cross section and σ_{SC}^{coh} is the coherent scattering cross section of the standard element Z' at energy of $E' = (Z'/Z)^{1/3} E$.

In the equation, parameter b is proposed to be 0.5 and the standard element is Oxygen [21]. When linear attenuation coefficients are measured for two energies E_1 and E_2, one can extract effective atomic number Z_{eff} and electron density ρ_e y solving following equations;

$$Z^4 = \frac{\mu(E_2)G(Z,E_1) - \mu(E_1)G(Z,E_2)}{\mu(E_1)G(Z,E_2) - \mu(E_2)G(Z,E_1)}, \quad (2)$$

$$\rho_e = \frac{\mu(E_1)F(Z,E_2) - \mu(E_2)F(Z,E_1)}{F(Z,E_2)G(Z,E_1) - F(Z,E_1)G(Z,E_2)}, \quad (3)$$

The effective atomic number Z_{eff} is defined for a compound or a mixture as;

$$Z_{\text{eff}} = \left(\sum_i q_i Z_i^k\right)^{1/k}, \quad (4)$$

where q_i is the fractional electron content of i th element in the compound or the mixture and the parameter $k = 4$.

This dual-energy method is limited by the K-edge energy of a material because the equation (1) can not be applied below K-edge energy. We will operate the compact hard X-ray source as electron beam energy is 35 MeV and the wavelength of the laser is 1064 nm and 532 nm. In this situation, the maximum X-ray energies are 21.9 keV and 43.8 keV. The X-ray energy should allow us to identify elements up to $Z = 38$. Energy spread $\Delta E/E$ of the monochromatic hard X-ray generated by SR is in the order of 10^{-4}, but the energy spread in the compact hard X-ray source is expected to be 1 to 10 % (rms), which is dependent on the collimator angle [5]. The energy spread of the X-ray is negligible with SR light sources, but in the case of the Compton scattering X-ray source the energy spread affects the accuracy of atomic number identification.

To examine the applicability of the dual-energy X-ray CT with the compact hard X-ray source, we have performed a numerical simulation for low to medium Z elements ($Z \leq 38$). We adopt linear attenuation coefficients of materials listed in the photon cross section database [23] in the simulation. The geometry of the dual-energy X-ray CT system in the simulation is shown in Fig. 6. We assumed a point light source and the thickness and width of the sample is 1 mm and 20 mm, respectively. In this case, X-ray photons are collimated at 3.3 mrad. It is noted that the X-ray energy of the Compton scattering X-ray source is dependent on the scattering angle. Thus, the X-ray energy is unique on each pixel of the 2D detector. When pixel size is 0.5 mm, $\Delta\theta$ is less than 0.2 mrad and the energy spread in the pixel is in the order of 0.1 %. This small energy spread in the pixel should improve accuracy of the dual-energy analysis. Effective atomic numbers obtained by the linear attenuation coefficients for two energies on each pixel of the detector are shown in Fig. 7(a). Average of Z_{eff} the is plotted as a function of atomic number of sample Z_{samp} in Fig. 7(b). The accuracy of estimated effective atomic number $\Delta Z/Z$ is less than 3 % (rms) except for $Z = 25$ and 33. We confirmed that atomic number can be identified up to $Z = 38$ using 21.9 keV and 43.8 keVX-rays with enough accuracy even energy spread of the monochromatic X-ray is 1 to 10 %. The limitation of this method is due to K-edge energy of the material.

Fig.6: Schematic drawing of the dual-energy X-ray CT system using the compact monochromatic hard X-ray source.

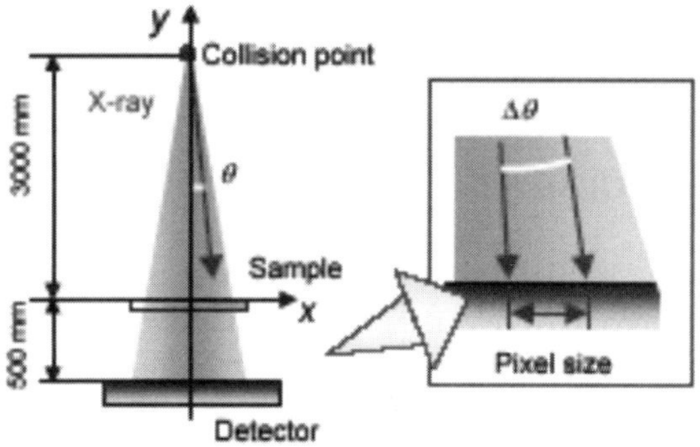

Fig.7: Schematic drawing of the dual-energy X-ray CT system using the compact monochromatic hard X-ray source.

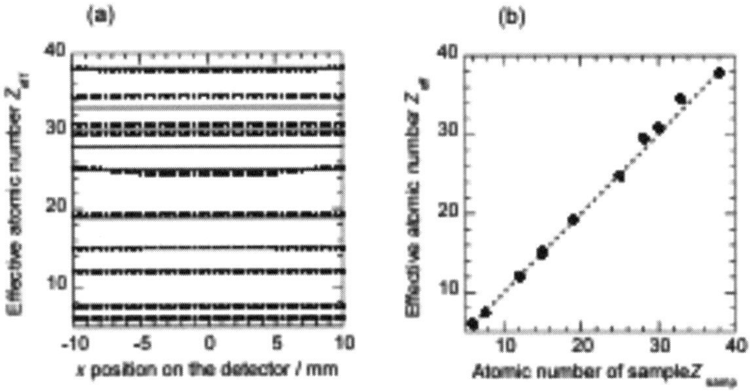

6. CONCLUSION

We are developing the compact X-ray source by laser-electron collision based on the X-band linac for medicine. To realize a remarkably compact system, we adopt the X-band system and commercial Q-switch laser. The X-ray yield by the electron beam and Q-switch Nd:YAG laser of 2.5 J/10 ns is 10^7 photons/RF-pulse (10^8 photons/sec in 10 pps). We also design to adopt a technique of laser circulation to increase the X-ray yield up to 10^8 photons/pulse (10^9 photons/s).

Final target of this study is the integrated system for medicine. A variety of medical applications using monochromatic tunable hard X-rays, such as the dual energy X-ray CT, the subtract X-ray CT and the angrography, will start in 2006.

ACKNOWLEDGEMENT

This study is performed in the national project of "Development of Advanced Compact Accelerators" of the Ministry of Science, Technology, Education, Sports and Atrs (MEXT), hosted by the National Institute for radiological Science, in Japan, and also supported by Research Program on Development of Innovative Technology (#0494) in Japan Science and Technology Agency of MEXT.

REFERENCES

[1] E. Rubenstein, et al., *Proc. National Academy Sci. USA* **83**, 24, p.9724(1986).
[2] S. Ohtsuka, et al., *British Journal of Radiology* **72**, p.24(1999).
[3] M Torikoshi, et al., *Opt.* **6** 371(2000).
[4] K. Dobashi, et al., *Proc. 8th European Particle Accelerator Conference in Paris*, France, 667(2002).
[5] K. Dobashi, et al., *Jpn. J. Appi. Phys.* 44 **4A** pp.1999-2005 (2005).
[6] A. Fukasawa, et al., *International Journal of Applied Electromagnetics and Mechanics* **14**, 221 (2001/2002)
[7] A. Fukasawa, et al., "A tunable monochromatic hard X-ray CT composed of an X-band linear accelerator and a Q-switched laser", *Nuclear Inst. and Methods B* 241 pp.921-925 (2005)
[8] M. Uesaka, et al., *Proc. XXI International LINAC Conference in Korea*, August 19 - 23, Gyeongju, Korea, 626(2002).
[9] E. Esarey, P. Sprangle, A. Ting, Nucl. Instr. and Meth. A 331, p.545 (1993).
[10] K.-J. Kim, S. Chattopadhyay, C.V.Shank, Nucl. Instr. and Meth. A 341, p.351 (1994).
[11] I.V.Pogorelsky et al. "Demonstration of 8×10^{18} photons/second peaks at 1.8 Å in a relativistic Thomson scattering experiment, Phys. Rev. ST-AB 3, 090702/1-8(2000).
[12] M.Uesaka et al. Nucl. Instr. and Meth. A 455, pp.90-98 (2000)
[13] M.Uesaka, Working Group D Summary, in this proceeding.
[14] H.Ogino et al. in this proceeding.
[15] M.Yorozu et al. Jpn.J.Appl.Phys., Vol.40, pp. 4228-4232 (2001).
[16] F.E.Carroll et al. Am.J.Rentgenol. 181, P.1197 (2003)
[17] W.J.Brown et al. Phys.Rev.Lett.Vol.7 060702 (2004)
[18] A.E. Vieks et al. Abstract of 21st Int. LINAC Conf., Korea, 2004, P.83.
[19] K.Hirano et al., "High Intense Multi-bunch Beam Generation by Photo-cathode RF Gun", (to be submitted to NIM)
[20] GLC report, *KEK report* 2003-7(2003).
[21] K.Hirata; An introduction to SAD (Strategic Accelerator Design), Second Advanced ICFA Beam Dynamics Workshop, CERN 88-04 (1988).
[22] Daphne F. Jackson and D. J. Hawkes, *Phys. Rep.* **70** 169 (1981).
[23] XCOM http://physics.nist/gov/XCOM

DESIGN OF A 2 kA, 30 fs RF-PHOTOINJECTOR FOR WATERBAG COMPRESSION

S.B. VAN DER GEER*, O.J. LUITEN
Eindhoven University of Technology, The Netherlands

M.J. DE LOOS
Pulsar Physics, The Netherlands

Because uniformly filled ellipsoidal 'waterbag' bunches have linear self-fields in all dimensions, they do not suffer from space-charge induced brightness degradation. This in turn allows very efficient longitudinal compression of high-brightness bunches at sub or mildly relativistic energies, a parameter regime inaccessible up to now due to detrimental effects of non-linear space-charge forces. To demonstrate the feasibility of this approach, we investigate ballistic bunching of 1 MeV, 100 pC waterbag electron bunches, created in a half-cell rf-photogun, by means of a two-cell booster-compressor. Detailed GPT simulations of this table-top set-up are presented, including realistic fields, 3D space-charge effects, path-length differences and image charges at the cathode. It is shown that with a single 10 MW S-band klystron and fields of 100 MV/m, 2 kA peak current is attainable with a pulse duration of only 30 fs at a transverse normalized emittance of 1.5 μm.

1. Introduction

As long as stochastic effects are negligible, Liouville's theorem dictates that the local phase-space density is constant along the particle trajectories during transport and acceleration. However, non-linear effects and correlations can easily degrade the overall bunch quality by orders of magnitude during the first stages of acceleration. Uniformly filled ellipsoidal 'waterbag' bunches have linear self-fields in all dimensions [1,2], and they therefore do not suffer from the deteriorating effects of non-linear space-charge forces. This in turn allows very efficient bunch manipulation of high-brightness beams at sub and mildly relativistic energies, a regime that formerly needed to be avoided because of space-charge induced brightness reduction. The novel concept of compression of ellipsoidal bunches puts 100 fs, 1 micron emittance electron bunches of kA peak current into the realm of small research groups, paving the way to a wide range of new and exciting table-top applications such as the generation of TeraHertz

* Corresponding author: s.b.van.der.geer@tue.nl

radiation [3], pulsed radiolysis for advanced material research [4], and pulsed electron diffraction experiments [5].

It was first suggested in [6] that waterbag electron bunches may be realized by 'blow-out' in the 'pancake' regime. A practical way to produce waterbag bunches is illuminating a metal cathode inside a sufficiently high accelerating field by an ultra-short photo-excitation pulse with a half-sphere transverse intensity profile, see [7]. In [8] it was first shown that waterbag bunches can be created in a 'standard' 1.5 cell rf-photogun, with a properly shaped 30 fs FWHM, 1 mm initial radius photo-excitation laser, taking into account the effects of realistic accelerating fields, space-charge forces, image charges and magnetic focusing. This resulted in a slightly egg-shaped bunch, also observed in [9]. However, despite this small deformation, the initial thermal emittance was fully recovered, see also [10], and the bunch showed an almost perfectly linear momentum-position phase-space correlation with a width in the order of only 10 fs. This suggests that a linear compression scheme can compress these bunches to multi-kA peak current with an rms emittance in the order of 1 μm.

The brightness conserving waterbag theory only applies if the bunches are accelerated in linear external fields, and if they have perfect shape. Both conditions are never perfectly met. Detrimental effects however can be minimized, as we show in this paper by detailed GPT simulation results of a relatively simple set-up, comparable in complexity to a single rf-photogun, thus proving the feasibility of longitudinal compression of waterbag bunches at 1 MeV. It is shown that a simple rearrangement of the cells of a 2.5 cell S-band rf-photogun, combined with proper transverse laser-shaping, peak currents of 2 kA at an emittance value of 1.5 μm are attainable with a FWHM pulse duration of only 30 fs.

2. Set-up

Shown in Figure 1 is the system under investigation. It consists of a half-cell S-band rf-cavity, illuminated by a femtoseceond uv photo-excitation laser with a half-sphere transverse intensity profile, a 2-cell S-band booster-compressor, four solenoids and a bucking coil. The set-up is smaller but analogous to the experiments described in [11].

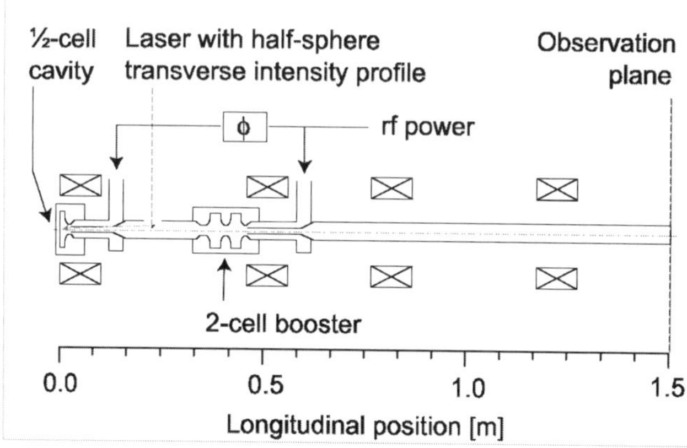

Figure 1: Schematic of the set-up under investigation.

To compress a bunch with length L over a distance d, the required velocity difference Δv between the head and tail particle is given by $\Delta v = v\,L/d$. From the definition of the Lorentz factor γ and the kinetic energy E of a particle $E = (\gamma - 1)\,m\,c^2$, it immediately follows that $\Delta v = \Delta E/(v\,\gamma^3\,m)$. The γ^3 factor in the denominator makes ballistic bunching particularly efficient at sub or mildly relativistic energies because far less energy spread is required for the same amount of compression. The amplitude of the electric accelerating field determines the total amount of charge that can be extracted. However, a high final energy is not desirable because of the γ^{-3} dependency mentioned before. To get the best of both worlds, we decided to start with a 100 MV/m S-band cavity only half-cell in length to limit the energy of the 100 pC bunch to be compressed to just 1 MeV, i.e. $\gamma \approx 3$. This radically breaks with the current paradigm in accelerator physics that only multi-ps electron bunches accelerated to sufficiently high energy, typically 10-100 MeV, can be compressed to the kA, 100 fs regime.

Because the waterbag shape prevents rms brightness deterioration, we use a simple drift space to let the bunch expand to the desired length of a few ps, sufficiently long for S-band compression. The proposed booster-compressor is a 2-cell 100 MV/m cavity, analogous to the last two cells of a 2.5-cell rf-photogun. It changes the slope in transverse momentum space while simultaneously increasing the energy to about 4 MeV. Compression is an intrinsically non-linear process because the relation between velocity and energy is not linear. Furthermore, when mildly relativistic particles gain or loose energy, they change velocity and spend less or more time in the cavity respectively. The phase of the rf-booster has been set such that at the point maximum compression, at z=1.5 m, both non-linear effects cancel each other to second order.

3. Simulation results

Detailed particle tracking simulations with the General Particle Tracer (GPT) code [12,13] form the basis of the predicted behavior of the set-up under investigation. The relativistic equations of motion of 10^5 sample particles are integrated through the combined fields of the rf-cavities, the solenoids and space-charge. This approach includes path-length differences, non-linear external fields, non-linear space-chare forces and image charges at the cathode. The integration steps of the embedded 5th order Runge-Kutta integrator are automatically chosen and expand from a few fs near the cathode region, into the ps regime in the cavities, to almost one ns in the drift spaces. The standing-wave field-pattern in both cavities is calculated with the Superfish [14] set of codes, and fed into GPT as cylindrically symmetric field-maps with 0.25 mm spacing in both the z and r-directions. Space-charge forces are calculated in 3D with a PIC approach, where technical details related to the multigrid solver tailor made for bunches with extreme and varying aspect ratio can be found in [15].

The bunch length, transverse size, and energy of a 100 pC bunch as function of longitudinal position are shown in Figure 2. Longitudinal compression by two orders of magnitude to 30 fs rms is reached at a final energy of 3.7 MeV. A transverse waist at the booster cavities is introduced to reduce the undesired effects of the non-linear entry and exit fields as much as possible. The reduction in energy at the entrance of the booster is to compensate for second-order effects in longitudinal phase-space. Detailed results at the longitudinal focus at z=1.5 m are shown in Figure 3. It is shown that a peak current of at least 2 kA and a FWHM bunch duration of only 30 fs can be realized by compression of 1 MeV ellipsoidal bunches in a set-up comparable to a single rf-photogun. Although the rms emittance of 1.5 μm is not spectacular, we would like to stress that it can be significantly improved when traded against pulse length. In fact, by only changing the magnetic field-profile slightly, it is possible to reach 0.7 μm, almost the initial thermal emittance, at a peak-current of 700 A.

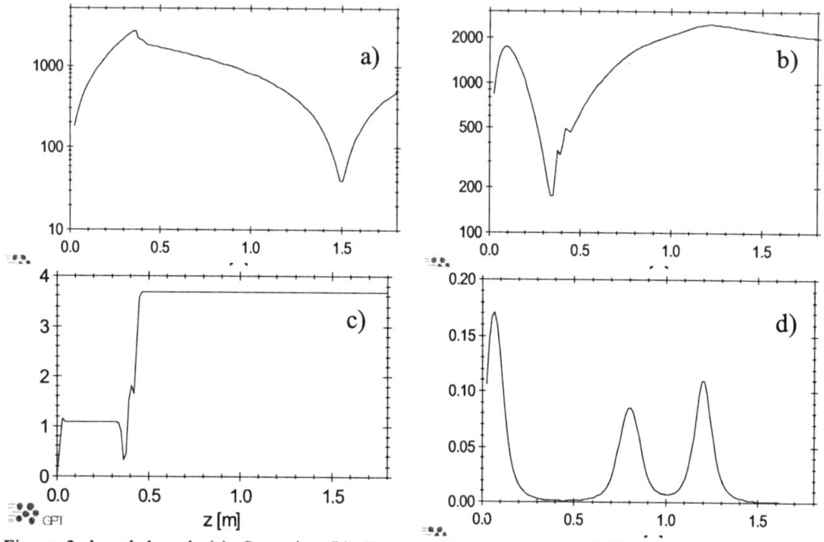

Figure 2: bunch-length (a), Spot-size (b), Energy (c) and Magnetic field on-axis (d) for a 100 pC bunch in the proposed set-up.

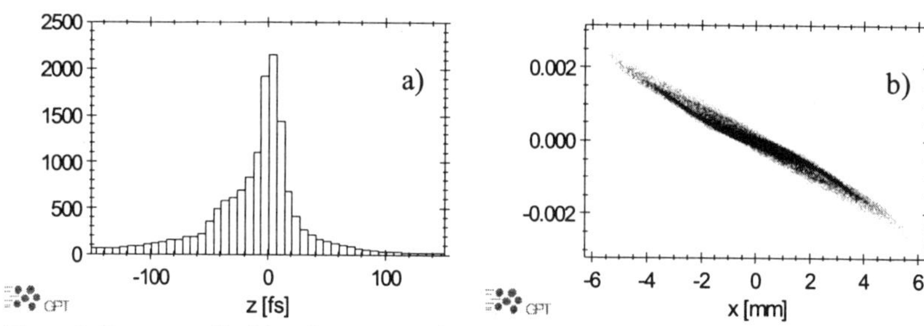

Figure 3: Current profile (a) and transverse phase-space (b) at the longitudinal focus at z=1.5 m. The normalized rms emittance corresponding to (b) is 1.5 μm.

4. Conclusion

Ideally, the brightness of uniformly filled ellipsoidal bunches does not deteriorate during acceleration and transport in linear fields. This fully eliminates the requirement that bunches need to be accelerated to highly relativistic energies before they can be compressed. In practice however, perfect waterbag bunches can not be created due to laser aberrations and image charges, external forces are never linear, and path-length differences need to be taken into consideration. Furthermore, once the ellipsoidal shape is distorted, non-linear space-charge forces come into play again which further deteriorates the rms brightness. Nevertheless, detailed GPT simulations of longitudinally compressed waterbag bunches predict that sub 100 fs, kA, high-brightness bunches at only a few MeV are feasible in a set-up comparable in size and complexity to a single rf-photogun.

References

1. O.D. Kellogg, *Foundations of potential theory*, Springer Verlag, (1929).
2. I.M. Kapchinskii and V.V. Vladimirskii, Proc. Int. Conf. High Energy Accelerators, CERN, Geneva, p. 274 (1959).
3. G.P. Gallerano and S. Biedron, *Overview of terahertz radiation sources*, Proceedings of the 2004 FEL Conference, p. 216-221, (2004).
4. Y. Muroya, T. Watanabe, G. Wu, X. Li, T. Kobayashi, J. Sugahara, T. Ueda, K. Yoshii, M. Uesaka, Y. Katsumura, *Design and development of a sub-picosecond pulse radiolysis system*, Radiation Physics and Chemistry **60**, (2001), p. 307.
5. B.J. Siwick, J.R. Dwyer, R.E. Jordan, R.J.D. Miller, *An Atomic-Level View of Melting Using Femtosecond Electron Diffraction*, Science 302, (2003) p. 1382.
6. L. Serafini, *AIP Conf. Proc.*, **413**(1), 321 (1997); L. Serafini, *AIP Conf. Proc.*, **279**, 645 (1993).
7. O.J. Luiten, S.B. van der Geer, M.J. de Loos, F.B. Kiewiet, and M.J. van der Wiel, *How to Realize Uniform Three-Dimensional Ellipsoidal Electron Bunches*, Physical Review Letters, **93**, 094802, (2004).
8. O.J. Luiten, S.B. van der Geer, M.J. de Loos, F.B. Kiewiet, M.J. van der Wiel, *Ideal waterbag electron bunches from an rf photogun*, Proc. 2004 Eur. Part. Acc. Conf., Luzern, Switzerland, (2004) p. 725.
9. J.B. Rosenzweig, A.M. Cook, R.J. England, M. Dunning, S.G. Anderson, and M. Ferrario, *Emittance compensation with dynamically optimized photoelectron beam profiles*, Nucl. Instr. and Meth. A, Vol. **557**, (2006) p. 87.
10. C. Limborg-Deprey, P.R. Bolton, *Optimum electron distributions for space charge dominated beams in photoinjectors*, Nucl. Instr. and Meth. A, Vol. **557**, (2006) p. 106.
11. S.G. Anderson, et al., *Velocity bunching of high-brightness electron beams*, Phys. Rev. ST Accel. Beams 8, 014401, (2005).
12. http://www.pulsar.nl/gpt
13. S.B. van der Geer, O.J. Luiten, M.J. de Loos, G. Poplau, U. van Rienen, *3D space-charge model for GPT simulations of high brightness electron bunches*, Institute of Physics Conference Series No. 175, p. 101, (2005).
14. J.H. Billen, L.M. Young, Poisson Superfish, Los Alamos National Laboratory, Report LA-UR-96-1834.
15. G. Pöplau, U. van Rienen, S.B. van der Geer, and M.J. de Loos, *Multigrid Algorithms for the Fast Calculation of Space-Charge Effects in Accelerator Design*, IEEE Trans. Magn., Vol **40**, 714 (2004).

PROPOSAL FOR A HIGH-BRIGHTNESS PULSED ELECTRON SOURCE*

M. ZOLOTOREV[1], E. D. COMMINS[2], S. HEIFETS[3] AND F. SANNIBALE[1]

[1]*Lawrence Berkeley National Laboratory, Berkeley, CA 94720, USA*
[2]*Physics Dept., University of California Berkeley, Berkeley CA 94720, USA*
[3] *Stanford Linear Accelerator Center, Stanford University, Stanford, CA 94309, USA*

We propose a novel scheme for a high-brightness pulsed electron source, which has the potential for many useful applications in electron microscopy, inverse photo-emission, low energy electron scattering experiments, and electron holography. A description of the proposed scheme is presented.

1. Introduction

We propose a novel scheme for a high-brightness pulsed electron source, which has the potential for many useful applications in electron microscopy, inverse photo-emission, low energy electron scattering experiments, and electron holography. The source makes use of Cs atoms in an atomic beam. Each cycle of the device begins with a laser pulse that excites a single Cs atom on average to a band of high-lying Rydberg nP states. The resulting valence electron Rydberg wave packet evolves in a nearly classical Kepler orbit. When the electron reaches apogee, an electric field pulse is applied that ionizes the atom and accelerates the electron away from its parent ion. The collection of electron wave packets thus generated in a train of cycles can occupy a phase volume near the quantum limit and it can possess very high brightness. Each wave packet can exhibit a considerable degree of coherence.

A proof of principle experiment is underway at the Lawrence Berkeley National Laboratory to demonstrate the essential features of the source.

* Work supported by the Director, Office of Science, of the U.S. Department of Energy under Contract No. DE-AC03-76SF00098.

2. Source Scheme Description

To explain the essential features of the source, we recall several basic quantities. In the rest-frame of a beam of electrons propagating in the z direction, the dimensionless differential phase volume dΓ is:

$$d\Gamma = \frac{1}{h^3} dxdp_x dydp_y dzdp_z = \frac{dxd\beta_x dyd\beta_y dzd\beta_z}{(2\pi)^3 \lambda_C^3} \quad (1)$$

where $\beta_{x,y,z} = v_{x,y,z}/c$ and c is the velocity of light, and $\lambda_C = \hbar/m_e c = 3.86 \times 10^{-11}$ cm is the electron Compton wavelength. Assuming that the electron beam is described by a six-dimensional Gaussian distribution in phase space, the total dimensionless phase volume Γ occupied by the electron beam may be expressed in terms of the transverse emittances $\varepsilon_{x,y}$ and longitudinal emittance ε_z as follows:

$$\Gamma = \frac{\varepsilon_x \varepsilon_y \varepsilon_z}{\lambda_C^3} \quad (2)$$

where

$$\varepsilon_x = \sqrt{\left(\langle x^2 \rangle - \langle x \rangle^2\right)\left(\langle \beta_x^2 \rangle - \langle x\beta_x \rangle^2\right)} \quad (3)$$

with similar definitions for ε_y and ε_z [1]. Let N_e be the actual number of electrons, and consider the ratio:

$$\delta_F = \frac{N_e}{\Gamma} = \lambda_C^3 \frac{N_e}{\varepsilon_x \varepsilon_y \varepsilon_z} = \lambda_C^3 B \quad (4)$$

where $B = N_e/\varepsilon_x \varepsilon_y \varepsilon_z$ is the “brightness”. The Pauli exclusion principle requires $\delta_F \leq 1$ for electrons of a given spin polarization; hence $\varepsilon_x \varepsilon_y \varepsilon_z \geq \lambda_C^3$. Typically, field emission electron guns of modern design achieve δ_F up to 5×10^{-6}, [2]. For the present electron source it appears possible to reach $\delta_F \approx 2 \times 10^{-2}$.

In our electron source (see Fig. 1) an atomic beam effusing from an oven at temperature $\approx$ 500 K is collimated to reduce the transverse beam temperature to $\approx$ 5 K. Also, laser excitation along the beam axis selects a narrow band of longitudinal velocities. Thus excited atoms have a thermal energy spread $\Delta E \approx 4 \times 10^{-4}$. If the valence electrons were then optically excited to ionization threshold and we could ignore subsequent space-charge heating, the electron energy spread would be $\delta E = (m_e/M_{Cs})\Delta E \approx 1.6\times10^{-9}$ eV. However space-charge interactions result in an electron temperature given by $kT \approx e^2 n_e^{1/3}$ where n_e is the electron density. For $n_e \approx 10^9$ cm^{-3} this would yield $\delta E \approx 10^{-4}$ eV. In the present proposal this undesirable heating is avoided by exciting one atom at a time on average (see Fig. 1) to a very high lying band of nP Rydberg states (for example $\bar{n} \approx 800$, $\Delta n \approx 50$), with 3 mutually perpendicular laser beams: [L1: CW, 852 nm ($6S_{1/2}$-$6P_{3/2}$)]; [L2: CW, 1.47 μm ($6P_{3/2}$-$7S_{1/2}$)]; [L3: pulsed, 777 nm ($7S_{1/2}$-nP)]. These intersect in a small interaction region [volume $\approx (10\ \mu m)^3$] midway between two plane parallel electrodes E1, E2 with separation ≈ 1 cm. The laser intensities and atom number density are chosen to give a single atom excitation probability per pulse ≈ 1.

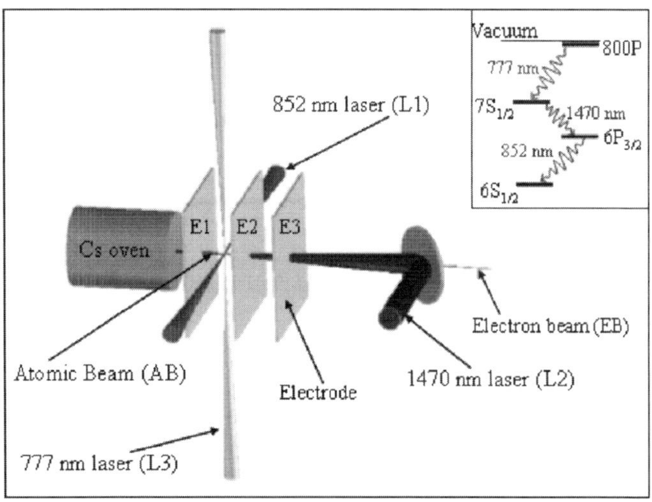

Figure 1. Schematic diagram of pulsed electron source, with insert showing relevant Cs energy levels (not to scale).

Each resulting valence-electron Rydberg wave packet describes a pendulum-like orbit that is very nearly classical, with a half-period $T = (\pi \hbar^3 \bar{n}^3)/(m_e e^4)$ ($T \approx 39$ ns for $\bar{n} \approx 800$). The frequency of the 777 nm laser pulse can be chirped so that the classical Kepler orbits for various n values in the vicinity of $n = 800$ all reach apogee simultaneously. The most favorable chirping parameters are found by optimizing a quantum mechanical calculation of laser excitation and subsequent time evolution of the Rydberg wave-packet. At apogee this wave packet fills a spherical shell centered on the remaining Cs ion, with mean shell radius $\langle R \rangle \approx 67$ μm and shell half-thickness $\Delta R \approx 6$ μm (see Fig.2), while $\Delta\beta_R$ reaches a minimum value such that $\Delta R \cdot \Delta\beta_R$ is very close to the uncertainty principle limit. (Note that $\Delta\beta_\theta$ is negligible compared to $\Delta\beta_R$).

The "intrinsic" transverse emittance at apogee is $\varepsilon_{x,y} \cong \left\langle \sin^2\theta \begin{pmatrix} \cos^2\phi \\ \sin^2\phi \end{pmatrix} \right\rangle \langle R \rangle \Delta\beta_R$. Here θ, ϕ are the usual angles in polar coordinates between the electron position vector **R** and the z axis. Spectroscopic observations show that for Cs the oscillator strengths for $7S\text{-}nP_{3/2}$ transitions are much larger than for $7S\text{-}nP_{1/2}$. Taking into account this spin-orbit effect [3, 4] and assuming that the 777 nm photons are linearly polarized along z, one finds that the probability density of the Rydberg wave packet is proportional to $1+3\cos^2\theta$. Thus $\left\langle \sin^2\theta \begin{pmatrix} \cos^2\phi \\ \sin^2\phi \end{pmatrix} \right\rangle = \frac{4}{15}$ and for optimum chirp conditions at $\bar{n} \approx 800$, $\varepsilon_{x,y} \cong 2.5 ƛ_C$.

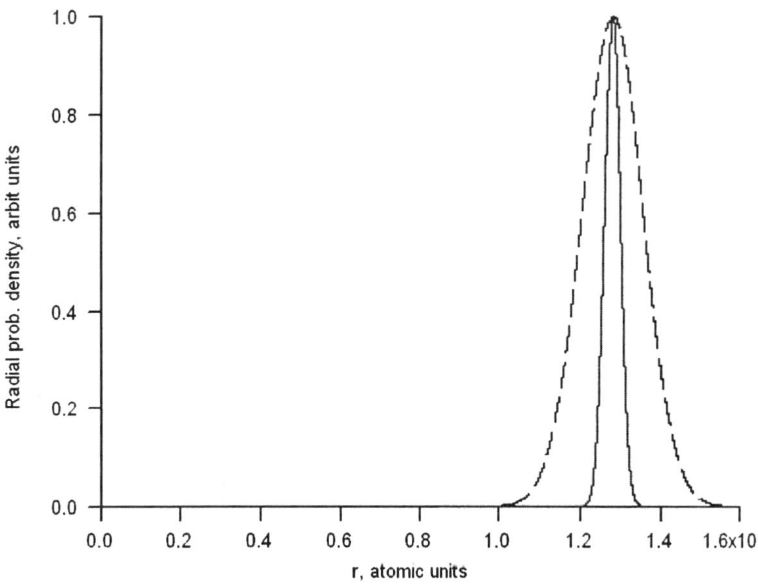

Figure 2. Radial probability distributions for Rydberg wave packet at apogee. Red solid curve: 777 nm laser pulse FWHM=2.35 ns, no chirp. Blue dashed curve: 777 nm laser pulse FWHM=8.2 ns, optimum chirp.

In addition appreciable phase coherence in each electron wave packet can be achieved. Thus in principle the proposed source could be employed for electron holography and other interferometric experiments.
At apogee, an electric field pulse is applied in the –z direction; it ionizes the atom and accelerates the electron in the +z direction. For a pulse peak = 100 V/cm and FWHM = 0.5 ns, the final electron energy is $E \approx 2$ eV, with $\Delta E \approx 10^{-4}$ eV. Several ns after the accelerating pulse, the electron reaches electrode E2 (Fig.1) and passes through a circular aperture A. The diameter of A is chosen to minimize aberrations and simultaneously to maintain reasonable field uniformity in the excitation region. To analyze the emittances of the electron bunch downstream from A we employ a Monte Carlo calculation to integrate the electron classical equations of motion in the acceleration pulse field and the Coulomb field of the parent ion, taking into account the apogee values for $\langle R \rangle$, ΔR, and $\Delta\beta_R$, the random value of θ at apogee, smearing effects due to the finite size of the interaction region, and the velocity distribution of the cesium atoms and the resulting transverse and longitudinal Doppler effects. We estimate that one can achieve emittances $\varepsilon_{x,y} \cong 3.0\lambda_C$, $\varepsilon_z = 4.8\lambda_C$, and a fractional energy spread $\Delta E/E \approx 4 \times 10^{-5}$.
For a Rydberg atom with principal quantum number n, the electric field required for spontaneous ionization is $E_I = e/\left(16\, a_0^2 n^4\right) \approx 8 \times 10^{-4}$ V/cm for $n = 800$. Such a field is generated by an electronic charge at a distance $s = 0.014$ cm. Therefore we require that when the next laser pulse occurs, the residual ion from the present pulse is at a distance $s' >> s$ from the interaction region. To achieve this, after the acceleration pulse another "ion clearing" electric field pulse in the –z direction is applied to remove the remaining Cs ion from the interaction region. Because the ion is massive, the duration of the clearing pulse and its amplitude must be quite large. After termination of the ion clearing pulse, another laser pulse is applied which begins the next cycle. For realistic ion clearing pulses, a cycle repetition rate ≈ 3MHz can be achieved.
If we are to achieve the emittances described above, it will be necessary to reduce stray electric fields in the interaction region to a level $E_S \leq 10^{-4}$ V/cm. This is a difficult but not impossible task, as has been demonstrated by the results of earlier experiments on Rydberg states [5]. Stray magnetic fields B can also cause undesirable perturbations that degrade brightness unless the electron radius of curvature is much greater than R. To achieve this, one must reduce the ambient magnetic field to a level ≤ 0.001 Gauss. This can be done readily by enclosing the interaction region in an appropriate magnetic shield. Care must

also be taken to avoid generation of photo-electrons by absorption of stray laser photons on electrode surfaces.

Collisions between the Rydberg atoms of interest and ground state atoms as well as molecules are unlikely to cause serious difficulties, even though the volume of an $\overline{n} \approx 800$ Rydberg atom is so large at apogee that for an atomic beam number density $n_0 \approx 10^{10}$ cm^{-3}, approximately 10^4 ground state atoms are contained within that volume. The cross-sections for collisions of the Rydberg atom with Cs_2 and with background gas molecules are undoubtedly large, but the number densities of these molecules will be so small (with proper design of the atomic beam source and in ultra-high vacuum) that they should not present serious problems. We are encouraged by the results of previous alkali Rydberg atom experiments done in rather rudimentary vacuum conditions, which achieved principal quantum numbers n ≈ 1000 [6].

3. Conclusions

The proposed source would be a first step toward a novel scanning electron microscope with current density on the sample of several kA/cm^2 and Angstrom resolution. Such resolution for a low energy microscope would be achieved for two main reasons: the requirements on chromatic aberration are minimized by extremely small energy spread, and spherical aberration can be compensated by axial symmetric lenses. In fact, the well defined time structure of the electron beam allows use of time- dependent fields that can be either focusing or defocusing, which implies the possibility of positive or negative spherical aberration. The source would also extend applications for electron beams to energy exchange with μeV accuracy in inelastic atomic and molecular scattering, and new ways of investigating chemical reactions and dynamics on a picosecond time scale using pump-probe techniques. Finally, the feasibility of phase coherence between successive electron wave packets opens the possibility of electron holography and other interferometric experiments.

A proof of principle experiment is underway at the Lawrence Berkeley National Laboratory to demonstrate the essential features of the source.

Acknowledgments

We thank J. Byrd, S. Chu, D. Colomb, P. Denes, G. Huang, Z. Hussain, J. Krupnik, G. Lebedev, S. Lidia, J. Oneill, D. Robin, K. Robinson, R. Schoenlein, R. Vogel, and W. Wan, who have provided very valuable discussions as well as scientific and technical support.

References

1. C. Lejeune, J. Aubert, Applied Charged Particle Optic, Advances in Electronics and Electron Physics, A. Septier Editor, Supp. 13 A (1980).
2. J. C. H. Spence, M. R. Howells, Ultramicroscopy **93** (2002) 213.
3. D. Norcross, Phys. Rev. A **7**, 606, 1973.
4. V H. Kaminski, J. Kessler, and K.J. Kollrath Phys. Rev. Lett. **45**, 1161, 1980.
5. J. Neukammer, et al., Phys. Rev. Lett. **59**, 2947, 1987.
6. M. T. Frey, S. B. Hill, K. A. Smith, and F. B. Dunning Phys. Rev. Lett. **75**, 810, 1995

Working Group 1

SUMMARY OF WORKING GROUP I: "ELECTRON SOURCES"

MASSIMO FERRARIO AND GIANCARLO GATTI

INFN-LNF
Via Enrico Fermi 40
Frascati (Roma), 00044, Italy

This report summarizes the presentations and discussions in the Working Group "Electron Sources" during the ICFA Advanced Accelerator Workshop on "The Physics and Applications of High-Brightness Electron Beams", held at the Ettore Majorana Foundation and Center for Scientific Culture, Erice, October 9-14, 2005. Latest developments towards a better understanding of high-brightness photo injectors physics and technology are reported.

1. Introduction

High brightness photo-injectors based on RF guns are still the main electron source for many advanced accelerator applications, ranging from high brilliance Free Electron Lasers (FEL) to Compton sources, plasma accelerator devices and to industrial and medical applications. The reliable RF guns operation achieved in the last decade, that lead for example to the proliferation of short wavelength SASE-FEL projects driven by a photo-injector, pushed forward the demands on beam quality produced by photo-injectors, asking for sub-micron normalized emittances and hundreds Ampere peak currents. The 16 presentations and the consequent formal/informal discussions in this Working Group covered a wide variety of issues aiming to a better understanding of the physics, the technology and the operation of electron photo-injectors. In these presentations, important experimental, analytical and numerical results have been discussed in support to new concepts concerning photo-cathode's quality, emission process, RF structure designs and beam dynamics. In the next paragraphs a brief summary of the presentations is reported.

2. Photo-cathodes

The first day of the working group has been entirely devoted to photo-cathodes issues, including cathode cleaning, QE and emission process studies, superconducting and needle cathodes.

Concerning polishing procedures on metal cathodes, D. H. Dowell reported results about Copper cathode polishing through a 1 keV hydrogen ion beam bombardment [1]. This technique allows to purify metals from different contaminants (e.g. C) that lower Quantum Efficiency (QE) of photo-cathodes. Measurements of the QE vs. wavelength, surface contaminants using x-ray photoelectron spectroscopy and surface roughness were performed at SLAC on a 1 cm in diameter copper sample. The results showed a significant increase in QE after cleaning with no effect on the surface roughness. The final results led to a very good agreement with the quantum yield calculated with the Fowler model. A possible implementation of H-ion cleaning on a RF-gun has been also discussed. With respect to H-plasma cleaning, Dowell said, H-ion technique is quick and can be made "in situ", while plasma seems to be advisable for a complete polishing before installation.

Results of a detailed characterization of plasma cleaning by low energy argon and hydrogen ion plasma on a Copper photocathode, were presented by D. T. Palmer [2]. The discussed experimental setup allows to perform sample polishing and after-process analysis in the same chamber. The Copper sample was prepared with diamond paste cleaning and different ultrasonic baths to remove the main impurities, characterized by Atomic Force Microscope (AFM), QE measurements (~ 10^{-7} at 266 nm) and X-ray Photoelectron Spectroscopy (XPS). After introduction in the chamber the cathode was exposed to different polishing cycles, made up of thermal baking and argon ion plasma cleaning. At each step the sample was analyzed with XPS to control the percentage of impurities through cleaning, and measuring the improved QE. At the end, another AFM was performed to control the effects of the process on the surface. XPS showed that plasma cleaning is able to remove some of the residual element left by baking (C, Na, O) with increasing of the QE up to 2.7×10^{-4} at 266 nm wavelength.

J. Schmerge presented a study on temporal response of metal photo-cathodes, mainly focused on Cu and Mg [3]. As Schmerge pointed out, QE of a metal cathode is strongly dependent on the field at the cathode surface due to the Schottky effect. Since the field is time dependent the quantum efficiency is also time dependent. It results that the laser pulse shape used to generate electrons doesn't correspond to the electron bunch shape. In addition since the thermal emittance and quantum efficiency are related, the thermal emittance is also time dependent. To overcome this unwanted Schottky effect, taking place on the time scale of the incident laser pulse length, some laser beam shape pre-compensations has to be considered to ensure a flat-top electron pulse, for example with an optimized laser energy ramp. Using the standard "three steps" theoretical emission model, Schmerge compared also QE measurements with theoretical predictions. Some discrepancy between theory and measurements remains, but it is probably due to surface effects still not taken in consideration in the Schmerge model.

In addition to the dynamics of the Schottky effect upon the QE and the thermal emittance there is also the wavelength dependence of the cathode's dielectric constant which can affect electron emission, as reported by D. Dowell [4]. The typical drive laser has a small bandwidth and therefore the wavelength dependence can be ignored. However there are recent suggestions for producing electron bunches from very short (10's of fs) laser pulses to produce very linear space-charge forces, as in the case of the so called Blow Out regime, resulting in a larger bandwidth in the laser spectrum. In this case one must take in consideration the excitation of electronic surface states of the cathode, the so-called plasmon states, that can be seen as resonance frequencies in the dielectric constant. These phenomena change strongly the emission process, leading to unwanted extracted charge distributions.

An electron beam with a low bunch charge as well as a short laser pulse length (~5 pC, ~ 3 ps) allows to study the emission mechanisms with a good phase definition and with a very small influence of the space charge force. In this case measured emittance is dominated by RF and thermal contributions. RF contributions can be computed while thermal emittance can be only estimated via simplified theoretical models. In his two talks J. H. Han has discussed experimental and numerical studies performed at PITZ toward detailed understanding of the photo emission and secondary emission processes at the Cs_2Te cathode. Concerning thermal emittance Han showed that the ratio η between experimental results and theoretical predictions resulted equal to 1.47 [5]. Such a high η value may be related to measurement errors, or may be a real emittance contribution due to a non-uniform QE or mechanical roughness of the photo cathode. Further studies are under way at PITZ to clarify such high discrepancy. Also multipacting in the gun cavity changes the surface status of the cathodes producing serious damages. J. H. Han reported in his second talk [6] about experimental observations of secondary electron emission at PITZ and comparisons with beam dynamics simulations, performed with an upgraded version of ASTRA modeling the electron multiplication process. Multipacting takes place at a low RF gradient (~ 1MV/m), Han recalled, with a strong influence of the solenoid field configuration. During operation of the gun cavity, two multipacting peaks in the extracted current signals have been observed at the beginning and the end of the RF pulse, when the cavity voltage is close to the multipacting threshold. The rear multipacting peaks have been measured for various RF field amplitudes in the cavity and photocathode damages were observed due to a consequent insufficient vacuum environment. Han simulations show that the position and the field configuration of the solenoids can be properly determined in order to inhibit the multipacting, thus reducing cathode risks.

G. Gatti presented results obtained with a new production technique of Mg film photo-cathodes [7]. The goal was to reach a Mg uniform layer strongly adherent to substrate, resistant to high field and laser radiation. For this purpose, well known deposition techniques such sputtering have been studied, as well as new Pulsed Laser Ablated Deposited cathodes (PLAD) have been considered. PLAD technique should ensure a high purity and adherence of the deposited film. In addition gives the possibility of multilayer deposition within the same session and an excellent control on the layer growth (nm). Two different PLAD cathodes synthetized for different purposes were tested in a DC chamber at low field (1 MV/m) with a Nd:Yag 266 nm laser source. The first sample was made by Mg film (1 μm thickness) covered by a graphite protective layer (few nm), and it showed good results comparable to sputtered sample (QE = $3x10^{-4}$) after laser cleaning removal of C layer. The second sample was produced with improved deposition rate of PLAD technique (2.5 μm thick film). Different areas were polished with laser cleaning at different fluences. Inside each of the area the uniformity of emission was 12%. The SEM analysis showed drastic morphological changes with different laser fluences. The Mg film that was originally amorphous, gradually became crystalline growing up with the fluence.

J. Smedley presented a study on superconducting Niobium and Lead photo-cathodes [8]. A Nb cathode would be a very suitable option for Nb superconducting guns, but its poor QE would limit average current. A Pb cathode instead should ensure a higher QE, but its behaviour inside at cryogenic temperatures has not yet been studied in detail. At first comparison between the two superconductors based on theoretical model has been presented, looking at the parameters that generate a different QE prediction (e.g. density of states, electron and photon mean free path). Then QE measurements were conducted on amorphous Nb and Pb films on Nb substrate samples. A Nd:Yag laser (266 nm) and a Krf excimer laser (248 nm) were used during measurements of QE versus laser cleaning energy, and a deuterium lamp during QE versus wavelenght scan. All samples surfaces were controlled by SEM, before and after laser cleaning, to look at morphological changes and damage threshold of material. Concerning Nb the best results were obtained with mechanical polished samples after 0.3 mJ/mm^2 laser cleaning (QE < 10^{-4} at 266 nm). Pb samples were measured mounted on a cold finger and cooled at -170° C to test cathode behaviour at low temperatures. Best results were obtained with vacuum arc deposited Pb samples (QE = $3.5x10^{-4}$ at 266 nm). According to Smadley tests cold temperature doesn't affect performances of the cathode as long as vacuum environment is good, a very promising results towards a fully superconducting RF gun.

R. Ganter [9] and M. Pedrozzi [10] presented in two talks a promising injector concept, based on cathode field emission process, to be employed by the Low Emittance Gun (LEG) project at PSI. The goal of the project to obtain 10^{-2} μm thermal emittance from the cathode and to prevent space charge induced emittance growth with a strong acceleration immediately after the emitter with a high voltage pulsed diode. Two possible choices for this cathodes have been discussed: field emitter arrays (FEA) and single needle cathode electron sources. FEA can give a high current density (~ 10^{12} A/mm^2) while a needle cathode (ZrC, HfC) would have the possibility to temporally modulate the electron beam through laser pulse assisted emission. Ganter discussed simulations showing the possibility to reach a very low emittance in case of uniform emission from FEA (less than 10^{-2} μm). Some preliminary measurements made on commercial FEA (Mo tips) has been also reported. To overcome the main observed drawbacks: non-uniform emission and arc breakdown, a conditioning process on FEA has been developed at PSI and first FEA samples have been prepared and measured. About needle cathodes, Ganter discussed also some measurements conducted on a ZrC tip illuminated with a 1064 nm laser light, reaching 1 ns long electron bunches with 35 mA peak current. In order to preserve the extremely low emittance expected from such a source in the downstream linac, the pulse diode will be immediately followed by an RF gun like structure, operating at 1.5 GHz, as discussed by M. Pedrozzi. To minimize the RF emittance dilution two modes will be exited in the cavity (fundamental and third harmonic) in order to linearize the beam longitudinal phase space and to optimise the bunching process. Preliminary Parmela simulations discussed by Pedrozzi show in fact that the harmonic voltage can compensate the non linear RF effects induced by the accelerating mode, especially with low injection phases, preserving low emittance in the gun bunching regime and making this approach attractive for downstream velocity bunching schemes.

3. RF electron gun designs and operation

The second day of the working group was focussed on presentations and discussions about RF gun operations and possible improvements of gun designs, from the popular 1.6 cells RF gun design to the new hybrid gun concept.

The π-mode resonant frequency of the 1.6 cells SLAC/BNL/UCLA type RF gun is conventionally tuned using cylindrical copper tuning pieces that extend into the full cell cavity through holes in the side of the gun. A. Cook pointed out in his talk [11] that this design begins to fail in many versions of this gun design at high voltage levels, when the cavity undergoes electric breakdown in the vicinity of the tuners. In order to remove the tuners from the region of high electric field, thus mitigating this problem, Cook recommended to change the full cell geometry significantly. At UCLA they have investigated a method for accomplishing this, stretching the gun structure to tune the full cell frequency up by over 2 MHz. A dedicate device has been constructed to perform stretching and the modified photo-injector in an RF test bed has been tested. After this treatment they succeed in putting approximately 8.4 MW of RF power into the gun, an improvement over the ~ 4.5 MW routinely achieved with a similar gun using conventional tuning methods installed at the UCLA Neptune laboratory.

Not only the π-mode is excited by the RF generator in any multi-cells RF standing wave structure, also the lower order modes of the fundamental pass-band can store a significant amount of energy. Lower order modes coupling with the beam could be significant especially when a not fully relativistic beam is accelerated as the one propagating in an RF gun. J. Schmerge pointed out in his talk [12] that normally only the π-mode in a 1.6 cell gun is considered in RF gun simulations. However, due to the finite Q there is a small but measurable 0 mode (for a different proposed nomenclature see [17]) present even at steady state, reaching a peak field of the order of 10% of the accelerating peak field. The net result with both modes is a phase shift between cells several degrees less than 180 degrees. This effect may produce a correlated energy spread on the beam higher than expected not considering 0-mode interaction. Simulations using the code GPT reported by Schmerge show in fact that both space charge and the 0 mode contribute to the correlated energy spread. The simulation can reproduce the experimental results but it requires a 0 mode amplitude larger than is measured. The best agreement with the measured data occurs with a 0 mode amplitude 12% of the π mode amplitude instead of the measured 8%, and a laser phase of 20° instead of the measured 25°. The discrepancy might be due to the transient RF field contribution and is currently under investigation. The non-zero component of the 0 mode at the cell to cell iris may also contribute to the transverse emittance but with a minor impact on the beam quality compared to its effect on longitudinal phase space. Schmerge reported also that the LCLS gun is designed with a mode separation of 15 MHz instead of 3.5 MHz as in the GTF gun. With the larger mode separation in fact the 0 mode contribution is significantly suppressed and the adverse beam effects will not be present.

V. Yakimenko presented experimental results obtained at ATF of the measured beam brightness generated by unbalanced gun fields [13]. Unbalanced gun tuning allowed to push the gradient in the half cell as high as 150 MV/m, while a manufacturing imperfection limited the field gradient in the full cell at 110 MV/m. This allowed to further improve the beam quality, achieving an impressive record: 1 μm rms normalised emittance at 5.5 MeV with 0.8 nC charge in a 6.2 ps FWHM electron pulse long. Based on this results the design of the new 1.6 cell gun will allow independent control of the temperature in the full and half cells of the gun. The results reported by Yakimenko have opened the opportunity to conveniently study unbalanced gun tuning that may have very important applications for high average current - high beam load injectors.

A new class of photo-injector design interesting for scientific purposes and for industrial and medical applications, has been discussed by D. Alesini [14]. In his talk, he has illustrated the electromagnetic and beam dynamics design procedure of a hybrid standing/traveling wave structure, a promising device in terms of compactness, simplicity and efficiency, developed by an UCLA/INFN/University of Roma collaboration. In this device a standing wave 1.6 cell RF gun section is integrated with a downstream traveling wave structure through a coupling cell that feeds simultaneously the two sections. The design has been conducted using the 2D and 3D electromagnetic codes Superfish and HFSS. Alesini illustrated the non trivial problem encountered to couple and tune an hybrid SW/TW RF structure and the preliminary beam dynamics simulations showing the remarkable high brightness achieved with such a compact device.

4. Beam dynamics and beam manipulation

The last day of the working group has been devoted to presentations about beam dynamics studies and beam manipulation tools, followed by discussion and concluding remarks.

About beam dynamics studies C. X. Wang [15] discussed the "double-minimum" feature of emittance oscillation in the drift of a photo-injector, which has been used for optimizing the working point of split photo-injectors like LCLS, SPARC and DESY X-FEL injector. To better understand this feature, Wang has derived approximate analytical expression for the emittance and has compared them with HOMDYN simulations, obtaining very good agreement. Wang pointed out that the "double-minimum" emittance oscillation it is mainly due to beam spreading (arising from chromatic effects in the solenoid [18]), a generic feature of a space-charge dominated beam in drift space.

The design of an X-band accelerating section for linearizing the longitudinal beam phase space of the high brightness S-band photo-injector SPARC, has been presented by B. Spataro [16]. As pointed out by Spataro this device by reducing the electron bunch longitudinal emittance may lead to a significant improvement of velocity bunching performances in the SPARC linac. Multi-megawatts peak power tests made in the past at SLAC with Travelling Wave X-band structures, in the context of NLC studies, showed frequent internal discharges caused by field emission, producing serious damages of the cavity surface. To overcome the problem, instead of TW units, Spataro considered using Standing Wave structures. In fact input RF matching in SW structures is only possibile if the RF power distribution is unperturbed. Field emission discharges detune the SW cavity by several bandwidths, self protecting the structure by reflecting the input power. The nine cells structure design presented by Spataro, operating in the π standing wave mode, is fed by a central coupler and has been designed to obtain a 42 MV/m accelerating gradient. The 2D profile has been obtained using the electromagnetic codes Superfish and Oscar2D while the coupler has been designed using HFSS. Bead-pull measurements made on a copper prototype have been performed and very sutisfactoring results in terms of field unflattness (less than 1%) have been achieved.

5. Conclusions

New ideas have been proposed during the workshop to achieve ultra-high brightness beams, as particularly needed in SASE-FEL experiments, based on new concepts or improved photo-injector designs.
The most popular RF gun design is still the 1.6 cells split photo-injector that focused the discussion towards the next generation of similar devices. A consensus emerged during the workshop with the following list of requirements for sub-micron emittance photo-injector operation:

- spatial and temporal laser profile uniformity,
- normal incidence of laser pulse on cathode surface,
- cathode's cleaning availability,
- symmetric RF input coupler,
- removal of tuning rods for high voltage operation,
- 0 and π modes separation increased,
- highest possible gradient,
- solenoid design free from quadrupolar field components,
- careful beam matching with the linac.

References

1. D. H. Dowell, "H-beam cleaning of metal cathode", these proceedings.
2. D. T. Palmer, "Quantum efficiency and topography of heated and plasma-cleaned copper photo-cathode surfaces", these proceedings.
3. J. Schmerge, " Time dependent emission from metal cathodes", these proceedings.
4. D. H. Dowell, "Ultra-fast effects in metal cathodes", these proceedings.

5. J. H. Han, "Analysis of thermal emittance measurement at PITZ", these proceedings.
6. J. H. Han, "Multipacting at the photo-cathode in the RF gun cavity", these proceedings.
7. G. Gatti, "Quantum efficiency measurements of Mg films produced by Pulsed Laser Ablation Deposition for high brightness electron sources", these proceedings.
8. J. Smedley, "Superconducting photocathodes", these proceedings.
9. R. Ganter, "Low emittance gun concept", these proceedings.
10. M. Pedrozzi, "Post acceleration concept for the LEG project", these proceedings.
11. A. M. Cook, "Mitigation of RF gun breakdown by removal of tuning rods in high field regions", these proceedings.
12. J. Schmerge, "The 1.6 cell gun correlated energy spread dependence on π and 0 mode amplitude", these proceedings.
13. V. Yakimenko, "1.6 Cell RF Gun Optimization For Submicron Emittance Operation", these proceedings.
14. D. Alesini, "The hybrid TW-SW photoinjector", these proceedings.
15. C. Wang," Emittance oscillation in the drift of split photo-injectors", these proceedings.
16. B. Spataro, "Design and RF measurements of an X-band accelerating structure for the SARC project", these proceedings.
17. M. Ferrario and C. Ronsivalle, "Resonant modes in a 1.6 cells RF gun", these proceedings.
18. M. Ferrario et al., "Recent advances and novel ideas for high brightness electron beam production based on photo-injectors" Proceedings of the ICFA workshop on "The physics & applications of high brightness electron beams", Sardinia, Italy, July 2002.

DESIGN AND RF MEASUREMENTS OF AN X-BAND ACCELERATING STRUCTURE FOR THE SPARC PROJECT

D. Alesini[1], A. Bacci[3], A. Falone[2], M. Migliorati[1,2], A. Mostacci[1,2], F. Palpini[2], L. Palumbo[1,2], B. Spataro[1]

1) INFN-LNF Via E. Fermi 40, I-00044 Frascati, Italy

2) Università degli Studi di Roma "La Sapienza", Dip. Energetica, Via A. Scarpa 14, 00161, Roma, Italy

3) INFN- MI Via Celoria 16, I-20133 Milano, Italy

The paper presents the design of an X-band accelerating section for linearizing the longitudinal phase space in the Frascati Linac Coherent Light Source (SPARC). The nine cells structure, operating on the π standing wave mode, is fed by a central coupler and has been designed to obtain a 42 MV/m accelerating gradient. The 2D profile has been obtained using the electromagnetic codes Superfish and Oscar2D while the coupler has been designed using HFSS. Bead-pull measurements made on a copper prototype have been performed and the results are illustrated and compared with the numerical predictions.

1. Introduction

The SPARC [1] injector aims to produce a high brightness electron beam able to drive a self-amplified spontaneous free-electron laser (SASE-FEL) experiment in the green visible light and to develop an ultra-brillant beam photo-injector needed for the future SASE-FEL based X-ray sources. The SPARC photo-injector (Fig. 1) consists of 1.6 cell RF gun of the BNL/UCLA/SLAC type [2] operating at S-band with incorporated metallic photo-cathode (Cu or Mg). The gun generates a 6 MeV electron beam to be then accelerated in two S-band, travelling wave sections (SLAC type). The beam parameter list at the end of the injector is reported in Table 1.

In the SPARC phase II the first LINAC section, downstream the RF gun, will be used to study RF compression [3]. These tests of RF compression assume great relevance in the SPARC R&D program [4] since the general layout for SPARX [5] foresees the use of a mixed compression scheme: RF compression in the photo-injector up to 500 A and one single stage of magnetic compression at 1 GeV up to the final peak current of 1 kA. In this case the use of the X-Band structure operating at 11.424 GHz is required to compensate the non-linearity distortions due to the RF curvature during acceleration and compression [6].

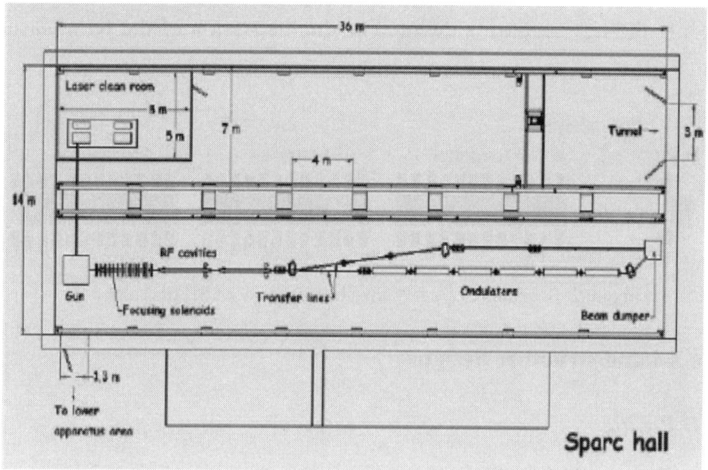

Figure 1. SPARC photo-injector layout.

Table 1. Beam parameter list at the end of the SPARC injector.

Electron Beam energy [MeV]	155
Bunch charge [nC]	1.1
Repetition rate [Hz]	10
Cathode peak field [MV/m]	120
Peak solenoid field [T]	0.273
Photocathode spot size [mm]	1.13
Laser pulse duration [ps]	10
Bunch energy @ gun exit [MeV]	5.6
Bunch peak current @ linac exit [A]	100
RMS normalized transverse emittance @ linac exit [mm mrad]	<2
RMS slice normalized transv. emit. @ linac exit (300 μm) [mm mrad]	<1
RMS longitudinal emittance @ linac exit [deg keV]	1000
RMS energy spread [%]	0.2
RMS bunch length @ LINAC exit [mm]	1

The schematic layout is reported in Fig 2. The X-Band structure, designed [7] to obtain 42 MV/m accelerating gradient, is a nine cells π-mode structure fed by a central coupler. The fourth harmonic of 2.856 GHz (instead of the third harmonic) has been chosen for space availability reasons and also because the technologies in the X-Band power sources and modulators have been already developed at KEK, SLAC and Varian [8-14] for future linear collider projects. The design of the accelerating section is described in sec. 2. In the third paragraph of this paper we describe the copper prototype and the measurement setup while the last section is devoted to the discussion of the RF measurement results.

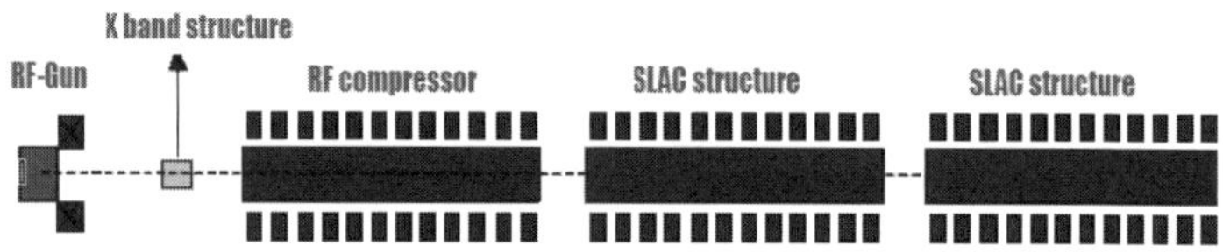

Figure 2. Schematic layout of the photo-injector of SPARC phase II.

2. The X band structure design

2.1. *2D Profile*

The detailed analysis of the structure design without coupler is reported elsewhere [7]. The single cell dimensions are reported in Fig. 3 and have been determined with the 2D electromagnetic (e.m.) codes SUPERFISH [15] and Oscar-2D [16]. In particular the beam aperture diameter (r_2) has been fixed by beam dynamics requirements and the outer diameter (r) has been chosen to tune the resonance of the π-mode at 11.424 GHz.

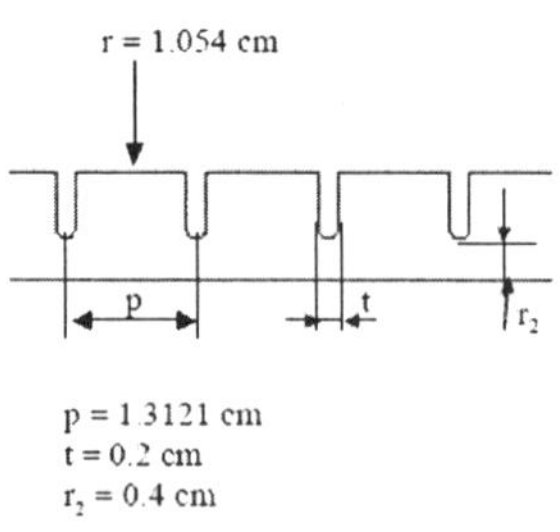

Figure 3. Single cell dimensions of the π-mode X-band structure

The complete structure (without coupler) is reported in Fig. 4. The choice of 9 cells has been done to achieve a total accelerating voltage V=5 MV with 3 MW of peak input power. The two cells connected to the beam pipe tube have different outer dimensions in order to compensate the detuning of the resonant frequency due to the presence of the beam pipe itself. No dedicated dampers of the parasitic higher order modes have been adopted since the X-Band structure operates on single bunch. The longitudinal electric filed on axis is reported in Fig. 5 while the cavity parameters are reported in Table 2.

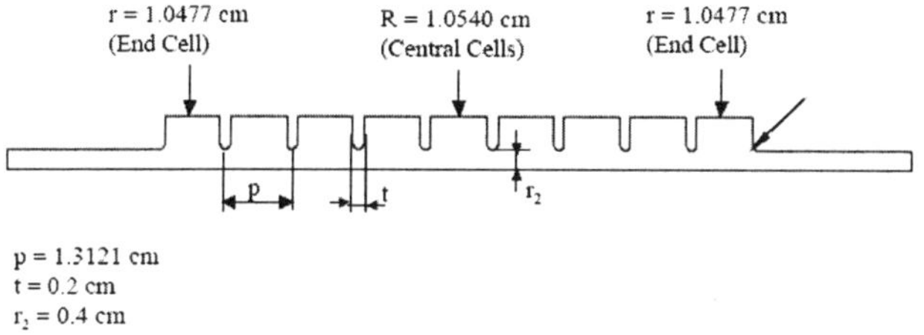

Figure 4. Complete π-mode X-band structure without coupler.

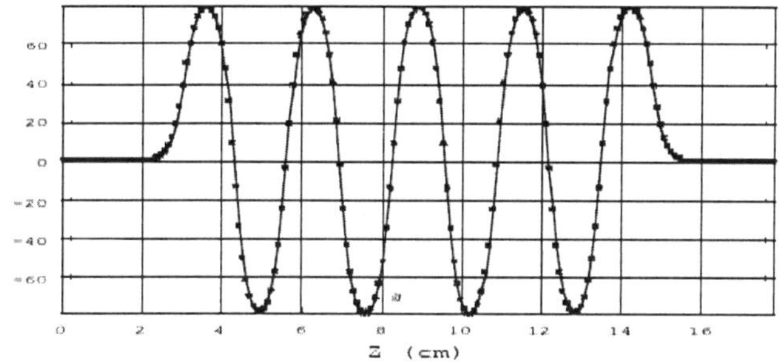

Figure 5. Longitudinal electric filed on axis (Superfish and Oscar 2D calculations.

Table 2. X-band cavity parameters (Superfish and Oscar 2D calculations).

Frequency [GHz]	11.424
Cells length [mm]	13.12
Number of cells 9	
Beam holes radius r [mm]	4
Iris Thickness t [mm]	2
Transit time factor T	0.731
Quality factor Q	8400
R/Q [Ω/ m]	9165
Coupling coefficient K [%]	2.42
Peak input power P [MW]	2.7
Duty cycle	10-4
Repetition frequency [Hz]	50
Average power dissipation Pd [W]	270
Peak axial electric field [MV/m]	57.5
Kilpatrick factor	1.197
Peak surface electric field [MV/m]	105

2.2. *Coupler Design*

Concerning the coupler design, it has been decided to feed the cavity in the central cell in order to not excite the mode 8/9 π that has the frequency nearest to the π mode and no field in the central cell. Therefore with a central coupler we have a much greater separation of the modes and, therefore, the working mode is less perturbed by the closest one. To reduce the coupler window dimensions we have decided to taper the smaller dimension ($10.16\ mm$) of the standard X-band waveguide ($10.16\ mm \times 22.86\ mm$) to the dimension of 4 mm (Fig. 6). The coupling cell is sketched in Fig. 7.

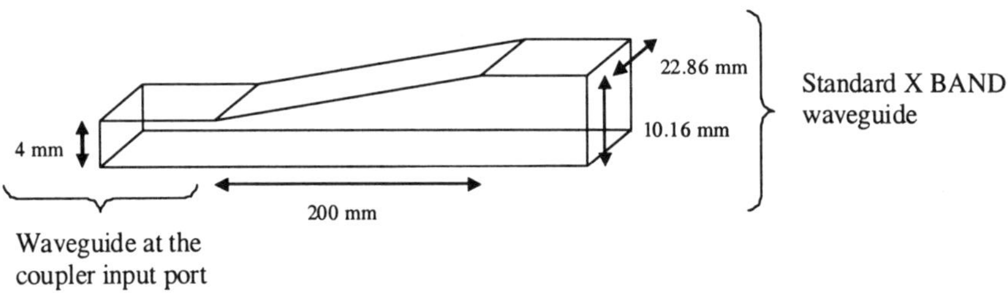

Figure 6. Tapering between the standard X-Band waveguide and the coupler input port.

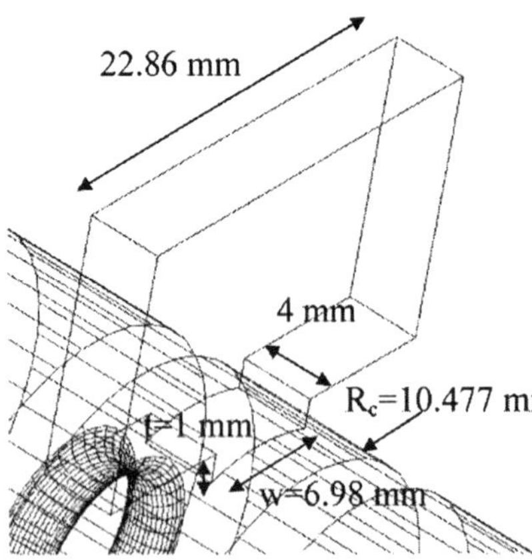

Figure 7. Sketch of the coupling cell.

The dimension of the coupler window (w) and of the central cell radius (R_c) have been tuned in order to obtain simultaneously a coupling coefficient $\beta = 1$, a resonant frequency of the whole system (cells+coupler) equal to 11.424 GHz and to preserve a good field flatness. This has been done using the electro-magnetic simulation code HFSS [17] by following two steps:

1. We have simulated a single cell with coupler as shown in Fig. 8. To get a better accuracy results we have simulated only one quarter of the structure with the proper boundary conditions. The dimensions of the coupler window and cell radius have been tuned in order to obtain a coupling coefficient $\beta = 9$ (exactly 9 times the coupling coefficient that we would reach with the complete structure) and a cell resonant frequency equal to 11.424 GHz;
2. We have simulated the complete structure (one quarter of the structure as shown in Fig. 9) with the dimensions found in the previous case and we have slightly adjusted the dimensions of the coupler cell and window to obtain a good field flatness at the resonant frequency of 11.424 with $\beta = 1$.

The final cell coupler dimensions are reported in Fig. 7 and the obtained reflection coefficient at the coupler port and the electric field on axis are reported in Fig. 10. By fitting the reflection coefficient at the input coupler port with the formula:

$$|S_{11}| = \sqrt{\frac{\left(\frac{1-\beta}{1+\beta}\right)^2 + (Q_L\delta)^2}{1+(Q_L\delta)^2}} \quad \text{with} \quad \delta = \frac{\omega}{\omega_{res}} - \frac{\omega_{res}}{\omega} \quad Q_L = \mathrm{Q_{EXT}}\frac{\beta}{1+\beta},$$

we have obtained the external quality factor Q_{EXT}=7900 and the coupling coefficient $\beta = 1.09$. The values of the simulated Q_{EXT} and field flatness could be optimized after measurement results.

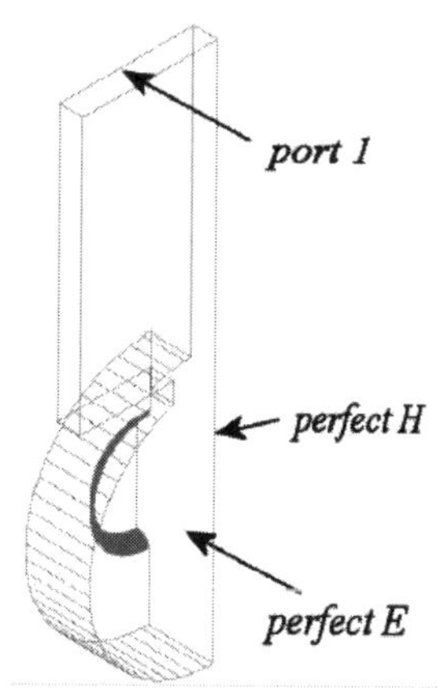

Figure 8. Single cell with coupler simulated by HFSS.

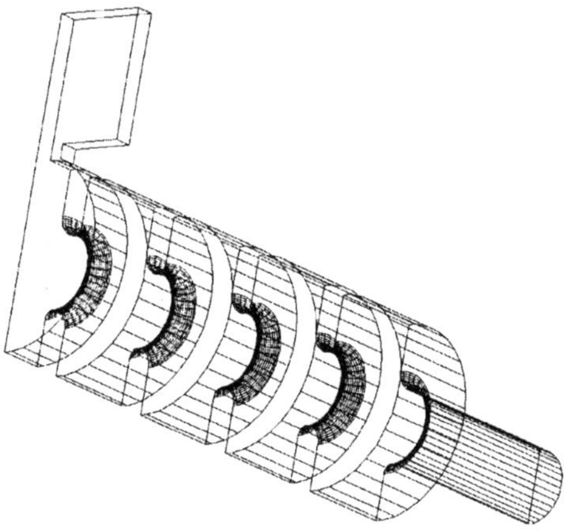

Figure 9. Complete structure with coupler simulated by HFSS

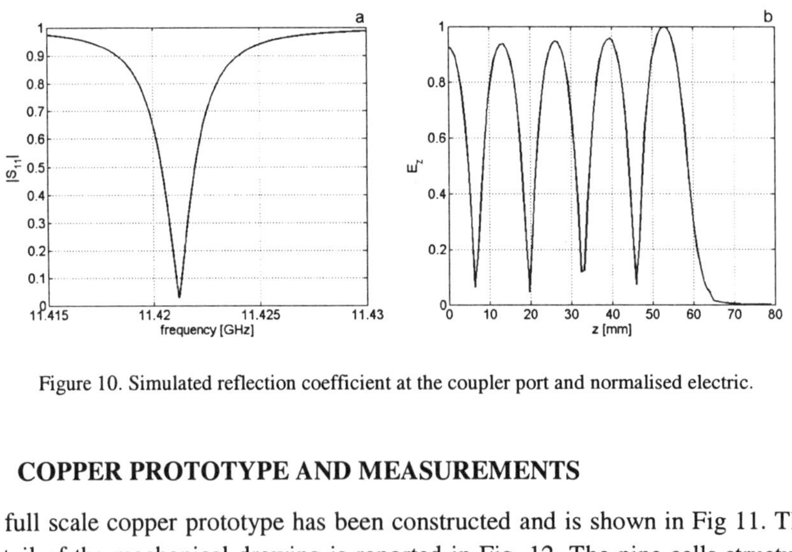

Figure 10. Simulated reflection coefficient at the coupler port and normalised electric.

3. COPPER PROTOTYPE AND MEASUREMENTS

A full scale copper prototype has been constructed and is shown in Fig 11. The detail of the mechanical drawing is reported in Fig. 12. The nine cells structure has been designed for brazing, but the RF tests refer to a mechanically joined structure. The material used to build this prototype is oxygen free copper. The structure has been realised by mechanical machining with a numerically controlled lathe and the obtained precision is below ±0.01 mm, while the surface roughness is not worst than 0.4 μR_a. The surface finishing was obtained directly by mechanical machining with custom cutting tools, avoiding any polishing technique and only silicon and sulphur free cutting fluid was used. The final machining was done at constant temperature in order to guarantee as much as possible the uniformity of the mechanical dimension of the cells. After machining, a standard cleaning procedure was performed using an alkaline solution at 3% at 50 °C followed by a rinse in tap water and a final rinse in distilled water followed by a chemical cleaning with citric acid solution and a rinse in tap water and finally a rinse in distilled water. Then the pieces were dried in a dust free oven. Each cell dimension has been checked with a quality control test.

The assembling procedure foresees the joining of the nine cells using two stainless steel disks used to press the structure by means of three 8 mm diameter copper rods. A torque of 5 Nm, corresponding to a pressure of roughly 80 N/mm^2, was applied. To feed the structure two lateral small antennas are placed.

Figure 11. Copper prototype of the X band structure.

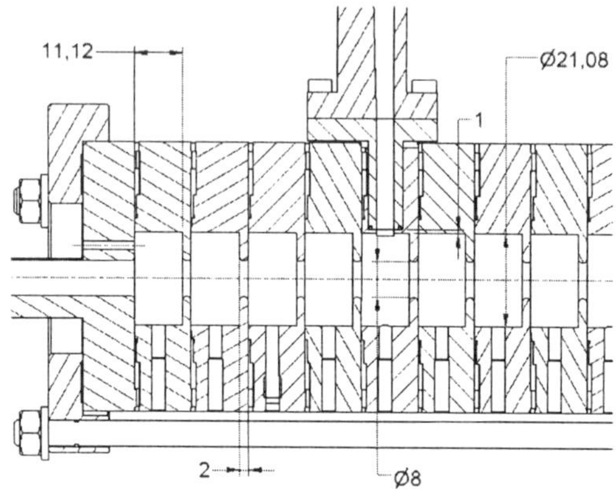

Figure 12. Detail of the mechanical drawing.

Two types of different measurements have been carried out: transmission (or reflection) scattering coefficient measurement (between the two antennas or between the antennas and the central coupler) and bead pull measurements.

With the first type of measurements we have found the resonant frequency, the β coefficient at the input ports and the unloaded or external quality factors of π (or other) mode(s). With the second type of measurements we have found the longitudinal electric field on axis and we have calculated the shunt impedance of the structure.

The complete measurement setup is shown in Fig. 13. The PC controls both the network analyzer Agilent N5230A (interfaced by a GPIB Ethernet device) and the control circuit of the steeping motor through LABVIEW [18]. The nylon wire is kept straight by a 75 g weight.

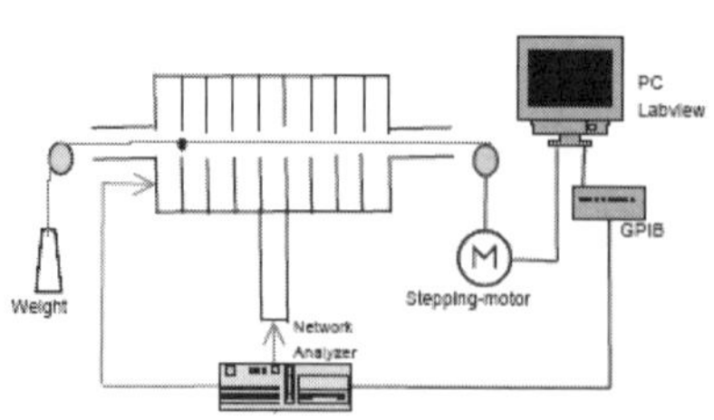

a)

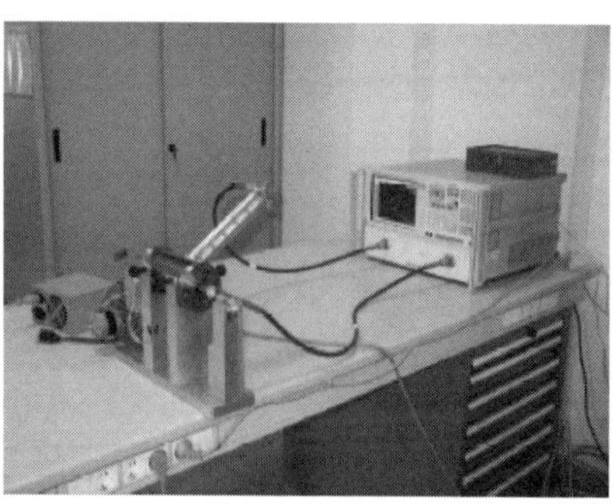

b)

Figure 13. Picture of the measurement setup (b) and conceptual scheme (a).

4. Measurement results

4.1. *Resonant frequency and quality factors measurements results*

The transmission coefficient between the two small antennas and between the antenna and the central coupler are reported in Figs 14a and 14b. As previously observed, we can excite only five over nine possible modes by the central coupler because we impose a non-zero field in the central cell. On the contrary with the two antennas we can excite all the possible modes. The measured dispersion curve, compared with the one obtained from HFSS and SUPERFISH, is reported in Fig 15.

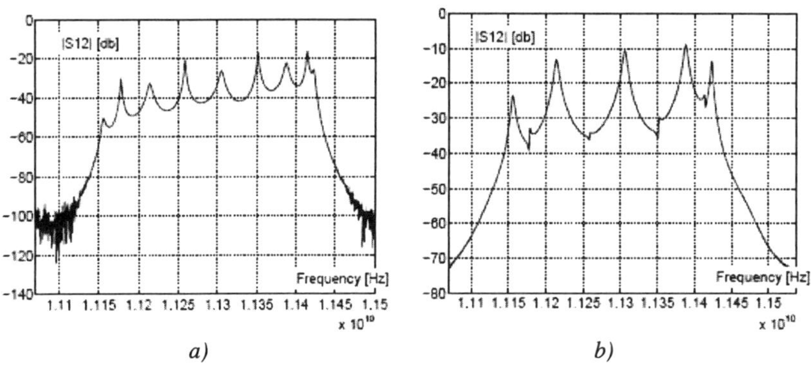

Figure 14. Transmission coefficients between the two small antennas (a) and between the antenna and the central coupler (b).

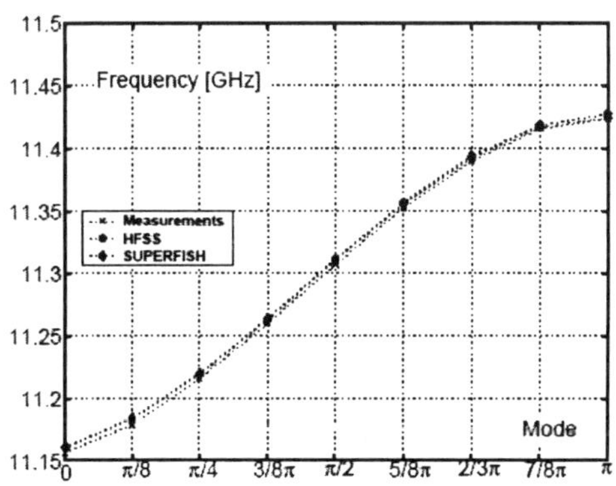

Figure 15. Measured dispersion curve, compared with the one obtained from HFSS and SUPERFISH.

The quality factor of the resonance has been measured as a function of pressure realized by the rods. The results are reported in Fig 16. As expected the Q factor increases with the increasing pressure and, with the maximum pressure, we have obtained the result reported in Table 3 and compared with the numerical one. These results prove that even if the structure is not brazed the electric contact is good enough to have a Q-factor of the resonance comparable with the numerical one. Concerning the external quality factor it has been found the value Q_{EXT}=8000 in agreement with the simulation results.

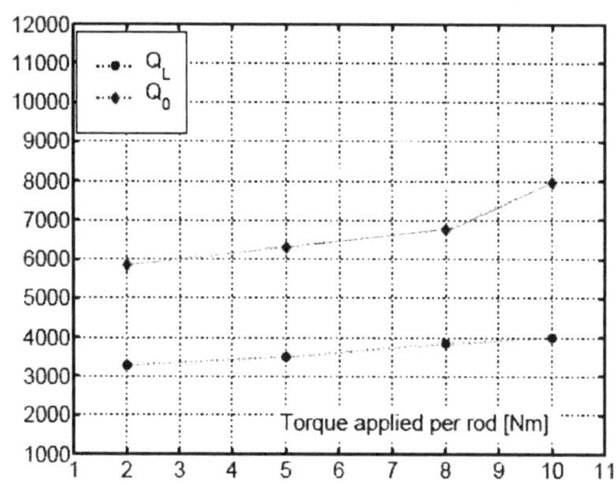

Figure 16. Measured quality factor of the mode resonance as a function of pressure realized by the rods.

Table 3. Unloaded quality factor measured with the maximum pressure and compared with the numerical results.

HFSS	SUPERFISH	OSCAR 2D	Measurements
Q0 8500	8070	8413	7960

4.2. *Electric field measurement results*

The bead-pull technique that we have used is based on the Slater theorem and is widely discussed elsewhere [19]. The small perturbing object inside the cavity induces a frequency shift that is related to the variation of the electric and magnetic energy stored in the cavity in the point where the object is located according to the formula:

$$\frac{\omega_p - \omega_0}{\omega_0} = \frac{1}{4U} \int_{\Delta V} \left(\mu_0 \left| \vec{H} \right|^2 - \varepsilon_0 \left| \vec{E} \right|^2 \right) dV$$

Considering the accelerating mode and the small transverse dimensions of the perturbing object (of the order of 0.1 mm) it is possible to find the longitudinal electric field on axis by the formula:

$$\frac{\omega_p - \omega_0}{\omega_0} = -\varepsilon_0 k_e \Delta V \frac{E_z^2}{4U} = \frac{1}{2Q_L} \tan(\phi(f_0))$$

where k_e is the form factor of the object, ΔV is the perturbing object volume, U the total energy stored in the cavity, Q_L is the loaded quality factor of the resonance and $\phi(f_0)$ is the phase of the transmission coefficient between two ports coupled with the field in the cavity. The latest expression have been written because it is more convenient to measure the phase shift of the resonant frequency instead of $\Delta\omega$ [20]. Moreover, for $\phi(f_0)$ reasonably small (e.g. smaller than 30°), $\tan(\phi)$ is approximately equal to ϕ and this has some practical advantages. The perturbing objects are cylinders made of copper with a radius of 0.09 mm and length of 0.7 mm and 0.9 mm (measured with a digital callipers with an uncertainty of $10\mu m$).

A careful optimization of the measurement setup has been done in order to minimize the systematic errors and to better understand the uncertainty of the measurements. The most important reasons of induced errors in the measurements were:

1. the effect of wire nylon that induced an unwanted perturbation in the frequency measurement. To cancel this systematic error different measurements have been done with different nylon wire diameters. A typical phase resonance measurement with different wire diameters is reported in Fig. 17. The best results have been obtained with the smallest wire and the final small systematic error has been completely cancelled considering the frequency shift with and without the perturbing object. As example in Fig. 18 it is reported the results of phase shift considering different pieces of nylon wire with the smallest diameter of 0.083 mm without perturbing object;
2. the effect of the drops of glue used to fix the perturbing object. It gave a perturbation in the frequency measurement of the order of 10% of the total frequency variation. In particular the frequency variation at the center of the irises didn't go to zero. To take into account this effect different measurements have been done in order to subtract the systematic perturbation to the measure as shown in Fig. 19;
3. the effect of jitter in the longitudinal coordinate between different measurements and within a single measure. This problem, given by the stepping motor, has been reduced using a weight smaller than 75 g, and

correcting off-line the measurement results. As example in Fig. 20 it is reported the result of two different measurements.

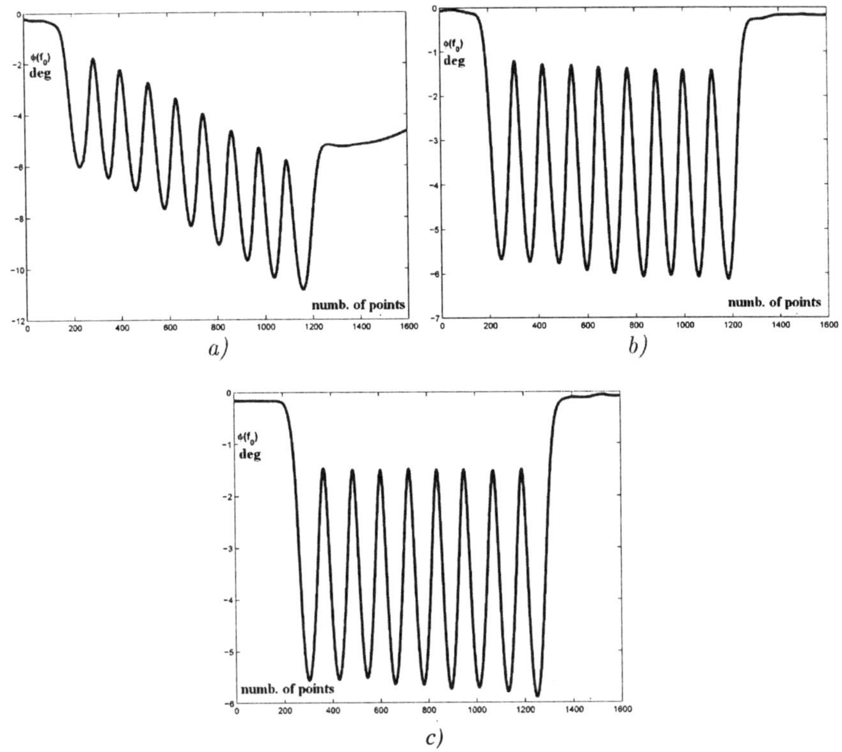

Figure 17: Phase of resonance measurement with different wire diameters: a) 0.18mm; b) 0.148mm; c) 0.083mm.

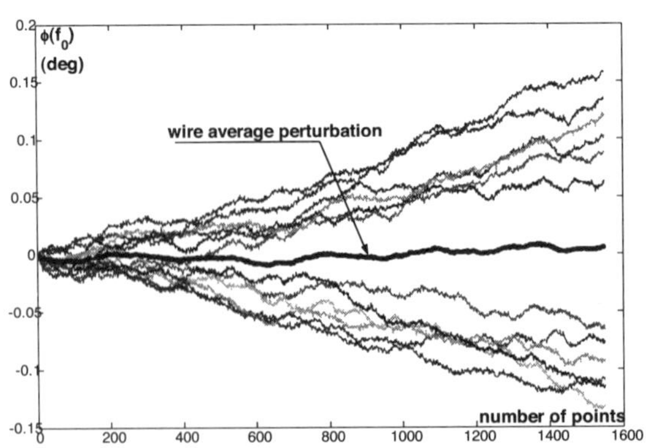

Figure 18: Resonance phase shift considering different pieces of nylon wire with the smallest diameter of 0.083 mm.

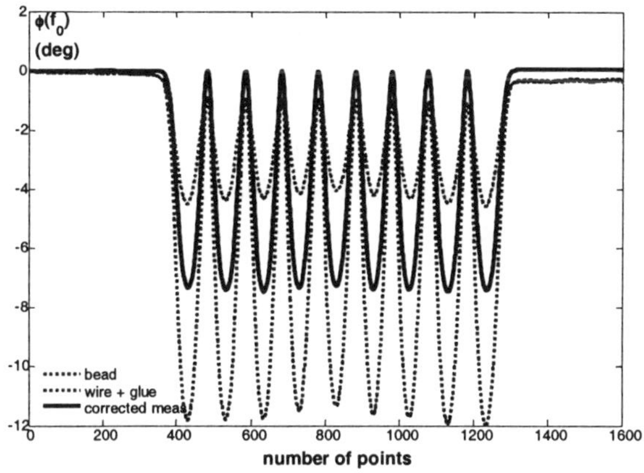

Figure 19. Phase measurement taking into account the effect of the wire and of the glue.

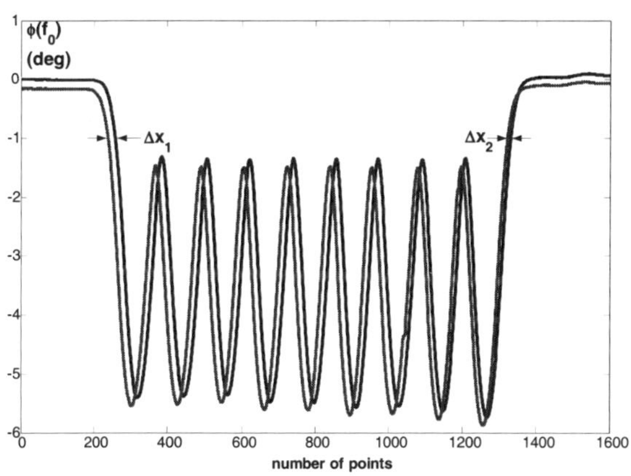

Figure 20. Jitter between two different measurements.

To calculate the R/Q of the structure it has been necessary to determine the form-factor of the different perturbing objects. In literature there are many analytical formulas for the determination of the form-factor for ellipsoid geometries [21]. The used perturbed objects are far to be perfect ellipsoid and we have preferred to determine the form factor through experimental calibration [22]. Using Slater's theorem we can calibrate the form factor comparing the perturbation induced by the perturbing object in a cavity with known field. For this purpose, it has been used a pill-box cavity working at

1.91 GHz on the TM $_{010}$ mode. Using different resonant mode of the pillbox cavity we have also checked that the form-factor does not depend on the frequency. The results of this calibration are reported in Table 4 where the obtained form factors of two objects with different length are reported [23].

The measured longitudinal electric filed on axis, after the tuning procedure, is plotted in Fig. 21. The reached field-flatness is of the order of 1% at the nominal resonant frequency of 11.424GHz. The calculated R_{sh}/Qs normalized to the cavity length are reported in Table 4 and they are in a very good agreement with the simulation results.

Table 4. Form factors obtained by calibration of two objects with different length.

Object length [mm]	Form factor ke	R/Q [Ω/m]
0.7	25.860± 0.031	9480± 96
0.9	28.400± 0.027	9367± 211

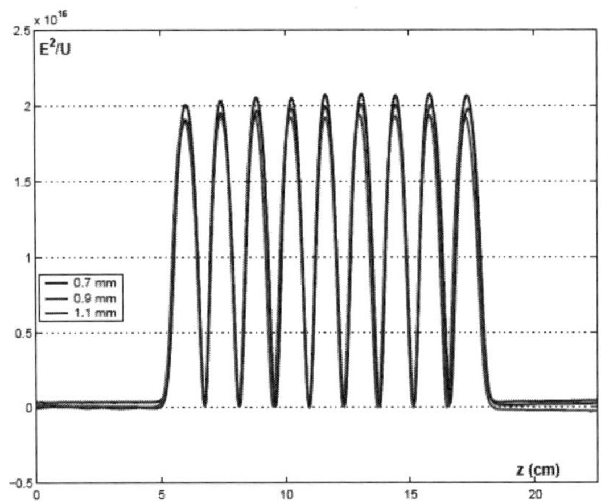

Figure 21. Measured longitudinal electric field on axis.

5. Conclusions

In the paper we have presented the design of the X-band accelerating section for linearizing the longitudinal phase space in the Frascati Linac Coherent Light Source (SPARC). The structure, operating on the π standing wave mode, is a 9 cells structure fed by a central coupler and has been designed to obtain a 5 MV accelerating voltage. Quality factors, resonant frequency and electric field measurements have been done on a copper prototype. Even if the prototype is not brazed the reached quality factors are very close to the theoretical ones. Concerning the electric field measurement with the bead pull technique we have discussed the most important reasons of induced errors and how to cancel these effects. In particular the effect of wire nylon, drops of glue used to fix the perturbing object and the effect of jitter in the longitudinal coordinate have been discussed. To calculate the R/Q factors the perturbing objects have been calibrated with a pill-box cavity and the final results are in very good agreement with the numerical ones.

Brazing tests are now in progress in the LNF for the construction of the final device. More details on this subject can be found in [24, 25]

Acknowledgements

The authors are grateful to P. Chimenti, V. Chimenti, A. Clozza and V. Lollo for their help in the mechanical design of the cavity prototype

References

1. Technical Design Report for the SPARC Advanced Photo-Injector. Technical report, LNF-INFN, ROMA, (2003).
2. D.T. Palmer, The next generation photoinjector, PhD.Thesis, Stanford University.
3. M. Ferrario et al., Homdyn Study for the LCLS RF Photoinjector , SLAC-PUB 8400.
4. D. Alesini et al., Conceptual Design of a High-Brightness Linac for Soft X-ray SASE-FEL Source, these proceedings.
5. D. Alesini et al. (The SPARX study group), Conceptual design of a high brightness linac for soft X-ray SASE-FEL source, NIM-A, 507 (2003), pp. 502-506.
6. P. Emma X-band RF harmonic compensation for the linear bunch compression in the LCLS. SLAC, LCLS-TN-01-1 2001.
7. A. Bacci et al., An X-band structure for a longitudinal emittance correction at SPARC. SIS-Pubblicazioni, LNF-INFN, LNF 03/008(R), (2003).
8. K. Takata, Proc. First Workshop on Japan Linear Collider (JLC),KEK, Oct. 24-25, 1989.
9. J.W. Wang and G.A. Loew, “Measurements of ultimate Accelerating Gradients in the SLAC Disk-loaded Structure,” 1985 PAC, Vancouver, B.C., May 1985, SLAC-PUB-3597, March 1985.
10. J. W.Wang, V. Nguyen-Tuong and G.A. Loew, “RF Breakdown Studies in a SLAC Diskloaded Structure,” Proceedings of the 1986 Linear Accelerator Conference, Stanford, Ca, June 1986, SLAC-PUB-3940, April 1986.

11. J.W. Wang and G.A. Loew,"Progress Report on New RF Breakdown Studies in an S-band Structure at SLAC,"presented at the 1987 PAC,Washington,D.C.,March 1987, SLAC-PUB- 4247,February 1987.
12. E. Tanabe,J. W. Wang and G.A. Loew, "Voltage Breakdown at X-band and C-band Frequencies, "Proceedings of the 1986 Linear Accelerator Conference, Stanford, Ca, June 1986.
13. J. W. Wang, "RF Properties of Periodic Accelerating structures for Linear Colliders", SLAC Report-339, July 1989.
14. J.R. Rees, "A Perturbation Approach to Calculating the behavior of Multi-cell Radiofrequency Accelerating Structures" PEP-255, Stanford Linear Accelerator Centre (1976).
15. Poisson Superfish, James H. Billen and Lloyd M. Young, software produced under U.S. Government by Los Alamos National Laboratory, Particle Accelerators 7 (4), 213-222 (1976).
16. P. Fernandes and R. Parodi, "LALAGE – A Computer Program to Calculate the TM Modes of Cylindrically Symmetrical Multicell Resonant Structures", PAC 1982,Vol. 12, pp. 131- 137.
17.]http://www.ansoft.com
18. http://www.ni.com/labview/
19. J.C.Slater L.C.Maier, Field Strength Measurements in Resonant Cavities, Journal of Applied Physics, 23 No.1, 1952.
20. Measurement of resonant frequency and quality factor of microwave resonators Journal of Applied Physics,84 No.6,1998.
21. G.Dome F.Caspers Precise Perturbation measurement of Resonant Cavities and Higher Mode Identification, Cern SPS/85-46,(1984).
22. J.C.Slater Microwave Electronics, Dover Publication (1969).
23. International Organization for Standardization et al., Guide to the expression of uncertainty in measurement. (ISO, Geneva, 1995).
24. P.Chimenti et al.,"First Results on Vacuum Brazing of RF 11 GHz Accelerating Structure at Frascati Laboratory", LNF-INFN,LNF-05/22(IR), November 14, 2005
25. P. Chimenti et al., "An Investigation on Electroforming Procedures for R.F. 11 GHz Linear Accelerating Structures at Frascati Laboratory", LNF-INFN,LNF-05/23(IR), November 14, 2005.

MITIGATION OF RF GUN BREAKDOWN BY REMOVAL OF TUNING RODS IN HIGH FIELD REGIONS *

A. M. COOK, J. B. ROSENZWEIG, M. DUNNING, P. FRIGOLA, AND K. SERRATTO

Department of Physics and Astronomy, University of California, Los Angeles, Knudsen Hall 3-174a,
Los Angeles, CA 90095-1547, USA
E-mail: alancook@physics.ucla.edu

The π-mode resonant frequency of the 1.6 cell SLAC/BNL/UCLA style RF photoinjector electron gun is conventionally tuned using cylindrical copper tuning pieces that extend into the full-cell cavity through holes in the side of the gun. This design begins to fail in many versions of this popular gun design at higher voltage levels, when the cavity undergoes electric breakdown in the vicinity of the tuners. In order to remove the tuners from the region of high electric field, mitigating this problem, one must change the full cell geometry significantly. We have investigated a method for accomplishing this, in which we stretch the gun structure to tune the resonant frequency up by over 2 MHz. We constructed a device to perform this stretching and tested the modified photoinjector in an RF test bed. We succeeded in putting approximately 8.4 MW of RF power into the gun, an improvement over the 4 MW routinely achieved with a similar gun using conventional tuning methods installed at the UCLA Neptune laboratory. Recent results in testing this gun with a magnesium cathode insert are reported as well.

1. Introduction

1.1. *The 1.6-Cell RF Photoinjector*

The 1.6-cell S-band RF photoinjector design is currently used in many high-brightness electron beam laboratories. This electron beam source consists of a cylindrical copper cavity divided by an iris into a *full-cell*, $\lambda/2$ in length, and a *half-cell*, $0.6\lambda/2$ in length. RF power is generated by a klystron and fed into the full-cell through a copper waveguide, driving a two-mode standing wave in the multi-cell structure via cell-to-cell coupling[1].

*This work is supported by the United States Department of Energy.

A circular cathode plate making up one end of the half-cell produces electrons when struck by a drive laser. These electrons are accelerated by the RF electric field along the axis of the cavity through both cells and out of the beam port. The accelerating mode resonates at an S-band frequency of 2856 MHz and is called the π-mode, as the fields in the cell centers are π radians out of phase with each other. This means that the longitudinal on-axis electric fields in each cell point in opposite directions, with a null at the center of the coupling iris. If the electrons are "injected" at the proper point in the RF phase, field oscillations will be such that the electrons always see an accelerating field as they pass between the cells. Thus the voltage experienced by the electrons across the gun is the sum of the voltages across each cell. The zero-mode, in which the fields are in phase, corresponds to effective decoupling of the cells and has a frequency equal to that of a single isolated cell.

1.2. *Frequency Tuning*

Prior to beam operation the photoinjector needs to be tuned in order to precisely match the π-mode frequency with that of the input RF power. The cavity geometry is constructed so that the frequencies of each cell, in an isolated pillbox model, combine via cell-to-cell coupling for a π-mode resonance at 2856 MHz according to the formulas[2,3]

$$
\begin{aligned}
f_1 &= f_\pi\left(1 - \frac{\kappa\sqrt{r}}{2}\right) \\
f_2 &= f_\pi\left(1 - \frac{\kappa}{2\sqrt{r}}\right)
\end{aligned}
\tag{1}
$$

where f_1 and f_2 are the respective frequencies of the half and full cells, f_π is the desired π-mode frequency, κ is the cell-to-cell coupling factor, and r is the ratio of the cell volumes. In practice, due to finite machining tolerances, the cavity must be tuned after construction to get a precise match to the correct frequency.

The resonant frequencies of the two cells can be tuned independently to set the correct field balance and π-mode frequency. This is implemented in practice by connecting a network analyzer via a waveguide coupler to the full-cell and monitoring f_π while adjusting the full and half cell frequencies. In this configuration the cavity is a one-port system, and the position of the π-mode frequency can be seen with an S_{11} measurement. Viewing the cavity as two capacitively-coupled LC circuits, tuning is accomplished by adjusting the inductances or capacitances of each circuit. The half-cell is tuned capacitively by forcefully deforming the cathode plate using a bolt threaded through a flange and into the back of the cathode. The full-cell is conventionally tuned inductively by inserting cylindrical copper tuning pieces into the cell through holes in the cavity walls. Finally, the π-mode position is adjusted *in situ* by changing the ambient temperature. This effectively provides three "tuning knobs" — the two individual cell frequencies and temperature.

The physics behind the individual cell tuning is described by Slater's theorem[8]. The theorem relates the resonant frequency perturbation to the energy stored in a small perturbative volume excluded from the cavity. It can be stated as

$$\frac{\delta\omega}{\omega} = \frac{\delta U}{U} = \frac{1}{2U}\int \left(\mu_0 \mathbf{H}^2 - \epsilon_0 \mathbf{E}^2\right) dV\,, \tag{2}$$

where ω is the resonant frequency, U is the energy stored in the cavity fields, $\delta\omega$ and δU are the perturbations to these quantities, and the integral is carried out over the perturbative excluded volume V. In the situation considered here the cylindrical copper tuning pieces extended into the full-cell and the deforming of the cathode in the half-cell represent volumes excluded from the cavity, as the fields are zero and therefore can store no energy inside the conductor.

2. Tuner Breakdown

2.1. *Gun Performance Limits*

It is a somewhat common problem that the limit on achievable cavity voltage levels in such photoinjectors is much lower than expected, resulting in disappointing beam production. In these cases the cavity often performs normally during initial RF conditioning, accommodating satisfactory input power levels of 7 MW or more. At some point arcing occurs, apparently causing catastrophic damage, and thereafter the cavity is limited to much lower power levels. Some examples are the 1.6-cell RF guns at the UCLA Neptune and LLNL PLEIADES laboratories, with post-damage upper input power limits of 4.5 MW and less than 4 MW respectively.

There is *in situ* evidence pointing to the tuning rods in the full-cell as the cause of these crippling electric breakdown problems. The tuning rods are cylindrical copper pieces that extend into the full-cell from outside of the gun through holes in the cavity wall. They are mounted on mechanical actuators that allow adjustment. A metal spring encircles the tuning rod several millimeters from its end in order to maintain contact between the copper piece and the wall of the hole it sits in, providing an electrical seal. Darkening and roughening of the copper appear on the tuning piece after continued use, indicating that much arcing has taken place there.

2.2. *Multipactoring and Gap Fields*

We present some simple calculations showing that it is reasonable to suspect arcing in the vicinity of the tuning pieces. Refer to the diagrams shown in Figure 1.

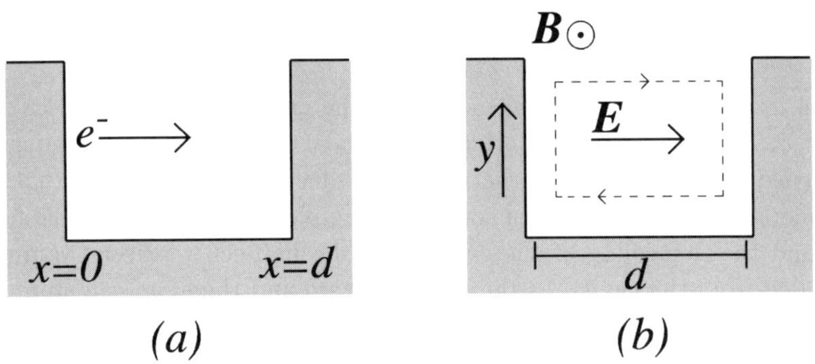

Figure 1. (a) Diagram for calculation of multipactoring threshold field in tuner gap. (b) Diagram for calculation of field present in tuner gap. Both diagrams represent the gap between the tuning piece and the hole wall, with the lower boundary representing the spring encircling the tuning piece.

Arcing in RF structures is often caused by multipactoring[4]. This is a phenomenon in which an electron strikes a metal surface, causing the emission of secondary electrons which then strike another surface, and so on in a chain reaction until there is a significant amount of charge involved. As shown in Fig. 1a, consider an electron born on one side of the gap between the tuning piece and the hole wall, at position $x = 0$ with velocity $\dot{x} = 0$. The one-dimensional equation of motion for this electron in a sinusoidally-varying electric field is

$$m\ddot{x} = eE_1 \sin \omega t \,. \tag{3}$$

Integrating both sides twice with respect to t and dividing by m yields

$$x = -\frac{eE_1}{m\omega^2} \sin \omega t + C_1 t + C_2 \,. \tag{4}$$

After applying the initial conditions $x = 0$ and $\dot{x} = 0$ we have

$$x = \frac{eE_1}{m\omega^2}(\omega t - \sin\omega t)\,. \tag{5}$$

Conditions are ideal for multipactoring if the applied electric field reverses direction at the moment the electron strikes the wall and secondary electrons are emitted, so that they are accelerated back toward the other side of the gap. This situation corresponds to $\omega t = \pi$ at $x = d$. Subsituting these into Eq. (5) gives

$$\begin{aligned} eE_1 &= \frac{m\omega^2 d}{\pi} \\ &= \frac{mc^2\omega^2 d}{c^2\pi} \\ &= \frac{mc^2k^2 d}{\pi} \\ &= mc^2\frac{4\pi^2 d}{\lambda^2}\,. \end{aligned} \tag{6}$$

Solving for E_1 and making the substitions $\lambda = 10.49$ cm (S-band) and $mc^2 = 0.511$ MeV we get

$$E_1 = 0.58 \text{ MV/m} \times d \text{ (in mm)} \tag{7}$$

for the electric field amplitude ideal for multipactoring in terms of the gap width d in millimeters. In our case d is on the order of one millimeter, so we would expect to see multipactoring occuring at around $E_1 = 0.58$ MV/m.

We wish to compare this to the electric field present in the tuner gap. Fields will be present in a small gap between tuning piece and hole wall but not in a larger hole left in the absence of a tuning piece, due to the activity of surface currents. Writing impedance in terms of capacitance and inductance as $Z = i\omega L + \frac{1}{i\omega C}$ illustrates that surface current traveling on the wall of the full-cell and coming across the tuner hole will see a smaller impedance entering a small (~ 1 mm) gap than it will entering a large hole. Therefore from a parallel impedance standpoint, the current will tend to flow into a narrow gap, whereas it will tend to flow around a larger hole. This suggests that retracting the tuners from the full-cell and back from the mouth of the tuner hole will result in the absence of fields in the holes.

We can do a simple calculation to find the field inside the gap when the tuning piece is present using a pillbox model to represent the full-cell. We will use this to argue that the fields present in the gap near the bottom are comparable with the multipactoring threshold. The well-known solutions[5] for the field components of the TM_{010} mode in a cylindrical conducting cavity are

$$E_z = E_0 J_0(kr)\cos\omega t$$
$$B_\phi = -\frac{E_0}{c} J_1(kr)\sin\omega t \tag{8}$$

where r is the radial coordinate inside the cavity and J_0 and J_1 are Bessel functions. At the wall of the cavity $kr = 2.405$ and so we have

$$B_\phi = \frac{E_0}{c} J_1(2.405)\sin\omega t \approx 0.52\frac{E_0}{c}\sin\omega t \tag{9}$$

for the pillbox field component B_ϕ at the mouth of the tuner gap in terms of the maximum electric field amplitude E_0 in the cylindrical cavity.

Inside the gap the magnetic field **B** is proportional[6] to the surface current **K**, which is governed by the transmission line equation

$$\frac{\partial^2 \mathbf{K}}{\partial \zeta^2} - \frac{1}{c^2}\frac{\partial^2 \mathbf{K}}{\partial t^2} = 0\,. \tag{10}$$

Referring to the diagram in Figure 1b, y is the vertical position in the gap, with $y = 0$ corresponding to the bottom of the gap and $y = h$ corresponding to the top. We use the coordinate $\zeta \equiv h - y$ for this part of the analysis in order to correctly represent the variation of **K** along the gap. Using the fact that **K** varies sinusoidally with time, we rewrite Eq. (10) as

$$\frac{\partial^2 \mathbf{K}}{\partial \zeta^2} + \frac{\omega^2}{c^2}\mathbf{K} = 0\,. \tag{11}$$

The simple solution for the surface current magnitude is

$$K = K_0 \cos\frac{\omega}{c}\zeta\,, \tag{12}$$

and thus the field **B** carries the same sinusoidal variation in ζ,

$$B = B_\phi \cos\frac{\omega}{c}\zeta\,. \tag{13}$$

Consider a line integral of the electric field in the gap around the closed rectangular path indicated by a dotted line. Using Maxwell's equation $\nabla \times \mathbf{E} = -\frac{\partial \mathbf{B}}{\partial t}$ and Stokes' theorem, we can write

$$\oint \mathbf{E}\cdot d\ell = -\int\!\!\int \frac{\partial \mathbf{B}}{\partial t}\cdot d\mathbf{A}\,. \tag{14}$$

To evaluate the integral on the left hand side of Eq. (14), we let the integration path expand to coincide with the conducting boundaries of the gap and note that the contribution to the integral along the left and right sides of the path is zero since $\mathbf{E}$ and $d\boldsymbol{\ell}$ are perpendicular there. In addition, the contribution along the lower side of the path is zero because $\mathbf{E}$ is parallel to the conducting boundary and thus zero there. The upper portion of the path does contribute, as the field here is nonzero and along the direction of $d\boldsymbol{\ell}$. Summing the contributions yields

$$E\ d = -\int\int \frac{\partial \mathbf{B}}{\partial t} \cdot d\mathbf{A}\,. \tag{15}$$

To evaluate the integral on the right hand side of Eq. (15), we carry out the dot product and substitute the functional form of $\mathbf{B}$ in the integrand. This gives

$$\begin{aligned} E\ d &= \int\int \frac{\partial}{\partial t}\frac{0.52\ E_0}{c} \sin\omega t \cos\left(\frac{\omega}{c}y'\right)\ dx'dy' \\ &= \int\int 0.52 E_0 k \cos\omega t \cos ky'\ dx'dy' \\ &= 0.52 E_0 \cos\omega t \sin ky\ d\,. \end{aligned} \tag{16}$$

We are interested in the amplitude of the field and not temporal oscillations. Suppressing the time dependence and solving for E at the top of the gap yields

$$E = 0.52 E_0 \sin kh\,. \tag{17}$$

Substituting a common peak field value $E_0 = 80$ MV/m, cavity radius $a = 4.2$ cm, gap depth $h \approx 0.1\ a = 4.2$ mm, and known pillbox cavity value $k = \frac{2.405}{a}$ into Eq. (17) gives

$$E \approx 9.9\ \text{MV/m} \tag{18}$$

for the electric field amplitude at the mouth of the tuner gap. This amplitude decreases to zero at the bottom of the gap approximately linearly, as a sine curve in the small-angle limit, meaning that near the bottom there is a field value equal to the multipactoring threshold value of ≈ 0.58 MV/m previously quoted. Thus our calculation of the electric field in the gap agrees with our examination of multipactoring conditions, and we conclude that we can expect arcing in lower parts of the tuner gap, coinciding with *in situ* evidence seen there.

If the depth h of the gap is equal to $\lambda/4$, the cosine argument in Eq. (12) becomes $\pi/2$ at the bottom of the gap and thus $\mathbf{K}$ is zero there. This is desirable if one does not wish to use a spring that is rated for high current. Deliberately placing the spring so that the gap depth is $\lambda/4$ in order to minimize the current in the spring is called a *quarter wave choke* and is often employed in electrical engineering applications. This is not useful for many resonating cavity applications, such as this case where $\lambda/4$ in S-band is much larger than the gap depth would normally be and according to Eq. (17) would result in unacceptably large electric fields near the mouth of the gap.

3. Solution

3.1. *Loss of Tuning Redundancy*

In order to mitigate the problem of breakdown in the vicinity of the tuners it is necessary to remove the tuners from the full-cell. This lowers the π-mode resonant frequency by approximately 2 MHz for a given temperature and mode separation and also removes one of the "knobs" available for tuning of the gun. Tuning can still be accomplished by adjusting the half-cell frequency through cathode deformation until the field balance is correct, and then adjusting the position of the π-mode by changing the temperature. Temperature tuning changes frequency by 44 KHz per degree Celsius, which means that a 45 degree temperature decrease is required to make up the 2 MHz lost by removing the tuners. This is a very undesirable situation from an experimental standpoint, because such drastic cooling is difficult in practice.

More recent gun designs bear this in mind, with the full-cell geometry such that the frequency rests at an ideal spot with the tuning pieces fully retracted. An alternative solution, and the one explored in this paper, is to physically and semi-permanently deform the full-cell itself in lieu of the tuning rods. This allows tuning to be done in the absence of the tuners as detailed in the preceding paragraph, but the necessary temperature variation is no longer extreme and is quite realizable.

3.2. *SUPERFISH Simulation*

The simulation code SUPERFISH[9] calculates the fields and frequency associated with particular modes resonating in a conducting cavity with symmetry and geometry specified by the user. The code was used in a preliminary determination of how to deform the full-cell geometry to achieve the desired frequency change, indicating that moving the downstream wall of the cell by about 0.015 cm would produce the necessary results. This served as verification that the amount of stretching needed was reasonably acheivable.

3.3. *Hardware Design*

The full-cell was stretched using a custom mechanical device designed and fabricated at UCLA. The device consists of three metal plates separated by stainless steel rods, and a large steel threaded bolt to apply the stretching force. Figure 2 shows a photograph of the stretcher device installed on the gun, illustrating the assembly. The stretcher device itself is artificially highlighted to appear very bright in the photograph so it can be easily visually distinguished from the body of the gun.

Figure 2. Stretcher device installed on former UCLA Neptune 1.6-cell RF photoinjector electron gun, highlighted for easy distinction from gun body.

The gun is clamped between two plates with the beam port extending through the upper plate. A third plate is raised above the beam port. A threaded rod extends through the third plate and is threaded through a flange bolted on to the beam port, with a nut tightened on the rod so that it doesn't move with respect to the beam port flange. The force of the pull is solely on the beam port flange, such that the downstream wall of the full-cell is moved during stretching.

Stretching is accomplished by turning a nut located on the threaded rod above the third plate. The π-mode frequency is monitored on a network analyzer connected through a waveguide coupler during stretching. An iterative process of stretching past the desired frequency and then releasing the bolt tension so that the full-cell relaxes is performed until the cell relaxes to the correct frequency under no tension.

3.4. *High-Power Conditioning*

After stretching, the tuning was finished using the cathode tuning bolt, with further frequency adjustment to be made by temperature variation during operation. A bead drop measurement, in which a metal bead attached to a dielectric fiber is lowered into the cavity and through both cells while monitoring the π-mode frequency, was done to check the field balance between the cells. The bead perturbs the fields and thus the frequency in the cavity[7]. The measurement showed an approximate 40 KHz difference in the frequency perturbations when the bead is positioned in the centers of each of the two cells. This is the number one expects to measure when the fields are properly balanced. The difference arises because of the dielectric fiber, which extends through most of the cavity when the bead is in the half-cell, but extends only partially into the cavity when the bead is in the full-cell. This difference is not present in a bead *pull* measurement, in which the fiber always extends through the entire cavity. The cell-to-cell coupling, measured with a network analyzer, was excellent, with a β value of 1.007. These measurements served to verify that the structural distortions from the stretching were not significant, so the gun should still function in the proper manner.

Following stretching and final tuning, the gun underwent RF conditioning with an unpolished copper cathode installed. RF power was put into the cavity, as it would be during beam operation, and increased slowly until a high operating level was reached. As mentioned previously, that operating level is around 4.5 MW in a similar gun operating in the UCLA Neptune laboratory, with the tuning pieces in use. Removal of the tuners in our test gun allowed a level of 8.4 MW to be reached, a marked improvement. Conditioning was performed again using a polished magnesium cathode with similar success, reaching a level of approximately 9 – 10 MW. This improvement over the copper cathode case is most likely due to the polishing. In both cases the gun responded extremely well to conditioning.

4. Summary

The 1.6 cell SLAC/BNL/UCLA style RF photoinjector electron gun is capable of operating at significantly higher levels of input RF power when the full-cell tuning pieces are removed from the cavity. Simple calculations and *in situ* evidence suggest that arcing occurs in the vicinity of the tuners when they are not fully retracted. This is supported by the increase in power accomodated by the gun after removal of the tuners.

An RF gun no longer in use in a linac system and known to have an unreasonably low RF power capacity was used as a test-bed for this study. After using a mechanical device to deform the cavity geometry so that frequency tuning could be accomplished without use of the full-cell tuning pieces, the cavity was conditioned with RF power. With an unpolished copper cathode installed, the gun was able to accomodate approximately 8.4 MW of input power. Repeating with a polished magnesium cathode, about 9 – 10 MW of power was achieved. These are drastic improvements over previously achievable power levels of 4 – 4.5 MW in both the tested gun and guns currently in use at UCLA Neptune and LLNL PLEIADES laboratories. More recent photoinjectors are being designed to be used with the tuners completely retracted, an attribute that allows operation at much higher power levels, approximatley 15 MW in the case of ORION at SLAC. The results of this study are a successful verification that full-cell tuning rods negatively affect RF photoinjector performance.

Acknowledgments

This work was made possible by funding from the United States Department of Energy. The authors are grateful to colleagues at the Particle Beam Physics Laboratory at UCLA for collaboration and support.

References

1. J. B. Rosenzweig, *Fundamentals of Beam Physics*, 193–197 (2003).
2. S. J. Park et al., *2004 FEL Conf. Proc. (Trieste)*, p 179 (2004).
3. R. Bossart et al., *Proc. 3rd Eur. Particle Accl. Conf. EPAC-92 (Berlin)* p 1026 (1992).
4. A. J. Hatch and H. B. Williams, *Phys. Rev.* **112**, 681 (1958).
5. T. Wangler, *RF Linear Accelerators*, 28 (1998).
6. J. D. Jackson, *Classical Electrodynamics 3rd Ed.*, 352–353 (1999).
7. L. C. Maier, Jr. and J. C. Slater, *J. Appl. Phys.* **23**, 68 (1952).
8. J. C. Slater, *Microwave Electronics* (D. Van Nostrand Company, Inc., New York, 1950), p.81.

9. J. H. Billen and L. M. Young, "POISSON SUPERFISH," Los Alamos National Laboratory report LA-UR-96-1834 (revision March 14, 2000).

MEASUREMENTS OF QUANTUM EFFICIENCY OF MG FILMS PRODUCED BY PULSED LASER ABLATION DEPOSITION FOR APPLICATION TO BRIGHT ELECTRON SOURCES.

G.Gatti, L.Cultrera, F.Tazzioli, C.Vicario
INFN-LNF

A.Fiori
University of Rome "Tor Vergata", Chemical Science Department

A. Perrone, C. Ristoscu*
University of Lecce, Physic Department and INFN
**Laser Department, National Institute for Laser, Plasma and Radiation Physics, Bucharest*

Abstract

We report the results of preliminary tests on Mg film photocathodes deposited by pulsed laser ablation deposition on various substrates. A comparison with other deposition techniques is presented. We illustrate briefly the technique of deposition of thick films (up to several microns). Scanning electron microscope analysis of the films before and after laser cleaning has been performed to characterize the changes occurred on the surface. Measurements of the distribution and integrated value of quantum efficiency over the illuminated surface are reported. Plans for the deposition on the back flange of an RF gun are illustrated.

1. Introduction

1.1. *Requirements on high brightness photo- injector cathodes.*

The main requirements on photocathodes for today's advanced high brightness injectors are emission uniformity over the illuminated surface, high Quantum Efficiency (QE) in near UV region (e.g. 266 nm), fast response (less than 1 ps), low dark current, ruggedness [1].

Semiconductors (e.g. Cs_2Te) offer the highest QE (up to 10% at 266 nm) but they require an extreme UHV environment, both during preparation and operation time [2], moreover their response is slow (order of picoseconds). On the other side metals are rugged and offer the fastest response, but their QE is lower. The most commonly used metal is Cu ($10^{-5}<QE<10^{-4}$ at 266 nm), but the highest efficiency is offered by Mg ($10^{-4}<QE<10^{-3}$ at 266 nm) but bulk Mg inserts have presented problems of non uniformity and poor adhesion to the host Cu substrate [3]. New proposals and studies about new photocathodes able to overcome these problems [4] are in progress, though they still present different drawbacks.

Mg films of few microns could represent a better solution with respect to Cu and bulk Mg. They show a more uniform and pure surface, moreover they can withstand high electric fields without excessive dark current. Mg films could represent an alternative to Cu when a higher current is desired. Mg films

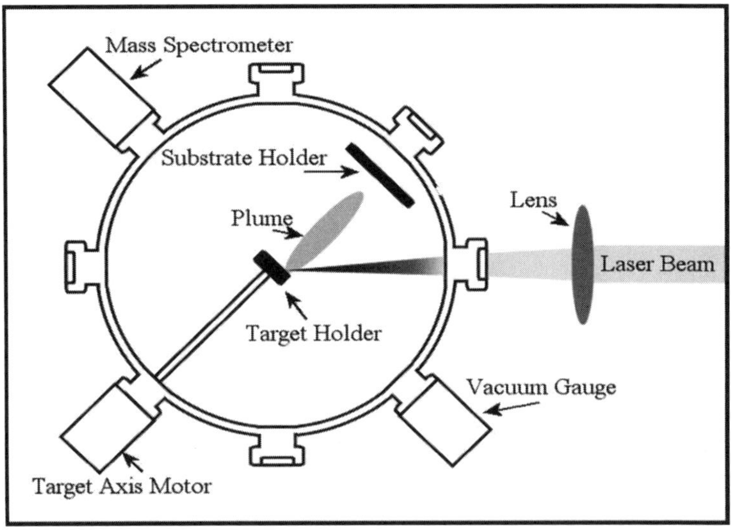

Figure 1. Schematics of the PLAD system

obtained by magnetron sputtering on Cu substrate have already been studied with promising results [3]. Tests in a RF gun were also performed, but arcing during the initial RF conditioning damaged the cathode [5]. Then it was decided to try an improved bulk welding technique and no further developments were made on Mg films. The poor performance could be attributed to inadequate quality of the film, especially regarding uniformity and adhesion to the substrate. It is then worthwhile to try alternative film deposition techniques, as the Pulsed Laser Ablation Deposition (PLAD), never used before for cathode deposition purposes. This technique promises better adhesion, uniformity and freedom from spurious composites.

1.2. *The PLAD sample preparation technique.*

The principle of the PLAD technique is the following. A powerful laser beam ablates Mg from the target, generating a plume of energetic particles (tens of eV) that impinge on the substrate placed at a suitable distance (see Figure 1). The energy of the impinging particles, higher by about an order of magnitude than in the sputtering technique, ensures a better adhesion and compactness of the film. The films were obtained by deposition of material ablated by a 30 ns pulsed laser beam of 308 nm wavelength (Lambda Physics LPX 315i XeCl excimer laser). In Figure 1 we show a sketch of the sample deposition process inside the related apparatus. The target where the laser beam was focused rotated at a frequency of 1 Hz to ensure a uniform target surface, hence better quality of the plume. The impinging angle of the light beam with respect to the target surface was 45 degrees. A quadrupole mass spectrometer controlled the quality of the vacuum before, after and during deposition together with quality of ablated material.

Two sets of samples were prepared: a thin one with a Graphite protective layer and a thicker one without protection. In the former the target was arranged as a disk with a 12 mm diameter graphite inner core fitted within a thick ring of high purity Mg (25 mm external diameter). This special target configuration allowed us to perform the deposition of the Mg film and in succession (within a unique process) of the protective graphite coating by moving the laser beam on different areas of the target. For the growth of thicker films without graphite covering layer the deposition process was performed ablating pure Mg target in low pressure of He (5 Pa). The presence of the background gases influence the laser induced plasma expansion dynamics. Due to the peculiar anisotropic distribution of velocities of the plume the main effect is the transverse confinement of the expanding ablated material [6]. This results in a strong increase of the deposition rate.

Moreover, the characteristics of the deposited films depend on the ablation rate, the plume deflection and the morphological changes of target surface created by prolonged laser irradiation. The right set of parameters for a thick and adherent film is chosen via systematic studies of Mg laser ablation process [7]. We have sampled out for discussion two of these films (samples 1 and 2) that have shown better uniformities. The experimental parameters used for such depositions are reported in Tables 1 and 2. Before each successive deposition run, the target surfaces were cleaned by applying a series of pulses. During these cleaning processes, a shutter is interposed between the target and the Cu substrate. A drawback of the PLAD technique is the formation, on the film surface, of droplets having a diameter of the order of one micron. They can in principle be avoided using different methods that frequently suffer of low deposition rates [8-11].

Target	Mg-C
Substrate	Cu
Laser fluence	10 J/cm^2
Target-Substrate distance	4.5 cm
Base pressure	10^{-6} Pa
Laser pulses • Mg cleaning • Mg deposition • C cleaning • C deposition	 5000 30000 2000 3000
Mg thickness	200 nm
C thickness	20 nm

Table 1. Sample 1 with Graphite protecting layer deposition settings.

Target	Mg
Substrate	Si (100)
Laser fluence	10 J/cm^2
Target-Substrate distance	3.5 cm
Base pressure	5 Pa of He
Laser pulses • Mg cleaning • Mg deposition	 5000 50000
Mg thickness	2500 nm

Table 2. Sample 2 layer deposition settings.

2. Experimental apparatus

The QE measurement and laser cleaning apparatus is made up of a test HV chamber, at 10^{-6} Pa background pressure, containing a vacuum diode; the film to be tested constitutes the cathode. An UV 266 nm laser pulse, 30 ps duration, from a mode-locked frequency quadrupled Nd:Yag laser, is injected through a quartz window and excites the cathode. Accelerating electric fields up to 1 MV/m can be applied to the diode to extract the electrons. The laser beam crosses the anode through a fine copper grid and impinges perpendicularly on the cathode (sketch of the cathode holder shown in Figure 2). The emitted charge is sent via a coaxial cable to the input of a high speed oscilloscope or to a high sensitivity charge amplifier. A scheme of the apparatus is shown in Figure 3.

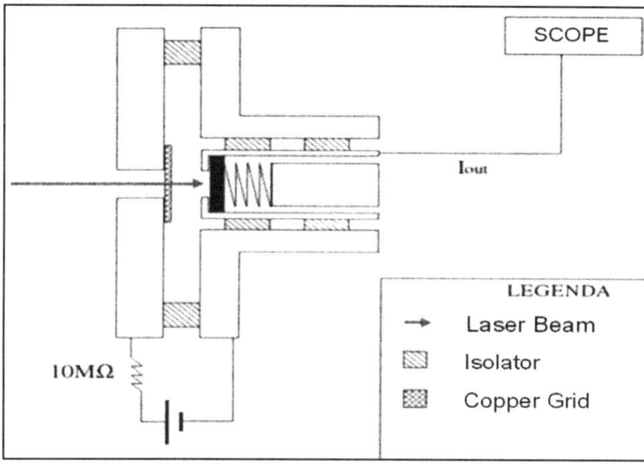

Figure 2. Sketch of the cathode holder

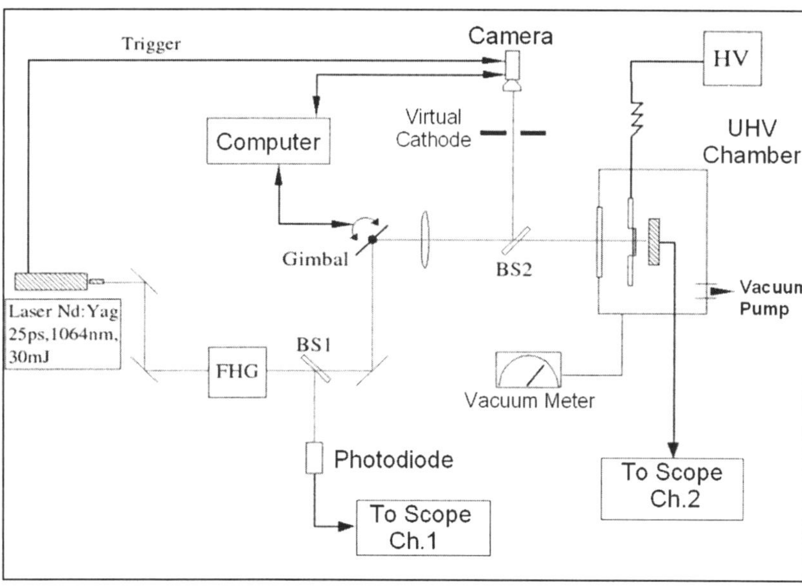

Figure 3. Layout of the experimental setup

2.1. *Measurement procedure and results*

The QE of the samples was measured before and after the laser cleaning operation. An area of 1 mm^2 on sample 1 was cleaned with a focused beam of high power density (4 GW/cm^2, 1200 shots). The beam was then defocused and the diameter was enlarged to cover the cleaned area for QE measurements. Starting yeld was poor, due to quadratic response of the graphite layer at 266nm; after removal of the coating by laser cleaning, a good linear response due to Mg with QE=3x10^{-4} was measured (Figure 4).

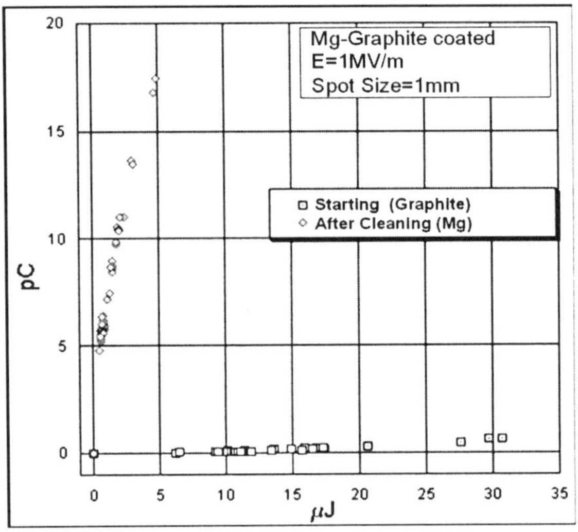

Figure 4. Sample 1: emitted charge vs. laser energy first and after laser cleaning

Sample 2 deposition was not optimized for maximum QE, but to obtain a large thickness, with the help of He gas background atmosphere. On this sample we performed two measurements.

1- The measurement of emission yield after a moderate cleaning (1200 shots at 1 GW/cm^2 on 1 mm^2 area. The final global QE obtained was about 2,5x10^{-5}. Fig. 5 shows the emission curve.

2- The measurement of average yield and density distribution over a smaller area of 0.5 mm^2 after a strong cleaning (100 shots at 20 GW/cm^2). The QE was measured on a matrix of 9 equidistant spots of 80 micron diameter. The average QE was 9x10^{-5} with a rms spread of 12%. The distribution is shown in Figure 6. The emission spread can be reduced by optimizing the quality of the film and the cleaning technique.

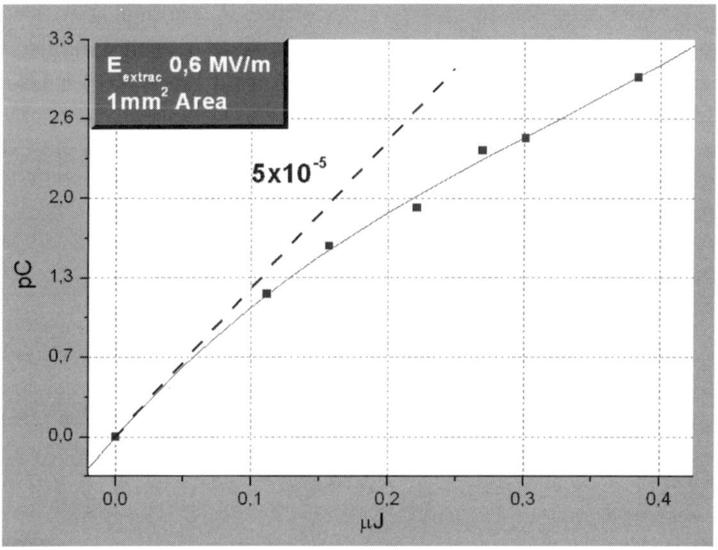

Figure 5. Sample 2: emission response after normal laser cleaning

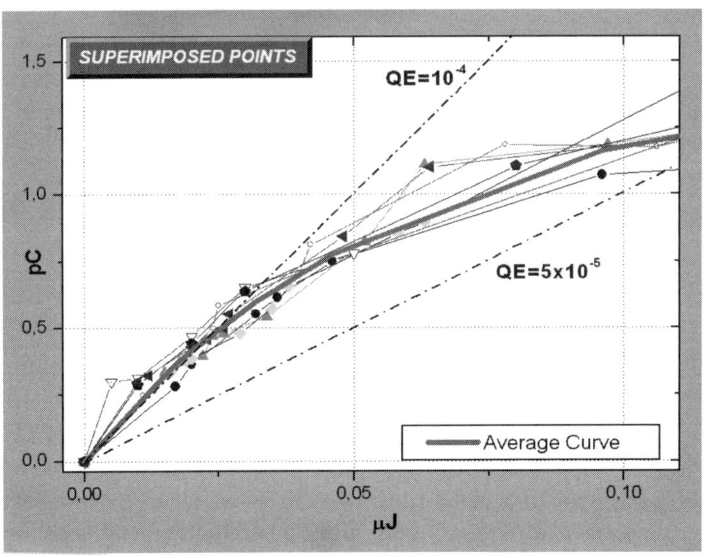

Figure 6. Sample 2: superimposed response of the 9 points after strong cleaning

SEM analysis of the surface was performed before and after irradiation. SEM micrographs in Figure 7 show that the starting PLAD Mg film morphology is amorphous. Under very intense irradiation, laser annealing process takes place and further on, recrystallization of Mg coating seems to take place.

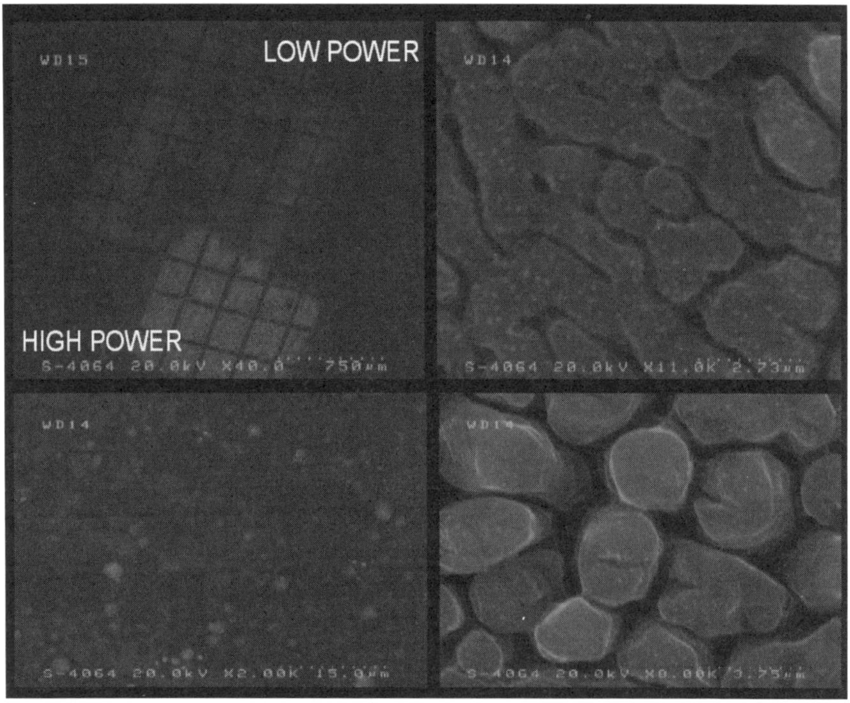

Figure 7. SEM of sample 2 after measurements. In top-left angle the two irradiated areas. In counterclockwise direction, details of the areas. Low power cleaning zone (amorphous structure with droplets), strong irradiated area (crystal structure), middle zone (in the shadow of the grid) forming crystals

2.1.1. *Sputtered films.*

We have acquired sputtered films from industry (Koral Labs) to establish a term of comparison with PLAD ones in the same testing conditions. The films we tested were 10 microns thick and were shipped to us in pressurized envelopes under Nitrogen. We have measured them as received, without any polishing. It was easy to remove the thin layer of oxides and other compounds present on the film surface. The maximum QE we measured was $1x10^{-4}$. The morphology of the films, analyzed with an AFM microscope, seemed to be crystalline with well ordered columnar structure. Resistance to laser radiation was good: after bombardment with about 1E4 shots at 500 μJ/mm^2 laser energy density, the ablated depth was less than 1 micron. The adherence to the substrate, after some operations of removal and re-deposition, looks satisfactory. The construction of the real cathode by deposition on the gun back-flange through masks, looks straightforward. We therefore think that this technique constitutes still a noteworthy competitor with PLAD.

3. Conclusions

Mg films deposited by PLAD on Cu substrates offer a viable alternative to bulk cathodes. The advantages are uniformity, high purity, and the possibility of co-depositing thin (tens of nm) protective layers, good adherence to the substrate. At low field the QE of few 10^{-4} at 266 nm has been demonstrated. An easily removable protective layer of graphite can prevent the oxidation of the surface during transportation and installation. This is important for thin films (less than half a micron) because the oxide can penetrate as deep as 200 nm. Thick films (several microns) can be deposited with the aid of inert gas atmosphere. Thicker films consent the removal of the oxide layer and do not require protection. The improvement of deposition of thick films is in course. The set up for the deposition on the back flange of a RF gun is being prepared. The final goal is to test the film in the real environment of an RF gun, to clarify the issues of dark current and resistance to arcing during RF conditioning.

Acknowledgments

This work has been partially supported by the EU Commission in the sixth framework program, contract no. 011935 – EUROFEL.

References

1. TDR for SPARC photo-injector, *http://www.lnf.infn.it/acceleratori/sparc.*
2. G. Suberlucq, *Proc. EPAC conference*, 64 (2004).

3. T. Srinivasan-Rao, J. Schill, I. Ben Zvi , M. Woodle, *Rev. Sci. Inst.* **69**, 2292 (1998).

4. X. Chang, Ben Zvi et al., *Proc. PAC conference*, 2251 (2005)

5. P.E. Dyer, A. Issa and P.H. Key, *Appl. Phys. Lett.* **57**, 186 (1990)

6. L. Cultrera, A. Pereira, C. Ristoscu, A. Clozza, F. Tazzioli, C. Vicario, *Appl. Surf. Sci.* **248**, 397 (2005)

7. R. K. Singh, D. Bhattacharya, J. Narayan, *Appl. Phys. Lett.* **61**, 483 (1992)

8. R. J. Kennedy, *Thin Solid Films* **214** 223 (1992)

9. H. B. Lu, S. F. Xu, Y. J. Tian, D. F. Cui, Z. H. Chen, Y. Z. Zhang, L. Li, G. Z. Yang, *J. Supercond.* **6**, 335 (1993)

10. D. Lubben, S. Barnett, K. Suzuki, S. Gorbatikin, J. Green, *J. Vac. Sci. Technol. B* **3**, 968 (1985)

11. X. J. Wang, M. Babzien, R. Malone, Z. Wu, *Proc. LINAC Conference*, 142 (2002).

THE S-BAND 1.6 CELL RF GUN CORRELATED ENERGY SPREAD DEPENDENCE ON π AND 0 MODE RELATIVE AMPLITUDE

J.F SCHMERGE, J. CASTRO, J.E. CLENDENIN, D.H. DOWELL, S.M. GIERMAN AND H. LOOS

SLAC, Stanford University, 2575 Sand Hill Rd
Menlo Park, CA 94025, USA

The π mode or accelerating mode in a 1.6 cell rf gun is normally the only mode considered in rf gun simulations. However, due to the finite Q there is a small but measurable 0 mode present even at steady state. The π mode by definition has a 180°phase shift between cells but this phase shift for the total field is several degrees different. This results in a correlated energy spread exiting the gun. A comparison of simulation and experiment will be shown.

1. Motivation

1.1. *Introduction*

Longitudinal phase space measurements performed at the SLAC Gun Test Facility (GTF) have shown a large correlated energy spread exiting the gun not predicted by simulations[1]. Previous simulations only include the π mode which is the dominant rf mode in the 1.6 cell gun. However, due to the finite Q there is a small but non-zero component of the 0 mode present in the gun at steady state. The total field has phase shift between cells that is approximately 7° off the 180° for a pure π mode case. The phase shift produces a correlated energy spread at the gun exit. This paper describes frequency domain measurements on the gun to characterize the π and 0 mode components as well as comparison of the measured longitudinal phase space with simulations including the 0 mode.

The phase shift of the total field with respect to the π mode also must be accounted for when measuring the phase in the rf gun. Typically the gun phase is determined by measuring the emitted charge as a function of phase[2]. The 0° phase point is defined as the phase where the gun starts to emit. This phase is measured with respect to the total field but is often used in the simulation as the phase with respect to the π mode. For the GTF gun the phase offset between the total field and the π mode in the half cell is approximately 5°.

1.2. *Phase Space Measurements*

Longitudinal phase space is measured at the GTF by measuring the energy spectrum as a function of linac phase[1]. The total energy is measured simultaneously with energy spread in order to determine the absolute phase. Maximum energy is defined as the 0° reference point. Figure 1 shows the results of a typical measurement and the resulting phase space distribution for a charge of 15 pC and a peak current of approximately 10 A. The longitudinal Twiss parameters can be determined at the linac entrance by fitting the measured energy spread as a function of linac phase just as the transverse Twiss parameters are determined from fitting measured beam sizes as a function of quadrupole focal length. Alternatively the measured energy spectrum can be fit using tomographic reconstruction to determine the actual phase space distribution[3].

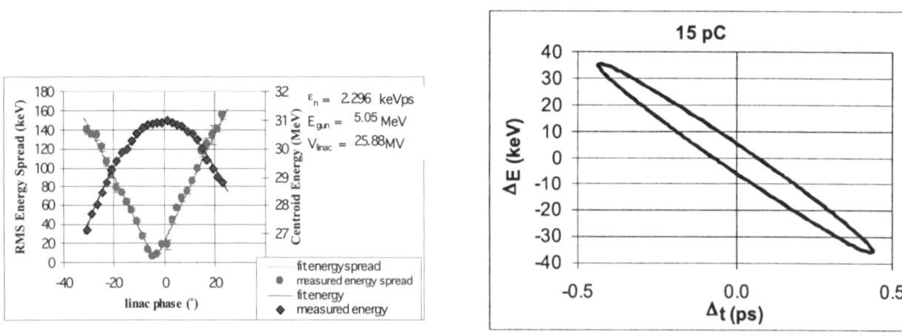

Figure 1. Measured energy and energy spread is plotted on the left along with the fit. The phase is defined to be 0 at the maximum energy. The Twiss parameters from the fit can be used to generate the rms phase space ellipse plotted on the right. This is the distribution approximately 75 cm downstream of the gun at the linac entrance. The beam is chirped with approximately -80 keV/ps.

The measured rms energy spread at the linac entrance is 36 keV while the simulation predicts approximately 10 keV when the applied rf field consists of only the π mode. The energy spread is clearly dominated by the correlated energy spread. The source of this chirp must either be an applied time varying field or a beam induced field. While the chirp does depend on charge a substantial chirp is present at all charges indicating one component of the source must be an applied field. The only applied fields between the cathode and the linac are the rf field in the gun and the DC magnetic field from an emittance compensating solenoid. Thus the rf field is the only time dependent field that could produce the measured chirp shown in figure 1.

Thermionic rf guns produce chirped beams as shown in figure 1 by using a ratio of full cell to half cell fields greater than one[4]. The field ratio was measured to be approximately one during cold testing of the GTF gun prior to installation. It was theorized that the gun tuning may have changed during multiple cathode changes or that the gun field distribution varied as the rf power was increased. Thus the gun was removed and carefully characterized.

2. Gun Characterization

Initially the gun had a single rf probe monitoring the field outside the full cell. The rf probe is a coaxial line with the center conductor extended past the outer conductor. The full cell had two identical holes located 180° apart one of which was used for the rf waveguide feed and the second was used as a vacuum pump out port. The probe is located in the pump out section. A second probe monitoring the half cell field was added outside one of the two laser ports. Each probe has a different response due to the different field strength at the probe position. The probe response was assumed independent of frequency over the few MHz range of the experiment. A cross section of the rf gun showing the location of the two probes is shown in Figure 2.

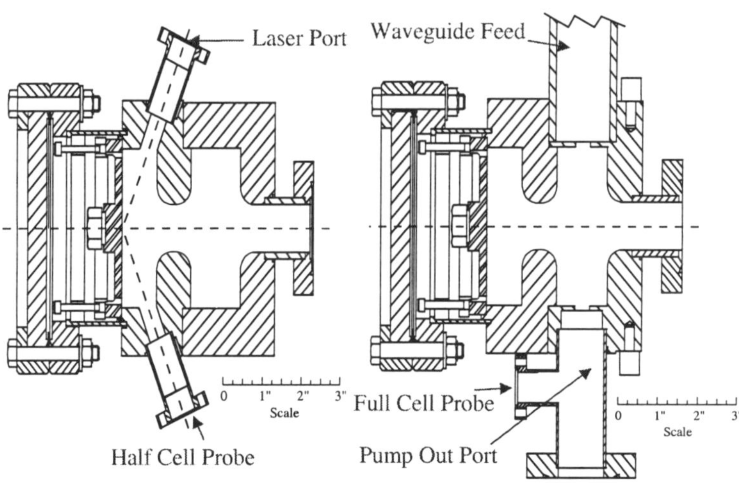

Figure 2. A cross section of the 1.6 cell photo-cathode gun is shown. The figure on the left shows the two grazing incidence laser ports, one of which was used to install an rf probe. The figure on the right is rotated 45° about the beam axis with respect to the first figure and it shows the waveguide feed and pump out port. The full cell probe is located in the pump out port region.

The gun was characterized in the frequency domain using a network analyzer with port 1 connected to the waveguide feed and port 2 to either the full cell or half cell probe. The measured transmission normalized amplitude and phase is plotted in Figure 3 for both probes. The constant and linear terms of the measured phase were modified to correct the phase reference point of the network analyzer and path length differences. Figure 3 shows the mode at 2856 MHz has a 180° phase shift between cells while the remaining mode has a 0° phase shift. These are by definition the π and 0 modes respectively.

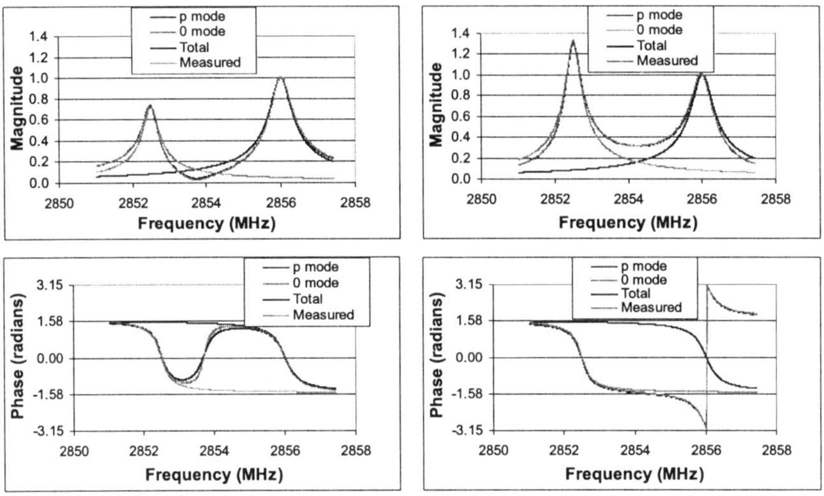

Full Cell Probe   Half Cell Probe

Figure 3. The S_{12} parameter normalized amplitude and phase with port 2 connected to the full cell on the left and the half cell on the right. The measured field is plotted in red and the fit in black. The fit is composed of two second order bandpass filters which are plotted in blue and turquoise.

A model of the two cell gun composed of two bandpass filters was used to fit the data with a least square error fitting routine. The frequency response of the model is given in equation 1 where f_{rf}, f_π and f_0 are the excitation frequency, the π mode frequency and the 0 mode frequency respectively. Q_π and Q_0 are the loaded Qs and A_π and A_0 are the normalized mode amplitude which depends on z. The model parameters fit to the data are listed in Table 1. The time constant for each mode is defined in equation 2. The reflection data or S_{11} parameter was also measured to determine the rf coupling coefficient β.

$$A_{ss}e^{j\theta_{ss}}\left(f_{rf},z\right)=\left[\frac{jA_\pi\left(z\right)\frac{f_\pi}{Q_\pi}f_{rf}}{f_\pi^2-f_{rf}^2+j\frac{f_\pi}{Q_\pi}f_{rf}}+\frac{jA_0\left(z\right)\frac{f_0}{Q_0}f_{rf}}{f_0^2-f_{rf}^2+j\frac{f_0}{Q_0}f_{rf}}\right] \tag{1}$$

$$\tau_{\pi,0}=\frac{Q_{\pi,0}}{\pi f_{\pi,0}} \tag{2}$$

Due to the different probe responses, the frequency response measurement can not determine the ratio of the field in the full cell to the half cell. A bead drop measurement was used to determine the field amplitude versus position. The bead data indicate the π mode has a full cell to half cell field ratio of 1 ± 0.02 and is used to set the normalized amplitude of the π mode in the full and half cell to +1 and -1 respectively. With this definition of normalized field, the probe response is -44.1 dB for the full cell probe and -53.0 dB for the half cell probe.

Table 1. The measured rf parameters for the π and 0 mode.

Parameter	π mode	0 mode
A ($_{\text{full cell}}$)	+1	+0.72
A ($_{\text{half cell}}$)	-1	+1.31
f_0 (MHz)	2856.03	2852.53
Q	5160	6980
β	1.34	0.68
τ (ns)	576	779

After characterization the gun was re-installed and operated. The rf probes were used to measure the field in each cell at full power. Figure 4 shows the measured fields in each cell as a function of time. Approximately 6 MW of power is required to reach a field of 90 MV/m in the gun. As the plot indicates the ratio of the field is roughly independent of time. The time shift between the two probe signals is due to cable length differences.

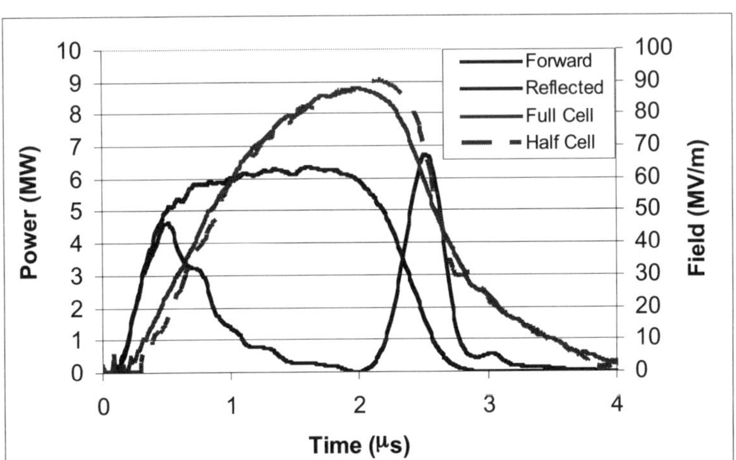

Figure 4. The measured amplitude of the full cell and half cell probes along with the incident and reflected rf power signals. The data was collected using a crystal detector.

The peak field measured by the two probes agrees to within approximately 2%. Thus the on axis field at full power and low power are identical and a field ratio >1 is not the source of the correlated energy spread.

3. 0 Mode Contribution

Next, the 0 mode component of the total field was studied to determine if it could produce a correlated energy spread. The total field in the gun is given by the sum of the fields from the two modes. The total field can be calculated from the model parameters listed in table 1. For a step function excitation the field is given in equation 3. A_{ss} and θ_{ss} are the steady state amplitude and phase from equation 1. The effect of the two transient terms are assumed small due to the exponential decay and are not considered in this paper.

$$V(t) = A_{ss}\cos\left(\omega_{RF}t + \theta_{ss}\right) + B_{\pi}e^{-\frac{t}{\tau_{\pi}}}\cos\left(\omega_{\pi}t + \phi_{\pi}\right) + B_{0}e^{-\frac{t}{\tau_{0}}}\cos\left(\omega_{0}t + \phi_{0}\right) \quad (3)$$

The relative amplitude and phase for each mode in each cell has been computed and is listed in table 2 for an excitation frequency of 2856.00 MHz. Since the 0 mode is nearly 90° out of phase with the π mode, the total field amplitude is approximately equal to the π mode amplitude but the total field phase is changed by several degrees. The effect is greatest in the half cell where the 0 mode amplitude is the largest. The phase advance between cells is nearly 7° larger for the total field than the π mode alone.

Table 2. The calculated amplitude and phase for the π , 0 mode and total field in both the full and half cell. The phase is measured with respect to the incident field.

Field	Full Cell	Half Cell
π mode	1∠0°	1∠180°
0 mode	0.044∠-86.7°	0.076∠-86.7°
Total ($A_{ss}\angle\theta_{ss}$)	1.003∠-2.5°	0.998∠184.4°

Figure 5 is a GPT simulation of the energy spread as a function of linac phase where the results of four different simulations are shown demonstrating the effect on the energy spread due to both the 0 mode and space charge. The green curve is a simulation excluding both the 0 mode and space charge. Including the 0 mode but not space charge results in the blue curve. This shifts the curve closer to the experimental data but it still does not reproduce the experimental results. The pink curve includes space charge but not the 0 mode and once again it does not match the experimental results. The red curve includes the 0 mode and space charge and actually matches the experiment fairly well. However, in order to get good agreement with the experiment the 0 mode amplitude in the half cell was increased to 12% of the π mode instead of the measured 8% and the laser phase was reduced to 20° with respect to the π mode field instead of the measured 24°. The discrepancy may be due to the presence of transient field components and measurements are in progress to determine the contribution of the transient rf fields to the correlated energy spread. The results indicate space charge does contribute to the correlated energy spread. Both space charge and the 0 mode must be included in the simulations even at the relatively low peak current of 10 A in order to reasonably reproduce the experimental results.

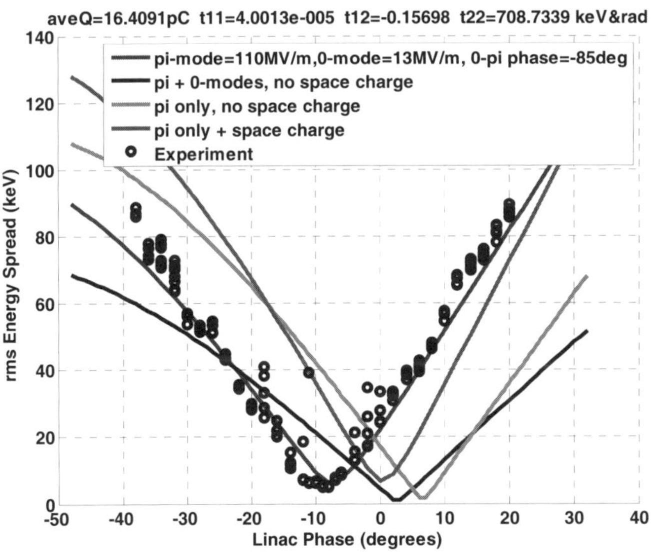

Figure 5. The measured energy spread versus linac phase is plotted as open circles along with the results of four GPT simulations. The green curve is the result of the simulation with the 0 mode and space charge excluded. The blue curve includes the 0 mode but not space charge and the pink curve includes space charge but not the 0. The red curve includes both space charge and the 0 mode.

4. Conclusions

A large correlated energy spread is produced at the gun exit. The correlated energy spread is not due to a full cell to half cell field ratio greater than one.

The field ratio was measured to be nearly one at both low power with a bead drop and at full power with calibrated field probes in each cell.

A model of the two cell gun was fit to the frequency domain measurements. Based on the fit parameters the steady state field due to the π and 0 modes could be computed along with the total field. The total field has a phase advance between cells of 187° instead of the 180° degrees for the π mode. Also, in the half cell the phase difference between the π mode and the total field is approximately 5°. Measurements of the laser phase determine the phase of the laser with respect to the total field and not the π mode. The correct laser phase must be included in the simulation to predict the correct energy spread.

Simulations using the code GPT show that both space charge and the 0 mode contribute to the correlated energy spread. The simulation can reproduce the experimental results but it requires a 0 mode amplitude larger than is measured. The best agreement with the measured data occurs with a 0 mode amplitude 12% of the π mode amplitude instead of the measured 8% and a laser phase of 20° instead of the measured 25°. The discrepancy might be due to the transient rf field contribution and is currently under investigation.

Simulations should include the 0 mode component of the gun field in order to accurately compute the longitudinal phase space. The non-zero component of the 0 mode at the cell to cell iris may also contribute to the transverse emittance. The LCLS gun is designed with a mode separation of 15 MHz instead of 3.5 MHz as in the GTF gun. Thus the LCLS gun will have a phase advance between cells of 181°. The 0 mode contribution is significantly suppressed with the larger mode separation and the adverse beam effects will not be present.

Acknowledgments

This work supported by the U.S. Department of Energy under contract number DE-AC02-76SF00515.

References

1. D.H. Dowell et al, "Longitudinal emittance measurements at the SLAC gun test facility", NIM Vol A507, pp. 331-334, 2003.
2. J.F. Schmerge et al, "RF gun photo-emission model for metal cathodes including time dependent emission", to be published in these Proceedings.
3. H. Loos, et al, "Longitudinal phase space tomography at the SLAC gun test facility and the BNL DUV-FEL ", NIM Vol A528, pp. 189-193, 2004.
4. M. Borland, "A high-brightness thermionic microwave electron gun", SLAC Report 402, 1991. http://www.slac.stanford.edu/pubs/slacreports/slac-r-402.html

RF GUN PHOTO-EMISSION MODEL FOR METAL CATHODES INCLUDING TIME DEPENDENT EMISSION

J.F. SCHMERGE, J.E. CLENDENIN, D.H. DOWELL, AND S.M. GIERMAN

SLAC, Stanford University, 2575 Sand Hill Rd
Menlo Park, CA 94025, USA

The quantum efficiency from a metal cathode is strongly dependent on the field at the cathode due to the Schottky effect. Since the field is time dependent the quantum efficiency is also time dependent. Thus the laser pulse shape used to generate electrons in a photocathode rf gun is not the same as the electron bunch shape. In addition since the thermal emittance and quantum efficiency are related, the thermal emittance is also time dependent.

1. Motivation

1.1. *Introduction*

The emission from a photo-cathode rf gun depends on both the rf field and the incident laser pulse. In metal cathodes the rf field and image charges affects emission through the Schottky effect. All of these fields have fast time dependencies which must be accounted for to get an accurate model of the emitted electron pulse.

This paper describes the main time dependent effects and provides simple mathematical models to predict experimentally observed phenomena. The results apply to metal cathodes where the Schottky effect leads to an effective lowering of the work function. For those applications which require a flat top temporal electron beam, the time dependent cathode emission will lead to the use of drive laser pulses with asymmetric temporal shapes. In addition to the time dependent quantum efficiency (QE), it will be shown that the same time dependent fields result in a time dependent thermal emittance.

1.2. *Measurements*

Typically one of the first measurements performed while commissioning a photo-cathode gun is the measurement of charge vs laser phase. This type of measurement demonstrates the time dependent emission process due to the Schottky effect. All measurements reported here are conducted with a transverse flat-top laser profile.

A typical measurement performed at the SLAC Gun Test Facility (GTF) is shown in Figure 1 where the measured QE is plotted as a function of the laser phase for two different metal cathodes[1]. The QE is defined as the ratio of the number of electrons emitted to the number of incident photons. The charge is measured on a Faraday cup approximately 80 cm from the gun exit with a solenoid immediately downstream of the gun that has been adjusted to maximize the collected charge. Dark current is subtracted in the measurement to determine the photo emitted charge contribution. The laser energy is measured with a joule meter monitoring the energy downstream of a 1% pickoff.

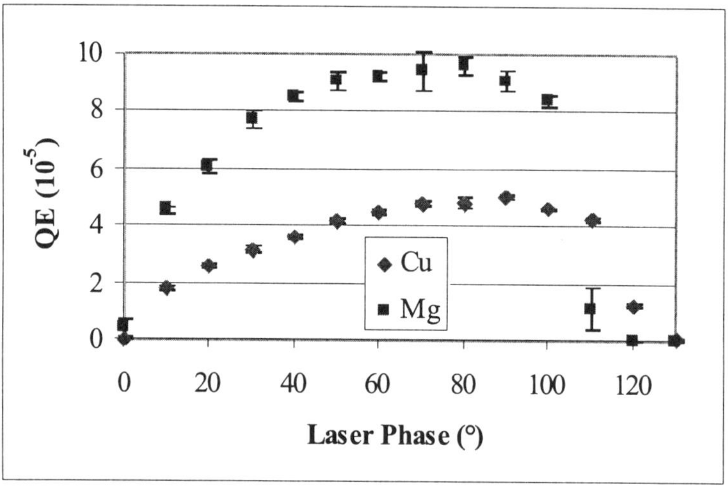

Figure 1. The measured QE as a function of laser phase for both Cu and Mg cathodes. A laser phase of 0° is defined as the zero crossing of the rf field at the cathode. The peak field at the cathode was 110 MV/m for the Cu cathode and 90 MV/m for the Mg cathode.

The Schottky scan in Figure 1 clearly demonstrates the importance of the rf field in the emission process. Equally important but not shown in Figure 1 is the effect of image charge. To demonstrate this effect the charge is measured as a function of laser energy. Figure 2 shows the result of such a measurement performed on the same Cu cathode but at a later time. Here the QE can be seen to be laser energy dependent. The linear fit is plotted and results in a QE of 6.6 10^{-5} which is roughly a factor of two higher than the QE shown in Figure 1. Other measurements on this and other Cu cathodes at the GTF have resulted in QEs ranging from 0.6-6 10^{-5} for an applied field of 120 MV/m and 30° laser phase. The reason for the variation of the QE has not been determined but it is suspected due to surface contaminants.

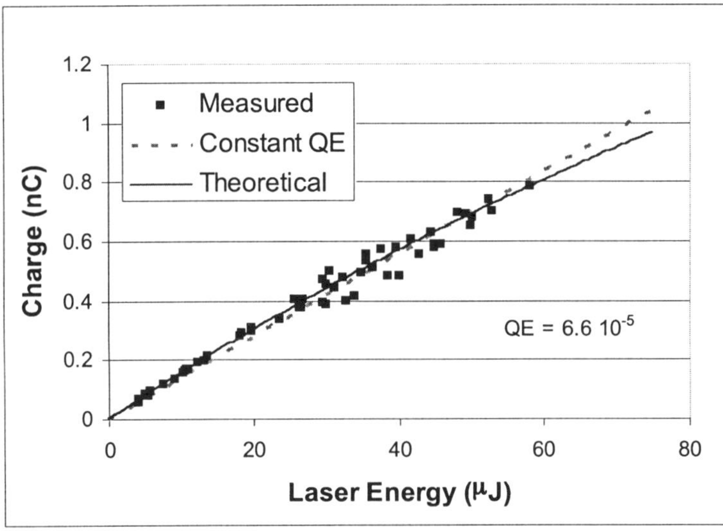

Figure 2. Measured charge as a function of laser energy for a Cu cathode at a laser phase of 40° and a peak field at the cathode of 110 MV/m. The dashed line is the linear fit to the measured data and the solid line is the fit using equation 1 which includes the effect of the image charge reducing the total field at the cathode. Due to the Schottky effect this leads to a reduction in QE for the tail of the pulse and thus the measured curve is not linear but rolls over as the laser energy is increased.

However, the linear fit always underestimates the charge for low laser energies and overestimates the charge for the highest laser energies. A better fit can be found by including the time dependent effects in a model that can accurately reproduce both experiments shown above.

2. Model

Typically, the QE from a metal cathode is assumed to behave as given in Equation 1[2] where the work function reduction due to the Schottky effect is defined in equation 2. $\Phi_{cathode}$ is the cathode work function with no applied field, hν is the laser photon energy and η_0 is a proportionality constant that can be determined from the slope of the charge vs laser energy plot at low charges.

The total field at the cathode is the superposition of the applied rf field and the field from the image charge where θ_{laser} is the laser phase and r is the laser beam radius at the cathode. Models for field emission include a field enhancement factor to account for geometrically enhanced surface fields at localized points. However, this model for photo-emission does not include a field enhancement factor since the current is emitted over a large area and the average field should be equal to the applied field.

It is assumed in the derivation of equation 2 that the electron beam is a pancake beam and thus the space charge field is nearly independent of transverse position. If the space charge field depends on position then the QE will depend on both space and time[3,4]. However, in typical rf guns the head of the beam gains a small amount of energy over the duration of the laser pulse and moves only 10's of microns. Therefore, the large aspect ratio assumed for the pancake beam is nearly always satisfied.

For an applied field of 110 MV/m, the total field decreases to zero and suppresses further emission when the total charge exceeds 1.5 nC assuming a 2 mm diameter laser pulse. This is the space charge limit. The total charge is calculated by integrating the QE as shown in equation 3 where I_{laser} is the temporal power profile of the laser. The fit shown in figure 2 was produced by numerically integrating equations 1-3. Likewise equations 1-3 can be used to fit the data in figure 1. While the model fits the functional dependence on laser phase, the value of η_0 is different for the data in the two figures.

$$QE(t) = \eta_0 \left[h\nu - \left(\Phi_{cathode} - \Delta\Phi(t) \right) \right]^2 \tag{1}$$

$$\Delta\Phi(t) = \sqrt{\frac{e^3 \left(E_{RF} \sin(\omega_{RF} t + \theta_{laser}) - \dfrac{Q(t)}{\pi r_{cathode}^2 \varepsilon_0} \right)}{4\pi\varepsilon_0}} \tag{2}$$

$$Q(t) = \int_{-\infty}^{t} e \frac{I_{laser}(\tau)}{h\nu} QE(\tau) d\tau \tag{3}$$

However, equations 1-3 do not give a hint as to the theoretical maximum value of the QE or how the thermal emittance is related to the QE. A model that includes the electron energy distribution is needed to determine these parameters. In addition it was desired to consider why the QE varies with laser incidence angle and polarization. Thus a simplified model was developed to describe the emission process.

The basic model described below includes the following assumptions:

1. Electrons are emitted from the bulk material
2. Only electrons with sufficient momentum to overcome the work function surface barrier are emitted
3. Electrons in the cathode material exhibit a Fermi-Dirac electron energy distribution

4. The laser pulse bandwidth is ignored (except when investigating laser chirp effects)
5. Single photon absorption
6. The mean free path due to electron-electron scattering is assumed much longer than the optical skin depth.
7. Cathode has a flat planar surface
8. No surface effects except the laser beam reflection are included
9. No polarization effect other than reflectance

Electrons are assumed in thermal equilibrium until a photon is absorbed which raises the electrons kinetic energy. Since the mean free path is assumed infinite, the electrons that absorb a photon are assumed to escape the cathode if they reach the cathode surface and the longitudinal momentum is sufficient to overcome the barrier. The criterion for emission is given in equation 4 where p_z is the momentum perpendicular to the surface and E_b is the true barrier energy as defined in equation 6[5]. This requirement leads to an upper limit on the angle with respect to the surface normal, Φ_{max}, that an electron can approach the surface and still escape as given in equation 5 where E is the electron energy prior to absorption. This maximum angle is 10.9° for Cu and 19.7° for Mg assuming a field at the cathode of 120 MV/m and a 263 nm photon. Figure 3 is an energy level diagram for both Cu and Mg showing the barrier energy and the minimum required energy for an electron to absorb a photon and emit as defined in equation 7.

$$p_z \geq \sqrt{2mE_b} \tag{4}$$

$$\cos\Phi_{\max} = \sqrt{\frac{E_b}{E + h\nu}} \tag{5}$$

$$E_b \equiv E_F + \Phi_{cathode} - \Delta\Phi \tag{6}$$

$$E_{\min} \equiv E_b - h\nu \tag{7}$$

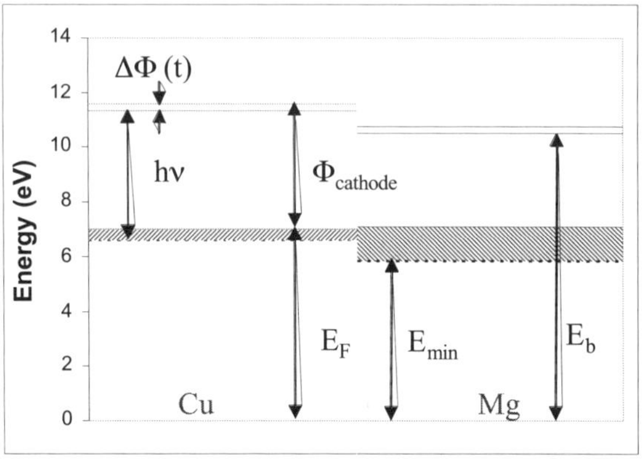

Figure 3. An Energy level diagram for Cu is shown on the left and Mg on the right. The lowest solid line is the Fermi Energy and for T=0 K is the uppermost electron energy state in the metal. The upper solid line is the work function plus the Fermi level and the line immediately below it represents the Schottky effect reduction. This is the real electron barrier the electron must overcome to be emitted. The cross hatched area represents the electrons with sufficient energy to escape.

Since e-e scattering is effectively ignored, the depth below the cathode surface at which the electrons absorb a photon is irrelevant in this model. Of course the number of photons penetrating into the metal exponentially decays with the skin depth and the deeper the electron are generated the less likely they will reach the surface with sufficient energy to escape due to electron-electron scattering. However, from the simple model assuming emission is independent of depth, a reasonable account of photo-emission can be established.

2.1. *Quantum Efficiency*

The QE is then the number of electrons that could escape divided by the total number of electrons summed over energy and the solid angle times the fraction of laser energy actually absorbed in the cathode. With this in mind and the physical model described above the quantum efficiency can be described as the integral shown in equation 8 where $N_{electron}$ is the electron energy distribution which is assumed a Fermi-Dirac distribution and R is the power reflectance of the laser which is angle and polarization dependent..

$$QE = (1-R)\frac{\int\limits_{E_{\min}}^{E_b}\int\limits_{0}^{\Phi_{\max}}\int\limits_{0}^{2\pi} N_{electron}\sin\Phi d\theta d\Phi dE}{\int\limits_{0}^{E_b}\int\limits_{0}^{\pi}\int\limits_{0}^{2\pi} N_{electron}\sin\Phi d\theta d\Phi dE} \tag{8}$$

An analytic solution for equation 8 can be found if we assume the temperature is 0 K. The result is given in equation 9 with E_k defined in equation 10. Equation 8 must be numerically integrated or solved approximately for non-zero temperatures. Higher temperatures lead to higher QEs since more electrons have sufficient energy to participate in the emission process. However, it typically requires very high temperatures to produce order of magnitude increases in the QE except when the photon energy is on the same order as the work function as seen in Figure 4. The theoretical QE is plotted as a function of laser wavelength in Figure 4 for Cu along with the measured QE with near zero applied field[6]. The theoretical QE is nearly a factor of three greater than the measured value. This is the same factor that the theoretical QE is reduced if the finite mean free path is included in the model[7].

$$QE = (1-R)\frac{\left(1-2\left(\frac{E_b}{E_k}\right)^{1/2}+\left(\frac{E_b}{E_k}\right)\right)}{2\left(\frac{E_F}{E_k}\right)} \tag{9}$$

$$E_k \equiv E_F + h\nu \tag{10}$$

This expression gives 16 10^5 and 21 10^5 for the QE of Cu and Mg respectively at room temperature under identical conditions as the experimental results in Figure 1 at 30° laser phase. Although the work function of Mg is nearly 1 eV lower than Cu, the QE is not significantly higher due to the much higher reflectivity at 263 nm. The Cu reflectivity is only 34%[8] while Mg is approximately 92%[9]. Our model QE values are a factor of 6 and 3 higher for the Cu and Mg measurements in figure 1 and 3 times higher than the QE shown in figure 2. Most of this discrepancy can be attributed to assuming an infinite mean free path as discussed above. The rest is assumed due to localized contaminants on the surface[6] which creates a spatially varying work function and surface reflectivity. Methods of cleaning the cathode in situ to raise the measured QE to the theoretical value are under investigation.

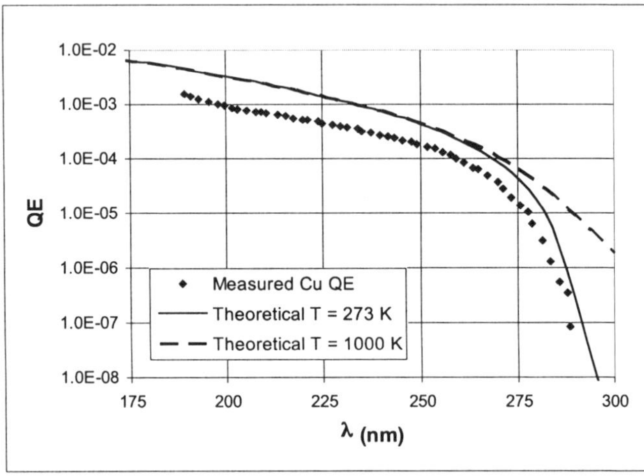

Figure 4. The QE as a function of laser wavelength is plotted for a Cu cathode at two different temperatures. Diamonds points are measured data for a Cu cathode after cleaning with 1 keV H^+ ions.

Interestingly equation 9 can be reduced to equation 1 where an approximate solution for η_0 can be derived. The expression for η_0 is a function of E_F, hν, $\Phi_{cathode}$ and $\Delta\Phi$ but is not reported here since it is cumbersome and not particularly useful. However, the model has the same functional dependence as equation 1 and the experiments, but the predicted value of the QE is larger than is measured at the GTF.

The QE is time dependent due to the Schottky effect. As the field at the cathode increases, the barrier energy decreases and the QE will increase because more electrons in the metal have sufficient energy to emit. Likewise the QE will decrease if the field at the cathode is decreased. For typical parameters at the GTF it requires approximately a 15% change in the cathode field to change the QE by 10%.

The physical model also provides some insight into the response time of the cathode. Electrons that absorb a photon near the surface can be emitted immediately while electrons that absorb photons below the surface must travel to the surface before escaping. The response time constant is estimated in equation 11 where δ is the optical skin depth. For Cu at 263 nm the time constant is approximately 6.0 fs and for Mg the time constant is only 4.8 fs due to the shorter skin depth.

$$\tau \approx \frac{\delta}{\sqrt{\frac{2E_k}{m}}} \tag{11}$$

Looking at figure 4 it appears that one can achieve much higher QEs by using a photon energy that significantly exceeds the cathode work function. While this is true it only tells part of the story since the thermal emittance and QE are related. The reason the QE rises as the wavelength is decreased is that an increased number of electrons have sufficient energy to escape from the metal. This also leads to larger amounts of excess energy once they are emitted or in other words a larger thermal emittance.

2.2. *Thermal Emittance*

The thermal emittance is a measure of the transverse energy of the electrons at the cathode surface. The definition of normalized thermal emittance is given in equation 12. Assuming the electron momentum is independent of position (correlated term is zero) on the cathode, the thermal emittance can be expressed as shown in equation 13 where the sum is performed only over the emitted electrons. The electron density distribution on the cathode, f(ρ,θ), is identical to the laser photon density distribution for a cathode with uniform QE.

$$\varepsilon_{nthermal} \equiv \frac{1}{mc}\sqrt{\langle x^2\rangle\langle p_x^2\rangle - \langle xp_x\rangle^2} \tag{12}$$

$$\varepsilon_{nthermal} \equiv \frac{1}{mc}\sqrt{\frac{\int_0^{r_{cathode}}\int_0^{2\pi} f(\rho,\theta)\rho^3\cos^2\theta d\theta dr}{\int_0^{r_{cathode}}\int_0^{2\pi} f(\rho,\theta)\rho d\theta dr}\frac{\int_{E_{min}}^{E_b}\int_0^{\Phi_{max}}\int_0^{2\pi} N_{electron}p_x^2\sin\Phi d\theta d\Phi dE}{\int_{E_{min}}^{E_b}\int_0^{\Phi_{max}}\int_0^{2\pi} N_{electron}\sin\Phi d\theta d\Phi dE}} \tag{13}$$

For a flat-top laser profile and a temperature of 0 K the thermal emittance can be expressed as shown in equation 14. Non-zero temperatures require approximate or numerical solutions. The thermal emittance as a function of laser wavelength is plotted in Figure 5 for a 1 mm flat-top pulse. Temperature effects are significant when the photon energy is greater than the work function. The thermal emittance increases as the temperature is increased due to the additional energy the electrons possess after emission. Likewise decreasing the work function will increase the thermal emittance. Thus the thermal emittance for Cu and Mg are calculated to be 0.25 μm/mm and 0.46 μm/mm respectively using a 263 nm photon.

$$\varepsilon_{nthermal} = \frac{r_{cathode}}{2}\sqrt{\frac{E_k}{mc^2}}\sqrt{\frac{\left(1-2\left(E_b/E_k\right)^{1/2}+2\left(E_b/E_k\right)^{3/2}-\left(E_b/E_k\right)^{2}\right)}{3\left(1-2\left(E_b/E_k\right)^{1/2}+\left(E_b/E_k\right)\right)}} \tag{14}$$

The theoretical thermal emittance is roughly a factor of two less than is typically measured[1,10,11]. Scattering will decrease the theoretical emittance due to a decrease in the average energy of the emitted electrons. The discrepancy could be due to acceleration from the rf field inside the metal before emission, surface roughness, non-uniform emission or other effects. Surface roughness can increase the thermal emittance since the average transverse momentum of an emitted electron is increased because the emitted electrons are peaked normal to the tilted surface. In addition, the rf field which is also normal to the tilted surface will add transverse energy.

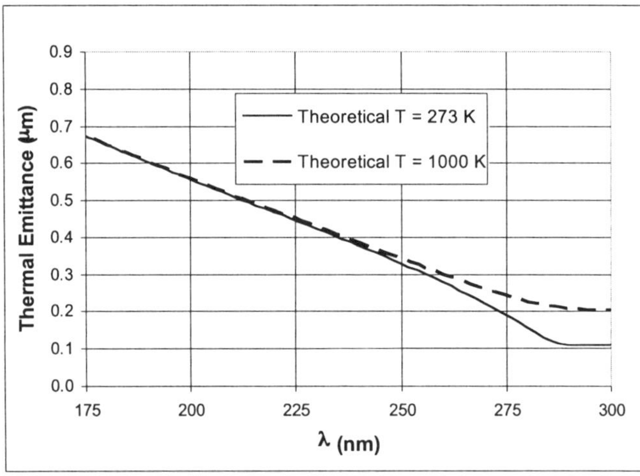

Figure 5. The thermal emittance as a function of laser wavelength is plotted for a Cu cathode at two different temperatures.

It should be clear from equations 9 and 14 that the thermal emittance and QE are related. One can combine figures 4 and 5 to give the thermal emittance as a function of QE. Figure 6 shows the cost in thermal emittance for increases in QE. Increasing the QE from 10^{-4} to 10^{-3} nearly doubles the thermal emittance. Based on these calculations the LCLS will shift the Ti:sapphire laser wavelength from 790 nm to 765 nm and use the third harmonic at 255 nm. This increases the expected QE by more than a factor of two while only increasing the predicted thermal emittance by 20%.

The thermal emittance is also clearly time dependent. If the field at the cathode increases the barrier energy decreases and the thermal emittance increases. Likewise if the field decreases the thermal emittance decreases. The percent change in thermal emittance is not as dramatic as the QE variation but depending on the pulse length and charge it can be significant. For typical parameters at the GTF a 70% change in field is required to produce a 10% change in the thermal emittance.

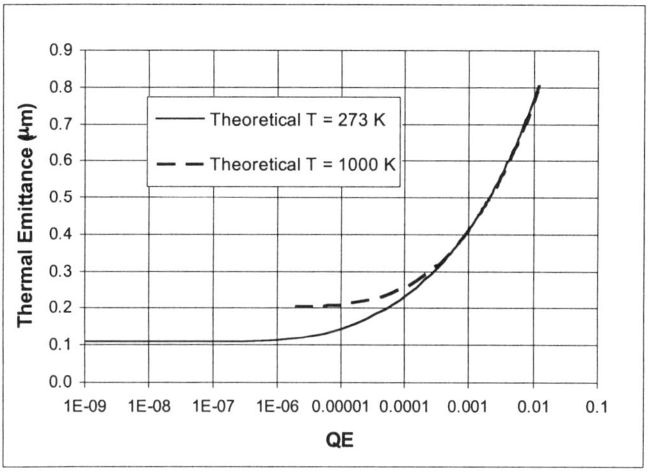

Figure 6. The thermal emittance as a function of QE is plotted for a Cu cathode at two different temperatures.

3. Laser and Electron Pulse Shape

Since the barrier energy is time dependent due to the Schottky effect, the QE and the thermal emittance are both time dependent. Of course the QE variation is more dramatic as is evident from figures 4 and 5.

When operating well below the space charge limit the head of the beam sees a lower field at the cathode than the tail due to the rising rf field. Thus the barrier energy is higher at the head and the QE and thermal emittance are smaller. When operating near the space charge limit just the reverse occurs since the field at the cathode for the tail is nearly zero. It is possible to operate in the middle where the decrease in the space charge field nearly cancels the rise in the rf field. At this point the Schottky effect is nearly constant and so are the QE and the thermal emittance. However, this is only valid at a specific charge and for a specific pulse shape which depends on the rf field, laser phase and beam radius.

Typically, the Schottky effect is not included in rf gun simulations and the electron beam pulse reproduces the drive laser pulse. In the LCLS where a very low emittance beam is required, the electron pulse shape helps determine the final transverse emittance due to space charge forces. It is assumed in the simulations that the electron beam will have the same temporal shape as the laser pulse that generates the beam. However, as shown earlier the Schottky effect has a large effect on emission from metal cathodes. Thus both the transverse and longitudinal phase space distributions will be affected.

Equations 1-3 are numerically integrated to determine the electron beam shape for a given laser pulse shape. Figure 7 shows the resulting electron beam pulse shape with a flat-top laser pulse and also the required laser pulse shape to generate a flat-top electron beam shape for the nominal LCLS case. With a flat top laser pulse the electron beam has about a -1 A/ps slope which is highly dependent on the beam parameters. For example a decrease in beam radius from 1.2 to 1 mm would change the slope to approximately -2 A/ps. The electron beam slope can be corrected by using a laser with an asymmetric pulse shape to compensate for the decrease in QE. A laser with an approximately 10% increase over the pulse will produce a flat electron beam. Of course the slope would need to increase to about 20% if the laser diameter were decreased to 1 mm.

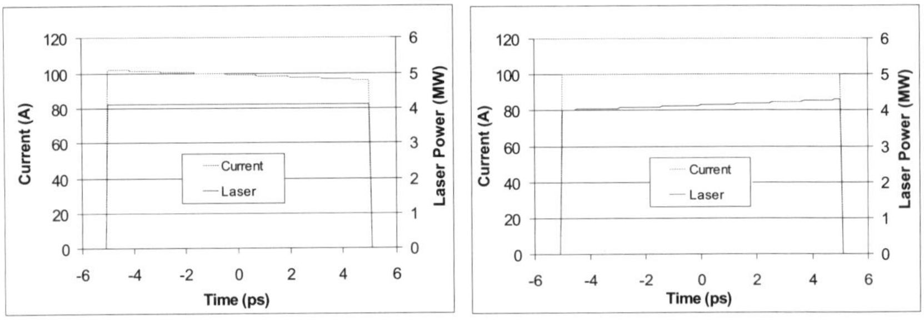

Figure 7. The electron beam pulse for a flat-top laser pulse is shown on the left and the required laser pulse shape to produce a flat-top electron beam is shown on the right. The nominal LCLS parameters include a laser wavelength of 255 nm, 120 MV/m rf amplitude, 30° laser phase, total charge of 1 nC and laser beam radius of 1.2 mm.

As mentioned earlier it is possible to produce a flat electron beam with a flat laser pulse but this only occurs for a specific set of parameters. Changing any single parameter leads to a time dependent QE. Another parameter which can be manipulated is the laser beam chirp. Changing the laser wavelength along the pulse can flatten the QE. Although one can not simultaneously compensate both the QE and thermal emittance with laser chirp exactly, it is possible to reduce the variation for both QE and thermal emittance to the order of 2% or less. A laser chirp of -0.1 nm/ps is sufficient to generate a flat electron pulse with a flat laser pulse for the nominal LCLS case.

4. Conclusions

Time dependent fields at the cathode lead to time dependent QE and thermal emittance through the Schottky effect. A simple model that includes the time dependent fields can be used to predict the QE and thermal emittance. The model overestimates the quantum efficiency most likely because the model does not properly include e-e scattering in the metal and also does not account for surface contaminants which may decrease the QE at localized positions. The model underestimates the thermal emittance possibly because it does not include acceleration prior to emission and surface roughness.

Due to the time dependent QE, the electron pulse shape is not identical to the laser pulse shape. Thus the laser pulse shape must be shaped to produce the desired electron pulse shape. If the charge is much less than the space charge limit the laser power will need to decrease during the pulse to compensate for the increase in QE. If the charge is close to the space charge limit the laser power will need to increase as time increases. When the space charge field temporal variation cancels the applied rf field temporal variation the QE is independent of time.

The charge, rf field amplitude, laser phase, or beam diameter will all affect the final electron pulse shape. In addition the laser beam wavelength affects the QE so the chirp of the laser is also important. However, the Schottky effect is typically not included in electron beam simulations so the electron beam shape at the cathode is incorrect. Thus the longitudinal space charge term will not be properly calculated by the codes and the beam will evolve differently from the simulation. Thus for those experiments that rely on simulations to design a beam-line that delivers a desired phase space distribution, the Schottky effect with a metal cathode should be included.

Acknowledgments

This work supported by the U.S. Department of Energy under contract number DE-AC02-76SF00515.

References

1. J.F. Schmerge et al, Proceedings of the 2004 FEL Conference, pp. 205-208, 2004 and SLAC-PUB-10763, 2004.
2. T. Srinivasan-Rao, J. Fischer and T. Tsang, Journal of Applied Physics, Vol. 69, no. 5, pp 3291-3296, 1991.
3. J. Rosenzweig et al, Nuclear Instruments and Methods Vol. A341, pp. 379-385, 1994.
4. J.L. Adamski et al, SPIE Vol. 2988, pp 158-169, 1997.
5. M. Cardona and L. Ley, Photoemission in Solids I, Springer-Verlag, pp 22-23, 1978.
6. D.H. Dowell et al, to be published in the 2005 FEL Conference Proceedings, Stanford, CA, August, 2005.
7. J.Smedley, to be published in These Proceedings.
8. M.J. Weber, "Handbook of Optical Materials", pg 317, 2002.
9. E.D. Palik, "Handbook of Optical Constants of Solids III", pg 239, 1998.

10. W.S. Graves et al, Proceedings of the 2001 Particle Accelerator Conference, pp. 2227-2229, 2001.
11. J. Yang et al, Proceedings of the 2002 European Particle Accelerator Conference, pp. 1828-1830, 2002.

SUPERCONDUCTING PHOTOCATHODES

J. SMEDLEY, T. RAO, J. WARREN
Brookhaven National Laboratory, Upton, New York

J. SEKUTOWICZ
DESY, Hamburg, Germany

J. LANGNER, P. STRZYŻEWSKI
The Andrzej Soltan Institute for Nuclear Studies, Swierk, Poland

R. LEFFERTS, A. LIPSKI
State University of New York at Stony Brook

We present the results of our investigation of lead and niobium as suitable photocathode materials for superconducting RF injectors. Quantum efficiencies (QE) have been measured for a range of incident photon energies and a variety of cathode preparation methods, including various lead plating techniques on a niobium substrate. The effects of operating at ambient and cryogenic temperatures and different vacuum levels on the cathode QE have also been studied.

1. Introduction

Several projects now being considered require injectors capable of delivering significant average current (>1 mA) with CW operation [1-4]. Some of these applications also require significant peak current, and would benefit from high (20 MV/m+) accelerating gradients. For these applications, a photoinjector based on superconducting technology is a leading contender.

Designing a superconducting photoinjector entails many challenges. As with any photoinjector, the choice of cathode material is of critical importance. The cryogenic environment of the superconducting cavity places additional considerations on the cathode choice compared to room-temperature copper cavities. Any cathode that is not itself a superconductor must be thermally isolated from the cavity, and must be mounted in such a way as to minimize RF loss in the cavity. A solution to these isolation problems has been proposed and investigated by D. Jensen at Rossendorf [5], using an RF choke-joint on the back wall of the cavity. This is an attractive solution, as it potentially enables the use of nearly the entire range of cathodes developed for room temperature cavities. Chemical contamination of the superconducting cavity may also be a problem for some cathode materials. To address this concern, a group at Brookhaven National Laboratory [6] is investigating the use of a diamond capsule to enclose the cathode, with the diamond acting both as an electron multiplier (via secondary emission) and as a barrier to prevent contamination of the cavity by the cathode and contamination of the cathode by the cavity.

Another option is to use a superconductor as the cathode. This eliminates the need for isolation and simplifies the overall design. These cathodes can be expected to exhibit the benefits and drawbacks of traditional metal cathodes – low quantum efficiency ($<10^{-3}$) and nearly unlimited operating life. The niobium that comprises the cavity walls is the obvious first choice, and a cavity utilizing a niobium cathode has been built and tested at BNL [7]. The quantum efficiency (QE) of niobium was found to be poor ($<10^{-5}$), even for a metal [8]. For this reason, this paper investigates the photoemission properties of lead. Lead is a type I superconductor commonly used in cavities for ion accelerators, with a critical temperature of 7.9 K and a critical magnetic field of 54mT [9]. This low critical field would limit the accelerating gradient achievable in a cavity constructed entirely of lead. For this reason, we focus primarily on coating a niobium substrate with lead - we hope to coat only the cathode region of a niobium cavity with lead. As the magnetic field on the beam axis is typically small, we were not expecting the maximum electric field in the cavity to be impacted by the lead coating. This has been verified in recent tests at Jefferson Laboratory [10].

The primary focus of this paper is the experimental measurement of the quantum efficiency of various lead coatings, both at room and cryogenic temperatures. A theoretical investigation of the QE of lead using Spicer's three-step model of photoemission [11] has been performed, and will be reported in the companion paper to this article [12]. Portions of the experimental section were previously reported [13,14], and are included here for completeness.

2. Preparation of Cathodes

Five distinct types of lead cathodes were investigated in this work - four methods of plating lead along with a solid, polished lead sample. In all cases the cathodes/substrates are 9 mm in diameter and 8 mm thick.

Electroplated samples were prepared at Stony Brook University, on both copper and niobium substrates. The polished cathodes were electroplated with lead by a procedure developed for use on superconducting cavities for heavy ion accelerators [15]. The plating solutions are based on methane-sulfonic acid chemistry and are created from commercially available products [16]. Each copper cathode was immersed in plating solution and flashed four times for several seconds with 10 mA/cm^2 plating current to prepare the surface. The current was then lowered to 2 mA/cm^2, corresponding to a plating rate of 10 μm/hour, and a plating thickness of 8 μm of lead. The cathodes are then rinsed with de-ionized water and dried with a stream of nitrogen gas. The procedure for plating niobium is identical to that for copper, except that the coating thickness is 2 μm.

The bulk lead cathode was prepared from cylindrical stock obtained from Goodfellow (99.95% purity). The surface was mechanically polished with Beuhler diamond polishing compounds. Nine, six and one micron polishing compounds were used, resulting in a finish with scratches on the order of one micron wide. Scanning electron microscope (SEM) images taken after the photoemission measurements revealed diamond inclusions covering roughly 5% of the cathode surface, likely imbedded during polishing (see fig. 2).

The vacuum-deposited (evaporated) sample was prepared at SBU in a vacuum evaporator, with a background pressure of 5 μTorr and a deposition time of 9 minutes. The coating thickness is 6 μm. A polished copper cathode was used as a substrate. Subsequent evaporations have been prepared on niobium substrates, but have not yet been measured.

A sample was covered by means of the magnetron sputtering technique at the Soltan Institute. Deposition of lead was performed using a cylindrical cathode, in the presence of argon under 3 mtorr of pressure. The Nb substrate was placed at a distance of 3 cm from the sputtered cathode. The current discharge was set at a value of 100 mA. Such conditions ensure stable discharge without any surface melting of the cathode. The obtained film was smooth and macrodroplet-free. Total deposition time was 60 minutes and the estimated thickness of the layer was 4 μm.

The arc-deposited sample was prepared at the Soltan Institute by means of arc discharge with a planar cathode in UHV conditions. This process is characterized by a high ionization ratio of metallic plasma, higher energy of ions in comparison with the magnetron sputtering technique and also a higher purity of the deposition process due to the absence of a working gas. A discharge current of 25 A allowed stable arc operation (typical value for niobium is 100 A). The deposition process was performed with an Aksenov-type magnetic filter connected to the plasma source in order to eliminate macrodroplets. The coating thickness is 1 μm on a niobium substrate.

3. Laser Cleaning

Laser cleaning with 248 nm light was used to improve the QE of all of the samples measured. The 248 nm light was provided by a KrF excimer (GAM Laser EX5), with a pulse duration of 10 ns and a repetition rate of 20 Hz, with the exception of the initial test, for which a Lambda-Physik LPX105 was used, with a 20ns pulse duration and a repetition rate of 10Hz. The change in surface morphology induced by various energy densities of 248 nm light was determined using an electroplated lead sample (shown in fig. 1). After irradiation, the cathode was removed from the vacuum and observed with an optical microscope and a SEM. The SEM was used to perform an analysis of the surface composition via X-ray florescence, as well as the surface structure. The first observable change in the surface morphology (fig. 1c) occurred at an energy density of 0.26 mJ/mm^2. An energy density of 1.8 mJ/mm^2 was found to locally melt the coating and expose the substrate – the dark areas in fig. 1f correspond to the copper substrate. Based on these results, an energy density of 0.2 mJ/mm^2 was chosen as the starting point for the cleaning.

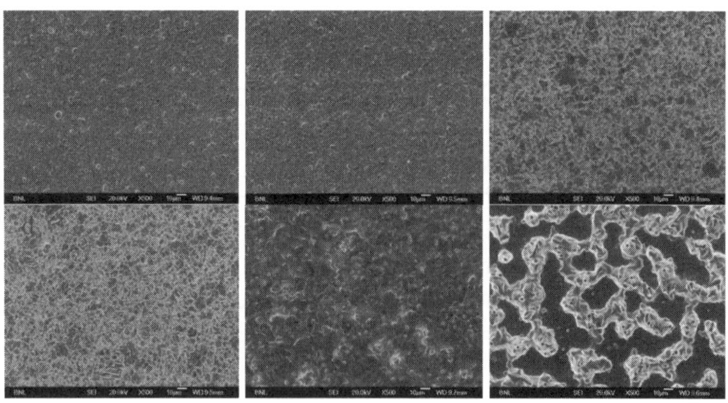

Figure 1. Surface structure of a lead coated cathode after 248 nm laser irradiation: (a) no laser (b) 0.11 (c) 0.26 (d) 0.52 (e) 1.1 (f) 1.8 mJ/mm^2. Scale: – is 10μm.

The surfaces were each irradiated for ~10 minutes, with the laser operating at 20 Hz. To avoid alignment difficulties, the region of the cathode exposed to the cleaning beam was significantly larger than the measurement area. For the solid lead, evaporated and electroplated samples, the cathode was cleaned a second time at an energy density of 0.4 mJ/mm^2. In both cases, little improvement was observed in the QE due to this increase. The samples were analyzed with an SEM after removal from the measurement system. Figure 2 shows the surface structure for four of the cathodes before cleaning. Figure 3

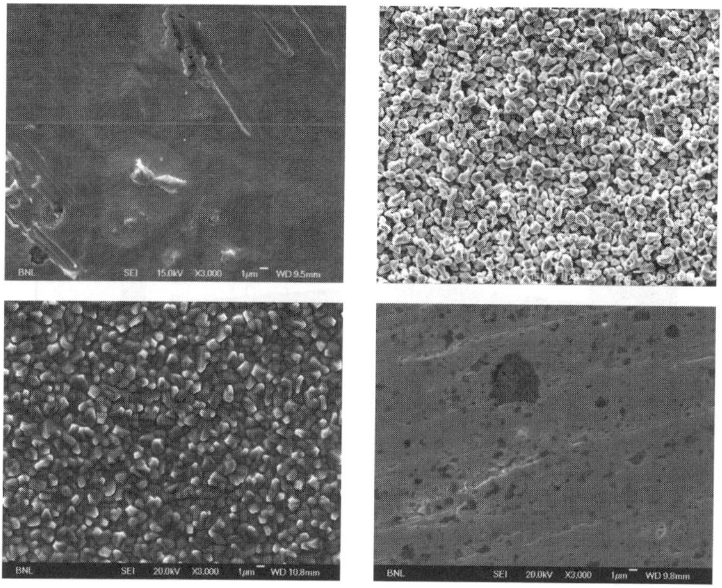

Figure 2. Lead cathode surface morphology prior to laser cleaning. (a) Arc-Deposited (b) Sputtered (c) Evaporated (d) Polished Solid Lead. Scale: – is 1μm.

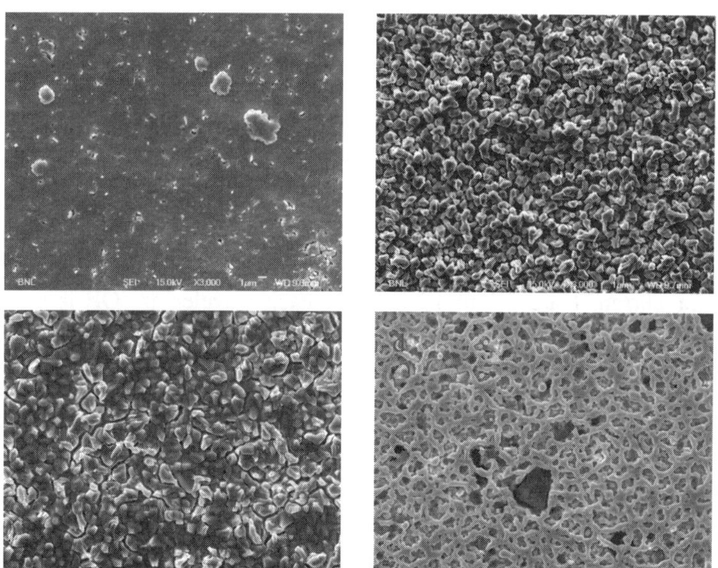

Figure 3. Lead cathode surface morphology after irradiation with 0.2 mJ/mm^2. (a) Arc-Deposited (b) Sputtered (c) Evaporated (d) Polished Solid Lead. Scale: – is 1μm.

shows the surface structure after cleaning. In all four cases, the X-ray fluorescence spectrum in the SEM showed only lead, indicating that the coating was intact.

4. Photoemission measurement

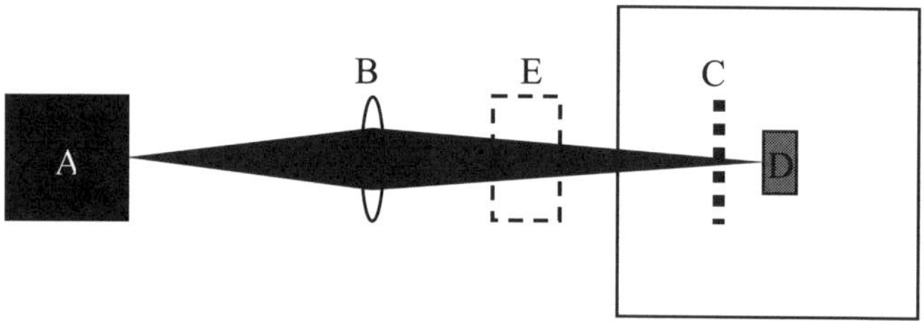

Figure 4. Schematic of experimental arrangement, showing: (A) Monochromator, (B) Lens, (C) Anode Grid, (D) Cathode and (E) Location of Power Meter.

A schematic of the photoemission measurement system is shown in fig. 4. A deuterium light source (Ocean Optics DH-2000-S-DUV) is fiber-coupled to a monochromator with a 300 micron exit slit. The desired wavelength λ is selected by the dial on the monochromator (Edmund DCM1-01). The output bandwidth is 2 nm, measured with an Ocean Optics HR2000 spectrometer. A fused silica lens is used to focus the light on the cathode through a vacuum window and the anode mesh. The output of the monochromator is measured for each wavelength before and after each QE measurement, at a point after the lens but prior to the vacuum window, using a power meter (Newport 918-UV). For the QE measurement, the anode is held at a positive voltage, and the current is measured leaving the cathode by a picoammeter (Keithley 487). The optical transmission of the vacuum window and the mesh are calibrated separately for each wavelength. Typical values for optical power P (in a 2 nm band) are 10-100 nW, and typical values for current I are 0.1 to 10 pA. For each wavelength, QE is calculated using the formula $QE=I\,(hc)/(P\,\lambda)$.

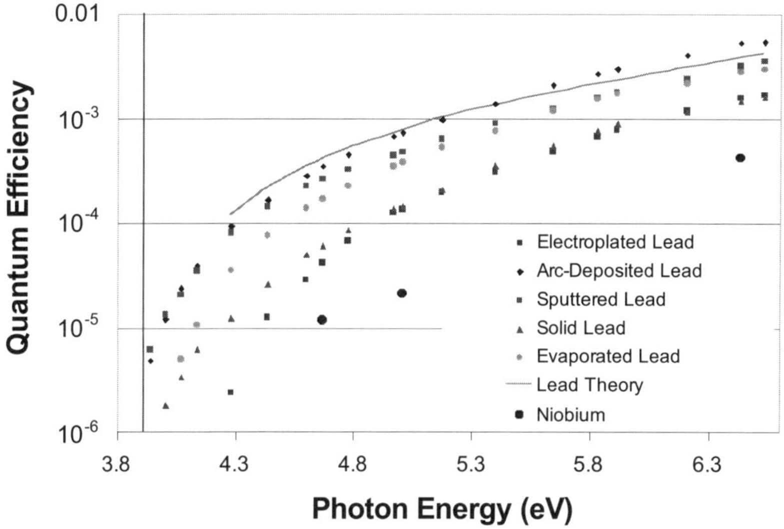

Figure 5. QE of various laser cleaned lead cathodes for photon energies between 3.9 and 6.5 eV. The vertical line represents the expected work function accounting for the Schottky effect (3.91 eV).

The quantum efficiency of the five lead cathodes is shown in fig. 5 as a function of incident photon energy. The theory line in the plot comes from the three-step model of photoemission [11,12]. The electron density of states was obtained from the NRL structures database [17]. The optical constants for lead and the "theoretical" work function were obtained from the literature [18,19]. The niobium data is included for reference [8].

Figure 6 shows the data for sputtered lead plotted as $(QE)^{1/2}$ vs. photon energy. This plot should be approximately a straight line, with a horizontal-intercept equal to the material work function. Data for both an electrode bias of 1 MV/m (1 kV across 1 mm) and 5 MV/m are shown. The Schottky effect (the reduction of the material work function due to the applied field) is evidenced by the lower work function for the 5 MV/m data. The work functions have been calculated using the method of Fowler [20,21]. No attempt was made to account for the 2 nm bandwidth of the lamp light. This introduces an uncertainty of ±0.04 eV on the values for the work function. Table 1 shows the best fit work function for each cathode, cleaning energy and bias field, along with the QE at 213 nm and 193 nm. For some cathodes, the QE prior to cleaning was too low to be reliably measured; in these cases, only the QE after cleaning is shown.

Table 1. Work function (Φ) and QE of cathodes, including Cleaning Energy Density (C.E.D.) of 248 nm light used, and bias field applied between the cathode and anode. A C.E.D. of 0 represents the value prior to laser cleaning.

Cathode	C. E. D. mJ/mm^2	Field MV/m	Φ(eV)	QE (x10^{-4}) 213 nm	QE (x10^{-4}) 193 nm
Solid	0	1	4.52	0.49	2.6
	0.18	1	4.19	5.5	13.1
	0.4	1	3.93	7.8	15.3
	0.86	1	4.02	5.6	12.8
Evaporated	0	1	4.22	0.65	2.9
	0	5	4.17	0.70	3.1
	0.21	1	3.97	15.6	28.4
	0.21	5	3.84	14.2	27.9
	0.37	1	3.92	13.3	25.4
	0.37	5	3.86	15.3	30.4
Arc	0.21	1	3.88	27.2	54.1
Sputtered	0	1	4.21	0.10	0.22
	0.23	1	3.83	16.0	32.6
	0.23	5	3.71	17.9	36.4
Electroplated	0.22	1	4.20	6.8	16.0
	0.37	1	4.11	7.1	16.3

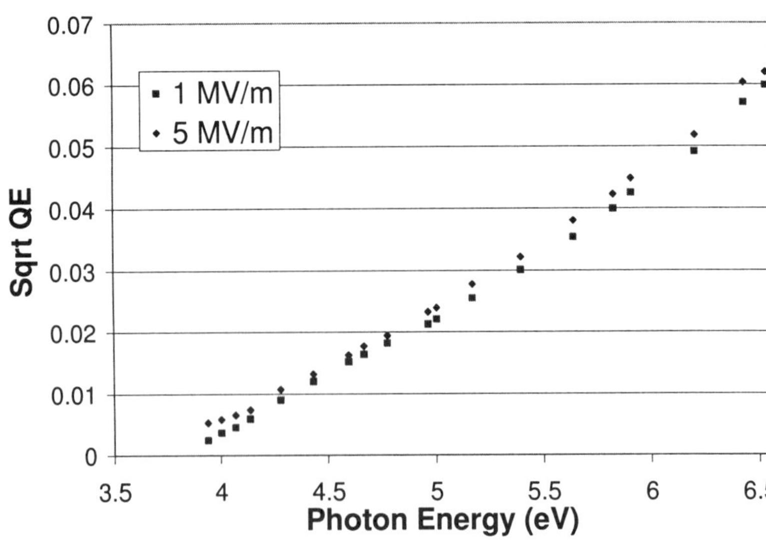

Figure 6. (QE)$^{1/2}$ vs. photon energy for sputtered lead.

5. Photoemission at Cryogenic Temperatures

A superconducting lead cathode will be used at cryogenic temperatures in a real injector. Tests with an all-niobium injector noted a QE at cryogenic temperatures that was inferior to the room-temperature QE [7]. For this reason, it was instructive to investigate the QE of lead cathodes at cryogenic temperatures. This was accomplished by mounting the cathode on a vacuum cold-finger attached to a LN_2 dewar. The cathode was electrically isolated from the thermal mass of the cold-finger by a ceramic standoff, but in adequate thermal contact. A thermocouple on the thermal mass was used monitor the cathode temperature. The cold-finger system has a 5 mm separation between the cathode and anode. A bias of 5 kV was used, yielding a field of 1 MV/m. The arrangement for the experiment was otherwise identical to that described above.

Table 2. QE and work function for arc-deposited sample at cryogenic temperature, for both good and poor vacuum conditions.

	Temp. C	Vacuum nTorr	Φ(eV)	QE ($x10^{-4}$) 213 nm	193 nm
Initial	20	8	4.57	0.6	2.4
Cleaned 0.23 mJ/mm^2	20	8	4.00	10.1	26.0
Cold	-169	6	4.01	9.6	23.2
Poor Vac	20	1300	4.04	8.6	23.5
Cold	-169	270	4.45	3.0	8.5
Cleaned 0.22 mJ/mm^2	-169	270	4.06	10.0	24.4
2 hrs later	-169	270	4.26	4.6	11.1

Both electroplated and arc-deposited samples were studied in this apparatus. The QE of each cathode was determined at room temperature, in good vacuum (8 nTorr), before and after laser cleaning. The cathode was then cooled to -169C over the course of half an hour. Once the temperature stabilized, the cathode QE was measured again. The cathode was then laser cleaned, while still cold. The QE was measured immediately after cleaning, and every half-hour afterwards. For the case of the good vacuum, the cold QE did not vary much from the room-temperature value. However, when the above protocol was repeated for "poor" vacuum conditions (1.3 μ Torr), the QE was found to degrade significantly when the cathode was cooled. These vacuum conditions were achieved by backfilling the cavity with high-purity N_2, then pumping with only the turbo pump, leaving the ion pump valve closed. The room temperature QE could be restored by laser cleaning, but the QE would continue to degrade on the scale of hours after the cleaning. Table 2 shows the vacuum level, QE and calculated work function for the arc-deposited cathode for each of the steps. In each case the QE was measured over a wide range of wavelengths, however only the QE @ 213 nm and 193 nm shown in the table. Note that, for the poor vacuum case, the act of cooling the cathode reduces the vacuum pressure in the cell considerably, suggesting that the cold finger is acting as a cryopump. This effect is likely the cause of the degradation of the QE in the cold, poor vacuum case – contaminants are being trapped onto the cathode. It should be noted that this is a second arc-deposited cathode, not the cathode described in section 4. This cathode has a slightly higher work function, and a slightly lower QE than the previous cathode. The results for the electroplated cathode cryogenic test were similar to those of the arc deposited, and are not shown.

6. Conclusion

Lead appears to be an attractive option for moderate average current sources. The best measured QE (arc-deposited) would require 2.1 W @ 213 nm to generate 1 mA. The optimal choice for lead coating method seems to be arc-deposition, as it provides a smooth surface (figs. 2a & 3a) and good QE. Electroplating is also an option, although the electroplated samples we tested seemed to have a work function significantly higher than the other lead samples, and correspondingly lower QE. The surface finishes of the evaporated and sputtered samples are likely too rough to use in an injector, at least at high field.

The three-step model of photoemission seems to well predict the QE of lead, especially given that no free parameters are used in the model. The work functions of the lead samples agree with the expected value, with the exception of the electroplated sample. Modest laser cleaning energy densities (~0.2 mJ/mm^2) are sufficient to achieve the maximum QE without damage to the coating.

A niobium cavity has been constructed at DESY. This cavity has undergone cold tests, and is awaiting coating of the cathode region with lead via arc-deposition. This cavity will then undergo a second round of cold tests to confirm the RF performance, after which QE measurements will be performed.

Acknowledgments

The authors would like to thank J. Walsh for his expert technical assistance and V. Radeka for his support. This manuscript has been authored by Brookhaven Science Associates, LLC under Contract No. DE-AC02-98CH10886 with the U.S. Department of Energy. The United States Government retains, and the publisher, by accepting the article for publication, acknowledges, a world-wide license to publish or reproduce the published form of this manuscript, or allow others to do so, for the United States Government purposes.

References

1. J. Sekutowicz et al., *Phys. Rev. ST Accel. Beams* **8**, 010701 (2005).
2. I. Ben-Zvi et al., *Proceedings of 2005 Particle Accelerator Conference, Knoxville, Tennessee,* 1150 & 2741 (2005).
3. V.N. Litvinenko et al., *Proceedings of 2005 Particle Accelerator Conference, Knoxville, Tennessee,* 2768 (2005).
4. Light sources, such as the operating FEL at Jefferson Laboratory (www.jlab.org/FEL), and the proposed light sources at Cornell (erl.chess.cornell.edu) and Daresbury (www.4gls.ac.uk)
5. D. Janssen et al., Nuc. Inst. and Meth. in Phy. Res. **A507,** 314 (2003).
6. X. Chang et al., *Proceedings of 2005 Particle Accelerator Conference, Knoxville, Tennessee,* 2251 (2005).
7. T. Rao et al., *Proceedings of 2005 Particle Accelerator Conference, Knoxville, Tennessee,* 2556 (2005).
8. J. Smedley, T. Rao and Q. Zhao, *J. Appl. Phys.* **98**, 043111 (2005)
9. Peter Schmuser, *Superconductivity* (www.desy.de/~pschmues).
10. P. Kneisel, J. Sekutowicz, R. Lefferts and A. Lipski., *Proceedings of 2005 Particle Accelerator Conference, Knoxville, Tennessee,* 2956 (2005).
11. W.F. Krolikowski and W.E. Spicer, *Phys. Rev.* **185**, 882 (1969)
12. J. Smedley et al., *Phys. Rev. ST Accel. Beams*, Submitted to Special Section for Workshop on The Physics and Applications of High Brightness Electron Beams (2006)
13. J. Smedley et al., *Proceedings of 2005 Particle Accelerator Conference, Knoxville, Tennessee,* 2598 (2005).
14. J. Smedley et al., *Proceedings of EPAC 2004, Lucerne, Switzerland,* 1126 (2004).
15. J.W. Noe, *Proceeding of the Heavy Ion Accelerator Technology, 8th Intr. Conf.,* 192 (1999)
16. LeaRonal Solderon from Shipely Company, LLC, Marborough Ma 01752
17. cst-www.nrl.navy.mil/ElectronicStructureDatabase
18. C. Norris and L. Wallden, *J. Phys. F: Metal Phys.* **2**, 180 (1972).
19. H.G. Liljenvall, A.G. Mathewson and H.P. Myers *Philos. Mag.* **22**, 243 (1970).
20. D. H. Dowell, F. K. King, R. E. Kirby and J. F. Schmerge, *Contributed to 27th International Free Electron Laser Conference (FEL 2005), 21-26 Aug 2005, Stanford, California.*
21. R.H. Fowler, *Phys. Rev.* **38**, 45 (1931).

Working Group 2

SUMMARY OF WORKING GROUP 2: DIAGNOSTICS AND BEAM MANIPULATION

G. TRAVISH†

UCLA, Department of Physics and Astronomy
Los Angeles, CA 90095, USA.

The efforts of the working group on diagnostics and beam manipulation held at the ICFA sponsored workshop, "The Physics and Applications of High Brightness Beams" are presented.

1. Overview

Working Group 2 was charged with assessing the state of diagnostics and beam manipulation techniques in high brightness beams. Through a series of some 12 talks and vigorous discussions, a picture emerged of the substantial progress made in high brightness beam metrology and control since prior such workshops (i.e. the 2003 workshop by the same name held in Sardinia). While no formal attempt was made to compare the progress, and indeed the working group participants were not all the same, the great strides made were clear. Below I will give some highlights of the talks which are meant solely to encourage review of the actual papers and presentations by the participating authors.

In the next section I list the working group participants, and following that is a summary of the presentations. I conclude with a few comments and impressions from the working group and workshop as a whole.

2. Participants

Twenty-seven (27) people signed in to the working group and participated for at least a portion of the sessions (see Table 1). Special recognition goes to Daniele Filippetto who was the working group secretary and without whom much of this information would be lost, and to Thorsten Kamps who acted as chair while I gave my talk in another working group. A number of people sat in on multiple groups, and it is possible that some of those are not listed here. It is a testament to the excitement generated at the workshop that both many participants

† travish@physics.ucla.edu

attended all sessions of the working group, and that many other participants "cross-pollinated" with the other groups.

Table 1: The list of signed participants in Working Group 2.

NAME	SURNAME	INSTITUTION
Scott	Anderson	LLNL LIVERMORE
Gerard	Andonian	UCLA
Manuela	Boscolo	LNF-INFN
Francesco	Broggi	INFN-LASA
Les	Dallin	CANADIAN LIGHT SOURCE
Mike	Dunning	UCLA
R Joel	England	UCLA
Mario	Ferianis	SINCROTONE TRIESTE
Daniele	Filippetto	LNF-INFN
Dario	Giove	INFN MILANO
Mark	Hogan	SLAC
Katja	Honkavaara	UNI-HH
Thorsten	Kamps	BESSY BERLIN
Sergiy	Khodyachykh	DESY
Patrick	Krejcik	SLAC
Joseph P.	Lidestri	COLUMBIA U.
Anne	Oppelt	DESY
Lenny	Rivkin	PSI SWITZERLAND
Jorg	Rossbach	UNIVERSITY- HAMBURG/ DESY
Fernando	Sannibale	LBL BERKELEY
Siegfried	Schreiber	DESY
Todd	Smith	Stanford U.
Lazar	Staykov	DESY
Gil	Travish	UCLA
Rao	Triveni	BNL UPTON USA
Alexander	Tron	LEBEDEV PHYSICAL INSTITUTE
Max	Zolotorev	LBL BERKELEY

3. Presentations

During the three working days (Monday, Tuesday and Thursday), a total of 12 full presentations were heard. This relatively small number of talks allowed for extensive discussion both prompted by the talks and participant initiated.

An example of an extended discussion held across multiple sessions was the topic of "formation length" and the related issues of "near field" versus "far field" effects. It seemed clear that an entire workshop could be held on these topics. And, it is clearly far outside the scope of this brief summary to explain the subject. However, it may be interesting to the reader to note that these discussions continue today via exchange of emails and in other forums.

A list of speakers and talks, broken down by presentation session, are given next. This summary is followed by highlights of the individual talks.

3.1. *Monday:*

G. Andonian
Compression Studies at the ATF with the BNL/UCLA Chicane

T. Kamps
Laserwire Based Beam Profile Monitor

A. Tron
New principles in photochronography of femtosecond resolution

3.2. *Tuesday:*

M. Boscolo
SPARC e-meter @PITZ: simulation studies compared to measurements

M. Dunning
Overview of the LCLS Single-Shot Relative Bunch Length Monitor System

R. J England
Beam Shaping and Permanent Magnet Quadrupole Focusing with Applications to the Plasma Wakefield Accelerator

D. Filippetto
The Sparc Movable Emittance Meter: Commissioning And First Operation At Pitz

K. Honkavaara
Measurements of the Transverse Emittance at the VUV-FEL at DESY

3.3. *Thursday:*

J. Rosenzweig
Optimum Beam Creation in Photoinjectors Using Space-charge Expansion II: Experiment

A. Sakumi
Synchronization between laser and electron beam at photocathode RF gun

T. Smith
Temporal Diagnostic with 10fs Resolution

L. Staykov
Emittance Measurements -- to the point

4. Presentation Highlights

After opening comments by myself, G. Andonian (UCLA) presented recent results from the UCLA-BNL bunch compressor at the ATF. Beyond commissioning of the device, a focus of this beam manipulation experiment was to gather data on coherent edge radiation (CER). Early results from the experiment gathered data from the boundary between the final two magnets and in the < 50 μm wavelength range. Using the already installed cooled bolometer, filters and polarizer, the team had gathered initial data and planned to gather further data to compare with careful simulations. An understanding of the microscopic beam physics and CER features, in general, is the stated goal.

T. Kamps (BESSY) followed with a presentation on a beam profile monitor that used the "laser wire" technique. The LBBD collaboration, of which the speaker is a part, intends to produce a high-precision non-invasive beam profile monitor for linear colliders. Apparently, other methods (OTR, wirescanners, etc.) are of insufficient resolution for colliders such as TESLA. The operating scheme was presented, as was a summary of the work of the three years of R&D and two years of commissioning and testing. Scan speed has been improved to 30 seconds over this time. Next steps include upgrades of the laser and installation of a second scanning dimension.

A novel method of bunch length measurement was presented by A. Tron (Lebedev). His technique is a type of streak camera that uses an RF cavity to directly modulate the electron beam energy which is then measured in a bend spectrometer. There was some discussion as to whether the method would, in fact, have better resolution. The speaker claims to have performed extensive simulations with space charge effects accounted for. Regardless of the viability of the method, it was clear that the working group members are desirous of such a device that can offer streak-camera like flexibility yet have 10 fs range resolution.

The Monday session concluded with general discussions and a first attempt to delineate the various diagnostics (more on this later). After a foggy evening, fine Sicilian meals, and a crisp morning, the Tuesday session began with M. Boscolo (INFN/PITZ) discussing simulations of machine studies at PITZ1.5. These PARMELA simulations included real laser profiles, and careful consideration of the solenoid field. The simulations still made assumptions such as radial symmetry, but managed to find good agreement on parameters such as energy spread (spectrum) and beam sizes and emittances as a function of solenoid current. The speaker pointed out that some machine parameters were still not well known, again pointing to the need for further diagnostics.

M. Dunning (UCLA) presented an overview of the UCLA project to build a bunch length monitor for the LCLS. The LCLS requires numerous feedback systems including one based on longitudinal beam parameters in order to maintain stable operation. A relative bunch length monitor using the radiation from the final bend magnet chicane is planned. The system must be non-invasive, highly reliable (maintenance free), run at up to 120Hz, and have <2% drift over 24 hours with better than 2% (of the nominal bunch length) resolution. At the time of the presentation, a single shot spectrometer looking at the combined CSR and CER was under design.

The challenging LCLS diagnostic presentation was followed by a beam manipulation talk given by R. J. England (UCLA). This presentation concerned itself with the production of a bunch whose profile is optimized for plasma wakefield accelerator schemes. Namely, a longitudinally ramped pulse produced through a dogleg with sextupole elements. As part of this work, an x-band standing-wave deflecting cavity is being produced to diagnose the longitudinal profile with high resolution. In addition, the speaker presented work on permanent magnet quadrupoles which allow for matching of the beam into a subsequent plasma experiment.

A second presentation involving the PITZ system was given by D. Filippetto (INFN SPARC). This work centered on a longitudinally positionable emittance measurement system that employs slit masks and a long bellows. The effort is concerned with measuring the emittance "bounce" seen in simulations of typical separated photoinjector-linac arrangements (the so-called "Ferrario Scenario"). In order to achieve this goal, the system has to reduce measurement error well below the typical 20%+. Error minimization will occur through higher resolution digitization, very stable beam parameters and averaging over large number of measurements.

Another talk on emittance measurements was presented by K. Honkavaara (DESY). Rather than a single system, this work used multiple methods to measure both the projected and slice emittances at various point along DESY's VUV-FEL. The techniques used included four beam-size monitors (at two distinct locations), quadrupole scans (at multiple locations), a slice measurement using a deflecting cavity and quadrupole scan, and finally seven wire-scanner stations. Extensive effort was put into the measurements including image analysis; careful background subtraction algorithms; fitting of the emittance data; and error estimation. The reproducibility of the measurements was very good and has even lead to improvements in the machine stability (through discovered hardware limitations). With this talk, and further discussions, the Tuesday session was closed.

Wednesday was spent diagnosing ancient ruins and taking in the beauty of Sicily. Refreshed by the break, Thursday's session began with a talk by the workshop Chair, J. Rosenzweig (UCLA), on a means of creating an optimal beam in a photoinjector using space charge expansion. This talk was actually the second part, describing the experimental work, of a plenary session presentation. Here, the speaker described an upcoming experiment to be carried out at the SPARC photoinjector using an 80fs Gaussian laser pulse to produce an ellipsoidal beam through "blowout". A first effort on laser shaping was shown including use of a "cut Gaussian" to approximate the desired radial profile. Diagnosis of the low energy beam profile is to be done through multiple methods: steak camera measurements of Cerenkov radiation, a novel laser gated system, and cross correlation of CTR radiation. At higher energy, additional methods are to be used including a deflecting cavity. At low energy however, generation of the needed photons for the streak camera demands using an aerogel based cell which has been designed and tested in the beamline.

A. Sakumi (KEK) presented work on synchronization of a probe laser and electron beam to be used as part of a pulse radiolysis system. Beyond the usual drive laser timing stabilization, the effort included laboratory temperature control to <0.1 degree and local heat source reduction around critical components. In the future, a new feedback system is planned as they require stability over an entire day of operation.

T. Smith (Stanford) presented a temporal diagnostic with 10 fs resolution in use at the Stanford Picosecond FEL Center. The speaker explained how the usual "zero-phasing" linac technique required a very large R_{65} in order to achieve high resolution (and hence the linac requires a high total gradient). Instead, the Stanford team used a chicane to produce a typical R_{56} and then sent this beam through a linac with a modest R_{65}. The resulting phase space rotation allows for temporal diagnosis of the beam through a standard bend spectrometer.

After the above series of diverse topics, it was time to conclude with a return to emittance measurements: L. Staykov (DESY) presented his work on a single slit emittance measurement technique. In the full system, the slit is scanned across the beam profile with x-y translators and angle tuned to the beam path using a goniometer. The speaker showed calculations of the beamlet width as a function of slit width and slit-to-screen drift, including space charge. This analysis indicated an optimal width of 10μm with a 1.5m drift. The system is under construction at the time of the presentation.

5. Conclusions

This working groups' efforts are best measured by the future results of the initial efforts presented in the following papers. The presentations and attendant discussions have no doubt stimulated many new ideas and refinements to existing efforts.

An attempt was made to tabulate existing diagnostic methods. While this exercise is not without merit, the diversity of diagnostics makes such documentation less than illuminating.

A few general truisms were reinforced by the workgroup discussions: diagnostics are more critical that ever; multiple approaches are necessary to measure even simple beam parameters; and, the working group leader (and possibly most everyone else) really have no idea what "formation length" means.

OBSERVATION OF COHERENT EDGE RADIATION EMITTED BY A 100 FEMTOSECOND COMPRESSED ELECTRON BEAM

G. ANDONIAN, M. DUNNING, E. HEMSING, J.B. ROSENZWEIG, A. COOK, A. MUROKH, S. REICHE

Particle Beam Physics Laboratory
Department of Physics and Astronomy,
University of California Los Angeles,
Los Angeles, CA 90095, USA

M. BABZIEN, I. BEN-ZVI, K. KUSCHE, V. YAKIMENKO

Accelerator Test Facility,
Brookhaven National Laboratory,
Upton, NY 11973, USA

D. ALESINI, L. PALUMBO, C. VICARIO

Istituto Nazionale di Fisica Nucleare
Laboratori Nazionali di Frascati
Via E. Fermi 15
Frascati (RM) 00046, Italy

A chicane compressor developed by UCLA for the production of ultra-short, 60 MeV electron beams at the Brookhaven National Laboratory Accelerator Test Facility has been commissioned, and initial beam physics experiments have been performed. These measurements have established the compression of electron beams to the 100 femtosecond (1 kA peak current) regime, via coherent transition radiation (CTR) based measurements. Investigations of coherent edge radiation (CER) include signatures that differentiate it from coherent synchrotron radiation (CSR), such as polarization and far-field angular distribution. Additionally, the radiation wavelength spectrum is determined from autocorrelation measurements. Radiation properties are compared to detailed start-to-end simulations derived from PARMELA and QUINDI (a Lienard-Wiechert code developed at UCLA). Plans for future experiments which further explore the observed wavelength spectra are presented.

1. Introduction

The chicane compressor developed by the UCLA Particle Beam Physics Laboratory (PBPL) was designed with numerous goals in mind. The compressor would be used to enhance the peak current of the electron beam needed to drive the existing free-electron laser (FEL) to saturation [1]. Moreover, the compressor was built to extend the capabilities of the Accelerator Test Facility (ATF) at Brookhaven National Laboratory (BNL).

The chicane compressor has been commissioned and fundamental beam physics experiments have been performed. In previous applications, the compressed beam was used for initial plasma wakefield acceleration experiments. In this paper, we report the basic beam physics experiments performed with the compressed beam as well as analytic studies on the consequent radiative processes. These compression experiments have focused on the creation of very short electron beam pulses, as short as 30 μm (or 100 femtoseconds rms), measurement of the longitudinal phase space characteristics of the bunches produced, and the subsequent initial measurement of *coherent edge radiation* (CER) emitted from such short beams.

Parametric studies of the bunch properties as a function of RF phase have been carried out. The ultra-short bunch longitudinal profile is measured using coherent transition radiation (CTR) interferometry, and analyzed by both time-domain fitting, and Kramers-Kronig-based reconstruction of the profile itself. The particular geometry of the vacuum vessel of the UCLA-ATF chicane, in which a radiation port allows direct viewing of the beam path as it leaves the third and enters the fourth magnet (Fig. 1), gives an ideal scenario for observing CER, in which radiation from the two adjacent edges add constructively.

The emitted CER power has been examined for its dependence on degree of compression, and its spectrum for its unique signatures in polarization, far-field distribution. The wavelength distribution, concentrated in the THz region, was examined using auto-correlation, and temporal reconstructions of the radiation pulse developed. These attributes of the radiation are benchmarked against start-to-end simulations of the beam dynamics and electromagnetic radiation with striking agreement. Based upon the comparison of experimental and simulation results, the application of CER for a non-destructive bunch length monitor are mentioned.

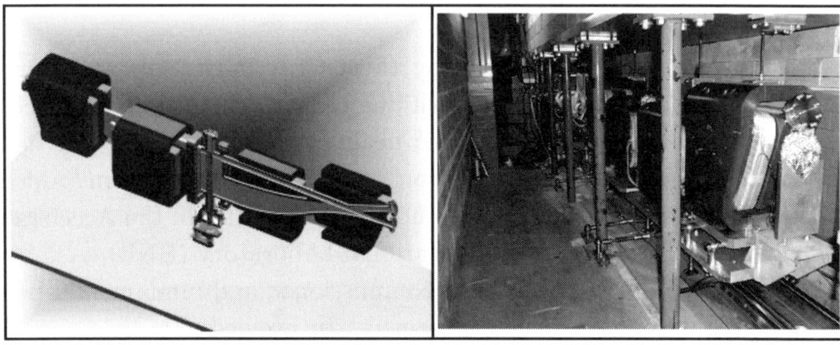

Figure 1. A computer aided draft of the chicane compressor (left) and a photograph of the chicane compressor during installation along the high-energy line of the ATF (right).

2. Experiment Description

2.1. *Chicane Compressor*

The chicane compressor was conceived of, designed and fabricated by the PBPL group at UCLA in close collaboration with the engineering staff of the ATF. The chicane compressor is installed in the high-energy beam line (H-line) of the ATF. The compressor consists of four dipole magnets oriented in a chicane layout (Fig. 1). Each magnet has a field of 2 kGauss and a gap of 2.1 cm. The design bend angle is 20 degrees.

The strong natural vertical-focusing of the magnets that would lead to excessive charge density in the chicane is compensated by the addition of initial and final horizontally focusing edge angles. The simulation tool, `AMPERES`, a fully 3D magnetostatic code, was used to model the chicane magnets. The field quality was rigorously bench tested and documented at UCLA [3]. Each dipole magnet has fine trim coils that are used to correct beam trajectories that arise from remnant magnetic fields when the chicane is powered off.

A custom vacuum vessel was designed and constructed at UCLA and incorporated into the chicane design considerations. The vacuum vessel mates to pre-existing vacuum pipes at the ATF. The vessel consists of a straight through pipe used for normal (non-compressed beam) operations, and a bend pipe for compressor operations. CER and CSR measurements are made through a dedicated radiation extraction port that is incorporated into the vacuum vessel. A two-stage beam profile monitor was designed to give the availability to intercept the beam in both modes of operation (compression on or off). A seven meter radiation transport originating from the CER port delivers the radiation to the a diagnostic station.

The chicane compressor power supplies and safety interlocks were designed by the ATF. Numerous considerations, including radiation and electrical safety, and other logistical concerns, such as installation and alignment in the high-energy line, were addressed prior to commissioning.

2.2. *Diagnostics*

Radiation exits the chicane compressor through a fused silica vacuum window located on the extraction port. The radiation is guided via adjustable, gold-caoted mirrors through metallic pipes (50 mm diameter) to a diagnostic station located external to the radiation tunnel. The entire length of the radiation transport line is approximately 7 m. A Picarin lens is positioned one focal length (1.25 m) from the approximate radiation source (between the third and fourth dipoles) which provides a point-to-parallel transport configuration. The transport line terminates in a diagnostic station, where both a bolometer and interferometer are used to conduct measurements on the emitted radiation.

2.2.1. *Silicon Bolometer*

In order to observe the effects of CER, a cold bolometer is used to detect the long-wavelength photons efficiently. The bolometer is manufactured by IR Labs (Model No. HDL-5) [4]. It is cooled with liquid helium and has built in filter wheel, loaded with specified filters that have cut-on wavelengths (low-pass frequencies) of 13, 27, 45, 103, and 285 μm. A wedged polyethylene window protects the Winston cone and detector of the bolometer.

2.2.2. *Interferometer (CER Spectral Measurements)*

The radiation spectrum is measured at the output of a Michelson-type interferometer as a function of bolometer voltage. The interferometer is optimized for the 15 μm to 1 mm wavelength range, and has a translatable mirror along one orthogonal leg with 1 μm spatial resolution [5]. The spectral measurement is described in the upcoming sections.

3. Overview of CER

Radiation exiting the chicane contains features of both synchrotron and edge radiation due to the measurement position and the magnet geome-

Figure 2. A photograph of the silicon bolometer (center right) adjacent to the diagnostic stand with focusing parabolic mirror (center left).

try. For electrons emitting under the same radiation process, the far-field intensity distribution is expressed as

$$I(\omega) = I_0(\omega)[N_e + N_e(N_e - 1)F(\omega)], \tag{1}$$

where $I_0(\omega)$ is the single electron intensity distribution and $F(\omega)$ is the bunch form factor. For wavelengths longer than the bunch length (~30 μm), each type of radiation is coherently enhanced by a factor of N_e ($\sim 10^9$ for given parameters), the number of electrons in a bunch.

The edge radiation emitted from electrons entering and exiting the edges of bend magnets is expected to have greater intensity at longer wavelengths than synchrotron radiation [7]. The characteristics of CER also depend on the topology of the field gradient. In the *zero-edge length model* [8], the field of the bending magnet is approximated by a step function. The resultant radiation is radially polarized, with a cylindrically symmetric spatial distribution characterized by a null on the straight section axis and maxima at $\theta \sim 1/\gamma$, analogous to transition radiation. When the magnetic edge is considered finite, the resulting radiation is no longer cylindrically symmetric or completely radially polarized. Analytic finite edge length models result in complicated expressions, requiring numerical calculations [9].

3.1. *Simulations*

In order to model the experiment, a start-to-end simulation suite has been developed at UCLA. The beam generation and acceleration is studied with the code PARMELA and the beam interaction through transport and compression are modeled with ELEGANT. The expected performance of the compressor, according to the simulation suite, is predicted to compress a 60 MeV, 55 A electron beam to a 1.6 kA peak. A normalized emittance growth in the bend plane from 1.5 mm-mrad to 6 mm-mrad is expected to accompany this compression.

Further, the parallel-computing code QUINDI was developed to perform the radiative studies of the compressed bunches at the ATF. The program avoids the sequential approach of a magnetic lattice, relying instead on an object-driven description of magnetic elements. Magnets with arbitrary shapes of convex polygons and user defined fringe fields are permitted. The observed radiation field is calculated on a user-defined plane based on the acceleration field of the Lienard-Wiechert potentials.

4. Performance Studies

Many compressor measurements have been performed at the ATF since 2002. The initial measurements concentrated on quadrupole-based transverse phase space tomography, which showed phase space bifurcation at 60 MeV [10]. The transverse bifurcation was accompanied by dramatic breakup in the momentum spectrum, a direct effect of the collective fields (Fig. 4).

4.1. *Bunch Length Measurement*

In order to benchmark CER measurements to other measurements as well as the start-to-end model, a reliable measurement of the pulse length was conducted. The pulse length is measured with the use of a *coherent transition radiation* (CTR) interferometer. Performed in tandem with energy and spectral measurements, the interferograms characterize the longitudinal phase space of the compressed beam in great detail. The CTR interferometer is a simple, compact design using thin partially reflective beam splitters, which has a greater resolution than the standard polarizing scheme [5]. A focusing parabolic mirror is used to direct the CTR emitted from a retractable foil downstream of the chicane compressor. Golay cells are used to detect the CTR signals.

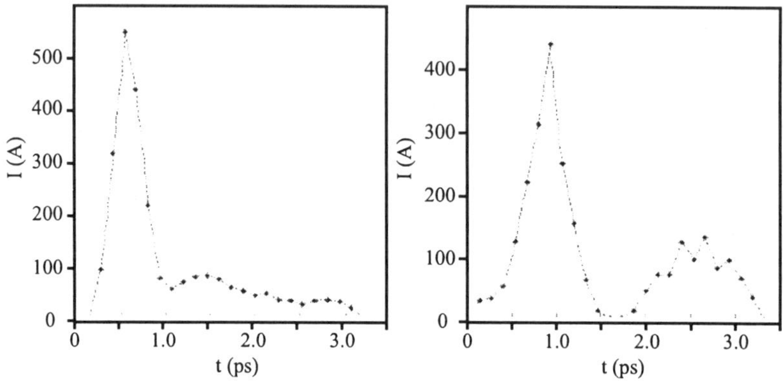

Figure 3. The reconstructed longitudinal beam profiles using the Kramers-Kronig analysis of CTR interferograms with optimum compression (left) and over compression (right).

The results of the CTR interferometry suggest that the beam is compressed to a bunch length of ~30 μm. The Kramers-Kronig analysis is used to reconstruct the time-domain bunch profile (Fig. 3). A characteristic one-sided distribution is seen at optimal compression. The beam displays distinct splitting in the time domain profile with over-compression.

4.2. *Momentum Spectrum*

Commissioning studies of the compressor included the observation of the momentum spectrum of the compressed electron beam. Initial coherent transition radiation (CTR) interferograms confirm that the electron bunch length was shortened to 30 μm RMS. The momentum spectrum of the ultra-short beam was observed on a diagnostic screen located just after a 20° bend dipole (spectrometer). Indeed, the beam exhibited bifurcation; longitudinal beam break up is a footprint of strong beam compression (Fig. 4).

5. Coherent Edge Radiation Measurements

5.1. *Frequency Spectrum*

Spectral measurements of the emitted chicane radiation were conducted at the ATF. Interferograms were obtained from scans with the Michelson-type interferometer to provide information about the spectral content. This data was then used to reconstruct the longitudinal charge distribution of the beam.

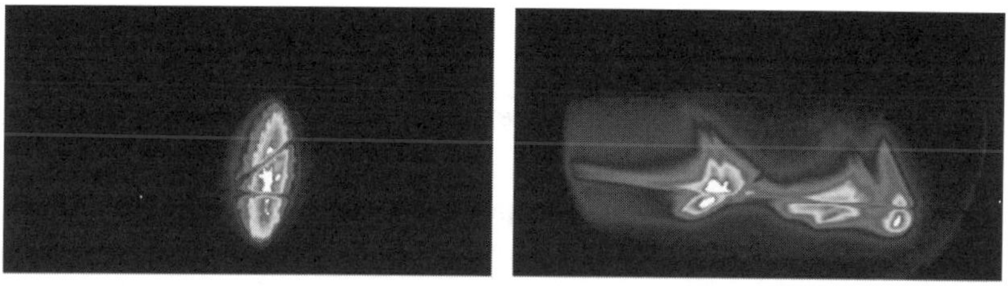

Figure 4. Electron beam images in the ATF spectrometer - the bend plane is horizontal. The uncompressed beam (left) at near minimum energy spread and 9 degrees forward of RF crest, and the beam at maximum compression (right), 19 degrees forward of crest. Bifurcation is evident in the compressed beam case.

Interferometer signal amplitudes from the output of the silicon bolometer were recorded for N shots per mirror position. An averaged, normalized interferogram with the corresponding measured electron beam charge is shown in Fig. 5a. Each point on the interferogram records the mean value and the error bars depict the standard deviation.

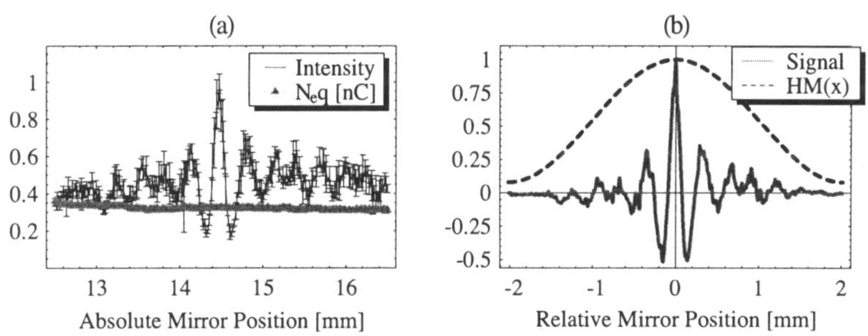

Figure 5. (a) Normalized, averaged signal amplitudes from interferometer scans of chicane radiation. (b) Apodized signal via Hamming function.

For an accurate spectral reconstruction with a discrete inverse Fourier transform (DIFT), the non-zero offset of the interferogram is removed and the signal is multiplied by an apodization function. Fig. 5b shows the results of apodization by a Hamming function $\mathrm{HM}(x) = 0.54 + 0.46\cos(\frac{\pi x}{a})$, where $2a = 4$ mm is the length of the interferometer scan centered on the maximum peak. This standard procedure removes artificial low-frequency components and corrects for the spurious tails that arise from the finite sum in the DIFT [12],

$$\tilde{I}_k = \frac{1}{n}\mathrm{Re}\sum_{j=1}^{n} I_j e^{2\pi i(k-1)(j-1)/n}, \tag{2}$$

where n is the number of equally spaced interferometer mirror positions recorded over a finite distance $[ct_{min}, ct_{max}]$. This yields $n/2$ non-repeating intensity values, I_k, up to the maximum frequency, $f_{max} = n/[2(t_{max} - t_{min})]$ where I_j is the jth intensity point in the modified interferogram.

Fig. 6(a) displays the resulting spectrum of the apodized signal. The peak at $f_{peak} \approx 0.5$ THz coincides with the dominant frequency given by results from `QUINDI` simulations, which show the simulated spectral distribution is dominated by frequencies 0.2 THz $\leq f \leq$ 1.5 THz.

5.1.1. *Time-domain Profile Reconstruction*

The forward far-field radiation intensity spectrum produced by the bunch is given by Eq. 1, where the longitudinal form factor $F(\omega)$ is

$$F(\omega) = \left| \int \hat{\rho}(z) e^{i\omega z/c} dz \right|^2, \tag{3}$$

with $\hat{\rho}(z) = \rho(z)/N_e q$, the normalized longitudinal charge density.Following a minimal-phase reconstruction technique , a discretized Kramers-Kronig relation is used to extract $\hat{\rho}(z)$ [14]. The minimal phase $\psi(\omega_i)$ from $\int_0^\infty \hat{\rho}(z)e^{-i\omega_i z/c}dz = \xi(\omega_i)e^{i\psi(\omega_i)}$ is

$$\psi(\omega_i) = \frac{2\omega_i}{\pi}\sum_{j\neq i}^{N_{max}} \frac{\ln[\xi(\omega_j)/\xi(\omega_i)]}{\omega_j^2 - \omega_i^2}\Delta\omega, \tag{4}$$

with $\xi^2(\omega_i) = F(\omega_i)$ and $\Delta\omega$ is the resolution from the spectrum. The normalized profile $\hat{\rho}(z)$ has the form

$$\hat{\rho}(z) = \frac{1}{\pi c}\sum_{i=0}^{N_{max}} \xi(\omega_i)\cos\left[\psi(\omega_i) - \frac{\omega_i z}{c}\right]\Delta\omega. \tag{5}$$

The bunch distribution is calculated from the form factor, determined by fitting well-behaved asymptotes to the measured normalized spectrum. normalized by $I_0(\omega)$. In principle $N_{max} \geq n/2$ since $F(\omega)$ vanishes for $\omega \to \infty$. From this expression it is clear that a reconstruction of the longitudinal profile proceeds from the structure of the form factor, which is given by fitting appropriate asymptotes to the modified spectrum such that

$F(\omega \to \infty) = 0$ and $F(\omega \to 0) = 1$. Since the high frequency asymptotic behavior of the form factor goes like an inverse power-law[14] for high frequencies such that $F(\omega \to \infty) = 0$, and because the dominant contribution to Eqn(4) is for $\omega_i \leq 2\pi f_{max}$, the high frequency contributions to Eqn(5) for $\omega_i \gg 2\pi f_{max}$ are negligible. As a result, the upper limit N_{max} can be much larger that $n/2$. In practice, this high frequency limit is set by the condition that the phase is changed only negligibly for increased values of N_{max} and it's implementation is supported by expectations from theory, simulation (Fig. 6(a) – dashed curve) and from the experimental limitation that the spectrum is attenuated for high frequencies (discussed below). Results of this reconstruction are shown in Fig. 6(b). As expected from simulation and theory for the chicane compressor, the bifurcated longitudinal bunch profile exhibits a shortened head of length $\sim$ 30 μm at FWHM. The values for current are scaled such that the integral over the entire bunch length yields to the average measured charge of $N_e q \approx 330$ pC. Due to experimental limitations (discussed below) and numerical fitting of the form factor, this reconstruction technique is useful as an approximation of the true bunch distribution. Nevertheless, the reconstructed profile shows general agreement with predictions from PARMELA simulations in the overall structure of the asymmetric bunch and the compressed head of $\sim$ 30 μm FWHM.

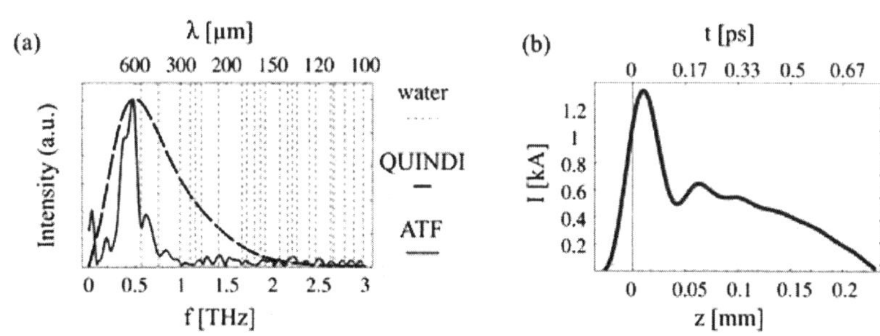

Figure 6. (a) Spectra from apodized interferogram and from simulation. Prominent water absorption frequencies are shown as vertical dotted lines. (b) Minimal phase Kramers-Kronig bunch reconstruction for the measured spectrum.

5.1.2. *Experimental Limitations on Spectral Analysis*

The simulated and measured spectra show that the radiation is dominated by frequencies below 1.5 THz, however, several experimental factors contribute to spectral filtering which affect the measured frequency distribution and the longitudinal bunch reconstruction. The primary experimental artifact near the dominant frequency band is the selective filtering from water absorption (due to high levels of humidity encountered through the radiation transport on the dates of data acquisition). This effect is exemplified by strong absorption troughs located near $f = 0.57$ and 0.75 THz. Significant absorption frequencies are plotted in Fig. 6(a) for a 7 m travel path [15]. The spectrum is also affected by the radiation transport line, which acts as a high pass filter due to its inherent finite apertures and acceptance angles. Further, the fused silica vacuum port window maintains a transmission coefficient that slowly decreases for frequencies greater than 0.3 THz, but rapidly approaches zero for frequencies greater than 6 THz. Consequently, the observed spectrum suffers from attenuation that increases steadily with frequency up to 3 THz, where-after the window becomes largely opaque near these frequencies.

5.2. *Polarization*

The transverse radiation profile and polarization were measured in an effort to observe the distinctions between CER and CSR. Fig. 7 shows the measured intensity (normalized to maximum) as a function of polarizer angle. The wire grid polarizer was mounted at the end of the transport and rotated in 15° increments. The focused signal intensity was measured with the bolometer. The linearly polarized component of radiation (sinusoidal) is consistent with that expected from synchrotron radiation. The non-linearly polarized component, which introduces a vertical offset, is a clear signature of edge radiation. Analytic studies show that pure synchrotron radiation would give a ratio of approximately 7:1 between the maximum and minimum signal [16], while the observed ratio is approximately 4:1. `QUINDI` results are consistent with the measured data (Fig. 7).

5.3. *Transverse Spatial Distribution*

The transverse far-field spatial intensity distribution of the emitted radiation was measured by scanning a small iris (3 mm diameter) in a 19 mm $\times$ 23 mm rectangular array of discrete points (Δd =3.8 mm). The

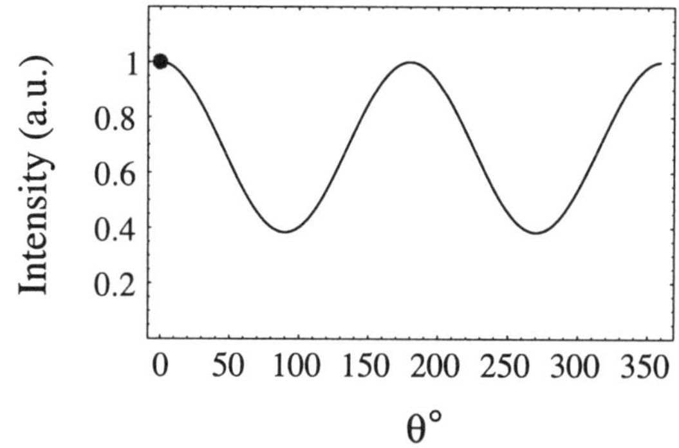

Figure 7. Normalized polarization intensity of the chicane radiation. Solid line is from QUINDI, dots are measured at the bolometer. The zero angle is arbitrary.

iris was transversely scanned across the transport tube exit and the signal was focused into the bolometer with the parabolic mirror. Iris scans were performed for varying radiation polarizations with the aforementioned wire-grid polarizer (Fig. 8).

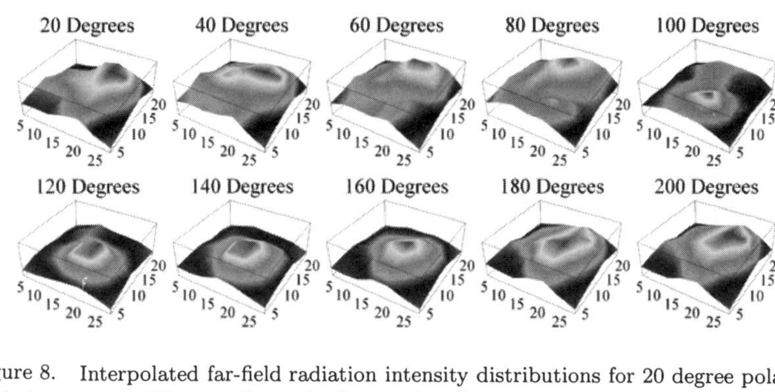

Figure 8. Interpolated far-field radiation intensity distributions for 20 degree polarizer angle increments. The x and y scales have units of mm.

The results of the iris scan measurement were compared to a similar study performed with the start-to-end simulation suite. In fact, QUINDI results (Fig. 9) show striking similarities with the experimental data. The overall presence of strong intensity peaks (σ-polarization) due to in-bend-plane radiation is complemented by the existence of out-of-bend-plane radiation (π-polarization) as evidenced by multiple, lower-intensity, peaks as a function of polarization angle.

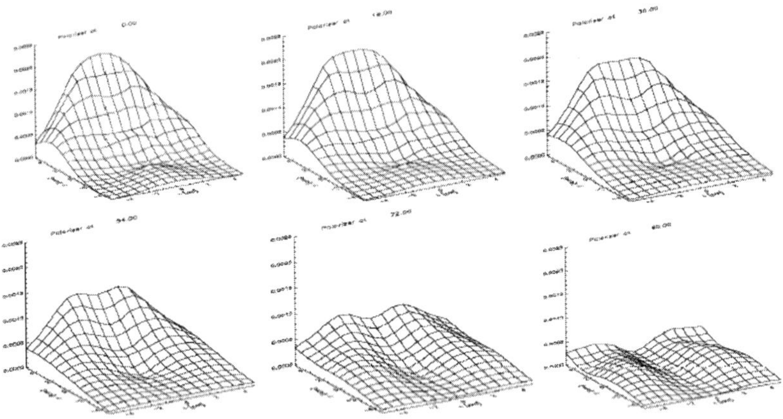

Figure 9. QUINDI results for the far-field angular distribution of the chicane radiation as a function of polarization angle.

6. Conclusions

The unique geometry of the compressor, in which adjacent magnet edges constructively interfere in producing edge radiation, has further allowed exploration of a new phenomena – short wavelength coherent edge radiation. The chicane radiation measurements conducted at the ATF display striking signatures of CER. The transverse spatial distributions show patterns consistent with QUINDI simulations, notably strong intensity peaks shifting into multiple peaks for varying degrees of polarization with identifiable nulls in the distribution on-axis. The non linearly polarized component of radiation observed as on offset to the sinusoidal in the overall polarization curve is a clear indication of prominent edge radiation.

Future plans to improve the measurements include radiation transport modifications such as replacement of the fused silica chicane radiation port window with a diamond or z-cut crystalline quartz window; this will greatly improve the spectral range of the measurements. In addition, enclosing and flushing the transport line with dry nitrogen will mitigate the effects of water absorption lines in the measured spectrum.

Acknowledgments

This work supported by DOE Grant DE-FG02-98ER45693 and NSF Grant PHY-0355048.

References

1. G. Andonian, et al., Proceedings of the 2003 Particle Accelerator Conference, 944, (2003)
2. T. E. May, R. A. Bosch and R. L. Julian, Proceedings of the 1999 Particle Accelerator Conference, 2394, (1999).
3. R. Agustsson, M.S. Thesis, UCLA (2003)
4. http://www.irlabs.com.
5. http://www.radiabeam.com.
6. M. Barranco and J. R. Buchler, *Phys. Rev. Lett.* **C34**, 1729 (1980).
7. O. V. Chubar and N. V. Smolyakov, *J. Optics (Paris)* **24**, 117 (1993).
8. R. A. Bosch, Il Nuovo Cimento **20D**, 483 (1998).
9. O. V. Chubar and N. V. Smolyakov, Proceedings of the 1993 Particle Accelerator Conference, 1626, (1993).
10. F. Zhou, et al., Proceedings of the 2005 Particle Accelerator Conference, 2263, (2005)
11. L.M. Young, J.H. Billen, Tech. Report LA-UR-96-1835, Accelerator Code Group, LANL, USA, (rev. 2000).
12. J. F. Rabolt and R. Bellar, *Applied Spectroscopy* **35**, 132, 1981.
13. M. Borland, Tech. Report LS-287, Advanced Photon Source, ANL, USA (2000).
14. R. Lai and A.J. Sievers, *Phys. Rev. E* **50**, 5 (1994).
15. P.U. Jepsen, *private communication.*
16. J.D. Jackson, *Classical Electrodynamics*, Wiley, New York, 1999.
17. L. Fröhlich, *DESY-THESIS 2005-011*, 2005.

PARMELA SIMULATIONS FOR PITZ: FIRST MACHINE STUDIES AND INTERPRETATION OF MEASUREMENTS

M. BOSCOLO, M.FERRARIO
INFN-LNF, Via E. Fermi 40, 00044 Frascati (Roma), Italy

S. CIALDI
INFN-Milano, Via Celoria 16, 20133 Milano, Italy

M. KRASILNIKOV, A. OPPELT
DESY, 15738 Zeuthen, Germany

First comparisons between measurements and simulations are discussed for the PITZ1.5 facility. In particular, dedicated PARMELA simulations performed using the realistic laser pulse hitting the cathode are compared to first emittance measurements performed with the SPARC e-meter and with the EMSY system. Measured energy spectra are compared to simulations, as well. It is discussed the overall good agreement between calculations and measurements.

1. Introduction

In summer 2005 the PITZ1.5 test facility at DESY in Zeuthen has started its commissioning phase and it is used essentially for beam dynamics studies. It consists of an intermediate setup (see figure 1) between phase 1 and phase 2 [1].

Phase 1 of PITZ has successfully concluded in November 2003 with the full characterization of a gun that has been installed and is currently in operation at the VUV-FEL. The gun was then replaced and PITZ1 run for the whole 2004 [2]. Since December 2004 the photo injector has been upgraded. The diagnostics beamline has been significantly modified, together with the installation of a normal conducting booster cavity and an extended water cooling system.

As the new 10 MW klystron has been installed in June-July 2005, PITZ1.5 has started its operations. The gun is being conditioned towards higher peak and average power with the goal of 60MV/m. The booster cavity is a normal conducting Tesla prototype cavity that accelerates the beam up to about 13MeV. Commissioning of new diagnostic components has been done at this first stage of operations. One of these new features is the SPARC emittance meter device [3], temporarily installed at PITZ.

First emittance measurements have been done with and without booster, i.e. at high and low energy for different bunch charge configurations. The measurements based on the slit mask techniques (e-meter) have been compared to the data taken with the PITZ emittance measurement system (EMSY) [1].

In this paper emittance measurements are compared to dedicated beam dynamics simulations performed with the PARMELA [4] code. Energy spectra are also measured and compared to simulations, as well.

An overall good agreement is found between measurements and simulations, as discussed in sections 3 and 4.

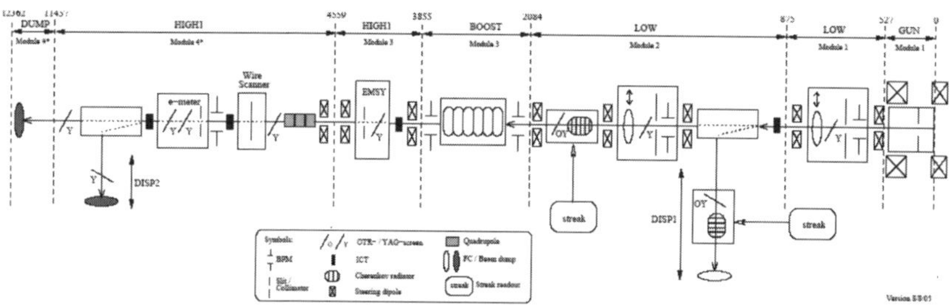

Figure 1. Scheme of the PITZ1.5 facility.

2. Simulations in nominal machine conditions

Table 1. Beam parameters for the PITZ1.5 injector.

Bunch charge [nC]	Up to 1
Temporal bunch length [ps]	20
Laser pulse rise/decay time [ps]	2
Rms transverse size [mm]	0.5
Thermal emittance [μm]	0.6
Energy @gun exit [MeV]	5.1
Energy @booster exit [MeV]	13

PARMELA simulations with a nominal 20 ps flat-top laser pulse length have been performed as a first step. Parameters used for these sets of simulations are listed in table 1. The gun and booster 2-D RF field maps for the PARMELA simulations have been calculated [5] from the electromagnetic code SUPERFISH [6].

Gun solenoid scans have been performed without accelerating cavity to find the best theoretical magnetic value B_{gun} for the emittance compensation scheme

[7] in a low charge configuration. A bunch charge of 0.23 nC has been considered for these studies.

The best beam behavior according to the theory [7] is found for a B_{gun} of 0.18 T. In fact, for this magnetic value transverse projected emittance shows two relative minima and the longitudinal position of the accelerating structure corresponds to the relative emittance maximum, as it should be from theory. Horizontal emittance and beam envelope along the beamline without and with booster are plotted in figure 2: pink and red curves are for emittance, blue and light blue curves for envelope without and with booster, respectively. A horizontal projected emittance of 1 mm mrad is calculated at the end of the beamline for the low charge configuration with these nominal parameters.

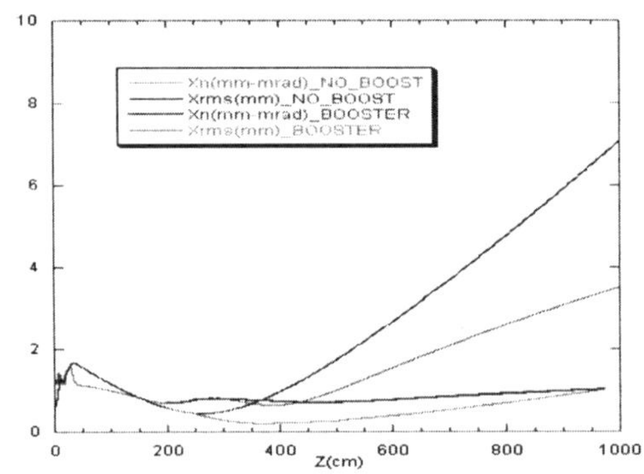

Figure 2. PARMELA simulations for a pulse length of 20ps flat-top, Q=0.23nC, transverse rms spot size of 0.5mm and B=0.18T with (red and light blue curves) and without acceleration (pink and blue curves).

3. Transverse Emittance and Beam size

Two sets of measurements have been selected for a comparison between data and simulations, one at low current (Q=0.5 nC) the other at high current (Q=1 nC). In both machine configurations transverse emittance and beam envelope have been evaluated for different values of the solenoid magnetic strength. Realistic beam pulses have been used for simulations.

3.1. *High bunch charge configuration*

Horizontal and vertical emittance and rms beam sizes have been measured at the EMSY position, at about 4.3 m from the cathode, for different gun solenoid magnetic strengths. The magnetic strength is determined from the measurement of the current powering the coil $I_{gun}[A]$, knowing the scaling factor between the current and magnetic strength. The agreement between PARMELA simulations and measurements is quite good, as discussed below.

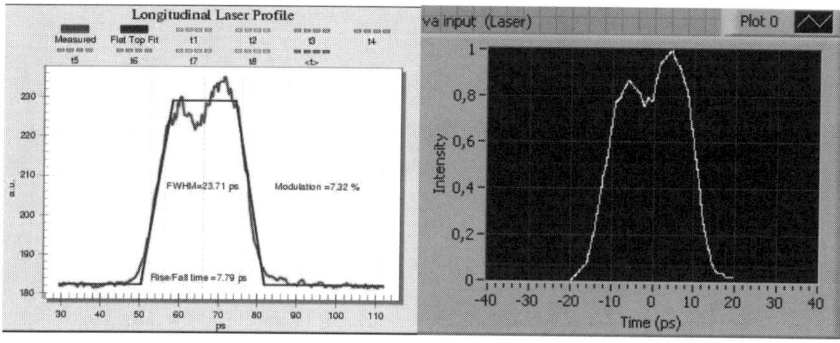

Figure 3. Left plot: Laser longitudinal profile during emittance measurements at EMSY , with the 1nC configuration. Right plot: Realistic beam pulse used for simulations.

The chosen measurements have been taken on 28-29/09/2005. During this data taking the laser temporal profile hitting the cathode was the one reported in left plot of figure 3. Its characteristics are: FWHM=23.71 ps, rise time (r.t.) of 7.79 ps with a ripple modulation on the flat top of 7.3%, but essentially two peaks can be clearly observed from the plot.

The initial longitudinal bunch shape for PARMELA simulations is taken identical to the real laser longitudinal profile (see right plot of figure 3); laser and beam pulse are assumed to be transversely uniform.

Measurements compared to simulations are reported in figure 4. Upper plot shows the horizontal and vertical beam sizes, lower plot the transverse emittance versus the gun solenoid current. Red and blue curves are the horizontal and vertical simulated values, respectively; black and green curves are the corresponding measured values. Increasing by 10% the scaling factor between I_{gun}[A] and the magnetic strength the agreement with simulations becomes quite good, as it appears from figure 4. The same effect has been observed in other analysis. The source of this mismatch is still under investigation.

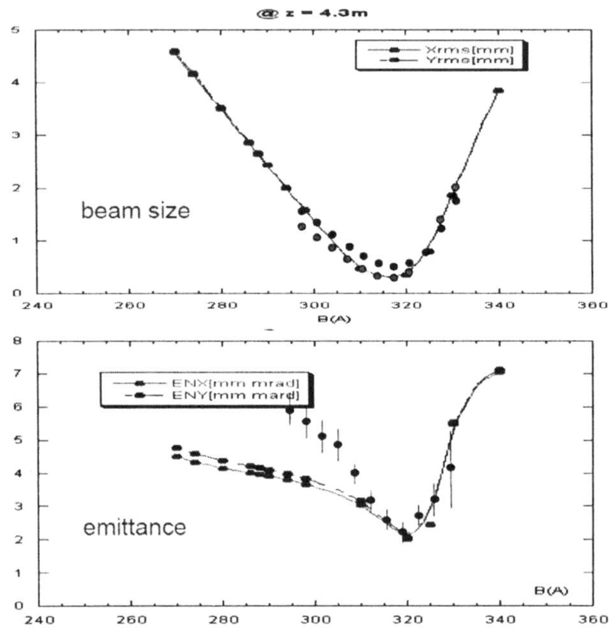

Figure 4. Beam size (upper plot) and transverse emittance (lower plot) measurements compared to simulations. Red and blue curves represent simulations for horizontal and vertical planes, respectively. Black and green dots represent measurements for horizontal and vertical planes, respectively. A 10% mismatch between the solenoid current and its magnetic strength is considered.

3.2. *Low bunch charge configuration*

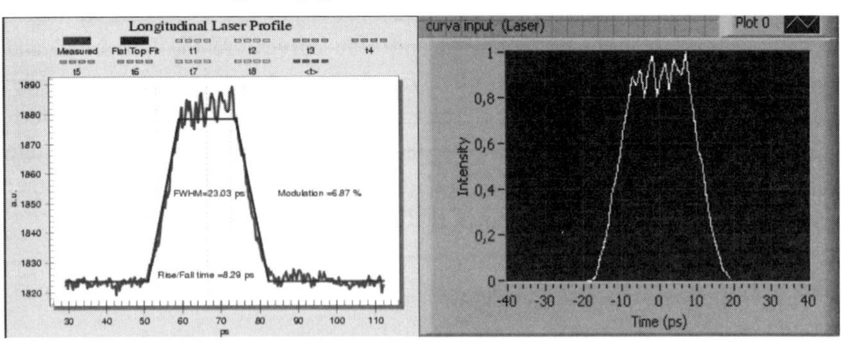

Figure 5. Left plot: Laser longitudinal profile during measurements at the e-meter with the 0.5nC machine setup. Right plot: corresponding realistic beam pulse used for simulations.

A second set of measurements taken on 29/09/2005 has been analyzed and compared to simulations. The machine was tuned for a bunch current of

Q=0.5nC and beam envelope and emittance were measured at the e-meter, at about ~7.7 m from the cathode.

The laser profile corresponding to these measurements is reported in left plot of figure 5: FWHM is 23.03 ps, r.t. is 8.29 ps and there are some high frequency ripples in the flat top with amplitude of 6.9%. This is the temporal distribution taken for the electron beam at the cathode used for simulations (right plot figure 5).

Comparison between simulations and measurements are reported in figure 6, where a 10% mismatch between solenoid current and magnetic value is taken into account, as for the studies performed at high current. Agreement with simulations is in this case not as good as for the previous one. Anyhow, especially emittance behavior is reproduced by calculations. More details on these measurements are discussed in [8].

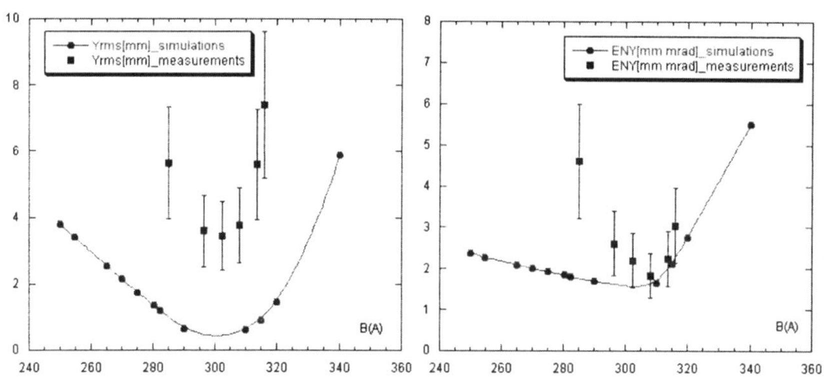

Figure 6: Vertical beam size (left plot) and projected emittance (right plot) at e-meter. Red and blue curves are simulated and measured values, respectively. A 10% mismatch between the solenoid current and magnetic strength is taken into account.

4. Energy Spectrum

The energy distributions are routinely measured at PITZ1.5 to tune the machine both after the gun and at the end of the beamline. Injection phase φ_{inj}, as well as booster RF phase $\varphi_{booster}$, are set to give the highest beam energy and the lowest energy spread, for a fixed accelerating voltage. This corresponds to φ_{inj}=40° and $\varphi_{booster}$=78°.

We discuss in the following the measurement of the longitudinal momentum distribution performed together with the emittance measurements already analyzed in section 3.1. Left plot of figure 7 shows the measured longitudinal momentum distribution at the two optimized phases; with average and rms values P_{mean}=13.0488 MeV, and P_{rms}=0.03183 MeV, respectively.

A double peak structure can be clearly observed. Decreasing by 3° the injection phase the two peaks structure disappear (right plot of figure 7). The results obtained using PARMELA are shown in left plot of figure 8 for the simulated energy distribution at optimal φ_{inj} =40° and in the right plot for a shift of φ_{inj} by -3°. We found the same effect observed in the measurements.

In addition, there is a good agreement between simulations and measurements on the energy spread, as well as on the distance between the two peaks and on their dependence on φ_{inj}.

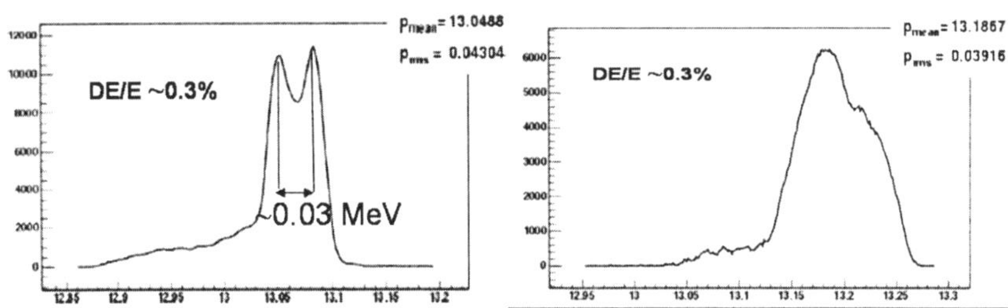

Figure 7: Measured longitudinal momentum distribution at the end of the injector, for optimized φ_{inj} and $\varphi_{booster}$ on left plot and for φ_{inj} = -3° on right plot.

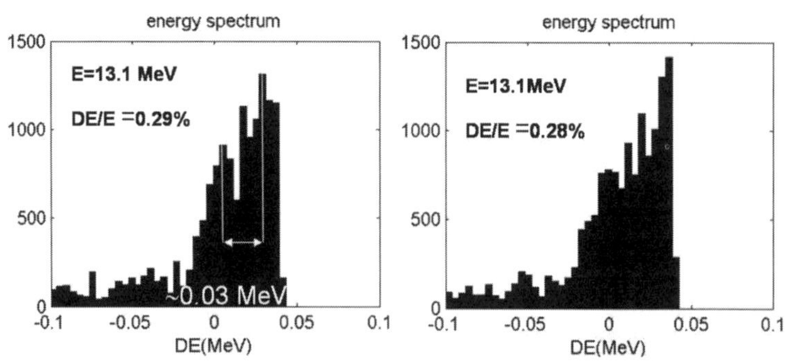

Figure 8: Simulated energy distribution at the end of beamline, for optimal φ_{inj}=40° and $\varphi_{booster}$=78°on left plot and for φ_{inj} = -3° on right plot.

The qualitative interpretation of this measurement is in agreement with the simulation studies carried on for SPARC and briefly summarized in the following.

The two peaks in the energy distribution reported in left plot of figures 7 and 8 can be explained by the argument that temporal oscillations transform into energy oscillations [9]. In fact, in this particular measurement the initial longitudinal distribution shows a double peak behavior (figure 3). As the beam goes through the gun and drifts the two longitudinal peaks transform into energy distribution peaks, as found for the SPARC injector [10].

PARMELA simulation studies have been carried on for SPARC taking into account an amplitude modulation in the longitudinal profile. They indicate that at the entrance of the first acceleration structure the beam has lost temporal ripples which have converted into energy ripples through a fractional plasma oscillation, as long as the modulation is within 30% of the total distribution. It is a space-charge induced compensation process. These energy ripples are then suppressed in the first traveling wave cavity during the acceleration process, so that energy spread at the end of the linac is not affected by the initial pulse shape.

More detailed studies on laser pulse shaping effects [11] in SPARC have shown that the value of the emittance remains substantially constant increasing the amplitude of the oscillations, as long as the frequency is sufficiently high. The slope of the emittance-amplitude line decreases with the increase of the frequency.

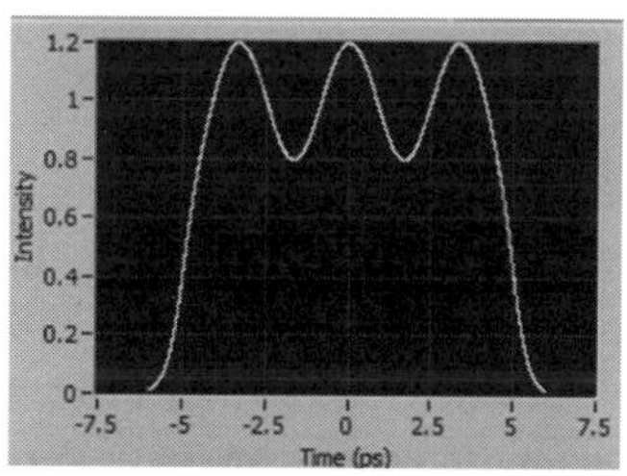

Figure 9: Initial beam pulse for a simulation for a SPARC case: FWHM=10ps, r.t.=1ps, low frequency modulation in the flat-top with 30% amplitude.

Moreover, the measurement performed at PITZ1.5 has been simulated for the SPARC case with a longitudinal profile of the initial beam pulse having a peaked structure. The dependence of energy spectrum on phase injection and on initial pulse shaping has been analyzed.

If we consider an initial pulse as the one in figure 9: FWHM=10ps, r.t.=1ps and a low frequency modulation in the flat-top with an amplitude of 30% we observe also in this case that the energy spectrum is very sensitive to the injection phase. Upper and lower plots of figure 10 show the energy distribution at the exit of the second accelerating cavity at z=4.6m from the cathode and for a beam energy of E=80.8 MeV for the optimal φ_{inj}=32°and for a -3° shift from this value. Different spectra are observed, but energy spread is not affected by the initial pulse shape. Energy spread is 0.21% in the first case, as for the nominal profile, and 0.37% with the -3° shift.

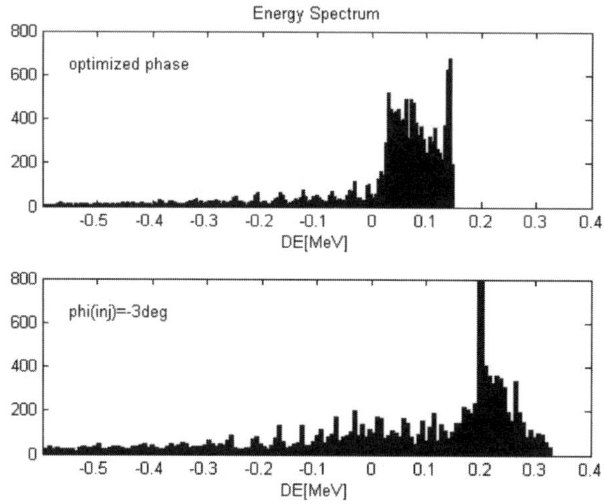

Figure 10: Simulated energy spectrum at the exit of the second cavity at ~80 MeV for a beam pulse as in figure 9. Upper plot: for optimal φ_{inj}=32° and $\varphi_{booster}$ on crest; lower plot: φ_{inj} = -3°.

5. Conclusions

First PARMELA simulations have been performed for the PITZ1.5 facility and been compared to first beam measurements, like transverse emittance, envelope and energy.

There are still some problems due to the limited knowledge of machine parameters, as the gun solenoid magnetic strength versus its current. However, an overall good agreement between simulations and measurements of energy, beam size and emittance has been found.

Acknowledgments

We acknowledge useful discussions with D. Alesini, F. Marcellini, C. Ronsivalle, V. Fusco and with the entire diagnostic group that operated with the SPARC e-meter system at the PITZ facility.
This work has been partially supported by the EU Commission in the sixth framework program, contract n. 011935 EUROFEL.

References

1. A. Oppelt et al., *Status and First Results from the upgraded PITZ Facility,* **FEL05**, S. Francisco, USA (2005).

2. A. Oppelt et al., *The Photo Injector Test Facility at DESY Zeuthen: Results of the first phase,* **LINAC 2004**, Lübeck, Germany (2004).
F.Stephan et al., *Recent Results and perspectives of the low emittance photoinjector at PITZ,* **FEL 2004**, Trieste, Italy (2004).
V.Miltchev et al., *Modelling the Transverse Phase Space and Core Emittance Studies at PITZ,* **FEL05**, S. Francisco, USA (2005).
3. L. Catani et al., *Commissioning of the SPARC movable emittance meter and its first Operation at PITZ,* **FEL05**, S. Francisco, USA (2005).
4. L.M.Young, *PARMELA*, Los Alamos National Laboratory report LA-UR-96-1835, (revision May 23, 2005).
5. D. Alesini and F. Marcellini, *private communications.*
6. J. H. Billen, L.M.Young, *POISSON SUPERFISH*, Los Alamos National Laboratory report LA-UR-96-1834, (revision January 5, 2005).
7. M. Ferrario et al., *Homdyn Study for the LCLS Photoinjector,* **SLAC-PUB-8400** (2000).
8. D. Filippetto et al., *these proceedings.*
9. M. Ferrario et al., *Beam Dynamics Studies for the SPARC Project,* **PAC03**, Portland, Oregon, USA (2003).
10. D. Alesini et al, *Status of the SPARC Project,* **PAC05**, Knoxville, Tennessee, USA (2005).
11. M. Boscolo et al., *Preliminary Results on Beam Dynamics of Laser Pulse Shaping Effects in SPARC,* **PAC05**, Knoxville, Tennessee, USA (2005).

THE LCLS SINGLE-SHOT RELATIVE BUNCH LENGTH MONITOR SYSTEM

M.P. DUNNING*, G. TRAVISH, A. COHEN, P. FRIGOLA, S. REICHE, J. ROSENZWEIG

Particle Beam Physics Laboratory
Department of Physics and Astronomy
University of California Los Angeles
Los Angeles, CA 90095, USA

The Particle Beam Physics Lab (PBPL) at UCLA is developing a single-shot relative bunch length monitor system for the Linac Coherent Light Source (LCLS) X-ray Free Electron Laser at SLAC. To ensure high-quality lasing, feedback systems are needed. The bunch length monitor system will be a crucial part of the linac longitudinal feedback system, which is necessary to keep the final peak beam current stable. The system will consist of two single-channel detector arrangements, one after each bunch compressor, and will use the coherent synchrotron radiation (CSR) and coherent edge radiation (CER) from bunch compressor chicane dipole magnets to monitor the bunch length. Details of the system are presented, with an emphasis on the device for bunch compressor one (BC1), and the system is compared to alternatives. Plans for testing are presented as well.

1. Introduction

1.1. *Background: The LCLS*

The LCLS, when operational in 2009, will be the world's first X-ray Free Electron Laser (FEL), producing pulses of up to 8 keV photons at up to 120 Hz[1]. The electron beam is born in a 1.6 cell S-band photoinjector, undergoes two stages of bunch compression, and is fed into a 100 meter undulator where the X-ray pulses are created. The resulting ultrashort (sub 200 fs) X-ray pulses will be directed into two experimental halls which will house six experimental stations. The LCLS will utilize the last kilometer of the SLAC linac. Table 1 below lists some relevant parameters for the LCLS, for the two different electron beam charge scenarios.

*mdunning@physics.ucla.edu

Table 1. Relevant LCLS parameters for the two different electron beam charge scenarios. BC1: Bunch Compressor 1, BC2: Bunch Compressor 2.

	1.0 nC	0.2 nC	
Nominal electron energy, BC1	.25	.25	GeV
Nominal electron energy, BC2	4.3	4.3	GeV
Final peak current	3400	2500	A
Nominal RMS bunch length, BC1	200	60	μm
Nominal RMS bunch length, BC2	20	8	μm
Nominal RMS bunch duration, BC1	667	200	fs
Nominal RMS bunch duration, BC2	67	27	fs
Max single bunch repetition rate	120	120	Hz

1.2. *The Problem*

High quality FEL lasing demands tight beam parameters; the LCLS will have some of the strictest tolerances to date. To ensure that these tolerances are not exceeded, feedback systems are needed[4], along with appropriate diagnostics. The bunch length monitor (BLM) system will be a crucial part of the longitudinal feedback system, which is based on measuring bunch length and energy, and adjusting the linac phase and voltage, respectively. The main requirements of the BLM system are[2]:

- Relative bunch length needed- not absolute bunch length
- Two separate bunch length monitors needed- one after each bunch compressor (see Fig. 1 below)
- Must be single-shot
- Must be non-invasive
- Must be maintenance free for several days (months)
- Must include possibility to run at 120 Hz
- Single-shot measurement resolution: 1-2% of nominal bunch length
- Long term signal drift: <2% over no less than 24 hours

1.3. *Some solutions*

There are a number of devices that can be used to measure bunch length shot-to-shot; most notable are streak cameras, interferometers, electro-optic methods, deflecting cavities, spectrometers, and various single detector arrangements. Each of these diagnostics was considered as a possible candi-

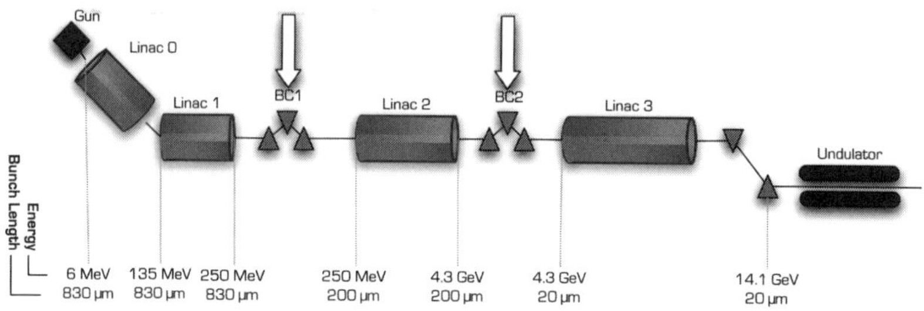

Figure 1. Locations of the bunch length monitors in the LCLS beamline.

date for the BLM system, and is briefly described below.

2. Possible Single-Shot Bunch Length Diagnostics

2.1. *Streak Cameras*

Femtosecond streak cameras are an established method to measure bunch length, in that they are single-shot and have a fairly wide dynamic range. However, they can be ruled out immediately for the present application due to their insufficient temporal resolution (currently 200 fs at best[a]) and the problem of trigger jitter[14].

2.2. *Interferometers*

Interferometers can be single-shot, have a fairly high temporal (frequency) resolution[20], and are compact, but were ruled out because of their narrow dynamic range and because single-shot devices were unproven.

2.3. *Electro-Optic Methods*

Electro-optic (EO) methods are single-shot and may have sufficient temporal resolution, but are not mature at this point. Furthermore, the technique may be invasive, due to fact that the electron beam travels very near to the electro-optic crystal[15]; induced wakefields will slightly influence beam properties[16]. EO methods also require an expensive femtosecond laser, and would probably be cost-prohibitive.

[a]Hamamatsu FESCA-200, http://sales.hamamatsu.com/

2.4. *Deflecting Cavities*

RF deflecting cavities can measure bunch length shot-to-shot at femtosecond levels[17,18], but are invasive, in that they destroy the measured shot.

2.5. *Spectrometers*

The most robust device with sufficient temporal resolution appears to be the spectrometer (polychromator)[11]. The frequency spectrum of the radiation can be obtained directly from the voltage output of the detectors used in the device. Through a simple analysis one can deduce a single value which is proportional (or inversely proportional) to the electron bunch length over a fairly wide dynamic range. The spectral information and hence, the resolution, is mainly limited by the number of detectors (channels). Such a device can accommodate changing operating modes, as the fitting function to the spectrum can be adjusted.

For example, a grating polychromator can be constructed for bunch compressor 1 (BC1) employing a 20 channel linear array of cryogenically cooled hot electron bolometers[12]. Such a system offers details about the bunch longitudinal geometry, and allows for diagnosis of and recovery from non-ideal conditions. This type of device was initially proposed for the LCLS BLM system, but was rejected due to cost and the maintenance involved with the cryogenic system (the system would have required liquid Helium refills approximately every 10 weeks and more frequent liquid Nitrogen refills).

2.6. *Single-Detector Arrangements*

A simple single-detector arrangement can also be constructed to measure a signal which varies approximately linearly with the bunch length over a reasonable dynamic range. The resulting time and frequency integrated signal can be used to monitor the relative bunch length. This type of device is robust and inexpensive. Measurement of the integral of the radiated signal from a compressed beam is a well established method of tuning a chicane to a peak compression setting. This method has been used by at least one group for the purpose of RF feedback, albeit at L-band and with more relaxed tolerances.[13]

Simplicity and cost drove the decision to use this type of system, however it is unclear at this time whether this device will achieve the desired resolution ($\approx$2% of nominal bunch length) and dynamic range ($\approx$2x smaller/larger than nominal bunch length). Rigorously testing the device under quasi-LCLS conditions (see Sec. 5)will determine if the device is suitable.

2.6.1. *Detectors*

A number of different radiation detectors can be used for this device. Among them are diode detectors, Golay cell detectors, and pyroelectric detectors. An ideal detector would have the following properties:

- Flat, broad frequency response
- Blackbody limited detectivity
- Fast response
- Solid state construction

Pyroelectric detectors were chosen because of their flat frequency response, detectivity, and speed (several hundred Hz is common). Diode detectors were ruled out because of their complex frequency response, and Golay cells because of their low speed (typically 25 Hz max).

3. Proposed Design

The proposed single-detector design for the BC1 BLM can be seen in Fig. 2 below.

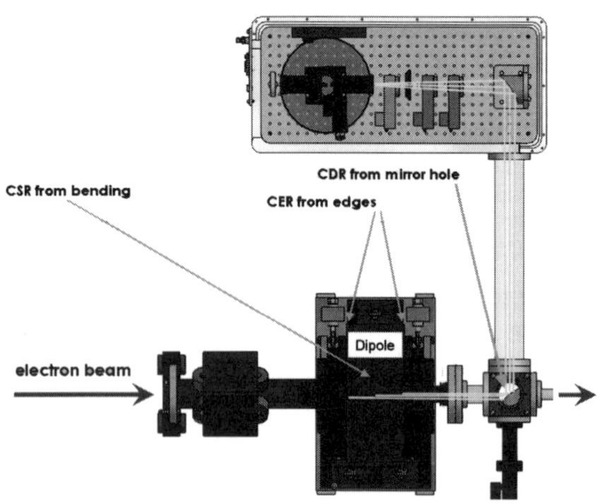

Figure 2. Top view of the BC1 BLM, showing the sources of CSR, CER, and CDR.

The radiation from chicane dipole magnets will travel out the chicane exit pipe (there is no separate radiation port in the chicane) and will be deflected out of the beamline with an insertable mirror, which contains a hole for the electron beam to pass through. The radiation will then travel out of the beamline through a diamond window and into a nitrogen purged diagnostic chamber. In the diagnostic chamber, detailed in Fig. 3 below, the radiation will be focused and turned by an off-axis paraboloidal mirror; travel through bandpass and amplitude filters; a polarizer, and an iris; and, into a suitable detector (Coherent J3-05 pyroelectric shown here).

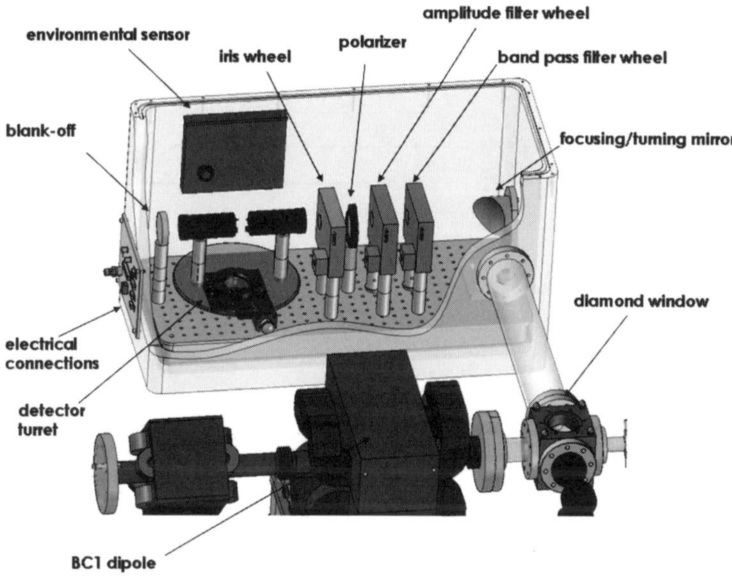

Figure 3. The proposed single-detector design.

The second detector is for background measurement and subtraction. One or more different detectors can be mounted on the rotatable platform. An environmental sensor will monitor temperature and humidity.

4. Detected Radiation

4.1. *CSR*

The CSR from bunch compressor chicane magnets will be used to monitor the relative bunch length. The coherent synchrotron radiated power (per unit frequency, per unit solid angle) can be written as[5,6,7]

$$P_{coh} = N^2 f(\omega) P_0 \;, \tag{1}$$

where P_0 is the power radiated by a single electron, N is the number of electrons in the bunch, and $f(\omega)$, the "form factor" is defined here as

$$f(\omega) \equiv \left| \int S(z) e^{i\frac{2\pi}{\lambda}z} dz \right|^2 , \tag{2}$$

(considering only the longitudinal dimension) where $S(z)$ is the longitudinal bunch distribution and $\lambda = \frac{2\pi c}{\omega}$ is the wavelength of interest. The radiated power is coherently enhanced, and substantial, for $\lambda \gtrsim \sigma_z$. Figure 4 shows $f(\omega)$ and the total predicted CSR power for the parabolic bunch distribution after BC1.

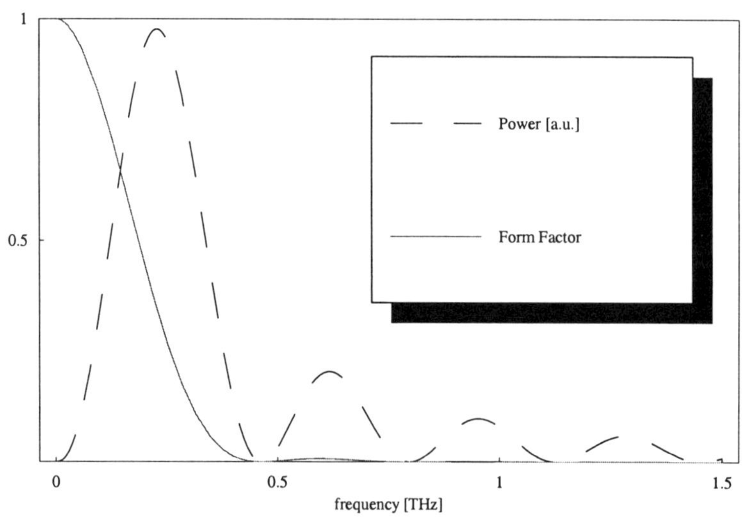

Figure 4. Form factor $f(\omega)$ and total predicted CSR power for the bunch distribution after BC1.

It can be seen from Fig. 4 that for the smooth parabolic bunch profile, most of the useful information is in the 0-0.5 THz range.

4.2. *CER*

Coherent radiation from the edges of bending magnets will also be substantial for BC1, for wavelengths longer than the bunch length[8]. The contribution from CER appears to be comparable to the contribution from CSR, and will be studied further using numerical tools.

4.3. *CDR*

The radiation will be deflected out of the beamline with a mirror. The mirror must have a hole for the electron beam to pass through, and in doing so, the relativistic electrons will emit diffraction radiation. This radiation, similar to transition radiation[19] also is coherently enhanced for wavelengths comparable to or longer than the bunch length[10]. The contribution from CDR appears to be unavoidable with the present design, but may be a useful source of radiation; initial calculations show that the CDR is strong.

Use of CDR presents some advantages over CSR/CER; the "radiator" location and strength can be chosen by selecting the placement of the mirror and the aperture size. Furthermore, because the distance from the mirror to the vacuum window can be made short, a large fraction of the radiation can be collected. The contribution from CDR is being investigated further.

4.4. *Comments*

Simulations of the detected radiation (CSR, CER, CDR) are currently underway using QUINDI, a numerical code which models spontaneous radiation in bend systems. However, the code is still under development, and diffraction radiation has not yet been included. It was originally planned to use TREDI and a locally developed post-processor, FieldEye, for the numerical simulations, but this approach proved to be inadequate to model the system.

Incoherent synchrotron radiation will not likely be a significant source of noise, but is being investigated further. The effect of vacuum chamber cutoff/shielding on the detected radiation is currently being investigated as well[21,22].

5. Testing

Testing is planned at the APS facility at Argonne National Lab, as the LEUTL bunch compressor is most similar to the LCLS BC1. The beam energy, energy spread, charge, and bunch length can closely mimic the LCLS BC1 parameters. Effects of noise, dynamic range in charge and bunch length, and simulation results can be tested; however, the use of the device as an LCLS feedback system can not likely be tested in any comparable way.

6. Summary

The LCLS bunch length monitor systems, part of the longitudinal feedback system, will use the coherent radiation from chicane magnets to monitor the relative bunch length. The BC1 BLM has been designed, component selection is almost complete, and the device will soon be built. It will first be bench-tested at UCLA, then installed and tested at the APS, and finally installed at SLAC for BC1 commissioning in late 2006. The BC2 BLM system will be installed for BC2 commissioning in late 2007.

7. Acknowledgements

The authors would like to thank J. Wu, P. Emma, E. Bong, and T. Borden of SLAC; B. Berg, and J. Stein of ANL; and O. Grimm of DESY.

References

1. P. Emma, *et al.*, *SLAC-R-593*, 2002.
2. P. Emma, D. Schultz, E. Bong, D. Marsh and P. Krejcik, *LCLS PRD #1.3-018*, (2005).
3. J. Wu, P. Emma, Z. Huang, *Proc. of 2005 Part. Acc. Conf.*, 428 (2005).
4. J. Wu, P. Emma, L. Hendrickson, *Proc. of 2005 Part. Acc. Conf.*, 1156 (2005).
5. J. S. Nodvick and D. S. Saxon, *Phys. Rev.* **96**, 180 (1954).
6. T. Nakazato, *et al.*, *Phys. Rev. Lett.* **63**, 1245 (1989).
7. C. J. Hirschmugl, *et al.*, *Phys. Rev. A* **44**, 1316 (1991).
8. R. A. Bosch, *Il Nuovo Cimento* **20D**, 483 (1998).
9. O. V. Chubar and N. V. Smolyakov, *J. Optics (Paris)* **24**, 117 (1993).
10. Y. Shibata, *et al.*, *Phys. Rev. E* **52**, 6787 (1995).
11. T. Watanabe, *et al.*, *Nuc. Instrum. and Meth. A* **480**, 315 (2002).
12. A. Cavallo, *et al.*, *Rev. Sci. Instrum.* **59**, 889 (1988).
13. O. Grimm, private communications.
14. M. Uesaka, *et al.*, *Nuc. Instrum. and Meth. A* **406**, 371 (1998).
15. P. Bolton, *et al.*, *SLAC-PUB-9529*, 2002.
16. G. Berden, *et al.*, *Proc. of 2004 FEL Conf.*, 343 (2004).
17. L. Fröhlich, O. Grimm, *et al.*, *Proc. of 2004 FEL Conf.*, 395 (2004).
18. R. J. England, *et al.*, *Proc. of 2005 Part. Acc. Conf.*, 2627 (2005).
19. C. Settakorn, *Proc. of 2nd Asian Part. Acc. Conf.*, 514 (2001).
20. P. Muggli, M. J. Hogan, *et al.*, it SLAC-PUB-11189, 2005.
21. M. Dohlus and T. Limberg, *Nuc. Instrum. and Meth. A* **407**, 278 (1998).
22. R. L. Warnock, *SLAC-PUB-5523*, 1991.

BEAM SHAPING AND PERMANENT MAGNET QUADRUPOLE FOCUSING WITH APPLICATIONS TO THE PLASMA WAKEFIELD ACCELERATOR*

R. J. ENGLAND, J. B. ROSENZWEIG, G. TRAVISH, A. DOYURAN, O. WILLIAMS, AND B. O'SHEA

University of California, Los Angeles
Department of Phyics and Astronomy
Los Angeles, CA 90095, USA

D. ALESINI

Laboratori Nazionali di Frascati
Rome, Italy

It has recently been proposed to use a dispersionless translating section (dogleg) with sextupole correction magnets as a bunch compressor to create longitudinally shaped (linearly ramped) electron bunches. We discuss the experiment soon to be underway at the UCLA Neptune Linear Accelerator Laboratory to test this technique with the 300 pC, 13 MeV electron bunches produced by the Neptune S-Band photoinjector. The experiment will utilize a dipole-mode deflecting cavity, as a temporal diagnostic, and a final focus system of permanent magnet quadrupoles with field gradients of 110 T/m. We also discuss the potential scaling of this technique to bunches of high (i.e.>1nC) charge for the purpose of creating a suitable drive beam for the plasma wakefield accelerator, operating in the blowout regime.

1. Introduction

The blowout regime of the plasma wakefield accelerator (PWFA)[1] has been the recent subject of various experimental and theoretical investigations.[2,3,4] In this regime, a relativistic beam (the drive bunch) produces large-amplitude wake-fields in a plasma of density n_0 much less than the drive beam density n_b. The high charge of the drive bunch causes plasma electrons in its wake to be expelled from the beam path, producing a rarified ion column behind the bunch. The blowout regime is of consider-

*This work is supported by the department of energy under grant number xxx.

able interest due to the fact that the induced wake-fields in the ion column exhibit linear focusing and accelerating forces, making it a well-behaved regime for the acceleration of either a low-charge witness bunch (injected in the wake of the drive beam) or the tail of the drive beam itself. Because plasmas can support very high electromagnetic fields, the accelerating gradients in this scheme can in principle be on the order of tens of GV/m, which is more than an order of magnitude improvement over traditional RF-based accelerating structures.

The current profile of the drive beam has been shown by analytical models [5,6] and by 2D simulations [7] to be optimized in this type of scenario by a triangular "ramp", with the beam current rising linearly from head to tail, followed by a sharp drop to zero at the tail. The transformer ratio (a figure of merit for the PWFA obtained by dividing the peak accelerating field behind the bunch by the deccelerating field within the bunch) for such a drive bunch in a plasma of density n_0 is given by $R = k_p L$, where $k_p = \sqrt{4\pi n_0 e^2/m_e c^2}$ is the plasma wavenumber and L is the length of the drive bunch 'ramp' [5]. Consequently, the value of the transformer ratio can in principle exceed the maximal value ($R = 2$) for a symmetric bunch profile, so long as the length of the ramp exceeds two plasma skin depths ($L > 2k_p^{-1}$).

A beam dynamics experiment designed to generate relativistic bunches whose current profile approximates the ideal 'ramped' bunches is presently underway at the UCLA Neptune linear accelerator laboratory. [8] The scheme proposed for this experiment takes advantage of the RF curvature imposed upon the longitudinal phase space distribution of the bunch when it is injected into the accelerating section behind the crest of the accelerating field. A sextupole-corrected dispersionless translating section (or dogleg) is used to partially compress the resulting bunch (which has a positive chirp in energy vs. longiudinal coordinate z), producing a final beam with a current distribution resembling a triangular ramp. Simulation results of the longitudinal phase space distribution and current profile for the Neptune experiment, using the particle tracking code PARMELA [9], are shown in Fig. 1. Part (a) shows the chirped beam generated by the photoinjector gun and linac. The final distribution, after passing through the dogleg, is shown in parts (b) and (c) with sextupole magnets off and on respectively. The use of sextupole correctors in part (c) to remove the nonlinear second-order correlation between energy and longitudinal position results in an almost purely linear compression of the initial chirped phase space distribution, producing a beam with a current profile that rises gradually from head to tail, followed by a sharp drop. This bunch shape approximates the optimal triangular ramp, and has been shown in 2D particle-in-cell simulations to be capable of producing transformer ratios greater than 2 when properly matched into a plasma.[8]

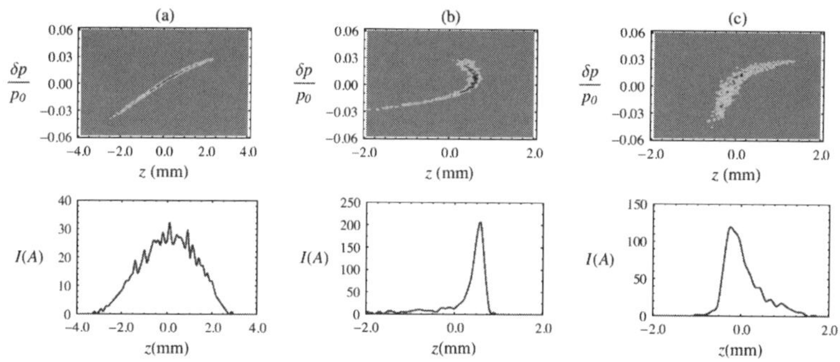

Figure 1. Simulated plots of longitudinal phase space and current profile for Neptune beam immediately after the accelerating section (a), and then after passing through the dogleg compressor with sextupoles turned off (b) and turned on (c).

To properly match such a beam into a plasma, we require that the betatron matching condition $\beta_r = \beta_{eq} = \sqrt{\gamma/(2\pi r_e n_0)}$ be satisfied. This requirement, combined with the previously mentioned condition on the bunch length L and the requirement for the blowout regime $n_b >> n_0$ (which we will interpret as $n_b > 4n_0$), we obtain a set of constraints on the plasma density, the drive beam RMS size σ_r, and the normalized emittance ϵ_N:

$$n_0 > n_{0,min} = \frac{mc^2}{\pi e^2 L} \tag{1}$$

$$\sigma_r < \sigma_{max} = \sqrt{\frac{Q/e}{4\pi n_0 \sigma_z}} \tag{2}$$

$$\epsilon_N < \epsilon_{N,max} = \gamma\beta \frac{\sigma_{max}^2}{\beta_{eq}} \tag{3}$$

In these relations, we have approximated the beam density by $n_b = Q/(e\pi\sigma_r^2\sigma_z)$ and have used the definition of the normalized emittance $\epsilon_N = \gamma\beta(\sigma_r^2/\beta_r)$. We can use these constraints to obtain the following expression for the minimum required beam brightness:

$$B > B_{min} = \frac{2cQ}{\epsilon_{N,max}^2 \sigma_z}. \tag{4}$$

Here we have used the definition for the transverse brightness $B = 2I/\epsilon_N^2$, where $I = en_b\beta c\pi\sigma_r^2$ is the beam current. These relations provide us with an estimate of the required beam parameters for successfully applying the bunch shaping technique that is the subject of our experiment to create an adequate drive beam for a PWFA. In Section 2, we will compare these calculated threshold values with the simulated beam parameters for the Neptune experiment. In Section 3 we will consider possibilities for future experiments, including the creation of a witness bunch and scaling the Neptune experiment to high charge (4 nC).

2. UCLA Neptune Experiment

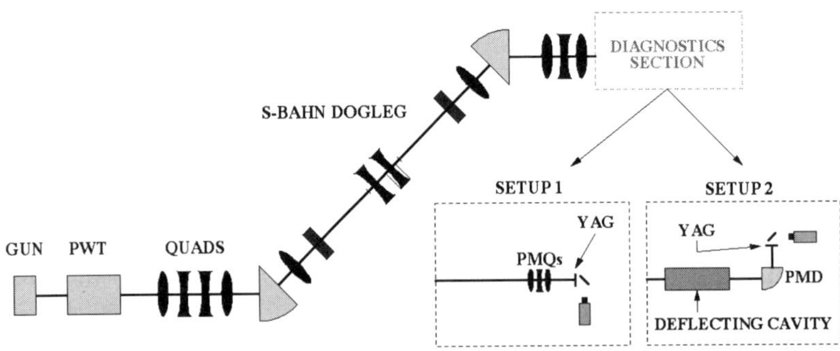

Figure 2. Cartoon graphic of the experimental beamline (not to scale). Blue lenses, red rectangles, and yellow wedges represent quadrupoles, sextupoles, and dipole magnets respectively. Two alternate setups are shown for the final diagnostics section.

2.1. *Experimental Overview*

A cartoon of the experimental beamline is shown in Fig. 2. Electron bunches with an RMS bunch length of 2.5 ps are generated in the 1.6 cell S-Band photoinjector gun and are then accelerated to an energy of 13 MeV by the plane wave transformer (PWT) accelerating cavity. The third and fourth dipole magnets of a chicane compressor are used as a 45-degree bend (yellow wedge) to divert the beam onto the dogleg section, which has been named $S - Bahn$, after a train system in Hamburg, Germany. The dogleg has a negative longitudinal dispersion (or R_{56}), which can be utilized to generate a partial compression of the beam as shown in Fig.

1 and thereby produce a ramp-shaped profile. A second pair of dipoles (second yellow wedge) bends the beam trajectory back by 45 degrees onto a path which runs parallel to but horizontally displaced from the original photoinjector beamline. This parallel beamline has a triplet of traditional electromagnetic quadrupole magnets followed by a diagnostic section. Two alternative diagnostic setups will be employed, as shown in the inset in Fig. 2. Setup 1 will consist of a triplet of permanent magnet quadrupoles followed by a Ce:YAG profile monitor, which will be used to obtain a high-brightness focus, as discussed in the Introduction. Setup 2 will employ a 9-cell X-Band dipole mode deflecting cavity, the final version of which is currently under construction. The deflecting cavity will serve as a temporal diagnostic to measure the current profile of the beam for comparison with simulation results such as those shown in Fig. 1. As the deflecting cavity diagnostic is the subject of a recent conference paper,[10] we will concentrate primarily on the details of Setup 1 in the following sections.

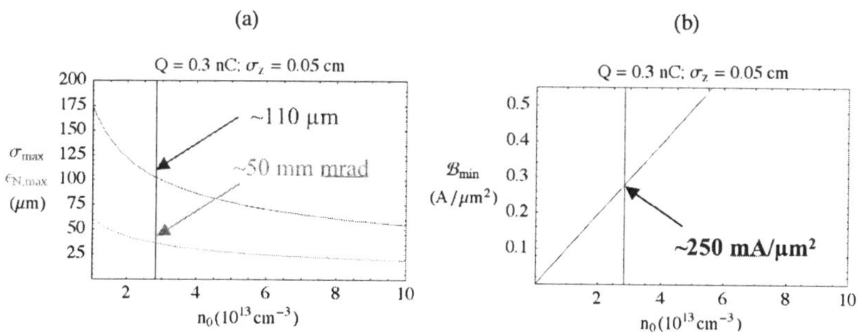

Figure 3. Plot of the constraints as given by Eqns. (1)-(4), with approximate maximal values for beam size and emittance, and minimum brightness.

Using the constraints on the beam parameters imposed by Eqs. (2)-(3), we can construct plots showing the maximum allowable beam size and emittance as a function of plasma density n_0 for a 0.5 mm long 300 pC beam. This is shown in Fig. 3(a). Combining this with the minimum density of $n_0 = 2.8 \times 10^{13} cm^{-3}$ required by Eq. (1) gives us estimated upper limits on RMS beam size and normalized emittance of 110 μm and 50 mm mrad respectively. A corresponding plot of minimum brightness using Eq. (4), shown in Fig. 3(b), gives a lower limit of 250 $mA/\mu m^2$. These limits are compared with simulation results in Section 2.3.

2.2. *Permanent Magnet Quadrupoles*

The permanent magnet quadrupoles (PMQs) to be used for the S-Bahn final focus shown in Setup 1 of Fig. 2 are of a hybrid iron and permanent magnet design originally developed for the nonlinear inverse Compton scattering experiment that is currently in progress at the Neptune laboratory.[11] The magnets, shown in Fig. 4, contain cubes of NdFeB, surrounded by an iron yoke which serves to close the magnetic circuit. Four hyperbolic pole faces constructed by wire electric discharge machining (EDM) are held against the NdFeB cubes in a quadrupole array around the geometric center by an aluminum keeper.

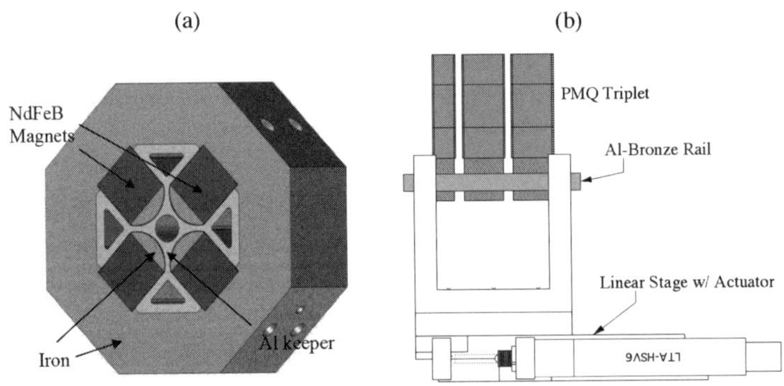

Figure 4. Drawing of hybrid permanent magnet quadrupole design (a), courtesy of A. Doyuran, *et al.*[12], and schematic of assembled triplet and stand (b).

Magnets of 1 cm and 2 cm lengths have been constructed, incorporating 4 and 8 NdFeB cubes respectively. The measured field strengths of the two types are similar (109 and 110 T/m respectively). The proposed configuration for the triplet is a single 1cm length defocusing PMQ, followed by a focusing and then a defocusing 2cm length PMQ. A side view of the assembly is shown in Fig. 4 (b) at a reduced scale.

2.3. *Simulations of Experimental Results*

The PMQ focusing system described in Section 2.2 was simulated initially using the matrix-based beam envelope code PowerTrace. Further studies were then done using the particle tracking code ELEGANT to model the dogleg and final focus sections.[12] The phase space coordinates for the particles used as the input for the ELEGANT simulation were generated by a UCLA-PARMELA simulation of the photoinjector and linac.

Simulated experimental values for energy E, charge Q, normalized emittances $\epsilon_{x,N}$ and $\epsilon_{y,N}$, RMS bunch length and transverse dimensions σ_t, σ_x, σ_y, and brightness B are given in Table 1. Initial values correspond to the beam parameters immediately after the accelerating section, and final values correspond to the final focus location of the permanent magnet quadrupole triplet of Setup 1. Final values represent design goals based upon the PARMELA and ELEGANT simulation results. The shape of the longitudinal phase space predicted by these simulations was shown previously in Fig. 1. The reduction in charge is a prediction based upon observed transportation losses in the beamline, and the emittance growth is due primarily to transverse nonlinear effects in the dogleg. Note that the predicted beam sizes, emittance values, and brightness fall roughly within the limits set by Eqs. (1)-(4), as discussed in Section 2.1, for applicability to plasma wake-field studies with large transformer ratios.

Table 1. Simulated Experimental Parameters

Parameter	*Units*	*Initial*	*Final*
E	MeV	13	13
Q	pC	300	240
$\epsilon_{x,N}$	mm mrad	5	41
$\epsilon_{y,N}$	mm mrad	5	15
σ_t	ps	2.5	1.8
σ_x	mm	1	0.130
σ_y	mm	1	0.057
B	mA/μm^2	7600	433

3. Future Directions

3.1. *Creation of a Witness Bunch*

In order for the ramped bunch mechanism described in Section 1 to represent a useful technology for the wake-field accelerator, it must be compatible with some feasible scheme for creating a witness bunch. The witness bunch would ideally be a bunch of much lower charge which trails behind the main drive bunch, and can therefore be accelerated by the wake-fields which are generated by it. One technique used in the past has been to accelerate the tail of the drive bunch itself. However, the mechanism described in Section 1 is designed to create a drive bunch with a sharp cutoff at the tail end, as seen in Fig. 1(c). We see, however, in Fig. 1(b), that with the sextupole correctors turned off, the nonlinear effects which are suppressed in 1(c) produce a significant lower-energy tail behind the bunch, but the ramped shape is lost. A potential solution would be to operate in a regime interme-

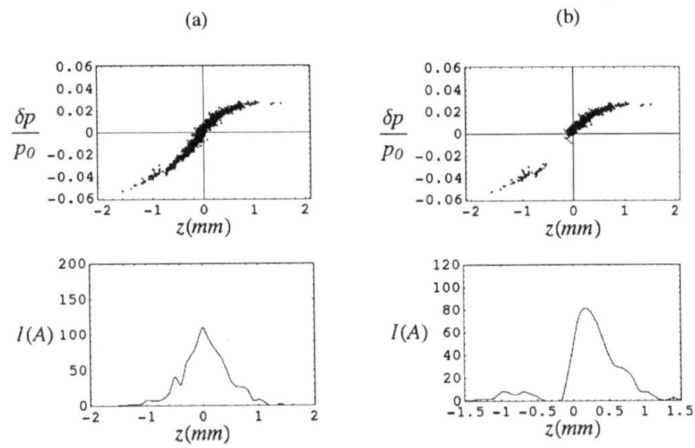

Figure 5. Simulation of undercorrected beam at exit of dogleg with collimator removed in (a) and inserted in (b), thereby producing a ramped drive beam followed by a low-charge witness bunch.

diate between the conditions represented in Fig. 2(b) and 2(c), where the sextupole magnets are turned on but at a lower field strength, producing a beam with a ramp at the front followed by a more tapered fall-off at the back. This situation is seen in Fig. 5(a), which shows the results of an ELEGANT simulation of the dogleg compressor. By inserting a 1cm wide collimator in the x-direction, at a location in the dogleg (corresponding to the position of the quadrupole before the final dipole in Fig. 2) where the horizontal dispersion is large and therefore there is a strong correlation between x and z, the tail of the beam can be truncated from the main body. As shown in Fig. 5(b) this results in a ramp-shaped primary bunch followed by a separate trailing bunch of lower charge. This scheme has the benefit of being relatively simple, requiring only the insertion of a collimator into the beamline. However, the resultant reduction in charge and horizontal truncation of the beam must be taken into consideration in the design of the downstream focusing optics.

3.2. *Scaling to High Charge: 4 nC*

Future upgrades to the Neptune laboratory, including a new drive laser oscillator, replacement of the photoinjector, photocathode laser cleaning, and higher RF power levels in the gun are expected to increase the bunch charge to as high as 4 nC. It is therefore of interest to consider how the bunch shaping mechanism described previously scales to higher charge. Preservation of the beam envelope under the emittance compensation mechanism in the gun and linac requires that the bunch dimensions at the cathode scale with charge as $Q^{1/3}$. This is accomplished by stretching the pulse length of the photocathode drive laser and expanding its transverse dimensions accordingly. Simulations of the Neptune photoinjector under this scaling in UCLA-PARMELA indicate a normalized emittance of $\epsilon_{x,N} = \epsilon_{y,N} = 25$ mm mrad at the exit of the linac with 4% transportation losses for an initial charge of 4 nC. Since the dogleg compression mechanism requires a chirped beam, the bunch was chirped in energy by setting the RF phase of the linac in the simulation to a value corresponding to an injection phase of 22 back of crest. Due to this chirp, the increase in bunch length due to the charge scaling resulted in a predicted 4.5% RMS energy spread, compared with 1.8% energy spread for the 300 pC case.

The final phase space coordinates of the particles in the PARMELA simulation were then used as the input for an ELEGANT simulation of the dogleg compressor. The results of these simulations indicated that due to the larger energy spread of the 4 nC beam, two undesirable effects became more pronounced: (1) distortion of the longitudinal phase space by third-order longitudinal dispersion (U_{5666} in transport notation) and (2) emittance growth due to horizontal second-order dispersion (T_{566}). The first effect results in the formation of a low-energy tail behind the beam. This is seen in Fig. 6(a). The tail can in principle be corrected by the use of octupole magnets. The second effect requires the use of sextupole magnets and is somewhat more difficult to remedy, due to the fact that, at least for the particular optical configuration of the Neptune dogleg, it is impossible to simultaneously eliminate both the horizontal and longitudinal second order dispersion (T_{166} and T_{566} respectively). Consequently, the sextupole magnets may be used to eliminate the second order horizontal dispersion, thereby improving the final emittance, but as a result the longi-

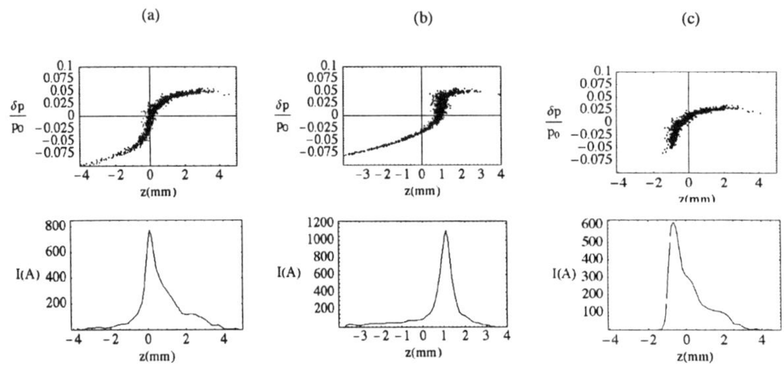

Figure 6. Simulation of longitudinal phase space and current profiles of 4nC beam at exit of dogleg compressor with (a) with sextupoles set to eliminate second-order longitudinal dispersion (T_{566}), (b) with sextupoles set to eliminate second-order horizontal dispersion (T_{166}), and (c) with collimator inserted to remove low-energy tail.

tudinal dispersion becomes nonzero and so the shape of the ramped profile is destroyed. This scenario is illustrated in Fig. 6(b). A solution which appears to solve both problems is to simply eliminate the tail in part (a) by collimating the beam. As it turns out, much of the emittance growth is due to the low-energy particles contained in this tail, and their removal improves the final emittance by a factor of two and restores the ramped profile, as seen in Fig. 6(c). By using a collimator of finite width, a small subset of the tail particles could be left as a witness bunch, making this technique compatible with the results of the previous section. To clarify these results, the simulated emittance and corresponding matrix element values are provided in Table 2.

Table 2. Simulated Parameters Corresponding to Fig. 6 (a), (b), and (c)

Parameter	*Units*	*(a)*	*(b)*	*(c)*
$\epsilon_{x,N}$	mm mrad	742	96	46
$\epsilon_{y,N}$	mm mrad	456	141	68
T_{166}	m	-0.26	0.00	-0.26
T_{266}	rad	-7.9	0.00	-7.9
T_{566}	m	-0.04	0.623	-0.04
U_{5666}	m	-2.44	-1.02	-2.44

It should be noted that the simulations above do not include transmission losses due to ordinary apertures of the beamline. And in fact, the simulated RMS beam sizes in the dogleg for the high-charge case are found to exceed the radius of the beam pipe. Consequently, although scaling the compressor to higher charge appears theoretically feasible, it would in practice be necessary to expand the apertures of the beamline, which would require a significant redesign of the beamline hardware.

4. Conclusions

We have described an experiment underway at the Neptune laboratory to create 300 pC electron bunches 1 to 2 ps in duration with a linearly ramped current profile suitable for use as a plasma wake-field drive beam. To obtain a final beam of sufficiently high brightness, a final focusing system has been constructed, incorporating a triplet of hybrid permanent magnet and iron quadrupoles. Simulation results using PARMELA and ELEGANT predict that the final transverse beam brightness and normalized emittances (450 mA/μm^2 and 100 μm respectively) should be within the limits required for generation of large-amplitude (transformer ratio >2) wake-fields. Simulations also indicate the feasibility of a proposed method for creating a witness beam, by undercompressing the beam slightly in the dogleg and severing the resulting low-energy tail from the main body of the beam using an insertable collimator. In addition, the feasibility of scaling the beam-shaping scheme to high charge (4 nC) was studied. Simulations of the high-charge case indicate that undesirable emittance growth and phase space distortion produced as a result of the larger energy spread can largely be corrected by truncation of low-energy particles using collimation. However, the larger beam size exceeds the apertures of the beamline and would most likely require significant redesign of the existing hardware.

References

1. J. B. Rosenzweig. Acceleration and focusing of electrons in two-dimensional nonlinear plasma wake fields. *Phys. Rev. A*, 44:R6189–6192, November 1991.
2. P. Muggli, B. E. Blue, C. E. Clayton, S. Deng, F. J. Decker, M. J. Hogan, C. Huang, R. Iverson, C. Joshi, T. C. Katsouleas, S. Lee, W. Lu, K. A. Marsh, W. B. Mori, C. L. O'Connell, P. Raimondi, R. Siemann, and D. Walz. Meter-Scale Plasma-Wakefield Accelerator Driven by a Matched Electron Beam. *Phys. Rev. Lett.*, 93:014802, June 2004.
3. N. Barov, J. B. Rosenzweig, M. E. Conde, W. Gai, and J. G. Power. Observation of plasma wakefield acceleration in the underdense regime. *Phys. Rev. ST-AB*, 3:011301, January 2000.

4. N. Barov, J. B. Rosenzweig, M. C. Thompson, and R. B. Yoder. Energy loss of a high-charge bunched electron beam in plasma: Analysis. *Phys. Rev. ST-AB*, 7:061301, June 2004.
5. K. L. F. Bane, Pisin Chen, and P. B. Wilson. On Collinear Wake Field Acceleration. Technical Report SLAC PUB-3662, Stanford Linear Accelerator Center, Stanford, CA 94305, April 1985.
6. T. Katsouleas. Physical Mechanisms in the Plasma Wake-Field Accelerator. *Phys. Rev. A*, 33:2056–2064, March 1986.
7. K. V. Lotov. Efficient operating mode of the plasma wakefield accelerator. *Physics of Plasmas*, 12:053105, May 2005.
8. R. J. England, J. B. Rosenzweig, G. Andonian, P. Musumeci, G. Travish, and R. Yoder. Sextupole Correction of the Longitudinal Transport of Relativistic Beams in Dispersionless Translating Sections. *Phys. Rev. ST-AB*, 8:12801, January 2005.
9. L. Young and J. Billen. PARMELA. Technical Report LA-UR-96-1835, Los Alamos National Laboratory, Los Alamos, NM, 1996.
10. R. J. England, B. O'Shea, J. B. Rosenzweig, G. Travish, and D. Alesini. X-band dipole mode deflecting cavity for the ucla neptune beamline. In *Proceedings of the 2005 Particle Accelerator Conference*, IEEE, page 2627, 2005.
11. A. Doyuran, O. Williams, R. J. England, C. Joshi, J. Lim, J. B. Rosenzweig, S. Tochitsky, and G. Travish. Investigation of x-ray harmonics of the polarized inverse compton scattering experiment at ucla. In *Proceedings of the 2005 Particle Accelerator Conference*, IEEE, pages 2303–2305, 2005.
12. M. Borland. ELEGANT: A Flexible SDDS-Compliant Code for Accelerator Simulation. Technical Report LS-287, Argonne National Laboratory Advanced Photon Source, Argonne, IL, September 2000.

COMMISSIONING OF THE SPARC MOVABLE EMITTANCE METER AND ITS FIRST OPERATION AT PITZ

D.Filippetto,_M.Castellano, G.Di Pirro,C.Vicario, INFN-LNF, Via E.Fermi 40, 00044 Frascati (Rome), Italy

L.Catani, E.Chiadroni, A.Cianchi, INFN-Roma2, Via della Ricerca Scientifica 1, 00133 Roma, Italy

M.v.Hartrott, BESSY, 12487 Berlin, Germany

A.Oppelt, F.Stephan, M.Krasilnikov, DESY Zeuthen, Platanenallee 6, 15738 Zeuthen, Germany.

For the SPARC Project a novel diagnostic device, called "Emittance-meter", has been conceived and constructed to perform a detailed study of the emittance compensation process in the SPARC photo-injector and to optimize the RF-gun and the accelerator working point. It consists of a movable emittance measurement system, based on the 1D pepper-pot method, installed between two long bellows with the possibility to scan a region 1.2 m long downstream the RF-gun. The construction of the device was completed in the first part of this year and a series of laboratory tests, to evaluate its performances, were carried out in Spring 2005. At the beginning of the summer the complete system was moved to DESY at Zeuthen to be installed on the Photo Injector Test Facility PITZ. After the commissioning it will used for measurements of the PITZ electron beam in the framework of a collaboration between the SPARC and PITZ Projects aiming on studies and operations with photo injectors.

1. Introduction

The "SPARC emittance meter" ("SPARC e-meter" shortly) is a diagnostic device designed for a detailed analysis of the beam dynamics in the region of the SPARC injector where a significant evolution of transverse emittance occurs.

Design of SPARC e-meter has been optimized with respect to the design parameters of the SPARC electron beam, namely energy, nominal transverse emittance and size, charge per bunch etc. [1].

The construction of the device was completed in the first part of this year while the beam studies of SPARC Injector with the e-meter are scheduled for the end of 2005.

The evidence of a temporal gap between the completion of the e-meter and its first scheduled operations at SPARC suggested the possibility of a temporary installation at DESY Zeuthen for a round of measurements of the PITZ electron beam in the framework of a collaboration aiming on studies and operations with photo injectors.

To permit the installation of the SPARC e-meter at PITZ we modified the original design of the system to ensure their mechanical compatibility. Mainly the table, being the support for the beamline, was modified by shortening its legs in such a way to allow its installation on the PITZ girder (Fig.1). Legs extenders have been prepared and they will be used to adjust the height of the "modified" e-meter to that needed for SPARC.

Before its shipment to DESY Zeuthen a number of test of the e-meter have been carried out to verify its performance and calibrate the components.

2. System overview

The SPARC e-meter consists of a moveable (in z, z being the distance from the cathode, measured along the accelerator axis) emittance measurement device allowing to measure the emittance in the range from about z=83 cm to z=210 cm.

The technique to measure the beam emittance and the lattice Twiss parameters, in both the horizontal and vertical planes, makes use of a double system of horizontal and vertical slits arrays. They select one or several "beamlets" by means of either a single slit moving transversely across the beam spot or a multi-slit mask intercepting the beam.

Because beam size and divergence strongly depend on the longitudinal position, the distance between the slits and the screen must be properly adjusted in order to optimize the accuracy of the measurement of beamlet profiles. A bellow is therefore interposed between the slit mask and the screen, allowing changing

Figure 1. 3D-rendered mechanical drawing of the SPARC e-meter. In the picture are shown one of the two alignment tools to be installed during alignment on top of the end flanges (grey cylinders) and the PITZ girder (green structure).

their distance from 20 to 40 cm, a range shown by simulations to produce the required accuracy under all explored scenarios.
Two 1.5 m long bellows allow two crosses, carrying the slit masks and the detector screen, to be moved along the region where the most relevant part of the emittance compensation process occurs.
The two long bellows are connected to two crosses, fixed on the e-meter support, that define the physical boundaries of the system. In the first of these two crosses a Farady cup, to measure the integrated beam charge, is installed. In the second one, downstream from the movable system, an aerogel radiator is installed to be used for an experiment on bunch length measurements using Cerenkov radiation.
The whole system is remotely controlled allowing the possibility to easily switch between different measurement configurations changing, for instance, relative distances between slits mask and screen or the longitudinal position of the movable measurement system.
In the single slit measurement the acquisition stores several images for each slit's beam cut, to minimize statistical errors. Distance between consecutive beam cuts can be set by the operator.
Alternatively it is possible to switch to single-shot measurement, just centering the multi-slit on the beam.
Signal intensity and CCD saturation can also be checked and adjusted by operator on line via gain, shutter and brightness controls.

2.1. *Tests and evaluation of performances*

2.1.1. *Slits Mask*

For the measurement of the emittance in the horizontal and vertical planes using the *1D pepper-pot method* we have two slit masks, mounted on two independent holders 90° with respect to each other. Two sets of slit masks have been prepared so far stacking single pieces made of 2 mm thick tungsten with a well defined profile. Once assembled these pieces form a mask with two single slits, 50μm and 100μm respectively, and an array of 7x50 μm slits separated by 500μm.

A preliminary analysis of the first prototype of the pepper-pot mask using a profile projector showed that agreement of measured widths, with respect to the nominal value, was better than 5μm for 7 over 9 of the slits produced, thus compatible with the needed tolerances.

Nevertheless, a more accurate analysis with an optical microscope evidenced that while the average width was within the tolerances, irregularities of the profile due to roughness of edges might locally exceed, in some cases, 10% of the nominal width.
An alternative method for the preparation of slits we investigated is the photo-chemical machining. Optical analysis of slits produced using this method showed higher uniformity of the slits and improved smoothness of edges consequence of the more precise etching process. It also eliminates irregularities produced by

machining, due to the mechanical stress of material (Fig.2).

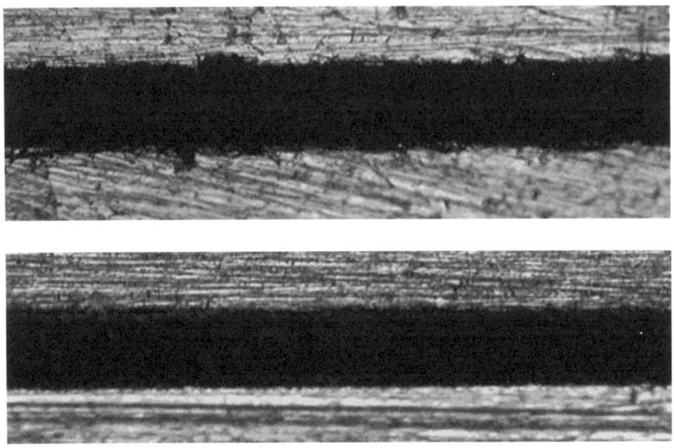

Figure 2. Optical microscope pictures of the single-slit pepper-pot obtained by mechanical machining (top) compared to single-slit produced by photo-chemical machining (bottom).

2.1.2. *Screens and imaging system*

The transverse distribution of low-charge beamlets emerging from the slit-mask needs to be measured with high accuracy.

It means that radiator screens, used for this purpose, need to have a linear response with beam charge in the range of few tenths of pC and they must guarantee a spatial resolution better than 20 μm.

Same performances are required to the imaging system that should not introduce any degradation to the figure above.

Doped-YAG radiators, either crystals or sintered screens, are good candidates because of their high resolution and efficiency.

For our application, we focused our attention on Ce:YAG radiators that we tested at the DAFNE Beam Test Facility.

We collect forward radiation emitted from Ce:YAG crystal with a mirror at 45° downstream the radiator. As result the radiator is observed at 90° with respect to the rear face thus minimizing the degradation of spatial resolution due to the non-negligible thickness of transparent crystal.

The performance of the Ce:YAG radiators has been compared to those of a Cr-oxide radiator used by our group in previous applications. Both radiators have been installed at the same diagnostic station in the DAFNE Beam Test

Facility and their performances measured under different beam conditions, within the range of values expected in the SPARC injector.

Analysis of results shows that performances of the Ce:YAG screen are superior: efficiency is a factor 2-3 higher than Cr-oxide and resolution is evidently better, as we can observe comparing the two pictures on Fig.3. At the same time we didn't report any evidence of a deviation from a linear correlation between the light yield from the two radiators varying the charge density of the beam in the range of values expected in the SPARC injector. This confirms a good linearity of the radiator with charge density in the range of values of interest.

Imaging system includes a digital CCD cameras (Basler 311f) and a 105mm "macro" type objective from SIGMA. In the current set-up, the magnification, and the correspondent resolution, has been chosen to better adapt to the PITZ beam size. Being the distance of CCD from the object (the radiator surface) 300 mm, we calculated a magnification of 1:1.7 and the resolution of the optical system has been measured to be 17,2 μm (Fig.4). It's worth to mention that tests of the imaging system previously made in our laboratory shown that a resolution better than 11 μm can be achieved.

The digital cameras are connected to the e-meter control system by means of their built-in firewire (IEEE1394) interface that can be used for both images read-out and control of the camera settings. In SPARC the digital cameras of the e-meter will be part of a larger system including those needed for other optical diagnostic stations. It will be based on the solutions developed for the TTF VUV-FEL large camera system [2].

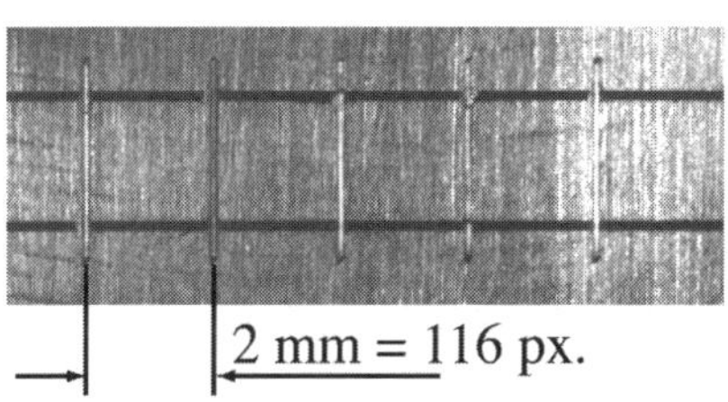

Figure 4. Image of the calibration screen used to check the resolution of the optical system.

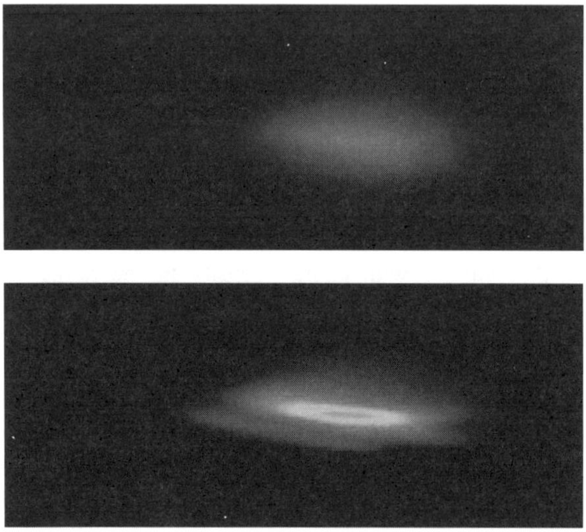

Figure 3. Electron beam imaging using Cr-oxide (top) and YAG-Ce doped (bottom) radiators. Electronic gain of the camera was approximately three times higher in the case of Cr-oxide screen to get comparable pixel values with the Ce:YAG.

2.1.3. *Motors and Actuators*

Motors and actuators are other important components of the SPARC e-meter. Linear actuators with stepper motors are used to control the insertion of screens and slits mask into the beamline. A differential encoder and a reference end switch guarantee reproducibility and accuracy of the movement, the latter being better than 2μm. This value is compatible with resolution needed for multi-shot measurement using a single slit to transversally scan the beam and collecting together the image from different positions.

More powerful brushless motors are used to move longitudinally the complete measurement system located between the two long bellows and to change the distance between the slits mask and the screen.

Absolute position of linear movements can be obtained by, or checked with, the value given by position transducer potentiometers.

The cross housing the two slits mask can be tilted around both x and y-axes to adjust the alignment of the slits with respect to the beam direction. Each stage is moved by a stepper motor having 200 steps per turn. μ-step movement is also possible (256 μ-steps per step).

Resolutions and accuracy of these movements have been tested and results obtained ($\Delta\theta < 2.5$ μrad) are compatible with the needed resolution of the rotational movement. Improvements are also possible by fine-tuning the assembly of mechanical components.

3. Operations at Pitz

The original design of the SPARC emittance meter has been partially modified during the construction to permit its installation in the Photo Injector Test Facility PITZ at DESY Zeuthen. The height of support was modified to allow installation on top of the PITZ girder. Furthermore we ensured mechanical compatibility between SPARC e-meter and PITZ vacuum beamline. Alignment tools and procedures have been jointly defined (Fig. 5).

The SPARC e-meter has been installed in the last section of the upgraded PITZ beamline, in the space located after the booster, before the electron beam spectrometer [3].

The SPARC e-meter has its own control and acquisition system. Two PCs have been installed in the PITZ control room to run control panels and measurements programs. Motors are controlled via CAN bus or RS232 serial interfaces. PCs running control panels communicate with motor controllers via network using a network serial-port server.

Installation of the SPARC e-meter was completed at beginning of July 2005 and commissioning started. Two rounds of commissioning and measurements shifts have been allocated for the SPARC e-meter in the PITZ shift plan for July and August 2005.

The first days of operations have been dedicated to the optimization of components and to fix minor problems.

In the last part of that two-weeks shift characterization of the e-meter with low energy beam started. We verified the reproducibility of measurements under

Figure 5. SPARC e-meter during installation at the Photo Injector Test Facility PITZ.

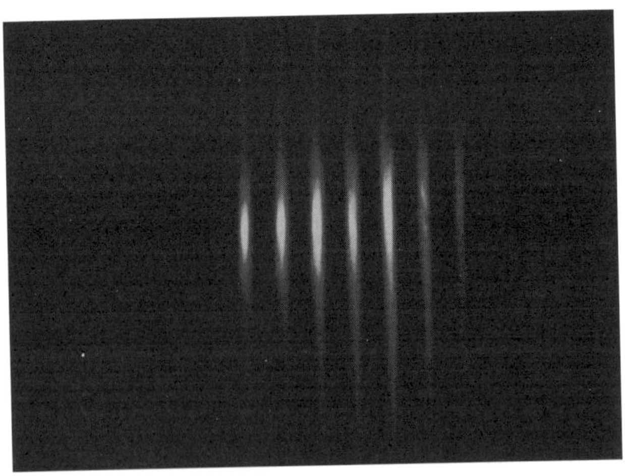

Figure 6: Picture at the YAG screen of the beam downstream the multi-slit mask.

different beam conditions and studied the strategies for the optimization of the measurements, e.g. adjust the distance between the slits mask and the screen

(Fig.6).

Although the transport of the beam is not yet optimized since commissioning of the upgraded PITZ beamline is still in progress, the measurements, yet preliminary, confirmed the value of transverse emittance expected and especially its dependence on the longitudinal position of the slits mask. As an example we report the measurements of the vertical emittance at three different positions along the e-meter (Fig.7). They show a constant growth of the emittance in the case with booster off and an almost constant value (with the measurement errors) for the higher energy beam accelerated by the booster.

4. Error analysis

Experience made with measurements at PITZ evidenced that a deeper look into error sources and their propagation is required. Since the purpose of this device is to measure emittance oscillations in the "*just after gun*" region, and the magnitude of these oscillations is expected to be small, we need very high accuracy on measured values. Moreover an accurate analysis of measurement errors, aiming their control and reduction is mandatory.

A first analysis shows two main kinds of error sources:

- instrumentation errors, i.e. mechanical tolerances, slit width, optical system's resolution and magnification, image digitalization, etc...
- errors due to possible beam fluctuations.

The first class of errors can be minimized increasing magnification up to 1:1 (but in this case the beam has to be small enough), improving image intensity resolution (bits per pixel), and taking into account, for each slit width, its measured value instead of the nominal one. We are now planning to improve the precision of the system in preparation for measurements on SPARC beam, reducing or removing all these instrumentation errors.

Being this a multi-shot measurement, the error due to either intensity and position fluctuation of the beam could be dominant. In fact by inserting the slit, any absolute reference with the beam is lost; so fluctuations can lead to explore different parts of the beam with the same slit from one shot to another, leading to big errors in measured parameters. One way to minimize these errors is to collect a large amount of data (images for each slit).

However to make good measurements, hard work on beam stabilization and optics has to be done. From RMS emittance formula qualitative error dependences from measured numbers errors and TWISS parameters are been found.

If one assumes that all relative errors of measured correlation matrix parameters can be considered as equal:

$$\Delta\langle x^2\rangle, \Delta\langle x'^2\rangle, \Delta\langle xx'\rangle = \delta\cdot\langle x^2\rangle, \delta\cdot\langle x'^2\rangle, \delta\cdot\langle xx'\rangle;$$

where δ takes into account instrumentation and fluctuation errors, what one can find by substitution is:

$$\varepsilon = \sqrt{\langle x^2\rangle\langle x'^2\rangle - \langle xx'\rangle^2};$$

$$\Delta\varepsilon = \sqrt{\frac{1}{(2\varepsilon)^2}\left[\left(\langle x^2\rangle\Delta\langle x^2\rangle\right)^2 + \left(\langle x'^2\rangle\Delta\langle x'^2\rangle\right)^2 + \left(2\langle xx'\rangle\Delta\langle xx'\rangle\right)^2\right]} = \ldots$$

$$\ldots = \varepsilon\delta\sqrt{\frac{1}{2} + \alpha^2 + \frac{3}{2}\alpha^4}$$

$$\boxed{\frac{\Delta\varepsilon}{\varepsilon} \propto \alpha^2\cdot\delta\,!!}$$

For typical values of $\delta \sim 0.02$ we need $|\alpha| < 3$ to have small enough errors. With these numbers in mind one can see from picture n.8 that in SPARC case we could have some problems measuring emittance at the beginning and at the end of E-meter region (gray rectangle).

Single shot multiple-slit measurements are a possible solution to this problem, minimizing errors from fluctuations: we'll find them directly on emittance values coming from same measurements in different times, but the uncertainty will not be amplified by propagation.

The bad side of multislit measurement is the fixed number of slits that can be used (7 in our case) with fixed distance from each other, and also the overlapping of signal from different slits which limits the distance between slits array and Yag optics, limiting the intrinsic resolution of system.

5. Conclusion

Construction of the SPARC Emittance-meter has been completed and it has been successfully commissioned at Photo Injector Test Facility PITZ in July 2005. Laboratory tests carried out at INFN-LNF before its shipment to DESY Zeuthen confirmed that expected performances have been achieved.

Although still preliminary, results of measurements of the emittance at the upgraded PITZ beamline have been obtained and they show a good agreement with those foreseen by simulations.

Operations with SPARC e-meter are in progress and they will continue until the device will be shipped back to Frascati, in early Fall 2005, to be used for the measurements at the SPARC photo injector.

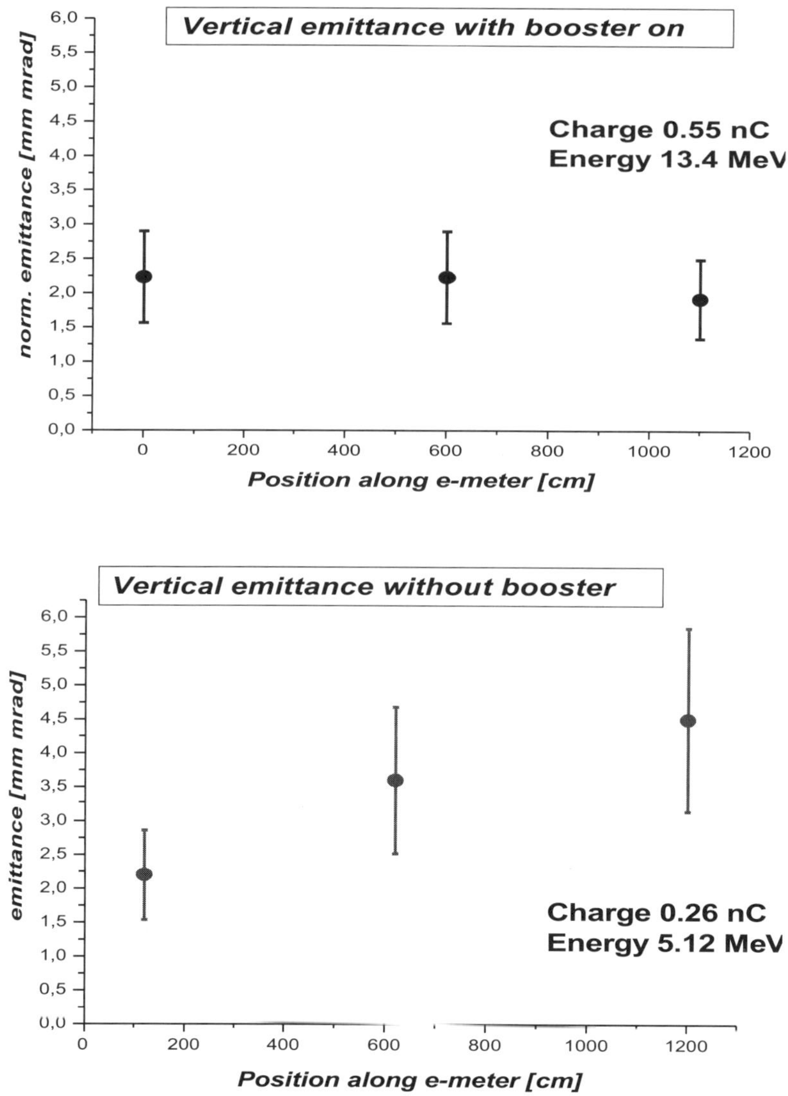

Figure 7: Preliminary result of vertical emittance measurement at PITZ with the SPARC e-meter under different conditions (booster off/on). Note that the three z-positions are not coincident in the two cases; also the beam charge is not identical.

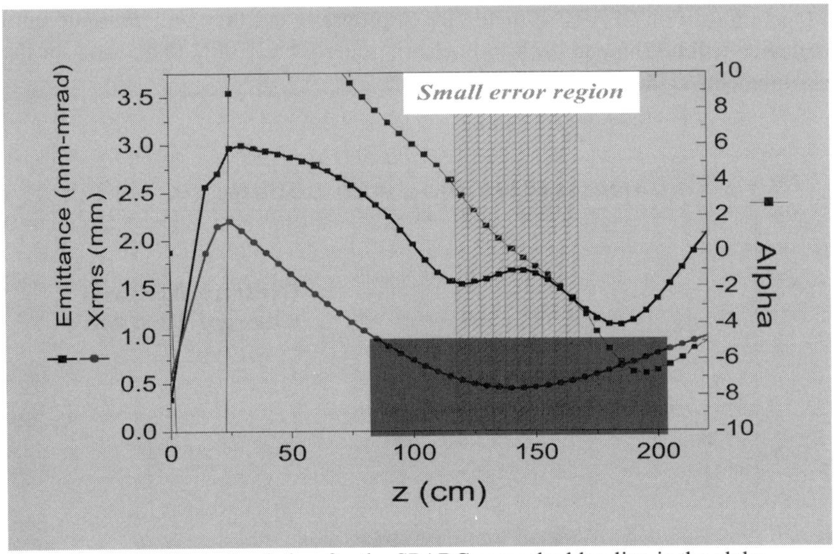

figure 8: PARMELA simulation for the SPARC case; the blue line is the alpha variation along first two meters of LINAC.

5.1.1. *Acknowledgements*

The authors whishes to thank V. Lollo, A.Clozza and C.Ligi for their contribution to the design and construction of the SPARC e-meter and F.Anelli and S. Fioravanti of INF-LNF for their work with motor controllers. We also thank L.V.Vu, A.Donat, F.Tonisch for the preparation and installation at PITZ, and J. Bähr, S. Khodyachykh, V. Miltchev, J. Rönsch and L. Staykov for the support during the shifts.

This work has been partially supported by the EU Commission in the sixth framework program, contract no. 011935 EUROFEL-DS1.

References

[1] A. Cianchi et.al., "Design study of a movable emittance meter device for the SPARC photoinjector", Proceedings of EPAC04, Lucern (Switzerland) 2004
[2] L. Catani et al., "A Large Distributed Digital Camera System for Accelerator Beam Diagnostics", Rev. Sci.Instr. 76, 073303 (2005)
[3] A.Oppelt et.al., "Status and First Results from the Upgraded PITZ Facility", FEL 2004 procedings.

EXPERIMENTAL TESTING OF DYNAMICALLY OPTIMIZED PHOTOELECTRON BEAMS*

J.B. ROSENZWEIG, A.M. COOK, M. DUNNING, R.J. ENGLAND

UCLA Dept. of Physics and Astronomy, 405 Hilgard Ave., Los Angeles, CA 90095, USA

P. MUSUMECI[†], M. BELLAVEGLIA[#], M. BOSCOLO[#], L. CATANI[#], A. CIANCHI[#], G. DI PIRRO[#], M. FERRARIO[#], D. FILLIPETTO[#], G. GATTI[#], L. PALUMBO[#], L. SERAFINI[£], C. VICARIO[#]

[#]*Istituto Nazionale di Fisica Nucleare, Laboratori Nazionale di Frascati, Frascati (RM) Italy,*

[†]*Istituto Nazionale di Fisica Nucleare, Sezione Roma 1 Roma (RM) Italy*

[£]*Istituto Nazionale di Fisica Nucleare, Sezione Milano Via Celoria Milano (MI) Italy*

Jet Propulsion Laboratory, California Institue of Technology, 4800 Oak Grove Drive Pasadena, CA 91101, USA

We discuss the design of and initial results from an experiment in space-charge dominated beam dynamics which explores a new regime of high-brightness electron beam generation at the SPARC (located at INFN-LNF, Frascati) photoinjector. The scheme under study employs the natural tendency in intense electron beams to configure themselves to produce a uniform density, giving a nearly ideal beam from the viewpoint of space charge-induced emittance. The experiments are aimed at testing the marriage of this idea with a related concept, emittance compensation, We show that the existing infrastructure at SPARC is nearly ideal for the proposed tests, and that this new regime of operating photoinjector may be the preferred method of obtaining highest brightness beams with lower energy spread. We discuss the design of the experiment, including developing of a novel time-dependent, aerogel-based imaging system. This system has been installed at SPARC, and first evidence for nearly uniformly filled ellipsoidal charge distributions recorded.

* This work is partially supported by the US Dept. of Energy under grant DE-FG03-92ER40693, the National Science Foundation under grant number, and by the EU Commission in the 6th framework program, contract no. 011935 EUROFEL-DS1.

1. Introduction

In order to obtain the highest brightness electron beams from photoinjectors, it is most common to rely on the *emittance compensation* process[1]. Optimization of this process demands that the transverse fields be as uniform, and linear (in radius *r*) as possible. The existing theoretical and experimental studies of emittance compensation have, to that end, assumed use of a uniform density electron beam, having a cylindrical shape. However, this shape produces space-charge fields near the beam head and tail that have pronounced nonlinear dependences on the spatial coordinates. These nonlinearities result in both transverse and longitudinal emittance growth.

It has been known for some time[2], however, that a uniform ellipsoidal density distribution yields space-charge fields that are linear in all dimensions (e.g. $E_x \propto x$, $E_z \propto z$). Under such conditions, it is conceivable that one may obtain essentially emittance growth-free dynamics. How to produce such a distribution has, until recently, remained an unanswered question.

In 1997, Serafini proposed the dynamic creation of an ellipsoidal bunch by launching an ultra-short, radially shaped beam[3], which then evolves through longitudinal expansion of differing radii in the beam to achieve the desired longitudinal shape. In this work, a 10's of femtosecond laser pulse with uniform time profile was assumed, which is not technically feasible — pulses this short are now a routine capability of the photocathode drive lasers, but not with such a restrictive profile. On the other had, it has recently been shown by Luiten, *et al.*,[4] that in obtaining the correct final ellipsoidal distribution, there is essentially no requirement on the shape of the initial laser pulse other than it be ultra-short (length τ_l much shorter than eventual beam length after space charge expansion). Thus such laser pulses are a natural, and technically achievable way of producing an ellipsoidal-shaped, nearly uniform density beam.

As the beam dynamics just after photoemission are qualitatively different in the traditional emittance compensation scenario and in the Luiten-Serafini scheme, it is not immediately apparent that one may successfully combine the two. The UCLA-SPARC collaboration has recently shown[5] that this marriage is indeed possible; further, the combination emittance compensation and dynamic creation of the ellipsoidal shaped beam produces results that in many ways are superior to those obtained in state-of-the-art designs. As the bunches produced are shorter than in such standard cases, very high brightness beam creation was shown to be possible.

The basic idea behind the Luiten-Serafini scheme is simple: the beam profile expands and deforms longitudinally to produce, in the final state, a uniformly filled ellipsoid of charge. In the process, phase space rearrangements occur which degrade the emittances — especially in the longitudinal dimension. In order to understand this process, to specify experimental requirements, and to identify experimental signatures associated with the process, we have analyzed the dynamics of space-charge-dominated beam expansion[5].

This analysis may be summarized in a few points:

First, the injected bunch surface charge density $\sigma_b = dQ_b / dA$ must not be too high, or image charge effects at the cathode distort the final pulse profile so that it is not ellipsoidal. This is quantified by the condition $\alpha \equiv 4\pi\sigma_b / E_0 << 1$.

Second, the beam must be much shorter than its eventual size in order to be able to ignore the details of the initial pulse profile, which is not a serious constraint. In practice 300 fs laser pulses (typical of the limitations of the SPARC photocathode drive laser after conversion to UV) excite roughly the same length electron bunch, which expands to around 4 psec in our example cases. The pulse length after expansion is estimated as $L_b \approx 2\pi\sigma_b m_e c^2 / E_0^2$.

The current density that is achieved after expansion, is $J_z = eE_0^2 / 4\pi m_e c$, a constant dependent only on the applied electric field E_0. All beams become uniform in density. To achieve the desired *ellipsoidal* beam shape, one must choose the initial surface current density distribution correctly, which implies that $\sigma_b(r) = \left(3Q / 2\pi a^2\right)\sqrt{1-(r/a)^2}$.

Several phenomena that do not occur in a standard geometry ($\sigma_b(r)$ =constant) are apparent from this analysis. In the Luiten-Serafini scheme, pulse length expansion is required, while in the standard scenario it is avoided. During this expansion, one has mixing of electrons between slices. This causes an initial fast increase of the longitudinal emittance, which is terminated by the transition of the space-charge field direction from mainly longitudinal to predominantly transverse as the beam accelerates. The "missing" region of transverse space-charge also differentiates this scenario — the geometry of the injected beam in the standard configuration has a length much longer than the radius, and transverse space-charge forces assert themselves nearly immediately, within a propagation length approximately equal to the beam radius.

While the analysis of the beam dynamics is useful, the central issue of joining this regime with emittance compensation must be explored with simulations. The initial simulations begun in Ref. 5 we have performed are in the context of the SPARC scenario, so that we may proceed directly to discussing the experimental tests of this new regime — now commonly known as the "blowout regime" — of the photoinjector there.

2. Blowout regime with emittance compensation: general study

We have performed initial UCLA PARMELA[6] simulations to explore the joining the Serafini-Luiten scheme with the optimized emittance compensation working point, of the SPARC injector at LNF. We assume that the gun (1.6 cell, 2856 MHz) and solenoid are the same, and run near to the standard conditions. Through trials, we have optimized the launch conditions of the beam. In order to have values of α which do not give excessive image charge effects, the beam charge is lowered and the beam radius is slightly enlarged. In the preliminary optimization, we launch a 0.33 nC beam with an initial longitudinal Gaussian distribution having σ_t =33 fs beam, and a radial Gaussian with σ_x =0.77 mm (cutoff at 1.8 σ). The gun is run with a peak on-axis gradient of 120 MV/m; the beam is launched at 33 degrees forward of crest. This is advanced in comparison to the nominal launch phase for a standard bunch, and serves to control the excessive beam energy-spread after the gun. The emittance compensation solenoid is run with peak field B_z= 2700 G, which is slightly below the standard scenario, as the beam has slightly lower energy exiting the gun. We note that the peak value of α in our case is 0.11, as opposed to 0.42 in the LCLS design.

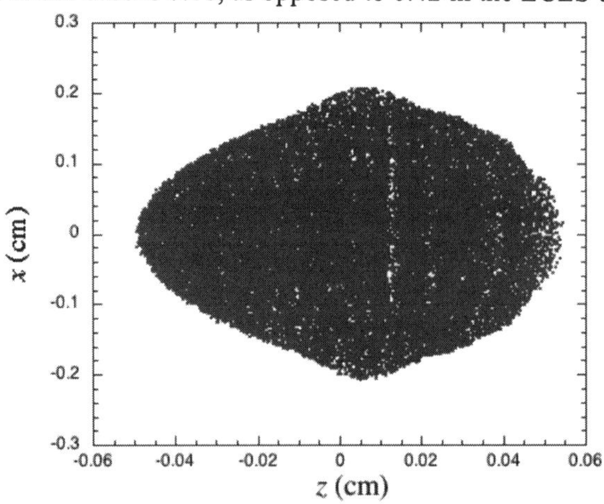

Figure 1. PARMELA simulation results, showing e- bunch (x,z) distribution 133 cm from cathode (6.3 MeV energy), before injection into first linac section, showing ellipsoidal beam boundary.

There is of course an initial transverse emittance growth which occurs during the reconfiguration of the bunch charge near the cathode, and subsequent growth which may occur due to the imperfections in the quasi-ellipsoidal distribution that is formed. It is these effects that are addressed by the emittance compensation process. Emittance compensation is accomplished in two steps: the focusing of the beam by the post-gun solenoid, and the matching of the beam in the first traveling wave linac section (3 m long, SLAC-type, 13.5 MV/m average acceleration), which has a 560 G solenoid field overlaid on it.

The formation of the quasi-ellipsoidal bunch is clearly shown in Fig. 1, which displays the bunch (x, z) distribution at a point 133 cm from the cathode, in the drift space after the gun and just preceding initial traveling wave linac section. Here the beam has 6.3 MeV mean energy, and its transverse dynamics are space charge-dominated. Thus one sees clearly the "inflated" ellipsoidal beam shape. As this shape is obtained purely through space-charge effects, the 6-dimensional transverse phase space is indeed close to the ideal Kapchinskii-Vladimirskii distribution. The final bunch length is 1.3 mm full width, corresponding to a peak current of 105 A. Thus even with one-third of the charge, this scheme should produce a higher current than obtained in simulations of the standard design.

Two notable defects are seen in the beam shape in Fig. 2. The first is the extension of the half-ellipsoid in the trailing part of the bunch as compared with the initial half. This asymmetry is caused by image-charge effects. This non-ideal behavior in fact gives the limit on α; when one attempts to launch a higher surface charge density, the bunch deformation from the desired symmetric ellipsoid produces poor emittance performance. The second notable feature is the existence of an anomalous ring at the outer radial edge of the beam. This part of the beam has low surface charge density and experiences radially fringing fields due to its edge location. Because of these effects, it does not experience enough longitudinal expansion to keep pace with the rest of the bunch, but instead has a moderate amount of radial expansion.

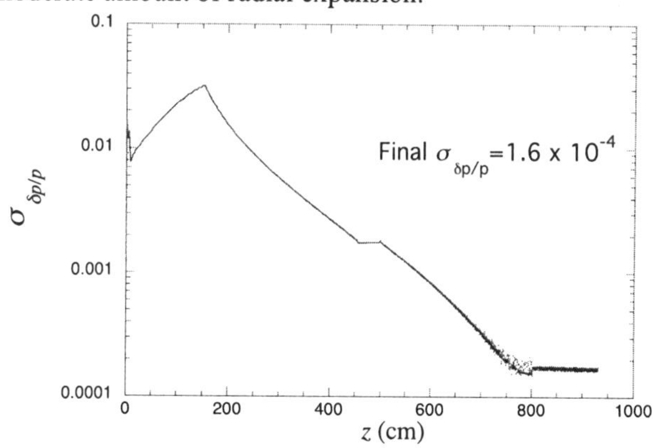

Figure 2. Evolution of $\sigma_{\delta p/p}$ in z for emittance compensation case, from PARMELA simulation.

As the longitudinal space charge during much of the acceleration is linear, and pulse length T is short, the longitudinal phase space is very compact. The evolution of the relative momentum spread $\sigma_{\delta p/p}$ in z is shown in Fig. 2. The final achieved rms value is $\sigma_{\delta p/p} = 1.6 \cdot 10^{-4}$, which is an order of magnitude smaller than that obtained in the standard LCLS-type (or SPARC-type) design.

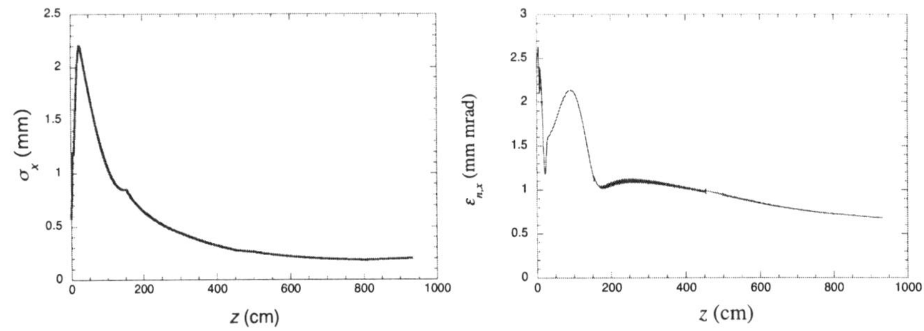

Figure 3. (left) The evolution of rms transverse beam size σ_x for emittance compensation case, fromPARMELA simulation. (right) Evolution of rms normalized emittance $\varepsilon_{n,x}$ for emittance compensation case.

The evolution of the rms transverse beam size σ_x, and the rms normalized emittance $\varepsilon_{n,x}$ are shown in Figs. 3. While the behavior of σ_x is similar in most respects to the standard design, with the beam approximately matched at linac entrance to the invariant envelope, the emittance behavior is not as familiar. In the standard LCLS design, $\varepsilon_{n,x}$ achieves a minimum value in the post-gun drift, rising to a local maximum at injection into the linac. The focusing and adiabatic damping of the motion in the linac then produce a monotonic decrease of $\varepsilon_{n,x}$ in z. In our case, the transverse space-charge and thus the plasma/emittance oscillations do not "turn on" until after the longitudinal expansion is well underway, thus delaying the emittance minimum in Fig. 3 to occur inside of the linac. In order to produce faster emittance oscillations in the linac to strongly diminish $\varepsilon_{n,x}$ before acceleration removes the plasma-dominated beam behavior, the solenoid field in the first linac section has been raised by 40% relative to the standard scenario. This ploy works well, as the final value (still slightly decreasing) of $\varepsilon_{n,x}$ at the end of the second linac (84.5 MeV energy) is 0.68 mm mrad. The thermal emittance at the cathode is 0.4 mm mrad, and so the space-charge induced emittance is well compensated.

After acceleration to higher energy (84.5 MeV), the beam is not space-charge dominated, and the (x,z) profile no longer ellipsoidal, as shown in Fig. 4. Nonetheless, the beam has excellent emittance, and maintains a current profile with shape $I(t) \propto \sqrt{1-(2t/T)^2}$. With a high initial current, and low intrinsic energy spread, this beam may be compressed further, with very high final peak current achievable. In Fig. 5, we show the resulting longitudinal phase space calculated by a further simulation, using Elegant[7] (with input obtained from PARMELA output), of post-acceleration running forward of crest, and then encountering a chicane. The distribution shown has a final rms bunch length $\sigma_z = 11\ \mu\text{m}$ (σ_t = 37 fsec), with a peak current of 4.5 kA. This beam, which has only 0.4% rms momentum spread, has obvious utility in ultra-short pulse FEL or inverse-Compton scattering experiments at SPARC.

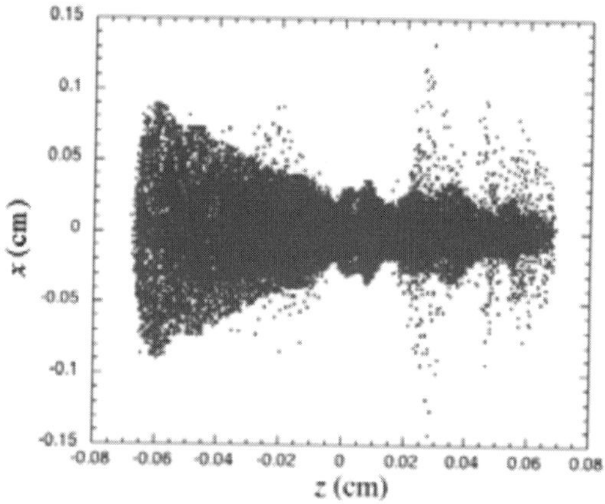

Figure 4. Electron bunch (x,z) distribution after second linac section (84.5 MeV energy), with ellipsoidal beam shape no longer apparent, from PARMELA simulation.

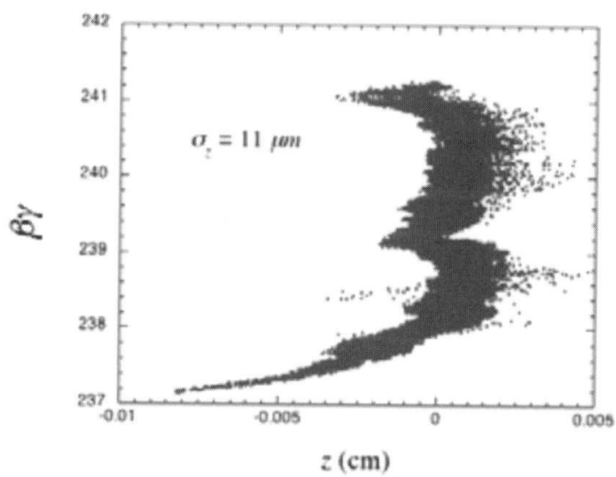

Figure 5. Longitudinal phase space after third (off-crest) linac section and chicane, showing compression of pulse to $\sigma_z = 11$ μm, from Elegant/PARMELA simulation.

3. Considerations for SPARC experiment

Several experimental scenarios have been investigated for possible tests of this new "blowout" regime of photoinjector operation, all at UCLA collaborating institutions. These laboratories include the PLEIADES injector, the ORION injector, and the SPARC injector. All have approximately the same gun design (fabricated at UCLA), and have traveling wave post acceleration linacs with solenoid focusing overlaid. All of these injectors possess lasers with ~100 fs pulse capability, and are adequate for studying the physics of this regime. The PLEIADES injector has been placed on hiatus. The ORION injector is a bit different than what we have shown above, in that it employs X-band linac sections; this is a notable but not serious difference, and thus one may consider its use. Finally, we note that the SPARC injector is the example we have employed here, and is thus apparently nearly optimum for exploring the physics illustrated in the preliminary simulations above.

3.1. *Laser and photocathode issues*

Before discussing the planned electron beam measurements, we first review some experimental considerations specific to the use of such short lasers. First, we note that it should be tailored, either by collimating and relay imaging ("cut-Gaussian", as in the simulations); even better, more sophisticated manipulations of the laser intensity may be accomplished by use of a deformable mirror[8]. Provisions need to be made for having a high resolution "virtual cathode" to monitor the laser intensity profile at the cathode.

The beam must also be fairly short. In first measurements at SPARC, we have found, through cross-correlation measurements, that the UV pulse is difficult to make shorter than 300 fs using 3rd harmonic generation with noncollinear mixing in the conversion crystals. In order to obtain even this result, we must give up UV energy, going from 1.5 mJ to 0.2 mJ. While this energy is still adequate for obtaining 0.33 nC of charge using a metal cathode (Cu or Mg), to go shorter may be difficult. In order to check the effect of this extra pulse length on the scheme, we have performed simulations analogous to the original exploratory calculation. Indications are that the added length in the laser pulse does not affect the final electron beam configuration at this level.

The laser transport in this case is also complicated by illumination at 70 degrees. The needed time-transverse correlation on the laser pulse is introduced by a blazed grating. This procedure should be accurately implemented for this experiment in particular, as small time errors may be important. Also, it should be noted that the grating is a dispersive element, and the effect of having a large laser bandwidth should also be examined.

The laser intensity needed for this scenario is a factor of 30-100 higher than in the LCLS case. The issue of laser damage at both the input window and the cathode surface has been examined, and found based on previous experience at UCLA and at the BNL ATF to not be worrisome. A large laser field at the cathode can also produce a previously unanticipated effect, that of "heating" the electrons through the induced wiggling motion. The degree to which the laser field causes emittance growth can be estimated as

$$\varepsilon_n \cong a_l \sigma_x \cong \frac{\lambda_l}{2\pi m_e c^2}\sqrt{Z_0 P_l}\,. \tag{1}$$

Here a_l is the peak normalized vector potential of the laser field, and P_l is the peak laser power (note the independence of the beam size σ_x). For the SPARC experiment we estimate that $\varepsilon_n \approx 0.04$ mm-mrad, so this effect, while physically interesting, is negligible in our case.

One must also examine the issue of cathode time response. For a metal such as Cu, the laser penetrates only a conduction electron plasma skin-depth $l_d \cong k_p^{-1} \cong \left[4\pi r_e n_e\right]^{-1/2} \approx 50$ nm. An excited photoelectron born in this thin region must leave essentially normal to the surface, and travels with velocity $v \cong c\left[2h\nu_l / m_e c^2\right]^{1/2} \approx 4.2\cdot 10^{-3} c$. The time constant associated with the escape of the photoelectrons is thus $\tau_e \cong l_d / v_e \approx 40$ fs, so we should not have significant lengthening in this experiment. On the other hand, L-band photoinjectors with lower gradients (e.g. TTF) typically use Cs_2Te high quantum efficiency (*QE*>1%) cathodes. The higher *QE* arises from the scattering of the photoelectrons off phonons in the semiconducting material, and thus the emission process may be one to two orders of magnitude longer. This must be kept in mind before considering extending this scheme to an L-band scenario.

3.2. *Experimental signatures and measurements*

The experimental signatures that one looks for in tests of this photoinjector operating regime may be delineated. In general terms, the complete compensation scheme presented will show good emittance performance, along with a higher current and shorter pulses. In addition, at low energy, the beam will have an ellipsoidal shape. This shape may be viewed trivially in *z*-projection by a standard profile monitor (scintillating crystal, optical transition radiator, etc.). In terms of the longitudinal profile, one may consider use of a streak camera, to observe the time dependence of the current, and spatially resolve the transverse direction, thus measuring for example, a (x,z) slice of the beam, which should give a uniformly filled ellipse. By scanning this slice in x position, one may reconstruct the entire ellipsoid. Streak cameras may have time resolution as low as 0.25 ps (in practice it may be a bit larger), which is adequate to resolve the beam in our example, which is longer than 4 ps full width. The *z*-projected transverse phase space (in one dimension) may be investigated at low energy using the multi-slit technique.

At higher energy, one may observe the final state of well-compensated emittance through quadrupole scanning[9] or transverse phase space tomography[10]. The ellipsoidal beam is not observable at this energy, as the beam transverse distribution is emittance, and not space-charge dominated, as shown in Fig. 4. The longitudinal distribution can be time-resolved at higher energy at the SPARC injector[11] using a fast RF sweeper with 30 fs resolution. One may also use longitudinal phase space tomography to observe the higher quality longitudinal phase space. In addition, at SPARC one may use a downstream compressor to investigate compression to the ultra-short bunch length illustrated in Fig. 5. This bunch length presents challenges in measurement, stressing both coherent radiation techniques, and RF deflectors.

In the SPARC experiment, we plan first to image the beam (time-integrated) at low energy (5-7 MeV) in the region after the gun, using a YAG detector. For time resolved measurements we will first convert the beam spatial information to photons with a prompt emitter. As transition radiation gives too weak of a signal at this energy, we will use a Cerenkov convertor (shown in Fig. 6). In order to have a manageably small-angle of emission we use aerogel, which has small index (n=1.005-1.02). At 5 MeV the Cerenkov emission threshold is reached for n=1.005; we may choose angles of emission from near zero at this threshold up to 9 degrees with the aerogels that are presently in-hand. The aerogels have been custom fabricated at the Jet Propulson Laboratory, using as a guide the mass/index relation $n = 1 + 2.1 \times 10^{-4} \cdot \rho\left(\mathrm{mg/cm^3}\right)$.

Figure 6. Cerenkov converter vacuum-tight holder, for containment of aerogel, with upstream entrance foil and downstream quartz window.

We presently are following the protocol of protecting the SPARC vacuum from out-gassing by the aerogel, developing an aerogel container as shown in Fig. 6. We have also studied the optical and vacuum properties of aerogel in order to optimize and simplify the design. The optical properties examined include the verification of the optical index and degree of scattering using green laser light (Fig. 7). These tests determined that the indices were as expected, and the scattering angles associated with green light were within the spreads already present due to Cerenkov and (scattered) beam angles.

At the same time, a program has been developed at UCLA to calculate, using a Monte Carlo approach the imaging and temporal resolution properties of the Cerenkov light from creation through transport to a final detector. As Cerenkov light has large angular spread (also enhanced by the entrance foil, with between 1 and 5 additional degrees of divergence from multiple scattering), and broad wavelength spectrum, these properties must be examined carefully to ensure that the measurement will give the desired results.

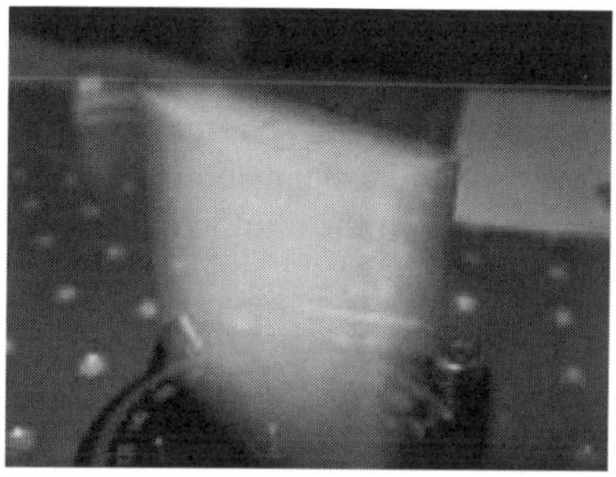

Figure 7. Testing of aerogel at UCLA, laser refraction determination of optical index.

The first step of the simulation is to provide electrons (typically 40,000) from PARMELA to GEANT, which simulates the scattering of the electrons in the entrance foil and generates a collection of Cerenkov photons in the aerogel. Optical transition radiation from the foil (weak) and very wide angle Cerenkov light generated in the output window (uncollected) are neglected in this simulation. The simulated distributions are filtered in wavelength — in practice this is accomplished with bandpass filtering in the downstream transport — and terminated at the output window. The photon distributions that result are then passed to a Mathematica-based, optical ray-tracing program, Rayica. Examples of these distributions are shown in Figs. 8 and 9.

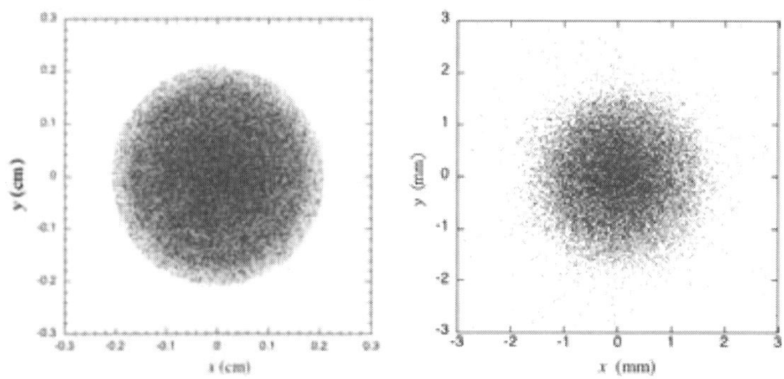

Figure 8. (Left) Electron beam distribution at aerogel holder input, from PARMELA; (right) photon distribution from GEANT at aerogel exit.

The GEANT simulations served to benchmark the analytical approximation to the number of photons obtained within a certain bandwidth $\Delta\omega/\omega$, that is $N_\gamma / N_{e-} \cong \alpha(2\pi/\lambda)\left[1-(\beta n)^{-2}\right](\Delta\omega/\omega)$, where α is the fine-structure constant. In Fig. 8, the transverse (*x*-*y*) distributions of electrons from PARMELA at the aerogel holder entrance and the analogous distribution of band-passed Cerenkov photons from GEANT are given. It is seen that the electron beam distribution is smeared by the "depth-of-field" of the photons, emitted at large angles, and at depths in the aerogel as large as 3.5 mm. To illustrate the full effect of this smearing on a time-resolved, transverse imaging measurement of the beam, we show in Fig. 9 the *t-r* distribution of photons at the aerogel exit. A clear degradation of the nearly ellipsoidal beam boundary that was displayed in the electron beam distribution is seen in the photon distribution.

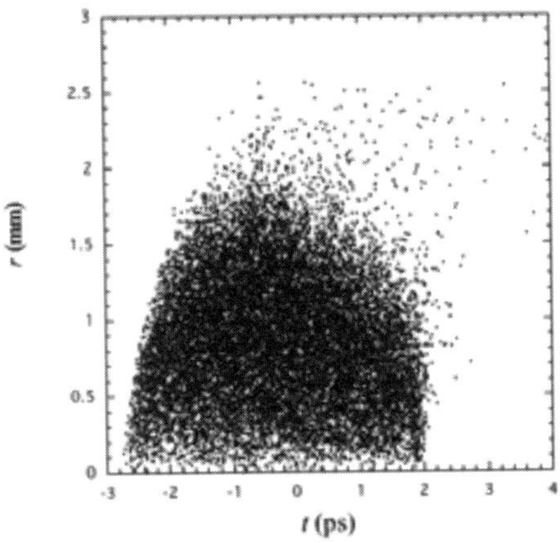

Figure 9. Photon distribution in radial position vs. time, from GEANT at aerogel exit.

In order to anticipate the quality of a time-resolved, transverse imaging measurement obtained with such Cerenkov photons, their transport to a detector has been simulated using Rayica. The transport system consists of an objective and focusing lens, with a field lens placed at the intermediate focal point located in between these lenses. The field lens serves to surpress "vignetting" of the imaged photons due to the simultaneous presence of large angles and a large initial field of Cerenkov photons.

Figure 10. Optical transport line, with 10 % bandpass filter centered on 500 nm. Aerogel index of refraction n = 1.018.

Two examples of Rayica simulations are displayed in Figs. 10 and 11. In Fig. 10, we show the initially considered scenario, concentrating on the visible (500 nm) spectrum, with a relatively large index (n=1.018) and 10% bandwidth. This case was chosen in order to have a large photon yield, but had two unfortunate attributes: the photons have such a large angle that they are collimated by the output window (as seen in Fig. 10), and the temporal dispersion is unacceptable large. The first effect is mitigated by obtaining smaller n aerogel, which has been enabled by JPL. Smaller n also ameliorates the second issue, which is further improved by working at longer wavelength. Such a case is shown in Fig. 11, which displays a simulation considering 800 nm central wavelength ($\Delta\omega/\omega$=5%), and n=1.0084. The results of a series of simulations at this preferred central wavelength is given in Table 1.

Figure 11. Optical transport line, with 5 % bandpass filter centered on 800 nm. Aerogel index of refraction n = 1.0084.

Table 1. Summary of results obtained from GEANT-Rayica Cerenkov creation and transport simulations, 40,000 electrons, 5% bandwidth centered on 800 nm.

Index of refraction	Number of photons	Surviving fraction at detector
1.00525	528	92.23 %
1.0063	1109	89.54 %
1.00735	1674	86.49 %
1.0084	2259	84.10 %

It should also be noted that recent tests of aerogel performance during vacuum pump-down at UCLA have indicated that quite good vacuum levels (few 10^{-8} Torr) are achievable at the pump nearest near the aerogel. Further, the aerogel is undamaged when pump-down proceeds slowly. The excellent pumping characteristics of aerogel, which has all voids simply connected, are well known from its use in space exploration missions. We can therefore consider, subject to further testing, the elimination of both the upstream foil and the downstream window. This would serve to clean up multiple scattering and spurious OTR and Cerenkov production from the initial photon distribution.

At SPARC there is a streak camera that is being reconditioned to have 2 ps FWHM time resolution, which is on the border of resolving for the bunch length of interest. This is handled in initial measurements by use of higher charges (above 1 nC), and thus longer beams, to test the longitudinal expansion dynamics of the beam.

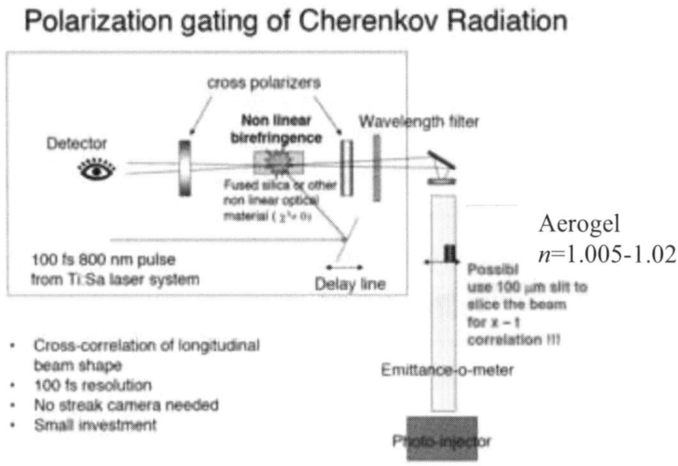

Figure 12. Schematic of multi-shot scanning, 100 fs resolution system based on polarization gating of nonlinear crystal. Horizontal collimation using slits allows scanning of transverse profile.

We must also consider alternative schemes based on ultra-fast gating. One such idea arising out of the SPARC program is summarized in Fig. 12, in which the Cerenkov light, appropriately limited in bandwidth, is injected into cross-polarizers, which nominally extinguish the light. Use of a crystal with nonlinear birefringence allows partial passage of the Cerenkov light when it is coincident with a strong gating laser pulse. This pulse will be derived from the 100 fs, 800 nm photocathode drive laser before frequency tripling.

It is instructive to review the polarization gating cross-correlation method in detail. The probe pulse is sent through crossed polarizers and a gate pulse is prepared with a polarization that is oriented at 45 degrees with respect to that of the probe pulse. The two pulses are then spatially overlapped in a piece of fused silica. In the fused silica, the gate pulse induces a birefringence through the electronic Kerr effect, a third-order optical nonlinearity, also known as the nonlinear refractive index. As a result, the fused silica acts as a wave plate while the gate pulse is present, rotating the probe pulse's polarization slightly, which allows some light to be transmitted through the analyzer. Because birefringence only occurs when the gate pulse is present, this geometry yields a cross-correlation of the probe pulse, if one measures the light energy transmitted through the analyzer as a function of optical delay between the two pulses.

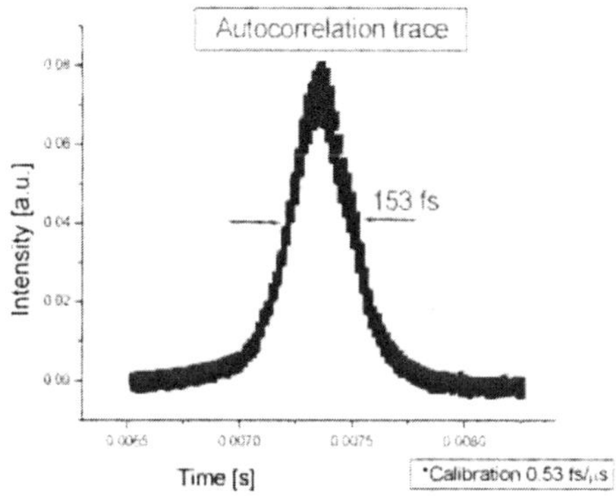

Figure 13. Example of autocorrelation of 100 fs IR beam in SPARC photoinjector laser system.

The amount of light transmitted depends on the intensity of the gate pulse following the relation $T = \sin(2.34\pi n_2 I(L/\lambda))$, where L is the length of the nonlinear material, n_2 is the nonlinear index of refraction, λ is the central wave-length of the filtered Cerenkov radiation and I is the intensity of the gate pulse. The appropriate level of energy in the gate pulse is easily obtained in the SPARC case by focusing a portion of the infrared beam that is the residual of the harmonic conversion. Further, we note that the harmonic conversion is done at short pulse length in this experiment. Such beam is very short and intense (> 20 mJ) and there are problems is transporting it to the interaction region g, as well as in converting the IR to UV without too much pulse lengthening.

The initial laser pulse length must be monitored in the experiment, both in IR — to know the resolution of the gating measurement — and in the UV, to obtain the initial conditions of the photoelectron beam at the cathode. In order to illustrate the technique of autocorrelation needed to measure laser pulse lengths of pulses obtained already in the SPARC laser system, we show in Fig. 13 an autocorrelation of the laser in the IR. This test gave an autocorrelation of 150 fs, which corresponds to a pulse length (FWHM) of 100 fs.

As it is envisioned, the gating measurement is based on scanning with a delay line, and is thus a multi-shot (~100) technique. If we further desire to scan the transverse distribution by collimating the electron beam with a slit, the number of data points to construct one (x,t) profile is quite large. Thus, even though the polarization gating scheme is inexpensive and promises good resolution, thought must still be given to a single-shot measurement. At SPARC, a single shot measurement, based on the RF deflector, will be available only at high energy, where the ellipsoidal beam is not observable. The 10 fs resolution of this time measurement should provide details of the dynamics not available through other methods, however.

3.3. *Recent simulation work*

The preliminary UCLA PARMELA simulations discussed above closely approximate the SPARC scenario, but do not provide an exact model yet for the experiments we propose. To this end, further simulations using UCLA PARMELA, Los Alamos PARMELA (at INFN-LNF), and TREDI (at ENEA) are being performed at present. We show the results of new simulations of the reference design from UCLA PARMELA in Figs. 14 and 15.

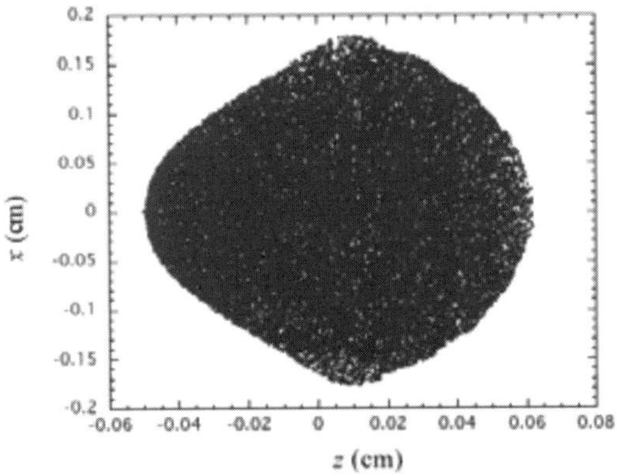

Figure 14. PARMELA simulation results, showing electron bunch (x,z) distribution 133 cm from cathode for case with 310 fs FWHM pulse length, taking into account possible lengthening mechanisms in harmonic crystal and cathode response.

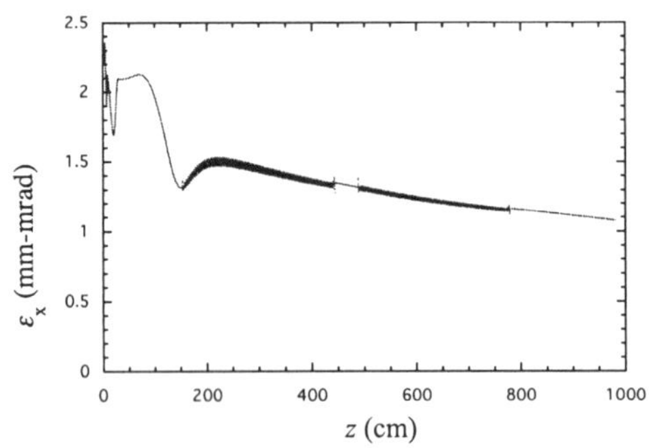

Figure 15. Emittance evolution for case shown in Fig. 14.

The first major difference introduced is the assumed lengthening of the injected beam, due to harmonic crystal and cathode response times, to 310 fs FWHM (4 times the case simulated above). As can be seen in Fig. 14, the overall ellipsoidal shape (shown at z=133 cm) is not notably changed from that shown in Fig. 1. The emittance compensation performance (Fig. 15) is acceptable, if a bit less effective than in the shorter launch beam case. We note however, that the working point has not been reoptimized in this case, as linac and solenoid settings have not been readjusted. Further optimization studies are being done to understand the baseline design for the experiment, as well as tuning/jitter performance curves for *e.g.* solenoid strength, linac phases, injection jitter, etc.

Further computational investigations planned include examination of velocity bunching using PARMELA; TREDI will also be used to model the effects of cathode emission nonuniformities and other 3D phenomena. It is also of crucial importance to evaluate the performance of a SASE FEL driven by dynamically optimized beams at SPARC.

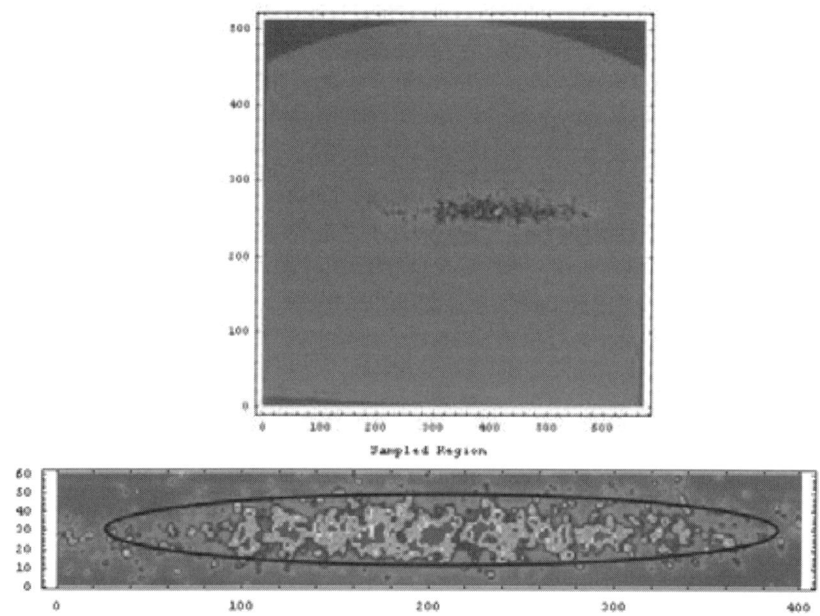

Figure 16. (top) Streak image after image correction from SPARC experiment, 150 psec total window in vertical dimension; (bottom) area of interest in streak image, with elliptical contour shown for analysis purposes.

4. First results

The first stage of experimentation on the blowout regime took place at LNF beginning at the end of March, 2006. These took place in the context of the commissioning of the RF photocathode gun alone. Upon startup, the UCLA-produced gun was conditioned quickly up to 11 MW, which produces 110 MV/m peak electric field, and 5.7 MeV electron beam. Here we describe preliminary measurements made with this system

With the laser reconfigured for short pulse (less than 0.5 psec FWHM), up to 1.6 nC of charge. While the laser was set to approximately the correct transverse size and profile shape, there remained considerable spatial fluctuations in both the laser and cathode quantum efficiency. Thus the conditions for observing the dynamic creation of nearly uniformly filled ellipsoidal charge distributions were not quite present; in fact, the emittance was not of equivalent quality to that obtained in standard operation. Nevertheless, impressive first data were obtained.

Initial measurements of the beam's longitudinal-transverse profile were made with aerogel with index n=1.008, with the Cerenkov radiator placed 2.4 m away from the cathode, downstream of the slit-based emittance measurement system. Streak camera images were obtained using the transport system described in the previous section. A complication arose from the correction applied to the streak image which corrects for the aberrations in the streak tube itself. A time-transverse correlation was introduced by this correction scheme which then had to be removed in analysis. Such a streak, after correction, is shown in Figure 16. This image displays the profile obtained from a bunch with charge of 700 pC. A large charge is preferred in this case in order to discern information at a time scale longer over the streak camera resolution; the 700 pC case is expected to have expansion of approximately 7 psec FWHM, well in excess of this 2 psec FHWM resolution.

Streak images obtained in the highest temporal resolution mode are inherently noisy; this condition is required in order to avoid space-charge induced pulse distortion inside of the streak tube. Thus in order to extract information from single shots concerning the streak image — which should represent the beam density distribution in an x-z slice in the midplane of the bunch — we have adopted a maximum likelihood analysis to test for different assumed types of beam distributions.

The x-z slice distributions we have tested for consistency with the data include: (1) a bi-Gaussian (thermal-type) distribution; (2) a uniformly filled ellipse (assumed arising from a parent uniformly filled ellipsoid); and (3) a nearly uniformly filled ellipse with a tail, which we choose to represent as a Fermi-Dirac distribution.

As all of the distributions assumed have contours of constant density that are elliptical, a systematic statistical approach is possible, in which we look at the total integrated intensity inside of ellipses of size varying from zero area to an area covering the entire streak image. These ellipses, an example of which is shown in Fig. 16, are all required to have the same aspect ratio, which is given by the intensity profile itself, $R = \sigma_x / v_s \sigma_t$ (v_s is the streak velocity, and σ_t = 3.45 psec for the streak in Fig. 16).

The bi-Gaussian distribution is given by

$$f(x,t) = f_0 \exp\left[-\frac{1}{2}\left(\left(\frac{x}{\sigma_x}\right)^2 + \left(\frac{v_s t}{v_s \sigma_t}\right)^2\right)\right] = f_0 \exp\left[-\frac{1}{2}A^2\right], \tag{2}$$

with normalized integrated signal inside of the contour of amplitude A

$$I_1(A) = 1 - \exp\left[-\frac{1}{2}A^2\right]. \tag{3}$$

Likewise, for a uniformly filled ellipse, one finds that the integrated signal inside of a contour of amplitude A is

$$I_2(A) = \begin{cases} A^2, & A \le 1 \\ 1, & A > 1 \end{cases}. \tag{4}$$

Finally, for a Fermi-Dirac function in which the density profile is assumed to have amplitude dependence $\left[\exp(\alpha(A - A_0)) + 1\right]^{-1}$, the integrated signal inside of a contour of amplitude A is

$$I_3(A) = A^2 - \frac{2}{\alpha^2}\ln\left[\frac{1+\exp(\alpha(A - A_0))}{1+\exp(-\alpha A_0)}\right] - \frac{2}{\alpha^2}\left[\text{Poly}\log\left[2, -\frac{\exp(\alpha(A - A_0))}{\exp(-\alpha A_0)}\right]\right]. \tag{5}$$

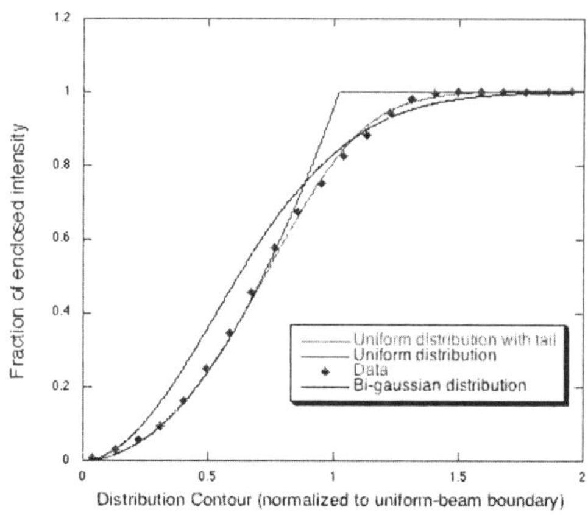

Figure 17. Analysis of streak data, with fraction of integrated intensity of data inside of elliptical contour shown. Best fit of data points to three models are shown: bi-gaussian distribution, uniform elliptical distribution, and Fermi-Dirac (uniform with tails) distribution.

With these functions in hand, we can fit to the data given in the streak images to determine the likelihood that one of the assumed three profiles is more likely than the others. Such an exercise has been performed for the streak given in Fig. 16, with the results shown in Fig. 17. It can be seen the bi-Gaussian hypothesis can be rejected as the least likely model. While the uniformly filled ellipsoid gives a good fit near the distribution center, it is not very accurate at the edge, where one expects strong deviations in any case from this model.

Finally, we note that the best fit obtained from the Fermi-Dirac model (using α and A as fit parameters), gives an excellent match to the data.

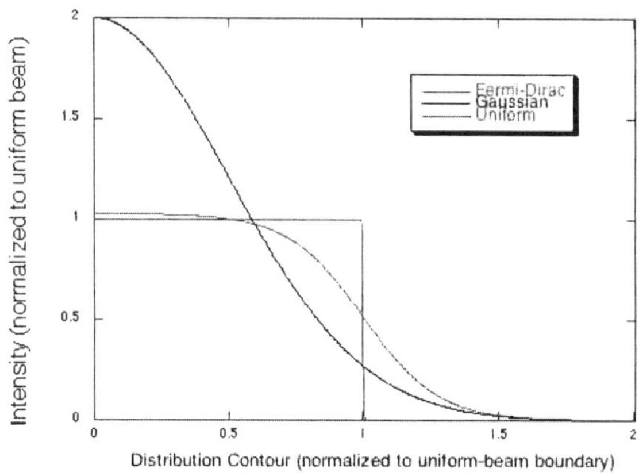

Figure 18. Analysis of streak data, with fraction of integrated intensity of data inside of elliptical contour shown. Best fit of data points to three models are shown: bi-gaussian distribution, uniform elliptical distribution, and Fermi-Dirac (uniform with tails) distribution.

The reconstruction of the distributions deduced from likelihood fits to the data in Fig. 17 are displayed in Fig. 18. It can be seen that the bi-Gaussian distribution is in large disagreement with the other two models, as it must be more peaked in the center — nearly a factor of two denser in our case. Note that the best fit, that of the Fermi-Dirac distribution, has a fall-off which can be attributed mainly to the approximately 2 psec full width half-maximum resolution of the streak camera, and thus expected physical erosion of the beam edges, as well as artifacts from the initial sub-psec beam pulse profiled are hidden in this measurement.

5. Conclusions

While the first measurements have established the soundness of the basic experimental approach, and given tantalizing first evidence for the formation of a nearly uniform ellipsoidal distribution, much more remains to be done. In order to more tightly control the initial conditions, it is necessary to correct the emission profile from the cathode. As of now, the SPARC injector is being modified to allow for near-axis, as opposed to 70 degree, injection. Additional improvements should result from use of laser cleaning of the cathode, as well a deformable mirror, thus making the emitted charge distribution more in line with expectation. In this scenario, we also expect improvement of the emittance obtained, to yield levels at or below those of the standard operating scenario.

After the next round of experiments, the SPARC injector will be completed with the addition of post-acceleration linacs and beam diagnostics (*e.g.* RF sweeper). In this fully mature experimental scheme, a complete test of the consistency of the Luiten-Serafini scheme with emittance compensation should be possible, using the large array of techniques described here. Further experiments will also emphasize the demonstration of high quality longitudinal phase space, and concomitant low energy spread as well as high compressibility.

Acknowledgments

The authors acknowledge useful interactions with Scott Anderson, Luca Giannessi, and Marcello Quattromini.

References

1. Luca Serafini and J.B. Rosenzweig Physical Review E 55, 7565 (1997).
2. I.M. Kapchinskii and V.V. Vladimirskii, in Proceedings of the International Conference on High Energy Accelerators, CERN, Geneva (Scientific Information Service CERN, Geneva, 1959), p. 274.
3. L. Serafini, AIP Conf. Proc. 413, 321 (1997).
4. O. J. Luiten, S. B. van der Geer, M. J. de Loos, F. B. Kiewiet, M. J. van der Wiel, Phys. Rev. Lett., 93, 094802-1 (2004).
5. "Emittance compensation with dynamically optimized photoelectron beam profiles" J. B. Rosenzweig, A.M Cook, R.J. England, M. Dunning, S.G. Anderson, Massimo Ferrario, Nucl. Instrum. Methods A, 57 87 (2006)
6. E. Colby, UCLA PhD Thesis, FERMILAB-THESIS-1997-03 (FNAL, 1997)
7. M. Borland, computer code ELEGANT, ANL, http://www.aps.anl.gov/asd/oag/oaghome.shtml.
8. http://www.okotech.com
9. X. Qiu et al., *Phys. Rev. Lett.* **76**, 3723 (1996).
10. V.Yakimenko, et al Phys. Rev. ST Accel Beams, 122801 (2003).
11. J.B. Rosenzweig *et al.*, "RF and Magnetic Measurements on the SPARC Photoinjector and Solenoid at UCLA", in Proc. 2005 Part. Accel. Conf. (IEEE, 2005).

SYNCHRONIZATION BETWEEN THE LASER AND ELECTRON BEAM IN A PHOTOCATHODE RF GUN*

A. SAKUMI, H. IIJIMA, M. UESAKA, K. YOSHII, T. UEDA, Y. MUROYA, A. FUKASAWA,

University of Tokyo, 2-22 Shirakata Shirane, Tokai, naka, 319-1188, Ibaraki, Japan

N. KUMAGAI, H. TOMIZAWA,

SPring-8, Japan

J. URAKAWA,

KEK, JAPAN †

The chemical reactions of hot or room temperature and/or critical water in a time-range of picosecond and sub-picoseconds have been carried out by the 18 MeV S-band linac and a Mg photocathode RF gun with the irradiation of third harmonic Ti: Sapphire laser, at Nuclear Engineering Research Laboratory (NERL), the University of Tokyo. Although this short bunch and 100 fs laser light are enough to perform the experiment of radiation chemistry in the time-range of sub-picoseconds, the total time-resolution become worse by the instability of synchronization between laser and radio frequency of linac. We found that the fluctuation of room temperature causes the instability, particularly the cycle of turning on/off of the air-conditioner. When we decrease the fraction of the room temperature within 0.1 degrees, the timing drift in an hour reaches 600 fs, closed to the timing jitter of 340fs than previous results.

1. introduction

A pulse radiolysis method is a useful and powerful technique for studying fast chemical reactions[1-11]. The intensive researches using the pulse radiolysis method have been carried out in aqueous solutions, organic liquids and polymeric systems[12-16]. Especially, the chemical reactions of hot or room temperature and/or critical water in a time-range of picoseconds and sub-picoseconds have been carried out by 18MeV S-Band linac with a BNL-gun IV-type(1.6cells)[17] and a Mg cathode of Spring8/KEK/U.Tokyo for three years[18]. The important factor for such as the fast radiation chemistry is not

* This work is supported by etc, etc.

† Work partially supported by KEK, .

only the pulse length of beam and laser but also the synchronization between the electron beam and laser. The synchronization is much influenced by external environments, especially for the fluctuation of laser-room temperature.

We have experimentally investigated the influence of this fluctuation to the synchronization. Furthermore we discuss active feedback system to achieve to be long-term stable, femto-seconds pulse radiolysis system.

2 Experimental setup

The experimental setup for sub-picoseconds pump- and probe-type radiation chemistry is shown in Fig. 1. The S-band linac, which provides an electron bunch as a pump-beam, consists of the photocathode RF gun, an accelerating tube and a chicane-type bunch compressor. Driven laser for RF gun and a probe-laser are generated from a Ti:Sapphire oscillator, which produces a laser light with a wavelength of 795 nm, an energy of 35mJ/pulse, a pulse duration of 300 ps and a repetition rate of 10 pps[18].

The laser and RF pulse of Klystron (and relevant gun and accelerator) are synchronized by using same master Oscillator. The frequency of master

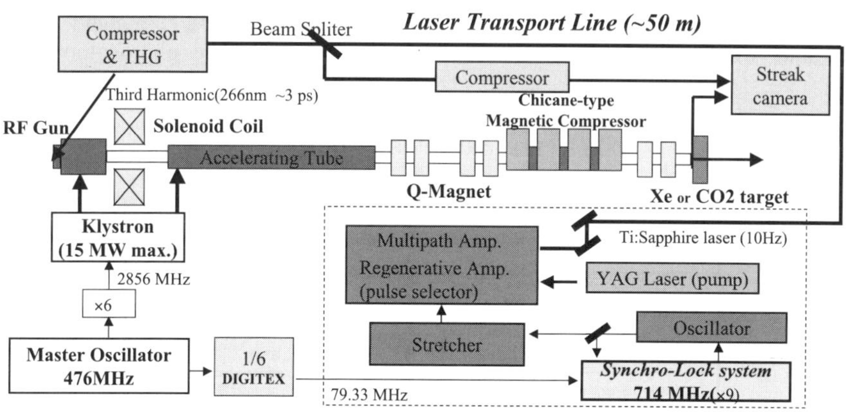

Fig. 1. Experimental setup for sub-picoseconds pump- and probe-type radiation chemistry

Oscillator is 476 MHz, so that we generate sixth harmonic(2856MHz) for the Klystron and sixth sub-harmonic(79.33MHz) for the laser.

The sub-harmonic generator (DIGITEX) also synchronizes detectors or other components. The laser Oscillator is synchronized with the frequency of 714MHz (9th harmonic) and the laser stretches from 100 fs to 300 ps. After stretcher the laser is amplified and retrieved with the frequency of 10 Hz. In order to achieve good synchronization, we use fs-Ti: Sapphire laser beams which separate to driven laser for RF gun and probe laser: A driven laser is compressed and irradiates to a third harmonic generator (THG), which is provided the third harmonic laser with a wavelength of 265 nm, an energy of ~100 mJ/pulse and pulse duration of a several picosecond. A spot size of the laser on the surface of the cathode is about 3 mm in diameter.
Another laser beam is also compressed to the time duration of 100 fs with the beam energy of several mJ for probing beam.

The performance of RF injector and the relevant laser and electron beam are shown in Table 1. The QE of the Mg cathode is achieve to 1.3×10^{-4}[19], corresponding to 1/10 as smaller as the expected value currently[20]. We think the reason that Mg cathode was exposed in air or moisture, unfortunately, though we kept it in helium gases immediately after diamond polishing. The beam is generated stably with the energy of 22 MeV, the charge of 1~2 nC(up to 3nC) /bunch and the bunch width of less than 1 ps. The normalized beam emittance are 26 pmm×mrad (horizontal) and 24 pmm×mrad(vertical)[20]. These bad emittance are thought to be caused by un-matching spot shape from oblique irradiation. The electron beam generated from the photocathode is

TABLE 1 PERFORMANCE OF RF INJECTOR

RF Injector			RF	
Cathode		Mg	Power	6.0 MW
Q.E.		1.3×10^{-4}	Pulse Duration	2 μsec
Charge		1nC/bunch	Repetition	10 Hz
		Up to 3nC/bunch	Laser	
Dark Current		800 pC/bunch	Driven Laser	Ti:Sapp., THG
Emittance	Horizontal	26 πmm × mrad	Laser Energy	100 μJ/pulse
	Vertical	24 πmm × mrad	Laser spot size	× 3mm
Bunch Duration		0.7 ps (1.5 nC, FWHM)		
Beam Energy		22 MeV		

focused by solenoid coil(1.0-1.8 kGauss), and a laser injection phase are optimized to charge-maximum.

After accelerating, the beam transports by sets of quadropole magnets and the electron bunch is compressed using the chicane type magnet (see Fig. 1). The compressed bunch goes through a chamber filled in Xe or CO2 gas, in which the electron bunch is emitted the Cherenkov light. The bunch duration is observed using a streak camera(FESCA, Hamamatsu Photonics Co.), which is measured the pulse duration of the Cherenkov light. The bunch duration is measured to be 0.7 ps (FWHM). The probe laser also irradiates to the streak camera after optical delay. The synchronization between the probe laser and the electron beam is measured from the relative time difference between the laser and the electron beam. Room temperature in laser room measures thermometer with a resolution of 0.1 degrees and the time evolution data are recorded in PC.

3. experimental results

The previous results of the time evolution of the synchronization between the probe laser and the electron beam (lower), and the relevant temperature of the laser room(upper) are shown in Fig.2[21]. We found that the frequency of the synchronization is good agreement with that of the temperature. It means the fluctuation of the room temperature has effect to the synchronization, especially much influence to laser Oscillator.

In order to decrease the fluctuation, we remove the local heat source near oscillator[22]. After re-construction, we can decrease the fluctuation within 0.1 degrees, limit of our thermometer (Fig. 3). Figure 3 also shows the Time evolution of the synchronization. We found quick and large components in the drift disappear. The jitter in one and a half hours reaches 600 fs(rms), which is closed to the jitter in short period (340 fs(rms)) than previous results(in Fig. 2). Nevertheless in more long period (ex. 1day), we still observe 6 ps drift (peak to peak) by temperature shift of outside in day and night. Our laser transport line is 50 m long[18], so that stretchy of the building might have bad influence to the synchronization.

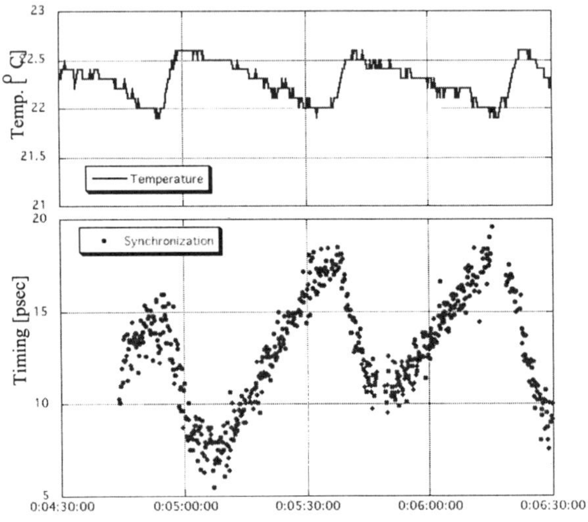

Fig.2 Previous results of the time evolution of the synchronization between the probe laser and the electron beam (lower), and the relevant temperature of the laser room(upper). The frequency of the synchronization is good agreement with that of the temperature.

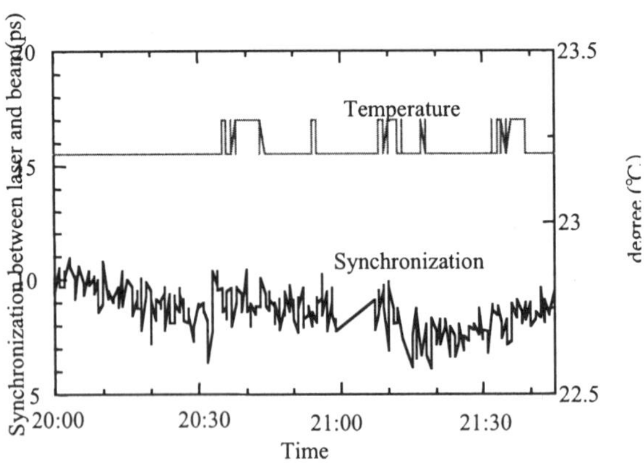

Fig.3 Time evolution of the synchronization and the relevant temperature. The fluctuation of the temperature decreased within 0.1 degrees,

and quick and large components in the drift disappears. The jitter in one and a half hours is 600 fs(rms), which is closed to the short time jitter of 340 fs(rms) than previous results(in Fig. 2).

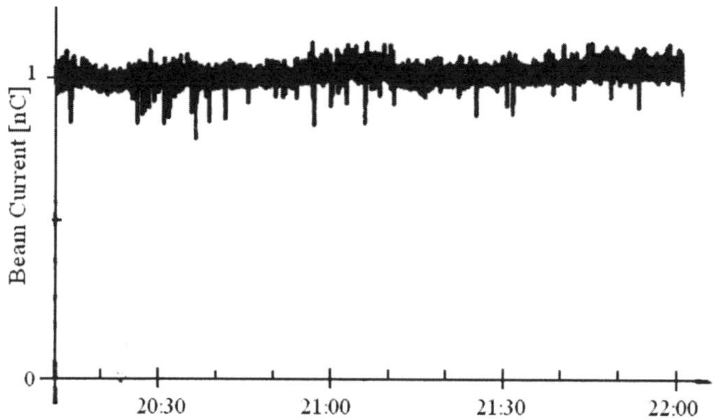

Fig.4 Stability of beam current during one and half hour.

Fig. 4 shows the electron beam stability, corresponding 4%(rms) during 1 hour (20:30-21:30).

4. discussion

We could observe a good synchronization between the pumping electron beam and the probe laser in hour, and we can also observe that the stability of the electron beam current is ~4%(rms) during 1 hour. These results are enough for one experiment of sub-picoseconds pulse radiolysis. However the experiments in long period are still not. When we change the accelerator RF phase slightly, synchronization can be adjusted. It means that the laser phase automatically goes out of whole alignment by laser's Synchro-Lock system. In order to avoid losing synchronization, we are developing the phase feed back system to fix the phase difference between accelerator and laser.

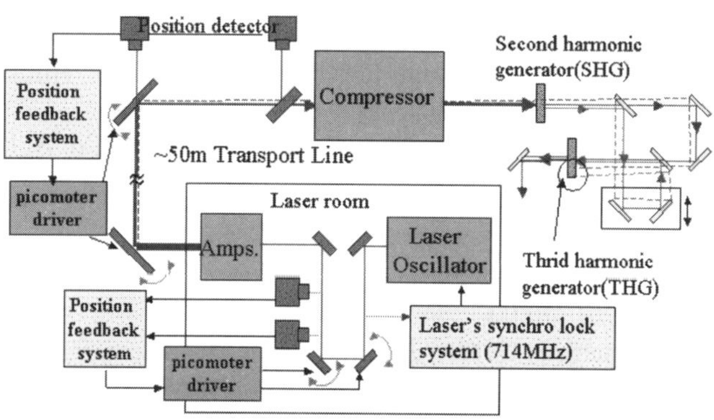

Fig.5 Diagram of the Position Feedback system

We have also aligned, manually, the beam point on the mirror in front of Compressor by CCD camera and mirror with computer controllable micro-motor. In an hour, we have hardly aligned, but need to align in one-day period. In order to obtain a good point stability, we are also developing the automatically feedback system. (Fig. 5).

5 Conculsion

We can reduce the fluctuation of room temperature within 0.1 degrees, so that we can observe a good synchronization of 600 fs between the pumping electron beam and the probe laser in an hour. It has the potential to do the experience of the pulse radiolysis experiment in an hour. In order to synchronize in long period as one-day, we are developing to new feedback system of phase matching and point stabilizing.

Acknowledgments

This is where one acknowledge funding bodies etc. Note that section numbers are not required for Acknowledgments, Appendix and References.

References

1. J.F.Wishart, D.G.Nocera, Adv. Chem. Ser. **254** 35 (1998).
2. J. Belloni, J. –L. Marignier, M. Gaillard, Cahiers Radiobiol **8** 25 (1998).
3. Y. Muroya, T. Watanabe, M. Uesaka, et al., *Proc. 1999 PAC*(New York City), 2602 (1999).
4. T. Kozawa, Y. Mizutani, M. Miki M, T. Yamamoto, S. Suemine, Y. Yoshida, S. Tagawa, Nucl. Instr. & Meth. A **440** 251 (2000).
5. Y. Aoki, J. F. Yang, M. Hirose, F. Sakai, A. Tsunemi, M. Yorozu, Y. Okada, A. Endo, X.J.Wang, I. Ben-Zvi, Nucl. Instr. Meth A **455** 99 (2000).
6. M. Washio, Y. Hama, S. Kashiwagi, R. Kuroda, T. Kobuki, T. Hirose, Ind. Soc. Rad. Photochem. 280 (2000).
7. T. H. Ghaddar, J. F. Wishart, J. P. Kirby, J. K. Whitesell, and M. A. Fox , J. Am. Chem. Soc. **123**, 12832 (2001).
8. Y. Muroya, M. Lin, T. Watanabe, G. Wu, T. Kobayashi, K.Yoshii, T. Ueda, M. Uesaka, Y. Katsumura, Nucl. Instr. & Meth. A **489** 554 (2002).
9. Y. Muroya, M. Lin, H. Iijima, T. Ueda, Y. Katsumura, Research on chmical intermediates, **31** 261 (2005).
10. J. Belloni, H. Monard, F. Gobert, J.-P. Larbre, A. Demarque, V. De Waele, I. Lampre, J.-L. Marignier, M. Mostafavi, J.C. Bourdon, M. Bernard, H.

Borie, T. Garvey, B. Jacquemard, B. Leblond, P. Lepercq, M. Omeich, M. Roch, J. Rodier, R. Roux, Nucl. Instr. & Meth. A **539** 527 (2005).

11. M. Uesaka, A. Sakumi, T. Hosokai, K. Kinoshita, N. Yamaoka, A. Zhidkov, T. Ohkubo, T. Ueda, Y. Muroya, Y. Katsumura, H. Iijima, H. Tomizawa, N. Kumagai, Nucl. Instr. & Meth. B **241** 880 (2006).
12. J.W. Boag, Nature **197,** 45 (1963).
13. J. W. Hunt, M. J. Bronskill, M. S. Matheson, et al., *Rev. Sci. Instrum.*, **41** 333 (1970).
14. L. Boilet, G. Burdzinski, G. Buntinx, C. Lefumeux, O. Poizat, J. Phys. Chems. A **105** 10271 (2001).
15. G.V. Buxton, C.R. Stuart, J. Chem. Soc. Faraday Trans. **91** 279(1995).
16. N. M. Dimitrijevic, K. Takahashi, D.M. Bartels, C.D. Jonah, J.Phys. Chem. A **105** 7236 (2001).
17. X.J. Wang, et al, *Nucl. Instr. Meth* **A483,** 82 (1996).
18. T. Kobayashi, Y. Muroya, T. Watanabe, T. Ueda, K. Yoshii, M. Uesaka, Y. Katsumura, K. Nakajima, X. Zhu, Int. J. Appl. Electromagnetics Mech. **14** 143 (2001).
19. H. Iijima, et al, *Proc. 2002 EPAC(Paris,France)*,1771 (2002).
20. T. Srinivasan-Rao et al. Rev. Sci. Instrum., **69**, 2292 (1998).
21. H. Iijima, M. Uesaka, F. Sakamoto, T. Ueda, N. Kumagai, L. Serafini, Comm. & Rev. paper **44** 5249 (2005).
22. H Tomizawa, H. Dewa, T. Taniuchi, A. Mizuno, T. Asaka, K. Yanagida, S. Suzuki, T. Kobayashi, H. Hanaki, F. Matsui, Proc. 2004 LINAC (Lübeck, Germany), 207 (2004).

METHOD OF BUNCH RADIATION PHOTOCHRONOGRAPHY WITH 10 FEMTOSECOND AND LESS RESOLUTION

ALEXANDER TRON†

Lebedev Physical Institute, Leninsky prospect 53
Moscow, 119991, Russia

IGOR MERINOV

Moscow Engineering Physics Institute, Kashirskoe shosse 31
Moscow, 115409, Russia

The success in creation of the proposed facilities, where the required electron bunch duration can be of about 100 fs and less, will be depend directly on the ability to measure the bunch longitudinal profile with resolution of the order of 10 fs and less. The only method for the bunch monitoring with the mentioned temporal resolution is the method of photochronography of the bunch incoherent radiation, for example, in the frequency range of visible light and at realizing streak camera with new principles of its operation. Results of novel type streak camera design and its investigation with photoelectron dynamics simulation taking into account space-charge effect are presented.

1. Introduction

Longitudinal profile monitoring of electron bunches or x-ray pulses with temporal resolution of the order of 10 fs is playing an increasingly critical role at present when development and creation of accelerator technologies of many applications are aimed at an electron bunch formation with its duration of about 200...100 fs and soft x-ray generation with much less duration [1, 2].

At the same time it is the well-known that a time convert streak camera is the only tool for monitoring the bunch radiation with resolution in femtosecond range, the resolution of which in the range of visible light does not exceed 300...200 fs [3], and in the range of soft x-ray, unfortunately, it is near 1 ps [4]

Hence, at present we need the measuring technique exceeding the reached temporal resolution by a factor of 10^2 at least in the case of the mentioned x-ray pulse monitoring. It means that it is necessary breakthrough in the measuring technique like a streak camera that could be made only on the base of new

† atron@sci.lebedev.ru.

principle of its operation. In the paper the results of streak camera design realizing new principles of operation [5] are considered and discussed.

2. Methods of bunch length monitoring

All well-known methods for short electron bunch length and shape monitoring with resolution in femtosecond range may be divided onto two groups: measurements of bunch radiation spectrum in frequency range of coherent radiation with further retrieving of the bunch shape assuming that the spectrum, emitted by a single electron in specific terms of device, is known exactly, - that is not so [6], and second and more reliable method bases on registration the bunch radiation intensity in the frequency range of incoherent radiation by means of technique like a streak camera.

Conventional streak camera has limitation in resolution, caused, mainly, by the longitudinal chromatic aberration of the camera' planar accelerating gap, magnitude of which is not less than 100 fs at very narrow initial electron energy spread, of about 0.1 eV, and at using the highest electric field (up to 10 kV/mm) at the photocathode surface in the considered camera type.

Space-charge effect is another reason caused limitation both in temporal resolution and in photoelectron bunch population that, as a rule, does not exceed 20…30 electrons in the cameras of the 300 fs – resolution.

It should be noted also the substantial feature of the conventional streak camera: the accelerating gap is plane-parallel configuration with photocathode surface field, restricted by small magnitude; time of flight of the electron bunch from the photocathode to the deflector can reach magnitudes of the order of 1 ns, during of which we have to keep the temporal structure of the photoelectron bunch in spite of space-charge effect.

3. New principles in time convert photochronography

New principle in the time covert photochronography, proposed in [5], consists in the following: the acceleration of photoelectrons and modulation of their dynamic variable in accord with the time of their escapement from photocathode must be performed simultaneously at the moment of this escapement and for the shortest time.

By means of combining the electrostatic accelerating field and rf- field, modulating electron on its longitudinal momentum, in a gap of resonator with internal conductor as a photocathode and taking the radius of the photocathode surface rather small (100…10 μm) one can enhance and localize the field near the surface of emitter so that the time of effective interaction between photoelectron and these fields will be about 1 ps. In the case many effects can not develop for short time, and resolution can reach 10 fs and much less.

Next term for the camera realization with the fs-resolution is identical conditions for all electrons starting from different points of emitter. It means that the modulating gap of the camera has to have appropriate symmetry. We will consider here the gap with spherical symmetry.

4. Streak camera realizing new principles of operation

In Figure 1 the possible scheme of the photoelectron camera, realizing new principle of its operation, is shown where the modulated photoelectrons in energy in the gap are analyzed by means of the spectrometer with uniform magnetic field. The rf-gap is formed by appropriate capacity gap of a quarter wave coaxial resonator with its internal conductor being under high voltage potential and ending by needle with a tip in the form semi sphere, covered by a photocathode material. Using the gap as a part of a RF-resonator we suppose that the camera can operate in sinchroscan or single shoot regime.

We note at once that at the accelerator beam measurement the rf-power for the camera resonator can be supplied from appropriate rf-system of an accelerator, so that, in the case there is no, in fact, a time jitter between the phase of the RF- field in the gap and the time of photoelectron escapement from the photocathode, and electronic part will be more simple in comparison with corresponding part of a conventional streak camera.

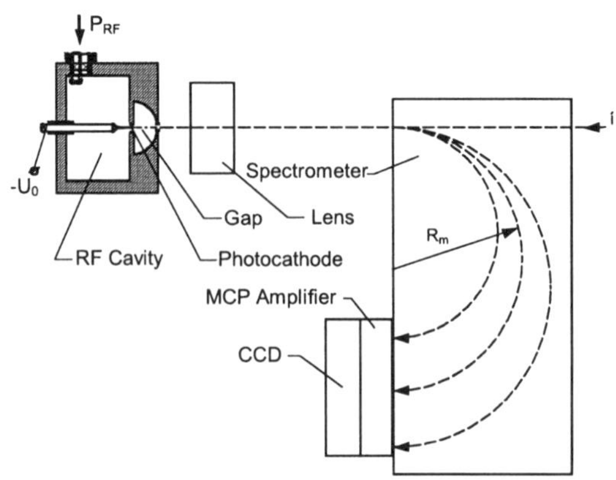

Figure 1. Scheme of streak camera with longitudinal modulation of photoelectrons in the gap of spherical configuration

4.1. *Accelerating gap*

In ideal case a radial electric field in the gap of spherical configuration can be presented by the well-known expression

$$E = E_0(1+m)\frac{R_0^2}{r^2}, \tag{1}$$

where $E_0 = -U_0/h$ – uniform field, when the voltage $-U_0$ is applied to the h-length planar gap; $1+m \equiv R/R_0$ – coefficient of enhancement of the field at the emitter relative to the uniform field; R_0 – radius of the emitter (internal conductor in spherical capacity gap) and R is a radius of anode.

Electrodynamic or modulating field in the gap we will consider in quasistatic approach when the expression for the field can be presented as a product of two functions: one of them depends on the time and another one – on coordinates. Temporal part of field we will write as $f_t = cos(\omega t + \varphi_0)$ where the phase φ_0 defines of the moment of the photoelectron escapement.

The modulating field is described in space by the same expression (1) with replacing U_0 on U.

In reality the electric field distribution in space, closed to the ideal distribution, is formed with a diaphragm as it is clear from the Figure 2 where equipotentials of electric field are shown.

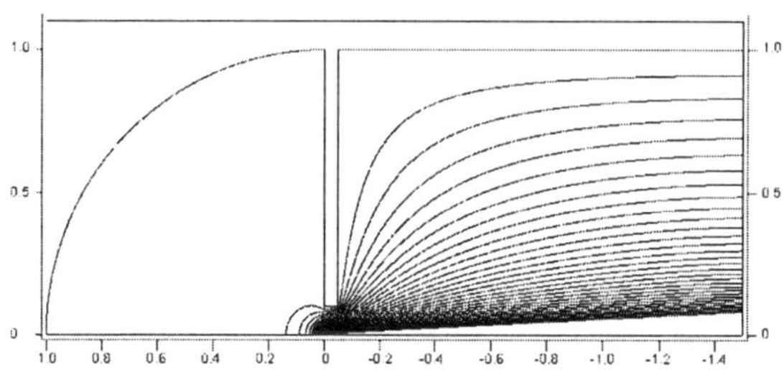

Figure 2. Equipotentials of electric field created by the charged needle shape electrode placed near conducting diaphragm

Axial field distribution in this system is determined by expression

$$E(z) = a_0 z^{-2}\,(1 + a_1 z + a_2 z^2 + a_3 z^3 + a_4 z^4) \tag{2}$$

Defining the field in units of kV/mm and magnitude of coordinate z in mm the values of coefficients in the expression (2) will be the following: $a_0 = 0{,}2$; $a_1 = 0{,}28152$; $a_2 = -\,0{,}167428$; $a_3 = 0{,}023894$; $a_4 = -\,0{,}00106$.

The expression (2) along with the mentioned above coefficients have been determined for the shown system with radius of emitter R_0 = 50μm, radius of anode R = 10 mm and for diameter of the diaphragm hole of 2 mm.

At the voltage U_0 = 4 kV the field on the emitter surface will be 81,087 kV/mm that is very close to the case for the ideal spherical gap. In Fig. 2 the shown boundary equipotential magnitude is 100 V at the emitter voltage 4 kV.

4.2. *Temporal resolution of the camera and its modulating gap*

The proposed method of high speed photochronography is based on two sequential separations of photoelectrons in accord with the time of their escaping from photocathode: first, in the phase space along the longitudinal component of the electron momentum using for that fast changeable field in the gap, and, second, in configurational space by means of electron spectrograph using stationary fields. Hense, the temporal resolution of the camera depends on both the resolution of the modulating gap in the phase space and the resolution of the electron spectrometer (operating in the regime of spectrograph).

4.2.1. Definition of resolution

The time resolution of the camera can be defined by the expression

$$\Delta t_c = \left| \frac{\Delta x}{\partial x / \partial t_0} \right|, \tag{1}$$

where Δx – width of line at the exit of spectrometer that corresponds to initial electron energy spread, magnitude of which in calculations has been taken from 0 to 0.5 eV: ΔW_0 = 0.5 eV (S-1 photocathode, Ag-O-Cs).

The time resolution of the modulating gap of resonator can be defined as

$$\Delta t_g = \left| \frac{\Delta P_g}{\partial P_g / \partial t_0} \right|, \tag{2}$$

where ΔP_g – momentum spread of electron at the gap exit corresponding to the initial electron energy spread.

These resolutions are connected through the expression

$$\Delta t_c = \Delta t_g \frac{\Delta P_{spectr} / P}{\Delta P_g / P}, \tag{3}$$

where the numerator of the quotient is the relative spectrometer resolution, and the denominator – relative momentum spread of the electrons at the gap exit. All magnitudes are determined for the same time of electron start. The quotient (3) should be never taken less than 1 because by taking the spectrometer resolution ΔP_{spectr} better then the width of the electron momentum line ΔP_g at the gap exit, corresponding to initial electron energy spread ΔW_0 = 0.5 eV, we can never increase temporal resolution of the camera.

4.2.2. Equation of motion

Investigation of longitudinal dynamics of the electrons in the gap, containing the fast changeable in time field, has to be carried out on the base of relativistic equation of motion that could be written in the form [7]

$$\frac{d\vec{v}}{dt} = -e\frac{c^2}{W_e}\gamma^{-1}\left\{\vec{E} + \left[\vec{v},\vec{B}\right] - \frac{\vec{v}}{c^2}\left(\vec{E},\vec{v}\right)\right\}, \tag{4}$$

where e, W_e, c and γ - are respectively an elementary charge, rest energy of an electron, speed of light in vacuum and Lorentz factor; v, E and B – are respectively velocity of the photoelectron, electric field and induction of magnetic field; t – is variable of integration.

We will consider the task in the assumption of ignoring an electrodynamics retardation, the fields are described in quasistationary approach.

By omitting the space-charge effect at first, there will be considered one-dimensional motion along the z-axis to determine the mentioned above resolutions of the gap and camera.

4.2.3. Temporal resolution dependences

Our investigations show that the gap resolution does not depend practicaly on the gap length $h = R - R_0$ within the range from our R = 10mm and up to 2 mm. The gap resolution mainly depends on the following parameters : R_0, U_0, U and frequency of the modulation field exitation f. Some results of these investigations for f = 3 GHz, R = 10 mm and for initial energy spread of the photoelectrons ΔW_0 = 0.5 eV are presented in Figs. 3 – 7.

It should be noted at once that in all cases, taken for consideration, the field on the photocathode surface does not exceed magnitude $1\cdot10^9$ V/m and in the case there will not be a problem with break-down in the gap after appropriate traning of it similar to prepareing a gun with thermal cathode.

The gap resolution more strongly depends on the photocathode radius. By reducing the radius of emitter up to several microns the gap resolution can reach 1fs and even 0.1fs at consideration of this task in the frame of relativistic classical mechanics, and such resolutions will correspond to the width of the electron momentum line $\Delta P_g / P$ close to $5\cdot10^{-5}$ or $1\cdot10^{-5}$, respectively

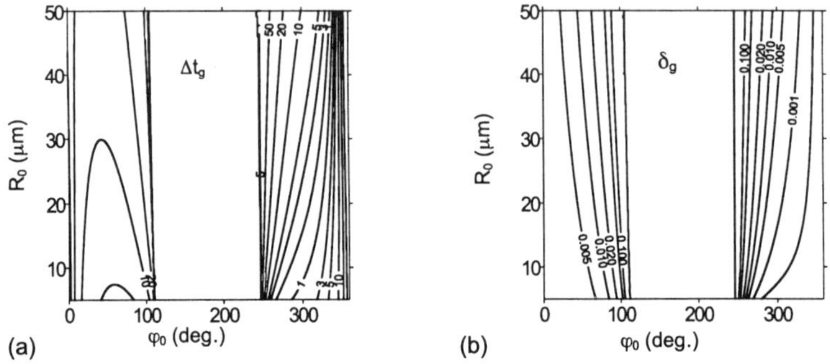

Figure 3. Isolines (in fs-units) of the gap resolution Δt_g (a) and the relative width of the electron momentum line δ_g in percents at the gap exit with parameters: h=10mm, U_0=4kV, U=10kV, - on the plane of variables R_0 and phase of the photoelectron escaping.

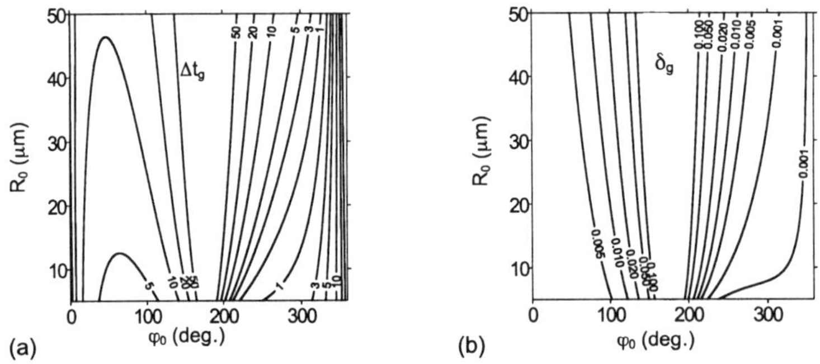

Figure 4. Isolines (in fs-units) of the gap resolution Δt_g (a) and the relative width of the electron momentum line δ_g in percents (b) at the gap exit with parameters: h=10mm, U_0=10kV, U=10kV, - on the plane of variables R_0 and phase of the photoelectron escaping.

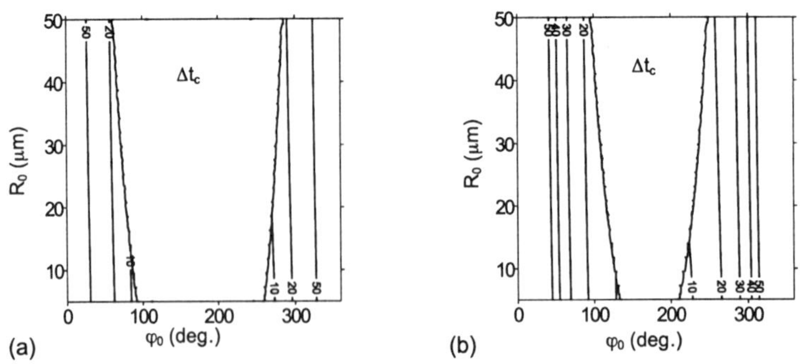

Figure 5. Isolines (in fs-units) of the camera resolution with gap parameter from Fig. 3 (a) and the with gap parameter from Fig. 4 (b) for constant spectrometer resolution of 0.02 %.

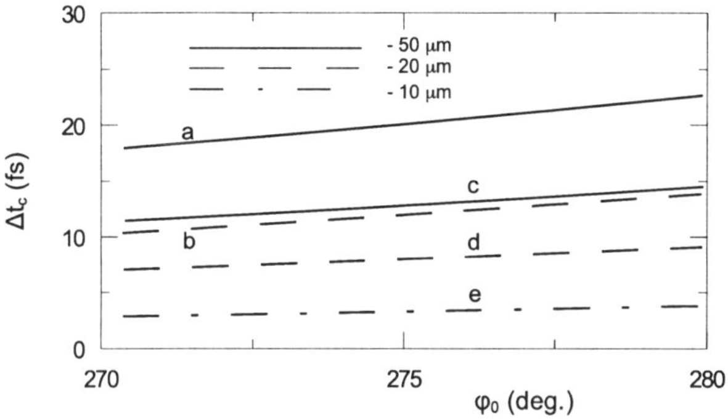

Figure 6. The camera resolution as a function of photoelectron escaping moment in degrees of the modulating field on the frequency of 3GHz for $U_0 = 4$ kV, $U = 10$ kV, $h = 10$ mm at $R_0 = 50$ μm, 20 μm, 10 μm and for the next spectrometer resolution: $\Delta P_{spectr}/P = 0.02$ % - ***a*** and ***b***; 0.01% - ***c*** and ***d***; 0.005% - ***e***. The curves e, d, a, c are shifted to the left by 5°, 10°, 15°, 25°, respectively.

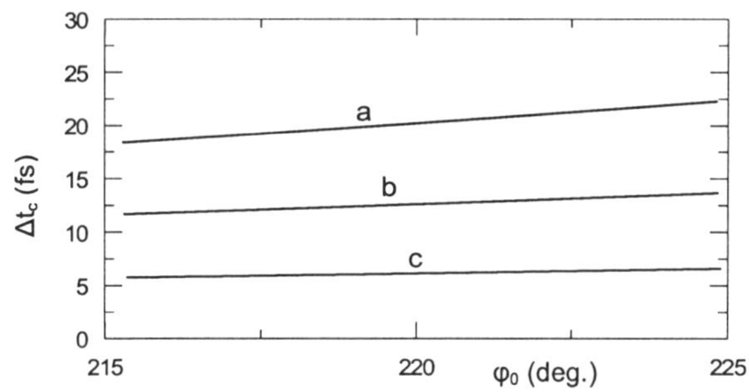

Figure 7. The camera resolution as a function of photoelectron escaping time, expressed in degrees of the modulating field, at the fields on the frequencies: 1.3 GHz (a), 3GHz (b), 11.4 GHz (c),- for $U_0 = 10$ kV, $U = 10$ kV, $h = 10$ mm and $R_0 = 20$ μm at the spectrometer resolution 0.02 %. The curves b and c are shifted to the left by 15° and 45°, respectively.

From the presented above dependences in Fig. 3-7 one can conclude that the camera resolution is strongly limited by the spectrometer resolution. At rather rough spectrometer resolution (0.02%) the resolution of the camera will be practically the same for the gaps with accelerating voltage of 4 kV or 10 kV. Nevertheless, at these voltages and the spectrometer resolution the camera can reach the10 fs - resolution and in the case we can use the emitter with any radius from 10 μm up to 50 μm. This freedom in option of parameters R_0 and U_0 allows to make further optimization concerning the spectrometer parameters and to mitigate the terms for space-charge effect.

5. X-ray streak camera realizing new principle of operation

Distinction of x-ray camera from the described above one will be mainly in material of the photocathode. For registration soft x-ray with quantum energy from 100 eV to 10 keV the gold photocathode is rather preferable because it is resistive to the air and many other medium. At the same time the initial energy spread in this case has FWHM of energy distribution equaled near 4 eV and quantum efficiency of 6% [8].

If we take the accelerating gap of planar configuration, as in a conventional camera, with its length of 1 mm and accelerating voltage of 10 kV we will have (in the case of gold photocathode) the magnitude of the longitudinal chromatic aberration of 0.7 ps.

It is quite different magnitude will be for the new type of streak camera, scheme of which is presented in Fig. 1, where the chromatic aberration is suppressed due to very high field on the photocathode surface and very small length of effective field, estimated by magnitude of several emitter radii. Moreover, by starting at the same moment and having different initial energy the photoelectron with smaller energy, being in the field increasing in time, can get additional energy in comparisons with other particle so that their initial energy spread can be decreased to the exit gap. It can be watch for definite phase range of their starts in the vicinity of 270 degrees where the resolution is the best.

In Figure 8 the x-ray camera resolution in the form of isolines in the fs-units is presented for the modulating gap with parameters: h = 10 mm, $U_0 = U$ = 10 kV, at frequency of the modulating field of 3 GHz and for constant spectrometer resolution equaled 0.01% for picture (a) and 0.02% - for (b).

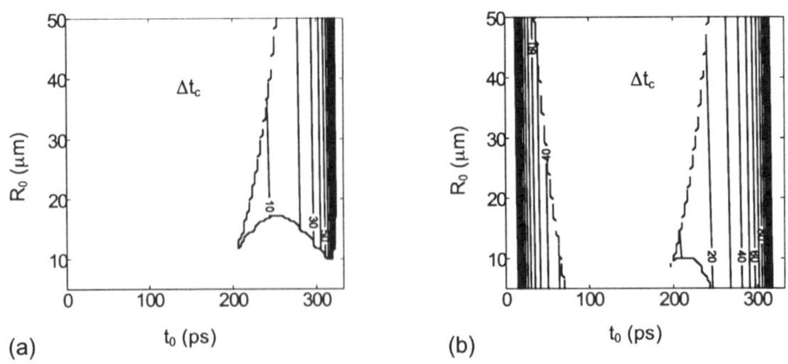

Figure 8. The camera resolution as a function of photoelectron escaping time for the modulating field on the frequencies 3 GHz, U_0 = 10 kV, U = 10 kV, h = 10 mm and at the spectrometer resolution $\Delta P_{spectr}/P$ = 0.01 %.- (a), $\Delta P_{spectr}/P$ = 0.02 %.- (b)

6. New type streak camera design

Proposed photoelectron camera, the scheme of which is outlined in Fig.1, has overall dimensions allowing it to be placed on the sheet of format A4.

The modulating gap is a capacity gap of a quarter wave coaxial resonator, schematically drawn in the same Fig.1. By taking diameters of internal and external conductors of about 3 mm and 30 mm, respectively, one can get the required amplitude of RF-voltage of 10kV on the gap at pulsed power consumption not more than 100 W on the frequency of 3 GHz.

One of the main units of the camera is spectrometer. There are several types for its implementation meeting requirements of the camera. At this stage of our investigations there has been chosen the simplest spectrometer with uniform magnetic field with semicircular trajectories of electrons in it. To minimize spherical aberration of his type spectrometer the lens is installed at the exit of the resonator.

Figure 9 makes clear operation of this electrostatic lens, where transformations of beam contour in the transverse phase space along optical channel are shown.

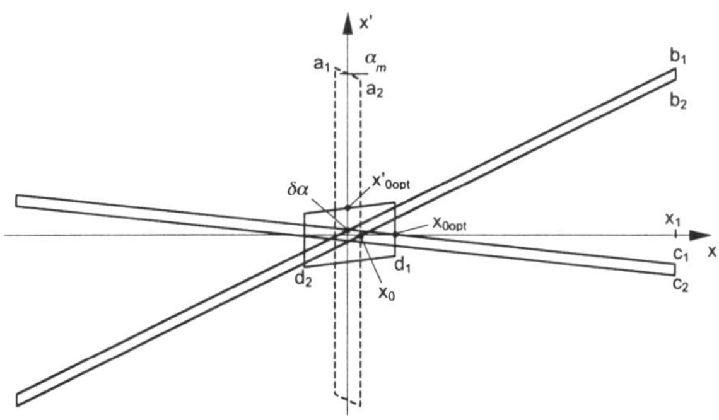

Figure 9. Photoelectron beam contour in the transverse phase space: b_1-b_2 – after the modulating gap; c_1-c_2 – after the lens; d_1-d_2 – at exit drift space, at entrance of spectrometer; a_1-a_2 –equivalent contour in the section coinciding with the center of the photocathode sphere. α_m – semiangle size of the anode hole, taken to be equaled 5°.

In a result of the channel optimization there were obtained the following parameters: x_{0opt} = 10.7 μm, x'_{0opt} = 8 mrad, resolved width of line at the exit magnet is 32μm, spectrometer resolution of 0.02% obtained at the trajectory radius of 80 mm and for electron source x_0 = 1μm, α_m = 5°. Focus length of lens is 9.145 mm, distance between the lens and magnet is about 100 mm. Emittance of the beam does not exceed 0.1 mm·mrad and that magnitude can be made much less in the considered system, based on a needle shape photocathode.

7. Conclusion

Considered method of electron bunch radiation photochronography, based on new principles of streak camera operation, allows to carry out measurement of the bunch shape with temporal resolution of 10 fs and much less.

Magnitude of this resolution is practically independent of frequency range of the investigated radiation and it can be used for the radiation registration in the range from visible light to x-ray with the same high temporal resolution.

There was proposed a simple scheme of the device realizing these new principles in high-speed photochronograhpy. It was shown that the mentioned high temporal resolution can be reach at rather low voltages, small RF-power consumption and all device can be placed on the sheet of format A4.

Acknowledgments

Authors thank O. Kozlov of JINR for his help in some investigations of field in the modulating gap.

References

1. Linac Coherent Light Source (LCLS) Conceptual Design Report, SLAC-R-593 (2002).
2. TESLA XFEL, Technical Design Report (Supplement), DESY 2002-167, TESLA-FEL 2002-09 (2002).
3. M.Ya. Schelev, *Physics – Uspekhi* **43** (9) 931 (2000).
4. Z. Chang et al., *Proc. of SPIE* **2869,** 971 (1998).
5. A.M. Tron, *Proc. of SPIE* **4948,** 141 (2003).
6. A.M. Tron, *Proc of PAC99 Conf.* 2190 (1999).
7. L.D. Landau, E.M. Livshits, *Field Theory. Theoretical Physics.* V.2, Fizmatgiz, Moscow, 1963.
8. B.L. Henke, J.A. Smith, *J.Appl.Phys.* **48,** No.5, 1852 (1977).

Working Group 3

"NEW CHALLENGES IN THEORY AND MODELING" SUMMARY FOR WORKING GROUP THREE

LUCA GIANNESSI

ENEA C.R. Frascati, Via E. Fermi 45, 00044 Frascati (Rome) Italy

Working group three was devoted to contributions on theory and modeling of devices and new ideas, aiming at increasing the beam brightness and at generating radiation at short wavelengths. The contributions presented span from the beam dynamics in radiofrequency accelerators, emittance compensation schemes and optimization of the beam brightness at low energy, analysis of the emittance dilution processes during acceleration and compression, laser plasma sources and the generation of radiation in FELs and other radiation sources. Analysis specifically oriented to the modeling of the beam and radiation dynamics were also presented.

1. Introduction

Working group three was devoted to the contributions on theory and modeling of devices and new ideas aiming at increasing the beam brightness, and at generating radiation at short wavelengths. About 20-25 participants joined the working group, with fluctuations due to the spotty participation of some participants from other working groups. A total of 16 talks during three days were presented, with a consistent amount of time devoted to discussion and active participation from all the participants. The presentations can be divided into the three main categories:

- Electron beam dynamics
- Laser plasma electron beam sources
- Free electron laser and radiation sources

The electron beam dynamics category includes seven contributions and general discussions encompassing several issues in conventional radio frequency (rf) accelerating devices, in photo-injector dynamics and in calculation and modeling of emittance dilution processes during acceleration and compression, as space charge and coherent synchrotron radiation effects. Two talks were presented in the laser plasma electron beam sources category and seven contributions attained to the FEL and radiation sources category, three of which were focused on recent calculations of FEL dynamics in the quantum regime.

2. Electron beam dynamics

In the first contribution **J. Rosenzweig** (UCLA Dept. of Physics and Astronomy) presented an overview of the beam dynamics in photo-injectors where the beam generation takes place on a space charge induced expansion ("blow up" regime). From a theoretical point of view this appears as a valid alternative to the classical "beer-can" uniform beam distribution, allowing the linearization of the space charge forces for an efficient emittance compensation scheme. Simulations done with Parmela[1] have pointed out that the emittance evolution is only slightly different than in the standard uniform beam scenario and that a final projected emittance of 0.7 mm-mrad with a peak current of 105A can be obtained.

V. Litvinienko (Collider Accelerator Department, Brookhaven National Laboratory, Upton, NY, USA) has analyzed a specific, critical component of electron re-circulated accelerators (ERL), the merger. Mergers used in presently operating ERL are not designed for very low emittance beams and therefore are not suitable in preserving emittances of the order of 1 mm-mrad. The concept of the "Zig-zag" merger based on an optimized geometry seems promising in solving some of the challenges presented by future ERL operating with super-bright intense electron beams.

The problem of simulating the coherent synchrotron radiation (CSR) effects has been addressed by **M. Migliorati** (Un. Roma "La Sapienza", Roma, Italy). The proposed method is based on the numerical solution of the Vlasov equation for the electron distribution based on the symmetric split technique[2]. This method is less sensitive to numerical "noise" introduced by macro-particles discretization of the charge density distribution, which is one of the issues in modeling CSR dynamics.

A description of the start to end strategy at the VUV-FEL at DESY [3] has been presented by **T. Limberg** (DESY, Hamburg). The code ASTRA[4] is used in the simulation of the space charge effects in the injector and in the linac, and code CSR-Track[5], which has been extensively described in the presentation, is used to model CSR effects. The data from the "LOLA" diagnostic device[6], which operates as a sophisticate streak camera, have been compared with the simulations (courtesy of M. Dohlus, DESY), showing an impressive agreement.

In the same frame of beam modeling and optimization **V. Fusco** (INFN LNF, Frascati, Italy) has presented the extension to the code HOMDYN [7] in order to include transverse wake field effects in accelerating structures. The effects of steering correctors combined with wakes has been presented in the specific case of the SPARC linac [8].

The transition between the space charge regime and the emittance dominated regime has been analyzed by **M. Ferrario** (INFN-LNF, Frascati (Rome) Italy). When dealing with high brightness beams the whole linac has to be designed as a emittance compensated photo-injector and step by step optimization is not always sufficient. The propagation close to the equilibrium solution allows the control of plasma oscillations "phase" and the "phase" leading to the lowest emittance can be tuned at the injector level with a "global" optimization.

C. Vaccarezza (INFN-LNF, Frascati (Rome) Italy) has analyzed the simulation results on laser pulse jitter and micro-bunching instability at SPARXINO [9] pointing out the apparent necessity of implementing a laser heater in order to suppress the micro-bunching instability gain.

3. Laser Plasma sources

Two contribution were presented in this category. The first was a brief overview of compression of externally injected electron beams in FLAT density plasmas and an analysis of bunch compression in "shaped" density plasmas, speaker **P. Tomassini** (ILIL IPCF-CNR and INFN Pisa 1, Pisa, Italy). Compression is required in order to increase the charge captured in the acceleration bucket and a shaped density plasma allows to separate the plasma wavelength and the phase velocity leading to the possibility of optimizing phase speed and wavelength for a given injected bunch. Simulations have shown a compression factor of 21 in 6 mm of acceleration with a final energy spread of about 4%rms and with a laser pulse energy of 0.75J.

In the second contribution the excellent activities on the generation of mono-energetic electrons at the University of Tokyo have been presented by **K. Kinoshita** (Nuclear Professional School, School of Engineering, University of Tokyo, Tokyo, Japan). The variable density plasma has been experimentally implemented in a two density staged device combining the advantages of high density plasma in the injection stage and the high energy and long de-phasing length of a low density plasma. The experiment has demonstrated the generation of a beam with 10 pC charge/shot, with an energy spread of about 10% at an energy of 12 MeV. The 2-staged acceleration using a gas-jet injector with capillary discharges is one of the most promising schemes to produce high quality electron bunches. The next step will be the further beam acceleration using capillary discharges.

4. FEL and radiation sources

An experimental demonstration of the super-radiance phenomenon in a single pass seeded FEL obtained at BNL-NSLS has been presented by **T. Watanabe** (Brookhaven National Laboratory, Upton, NY, US) . The DUV-FEL [10] at BNL is an ideal facility for laser-seeded FEL research. The laser beam driving the photo-injector which is naturally synchronized with the electron beam, has been used to seed the NISUS undulator with the resonance corresponding to the seed laser wavelength. The transition at saturation has been studied and FROG [11] traces of the radiation pulses after the FEL amplification have been recorded. Non-linear gain after saturation has been observed together with wide band multi-peak spectrum and pulse shortening from 150 fs down to 80 fs. The experimental data agreed well with GENESIS 1.3 [12] simulations and showed the typical energy scaling vs. the longitudinal coordinate along the undulator of the super-radiance phenomenon [13].

S. Reiche (UCLA, Los Angeles, US) has presented an overview of the numerical challenges for FEL simulations ranging from the undulator modeling issues, the harmonic cascade FEL simulations, the problem of quiet start and the phase space loading, the spontaneous radiation modeling, the influence of the wake-fields and CSR in undulators and magnetic chicanes. Most challenges arise in the time dependent calculations and in the large number of FEL types that have to be treated by simulations, as SASE, high harmonic generation, cascades ... In some cases the number of macro-particles is getting closer to the number of real electrons. The importance of a massive parallel approach has been pointed out, a thousand of nodes with 1 gigabyte of RAM each, would allow the simulation of more complicated undulators and with a more advanced solver.

Calculations of the wake-field undulator radiation were briefly presented by **A. Opanasenko** (NSC KIPT, UKRAINE).

R. Bonifacio (INFN Sez. Milano, Italy) has presented an overview of his recent results on the quantum free electron laser providing a preliminary list of parameters for pointing out quantum effects in a free electron laser experiment. This brief talk was a complement to the joint session contribution where the Free Electron Laser (FEL) equations, extended to the quantum domain describing the particle's motion by a Schrodinger equation and coupled to a self-consistent radiation field, were analyzed. The model depends on a single collective parameter $\overline{\rho}$ which represents the maximum number of photons emitted per particle. Quantum effects are observed when $\overline{\rho} < 1$, the classical model is recovered in the limit $\overline{\rho} \geq 1$.

The discussion on the quantum FEL regime continued with the talk of **A. Schiavi** (University of Rome La Sapienza, Rome, Italy) who presented the QFEL code, devoted to the simulations of FEL operating in the quantum regime.

In the same context the quantum effects in gain and startup of Free Electron Lasers were presented by **K. J. Kim** (ANL, Un. Chicago, US). With the formalism of the Wigner function satisfying the Quantum Vlasov – Maxwell equations, an expression to estimate the effects of a distribution of electrons with a finite spread in momentum on the quantum FEL dynamics has been derived.

Finally **J. Rosenzweig** (UCLA Dept. of Physics and Astronomy) has presented an analysis of the relation between the photon production and the acceptance bandwidth in a Compton backscattering source. As an example the number of photons in 1% bandwidth scattered by a 1J/1.5ps laser pulse at 800 nm is of the order of 0.5 photons per electron, i.e. about 3 10^9/nC showing the potentialities of this sources in providing reasonably high photon fluxes when combined with a low emittance (less than 5 mm-mrad) electron photo-injector.

5. Conclusions

On one side we may observe that many new ideas are coming in the field. The analysis of the “blow up” regime in photo-injectors or the enlightening discussion on the quantum FEL and the discussion of the experimental results at BNL on super-radiance, are examples in this direction. On the other side the Compton backscattering based sources or the laser plasma acceleration schemes, although may not be considered “new” ideas, are becoming more and more promising. These advances, these new proposals and new ideas continuously pose renovate demands and challenges, stimulating the community in further investigating the theory and modeling of fields and charged particles system dynamics.

Acknowledgments

I would like to thank all the speakers who presented their work in this working group and to acknowledge the contribution of all the participants who stimulated the discussion with their numerous and interesting questions. Finally I wish to acknowledge the special contribution of Angelo Schiavi for his help in the working group coordination and in the preparation of the working group summary.

References

1. Parmela: UCLA version: E. Colby, Ph.D. thesis, UCLA, 1997. ''Design, Construction and Testing of a Radiofrequency Electron Photoinjector for the Next Generation Linear Collider.'' LANL version: see e.g., http://laacg1.lanl.gov
2. 2.J.A. Fleick, J. R. Morris M. D. Feit, Appl. Phys 10, 129 (1976)
3. S. Schriber, "First Lasing at 32 nm of the VUV-FEL at Desy" in Proceedings of the 27th International Free Electron Laser Conference, www.JACoW.org, 12 (2005)
4. K. Floettmann, Astra User Manual, see http://www.desy.de/~mpyflo/Astra_dokumentation/
5. M.D. Dohlus, T. Limberg, "CSRtrack: Faster Calculation of 3-D CSR Effects" in proceedings of the 2004 FEL Conference, www.JACoW.org, 18-21 (2004)
6. M. Huning et al. "Observation of Femtosecond Bunch Length Using a Transverse Deflecting Structure" Proceedings of the 27th International Free Electron Laser Conference, www.JACoW.org, 538 (2005)
7. M. Ferrario et al., Part. Accel. **52**, 1 (1996).
8. A. Renieri et al. Status report on SPARC project, in Proc. of the 26th FEL conference, Trieste 2004 JACoW (Joint Accelerator Conference Website) at http://www.JACoW.org, p. 163 (2004)
9. M. Boscolo et al., "Beam Dynamics Studies for the Sparxino Linac*" in Proceedings of the 27th International Free Electron Laser Conference, www.JACoW.org, 67 (2005)
10. A. Doyuran et al., Phys. Rev. ST AB 7, 050701 (2004)
11. D. J. Kane and R. Trebino, IEEE J. Quantum Electron. **29**, 571 1993
12. S. Reiche, Nucl. Instrum. Methods Phys. Res., Sect. A **429**, 243 (1999).
13. R. Bonifacio, L. De Salvo, P. Pierini, and N. Piovella, Nucl. Instrum. Methods Phys. Res. A 296, 358 (1990).

RESONANT MODES IN A 1.6 CELLS RF GUN

MASSIMO FERRARIO
INFN-LNF, Via Enrico Fermi 40
Frascati, 00044, Italy

CONCETTA RONSIVALLE
ENEA,C.R. Frascati, Via E. Fermi,45
Frascati, 00044,,Italy

The SPARC photoinjector RF gun consists in the BNL/SLAC/UCLA 1.6 cell structure designed to resonate at 2856 MHz in the π mode. It will be demonstrated by a numerical modelization based on SUPERFISH code combined with the LC-circuit analysis that the two oscillating modes of the system usually indicated as 0-mode and π-mode (the operating mode) are in reality a π/3-mode-like and a π-mode-like. The consequences on the definition of the coupling coefficient and on the use of mode-separation based RF measurements are described.

1. The SPARC gun

The SPARC photoinjector gun consists in the BNL/SLAC/UCLA 1.6 cell structure designed to resonate at 2856 MHz in the π mode [1]. The electromagnetic field in the two cells is coupled through an aperture on the axis. The size of the aperture determines the coupling of electromagnetic power between the two cells. The two resonant modes (fig. 1) are usually identified by 0 and π modes, being the fields in phase in the low frequency mode and in phase opposition in the high frequency mode (fig.2).

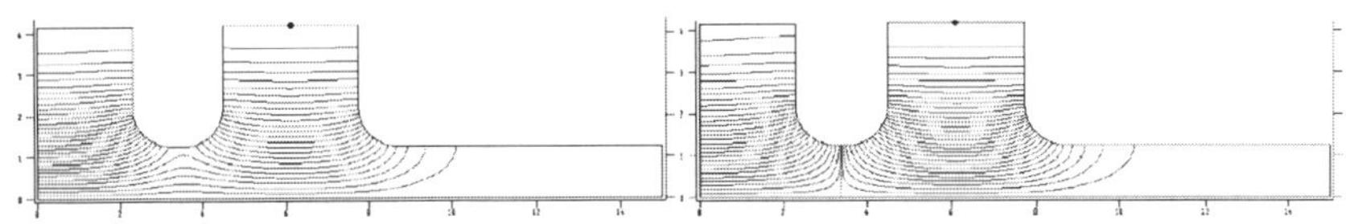

Figure 1. SUPERFISH computed contour lines of the two resonant modes in the SPARC gun: (left) low frequency mode usually identified as 0-mode, (right) high frequency mode usually identified as π-mode.

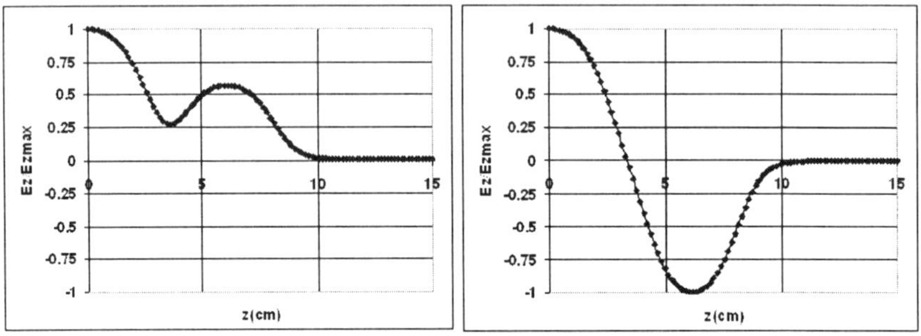

Figure 2. SUPERFISH computed on axis fields in the two resonant modes in the SPARC gun: (left) low frequency mode usually identified as 0-mode, (right) low frequency mode usually identified as π-mode.

However within the dispersion curve nomenclature [2] according with the coupled-circuit analog of capacitive coupled cavities (fig.3) both 0 and π modes exist only for half-cell terminations on both ends, considering as half-cells half length cells with conducting planes inserted in the center plane of symmetry,. In a lumped element model the resonant frequency for a cell is represented by $\omega_R = 1/\sqrt{(LC)}$ and the cell-to-cell coupling by $k=2C/C_a$. The dispersion relation for a such system is

$$\omega^2 = \omega_R^2 \cdot \left[1 + k \cdot (1 - cos\varphi)\right] \tag{1}$$

giving the frequency range over which an infinite chain of cells with cell frequency ω_R can oscillate for a particular mode distribution such as TM_{010}.

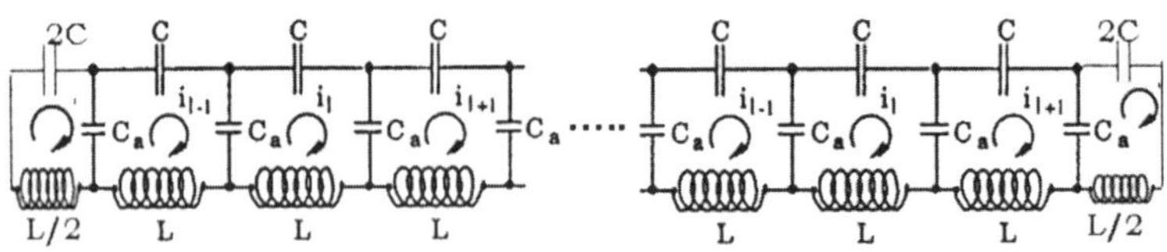

Figure 3. Equivalent circuit of a chain of capacitively coupled oscillators ended by half cells

The expression of the coupling coefficient in terms of the system bandwidth $\delta\omega = \omega_\pi - \omega_0$ is

$$k = \frac{\omega_\pi^2 - \omega_0^2}{2 \cdot \omega_0^2} \approx \frac{\delta\omega}{\omega_0} \tag{2}$$

and the field amplitudes in a finite chain with N cells with half-cell terminations are given by

$$Xn=A\cos(n\varphi) \qquad n=1,2....N \text{, with } \varphi=(q-1)\pi/(N-1) \quad q=1,2....N \qquad (3)$$

Different terminations and tuning conditions affect these results.
This is the case of the SPARC gun. In the following paragraph the effect of terminating and tuning in such system on the oscillation modes will be considered.

1.1. *Effect of termination in the oscillation modes of a two coupled cavity system*

Following the approach described in ref. [3] we can study the gun system starting from the first half cell by using SUPERFISH code with different boundary conditions. Two types of conditions can be imposed: Neumann ($E_t=0$, electric mirror, corresponding in the equivalent circuit to $i_{n+1}=i_n$) and Dirichlet ($H_t=0$, magnetic mirror, corresponding in the equivalent circuit to $i_{n+1}=-i_n$).
The Neumann-Neumann condition and Neumann-Dirichlet conditions on the half cell give respectively the frequency of the 0-mode and π mode of an infinite chain of capacitively coupled resonators described by the dispersion relationship (1).
Coupling the half cell to an equal frequency full cell, the Neumann-Neumann condition gives a 0-mode and a 2π/3-mode and Neumann-Dirichlet gives a π/3-mode and a π-mode. This modes identification can be immediately retrieved from the general solution (3) for a chain of N=4 oscillators ended by half cells, that can be obtained mirroring the half+full cell structure respect to the end plane.

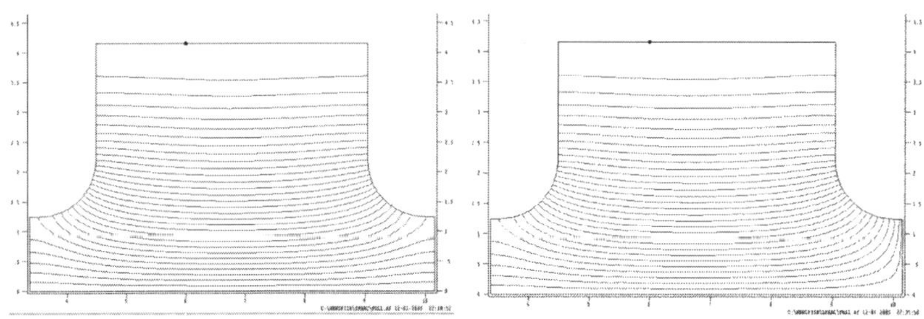

Figure 4. SUPERFISH computed contour lines in the full cell: (left) Neumann-Neumann condition, (right) Neumann-Dirichlet condition

The Neumann condition corresponds to closing the boundary by a metallic plate, while the Dirichlet condition is only a mathematical condition that is non-achievable and is equivalent to a perturbation of the full cell whose frequency rises from ω_R to $\omega_R \cdot \sqrt{(1+k)}$, i.e. of 2 MHz (fig.4). In fact in the first case the full cavity without losses can be described by a LC-circuit with an impedance $Z=j\omega L+1/j\omega C$, while the Dirichlet condition can be represented by the addition of a shunt capacitor C_a with a current $i_{n+1}=-i_n$ giving an impedance $Z= j\omega L+1/ j\omega C+2/ j\omega C_a$. This means that, in order to get the field flatness in π-mode in the physical situation with metallic plates on both ends, it is necessary to perturb the full cell rising its frequency (of 2 MHz) to the frequency value of the full cell closed by Neumann-Dirichlet conditions. The summary of the excited modes and relative frequencies in these different situations are reported in tables 1and 2.

Table 1. Modes frequency for different boundary conditions in the single cells

	Boundary-conditions: Neumann-Neumann	Boundary-conditions: Neumann-Dirichlet
Half cell	2851.6 MHz	2855.6 MHz
Full cell	2851.6 MHz	2853.6 MHz
Perturbed full cell	2853.6 MHz	

Table 2. Modes frequency for different boundary conditions in a system half cell+full cell

	Boundary-conditions: Neumann-Neumann	Boundary-conditions: Neumann-Dirichlet
Half cell + full cell	0-mode 2851.6 MHz 2π/3-mode 2854.6 MHz	π/3-mode 2852.6 MHz π-mode 2855.6 MHz
Half cell + perturbed full cell	π/3-mode-like 2852.6 MHz π-mode-like 2855.6 MHz	

The above frequency perturbation of the full cavity corresponds to shift the modes 0 and 2π/3 in a system composed by an half+full cell closed by metallic plates on both ends to the modes π/3 and π of an infinite chain of capacitively coupled oscillators. In the real situation the end full cell is not closed by a metal plate, since it is perturbed by a tube starting at the iris plane for the beam passage. This is a third type of boundary condition (open) that in terms of equivalent circuit corresponds to have $i_{n+1}=0$. The beam tube can be simulated in the lumped-constant model by a shunt capacitor C_a with the open condition

corresponding to an impedance $Z= j\omega L+1/\ j\omega C+2/\ j\omega C_a$. This gives a frequency increase of 1 MHz to the full cell frequency (that becomes $\omega_R \cdot \sqrt{(1+k/2)}$).
As a consequence in the coupled system composed by half cell+full cell+tube the 0 and π modes do not longer exist and the resonant frequencies of the whole structure correspond to $\pi/4$ and $3\pi/4$ modes. This modes identification can be immediately retrieved from the general solution for a chain of N=3 oscillators ended by full cells with open condition on both ends, that can be obtained mirroring the half+full cell+tube structure respect to the z=0 plane. For this kind of termination the field amplitudes are *sin-like* instead of *cos-like*[4]:

$$Xn=A\sin(n\varphi) \qquad n=1,2\ldots.N\text{ , with } \varphi=(q\pi)/(N+1) \quad q=1,2\ldots.N.$$

Since the tube increases the full cell frequency of 1 MHz, in order to get field-flatness in π-mode it is necessary to detune the full cavity of only 1 MHz by reducing the cavity radius (table 3). The field in the two modes before and after the full cell perturbation is shown in figures 5,6.

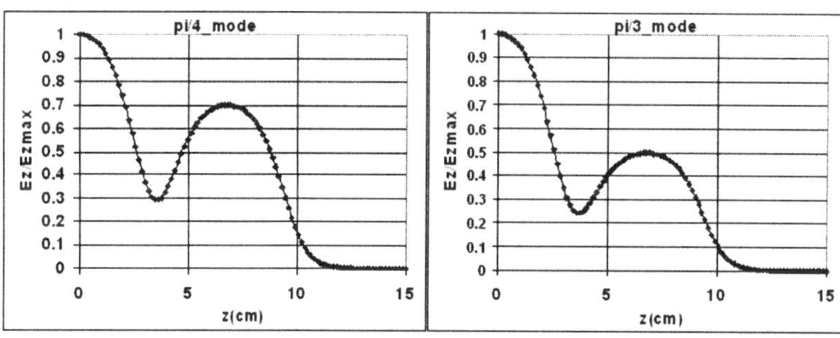

Figure 5. SUPERFISH computed fields in the half+full cell+tube in the low frequency mode: (left) before full cell radius perturbation, (right) after full cell radius perturbation

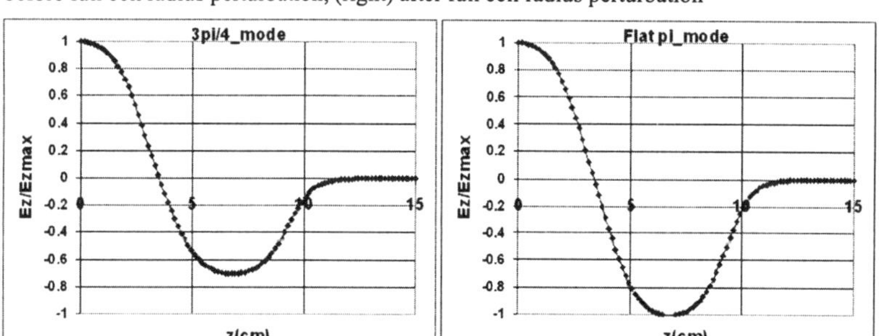

Figure 6. SUPERFISH computed fields in the half+full cell+tube in the high frequency mode: (left) before full cell radius perturbation, (right) after full cell radius perturbation

Table 3. Modes frequency for different tuning conditions in a system half cell+full cell+tube

	Boundary-conditions: Neumann-open
Full cell+tube	2852.6 MHz
Half cell + full cell+tube	π/4-mode 2852.15 MHz 3π/4-mode 2855.01 MHz
Perturbed full cell+tube	2853.6 MHz
Half cell + perturbed full cell +tube	π/3-mode-like 2852.6 MHz π-mode-like 2855.6 MHz

In conclusion in a half+full cell system two physical situations are possible:

1. End iris closed by metal plate (in this case equal frequencies give 0 and 2π/3 modes)
2. End iris with beam tube (equal frequencies give π/4 and 3π/4 modes)

In both cases the amount of correction can be derived from 0 and π modes of an infinite chain of capacitively coupled resonators that can be found by SUPERFISH calculations with different boundary conditions on a single half cell.

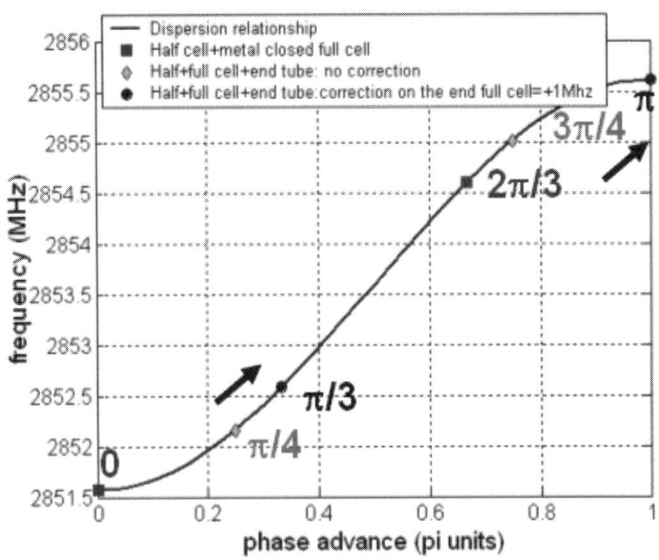

Figure 3. Modes shift due to the effect of terminating and tuning in our half+full cell system.

The amount of correction in the case #1 is $\Delta\omega \cong \delta\omega/2$ (2 MHz with our numbers) and in the case #2 is $\Delta\omega \cong \delta\omega/4$ (1 MHz with our numbers).
The correction is equivalent to induce a shift of $\pi/3$ in the modes 0 and $2\pi/3$ of the case 1 on the dispersion curve of an infinite chain of capacitively coupled cavities (fig.3), so that the resulting modes after correction become $\pi/3$-like and π-like.
The analysis has been carried out for a 1.5 cell-system, while the SPARC gun is a 1.6 cell system having a full cell shorter respect to the case examined above. However this does not change the previous conclusions. It only means that the radius of the full cell have been re-tuned in order to get the flat π mode and a small error is induced on the $\pi/3$ mode .

2. Practical consequences of the previous analysis

An uncorrect identification of the gun resonant modes as 0 and π modes can induce some errors in the definition of the coupling coefficient and in the mode-separation based techniques used for its measurement and for the electric field balancing in the high frequency mode.

2.1. *Coupling coefficient redefinition*

Following the previous analysis the correct definition of the coupling coefficient can be derived from the dispersion relationship (1) by using the measured frequencies in condition of flat π mode:

$$k = \frac{1}{2} \cdot \left[\frac{\omega_{\pi}^2 - \omega_{\pi/3}^2}{\omega_{\pi/3}^2 - \dfrac{\omega_{\pi}^2}{4}} \right] \tag{4}$$

The error induced on the k value if the measured frequencies f_1 and f_2 are attributed to 0 and π modes instead of $\pi/3$ and π is given by

$$\Delta k = \frac{f_2^2 - f_1^2}{2 \cdot f_1^2} - \frac{1}{2} \cdot \left[\frac{f_2^2 - f_1^2}{f_1^2 - \dfrac{f_2^2}{4}} \right] \tag{5}$$

giving a percentage error of

$$\frac{\Delta k}{k} = -\frac{1}{4} \cdot \left[\frac{f_2^2}{f_1^2} \right] \approx -\left(0.25 + \frac{\Delta f}{2 f_1} \right) \tag{6}$$

where $\Delta f = f_2 - f_1$ is the measured mode frequencies separation.
So the identification of the modes as 0 and π gives always an underestimation of the coupling coefficient > 25% attributing to the system bandwidth a smaller extension.

2.2. *Mode separation based RF measurements*

From the coupled circuit theory applied to two coupled cavities of different frequency it can be demonstrated [5] that, if the system supports the 0 and π modes, varying the frequency of one of the two cavities the quantity defined as

$$y = 0.5 \cdot (f_2^2 - f_1^2) / f_1^2 \tag{7}$$

is minimum and equal to the coupling coefficient k in correspondence to the minimum modes separation that occours when the two frequencies are made equal. The minimum mode separation corresponds also to equal amplitude fields in π mode and this method can be used to balance the fields in the π mode [6]. If this technique is directly applied to the gun for field balancing, assuming 0 and π modes, some errors occour, because the minimum modes frequency distance does not correspond exactly to a ratio $E_2/E_1=1$ (E_2=peak field in the cell#2,E_1=peak field in the cell#1). This is shown in figure 4 where the plots obtained by SUPERFISH simulation of the axial peak fields ratio vs mode separation for a two- cavity system composed by two half cells (where the 0 and π modes exist) and the SPARC gun are compared. One can observe that in the second case in correspondence to the minimum mode distance the fields ratio is no longer equal to 1 but is about 0.7 (equal frequency cavities giving a $3/4\pi$ mode). So only the bead-pull measurement can be used to achieve the required field flatness.

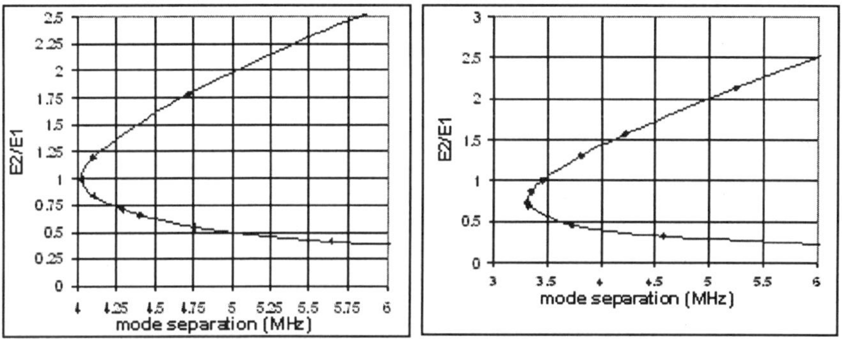

Figure 1. Ratio between the peak electric field in the cell#2 divided by the peak electric field in the cell#1 : (left) two half cells system (supporting 0 and π modes for equal frequencies), (right) SPARC gun (supporting π/3 and π modes-like for flat π mode)

As to the coupling coefficient measurement the parameter that must be measured is no longer given by the expression (7), but must be equal to

$$y = \frac{1}{2}\left[\frac{f_2^2 - f_1^2}{f_1^2 - \frac{f_2^2}{4}}\right] \tag{8}$$

Near to a minimum of modes separation this parameter is minimum and about equal to k. In this case (at difference of a 0-π mode system) in reality the minimum mode frequencies separation does not correspond exactly to the k value but the error that is done by assuming $k=y_{min}$ is only ~ 5%.

3. Conclusions

The analysis of the effect of tuning and terminating in the SPARC gun put in evidence that the usual identification of the two resonant modes as 0 and π is not correct, because within the dispersion nomenclature 0 and π modes exist only for half cell terminations. The cavities frequency is adjusted in order to get a flat π mode and from the resulting field amplitude ratio and resonant frequency in the two modes we found an equivalence between these two modes and the π/3 and π modes in the dispersion relationship of an infinite capacitively coupled chain of equal frequency oscillators.

It is not only a matter of nomenclature because the identification of the two modes as 0 and π modes can induce non negligible errors in the definition and measurement of the coupling coefficient and in a technique of field balancing based on the measure of mode separation for different perturbations of the frequency of one of the two cavities .

References

1. K.T. Mc Donald, "Design of the Laser-Driven RF Electron Gun for the BNL Accelerator Test Facility", *IEEE Transaction on Electron Devices*, **Vol. 35**, No.11,p. 2052
2. Nagle D.E,. Knapp, E.A.,Knapp B.C., "Coupled resonator model for standing wave accelerator tanks", *Rev. of Sci. Instr.* **Vol 38**, no. 11 (1967) 1583.
3. P. Fernandes, R. Parodi "On compensation of axial electric field unflatness in multicell accelerating structures", *Particle accelerators* (1984) **Vol. 14** pp. 185-199.
4. L. Picardi, C. Ronsivalle, B. Spataro "Numerical studies and measurements on the side-coupled drift tube linac (SCDTL) accelerating structure", *NIMB* **170** (2000) p. 219-229.
5. Rard de Leeuw "The accelerator injection chain of the elctron storage ring EUTERPE", *Thesis* (1996), p.43
6. D. T. Palmer, R.H. Miller, H. Winick, X.J. Wang, K. Batchelor, M. Woodle, I. Ben-Zvi, "Microwave Measurements of the BNL/SLAC/UCLA 1.6 Cell Photocathode RF gun", *Particle Accelerator Conference* (1995).

EMITTANCE DEGRADATION DUE TO WAKE FIELDS IN A HIGH BRIGHTNESS PHOTOINJECTOR*

M. FERRARIO, V. FUSCO†

INFN, Via E. Fermi 40,
00044 Frascati, Italia

M. MIGLIORATI, L. PALUMBO

La Sapienza University, Dip. di Energetica, Via A. Scarpa
Roma, Italia

Wake fields effects in addition to space charge forces may have an important impact during the emittance compensation process in a high brightness photo-injector. To study this effect we developed an upgraded version of the Homdyn code including off axis beam dynamics and wake fields.

Homdyn describes a bunch as a uniformly charged cylinder, divided in cylindrical slices; in the upgraded version each slice's centroid can be transversally displaced from the nominal axis thus inducing wake fields. When the bunch is short as compared to the beam pipe radius, wake fields for a single cavity are calculated using methods of diffraction theory; instead we use, for a periodic collection of cavities, an asymptotic wake field obtained numerically at SLAC and then fitted to a simple function.

As a first application we studied and verified a correction scheme for the SPARC photo-injector to control the bunch trajectory and angle at the entrance of the undulator. The correction scheme consists of a number of steering magnets and beam position monitors placed along the photo-injector. Two different steering approaches are analyzed and the emittance degradation is studied. The code demonstrates the steering positions and number do correct the bunch's orbit and angle and gives good results concerning the emittance degradation.

The emittance and energy spread degradation due to wake fields in the emittance meter experiment is also discussed.

---

* Work partially supported by the EU Commissioning in the sixth frame-work programme, contract no 011935 EURO-FEL-DS2

† valeria.fusco@lnf.infn.it

1. Introduction

Good performances of a free electron laser depend on the quality of the beam at the entrance of the undulator. To achieve short radiation wavelength and short gain length we need a high brightness beam that is a high current and a low emittance beam.

During the propagation of a beam trough a photo-injector, the beam may experience emittance growth. Emittance degradation is due to space charge and RF fields non linearties and longitudinal correlation along the bunch induced by transverse space charge, transverse RF fields and transverse wake fields.

Figure 1 represents the bunch in the phase space when the longitudinal correlation is induced along the bunch. The bunch has been divided into slices; each one is subject to a different transverse force, due to the space charge field, the RF field and the wake field, so each slice fills a different oriented area in the phase space assuming as a whole a fan shape. The consequence is that even if the emittance of each slice is preserved (if only linear forces are acting), the emittance over the whole bunch is not preserved.

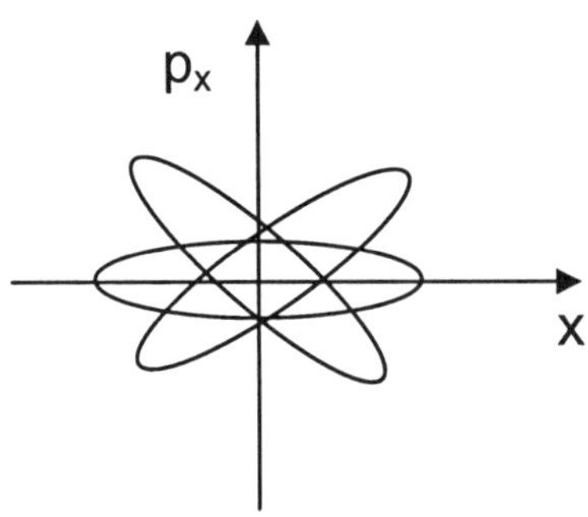

Figure 1. Fan shape of the emittance due to transverse fields induced by correlation along the bunch.

2. The Homdyn Model

Because wake fields may have an important impact during the space charge emittance compensation process [1], the Homdyn code has been developed to include off axis beam dynamics.

The Homdyn code [2] is a fast semi-analytical code whose main approximation consists in supposing the bunch as a uniformly charged cylinder. The cylinder is divided in slices whose evolution is described by differential equations for the envelopes and the centroids [3].

When the bunch is traveling on the axis each slice can evolve as in Figure 2a, that is after a time Δt each slice's radius and length can change under the space charge, the RF force, the solenoid force and eventually the length of the bunch can change also under the longitudinal wake field force. Anyway the slice's centroid can't be displaced from the axis and the bunch has to travel along the main axis.

The improvement is depicted in Figure 2b [4]: after a time Δt each slice's radius and length can change and moreover the slice's centroid can be displaced from the nominal axis.

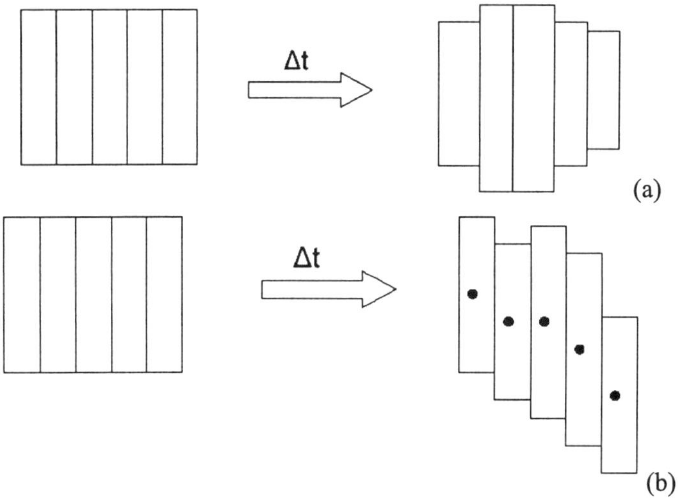

Figure 2. Multi-slice approximation in the Homdyn code: when the bunch is traveling on axis (a) and when it is traveling off axis (b).

The envelope equations for each slice of a bunch are:

$$\ddot{x}_s + \beta_s \gamma_s^{\,2} \dot{\beta}_s \dot{x}_s + (k_s^{rf} + k_s^{sol})^2 x_s = \frac{e}{\gamma_s^3 m} E_x^{sc}(\xi_s, A_s) + \left(\frac{4\varepsilon_n^{th}}{\gamma_s}\right)^2 \frac{1}{x_s^3}$$
$$\ddot{y}_s + \beta \gamma^2 \dot{\beta} \dot{y}_s + (k_s^{rf} + k_s^{sol})^2 y_s = \frac{e}{\gamma_s^3 m} E_y^{sc}(\xi_s, A_s) + \left(\frac{4\varepsilon_n^{th}}{\gamma_s}\right)^2 \frac{1}{y_s^3}. \quad (1)$$

where x_s and y_s are the single slice's radius, whilst the centroids motion for each slice traveling on the axis is simply represented by the following differential equation:

$$\ddot{z}_s = \dot{\beta}_s c$$
$$\dot{\beta}_s = \frac{e}{mc\gamma_s^3} E_z^{sc}(\xi_s, A_s) + \frac{e}{mc\gamma}\left[E_z^{rf}(z_{cs}, t) + E_{\|}^{w}(\xi_s)\right] \tag{2}$$

where e is the electron charge, m the electron rest mass and c the speed of light.

It is worth noting no off axis motion is included but only a longitudinal motion represented by z_s.
The Lorentz contraction factor for each slice is γ_s, being $\gamma_s=1/(1-\beta_s^2)^{1/2}$. The RF focusing gradient for each slice is[2]:

$$k_s^{rf} = \frac{e}{2\gamma_s m}\left(\frac{\partial E_z^{rf}}{\partial z} + \frac{\beta_s}{c}\frac{\partial E_z^{rf}}{\partial t}\right). \tag{3}$$

expressed through the linear expansion off-axis of the accelerating field $E_z^{rf}(z_s,t)$, the solenoid focusing gradient is:

$$k_s^{sol} = \left(\frac{eB_z(z_s)}{2\gamma_s m}\right)^2. \tag{4}$$

The transverse and longitudinal space charge is represented by the terms $E_x^{sc}(\xi_s,A_s)$, $E_y^{sc}(\xi_s,A_s)$ and $E_z^{sc}(\xi_s,A_s)$ respectively where $\xi_s=z_s-z_t$ and z_t is the position of the bunch's tail, A_s is the aspect ratio defined as $A_s=R/\gamma_s L_s$ where R is the radius of the bunch section [5] and L_s is the slice length. Finally in Eq. (2) the last term on the right hand side, $E_{\|}^{w}(\xi_s)$, represents the longitudinal wake fields (see section 3).
When the bunch is traveling off axis, the differential equations for the slice's centroids transverse motion are:

$$\ddot{x}_c + \beta\gamma^2\dot{\beta}\dot{x}_c + k_s^{rf}x_c = \frac{e}{\gamma^3 m}E_x^{sc}(\xi_s, A_s, d_{xc}) + \frac{e}{\gamma m}E_{\perp}^{w}(\xi_s, x_{cs})$$
$$\ddot{y}_c + \beta\gamma^2\dot{\beta}\dot{y}_c + k_s^{rf}y_c = \frac{e}{\gamma^3 m}E_y^{sc}(\xi_s, A_s, d_{yc}) + \frac{e}{\gamma m}E_{\perp}^{w}(\xi_s, y_{cs}) \tag{5}$$

As shown in Eq. (5) besides the RF field, each slice centroid can experience transverse wake fields, $E_\perp(\xi_s, x_{cs})$ (illustrated in section 3), and transverse space charge fields, $E_x^{sc}(\xi_s, A_s, d_{xc})$ and $E_y^{sc}(\xi_s, A_s, d_{yc})$, due to neighbor slices displaced d_{xc} or d_{yc} from the considered slice. It's worth noting Eq. (5), which is similar to Eq. (1), doesn't include the emittance term since there's no emittance preassure acting on the slice's centroid.

3. Wake Fields in the Homdyn Model

Let's consider a point charge Q (the leading charge) and a test charge q (the trailing charge) situated behind Q. If the charges are moving in an enclosed space and the conducting wall has a non zero resistivity or the devices have wall discontinuities, the electromagnetic field generated by the leading charge is scattered by the wall thus affecting the trailing charge [5]. These electromagnetic fields are called wake fields since, from causality, the leading charge cannot affect itself nor any charge in front of it but only charges lying behind it.

As a consequence an accelerator device whose shape is not uniform induces longitudinal wake fields and when the devices are misaligned respect to a nominal axis or the bunch is traveling off axis it induces transverse wake fields. Transverse wake fields contribute to emittance degradation.

We distinguish the case the bunch travels along a single cavity and a periodic collection of cavities.

3.1. *Single Cavity*

When the bunch's length σ is much smaller than the beam pipe radius a, $\sigma << a$, methods of diffraction theory [6] are used to calculate the impedance at high frequencies, $\omega >> c/a$, where c is the speed of light. The model suppose each structure as a pill box cavity, whose geometric dimensions are: a the beam pipe radius, b the cavity radius and g its length. When a bunch reaches the edge of the cavity, the electromagnetic field is just the one that would occur when a plane wave passes trough a hole; with this hypothesis it is possible to use the classical diffraction theory of optics to calculate the fields.

According to it, the longitudinal and transverse wake potential are respectively [7]:

$$W_{\parallel 0}(s) = \frac{Z_0 c}{\sqrt{2}\pi^2 a}\sqrt{\frac{g}{s}}$$
$$W_{\perp 0}(s) = \frac{2^{3/2} Z_0 c}{\pi^2 a^3}\sqrt{gs}. \tag{6}$$

where Z_0 is the characteristic impedance and s the longitudinal coordinate inside the bunch, being $s=0$ the bunch's head.

Eq. (6) are the fields produced by a point charge on a test charge at a distance s: they represent Green functions. One can calculate the longitudinal and transverse wake potential of an arbitrary distribution by making the convolution of the bunch distribution with the Green functions.

For a uniformly charged cylinder, as the one considered in Homdyn, we can obtain the wake potential and finally the wake fields to be inserted in Eq. (2) and Eq. (5) [8]:

$$E_{\parallel}^{w}(s) = \frac{q}{L}\frac{2}{\sqrt{2}\pi^2 a\varepsilon_0}\sqrt{\frac{s}{g}}$$
$$E_{\perp}^{w}(s,x) = \frac{q}{L}\frac{2^{5/2}}{3\pi^2 a^3 \varepsilon_0}\frac{s^{3/2}}{\sqrt{g}}x \tag{7}$$

where q is the bunch's charge, L its length and x is the bunch's head displacement from the nominal axis, having supposed the other slices' displacement do not differ much from the head.

It's worth noting that both the longitudinal and transverse wakes do not depend on the cavity radius b. Infact part of the diffracted field, generated when the leading edge of the bunch enters the cavity, will propagate in the cavity; if the bunch's rms length σ is shorter than the cavity radius b, then the geometrical condition $g<(b-a)^2/2\sigma$ is verified and the scattered field coming from the upper wall of the cavity will never reach the tail of the bunch itself: this is called cavity regime.

3.2. *Periodic Structures*

For a collection of cavities, Eq. (6) can't be used anymore because the wake fields, along the cavities, do not sum in phase and the result would be an overestimation of the effects. An asymptotic wake field, for a periodic collection of cavities of period p, obtained numerically at SLAC [9] and then fitted to a simple function is used instead. Such wake field is thus valid after a certain number of cavities given by [10]:

$$N_{crit} = \frac{a^2}{2g(\sigma + \frac{2a}{\gamma})} \tag{8}$$

The Green functions are:

$$\begin{aligned} W_{\|0}(s) &= \frac{Z_0 c}{\pi a^2} e^{-\sqrt{s/s_1}} \\ W_{\perp 0}(s) &= \frac{4 Z_0 c s_2}{\pi a^4} \left[1 - \left(1 + \sqrt{\frac{s}{s_2}} \right) e^{-\sqrt{s/s_2}} \right]. \end{aligned} \tag{9}$$

where:

$$\begin{aligned} s_1 &= 0.41 \frac{a^{1.8} g^{1.6}}{p^{2.4}} \\ s_2 &= 0.17 \frac{a^{1.79} g^{0.38}}{p^{1.17}}. \end{aligned} \tag{10}$$

Eq. (9) are valid when $s/p \leq 0.15$, $0.34 \leq a/p \leq 0.69$ and $0.54 \leq g/p \leq 0.89$. Such constraints are fulfilled for the SLAC cavities used in SPARC where $a=11.6\ mm$ $g=29.2\ mm$ and $p=35.0\ mm$.

Following the same procedure of a single cavity one can obtain [11] the longitudinal and transverse wake potential from the Green functions and thus the wake fields for a uniform charge distribution:

$$\begin{aligned} E_{\|}^{w^{in}}(s) &= \frac{2 Z_0 c s_1 q}{\pi a^2 L} \left[1 - e^{-\sqrt{s/s_1}} \left(1 + \sqrt{\frac{s}{s_1}} \right) \right] \\ E_{\perp}^{w^{in}}(s,x) &= \frac{4 Z_0 c s_2^2 q x}{\pi a^4 L} \left[-6 + \frac{s}{s_2} + 2 e^{-\sqrt{s/s_2}} \left(3 + 3\sqrt{\frac{s}{s_2}} + \frac{s}{s_2} \right) \right]. \end{aligned} \tag{11}$$

Eq. (11) can be inserted in Eq. (2) and Eq. (5). In this case the displacement x is, as for the single cavity, the offset of the first slice.

Anyway to obtain a better evaluation of the beam dynamics due to wake fields, the fields to be inserted in Eq. (2) and Eq. (5) can be obtained as the superposition of the wake fields generated by the single slices with x the single's slice offset. The wake fields generated outside a bunch, that is when $s>L$, are:

$$E_{\parallel}^{w^{out}}(s) = \frac{2Z_0 c s_1 q}{\pi a^2 L}\left[e^{-\sqrt{(s-L)/s_1}}\left(1+\sqrt{\frac{s-L}{s_1}}\right) - e^{-\sqrt{s/s_1}}\left(1+\sqrt{\frac{s}{s_1}}\right)\right]$$

$$E_{\perp}^{w^{out}}(s,x) = \frac{4Z_0 c s_2^2 q x}{\pi a^4 L}\left[\begin{array}{l}\frac{L}{s_2} + 2e^{-\sqrt{s/s_2}}\left(3+3\sqrt{\frac{s}{s_2}}+\frac{s}{s_2}\right)+ \\ -2e^{-\sqrt{(s-L)/s_2}}\left(3+3\sqrt{\frac{s-L}{s_2}}+\frac{s-L}{s_2}\right)\end{array}\right]. \tag{12}$$

As illustrated in Figure 3 each slice generates a longitudinal and a transverse wake field. The fields inside the slices (Eq. (11)) and outside the slices(Eq. (12)) sum up to give the whole field, the displacement x depending on the single slices offset. Thus the convoluted fields to be inserted in Eq. (2) and Eq. (5) are:

$$E_{\parallel}^{w}(s) = \sum_{1}^{N-N_s}\left[E_{\parallel\, i}^{w^{out}}(s) + E_{\parallel\, i=N_s}^{w^{in}}(s)\right] \tag{13}$$

$$E_{\perp}^{w}(s) = \sum_{1}^{N-N_s}\left[E_{\perp i}^{w^{out}}(s) + E_{\perp i=N_s}^{w^{in}}(s)\right] \tag{14}$$

where N_s is the considered slice number.

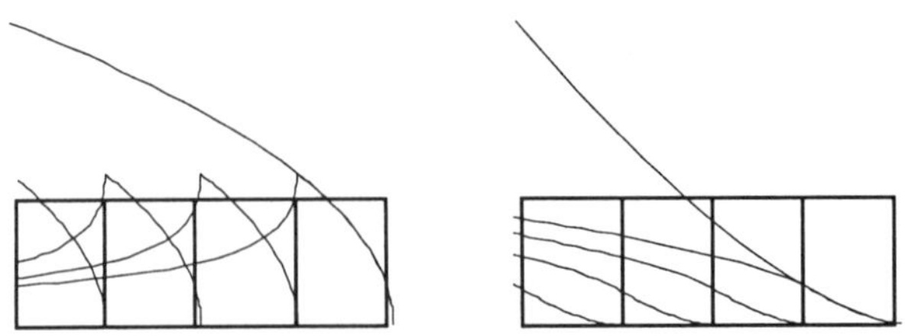

Figure 3. Longitudinal (left hand side) and transverse (right hand side) wake field generated by single slices.

3.3. *Energy spread and offset due to wake fields*

The energy spread and the offset computed by Homdyn when wake fields are switched on have been compared with the energy spread and the offset calculated analytically for a single cavity and a periodic structure. The comparison is good thus the simple expressions obtained give the possibility to evaluate quickly the energy spread and the offset.

The energy spread is defined in Homdyn as:

$$\frac{\Delta\gamma}{\gamma} = \frac{\sqrt{<(\gamma - <\gamma>)^2>}}{<\gamma>}. \tag{15}$$

In the case the space charge and the RF fields are switched off in the code, the energy spread is only due to the longitudinal wake field thus:

$$\frac{\Delta E}{E_0} = \sigma_E = \frac{e l_{str}\sqrt{<(E_{\|}^w(s) - <E_{\|}^w(s)>)^2>}}{E_0}. \tag{16}$$

where e is the electron charge, E_0 is the average energy, $E^w{}_{\|}(s)$ is the longitudinal wake field per unit length of the path of a single cavity (Eq. (7)) or a periodic structure (Eq. (11), Eq. (13) and Eq. (14)) and l_{str} is the length of the structure where wake fields are acting. From Eq. (16) we have

$$\sigma_E^2 = \frac{e^2 l_{str}^2}{E_0^2 q}\int_{-\infty}^{+\infty}\lambda(s)(E_{\|}^w(s) - <E_{\|}^w(s)>)^2 ds \tag{17}$$

and

$$<E_{\|}^w(s)> = \frac{1}{q}\int_{-\infty}^{+\infty}\lambda(s)E_{\|}^w(s)ds = k_{\|} \tag{18}$$

With $k_{\|}$ we introduce the average energy loss, the so called loss factor. In our treatment the loss factor is measured as V/m. Solving Eq. (17) we obtain:

$$\sigma_E^2 = \frac{e^2 l_{str}^2}{E_0^2}\left[\frac{1}{L}\int_0^L (E_{\|}^w(s))^2 ds - k_{\|}^2\right] \tag{19}$$

Introducing the fields for a single cavity or a periodic structure we can solve Eq. (19). For a single cavity we obtain a very simple expression here shown:

$$\sigma_E = \frac{1}{3}\left(\frac{Z_0 cq}{\pi^2 a}\right)\sqrt{\frac{1}{gL}}\frac{el_{str}}{E_0} \tag{20}$$

In the case of the transverse wake fields we can evaluate their effects considering the transverse offset given to the bunch. Using Eq. (5) in the time domain, we obtain :

$$< x_c > \cong \frac{k_\perp}{\gamma mc^2 / e}\frac{l_{str}^2}{2\beta^2} + x_c^{'}(0)z + x_c(0) \tag{21}$$

Where $x_c^{'}(0)$ and $x_c(0)$ are the initial angle and offset respectively and

$$k_\perp = \frac{1}{q}\int_{-\infty}^{+\infty}\lambda(s)E_\perp^w(s)ds \tag{22}$$

It's worth noting Eq. (21) has been derived supposing the offset of the slices is constant and equal to the offset of the first slice.
Inserting the transverse fields for a single cavity or a periodic structure we can solve Eq. (22). For a single cavity we have:

$$< x_c > = \frac{e}{\beta^2 \gamma mc^2}\frac{2^{7/2}}{15}\frac{Z_0 cqx}{\pi^2 a^3}\sqrt{\frac{L}{g}}\frac{z^2}{2} \tag{23}$$

when $x_c^{'}(0)=0$ and $x_c(0)=0$.

Such analytical calculations have been compared with the results computed by the Homdyn code for the energy spread and the transverse offset.

Table 1 summarizes the energy spread, $\Delta\gamma/\gamma$, and the offset, x_c, for a single cavity of the SLAC type, as computed by Homdyn and by the analytical calculation, versus the average energy, E_0, of the bunch.

Table 2, as for a single cavity, gives the energy spread and the offset for a periodic structure of the SLAC type, versus the average energy of the bunch as computed by Homdyn and by the analytical calculation.

Finally we evaluate the approximation made in Homdyn to calculate the effects of the transverse wake fields when we only consider the offset of the first slice compared to the case, described in the preceding section, when every slice offset is considered in Homdyn and the fields, both longitudinal and transverse, are obtained as a superposition of the fields generated by the single slices.

Table 1. Energy spread in percent, $\Delta\gamma/\gamma\%$, and offset, x_c, versus the bunch's average energy, E_0, for a single cavity obtained with the analytic calculation and as computed by the Homdyn code.

E_0 [MeV]	$\Delta\gamma/\gamma\%$ Homdyn	$\Delta\gamma/\gamma\%$ Analytic	x_c [mm] Homdyn	x_c [mm] Analytic
1	0.07	0.10	$6.3\ 10^{-4}$	$5.6\ 10^{-4}$
3	0.029	0.034	$2.1\ 10^{-4}$	$2.1\ 10^{-4}$
5	0.019	0.021	$1.4\ 10^{-4}$	$1.4\ 10^{-4}$
9	0.011	0.011	$0.87\ 10^{-4}$	$0.80\ 10^{-4}$
30	0.0035	0.0034	$0.28\ 10^{-4}$	$0.25\ 10^{-4}$
50	0.0021	0.0021	$0.17\ 10^{-4}$	$0.15\ 10^{-4}$
70	0.0015	0.0015	$0.12\ 10^{-4}$	$0.11\ 10^{-4}$

Table 2. Energy spread in percent, $\Delta\gamma/\gamma\%$, and offset, x_c, versus the bunch's average energy, E_0, for a periodic structure obtained with the analytic calculation and as computed by the Homdyn code.

E [MeV]	$\Delta\gamma/\gamma\%$ Homdyn	$\Delta\gamma/\gamma\%$ Analytic	x_c [mm] Homdyn	x_c [mm] Analytic
1	2.9	3.0	6.3	7.2
3	2.4	2.6	1.4	1.5
5	1.7	1.7	0.77	0.78
9	0.99	0.67	0.42	0.42
30	0.31	0.30	0.13	0.13
50	0.19	0.18	0.077	0.077
70	0.13	0.13	0.056	0.055

Table 3. Energy spread in percent, $\Delta\gamma/\gamma\%$, and offset, x_c, versus the bunch's average energy, E_0, for a periodic structure obtained with and without wakefields convolution.

E [MeV]	$\Delta\gamma/\gamma\%$ Noconv	$\Delta\gamma/\gamma\%$ Conv	x_c [mm] Noconv	x_c [mm] Conv
1	2.9	3.5	6.3	7.9
3	2.4	2.5	1.4	1.5
5	1.68	1.69	0.77	0.90
9	0.99	0.99	0.42	0.50
30	0.31	0.31	0.13	0.15
50	0.19	0.18	0.077	0.091
70	0.13	0.13	0.055	0.065

This gives an estimation of the mistake made in the analytical calculation considering only the first slice's offset. Thus Table 3 compares the energy spread and the offset when Homdyn is used in the convolution configuration or not. The table demonstrates there is a little difference for the energy spread case whilst the difference increases for the offset case.

4. Emittance Computation

The total rms normalized emittance is calculated in the code following the normalized rms emittance definition [12]:

$$\varepsilon_{nx} = (<(x-<x>)^2><(\beta\gamma x'-<\beta\gamma x'>)^2> \\ -<(x-<x>)(\beta\gamma x'-<\beta\gamma x'>)>^2)^{\frac{1}{2}}. \tag{24}$$

Such calculation is made following the code's approach which supposes the bunch divided in cylindrical slices.
Let's take S slices each containing M particles thus the whole bunch contains $N=M{\cdot}S$ particles. The media of Eq. (24) can be expanded in the following way:

$$<>=\frac{1}{N}\sum_{n=1}^{N}[\;]=\frac{1}{S\cdot M}\sum_{s=1}^{S}\sum_{m=1}^{M}[\;]=\frac{1}{S}\sum_{s=1}^{S}<>_s \tag{25}$$

For a uniform cylindrical charge distribution, the following relation holds:

$$<x^2>_s=\frac{X_s^2}{4} \tag{26}$$

and

$$<\frac{X_s^2}{4}>=\frac{1}{S}\sum_{s=1}^{S}\frac{X_s^2}{4} \tag{27}$$

where X_s is the slice radius.
Solving Eq. (24) we obtain the following results: when all the slices lie on the same axis, the emittance is only given by the 'envelope' emittance

$$\left(\varepsilon_n^e\right)^2=<\frac{X^2}{4}><\frac{p_X^2}{4}>-<\frac{Xp_X}{4}>^2 \tag{28}$$

where we omitted the symbol s for the slice and $p_X=\beta\gamma dX/dz$ is the transverse momentum.

If the slices do not lie on the same axis then the emittance is not simply given by the envelope emittance but it is given by the quadratic sum of the envelope emittance and two more terms, the 'centroids' emittance

$$\left(\varepsilon_n^{cent}\right)^2 = <(x_c - <x_c>)^2><(p_{xc} - <p_{xc}>)^2> + \\ - <(x_c - <x_c>)(p_{xc} - <p_{xc}>)>^2 \tag{29}$$

and the 'cross' emittance

$$\left(\varepsilon_n^{cross}\right)^2 = <\frac{X^2}{4}><(p_{xc} - <p_{xc}>)^2> + \\ + <\frac{p_X^2}{4}><(x_c - <x_c>)^2> + \\ - 2<\frac{Xp_X}{4}><(x_c - <x_c>)(p_{xc} - <p_{xc}>)> \tag{30}$$

Each slice's envelope and each slice's centroid can be represented in the phase space x, p_x and the spread of their distributions around their media represent the envelope emittance and the centroid emittance.
Let's suppose the bunch has a zero envelope emittance that is the slice's envelopes lie on a straight line as in Figure 4a and the centroid emittance is zero as well, being for example $p_{xc}=0$: this is represented in Figure 4b.

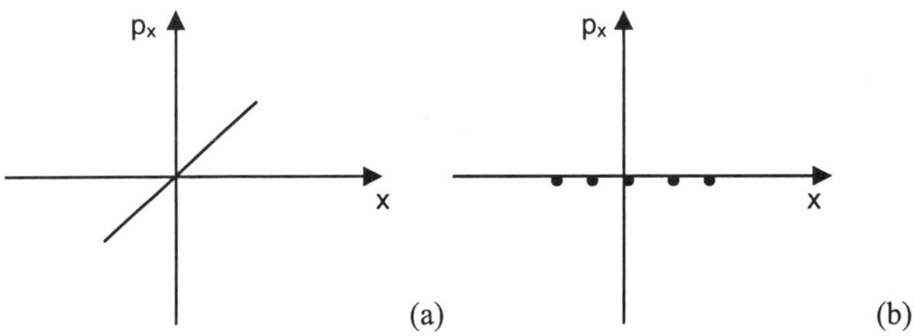

Figure 4.Slices' envelope lying (dots) on a straight line (a) and slices' centroids (dots) with $p_{xc}=0$

Nonetheless the cross emittance, thus the total emittance represented in Figure 5, can be different from zero, being from Eq. (30)

$$\left(\varepsilon_n^{cross}\right)^2 = + < \frac{p_X^2}{4} >< (x_c - < x_c >)^2 > \qquad (31)$$

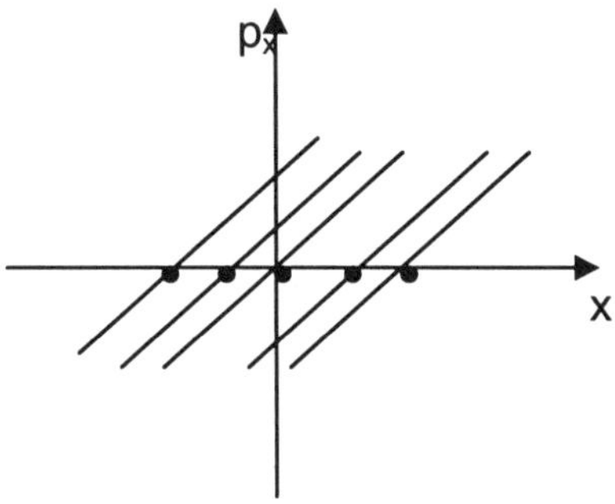

Figure 5. Cross emittance due to the distribution of the centroids and the envelopes of each slice.

The emittance complete expression of an off axis bunch has been inserted in the Homdyn code and it has been tested using the Parmela code [4].

5. Emittance Degradation Study in the SPARC Photo-injector

The SPARC project consists of a high brightness photo-injector for advanced accelerator experiments including a free electron laser. The whole device can be divided into two parts: the undulator devoted to the production of the radiation and the photo-injector whose aim is the production of a high quality electron beam to fulfill FEL requirements. Thus the Homdyn code, with its improvements on off axis beam dynamics has been used to study the bunch quality along the photo-injector until the entrance of the undulator.

The photo-injector consists of an electron source and accelerator sections.

The electron source for SPARC is a 1.6 cell RF gun operating at *2.856 GHz* with a high peak field of *120 MV/m*. The RF gun includes a metallic photo-cathode which is illuminated by a temporally flat, pico-second laser source. The system generates a *5.6 MeV* electron beam which is accelerated to *155 MeV* by three traveling wave sections of the SLAC type.

5.1. *The emittance meter experiment: emittance and energy spread degradation*

As a first application, we studied the emittance and energy spread degradation in the emittance meter experiment: a very accurate measure of the emittance evolution downstream the RF gun is required to evaluate if the compensation process is fulfilled.

The emittance meter is a movable device [13] whose aim is the emittance's measure in different positions along the nominal axis; it can be moved thank to a bellow structure, whose triangular shape, generates wake fields. Supposing the bellows as pill box cavities, Homdyn allows the analysis of the energy spread due to longitudinal wake fields introduced by the devices and emittance degradation due to transverse wake fields when the emittance-meter is off axis or the bunch is travellling off axis.

For an emittance-meter *1.5m* long, the obtained energy spread is shown in Figure 6.

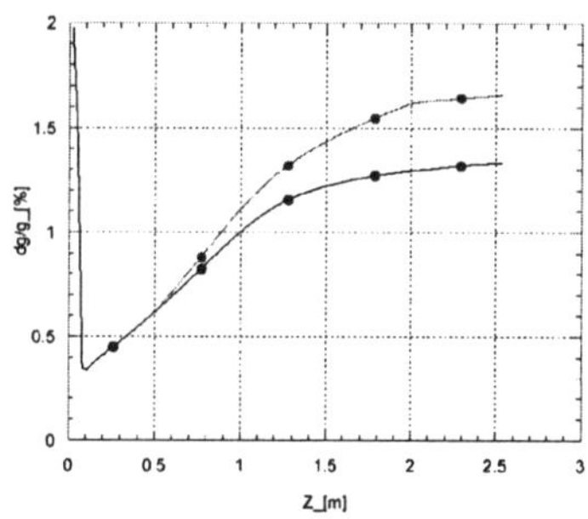

Figure 6. Energy spread versus position with (upper plot) the emittance-meter and without.

Figure 7 represents the emittance variation in percent versus the emittance-meter offset from the nominal axis.

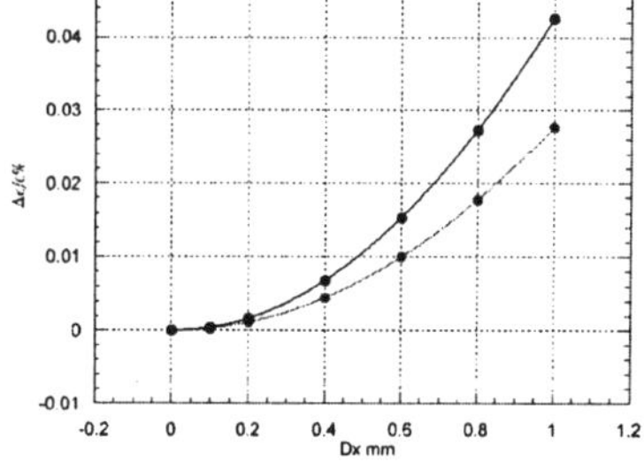

Figure 7. Emittance degradation in percent versus the emittance-meter offset Δx respect to the nominal axis, when $z=1.5m$ (booster position, upper plot) and $z=2m$ (at the end of the emittance meter).

5.2. *Misalignment correction scheme study in the SPARC photo-injector*

As for the emittance-meter, a bunch can travel off axis for different reasons: for example a laser pointing instability can directly generate the bunch off axis or the structures can be misaligned. In this case RF fields transverse components, space charge effects and transverse wake fields can further push the bunch off axis and increase the emittance. Thus the bunch has to be driven to the nominal trajectory with the help of steering magnets and beam position monitors (BPMs), placed along the accelerator structures.

To calculate the right angle in the vertical and horizontal plane that correct the trajectory, we use a transfer matrix between two points, for example the steering and the following BPMs or two adjacent steerings [3]

$$\begin{pmatrix} x \\ x' \\ y \\ y' \end{pmatrix}_2 = \begin{pmatrix} a & b & e & f \\ c & d & g & h \\ i & l & o & p \\ m & n & q & r \end{pmatrix} \begin{pmatrix} x \\ x' \\ y \\ y' \end{pmatrix}_1 \tag{32}$$

The matrix calculation is made up of two steps: the first step consists in the calculation of the matrix elements: we turn off the steering magnets, $x'=y'=0$, we obtain the nominal position on the following steering or BPM, x_{2n} and y_{2n}

$$\begin{cases} x_{2n} = ax_1 + ey_1 \\ y_{2n} = iy_1 + ox_1 \end{cases} \tag{33}$$

Turning on the horizontal steer we get the new positions x_{2k} and y_{2k}

$$\begin{cases} x_{2k} = x_{2n} + fy' \\ y_{2k} = y_{2n} + ly' \end{cases} \tag{34}$$

The second step determines the horizontal and vertical angle of the sterring, being known the matrix element and the position wished on the following steer or BPM:

$$\begin{cases} bx' + y' = x_2 \\ px' + ly' = y_2 \end{cases} \tag{35}$$

It's worth noting such matrix determines the horizontal and vertical angle for one steering. The general case of more than one steering is solved using the superposition principle thus adding the contribution of each steering coming before the ending point.

The above treatment includes the case the horizontal and vertical plane are coupled, as in a solenoid magnetic field.

Figure 8 is a scheme of the steering magnets and the beam position monitors along the SPARC photo-injector: a steering is positioned just outside the gun, two in each traveling wave structure; one more at the end of the linac, after the focusing quadrupoles and finally the last one just before the undulator. Outside the linac, beam position monitors are placed after every steering whilst they are placed after every two steerings inside the linac structure.

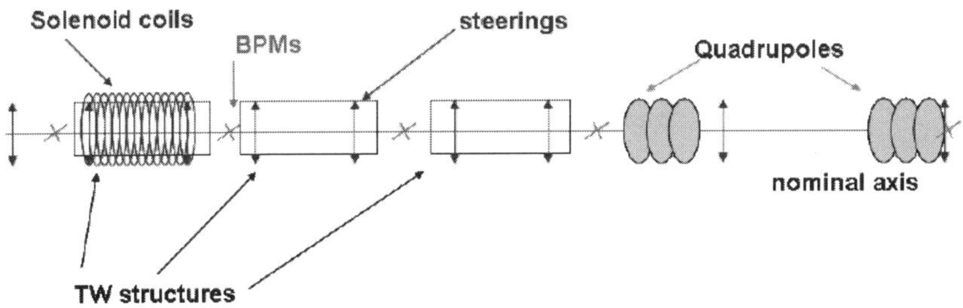

Figure 8. Steering magnets and beam position monitors positions in the SPARC photo-injector.

As a preliminary study of the emittance degradation, we analyze a case in which the bunch's centroid is driven further and further away from the nominal axis as shown in Figure 9.

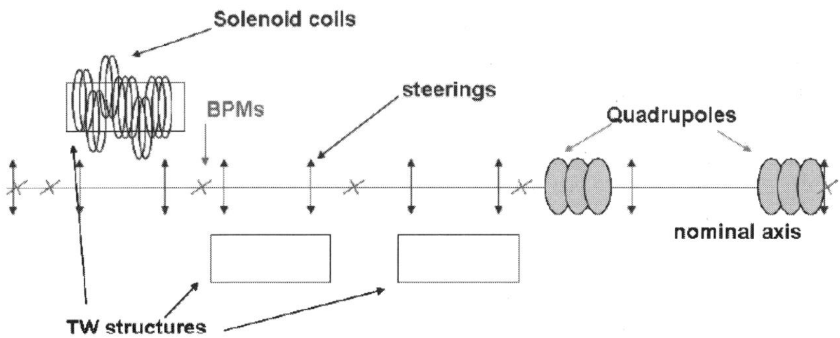

Figure 9.Traveling waves and solenoid coils misalignment in the SPARC photo-injector.

The bunch is generated on axis whilst the traveling waves can be transversally displaced with respect the nominal axis; besides, the thirteen coils forming the solenoid of the first traveling wave can be independently displaced as well: a study of different misaligned combinations of the solenoid coils shows the worst configuration as concern the offset induced on the bunch. The configuration is summarized in Table 4.

To calculate the right angle in the vertical and horizontal plane that corrects the trajectory, the transfer matrix between two points has been used [4].

We first correct the misaligned configuration to drive the centroid trajectory back to the nominal axis. We use the Homdyn code as if we were simulating a virtual experiment; this means we try to align the bunch on the BPMs after each steering and we read the centroid position in the BPMs. Therefore only one steering before each BPMs is used.

Table 4. Solenoid coils' and TWs offset respect to the nominal axis.

Device	Δx mm	Δy mm
Solenoid coil 1	0.0	0.1
Solenoid coils 2-3-4-5-6	0.1	0.0
Solenoid coils 7-8-9-10-11-12-13	0.0	-0.1
TW1	0.1	0.1
TW2	-0.1	-0.1
TW3	-0.1	-0.1

The centroid position is successfully driven back to the nominal axis, anyway Table 5 shows an emittance which is not perfectly compensated. The reason is that even if the centroids travels back to the nominal axis, the traveling wave structures are misaligned thus the transverse wake fields exist and cause an emittance's growth that should be minimized.

Table 5. Horizontal and transverse emittance with and without steering correction when the bunch's centroid is driven back to the nominal axis.

ε_n nominal	ε_{nx} steer off	ε_{nx} steer on	ε_{ny} steer off	ε_{ny} steer on
0.79 μm	2.95 μm	1.08 μm	1.12 μm	1.06 μm

Thus the technique used is the beam based alignment: it consists in forcing the bunch to undergo bumps and pass along the traveling wave axis as shown in Figure 10. Such technique uses all the steerings placed inside the linac instead of one per linac as in the previous case. We use again a transfer matrix to determine the steering angles.

Plots in Figure 11 represent the centroid trajectory with and without steerings and demonstrate that the centroid travels on the axis of each traveling wave structure as in Figure 10. Finally Table 6 demonstrates the emittance obtained is perfectly compensated.

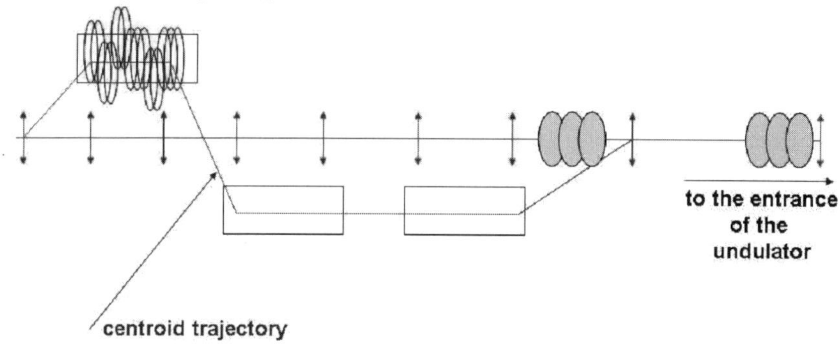

Figure 10.Centroid's trajectory along the linac with the beam based alignment technique.

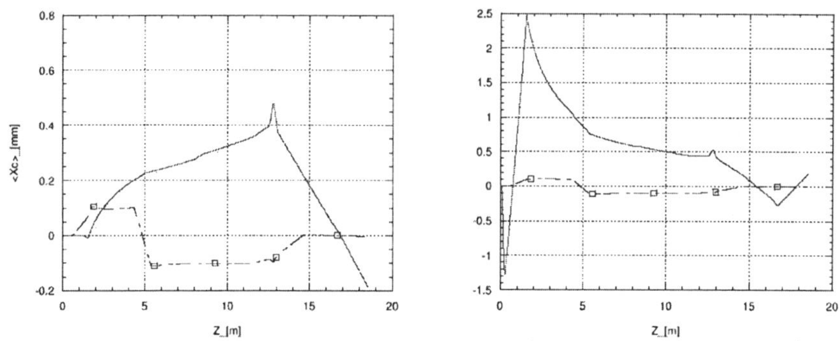

Figure 19.Horizontal and vertical centroid's trajectory along the linac using the bump technique with and without steering.

Table 6. Horizontal and transverse emittance with and without steering correction when the beam based alignment technique is used.

ε_n nominal	ε_{nx} steer off	ε_{nx} steer on	ε_{ny} steer off	ε_{ny} steer on
0.79 μm	2.95 μm	0.79 μm	1.12 μm	0.79 μm

6. Conclusions

The Homdyn model has been described together with its improvements: off axis beam dynamics has been included in the code allowing the study of wake fields. Infact wake fields, besides space charge and RF fields, may have an important impact in the emittance compensation process.

The analytical formula used for the wake fields has been described: wake fields for a single cavity, in the diffraction model limit, and wake fields for a periodic structure, for the SLAC type cavity, can be considered in the Homdyn computation.

Comparisons between Homdyn and analytical calculations, concerning the energy spread and the offset induced by wake fields, has been also discussed together with the description of the emittance, calculated using its statistical definition.

The code has been applied to the emittance-meter of the SPARC project to evaluate the energy spread and the emittance degradation induced by the bellow structure. Finally Homdyn allowed the study and the project of the steering magnets positions; the choice proposed is an example of emittance compensation process strongly influenced by wake fields.

References

1. M. Ferrario, M. Boscolo, V. Fusco et al., Proc. of The Physics and Application of High Brightness Electron Beams (2002).
2. M. Ferrario et al., Proc. of The Physics and Application of High Brightness Electron Beams (2000).
3. M. Reiser, Theory and design of charged particle beams, Wiley-Interscience Pub. (1994).
4. V. Fusco, Beam dynamics and collective effects in the Sparc project, phD thesis (2005).
5. L. Palumbo et al., Wake fields and impedante, Proc. of CAS (1993).
6. J. Lawson, Rutherford High Energy Laboratory Report, RHEL/M144 (1968).
7. K. Bane, M. Sands, Wake fields of very short bunches in an accelerating cavity(2001), Proc. Of Workshop on Impedance Beyond cut off (1987).
8. V. Fusco et al., Wake fields effects in the Sparc photo-injector, Proc. of EPAC (2004).
9. K. Bane, SLAC-PUB 9663, (2003).
10. R. B. Palmer, A qualitative study of wake fields for very short bunches, Tech. Rep. BNL (1987).
11. V. Fusco et al., Wake fields effects in the SPARC linac, Proc. of SPIE (2004).

12. J. Buon, Beam phase space and emittance, Proc. of CAS (1992).
13. A. Cianchi et al., Design study of a movable emittance meter device for the SPARC photo-injector, Proc of EPAC (2004).

SIMULATIONS OF COHERENT SYNCHROTRON RADIATION EFFECTS IN ELECTRON MACHINES*

M. MIGLIORATI, A. SCHIAVI

Dipartimento Di Energetica, Università di Roma "La Sapienza", P.le A. Moro, 5-00100 Roma, Italy

G. DATTOLI

ENEA, Unità Tecnico Scientifica Tecnologie Fisiche Avanzate, Via Enrico Fermi 45, Frascati, Rome, Italy

Coherent synchrotron radiation (CSR) generated by high intensity electron beams can be a source of undesirable effects limiting the performance of storage rings. The complexity of the physical mechanisms underlying the interplay between the electron beam and the CSR demands for reliable simulation codes. In the past, codes based on Lie algebraic techniques have been very efficient to treat transport problems in accelerators. The extension of these methods to the non linear case is ideally suited to treat wakefields - beam interaction. In this paper we report on the development of a numerical code, based on the solution of the Vlasov equation, which includes the non linear contribution due to wakefields. The proposed solution method exploits an algebraic technique that uses the exponential operators. We show that, in the case of CSR wakefields, the integration procedure is capable of reproducing the onset of an instability which leads to microbunching of the beam thus increasing the CSR at short wavelengths. In addition, considerations on the threshold of the instability for Gaussian bunches is also reported.

1. Introduction

Synchrotron radiation generated by high intensity electron beams in magnetic bending fields becomes coherent when the emission of electrons in a bunch is in phase. The consequence is a quadratic dependence of the emitted power on the number of electrons in the bunch[1]. The effect of this coherent emission may be the reason of a dilution of the beam qualities due to the interaction induced microbunching and to the self amplification of

*This work has been partially supported by the EU commission in the sixth framework programme, contract no. 011935 EUROFEL

the radiation determined by the CSR interaction[2].

The physical mechanisms underlying the CSR instability are not yet well understood. It is however clear that it can play an important role in storage rings where it can manifest itself as a burst of radiation in the terahertz region[3,4] with a consequence deterioration of the beam qualities, as e.g. undesired emittance growth in bunch compressors used in the new generation of FELs[5].

In this paper the effect of CSR on beam dynamics in a storage ring is studied by means of a new numerical code which solves the Vlasov equation using an algebraic technique, employing the well established evolution operator method commonly adopted in quantum mechanics[6]. The instability threshold for an initial Gaussian distribution perturbed with white noise is investigated using numerical simulations. Finally a scaling law for both the beam evolution and the microbunching effect are derived.

2. Evolution operators method

In the following we will consider the Liuville Vlasov evolution equations describing the dynamics of a given charged beam density distribution ρ, of the type

$$\frac{\partial}{\partial s}\rho = H\rho \qquad \text{with} \qquad \rho|_{s=0} = \rho_0 \tag{1}$$

where s refers to the propagation coordinate and ρ_0 denotes the initial distribution.

The operator H (Liuville Vlasov operator) contains the physical properties of the problem. It may be specified by simple differential operators when describing the bunch evolution through magnetic systems[7], or by integral operators when accounting for non linear problems associated with the effects of wakefields on the beam.

Whatever is the realization of H, the formal solution of eq. (1) can be written as

$$\rho = \exp(Hs)\rho_0 \tag{2}$$

if H is a not explicitly time-dependent operator.

Let us specify the operator H to study the longitudinal beam dynamics under the influence of wakefields. In this case, by neglecting the damping and the quantum fluctuations due to the synchrotron radiation, it can be cast in the form

$$H = \eta\varepsilon\frac{\partial}{\partial z} + \frac{Ne^2}{E_0}W(z,s)\frac{\partial}{\partial \varepsilon} \tag{3}$$

with η the slip factor, ε the relative energy variation from the nominal value E_0, z the longitudinal coordinate, N the number of particles in the bunch, e the electron charge, and $W(z, s)$ the longitudinal wake potential per unit length of the path. The wake potential is a convolution of the wakefield times the bunch distribution

$$W(z,s) = \int\int W_{||}(z-z')\rho(z',\varepsilon',s)dz'd\varepsilon' \tag{4}$$

In this specific case the formal solution of the Liuville Vlasov equation reads

$$\rho(z,\varepsilon,s) = \exp\left(\eta s\varepsilon\frac{\partial}{\partial z} + \frac{Ne^2}{E_0}\Omega(z,s)\frac{\partial}{\partial\varepsilon}\right)\rho_0(z,\varepsilon) \tag{5}$$

with

$$\Omega(z,s) = \int_0^s W(z,s')ds' \tag{6}$$

In writing the above solution, we are neglecting any contribution due to time ordering corrections, that arises whenever the operator H is explicitly time-dependent and does not commute with itself at different times. With these assumptions we neglect third order terms in δs when evaluating eq. (5)[8].

The exponential operator in eq. (5) consists of two non commuting parts. We decouple it using the split technique, i. e. we approximate the exponential operator for small step intervals δs as

$$\exp(A+B) = \exp\left(\frac{1}{2}A\right)\exp(B)\exp\left(\frac{1}{2}A\right) \tag{7}$$

It is therefore possible to calculate the action of the previous operator on the initial distribution function taking into account that

$$\exp\left(\alpha\frac{\partial}{\partial z}\right)\exp\left(\beta\frac{\partial}{\partial\varepsilon}\right)f(z,\varepsilon) = f(z+\alpha,\varepsilon+\beta) \tag{8}$$

The distribution function $\rho(z,\varepsilon,\delta s)$ at the position δs can then be expressed in terms of the distribution function $\rho(z,\varepsilon,0)$ at $s=0$ thus giving

$$\rho(z,\varepsilon,\delta s) = \rho\left[z + \eta\varepsilon\delta s + \frac{1}{2}\eta\delta s\frac{Ne^2}{E_0}\Omega\left(z+\frac{1}{2}\eta\varepsilon\delta s,\delta s\right),\right.$$
$$\left.\varepsilon + \frac{Ne^2}{E_0}\Omega\left(z+\frac{1}{2}\eta\varepsilon\delta s,\delta s\right),0\right] \tag{9}$$

The above equation allows to study the evolution of the distribution function in a storage ring due to the effects of wakefields. However, by neglecting the radiation damping and the quantum fluctuations, we are actually dealing with effects that have characteristic times shorter than the synchrotron damping time.

We can now introduce the CSR wakefield per unit length of the path that for the steady state radiation of an ultra relativistic particle in a long magnet is

$$W_{||}(z) = \frac{1}{4\pi\varepsilon_0}\frac{2}{R^{2/3}}\frac{1}{(3z)^{4/3}} \tag{10}$$

with ε_0 the dielectric constant of vacuum, and R the curvature radius of the bending magnet.

A simulation code (TEO[a]) has been developed to study the beam dynamics under the effect of the CSR by numerical implementation of eq. (9). The initial distribution is sampled on a uniform Cartesian grid. The simulation domain in phase space is monitored at runtime and dynamically expanded in order to follow the distortions induced by the beam dynamics. Throughout the simulation, the resolution of the mesh is kept constant in order to avoid spurious numerical noise due to remapping. The simulations are carried out with an adaptive algorithm for the step size which controls the accuracy of the computed solution.

As a result of the TEO code we show in fig. 1 the evolution of the normalized longitudinal distribution function $f(z) = \int_{-\infty}^{\infty} \rho(z,\varepsilon)d\varepsilon$ for different values of the longitudinal coordinate s. As the distribution function propagates inside the magnet, a microbunching develops and several peaks in the distribution appear.

The microbunching effect can also be seen in fig. 2 where we have represented the longitudinal phase space distribution corresponding to the final longitudinal position of the bunch.

It is important to note that the microbunching process does not appear if the initial distribution function is an unperturbed Gaussian distribution. This is an important feature of the code that allows to study the instability threshold and the coherent unstable modes more accurately than a tracking code that uses macroparticles. In fact, by solving the Vlasov equation, the distribution function does not contains the noise due to the macroparticles and its evolution can be very smooth. In order to excite an instability it is necessary to perturb the initial distribution, even if, below the threshold, any initial perturbation of the distribution function disappears during the

[a]Transport by Exponential Operators

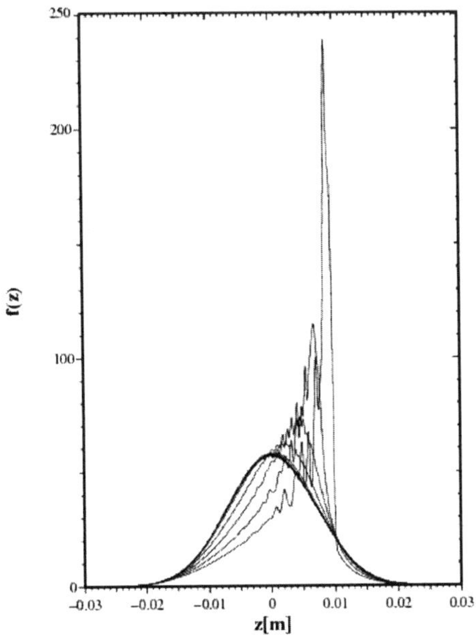

Figure 1. Evolution of an initially Gaussian perturbed longitudinal distribution function $f(z) = \int_{-\infty}^{\infty} \rho(z,\varepsilon)d\varepsilon$.

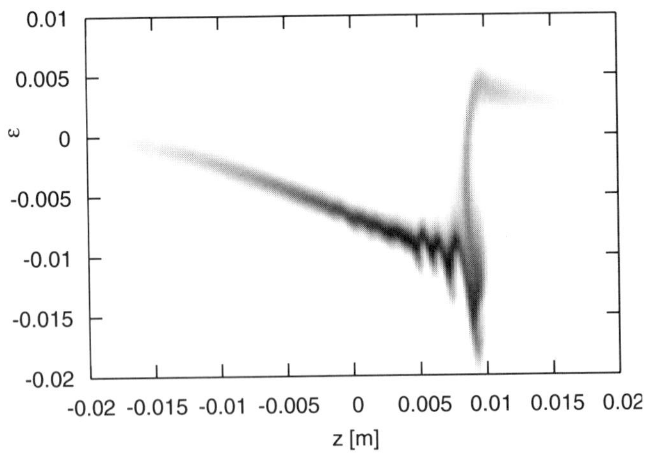

Figure 2. Phase space distribution at the final longitudinal position.

bunch evolution.

The difference with respect to a tracking code is shown in fig. 3 where we present the longitudinal distribution function obtained in the stable a) and unstable b) regime with both the TEO and a macroparticle tracking code. The smoothness of the distribution function is evident in the case of the Vlasov code.

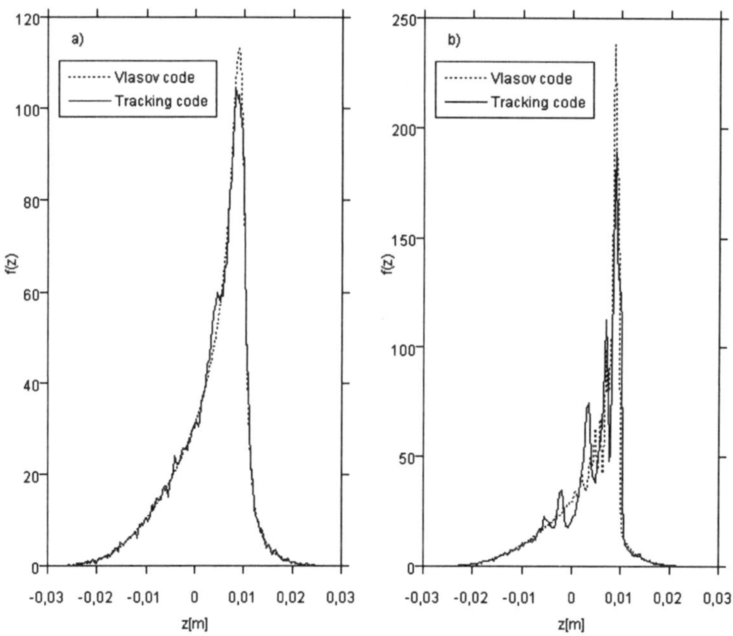

Figure 3. Normalized longitudinal distribution function in stable a) and unstable b) regime.

3. Microbunching effect

In order to study the effect of the CSR on the beam dynamics, simulations under different initial conditions should be performed. For example, in the case of a Gaussian bunch, the dependence of the microbunching threshold on bunch length, energy spread and other machine parameters has to be investigated. However, we will show that the Vlasov equation can be rescaled in such a way that two parameters only play a key role on the bunch evolution. To this aim we write eq. (1) with the operator H given by eq. (3)

$$\frac{\partial \rho}{\partial s} - \eta\varepsilon\frac{\partial \rho}{\partial z} - \frac{Ne^2}{E_0}W(z,s)\frac{\partial \rho}{\partial \varepsilon} = 0 \tag{11}$$

with $W(z,s)$ being the wake potential due to CSR that can be written as[9]

$$W(z,s) = \frac{1}{4\pi\varepsilon_0}\frac{2}{3^{1/3}}\frac{1}{R^{2/3}}\int_{-\infty}^{z}\frac{1}{(z-x)^{1/3}}\frac{\partial f(x,s)}{\partial x}dx \tag{12}$$

and

$$f(z,s) = \int_{-\infty}^{\infty}\rho(z,\varepsilon,s)d\varepsilon \tag{13}$$

It is now convenient to use the new phase space variables z' and ε' corresponding to z and ε normalized to the initial rms bunch length σ_{z0} and rms energy spread $\sigma_{\varepsilon 0}$ respectively. The normalized distribution function $\rho'(z',\varepsilon')$ is then

$$\rho'(z',\varepsilon') = \sigma_{z0}\sigma_{\varepsilon 0}\rho(z,\varepsilon) \tag{14}$$

By introducing the above quantities in eq. (11), we obtain

$$\frac{\partial \rho'}{\partial s'} - \frac{N'}{N}\varepsilon'\frac{\partial \rho'}{\partial z'} - F(z',s')\frac{\partial \rho'}{\partial \varepsilon'} = 0 \tag{15}$$

with

$$\begin{aligned} s' &= \frac{2Ne^2}{3^{1/3}4\pi\varepsilon_0 E_0 R^{2/3}\sigma_{\varepsilon 0}\sigma_{z0}^{4/3}}s \\ N' &= \frac{3^{1/3}4\pi\varepsilon_0\eta E_0 R^{2/3}\sigma_{\varepsilon 0}^2\sigma_{z0}^{1/3}}{2e^2} \\ F(z',s') &= \int_{-\infty}^{z'}\frac{1}{(z'-x)^{1/3}}\frac{\partial f'(x,s')}{\partial x}dx \end{aligned} \tag{16}$$

An important feature of the above equation is that $F(z',\varepsilon')$ is a non-dimensional form factor independent on the machine parameters. For an initial Gaussian bunch distribution it is independent on the bunch dimensions too and, as a consequence, the evolution of the distribution function depends only on the two parameters s' and N'/N. For two different Gaussian initial distribution functions, the beam dynamics of the normalized ρ' is exactly the same for the same values of s' and N'/N.

The above observations simplify the parametric study of the longitudinal evolution of a Gaussian bunch influenced by the CSR wake field. The variable s' is related to the distance the bunch has to travel in order to allow the microbunching effect to evolve itself, and the variable N'/N is related to the intensity of this effect. As an example, in fig. 4, we show the distribution function obtained with the parameters of Table 1 that give

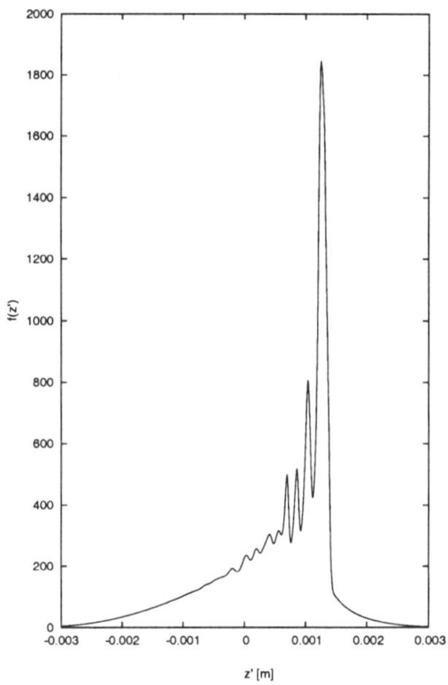

Figure 4. Normalized longitudinal distribution function corresponding to the two different sets of parameters indicated in Table 1.

Table 1.

N	$\sigma_{\varepsilon 0}$	σ_{z0} [m]	η	E_0 [GeV]	s [m]
5×10^{10}	7×10^{-4}	7×10^{-3} m	1.4×10^{-3}	1.5	1000 m
6.35×10^{10}	10^{-3}	10^{-3} m	5×10^{-3}	0.5	28 m

identical values of s' and N'/N.

Under these conditions, we performed simulations with an initial Gaussian distribution function, with different values of N'/N, and for a distance s' long enough to develop the microbunching effect when above threshold. The initial distribution function was perturbed with white noise such that the obtained threshold corresponded to the strongest unstable modes that could be excited. The results of the numerical simulations give a threshold of the microbunching effect of

$$N_{th} \simeq 91.69N' \simeq \frac{1}{12^{2/3}\pi^{5/2}}N' \tag{17}$$

The simulations show that above threshold there is not a single unstable mode, but a range of wavelengths of the order of a fraction of σ_{z0} that are unstable and change as the distribution function propagates in the machine.

It is important to underline that the threshold given by eq. (17) has been obtained by adding white noise to the initial distribution function. Simulations with a single mode of a given wavelength superimposed to the initial Gaussian distribution function showed that the mode can grow up even though $N < N_{th}$. This aspect of the problem is under active consideration.

4. Conclusions

The numerical code TEO for the longitudinal beam dynamics with the inclusion of the non linear contributions due to the wakefields has been developed. The numerical scheme is based on the solution of the Vlasov equation, and it exploits an algebraic technique that uses the exponential operators commonly adopted in quantum mechanics.

The code has been used to study the effect of the microbunching produced by the CSR in a storage ring. A comparison with a multiparticle tracking code showed that it is well suited for study threshold effects due to the very low numerical noise levels it introduces with respect to the tracking code. This feature allowed to study the microbunching threshold produced by CSR for an initial Gaussian bunch perturbed with white noise during the propagation in a storage ring.

The technique we have adopted preserves the simplecticity of the problem in a natural way and therefore no particle losses are expected due to numerical artifacts. Furthermore it is, to some extent, analytical and flexible enough to allow to introduce other effects on the beam due to transport elements or other kind of wakefields. This flexibility will be exploited to study the effect of CSR in a bunch compressor.

Acknowledgments

The Authors express their sincere appreciation to Prof. L. Palumbo for enlightening discussions and encouragements.

References

1. S. A. Kheifets and B. Zotter, CERN SL Report 95-43 (AP), (1995).
2. G. Stupakov and S. Heifets, *Phys. Rev. ST - Accel. Beams*, **5**, 054402 (2002).
3. J. M. Byrd, et al., *Phys. Rev. Lett.* **89**, n. 22, 224801 (2002).
4. U. Arp, et al., *Phys. Rev. ST - Accel. Beams*, **4**, 054401 (2001.)
5. S. Heifets, et al., *Phys. Rev. ST - Accel. Beams*, **5**, 064401 (2002).
6. J. Sanchez Mondragon and K. B. Wolf, "Lie methods in optics", Springer - Verlag, Berlin 1986.
7. A. J. Dragt, *AIP Conf.Proc*, **87**, pp. 147-313 (1982).
8. G. Dattoli, et al., "Study of coherent synchrotron radiation effects by means of a new simulation code base on the non-linear extension of the operator splitting method", submitted to publication.
9. G. Stupakov, *AIP Conf.Pro*, **468**, pp. 334-347 (1998).

QFEL: A NUMERICAL CODE FOR MULTI-DIMENSIONAL SIMULATION OF FREE ELECTRON LASERS IN THE QUANTUM REGIME.

A. SCHIAVI[1,2], N. PIOVELLA[3,4], G.R.M. ROBB[5] AND R. BONIFACIO[3]

[1] *Dipartimento di Energetica, Università di Roma "La Sapienza" , Via A.Scarpa 14, I-00161 Roma, Italy*
E-mail: angelo.schiavi@uniroma1.it

[2] *INFN-LNF, via E. Fermi 40, I-00044 Frascati, Roma, Italy*

[3] *INFN-Sezione di Milano, via Celoria 16, I-20133 Milano, Italy*

[4] *Dipartimento di Fisica, Università Degli Studi di Milano, via Celoria 16, I-20133 Milano, Italy*

[5] *SUPA, Department of Physics, University of Strathclyde, Glasgow, G4 0NG, Scotland, UK*

A new simulation tool has been developed for the simulation of the FEL equations in both the classical and quantum regimes to be applied to the investigation of proposed FEL models and to the exploration of parameter space for new experiments.

1. Introduction

Presently several projects are active in the research and development of new bright sources in the soft X-ray range using the free electron laser (FEL) mechanism [1,2,3]. Experiments are currently run or proposed to run in the self-amplified-spontaneous-emission (SASE) mode. One of the key disadvantages of SASE FEL is the large number of uncorrelated spikes in the pulse of emitted radiation, which also affects the spectrum of the FEL source [4]. A new regime of operation, called quantum SASE, could provide a dramatic improvement to the quality of the radiation extracted from an FEL [5]. For the realization of quantum SASE, the present paradigm of GeV electron bunches combined with a magnetic wiggler is not feasible due to the extremely long gainlength for this operation mode. It has been then proposed to test quantum SASE with an electromagnetic wiggler [6]. Detailed simulations of the quantum FEL model are necessary in order to address the feasibility of future experiments in this new regime.

2. The 1d model equations

The QFEL code is based on the one-dimensional quantum FEL model presented in [5]. The dynamics of the electron bunch is described by a Schrödinger equation for a matter-wave field Ψ interacting with a pendulum potential proportional to the dimensionless radiation field amplitude A. The evolution of the radiation field is determined by the Maxwell equation for the A field (in the slowly-varying envelope approximation, SVEA) interacting with the electron transverse current:

$$i\frac{\partial\Psi}{\partial\bar{z}} = -\frac{1}{2\bar{\rho}^{3/2}}\frac{\partial^2\Psi}{\partial\theta^2} - i(Ae^{i\theta} - A^*e^{-i\theta})\Psi \tag{1}$$

$$\frac{\partial A}{\partial\bar{z}} + \frac{\partial A}{\partial z_1} = \int_0^{2\pi}\frac{d\theta}{2\pi}|\Psi|^2e^{-i\theta} + i\bar{\delta}A \tag{2}$$

Here $\Psi(z_1, \bar{\delta}, \theta, \bar{z})$ is a function of z_1, the coordinate along the electron bunch moving at the resonant velocity, of $\bar{\delta}$, the detuning parameter, of θ, the electron phase, and of $\bar{z}$, the dimensionless wiggler length. The radiation amplitude $A(z_1, \bar{z})$ is a function of the relative electron coordinate z_1 and of the position along the wiggler $\bar{z}$. The electron position z_1 is expressed in units of cooperation length L_c, whereas the wiggler position $\bar{z}$ is in units of gainlength L_g. For the definition of all these lengths and parameters, refer to [5].

The quantum FEL parameter $\bar{\rho}$ represents the maximum number of photons emitted by a single electron as the bunch propagates through the wiggler. The value of $\bar{\rho}$ discriminates between the quantum and the classical regime. In the classical regime ($\bar{\rho} \gg 1$), every electron emits and absorbs many photons in passing through the wiggler. In the quantum regime instead ($\bar{\rho} < 1$), the electron has a small probability of emitting just one photon in the whole process. This eventually leads to a completely different behavior in the quantum regime, where the above equations describe a two-level system. For solving numerically the quantum FEL equations, we chose to expand the Ψ field into longitudinal momentum eigenstates c_n

$$\Psi(z_1,\bar{\delta},\theta,\bar{z})=\sum_n c_n(z_1,\bar{\delta},\bar{z})e^{in\theta}\,, \tag{3}$$

so that the equations to solve become

$$\frac{\partial c_n}{\partial \bar{z}}=-i\left(\frac{n^2}{2\bar{\rho}^{3/2}}+n\bar{\delta}\right)c_n-(Ac_{n-1}+A^*c_{n+1}) \tag{4}$$

$$\frac{\partial A}{\partial \bar{z}}+\frac{\partial A}{\partial z_1}=b(z_1,\bar{z}) \tag{5}$$

where the bunching factor b is computed using a weight function $G(\bar{\delta})$, which is a normalized Gaussian distribution centered on the mean detuning parameter $\bar{\delta}_0$:

$$b(z_1,\bar{z})=b_n\sum_n\int d\bar{\delta}\,G(\bar{\delta})c_nc^*_{n-1} \tag{6}$$

The normalization factor b_n takes into account the dimensionality of the simulation, as described later on in this paper.

3. Higher dimensions and transverse terms

The principal aim of developing the QFEL code was to extend the previous analysis and simulation work carried out in one longitudinal dimension by adding the transverse dimensions and introducing a description of the transverse FEL dynamics. As a consequence, the electron field $\Psi(z_1,\bar{x},\bar{y},\bar{\delta},\theta,\bar{z})$ and the radiation amplitude $A(z_1,\bar{x},\bar{y},\bar{z})$ acquired two transverse coordinates, namely $\bar{x}$ and $\bar{y}$, expressed in units of the initial width of the bunch in the transverse plane. The beam profile had a Gaussian profile along each transverse direction, with the possibility of simulating axis-symmetric beams ($\sigma_x=\sigma_y$), and beams with an oblate transverse profile ($\sigma_x\neq\sigma_y$). The effects due to beam emittance and to the profile of a laser wiggler will be discussed in a forthcoming work.

4. Numerical scheme

In order to integrate the quantum FEL equations, the phase space was sampled on a discrete grid and the dynamical variables, namely A and the c_n states, were evaluated on this grid. A combination of different finite difference (FD) schemes were adopted in an operator splitting algorithm.

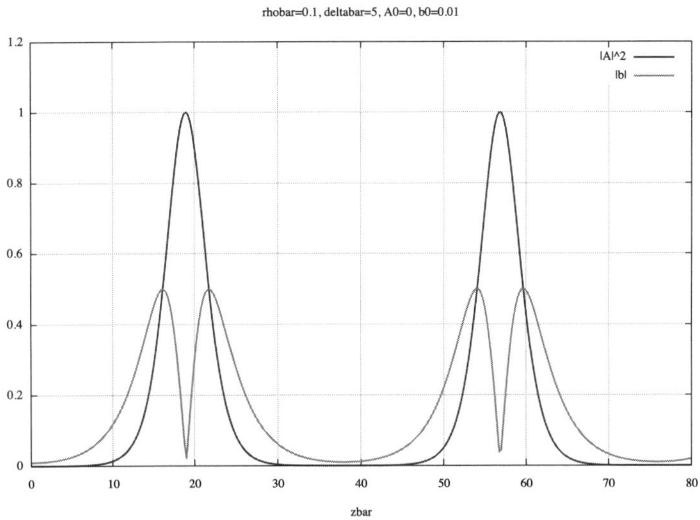

Figure 1. 1d steady-state simulation in the quantum regime ($\bar{\rho}$ =0.1): normalized radiation intensity $|A|^2$ and bunching factor b as a function of the position $\bar{z}$ along the wiggler.

An explicit FD scheme coupled to a Runge-Kutta stepper was used for integrating the interaction of the field and the electron bunch [8]. A cubic-interpolation propagation (CIP) method was implemented for solving the advection equation of the radiation field in the electron bunch rest frame [9]. The high dimensionality of the simulation phase space (up to $N = 5$) and the large number of unknowns (typically from 10^6 to 10^8) demanded for a parallel implementation of the code. To this end we adopted a fixed linear domain decomposition technique using a master-slaves paradigm. The whole code has been written in F90 using the MPI library.

5. Quantum 2d steady-state

To illustrate some typical effects studied using `QFEL`, we present here the 1d and 2d steady-state evolution in the quantum regime. These results were reproduced to test the high-gain FEL dynamics. It is important to note that for the simulation of SASE FEL the radiation propagation is an essential ingredient. The simulations were carried out for $\bar{\rho}$ =0.1 and with a mean detuning centered on the quantum resonance, i.e. $\bar{\delta}_0 = 5$. The initial field amplitude was set to zero, $A_0 = 0$, whilst the beam was initialized in a state with bunching $b_0 = 0.01$. Such a small value served as a seed to initiate the beam-radiation interaction. The one-dimensional case, see figure 1, shows the typical quantum FEL evolution for the steady-state regime, were the electrons swing back and forth from the initial momentum state c_0 to the first recoil state c_{-1}. Trace of this mechanism remains in the cyclic pulsing of the radiation intensity $|A|^2$. The peak value of the intensity is equal to 1 as implied by the quantum scaling used in the FEL equations.

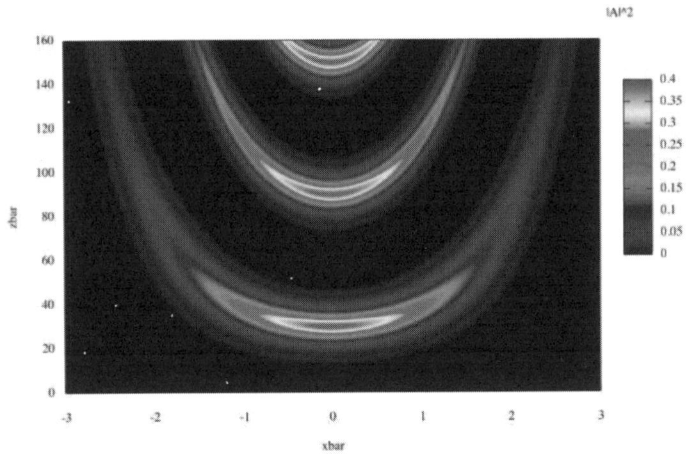

Figure 2. 2d steady-state simulation in the quantum regime ($\bar{\rho}$ =0.1): transverse radiation intensity $|A|^2$ streaked vertically along the wiggler position $\bar{z}$.

Figure 2 presents a 2d simulation for the same initial parameters, with the exception that the beam has now a finite transverse width. The figure shows the field intensity map streaked vertically along the wiggler. No transverse terms are switched on here, so that each part of the beam evolves independently from the others. The peak intensity is reached on axis ($\bar{x} = 0$) and the maximum value is reduced to 0.4. The gain length and the lethargy period are longer compared to the 1d case. This is because the bunching factor b, which is the source term for the radiation field, is a function of the local electron "density". Hence the wings of the beam effectively see a *reduced* bunching factor with respect to the center of the beam and evolve on a longer scalelength.

The effects of radiation diffraction were then included in the model by adding a term proportional to the transverse laplacian of the radiation field A:

$$\frac{\partial A}{\partial \bar{z}} + \frac{\partial A}{\partial z_1} - ia\nabla_{\perp}^2 A = b(z_1, \bar{x}, \bar{z}) \,, \tag{7}$$

where a is an adimensional parameter [7]. Figure 3 shows the effects of radiation diffraction: the intensity of the radiation reduces even more with respect to the 1d case, because the field can now escape in the transverse direction and is not further amplified by the interaction with the electrons.

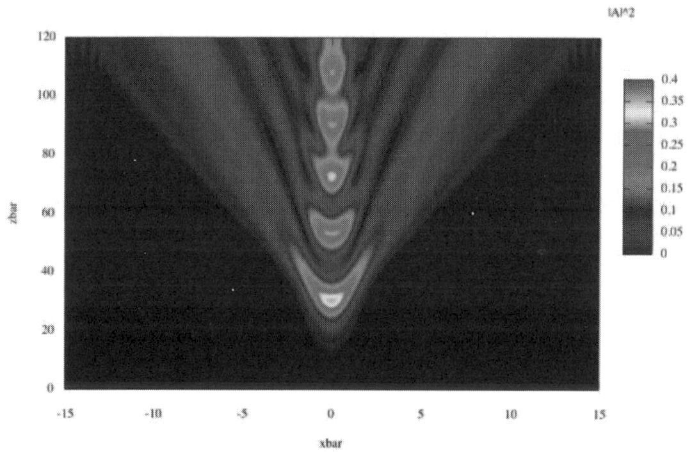

Figure 3. 2d steady-state simulation in the quantum regime ($\bar{\rho}$ =0.1) with radiation diffraction: transverse radiation intensity $|A|^2$ streaked vertically along the wiggler position $\bar{z}$.

6. Normalization and Scaling

The electron field is *slice-normalized* for each z_1 to the constant current profilc $I_0(z_1)$

$$\int d\bar{x} \int d\bar{y} \int d\bar{\delta} \int_0^{2\pi} \frac{d\theta}{2\pi} |\Psi(z_1, \bar{x}, \bar{y}, \bar{\delta}, \theta; \bar{z})|^2 = I_0(z_1) \,. \tag{8}$$

Since the Maxwell equation for A is quadratic in the electron field, it not possible to renormalize the Schrödinger equation for the electron beam without affecting the source term in the Maxwell equation. The dynamics of 1d, 2d and 3d models is intrinsically different due to the spreading out of the electron wavefunction, even in absence of transverse coupling terms. This leads to different gain and peak intensity values depending on the dimensionality of the simulation. In order to recover consistency between the three models, a renormalization factor b_n has been introduced in equation 6, which carries the information about the number of dimensions. The complete 3d model was taken as a reference, whereas the 1d e 2d models were rescaled in order to match the gain of the 3d equations. Effectively this corresponds to considering the 1d model as a description of just the inner core of the beam, whilst the 2d equations are those for a longitudinal slice of the beam. As a consequence the gain of the 1d model and of the central part of the 2d model are equal to the gain of the axial part of the beam in the fully 3d model. Clearly the total radiation intensity, integrated over the transverse directions, becomes smaller going from 3d to 1d, as just a part of the whole beam is considered. These observations are summarized in figure 4, which presents the log plot of the integrated field intensity for the three models when the renormalization factor b_n was introduced.

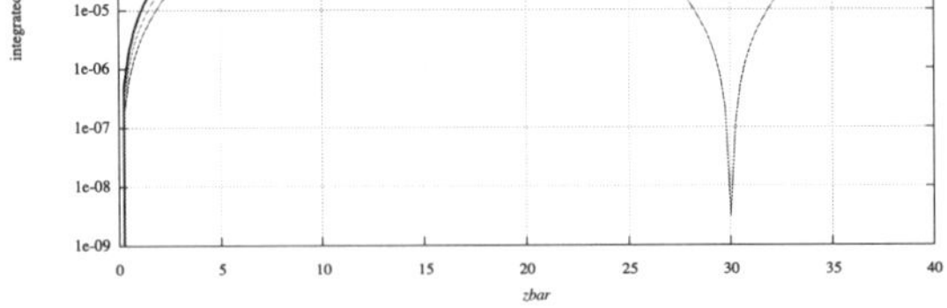

Figure 4. Comparison of the integrated radiation intensity as a function of position $\bar{z}$ along the wiggler for the 1d (red), 2d (green) and 3d models (blue).

7. SASE regime and grid-independent noise

One of the most fundamental requirements in addressing the accuracy and stability of a numerical simulation is that the solution to the dynamical equations is not dependent on the choice of a particular discretization of the integration domain. It is therefore common practice to refine the simulation mesh until the numerical solution stabilizes. This condition often represents the best compromise, balancing the accuracy of the computation and the resources (memory and CPU time) required to achieve it. Several criteria exist for monitoring the stability of a particular numerical scheme (e.g. time-step limiters for implicit or explicit FD schemes), but it is only through a trial-and-error process that the stability of the final solution can be addressed. It is important to note that in a non-linear system, as the one solved by `QFEL`, this process as to be repeated for every choice of the dynamical parameters (in our case the most sensible one is clearly the $\bar{\rho}$ parameter). In the simulation of the SASE operation, we included the radiation propagation, and the electron bunch was initialized with a small bunching (usually $b < 0.01$) and with a random phase $e^{i\phi}$. The phase $\phi(z_1, y, z)$ is a function of the point coordinate and is used to cancel correlations between the different parts of the electron beam. Instead of just sampling the random phase on the grid points, we developed an algorithm for generating a noise function in the Fourier-transformed space and remapping it onto the simulation grid. This allowed us to set a particular level of phase noise, controlling amplitude and waveform in a grid-independent way. It was then possible to check the stability of the simulation as described at the beginning of this section. Figure 5 presents the same pseudo-random phase mapped to a coarse mesh and to a finer one, where the grid has 5 times more points.

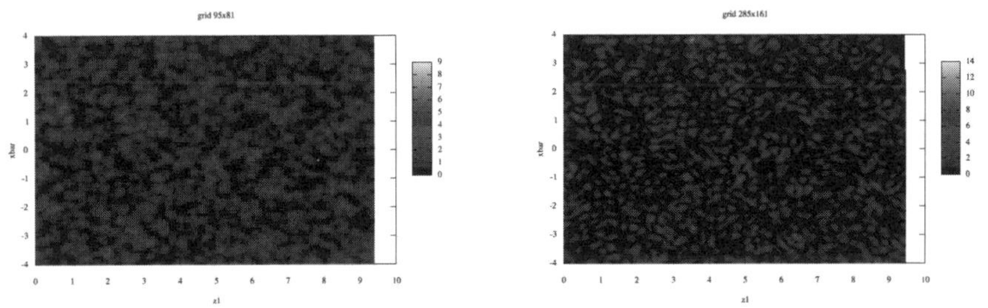

Figure 5. Random noise for a bidimensional simulation: coarse grid (left) and fine grid (right).

8. Conclusions

A new tool, the `QFEL` code, has been developed for studying quantum FEL systems in a full 3d spatial geometry. The choice of a 3d cartesian model over a 2d cylindrical geometry was motivated by the necessity of simulating the start-up from noise in the SASE operation mode. To this end, initializing a random annular phase, as in the 2d cylindrical case, would have introduced unwanted correlations far from the actual beam conditions. As a consequence, the model equations required up to 100 million grid points for the accurate description of the FEL dynamics. To tackle such a large number of unknowns, the numerical scheme was coded to run on a parallel computer using standard and efficient software solutions. The `QFEL` code can be easily run in 1, 2 or 3 spatial dimensions in a consistent way. Therefore feasibility studies of proposed experiments can be carried out in lower dimensions, which require less time and computing resources. Fine tuning of the parameters and exploration of the details of the full dynamical content of the model equations can then be obtained by going to higher dimensions, including transverse terms and propagation effects.

Acknowledgments

G.R.M. Robb acknowledges support from the Leverhulme Trust (grant F00273I).

References

1. R. Bonifacio, C. Pellegrini, L.Narducci, *Opt. Commun.* **50**, 373 (1984).
2. J. Andruszkow et al.,*Phys. Rev. Lett.* **85**, 385 (2000).
3. Linac Coherent Light Source (LCLS) Conceptual Design Report, SLAC Report No. SLAC-R-593, (2002).
4. R. Bonifacio, L. De Salvo, P. Pierini, N. Piovella and C. Pellegrini, *Phys. Rev. Lett.* **73**, 70 (1994).
5. R. Bonifacio, N. Piovella and G.R.M. Robb, *Nucl. Inst. and Meth* **A 543**, 645 (2005).
6. R. Bonifacio, *Nucl. Inst. and Meth* **A 546**, 634 (2005).
7. E.T. Scharlemann et al.,*Phys. Rev. Lett.* **54**, 1925 (1985).
8. W.H. Press and S.A. Teukolsky and others, *Numerical Recipes in C - The Art of Scientific Computing - Second Edition* , Cambridge University Press, USA (1994).
9. T. Yabe and T. Aoki, *Comp. Phys. Commun.* **66**, 219 (1991).

FIRST SIMULATIONS RESULTS ON LASER PULSE JITTER AND MICROBUNCHING INSTABILITY AT SPARXINO

M. BOSCOLO, M. FERRARIO, V. FUSCO, M. MIGLIORATI, L. PALUMBO, B. SPATARO, C. VACCAREZZA

INFN-LNF, via E. Fermi 40
Frascati, 00044, Italy

L. GIANNESSI, M. QUATTROMINI, C. RONSIVALLE

ENEA-CRE, via E. Fermi 45
Frascati, 00044, Italy

L. SERAFINI

Group, Laboratory, Address
City, State ZIP/Zone, Country

In the frame of the R&D activity foreseen for the first phase of the SPARX project, the SPARXINO test facility is meant to provide an ultra-high electron beam at 1.2-1.5 GeV, able to drive SASE and seeded FEL experiments in the 10-3 nm range. The proposed layout is analyzed regarding its sensitivity to laser pulse jitter effects and microbunching instability limitations. First simulation results are presented.

1. Introduction

The first phase of the SPARX project funded by MIUR, (Research Department of Italian Government) is an R&D activity focused on developing techniques and critical components for future X-ray FEL facilities. The activity consists of two main lines of research:

1. experimental tests on RF compression techniques by means of using the SPARC [1] high brightness Photoinjector presently under installation at Frascati INFN-LNF Laboratories,
2. exploring the feasibility of soft and hard X-rays sources in the SASE and seeding schemes in the SPARX test facility (SPARXINO), and upgrading in energy and brightness the existing Frascati 800 MeV LINAC working at present as injector system for the DAΦNE Φ-factory [2].

Other ongoing activities are the studies of high repetition rate S-band gun, high Quantum Efficiency cathodes, high gradient X-band RF accelerating structures and harmonic generation in gases [3].

The Sparxino test facility consists of an advanced high brightness photoinjector, followed by two Linac sections, Linac1 and Linac2 that drive the beam up to 500 MeV and 1.2 GeV respectively. A rectilinear RF compressor is integrated in the high brightness photoinjector, as proposed in [4], thus producing a 300-500 A beam in the early stage of the acceleration. After the Linac1 a magnetic chicane is used as a second stage of compression. A comparison between RF and magnetic compression technique is scheduled during the SPARC operation. In this paper two different RF compression schemes are analyzed regarding their sensitivity to the laser pulse and Linac1 phase jitter. In the first case a fourth harmonic RF cavity is employed at the exit of the RF gun, in the second one the cavity is placed at the entrance of the magnetic compressor, the effect on the jitter sensitivity and beam characteristics is described. Further preliminary simulation results on microbunching instability are presented.

2. The Sparxino test facility

A spectral range from 10 nm to 1 nm has been considered for the radiation. SASE-FEL's in this wavelength range require high brightness beam at the undulator entrance. In Table 1 a preliminary parameter list is reported for such a source, while in Fig. 1 the schematic layout is shown.

Table 1. Electron Beam parameters

Beam Energy	1.2	GeV
Peak current	1-2.5	kA
Emittance (average)	2	mm-mrad
Emittance (slice)	1	mm-mrad
Energy spread (correlated)	0.1	%

2.1. *The Photoinjector*

The injector design for the SPARXINO test facility is based on the SPARC high brightness photoinjector [1]. It considers a 1.1 nC bunch 10 ps long (flat top) with 1.1 mm radius, generated inside a 1.6-cell S-band RF gun of the same type of the BNL-SLAC-UCLA one [5] operating at 120 MV/m peak field equipped with an emittance compensating solenoid. Three standard SLAC type 3m TW structures each one embedded in a solenoid boost the beam up to 150

MeV. With a proper setting of the accelerating phase and solenoid field amplitude it is possible to increase the peak current preserving the beam transverse emittance. According to the simulation results the beam compression at low energy (<150 MeV), still in the space charge dominated regime, turns out to be feasible provided that a proper emittance compensation technique is adopted [4], a possibility that is not viable in a magnetic chicane. Moreover the propagation of a short bunch in the Linac1 reduces the potential emittance degradation caused by transverse wake fields, while a proper phasing of the linac can control the longitudinal wake fields. A systematic study based on PARMELA [6] simulations has been done in order to optimize the parameters that influence the compression [7], the results of computations show that peak currents up to kA level are achievable at the injector exit with a good control of the transverse and longitudinal emittance by means of a short SW section operating at 11424 MHz [8] placed before the first accelerating section. On the other side the results obtained by Ronsivalle et al [7] without the IV harmonic correction prior the RF compressor show that is anyway possible to reach good compression factors but paying for a heavy deformation of the longitudinal bunch distribution, a strong sensitivity to RF compressor phase, and a highly non linear longitudinal phase space.

Two cases are presented here to analyze the effects of this fourth harmonic correction on the beam jitter sensitivity. In the first photoinjector setting the correction is provided at the exit of the gun and a bunch average current of 450 A is obtained, i.e. a four times compression on the nominal SPARC beam, with a normalized rms emittance below 1 mm-mrad. The peak current, the transverse emittance and the axial magnetic field settings along the photoinjector length are shown in Fig. 2.

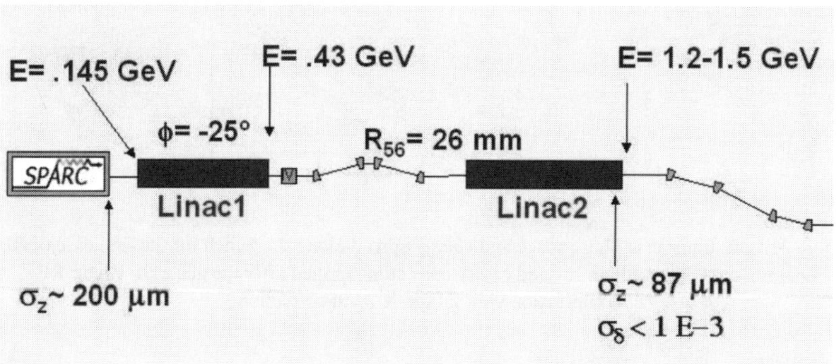

Figure 1. Schematic layout of the SPARXINO test facility.

For comparison a second case is presented here namely without the fourth harmonic correction applied after the gun; the obtained bunch average current is around 300 A, with a three times compression factor. In Fig. 3 the transverse phase space and the energy spread along the bunch are shown for the two cases as they appear at the photoinjector exit.

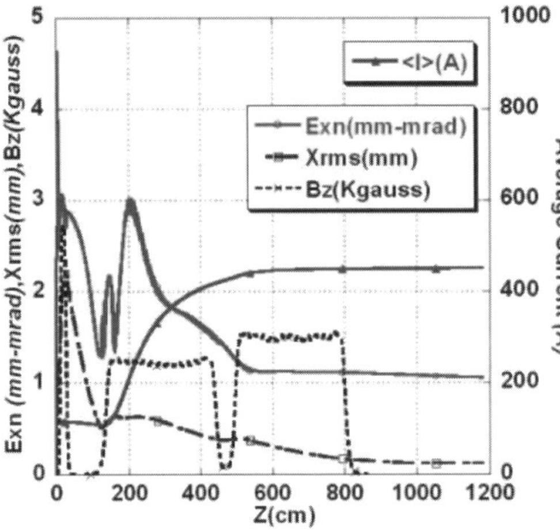

Figure 2: Average current, transverse emittance and envelope, axial magnetic field vs. z for a 450 A compressed beam along the photoinjector.

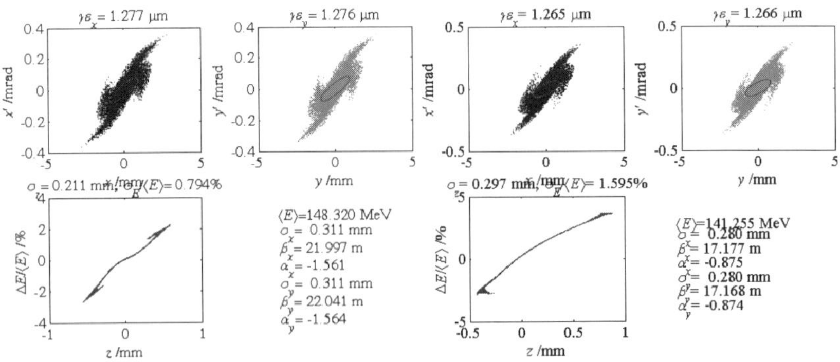

Figure 3: Left: transverse phase space and energy spread along the bunch for the first case of RF compressed beam with an X-band cavity correction applied after the gun exit. Right: RF compression without the X-band correction.

2.2. *The Linac*

The SPARXINO linac is based on the upgrade of the existing DAΦNE injection facility as reported in Fig. 4. The DAΦNE injection system is a 60 m long LINAC [9] equipped with 15 S-band (2.865 GHz) SLAC-type 3 m long accelerating structures driven by four 45 MW klystrons each followed by a SLED peak power doubling system. The Linac energy upgrade to 1.2 GeV can be achieved by upgrading the accelerating field of the existing units up to 25 MV/m. The Linac waveguide network must also be modified in order to supply two accelerating units per RF station. This system configuration requires four new 45 MW klystrons. More over the room availability for inserting four more SLAC-type sections will provide the possibility to reach a final energy of 1.5 GeV. In order to reach the current values required for the FEL application in the 3-5 nm range a second magnetic compression stage is required after the RF compressor. A magnetic chicane is placed for this purpose after the first linac L1 set to provide the needed correlated energy spread, while the second linac L2 helps to remove it before the injection into the undulator. The schematic drawing is shown in Fig. 1. For both the two presented working points, previously described, the beam propagation through the SPARXINO linac mostly occurs in a space charge dominated regime, namely at the exit of the photoinjector and after the magnetic compression. A detailed study has been devoted to the SPARXINO beam dynamics as reported in [10], from this analysis it turns out that the invariant envelope condition has to be applied at the entrance of the two linac sections, L1 and L2, to control and damp the emittance oscillation. The SPARXINO lattice is embedded in the existing DAΦNE LINAC and Transfer Lines lattice (Fig. 4). After the linac exit a 43 m dogleg is foreseen to inject the beam into the undulator.

2.3. *The bunch compressor*

The second compression stage consists in a magnetic chicane designed to provide a momentum compaction around $R_{56}\approx 26$ mm, and to minimize the emittance dilution due to the coherent synchrotron radiation. The chicane is made up of four rectangular dipoles and is placed in between the two Linac sections described above, where the beam energy is approximately 500 MeV. In Table 2 the parameters of the bunch compressor chicane, for the cases discussed here, are listed. At the exit of the Linac1 a second IV harmonic section provides the possibility of a longitudinal phase space correction prior the magnetic compression. The SPARXINO lattice and the optic functions are shown in Fig. 4.

Table 2. Bunch compressor chicane parameter list.

Parameter	Symbol	Unit	Value
Beam Energy	E	GeV	0.5
Initial rms bunch length	σ_{zi}	mm	210
Finale rms bunch length	σ_{zf}	mm	87
Rms total incoming energy spread	σ_{δ}	%	.48
RMS uncorrelated relative energy spread	$\sigma_{\delta u}$	10^{-5}	0.6
Momentum compaction	R_{56}	mm	-26
Second order momentum compaction	T_{566}	mm	42
Total chicane length	L_{total}	m	7
Floor length of each dipole magnet	L_B	m	0.20
Floor length of drift between 1st and last two dipoles	ΔL	m	3.0
Floor length of drift between center two dipoles	ΔL_c	m	0.20
Bend angle of each dipole	$\lvert\theta_B\rvert$	deg	3.68
Maximum dispersion	$\lvert\eta_{max}\rvert$	m	.21

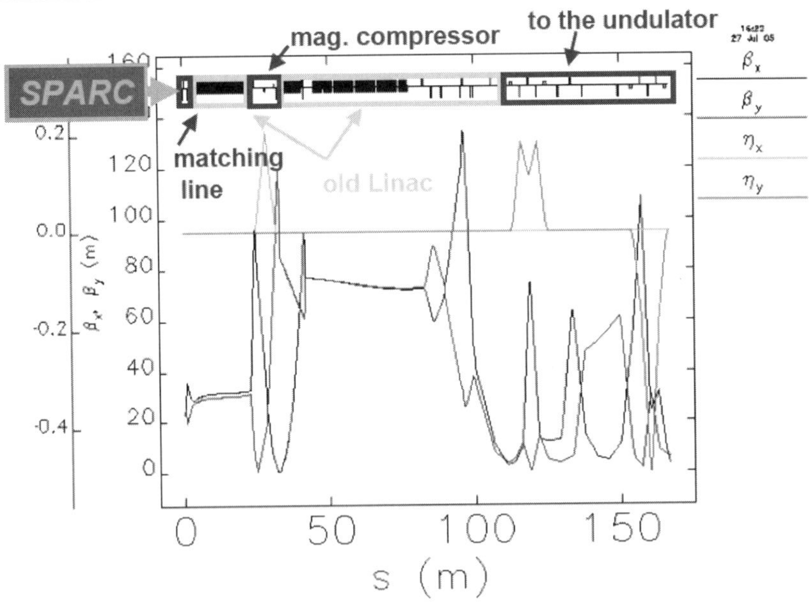

Fig. 4 Twiss parameters for the Sparxino Linac.

3. Phase jitter analysis

A phase jitter error $\varphi = \pm\ 1°$ has been simulated with Parmela and Elegant [11], for the laser pulse and for the Linac1 accelerating sections, in order to check the related sensitivity of the hybrid compression system for the two injector working points described above. The two cases have been previously optimized, with the code Litrack [12], to obtain the minimum sensitivity of the compressed beam properties to the phase error. In Fig. 5 the transverse slice emittance and the current distribution of the two reference cases are shown as tracked up to the end of the SPARXINO channel. The results of the phase jitter analysis are summarized in Table 3.

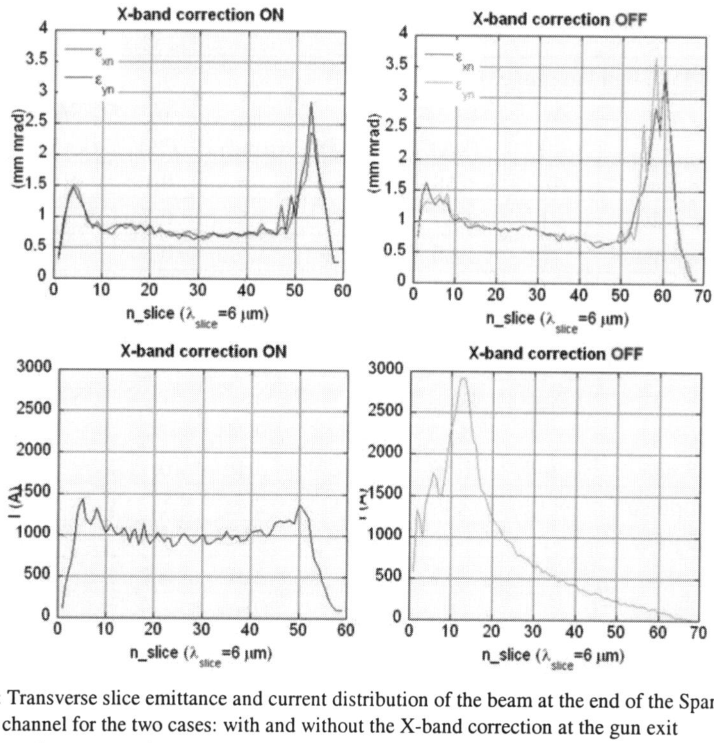

Figure 5: Transverse slice emittance and current distribution of the beam at the end of the Sparxino channel for the two cases: with and without the X-band correction at the gun exit

Several aspects have been considered for the two settings: the current distribution, the average peak current value, the transverse emittance and the brilliance. When the fourth harmonic correction is applied soon after the gun, quite a flat top current distribution is obtained at the exit of each compression stage and the final average peak current value, $I \approx 1$ kA, is almost unaffected by the applied phase error. On the other side only a factor two of magnetic compression is applicable to limit the emittance dilution due to CSR in the compressor chicane. When the longitudinal phase space correction is applied only after the Linac1, case two, the beam is more easily compressed, and a lower momentum compaction value, $R_{56} \approx 24.5$ mm, is needed to obtain the same average value for the peak current. Nevertheless a much higher sensitivity is evident to the one degree error phase considered for the laser pulse or for the Linac1 sections. It is straightforward that a compromise has to be pursued between the overall compression effectiveness and the phase jitter sensitivity in order to optimize the whole SPARXINO channel.

Table 3: Summary table of the laser pulse and first linac phase jitter effects with and without the X-band correction after the gun exit

Case 1: X-band at gun exit : **ON**						
	$\Delta\varphi$ (°)	$I_{pk\ av}$ (kA)	ε_x (μrad)	ε_y (μrad)	σ_δ (%)	SASE eff. (%) (λ=5nm, k=1.47)
Reference	0	1.68	2.5	1.5	0.1	86
Laser pulse phase jitter	+1	1.67	2.5	1.5	0.7	81
	-1	1.60	2.5	1.5	0.8	82
Linac 0 phase jitter	+1	1.94	2.7	1.6	0.7	83
	-1	1.38	2.3	1.4	0.8	86
Case 2: X-band at gun exit : **OFF**						
	$\Delta\varphi$ (°)	$I_{pk\ av}$ (kA)	ε_x (μrad)	ε_y (μrad)	σ_δ (%)	SASE eff. (%) (λ=5nm, k=1.45)
Reference	0	2.58	2.1	3.1	0.3	84
Laser pulse phase jitter	+1	4.49	2.0	2.4	1.7	79
	-1	1.88	3.0	6.4	1.5	85
Linac 0 phase jitter	+1	1.87	1.9	2.3	1.7	80
	-1	3.76	3.9	10.0	1.5	59

4. Microbunching instability analysis

A preliminary study has been carried out on the effect of the microbunching instability induced by collective self-fields such as longitudinal space charge, coherent synchrotron radiation and geometric wakefields. Numerical and theoretical studies together with the DUV-FEL experimental results have recently pointed out that the beam quality can be severely degraded by this type of instability and the Sparxino channel has been analyzed to evaluate the effects on the beam quality. As reported in [13] the microbunching gain of a bunch compressor due to its upstream LSC impedance for a Gaussian energy distribution can be written as:

$$G \approx \frac{I_0}{\gamma I_A}\left| k_f R_{56} \int_0^L ds \frac{4\pi Z(k_0;s)}{Z_0} \right| \exp\left(-\frac{1}{2} k_f^2 R_{56}^2 \sigma_\delta^2 \right), \tag{1}$$

where I_0 is the initial peak current, $I_A \approx 17$ kA is the Alfven current, γ is the electron energy in units of its rest mass, $k_f = k_0/(1+hR_{56})$ is the compressed wave number, for an initial modulation wavelength $\lambda = 2\pi/k_0$ and h representing the energy chirp, σ_δ is the rms relative energy spread, Z_0 =377 Ω and $Z(k_0;s)$ is the free-space LSC impedance per unit length given by [13]:

$$Z_{LSC}(k) = \frac{iZ_0}{\pi k r_b^2}\left[1 - \frac{k r_b}{\gamma} K_1\left(\frac{k r_b}{\gamma} \right) \right], \tag{2}$$

where r_b is the radius of the beam transverse cross section and K_1 is the modified Bessel function. The gain from (1) has been calculated as function of the density modulation wavelength for the SPARXINO linac up to the magnetic bunch compressor and plotted in Fig. 6 (red curve), an initial 3 keV uncorrelated energy spread was assumed. A numerical tracking has also been performed with the Elegant code for a 2 million macroparticle Gaussian distribution varying the

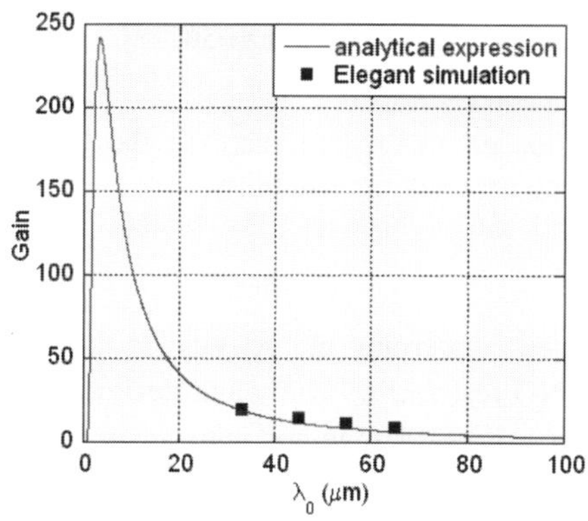

Fig. 6 Spectral gain function of the microbunching instability: the red curve is the result of the analytical expression (1), the blue dots are the results of a 2M particle tracking up to the BC2 compressor with the Elegant code.

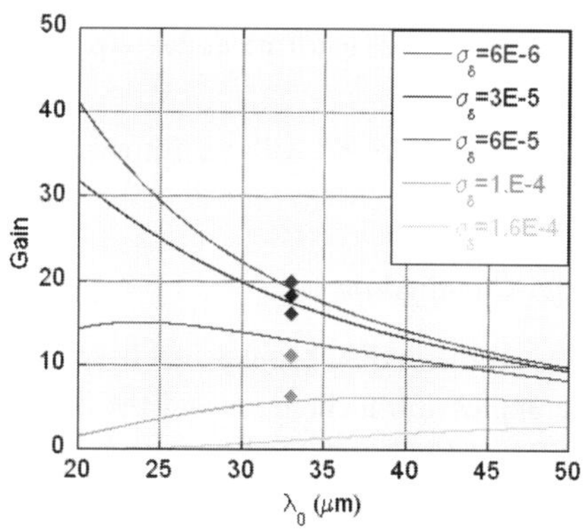

Fig. 7: Spectral gain function of the microbunching instability: the different curves are the result of the analytical expression (1), the dots are the results of a 2M particle tracking up to the BC2 compressor, varying the uncorrelated energy spread

initial density modulation wavelength, the results are also reported in Fig. 6 (blue dots). In Fig. 7 the analytical (curve) and numerical (dots) estimation is reported of the spectral gain function obtained varying the initial uncorrelated energy spread. The energy spread induced by the microbunching instability at the end of the Sparxino Linac is shown in Fig. 8, for a 2M particle Gaussian beam with initial modulation wavelength and amplitude λ_0~33 μm and A_0=5%. It is straightforward to notice how deep can be the effect of the microbunching instability on the beam quality.

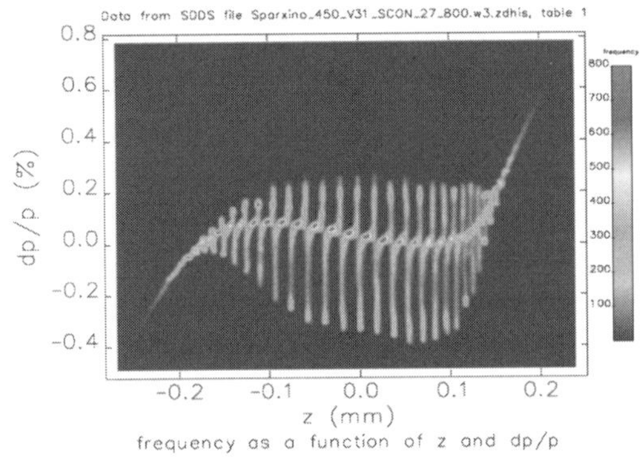

Fig. 8: Longitudinal phase space at the end of the Sparxino Linac for a 2M particle Gaussian beam with initial modulation wavelength and amplitude λ_0~33 μm and A_0=5%.

A "laser heater" will be foreseen as proposed in [13] to increase the beam uncorrelated energy spread and damp the microbunching instability.

5. Conclusions

The SPARX project has been presented concerning the activity on the RF compression techniques and the SPARXINO test facility study and design. The SPARXINO test facility layout has been presented together with an updated lattice design based on the invariant envelope condition. Two RF compressor working points have been chosen to be analyzed in detail and tested in the SPARC test facility presently under installation at Frascati. From the simulation results it comes out that the application of an X-band correction at the gun exit is effective in reducing the temporal jitter sensitivity of the beam while reducing the effectiveness of the following magnetic compression stage. First simulation results on the microbunching instability effects on the Sparxino linac have also been presented, indicating that to preserve the final beam quality an increase of the uncorrelated energy spread is needed to provide the required damping of the instability.

6. References

1. SPARC Project Team, Sparc Injector TDR, www.lnf.infn.it/acceleratori/sparc/.
2. G.Vignola *et al.*, "DAΦNE, The Frascati Phi-factory", Proc. of PAC 1993, Washington
3. L.Giannessi *et al.*, "Seeding High harmonic generation with laser harmonics produced in gases", FEL 2004, Trieste, Italy
4. L.Serafini and M. Ferrario, Velocity Bunching in PhotoInjectors , AIP CP 581, 2001, pag.87
5. D.T. Palmer, The next generation photoinjector, PhD. Thesis, Stanford University
6. Billen, "PARMELA", LA-UR-96-1835, 1996.
7. C.Ronsivalle *et al.*, " Optimization of RF compressor in the SPARX injector", Proc. of PAC 2005, Knoxville, (TN)
8. B. Spataro, *et al*, "Design and RF measurements of an X-band acc. structure", NIM A554:1-12, 2005.
9. S.Kulinski *et al.*, DAΦNE Tech. Note LC-4
10. M. Ferrario, *et al*, "Beam Dynamics around transition", these proceedings.
11. M. Borland, "ELEGANT" LS-287, ANL, Argonne, IL 60439, USA.
12. K.L.F.bane, P.Emma, "LiTrack: A fast longitudinal phase space tracking code with graphical interface", PAC2005, Knoxville, Tennessee, (2005) 4266
13. Z. Huang *et al*, PRSTAB 7, 074401.

Working Group 4

WORKING GROUP 4 SUMMARY: APPLICATIONS OF HIGH BRIGHTNESS BEAMS TO ADVANCED ACCELERATORS AND LIGHTS SOURCES

MITSURU UESAKA

Nuclear Professional School, School of Engineering, University of Tokyo,

2-22 Shirane-shirakata, Tokai, Naka, Ibaraki, Japan

E-mail: uesaka@nuclear.jp

ANDREA ROSSI

Department of Physics, INFN-Milan and University of Miran

Via Celoria 16, 20133 MILAN, Italy

We categorized 16 contributions into the three sub-fields. Those are 1. Compton scattering X-ray sources, 2. FEL and RF photoinjectors and 3. Plasma wakefield acceleratoion/innovative acceleration schemes. We performed a half day working group for each sub-field. The titles and summaries of the contributions appear in the article.

1. Introduction

The contributions to this working group are categolized into 1. Compton scattering X-ray sources, 2. FEL and RF photoinjectors and 3. Plasma wakefield acceleratoion/innovative acceleration schemes this time, based on the pre-communication among the chair-persons and working group leaders. The schedule is shown in Table 1. The number of the contributions in the three is written in Table 2. Finally, we discussed the current status of those innovative systems and the feasibility and breakthrough forward future accelerator systems.

Table 1: Schedule-Working Group D Applications of High Brightness Beams to Advanced Accelerators and Light Source

Sunday, Oct. 08	Monday, Oct. 09	Tuesday, Oct. 10	Thursday,Oct. 12	Friday, Oct. 13
23:00~24:00 Working Group Organization	14:00~17:00 Compton scattering X-ray sources	14:00~17:00 FEL and RF photoinjectors	14:00~17:00 Plasma wakefield acceleratoion /innovative acceleration schemes	10:00~10:40 Report on working group D

Table 2: Topics/ Issuers
Application to Advanced Accelerator.

Topics	Number of talks
Compton scattering X-ray sources	7
FEL and RF photoinjectors	5
Plasma wakefield acceleratoion /innovative acceleration schemes	4

2. Compton scattering X-ray sources

The list of the contributions are written as below.

1. Multi-scattering system to increase the X-ray yield and its medical applications
   Mitsuru Uesaka(Univ.Tokyo)
   "Medical Application of Multi-beams Compton Scattering Monochromatic Tunable Hard X-ray Sources"

Haruyuki Ogino(Univ.Tokyo)
"Laser Pulse Circulation System for Compact Monochromatic Hard-X-Ray Source"

2. Strong Focusing and Higher Harmonics Generation
Jae Lim(LLNL)
"A permanent –Magnet Quadrupole Final focusing Optics at PLEIADES Thomson X-ray Source"Oliver Williams(UCLA)
"Status of the Nonlinear Inverse Compton Scattering Experiment at UCLA"Rodney Yoder(Manhattan College)
"An Inverse Compton Scattering Radiation Source via Self-Guiding in a Plasma"Vitaly Yakimenko(BNL)
"Nonlinear Compton Scattering at BNL-ATF"

3. Theory and simulation
Vittoria Petrillo(Univ.Milan)
"Study of transverse effects in a collective Thomson Back-scattering Source of X-rays"

At first, it was declared by M.Uesaka that the Compton scattering X-ray source for medical use had entered the second generation of multi-scattering to increase the X-ray yield. As for the experimental beam physics, the higher harmonics has been observed at BNL-ATF (Vitaly Yakimenko) and is going to be measured at LLNL(Jae Lim) and UCLA(Oliver Williams). Especially, strong permanent focusing magnets are manufactured (UCLA) and used (LLNL) for the purpose. On the other hand, a new simulation is performed to take into account the collective effects for the interaction of longer laser and lower energy-electrons by V.Petrillo et al(Univ.Milan). Furthermore, a new scheme to use the plasma channel guided electron beam to strengthen the interaction and scattering between the laser and electron-beam and to finally increase X-ray yield is proposed by simulation by .R.Yoder (Mahatten College).

3. FEL and RF photoinjectors

The contributions are listed in the following.

1. FEL
Gerald Andonian (UCLA)
"Observation of Ultra-wide Bandwidth SASE FEL"

Takahiro Watanabe (BNL)

"Numerical simulation and experimental demonstration of seeded FEL at BNL-NSLS"
Kwang-Je Kim (ANL)
"Quantum Effects in Gain and Start-up of Free Electron Lasers—Wigner Function Approach "
William Graves (MIT)
"Longitudinal aspects of an ellipsoidal electron distribution and its effects on seeded FEL performance"

2. Application of RF Photoinjectors to Radiation Chemistry
Akira Sakumi (Univ.Tokyo)

"Review of RF photoinjectors for radiation chemistry in the world"Several new SASE FEL results were reported. Ultra-wide bandwidth (15%) SASE FEL has been observed at VISA, BNL-ATF (Andonian(UCLA)).Gain curve of seeded FEL and spatial profiles of harmonics have been experimentally demonstrated and 1 GW system design is shown. (T.Watanabe(BNL)). Then, an upgrade of the simulation was presented. Longitudinal studies of ellipsoidal/flattop bunches were done to achieve low energy spread in the whole bunch. Ellipsoidal bunch distribution produces linear correlation of energy and time and substantially improves over flat-top bunch (W.Graves(MIT)). Discussion on the QFEL was done by K.J.Kim(ANL). Concerning RF photoinjectors, the activity status of the RF photoinjectors for radiation chemistry in the world are introduced and reviewed. It is important because the radiation chemicists are the second-major users (A.Sakumi (Univ.Tokyo)). Suppression of the timing-jitter and long-term drift between electron- and laser-pulses are reported and discussed.

4. Plasma wakefield acceleratoion /innovative acceleration schemes

The contributions are shown below.

1.Plasma Wakefield Acceleration
Kenichi Kinoshita (Univ.Tokyo)
"Application of Laser Plasma Cathode to Radiation Chemistry, All Thomson Scattering and Electron Microscopy"

2. Innovative Acceleration SchemesGil.Travish (UCLA)
"Preliminary Results from the UCLA/SLAC Ultra-High Gradient Cerenkov Wakefield Accelerator Experiment"Rodney Yoder (Manhatten College)
"Coupling Laser Power into a Slab-Symmetric Acceleration Structure"

S. Dabagov (INFN)
"Channeling Projects at LNF"

Generation of monoenergetic electrons from a laser plasma cathode and the future applications for pulse radiolysis, all optical Thomson scattering X-ray generation and electron microscopy are presented by K.Kinoshita (Univ.Tokyo). Preliminary results from the UCLA/SLAC ultra-high gradient Cerenkov wakefield accelerator experiment for developing GV/m class dielectric accelerators was reported by G.Travish(UCLA). Coupling laser power into a slab-symmetric accelerator structure is numerically analysed by R.Yoder (Manhatten College). Finally, CHANNELING projects at LNF-From Crystal Undulators to Capillary Waveguides- is introduced by Sultan Dabagov (INFN). I would like to thank all contributors and participants for their cooperation, nice and vivid discussion. Applications of high brightness beams gives many important subjects to be solved, but also, Many Dreams!!. Thank you.

STUDY OF TRANSVERSE EFFECTS IN THE PRODUCTION OF X-RAYS WITH A FREE-ELECTRON LASER BASED ON AN OPTICAL UNDULATOR

A. BACCI, C. MAROLI, V. PETRILLO, L. SERAFINI

Dipartimento di Fisica dell'Università,INFN-Sezione di Milano, Via Celoria,16 20133 Milano (Italy)

M. FERRARIO

INFN-LNF, Via Fermi 40 00044 Frascati (RM), Italy

The interaction between high-brilliance electron beams and counter-propagating laser pulses produces X rays via Thomson back-scattering. If the laser source is long and intense enough, the electrons of the beam can bunch and a regime of collective effects can establish. In this case of dominating collective effects, the FEL instability can develop and the system behaves like a free-electron laser based on an optical undulator. Coherent X-rays can be irradiated, with a bandwidth very much thinner than that of the corresponding incoherent emission. The emittance of the electron beam and the distribution of the laser energy are the principal quantities that limit the growth of the X-ray signal. In this work we analyse with a 3-D code the transverse effects in the emission produced by a relativistic electron beam when it is under the action of an optical laser pulse and the X-ray spectra obtained. The scalings typical of the optical wiggler, characterized by very short gain lengths and overall time durations of the process make possible considerable emission also with emittance of the order of 1mm mrad.

1. *Introduction.*

A Thomson back-scattering set-up can be considered in principle as a source of intense X-ray pulses which is at the same time easily tunable and highly monochromatic. Due to recent technological developments in the production of high brilliance electron beams and high power CPA laser pulses, it is now even conceivable to make steps toward their practical realisation [1-6].

The radiation generated in the Thomson back-scattering is usually considered incoherent and calculated by summing at the collector the intensities of the fields produced in single processes by each electron [7-13]. If the laser pulse is long enough, however, collective effects can establish. The system in this range of parameters behaves therefore like a free-electron laser, where the static wiggler is substituted by the optical laser pulse[14-16].

From the point of view of the theoretical description of the process, the possibility of generate coherent X radiation can be demonstrated with the same set of one-dimensional equations that are used in the theory of high-gain free-electron laser amplifier [17-18]. However, many aspects of the process are connected with the finite transverse geometry of the electron beam and of the laser and, in order to give a quantitative evaluation of the radiation pulses it is obviously necessary to consider 3D equations.

In this paper, we write and solve a system of 3D equations that describes the coherent growth of the radiation in the laser-beam interaction. In particular, in section 2 we recall the derivation of such a system. A set of numerical results relevant to a case of practical interest, with a discussion of their importance will be given in section 3.

2.Equations of the model

We start from the Maxwell-Lorentz equations that describe both laser and collective electromagnetic fields and from the relativistic equations of motion for the electrons of the beam. The laser and collective fields are given in terms of the corresponding scalar and vector potentials in the Coulomb gauge.

We assume that the laser is circularly polarised with the following form of the vector potential $\mathbf{A}_L$ (the laser pulse propagates along the z-axis in the negative direction):

$$\mathbf{A}_L(xyzt) = \frac{a_{L0}}{\sqrt{2}}(g(xyzt)e^{-i(k_L z+\omega_L t)}\hat{\mathbf{e}} + cc) + O(\frac{\lambda_L}{w_0}) \tag{1}$$

where $\lambda_L = 2\pi/k_L$ is the laser wavelength, w_0 the laser spot size, $\omega_L = ck_L$ the angular frequency and $\hat{\mathbf{e}} = \frac{1}{\sqrt{2}}(\mathbf{e}_x + i\mathbf{e}_y)$. The envelope g(xyzt) is considered to be a slowly varying function of all variables xyz and t and is defined as a complex number with $|g(xyzt)| \leq 1$. In the case, for instance, of a laser pulse with a Gaussian transverse shape, the envelope has the form given for instance in Ref[19].

The quantity $z_0 = \pi w_0^2/4\lambda_L$ is the Rayleigh length of the laser, and w0 is its focal spot size. Notice that $\mathbf{A}_L$ is perpendicular to the z-axis up to terms of the order of λ_L/w_0, . Another interesting case is when the profile of the laser evolves starting by a step function and the pulse can be considered guided.

We make at this point the basic assumption that the collective scalar and vector potentials $\varphi(xyzt)$ and $\mathbf{A}(xyzt)$ have a slow dependence on the transverse space variables x and y, i.e., that they vary on a transverse scale L_T much greater than the radiation wavelength $\lambda = 2\pi/k$ and write

$$\mathbf{A}(xyzt) = A(xyzt)e^{i(kz-\omega t)}\hat{\mathbf{e}} + cc + O(\lambda / L_T) \qquad (2)$$

where ω=ck is the radiation angular frequency. The amplitude $A(xyzt)$ will therefore be considered as a slowly varying function of xyz and t and the collective potential $\mathbf{A}$ is perpendicular to the z-axis up to terms of the order of λ/L_T.

We write the relativistic equation of motion of each electron in the form

$$\frac{d\Pi_j(t)}{dt} = e\left[\nabla(\varphi - \beta_j(t)\cdot(\mathbf{A}_L + \mathbf{A}))\right]_{\mathbf{x}=\mathbf{r}_j(t)} \qquad (3)$$

with $\mathbf{r}_j(t)$ the position of the electron, $\Pi_j(t) = \pi_j(t) - (e/c)(\mathbf{A}_L + \mathbf{A})_{\mathbf{x}=\mathbf{r}_j(t)}$ the generalized momentum and $\pi_j(t) = mc\mathbf{p}_j(t)$ the mechanical momentum in which we use the definition

$$\mathbf{p}_j(t)=\gamma_j(t)\boldsymbol{\beta}_j(t) \qquad (4)$$

With some algebra, and neglecting all terms containing the collective scalar potential φ ("space charge" effects), one has

$$\frac{d\Pi_j(t)}{dt} = -\frac{e^2}{2mc^2\gamma_j(t)}\left[\nabla(\mathbf{A}_L + \mathbf{A})^2\right]_{\mathbf{x}=\mathbf{r}_j(t)} + \ldots.. \qquad (5)$$

where

$$(\mathbf{A}_L + \mathbf{A})^2 = a_{L0}^2\,|\,g\,|^2 + 2\,|\,A\,|^2 + 2\sqrt{2}a_{L0}\,\mathrm{Re}\,al(g^* A e^{i\theta}) \qquad (6)$$

$\theta=(k+k_L)z+c(k_L-k)t$ and the term $2|A|^2$ in the r.h.s. of (8) is usually omitted. The axial motion of the single particle is described by

$$\frac{dp_{jz}(t)}{dt} = -\frac{e^2}{2m^2c^3\gamma_j(t)}\Big\{ a_{L0}^2\left[\frac{\partial}{\partial z}|\,g\,|^2\right]_{\mathbf{x}=\mathbf{r}_j(t)} + . \qquad (7)$$
$$-2\sqrt{2}(k+k_L)a_{L0}\,\mathrm{Im}\left[(g^* A)_{\mathbf{x}=\mathbf{r}_j(t)}e^{i\theta_j(t)}\right] + \ldots..\Big\}$$

where the phase angles $\theta_j(t)$ of the particles in the combined laser plus collective field are given by

$$\theta_j(t) = (k_L+k)z_j(t)+c(k_L-k)t \qquad (8)$$

The first term in the r.h.s. of Eq.(7) gives the ponderomotive force exerted on the electron as it enters or leaves the laser pulse, while the second term is due to the action of all other electrons of the beam and is therefore responsible for collective effects and, in particular, for the FEL instability.
With similar hypotheses one obtains for the transverse motion

$$\frac{d\mathbf{p}_{j\perp}(t)}{dt} = -\frac{e^2}{2m^2c^3\gamma_j(t)}\{ a_{L0}^2(\nabla_\perp |g|^2)_{\mathbf{x}=\mathbf{r}_j(t)} + 2\sqrt{2}a_{L0}\,\mathrm{Real}\left[(\nabla_\perp (g^* A))_{\mathbf{x}=\mathbf{r}_j(t)}\, e^{i\theta_j(t)}\right] + ...\} \tag{9}$$

The first term to the right of (9) gives the ponderomotive focussing or defocussing actions due to the laser transverse gradients while the second term takes into account collective contributions to the transverse motion. The last point consists in the derivation of an approximate equation for the collective vector potential $\mathbf{A}(xyzt)$ directly from the Maxwell equation

$$(\frac{\partial^2}{\partial t^2} - c^2\nabla^2)\mathbf{A}(xyzt) = 4\pi c\mathbf{J}_b - c\frac{\partial}{\partial t}\nabla\varphi \tag{10}$$

where the beam current density is taken in its microscopic form, i.e., $\mathbf{J}_b = -ec\sum_j \beta_j(t)\delta(\mathbf{x}-\mathbf{r}_j(t))$, the integer j running from 1 to N the total number of electrons of the beam. By dropping the contribution from the scalar potential φ, remembering that $\mathbf{A}$ is to dominant order perpendicular to the z-axis, neglecting $\mathbf{A}$ with respect to $\mathbf{A}_\mathrm{L}$ in the driving term of this equation , and averaging over a specified volume we obtain the form

$$(\frac{\partial}{\partial t} + c\frac{\partial}{\partial z})A(xyzt) - i\frac{c}{2k}\nabla_\perp^2 A = -i\frac{\omega_b^2 a_{L0}}{2\sqrt{2}ck}\frac{1}{N_s}\sum_s \frac{g(\mathbf{r}_s(t),t)}{\gamma_s(t)}e^{-i\theta_s(t)} + ... \tag{11}$$

where $\omega_b^2 = 4\pi e^2 n_b/m$, $n_b = N/V_b$ being the average value of the beam volume density (V_b is the volume of the beam).
This equation, obtained for the static wiggler by Scharleman [20], shows the typical driving ("bunching") factor

$$b(xyzt) = \frac{1}{N_s}\sum_s \frac{g(\mathbf{r}_s(t),t)}{\gamma_s(t)}e^{-i\theta_s(t)} \tag{12}$$

The phase angles $\theta_s(t)$ have already been defined in (10), while the integer s in both equations (13) and (14) runs over all particles present at time t inside the volume chosen for the average.
Once restated using the traditional non dimensional variables $\bar{t} = 2\rho\omega_L t$ and $\bar{\mathbf{x}} = 2\rho k_L \mathbf{x}$, our basic euqation may be summarized as follows:

$$\frac{d}{d\bar{t}}\bar{\mathbf{r}}_j(\bar{t}) = \rho\frac{\mathbf{P}_j(\bar{t})}{\bar{\gamma}_j(\bar{t})} \tag{13}$$

$$\frac{d}{d\bar{t}}P_{jz}(\bar{t}) = -\frac{\bar{a}_{L0}^2}{2\rho\gamma_0^2}\frac{1}{\bar{\gamma}_j}\left[\frac{\partial}{\partial\bar{z}}|g|^2\right]_{\bar{\mathbf{x}}=\bar{\mathbf{r}}_j} - \frac{2}{\bar{\gamma}_j}\mathrm{Real}\left[(g^*\bar{A})_{\bar{\mathbf{x}}=\bar{\mathbf{r}}_j}e^{i\theta_j(\bar{t})}\right] + \ldots \tag{14}$$

$$\frac{d}{d\bar{t}}\mathbf{P}_{j\perp}(\bar{t}) = -\frac{\bar{a}_{L0}^2}{2\rho\gamma_0^2}\frac{1}{\bar{\gamma}_j}\left[\bar{\nabla}_\perp|g|^2\right]_{\bar{\mathbf{x}}=\bar{\mathbf{r}}_j} - \frac{4\frac{k_L}{k}\rho}{1+\frac{k_L}{k}}\frac{1}{\bar{\gamma}_j}\mathrm{Im}\left[(\nabla_\perp(g^*\bar{A}))_{\bar{\mathbf{x}}=\bar{\mathbf{r}}_j}e^{i\theta_j(\bar{t})}\right] + \ldots$$

(15)

$$\left(\frac{\partial}{\partial\bar{t}} + \frac{\partial}{\partial\bar{z}}\right)\bar{A}(\bar{\mathbf{x}},\bar{t}) - i\frac{k_L}{k}\rho\bar{\nabla}_\perp^2\bar{A} = b = \frac{1}{N_S}\sum_S\frac{g(\bar{\mathbf{r}}_S(\bar{t}),\bar{t})}{\bar{\gamma}_S(\bar{t})}e^{-i\theta_S(\bar{t})} + \ldots \tag{16}$$

with the phase angles written in non dimensional form as

$$\theta_j(\bar{t}) = \frac{k}{2\rho k_L}\left(\left(1+\frac{k_L}{k}\right)\bar{z}_j(\bar{t}) + \left(\frac{k_L}{k}-1\right)\bar{t}\right) \tag{17}$$

and

$$\bar{\gamma}_j^2 = 1 + \gamma_0^2\rho^2P_{jz}^2 + \bar{a}_{L0}^2(|g|^2)_{\bar{\mathbf{x}}=\bar{\mathbf{r}}_j(\bar{t})} + \ldots \tag{18}$$

In the preceding equations γ_0 is the average value of γ over all electrons of the beam at t=0,

$\bar{\gamma}_j = \gamma_j/\gamma_0$, $\mathbf{P}_j = \mathbf{p}_j/\gamma_0\rho$, where $\bar{a}_{L0} = \frac{e}{mc^2}a_{L0}$ is the laser parameter,

$$A = -i\left(\frac{\omega_b^2 a_{L0}}{4\sqrt{2}\omega\omega_L\gamma_0\rho}\right)\frac{e\bar{A}}{mc^2} \tag{19}$$

and the FEL parameter

$$\rho = \frac{1}{\gamma_0}\left(\frac{\omega_b^2 a_{L0}^2}{16\omega_L^2}(1+\frac{\omega_L}{\omega})\right)^{\frac{1}{3}} \tag{20}$$

The resonant frequency is given by $k \approx \frac{4\gamma_0^2 k_L}{1+a_{L0}^2}$.

3. Numerical result and discussion

We have developed a three-dimensional code that solves the set of equations (13)-(16), based on a fourth-order Runge-Kutta for the particles and on a finite-difference scheme for the radiation field.
An example is given by a beam with an energy of about 15 MeV, corresponding to $<\gamma>=30$, with a mean radius $\sigma_0=10$ micron, a total charge of 1 nC and a length $L_b=$ 120 micron. The laser pulse has 3 TW of power with a focal spot of about 50-100 μm so that the Pierce parameter ρ is 5 10^{-4}, and the quantum parameter q [21] is 0.5. With these values the gain length corresponds to about 100-200 micron and the appearance and the saturation of the collective effects (taking place in 7-12 gain lengths) are contained in 5 picoseconds, a time of the same order of the duration of the laser pulse.
The energy spread $\Delta\gamma/\gamma$ of the electron beam has been taken equal 1. 10^{-4} and the initial normalized transverse emittance has been varied from 0 up to 3.
In Figure 1 the growth of the collective potential amplitude evaluated in the middle of the bunch and averaged on the transverse section is shown in time (lower window), as well as the bunching factor (upper window). The parameters of this calculation are: a flat laser profile inside a region with $w_0=70$ micron, the laser power of 3 TW, a value of the initial emittance of 0,45 mm msec . In Fig 2 the transverse shape of the signal is shown in the phase after the saturation (t=5 psec). In Fig 3 the dependence of the maximum of $<|A|^2>$ on the transverse normalized emittance is shown for a gaussian laser pulse, while in Fig 4 the typical spectrum of the radiation is presented. As regards the spectrum, it has to be noted that the bandwidth is considerably larger than the 1d case. In fact, one can see from Fig 4 (obtained with $\varepsilon_N=1,11$ and $w_0=500$) that $\delta\omega/\omega=4\ 10^{-3}$ a factor 10 larger than ρ..

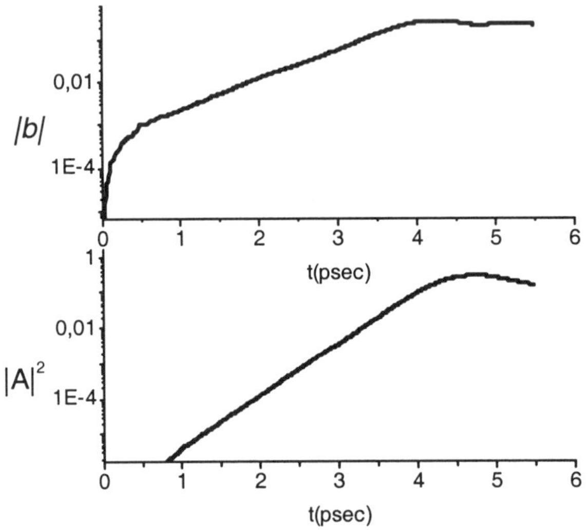

Fig. 1 |b| and $|A|^2(z_m)$ averaged on the transverse section vs t in psec for w_0= 70 μm, ε_N=0,45 mm mrad

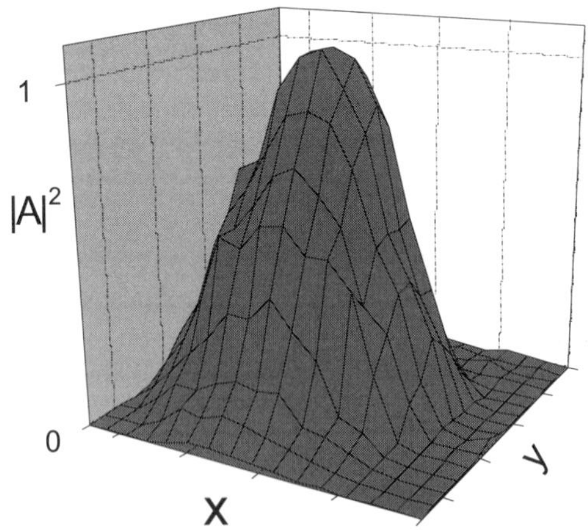

Fig 2: Transverse structure of |A|2 for the same case of Fig 1 at t=5 psec.

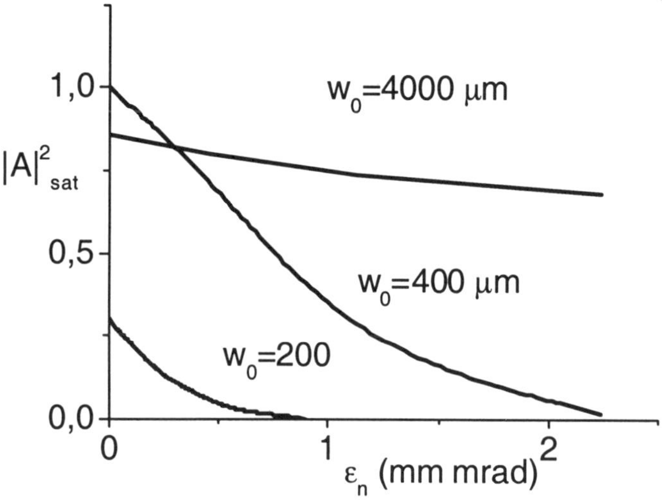

Fig 3 : Dependence of the peak value of |A|2 vs the normalized tranverse emittance for a Gaussian laser pulse

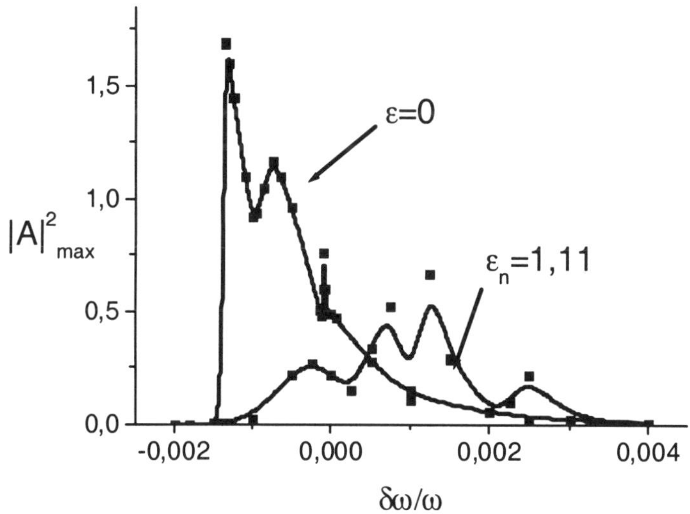

Fig 4 : Spectrum of the radiation versus dw/w for different emittances.

We have considerable emission also in violation of the Pellegrini criterion [22] for a static wiggler. In fact, the emittances considered largely exceed the value $\gamma\lambda/4\pi$, that in this case is 8,5 10-4 micron. On the other hand, on the basis of the fact that $Lg/Z_R=1.2\ 10^{-4}$, the criterion of Pellegrini can be rewritten in a generalized form for both static and optical undulators as

$$\varepsilon_N \leq \alpha\sqrt{Z_R / L_g}\,\lambda_R \gamma / 4\pi$$

with $\alpha=\delta\omega/\rho\omega$ and gives $\varepsilon_n<0.25$ a value closer to the limit shown by the numerical calculation.

References

1. R.W. Schoenlein, W. P. Leemans, A. H. Chin, P. Volfbeyn, T. E. Glover, P. Balling, M. Zolotorev, K. J. Kim, S. Chattopadhyay and C. V. Shank, *Science* **274**, 236 (1996)
2. W. P. Leemans, R. W. Schoenlein, P. Volfbeyn, A. H. Chin, T. E. Glover, P. Balling, M. Zolotorev, K. J. Kim, S. Chattopadhyay, and C. V. Shankl, *Phys. Rev. Lett.* **77**, 4182 (1996)
3. A. Ting, R. Fischer, A. Fisher, C. I. Moore, B. Hafizi, R. Elton, K. Krushelnick, R. Burris, S. Jadel, K. Evans, J. N. Weaver, P. Sprangle, M. Baine, and S. Ride: *Nucl. Instrum. and Meth. Phys Res. A*, **375** ABS68 (1996)
4. I. V. Pogorelsky, I. Ben-Zvi, T. Hirose, S. Kashiwagi, V. Yakimenko, K. Kusche, P. Siddons, J. Skaritka, T. Kumita, A. Tsunemi, T. Omori, J. Urakawa, M. Washio, K. Yokoya, T. Okugi, Y. Liu, P. He, and D. Cline, *Phys. Rev. S. T. Accel. Beam.,* **3**, 090702 (2000)
5. W. J. Brown, S. G. Anderson, C. P. J. Barty, S. M. Betts, R. Booth, J. K. Crane, R. R. Cross, D. N. Fittinghoff, D. J. Gibson, F. V. Hartemann, E. P. Hartouni, J. Kuba, G. P. Le Sage, D. R. Slaughter, A. M. Tremaine, A. J. Wootton, and P. T. Springer: *Phys. Rev. S. T. Accel. Beam.* **7** 060702 (2004)
6. F. Sakai, A. Endo, S. Ito, K. Takasago, Y. Okada, T. Yanagida, J. Yang, M. Yorozu *"Short pulse X-ray generation by laser-Thomson scattering"* Proceedings of Contributed papers. 2001 Particle Accelerator Conference, Chicago, (USA) 2001, (IEEE, Piscataway, NJ, 2001) Vol.4 p. 2695
7. D. J. Jackson: "*Classical Electrodynamics*" (John Wiley and Sons, NY, 1975) page 679
8. E. Esarey, S.K.Ride, P.Sprangle, *Phys.Rev. E* **48**, 3003 (1993)
9. P.Catravas, E. Esarey, W.P. Leemans, *Meas. Sci Technol* **12**, 1828 (2001)

10. J. Yang, M. Washio, A. Endo, T. Hori, *Nucl. Instr, Methods Phys. Res*. A **428**, 556 (1999)
11. S.Ride, E. Esarey, M. Baine, *Phys. Rev. E* , **52** 5425 (1995)
12. P. Tomassini, A. Giulietti, D. Giulietti, L.A. Gizzi, *Appl. Phys. B,* **80** 419-436 (2005)
13. K. Lee, B.H. Kim, D. Kim, *Physics of Plasmas* **12**, 043107 (2005)
14. J. Gea-Banacloche, G. T. Moore, R.R. Schlicher, M. O. Scully, H. Walther, *IEEE Journal of Quantum Electronics*, **QE-23**, 1558 (1987).
15. B.G.Danly, G.Bekefi, R.C.Davidson, R.J.Temkin,T.M.Tran,J.S.Wurtele, *IEEE Journ. of Quantum Electronics,* **QE-23**,103(1987).
16. J.C.Gallardo, R.C. Fernow, R.Palmer, C. Pellegrini, *IEEE Journal of Quantum Electronics* **24**, 1557-66 1988.
17. W.B. Colson, *Phys.Lett.* **59A**,187,(1976).
18. R. Bonifacio, C. Pellegrini, L. Narducci, *Opt. Commun*. **50**, 313 (1984)
19. M.V. Klein and T. E. Furtak: "*Optics*" Wiley and Son N.Y. (1987)
20. E.T. Sharlemann, *J. Appl. Phys* **58** (1985) 2154
21. C.B. Schroeder, C. Pellegrini and P. Chen, *Phys. Rev. E*, **64**, (2001) 056502
22. J.B. Murphy and C. Pellegrini, *J Opt.Soc. Am. B,* **2**, 259 (1985).

CHANNELING PROJECTS AT LNF: FROM CRYSTAL UNDULATORS TO CAPILLARY WAVEGUIDES

S. B. DABAGOV[1,2], M. FERRARIO[1], L. PALUMBO[1,3], AND L. SERAFINI[4]

[1)] *INFN - Laboratori Nazionali di Frascati*
Via E. Fermi 40, I-00044 Frascati(RM), Italy
E-mail: sultan.dabagov@lnf.infn.it

[2)] *RAS - P.N. Lebedev Physical Institute*
Leninsky Pr. 53, 119991 Moscow, Russia

[3)] *University of Rome "La Sapienza"*
P-le A. Moro 5, 00185 Rome, Italy

[4)] *INFN - Milan*
Milan, Italy

Frascati's National Laboratories (LNF INFN) are well known in the world for pioneering research in the particle interaction and synchrotron radiation physics fields. Good experience in designing accelerators, storage rings and beamlines for synchrotron radiation allows presently LNF to be in the frontier for the construction of new X-ray generation sources. This report is an introduction to new research activity "Coherent Scattering Phenomena for Radiations in Solids" started in Frascati within the approved projects SPARC, SPARX and PLASMON-X. The main purpose of the project is to develop research area for studying the channeling phenomena of charged and neutral particles in periodic solid structures.

1. Introduction

Interactions of different kinds of radiation in solids has been studied since scientific community started to operate with penetrating radiations. Among various research branches, coherent phenomena of radiation propagation into solids, as coherent bremsstrahlung, channeling and related effects in crystals, attract attention because it can create a strong flux of photons in broad energy intervals, starting from optical frequencies, via soft and hard X-ray, up to γ-ray frequencies. The latter makes this research field more and more attractive due to the importance of creating new powerful photon sources like the next generations of synchrotrons, free-electron lasers, Z-pinch machines, etc. Another approach to the problem of searching for intensive radiation sources is to increase the radiation density of existing X-ray and γ-sources that can be solved by means of novel optical elements, for instance capillary optics. Passage of charged particles through periodic systems and propagation of neutral particles (photons, neutrons) in capillary systems of micron or submicron channel sizes, even strongly different by nature, have much in common as both can be described within the united channeling theory. Moreover, the use of this analogy can much simplify researches in both areas.

Many intense X-ray sources are now available for scientific and technological applications. The recent ideas and the continuous developments bring X-ray sources, optics and detectors to performances that were almost unbelievable a few years ago. The new generation of powerful table-top devices for elemental and diffraction analysis, but in particular the many synchrotron radiation facilities all around the world that deliver bright and ultra-bright X-ray beams to hundreds of users investigating with different techniques the atomic structure of matter witness the almost exponential increase of X-ray brilliance. X-ray sources produced countless discoveries, ranging from atomic scale maps of proteins to structures of high-temperature superconducting materials. However, only now for the first time in decades, thanks also to the developments of new optical elements such as refractive optics, multilayers, polycapillary lenses, phase gratings, etc., X-ray scientists look at a new scenario that promise to be more exciting than ever. Researchers and funding agencies in Europe, USA and Japan are supporting the construction of new X-ray sources capable not only to generate the most brilliant sources but also to produce short X-ray pulses comparable with the dynamics of the chemical bonds. Actually, this class of machines, called free electron lasers (FELs), will be able to deliver powerful beams of coherent X-rays probably reshaping the landscape of X-ray science.

Since the middle of the last century Frascati's National Laboratories are well known in the world for pioneering research in the particle interaction and coherent radiation physics fields. Good experience in designing accelerators, storage rings and beamlines for synchrotron radiation allows presently the LNF to be in the frontier for the construction of new radiation sources, as FEL, Thomson scattering, coherent bremsstrahlung, channeling radiation, etc. In order to discuss new results, exchange experiences and ideas by various groups in the world, form the basis for future research activities within new collaborations, the International workshop on Charged and Neutral Channeling Phenomena - "Channeling 2004" was held in Frascati (November 2-6, 2004)[1].

2. History of coherent bremsstrahlung at Frascati

Perhaps the first idea on the processes of radiation emission by charged particles traversing solids were published by Williams in 1934. Then Ferretti developed semi-qualitative theory of the phenomena based on the Weiszacker-Williams method of the virtual quanta [2]. After that work we are finding a number of papers by Ter-Mikaelian, Purcell, Feinberg, Pomeranchuk, Dyson, and Überall devoted to theory of electromagnetic radiation by charged particles in matter (see [3,4,5] and Refs in). The full calculations of the cross sections and polarization within the Born approximation were first described in papers [6,7].

The obtained theoretical results stimulated the performance of experiments by means of the high-energy electron accelerators. However, first experimental results were in apparent contradiction with the theoretical ones. Indeed the measurements [8,9] showed an enhancement in the low region of the bremsstrahlung spectrum for a certain crystal orientation, while the theoretical expectation, was a minimum! This contradiction between theory and experimental results was due to the insufficient resolution of angle and energy measurements.

In 1959 Diambrini Pallazi proposed an experimental research program to be realized at the Frascati's 1 GeV electrosynchrotron. The final purpose was to develop a source of a polarized photon beam to be utilized for experiments in particle physics that were planned at the Frascati laboratories. The main problem to be solved was to obtain an energy and angular resolution higher than that used in the previous experiments [10].

The experiments at Frascati were first carried out by Bologna, Diambrini and Murtas, and later also by Barbiellini. These experiments in both the electron-positron pair production and bremsstrahlung showed the expected central minimum. However, having a qualitative agreement with the Überall's results, systematic difference in the results was also revealed [11,12].

A further experimental investigation carried out with a better angular resolution showed an unexpected fine structure in the angular dependence of the bremsstrahlung cross sections. It was realized that the approximation of continuous lattice planes used in the previous calculations was unsuitable for bremsstrahlung in the GeV-energy range, and new calculations, taking into account the actual structure of the lattice planes, are necessary. It became also clear that the bremsstrahlung spectra would have large peaks with high polarization. Detailed calculations and precise measurements on the discovered coherent bremsstrahlung and on the polarization have been carried out by the Frascati's group, finding good agreement between theory and experiment [13].

These authors also proposed a new method for measuring the polarization of very high energy photons by using a second crystal as analyzer [14,15,16,17]. The method was applied in almost all the electron accelerators around the world, namely at DESY, Cornell University (USA), Harvard University (USA, MIT), Yerevan Electro-Synchrotron Lab (Armenia), Tokyo Electro-Synchrotron Lab (Japan) [18,19,20,21].

3. SPARC/SPARX/PLASMON-X

The overall SPARC (Sorgente Pulsata Auto-amplificata di Radiazione Coerente) [22] project consists of 4 main lines of activity aiming at several goals: their common denominator is to explore the scientific and technological issues that set up the most crucial challenges on the way to the realization of a SASE-FEL based X-ray source, the SPARX proposal [23] ((Fig. 1)).

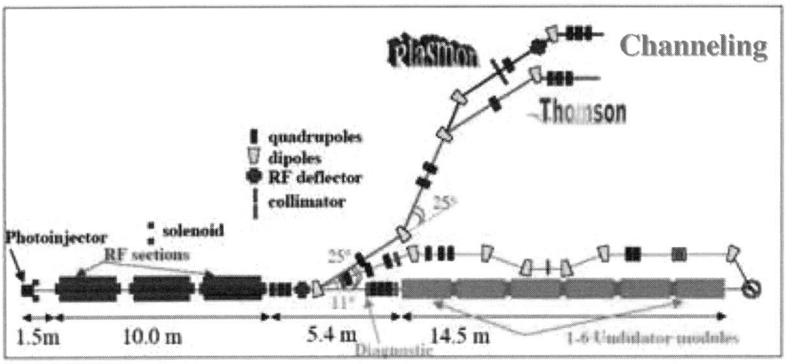

Figure 1. Schemes of different experimental layout disposition.

These are:
- 150 MeV Advanced Photo-Injector;

- SASE-FEL Visible-VUV Experiment;
- X-ray Optics/Monochromators;
- Soft X-ray table-top Source.

Since the performances of X-ray SASE-FEL's are critically dependent on the peak brightness of the electron beam delivered at the undulator entrance, investigations have been performing for two main issues - generation of the electron beam and bunch compression via magnetic and/or RF velocity bunching - by means of an advanced system delivering 150 MeV electrons, the minimum energy to avoid further emittance dilutions due to the time-dependent space charge effects.

In order to study the problems related to matching the beam into an undulator and keeping it well aligned to the radiation beam, as well as the generation of non-linear coherent higher harmonics, a SASE FEL experiment with the 150 MeV beam, using a segmented undulator with additional strong focusing, will be organized to observe FEL radiation at 530 nm and below. The X-ray FEL radiation will provide unique radiation beams to users in terms of peak brightness and pulse time duration (100 fs), posing at the same time severe challenges to the optics that is necessary to guide and handle such radiation. This project is pursuing also a vigorous R&D activity on the analysis of radiation-matter interactions in the spectral range typical of SASE X-ray FEL's (from 0.1 to 10 nm), as well as the design of new optics and monochromators compatible with these beams. In order to test these optics and to start the R&D on applications, the project undertakes an upgrade of the presently operated table-top source of X-rays at Politecnico Milano, delivering 10^7 soft X-ray photons in 10-20 fs pulses by means of high harmonic generation in a gas.

X-rays are presently utilized in many research and application fields, for instance, in imaging technique and in the analysis of industrial products. The SPARX source will be able to boost into excellence the actual applications and to steer research and application toward new objectives. There will be implications on many disciplines and innovative technologies will be developed in the field of material science, biology, medicine, non linear optics, x-ray microscopy; new methods will be developed in protein cristallography.

SPARX source ability to deliver an enormous amount of energy on small areas will make possible to study effects that, up to now, can only be theoretically predicted. In particular interesting results are expected in the field of photochemistry, with important industrial applications. High peak brightness and short pulse duration (a few fs) will be the main characteristics of the SPARX source. By using a 2.5 GeV linear electron accelerator and two magnetic undulators it is possible to obtain radiation at 10 nm and 1.5 nm. Exploitation of 3rd and 5th harmonics will allow emission in the range between 10 and 2 nm for the first beam line and between 1.5 and 0.3 nm for the second beam line.

The aim of the PLASMON-X project [24] is to provide the LNF with a world-class, high-power laser facility suitable for the development of an innovative, high-gradient acceleration technique based upon super-intense and ultra-short laser pulses, and X/γ-ray tunable sources using Thomson scattering of optical photons by energetic electrons. The facility will be built in close interaction with the SPARC project presently in progress.

The main purpose of the facility will consist in R&D activity aimed at the following objectives:

1. demonstration of high-gradient acceleration of electrons injected into electron plasma waves excited by ultra-short, high-power laser pulses;

2. development of a monochromatic and tunable X-ray source in the 20-1000 keV range, based upon Thomson scattering of laser pulses by relativistic electrons.

Both lines require very bright and short electron bursts together with very high power laser pulses of sub-100 fs duration. Studies at the forefront of research in this field consistently indicate that the power level and the focused intensity for a laser system appropriate for these studies are > 10 TW and $> 10^{19}$ W/cm^2, respectively.

The activity will be carried jointly by the INFN and CNR laboratories.

The R&D activity carried out at this facility will be open to some extent to external users, in particular for what concerns bio-medical oriented uses of the available sources for diagnose and therapy.

4. Comparison of spectral characteristics for channeling radiation, bremsstrahlung and Thomson scattering

Similar to bremsstrahlung (B) (brake radiation, Bethe-Heitler radiation) channeling radiation from relativistic electrons in crystals (ChR) has an angular divergence of about $\Delta\vartheta \simeq \gamma^{-1}$, where $\gamma = \frac{E}{m_0 c^2}$ is a Lorentz-factor, E is an electron energy. However, typical crystal thickness to generate these types of radiation varies due to the fact that there is big difference in the processes of multiple scattering. The latter results in strong enhancement for spectral and angular characteristics of ChR in comparison to B for $\gamma \gtrsim 10^3$ (see in [25,26,27,28]).

Let consider the expressions for angular densities of B. A maximum in the angular density of B $(dW/d\vartheta)_B$ is limited by the multiple scattering of electron in a matter. A characteristic thickness x_B at radiation within the angles γ^{-1} is defined by the expression for averaged angle of scattering (with logarithmic accuracy)

$$\frac{1}{\gamma^2} = \left(\frac{E_s^2}{E^2}\right)\left(\frac{x_B}{R}\right) \tag{1}$$

where R is the radiation length, $E_s = 21$ MeV. So we have a simple formula for the estimations

$$x_B = \frac{R}{1800} \tag{2}$$

from which it is easy to obtain x_B =56 μm for Si, 1.7 μm for W, and 73 μm. Then the energy emitted within the angles γ^{-1}can be estimated as

$$\Delta E = \left(\frac{dE}{dx}\right)\Delta x = \left(\frac{E}{R}\right)x_B = \frac{E}{1800} \tag{3}$$

Hence approximately for $\Delta\vartheta \simeq \pi\gamma^{-2}$ one writes

$$\left(\frac{dW}{d\vartheta}\right)_B \Delta\vartheta = \frac{E}{1800} \tag{4}$$

In the same way one can consider an oriented crystal, however we need additional information on energetic lost for the spontaneous radiation. The dependence of energy lost for the spontaneous radiation has a form of $dE/dx \sim E^2$. Now a characteristic length, on which the radiation happens in oriented crystal, is determined by a dechanneling length x_{Ch}. The emission of energy into the angles of γ^{-1} for the main harmonics, i.e. for the narrow spectral frequency range, takes place up to an optimal value of energy that is defined by the dipole character of radiation. At the planar channeling E_{opt} is close to 5-10 GeV, and at the axial channeling is about 1 GeV.

Let introduce a ratio of the angular densities of radiation by the expression

$$\eta_1 = \frac{(dW/d\vartheta)_{Ch}}{(dW/d\vartheta)_B} \tag{5}$$

Because of the angular fans, inside of which B and ChR concentrate, are equal, we obtain

$$\eta_1 \simeq \frac{(dE/dx)_{Ch}}{E/R}\frac{x_{Ch}}{x_B}, \tag{6}$$

where $x_{Ch} = \int \chi(x)\ dx$; $\chi(x)$ is a portion of the charged particles beam remained in the mode of channeling at the crystal thickness of x. For example, the dechanneling lengths for positrons (E_{e+}=200 MeV) approximately equal to 50 μm for a monocrystal Si (110), and 30 μm for diamond. The close values are obtained also for the case of planar positron channeling in W crystals (for positrons $x_{Ch} \approx x_{1/2}$ is a depth, at which a half of the particles captured by the crystal channels dechannels). For the axial channeling of positrons the dechanneling length increases by about one order of value. For electrons at the axial channeling in W we obtain $x_{1/2} \approx 10$ μm at E= 200 MeV (close values have been obtained also for Si). However, we have to underline that for the large depths (more than $x_{1/2}$) the rate of dechanneling essentially decreases. Hence there is an opinion to use the crystals of larger thickness than $x_{1/2}$. From the Eq.(6) it follows that at E= 200 MeV $\eta_1 \simeq 10 \div 10^2$.

Taking the problem of the spectral-angular characteristics of radiation, i.e.

$$\eta_2 = \frac{\left(d^2W/\left(d\vartheta\ d\omega\right)\right)_{Ch}}{\left(d^2W/\left(d\vartheta d\omega\right)\right)_B},$$

we will calculate the thickness, at which the ratio η_2 has a maximum. Approximately one can be written

$$\eta_2 = K_1 K_2\ , \tag{7}$$

where $K_1 = \frac{(dW/d\omega)_{Ch}}{(dW/d\omega)_B}$ describes a ratio of spectral densities obtained experimentally; $K_2 = \frac{x_{Ch}}{x_B}$. It is well known that the dechanneling length is proportional to energy of particle: $x_{Ch} \sim E$. Thus, if K_1 is a constant or changes a little, then η_2 increases with the growth of particle energy. As analysis shows, one expects a factor of $\eta_2 \approx 100 \div 300$ at $E \leq 1$ GeV.

Comparison for the Eqs.(6) and (7) speaks on the fact that the main features (advantages) of channeling radiation can be displayed studying the spectral and angular characteristics of radiation. Due to the fact that for the energy range $E = 0.1 \div 0.5$ GeV spontaneous radiation can be described in the dipole approximation we can conclude that the maximum radiated energy will be obtained for the maximal energy of channeled particles. Additionally to this fact we have to note that the crystal thickness should be optimized to be equal or more than the dechanneling length.

The study of spectral and angular distributions of ChR shows a possibility for creating a powerful source of narrow forwarded electromagnetic radiation. A number of emitted photons can be evaluated by the following simple expression

$$N_{Ch} \simeq \frac{1}{137} \frac{x_{Ch}}{\lambda_{Ch}} \gamma^2 \psi^2 , \tag{8}$$

where ψ is the critical angle for channeling (the Lindhard angle), for $E = 0.1 \div 10$ GeV $\gamma^2\psi^2 \approx 1$, $\lambda_{Ch} \simeq d/\psi$, d is the channel's width.

Meanwhile, for the bremsstrahlung we can estimate a number of photons as

$$N_B \simeq \frac{x_B}{R} \ln \frac{\omega_2}{\omega_1} \tag{9}$$

From the examination of two last relations we see that one channeled particle enables to emit photons, the number of which in one or two orders exceeds one for bremsstrahlung. The same ratio is obtained for the angular densities of ChR and B.

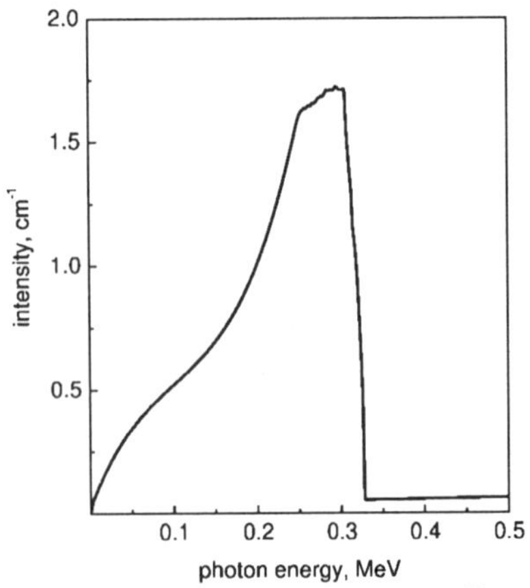

Figure 2. Intensity of ChR for relativistic positrons $E_{e+} = 200$ MeV in Si (110) /Maisheev/.

In Fig.2 one presents the dependence of ChR intensity versus emitted photon energy within the angles of γ^{-1} that was obtained by the simulation of planar channeling of positrons with energy $E_{e+} = 200$ MeV in Si (110)[29]. As seen a channeled positron can produce in the maximum more than one photon of about 300 keV per cm that exceeds at least in one order the number of bremsstrahlung photons.

There is another type of radiation that have to be considered as a competitor of ChR too, namely, coherent bremsstrahlung (CB). A braking particle radiated coherently in crystals is scattered as typical particle, in other words, for this event there is no a suppression of scattering as we observe in the case of channeling. That's why a factor of x_{Ch}/λ_{Ch} responsible for the sharp increase in the spectral-angular density of ChR, is absent for CB. For the channeled particles the radiation concentrates in the narrower spectral range, for which its density is essentially higher than for CB. Moreover, due to the suppression of multiple scattering, ChR is more forwarded than CB, and at last, an integral of ChR over the given range of frequencies and angles is significantly larger than coherent one.

This discrepancy can be explained by the various mechanisms of radiation: ChR regards to particle deflection by the averaged field of atomic rows and planes, while CB - to periodical collisions of particles with the atomic rows and planes that does not cause a meaningful deflection of particle from the quasi straight trajectory. Hence, they are characterized by various radiation frequencies and observed typically for the different crystal-beam orientations: pure CB is detected for the entrance angles that overcome the critical angles of channeling. However, for some specific cases of a beam transversing a crystal under the small angles both along the plane and the row, there is a possibility to observe simultaneously ChR and CB (or combination) but for the different frequency ranges.

Below we present several graphs of spectral distributions of electromagnetic radiation by relativistic electrons and positrons in oriented crystals [28]. The graphs for CB calculation and simulation are shown by dott-dashed line and thick line, correspondingly ; dashed line - for Bethe-Heitler radiation spectrum; α is the angle of the particle momentum in the transverse to the axis plane, ψ_c is the critical angle for axial channeling, ψ is the angle between particle momentum and crystal axis.

As seen from the Fig.3 in comparison to B we have enhance in the emission of photons for classical CB calculation as well as for the simulation. However, the results of simulations strongly varies from the calculations by the classical formulas. Taking into account the analysis above done, we

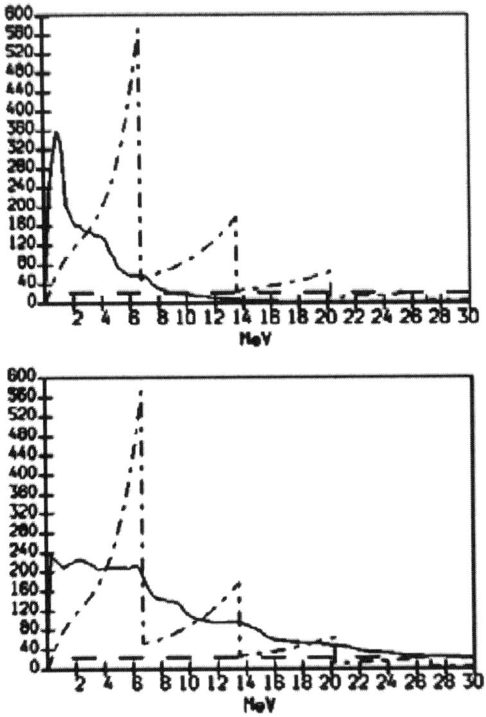

a.

b.

Figure 3. Spectra for coherent radiation of positrons (a) and electrons (b) with energy E_{e+}=0.5 GeV, moving in Si crystal at small angles to < 100 > axis ($\psi = \psi_c$, $\alpha = 0$) /Shul'ga et al/.

expect to have another gain factor in the radiation exit at the particle channeling.

Now we consider another source of radiation, namely, Thomson scattering (TS), in comparison to ChR. Even these two types of radiation are similar by their description, we have to note that the mechanisms of photon production are different. ChR is based on the scattering of relativistic particles (electrons and positrons) in the field of averaged potentials of crystal rows or planes, while in the case of TS we deal with laser photons scattered by relativistic electrons. In both cases we have to keep in mind different factors that can suppress significantly the photon emission: multiple incoherent scattering by crystal atoms for the channeling and mutual refractions in a "photon-electron" system for the TS.

As well known photon energy emitted by relativistic channeled particle

($\gamma \gg 1$) can be expressed by

$$\omega_{lab}^{ChR} \simeq \frac{2\gamma^2}{1+\vartheta^2\gamma^2}\omega_0^{ChR} \tag{10}$$

where ω_{lab} is the observed energy of ChR, ω_0 is the energy of particle oscillation that is defined by the interaction potential, i.e. crystal parameters, while in the case of TS the photon energy (radiated in the forward direction, $\vartheta \ll 1$) may vary between two extremes ($\vartheta = 0$ and $\vartheta = \pi$), thus

$$\omega_{lab}^{TS}\left\{\begin{array}{c}\vartheta=0\\ \vartheta=\pi/2\\ \vartheta=\pi\end{array}\right\} \simeq \left\{\begin{array}{c}1\\ 2\\ 4\end{array}\right\}\frac{\gamma^2}{1+\vartheta^2\gamma^2}\omega_0^{TS} \tag{11}$$

From this relations it follows that for these types of radiation we obtain the next dependences: $\omega_{lab}^{ChR} \propto \gamma^{3/2}$ and $\omega_{lab}^{TS} \propto \gamma^2$. For the electron/positron energy of 100-200 MeV the emitted radiation interval (frequencies) will be practically the same. So we have to compare total radiated powers for the both types of radiation. The number of photons emitted per unit of time at channeling increases with the increase of the electron/positron energy by the law $\propto \gamma^{1/2}$ whereas at TS this value is independent on the energy of projectile. However, in the same time the radiated power for the both cases $P \propto \gamma^2$. The fact that mathematical description for both ChR and TS is the same, allow us to introduce a simple comparison factor as

$$f \simeq \frac{\mathbf{A}_{Ch}^2}{\mathbf{A}_{TS}^2}\frac{L_{Ch}}{L_{TS}} \tag{12}$$

where $\mathbf{A}$ is the strength parameter of a field (crystal or laser) with the interaction length L (in the case of channeling the latter is defined as an effective length of channeling, i.e. $L_{Ch}(z) \simeq \int_0^z N_{ch}(z)\,dz$, is the integral of channeled particles at fixed depth, while for the TS L_{TS} - by the laser beam sizes and the mutual orientation). Analyzing the existing experimental data we can reduced the strength parameters for electron scattering in crystals

	Si ⟨110⟩	C ⟨100⟩	W ⟨111⟩
$\mathbf{A}_{Ch}^2$, eV/Å^3	~ 520	~ 580	~ 10000

and $\mathbf{A}_{TS}^2$ ~ 700 eV/Å^3 for the 10 TW laser with a beam diameter of 0.1 mm.

For 100-200 MeV electrons/positrons in the ideal situations when we have a very powerful laser beam, very high time and spatial synchronization, and without any destructive effects, the radiated power for the TS exceeds the power for channeling radiation. However, such a situation is very difficult to realize. If we consider more realistic case for the TS, namely, the laser power of 5 kW with a 0.1 mm beam diameter at wavelength of 1 μm, we can get ˜ 10^{-8} ph/e^- at $L_{TS} \approx 1$ cm. However, the exit of photons, for instance, from E_{e+}= 200 MeV positrons planar channeled in Si (110) of z = 0.1 mm thickness may reach ˜ 10^{-3} ph/e^+(some experiments indicate ˜ $5{\cdot}10^{-3}$ ph/e^- at planar channeling of 350 MeV electrons and ˜ $3{\cdot}10^{-2}$ ph/e^- at axial channeling in 170 μm Si, and photon energy with maximum for ω_{lab}^{ChR} ˜ 0.8 MeV). Going down by the photon energy to X-ray region we can mention an experimental results with ˜ 10^{-3} ph/e^- emission by E_{e-}= 100 MeV electrons channeled in 105 μm Si (110) (note that this value can be essentially increased by the use of thicker crystals).

Hence one should be emphasized that channeling radiation is really effective source of photons in very wide range of frequencies. In the X-ray interval of photon energies it will provide strong electromagnetic radiation higher than we get by bremsstrahlung, coherent bremsstrahlung and Thomson scattering. However, in the latter we have to underline that Thomson scattering radiation has a higher degree of monochromaticity and is free from incoherent background, which is always present at channeling.

5. X-Ray channeling

Propagation of x-ray and neutron through microcapillaries has been intensively investigated in the last 20 years [30,31] due to the high transmission efficiency of the radiation by the internal hollow part of microcapillaries, but also, due to the possibility to characterize new optical elements capable to deflect or bend intense radiation beams. These researches shown, in addition to the relevance of the capillary optical systems, that capillaries represent a fine tool to investigate fundamental dispersion features of x-ray radiation by a surface [32,33,34,35].

In microcapillaries the sizes of channels for radiation propagation are much larger than the effective wavelength of the radiation and the process dispersion is described within the framework of surface (superficial) channeling. Capillary optics manufacturing technology allows us to reach extremely small channel diameters, actually, the last (fifth) generation of these optics achieved submicron channel diameters. The last reports show

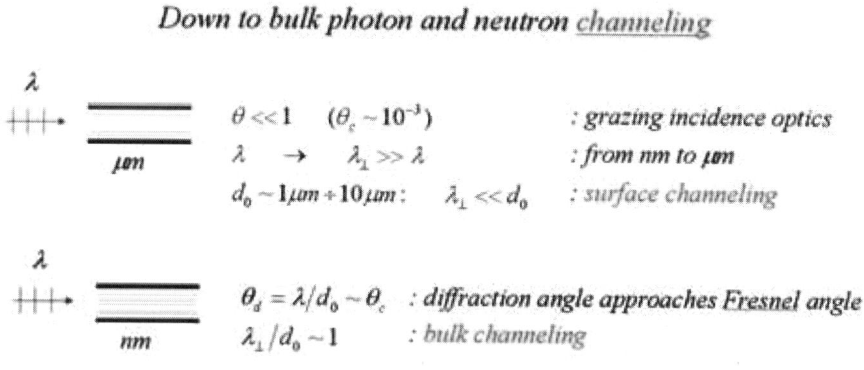

Figure 4. Changes in the propagation character of the radiation inside micro- and nanochannels.

that capillary optical elements with an inner channel diameter of a few tens of nanometers can be obtained with the existing technologies [36].

5.1. *Channeling in microcapillaries: Surface channeling*

As it has been shown recently, the propagation of X-ray photons through capillary systems exhibits a rather complex character [37]. Not all features shown in the experiments can be explained within the geometrical (ray) optics approximation [38,39,40]. On the contrary, the application of wave optics methods allows us to describe in details the processes of radiation spreading into capillaries.

The propagation of X radiation through capillary systems is mainly defined by its interaction with the inner channel walls. In the ideal case, when the boundary between hollow capillary and walls represents a smooth edge, the beam is split in two components: the mirror-reflected and refracted ones. The latter appears sharply suppressed in the case of total external reflection. The characteristics of scattering inside capillary structures can be evaluated from the solution of a wave propagation equation. In the first order approximation, without taking into account the roughness correction $\Delta\varepsilon(\mathbf{r}) = 0$ ($\Delta\varepsilon$ is the perturbation in dielectric permittivity induced by the presence of roughness), the wave equation in the transverse plane to the propagation direction reads

$$\left(\nabla_\perp^2 - k^2\delta(r_\perp) + k_\perp^2\right) E(r_\perp) = 0 \,, \tag{13}$$

where E is a radiation field function, and $\mathbf{k} \equiv (\mathbf{k}_{\|}, k_\perp)$ is a wave vector.

Due to the fact that the transverse wave vector may be presented as $k_\perp \approx k\theta$ under the grazing incidence ($\theta \ll 1$), an "effective interaction potential" is estimated by the expression

$$V_{eff}(\mathbf{r}_\perp) = k^2\left(\delta(\mathbf{r}_\perp) - \theta^2\right) =$$

$$= \begin{cases} -k^2\theta^2 \ , r_\perp < r_1 \\ k^2\left(\delta_0 - \theta^2\right) \ , r_\perp \geq r_1 \ , \end{cases} \tag{14}$$

where r_1 corresponds to the reflecting wall position. From the latter the phenomenon of total external reflection at $V_{eff} = 0$ follows, when $\theta \equiv \theta_c \simeq \sqrt{\delta_0}$ - the Fresnel's angle.

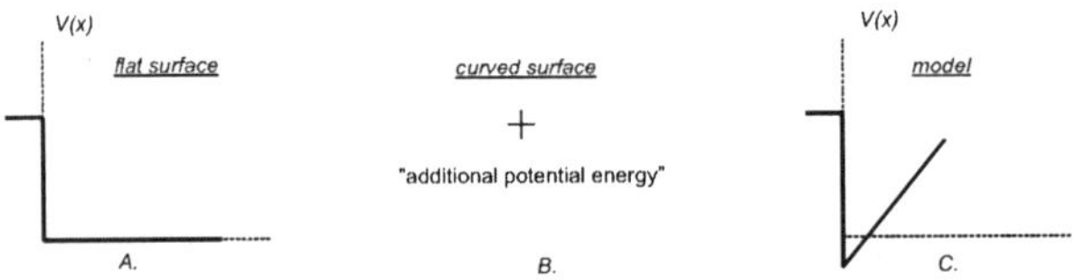

Figure 5. The change of the interaction potential between the flat surface (A.) and the curved one (B.). For simplicity, in calculations the "real potential" (B.) may be replaced by the model potential (C.).

When we introduce a curvature in the reflecting surface, the effective potential obtains an additional contribution. This term, that corresponds to the additional "potential energy", can be seen physically in the following way. Due to the reflecting surface curvature a photon receives an angular momentum $kr_{curv}\varphi$, where r_{curv} is the curvature radius of the photon trajectory, φ is an azimuthal angle in the cylindrical coordinates chosen along the direction of radiation propagation. The latter is supplied by the "centrifugal potential energy" $-2k^2 r_\perp / (r_{curv})$

$$V_{eff}(\mathbf{r}_\perp) = k^2\left(\delta(\mathbf{r}_\perp) - \theta^2 - 2\frac{r_\perp}{r_{curv}}\right). \tag{15}$$

The situation is explained in the scheme in Fig.5. Because of the variation in the spatial system parameters, the interaction potential has been changed from the step potential with the potential barrier of $k^2\delta_0$ to the well potential, with the depth and width defined by the channel characteristics.

In the following we briefly discuss a solution of the wave equation in the case of an ideal reflecting surface (i.e. without roughness), when the reflected beam is basically determined by the coherently scattered part of radiation ([41,42]). The evaluation of the wave equation with the boundary conditions of a capillary channel shows that X-radiation may be distributed over the bound state modes defined by the capillary channel potential (see below). It is important to underline here that the channel potential acts as an effective reflecting barrier, and then, an effective transmission of X-radiation by the hollow capillary tubes is observed. While the main portion of radiation undergoes incoherent diffuse scattering [43], the remaining contribution (usually small) is due to coherent scattering that represents a special phenomenon, extremely interesting to observe and clarify [44].

Let us estimate the upper limit of curvature radius $(r_{curv})_{\max}$ (which is defined by capillary/system of capillaries bending and/or by the radiation propagation trajectory), at which the wave behaviors are displayed under propagation of radiation in channels [32], by considering a photon with the wave vector $\mathbf{k}$ channeling into capillary with curvature radius $(r_{curv})_i$ (i-th trajectory). At small glancing angles, θ, the change of the longitudinal (along the propagation direction) wave vector, $k_{||}$, under reflection from the capillary wall is negligibly small; but mainly one changes the transverse wave vector, $k_{\perp}$,

$$k_{\perp} \simeq k\theta \ (\theta < \theta_c). \tag{16}$$

Correspondingly, from this relation it follows that the transverse wavelength will much exceed the longitudinal wavelength. This makes the interference effects observable even for very short wavelengths. Indeed,

$$\lambda_{\perp} = \lambda/\theta >> \lambda \tag{17}$$

quantum mechanical principles say that, in order to display the wave properties of a channeling photon, it is necessary that typical sizes of an "effective channel" δ_i, in which waves have been propagating, be comparable to the transverse wavelength, i.e. $\delta_i(\theta) \simeq \lambda_{\perp}(\theta)$ (Fig.6). Taken into account the fact that the term for δ_i is defined by the i-th layer curvature radius $(r_{curv})_i$ and the incidence angle θ, namely, $\delta_i \simeq (r_{curv})_i\theta^2/2$, this condition may be rewritten in the following form:

$$(r_{curv})_i\theta^3 \sim \lambda \ , \tag{18}$$

from which, for instance, we obtain $(r_{curv})_{\max} \sim 10\ cm$ for a photon of $\lambda \sim 1$ Å wavelength at glancing angle $\theta \simeq \theta_c/3$. So, from this simple estimate we can conclude that the relation (18) provides a specific dependence for surface bound state propagation of X-rays - surface channeling - along the curved surfaces (for instance, in μ-capillary systems) (see also [45]).

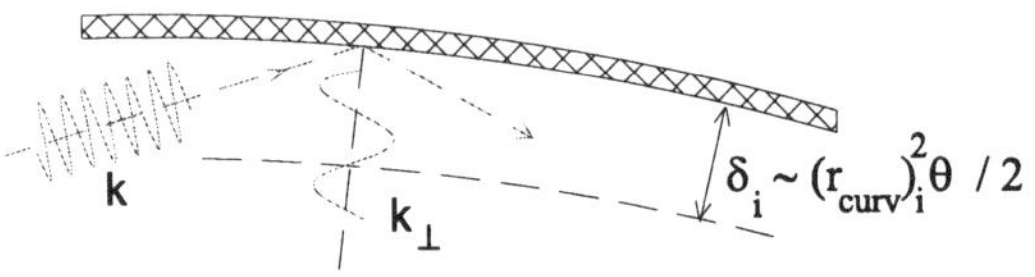

Figure 6. Illustration of X-ray reflection from the inner capillary surface. At glancing angles θ, when the cross size of a beam $\delta_i(\theta)$ becomes comparable with the transverse wavelength $\lambda_\perp(\theta)$, the radiation is grasped in a mode of surface channeling.

Since the waveguide is a hollow cylindrical tube, if the absorption is considered to be negligible, the interaction potential, in which a wave propagates, is determined by Eq.(15) with the radiation polarizability parameter $\delta_0 \simeq \theta_c^2$. Solving the wave equation in cylindrical coordinates $\mathbf{r} \equiv (r_\perp, \varphi, z)$ we are mainly interested in the surface propagation, $r_\perp \simeq r_1$, which, in fact, defines a wave guiding character inside the channel of the curvature radius r_1 ($r_\perp = r_1 - \rho$, $|\rho| \ll r_1$, along the trajectory for the given mode m)

$$E_n \simeq \sum_m C_m u_m(\rho)\; e^{i(k_\parallel z + n\varphi)} \;,$$

$$u_m(\rho) \propto \begin{cases} Ai\;(\rho + t_m) \;,\; \rho > 0 \\ \alpha Ai'(t_m)\, e^{\alpha\rho} \;,\; \rho < 0 \quad (\alpha > 0) \end{cases} \tag{19}$$

where $Ai\;(t)$ is the Airy function, satisfying an equation $Ai'(t) = t\;Ai\;(t)$ and decaying exponentially at $t > 0$, t_m is the m-th pole of the Airy function, and α is an inverse length characteristic of the capillary substance (the method used was described in detail in previous publications, see, for example, Ref. [46]). Evidently, these expressions are valid only for the lower-order modes and in the vicinity of a channel surface. The expression (19) characterizes the waves that propagate close to the waveguide wall, or in other words, the equation describes the grazing modal structure of the electromagnetic field inside a capillary (surface bound X-ray channeling states) (Fig.7).

The solution shows also that the wave functions are exponentially damped both inside the channel wall: $\rho < 0$ (as $e^{-2|\alpha||\rho|}$), and going from

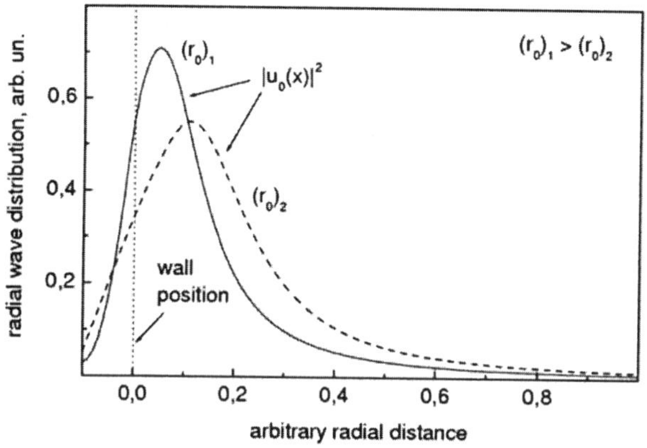

Figure 7. The radial distributions of the main bound mode of radiation inside a capillary channel for various channel diameters. The decrease of diameter ($2r_0$) results in a spatial displacement of the distribution away from the channel wall towards the center. The wall surface position is shown by the dotted line.

the wall towards the center: $\rho > 0$ (as the Airy function decays). It should be underlined here that the bound modal propagation in the hollow part of the system takes place without the wave front distortion. The analysis of these expressions allows us also to conclude that almost all radiation power is concentrated in the hollow region and, as a consequence, a small attenuation along the waveguide walls is observed.

As for the supported modes of the electromagnetic field, estimating a characteristic radial size of the main grazing mode ($m = 0$) results in

$$\overline{u}_0 \simeq 0.75 \left(\frac{\lambda^2 r_1}{\pi^2} \right)^{1/3} , \tag{20}$$

and we can conclude that the typical radial size $\overline{u}_0$ may significantly overcome the wavelength λ, whereas the curvature radius r_1 in the trajectory plane exceeds the inner channel radius, r_0: $\overline{u}_0 \gg \lambda$. For example, at the radiation wavelength $\lambda = 10$ Å an estimation of the characteristic size $\overline{u}_0$ in the extreme case, when $r_1 \simeq r_0$, gives us that $\overline{u}_0 \gtrsim 75$ Å for a capillary channel with the radius $r_0 = 10\ \mu$m.

5.2. *Channeling in nanocapillaries: Bulk channeling*

Designing and testing new natural systems composed by many channels capable to guide radiation have been continued since the mid 80's. The first manuscript on the new nano-crystalline structures [47] presently well-known as carbon nanotubes was published in 1991 [48]. The films of carbon nanotubes with specified geometry attracted attention of researches for the development of new electronic emitters, nanodevices, etc. [49]. The discovery of carbon nanotubes suggested also their use as collimators for hard x-ray radiation and charged particles beams or for channeling of charged particles, x-rays and neutrons [50,51,52,53]. Features of nanotube morphology, e.g., the presence of inner cavity, address the possibility to achieve an efficient transmission of x-ray, thermal neutron and charged particle beams. Actually, nanotubes can be considered as nanoscale capillaries. However, for carbon nanotubes, the effective wavelength of x-radiation, transverse wavelength $\lambda_\perp$, becomes comparable with the diameter of channels d and the description of the propagation as a reflection of photons from the internal surfaces of the tube walls is not consistent. Hence, instead of surface effects of radiation scattering we deal with bulk features of radiation interaction with the system [54]. We would like also to underline that the last advantages in capillary drawing technology have shown possibility to reach a nano scale for the capillary channels.

To understand the propagation nature and the scattering of soft x-rays inside channels of extremely small diameters is a nontrivial problem. Let us to start with simple approach. As mentioned above, a nanotube can be considered as a capillary of very small inner diameter and wall thickness. However, there is a strong difference between a typical capillary and a nanotube from the point of view of radiation propagation inside these structures. First of all, the dielectric parameter as a function of the distance from the center of capillary channel varies with a step-law from zero in the inner hollow cavity to a constant value defined by the substance of channel wall. On the contrary, for a nanotube channel a continuous change of the dielectric parameter value takes place. However, character of radiation propagation in a nanochannel is different from the propagation in a microchannel. It is associated to the large difference in channel diameters. Indeed, radiation transmission by a cavity is defined as the ratio between the transverse wavelength of the radiation and the channel diameter $\lambda_\perp/d$. When this ratio is quite small, $\lambda_\perp/d \ll 1$, we can neglect wave features in the propagation process in the framework of the ray optics approximation. The same approach is not valid if $\lambda_\perp/d \simeq 1$ e.g., when the diffraction angle becomes comparable to the Fresnel angle $\theta_c = \frac{\omega_0}{\omega}$(the critical angle of total external reflection), where $\omega_0 = \left(\frac{4\pi e^2}{m_0}N_e\right)^{1/2}$ is the plasmon energy, N_e,

e and m_0 are the electron density of matter, charge and mass of electron, respectively, ω is the quantum of radiation. This last condition occurs in nanosize channels and as a consequence nanochannels work as waveguides of radiation (see explanation in Fig.4).

The main characteristic of radiation propagation refers to the electron density $N_e(\vec{\mathbf{r}})$, which is position dependent. The use of typical atomic potential approximation procedure (for simplicity, a Molier approximation is chosen) and averaging of interaction potential as for continuous potential at channeling results in the following expression for a bent nanotube:

$$\langle N_e(r)\rangle \approx \frac{r_{curv} N_a Z}{\pi a^2} \sum_i \alpha_i \beta_i^2 \int_0^{\pi} d\theta\, K_0\left(\frac{\beta_i \rho}{a}\right), \tag{21}$$

where $\rho^2 = r^2 + r_{curv}^2 - 2 r r_{curv} \cos\theta$, r_{curv} is the radius of nanotube bending, N_a is the atomic density, a is the screening radius of atomic form-factor. Evaluation of this relation gives estimation for the plasmon energy dependence presented in Fig.8.

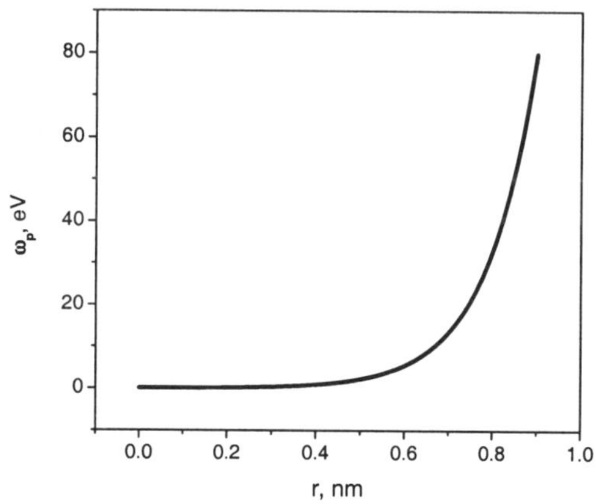

Figure 8. Plasmon energy distribution in the cross section of a nanotube

As expected, the dependence is a smooth continuous function crossing a nanotube cavity. So, propagation of polychromatic radiation through a straight nanotube will be accompanied by continuous energy filtering: the effective width of nanotube channel as wider as higher photon energy.

As follows from analysis of the main parameters that defines characteristics of radiation propagation through a nanotube; the transverse wavelength $\lambda_\perp$ of the radiation is independent by the projectile energy but is a function of the electron density of the nanotube. The density difference at the nanotube axis and inside the nanotube wall determines the transverse wavelength which is just a few nm.

However, channeling of radiation will be accompanied by strong tunneling of radiation through the channel walls. Its probability ($\propto \exp\left(-\frac{\Delta d}{\lambda_\perp}\right)$) is quite high for a single wall nanotube due to the small thickness of the wall. This effect is suppressed in multi wall nanotube as well as in typical capillaries with rather thick wall (single capillaries; in polycapillary case, thickness of the walls will be sufficiently less) . However, multilayer wall acts as periodic structure, and diffraction of radiation in a multilayer wall system can not be neglected.

5.3. *X-Ray channeling in carbon nanotube*

At present many techniques may produce layers of oriented nanotubes. Mostly aligned multiwall carbon nanotubes have been synthesized using zeolite-supported catalysts [56], lithographyc techniques [57,58], and mixtures of ferrocene and fullerene C_{60} or hydrocarbons compounds [59,60,61]. The nanotube structure and the orientation character change following the synthesis conditions. It should be underlined here that, nanotubes composed by one layer are often bent and this contributes to the no ideality of the film structure. A simple explanation of the formation of the nanotube structures is illustrated in Fig.9.

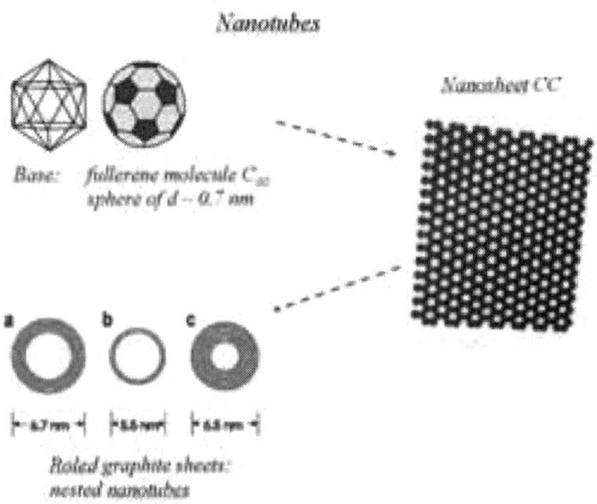

Figure 9. Evaluation of a carbon nanotube structure.

The material produced by thermal chemical vapor deposition, namely, a film of vertically aligned nanotubes on a Si substrate was characterized by a scanning electron microscope (SEM) with a resolution of ~ 10 nm. Fig.10 shows a SEM image of aligned carbon nanotube bundles deposited perpendicularly to the substrate surface. The mean length of nanotubes coincides with the thickness of a layer and is about 10 μm. The bundles are rather closely packed in the film and their density was estimated about 1000 pieces per μm^2. For a second sample the layers of the oriented nanotubes were destroyed by pressing nanotubes onto a substrate surface. Due to the fact that nanotubes in the layer were quite long and defective, in the second investigated sample, nanotubes were mainly parallel to the layer surface.

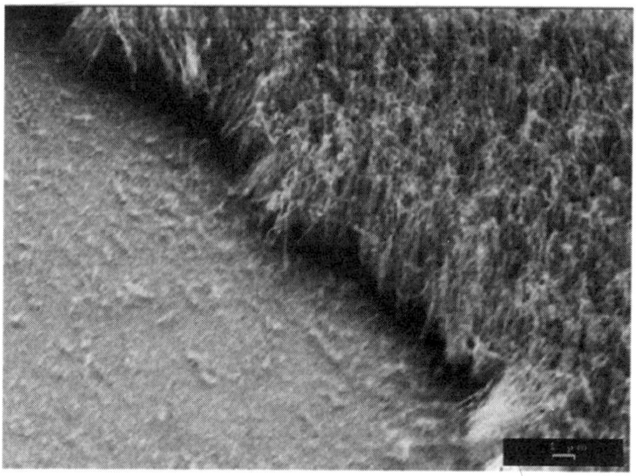

Figure 10. SEM image of a forest of the alligned carbon nanotubes.

As the TEM (transmission electron microscope) images show, multilayered carbon nanotubes, characterized by external diameter of $100 \div 200$Å, and diameter of an internal cavity of $50 \div 70$ Åmay support the growth of film structures that range from 5 up to 25 layers. Defects and bends of nanotubes are formed during growth, at destruction of a film or during the preparation of a sample for electron-microscopic measurements. In each layers the nanotubes are linked enough to form a sheaf of hundreds separate nanotubes, whose internal cavity is filled in part by metal iron.

From the SEM analysis we observed that nanotubes are most ordered in an internal part of a film. In the beginning of the growth process the silicon surface with deposited iron particles acts as the catalytic surface capable to support the alignment of carbon nanotubes. Then, a simultaneous growth does not allow tubes to be significantly bent. The external area of a film with thickness of $0.1 \div 0.2$ μm is probably the most disordered region due to the final contribution of the growing process, e.g., when the fullerene concentration is low.

The angular dependence of the x-ray fluorescence from nanotube films was measured using a laboratory spectrometer. Experimental layout was placed into the spectrometer chamber operating in a vacuum of $2 \cdot 10^{-6}$ Torr. The sample 8 mm long and about 2 mm wide was placed on the rotatable axis providing an angular range α, between the sample surface and the optical axis of spectrometer from 60° to 100°. The position of the copper anode of the x-ray tube and the entrance slit was fixed. The angle β between the radiation source (the anode surface is about 6 mm wide) and the rotating axis was 20°. A gas proportional counter with methane pressure of 0.2 atm was placed beyond the slit. A counter window was made from polypropylene film of 1 μm thickness. At anode voltage of 2.3 kV and current of 100 mA the maximal counter charge was $\sim$ 3000 impulses/sec. The total intensity at the maximum accounted for $\sim$ 60000 impulses.

X-ray fluorescence of a carbon atom does no show any angular dependence due to the localization and symmetry of the inner 1s-level. However, in crystals the probability of radiation emission is determined by the distribution of electrons in the valence band and may have a complex angular dependence. For example, in the emission x-ray spectra of graphite, the contribution of π and σ components is varied depending on angular orientation of the detector [62]. The shape of the C K_α-spectrum of graphite is continuously changed with a take-off angle of radiation and an integral intensity of a spectrum almost constant.

To detect effects due to channeling in carbon nanotubes the x-ray fluorescence from aligned nanotube film is compared to the emission of disordered nanotube film. The vertical orientation of carbon nanotubes in the second sample was destroyed by applying a pressure to the film. Indeed, after the treatment SEM analysis has shown an orientation almost parallel to the film surface. The angular dependence of the intensity of the x-ray fluorescence measured for both nanotube samples, aligned and disordered (curves 1 and 2, respectively) are compared in Fig.11. Intensity is normalized to the maximum at about 85°. All curves show a similar angular dependence increasing at angles from 60° up to 85°. The sample of aligned carbon nanotubes is characterized by additional small feature around 90°. For the disordered film the angular dependence of the X-ray fluorescence

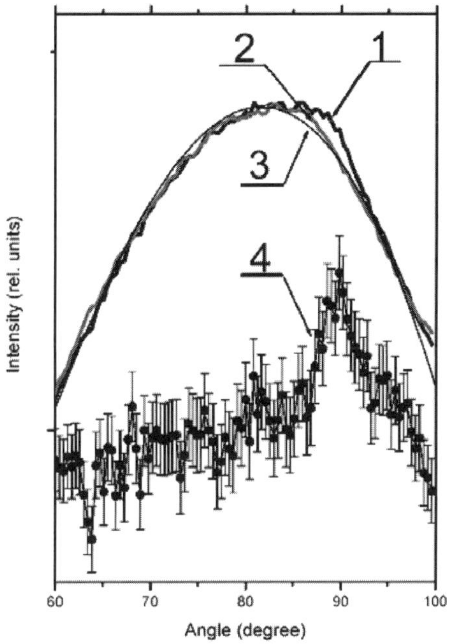

Figure 11. Angular dependence of X-ray emission from the aligned and chaotic nanotube arrangement in the carbon layer.

intensity can be described with the geometrical parameters of the incoming beam and the emitted fluorescent beams. The measured intensity fits the following expression

$$I = I_0 \sin\alpha \cos(\alpha - 20^\circ) \exp\left(\frac{-\eta \cos(\alpha - 20^\circ)}{\sin\alpha} \right) , \qquad (22)$$

where I_0 is the initial radiation intensity, η is the cumulative factor included the absorption and the density of the sample. The formula takes into account the change of the measured area with an angle α and the absorption of the incident radiation and fluorescence when the sample thickness is larger than the characteristic absorption depth of the exciting radiation. The behaviour (22) is represented by curve 3 in Fig.11. The angular behaviour of the radiation intensity is a strong function of η and using this parameter to fit curve 3, we obtain a maximum at 82°.

The experimental angular dependence of the fluorescence emission is in good agreement with simulations and doesn't show features in the angular range near to the normal. The excitation of carbon x-ray fluorescence in nanotubes occurs through all the depth of a layer. However, the contributions of the radiation emitted by the deeper layers tend to decrease. On the other hand, the absorption of C K_α-radiation in a carbon layer is negligible. Therefore the contribution to the spectrum is given by the entire sample.

The difference between curves 1 and 2, multiplied by five is represented by the curve 4 in Fig.11. The intensity of this curve is estimated to be about 0.4% of the total x-ray fluorescence intensity. Despite the large signal/noise ratio curve 4 exhibits a clear maximum at 90° with a width of about 6°. This feature can be interpreted as due to the interaction with carbon nanotubes oriented in the direction of the radiation. The width is then determined by the deviation of carbon nanotubes from the ideal vertical alignment. X-ray fluorescence from carbon nanotubes significantly misaligned, which mainly occurs near to the sample surface, contribute to the diffuse halo. The complex form of curve 4 is actually determined by a combination of different phenomena of dispersion of the radiation from oriented nanotubes.

Let us suppose that the angular distribution of x-ray fluorescence from carbon nanotubes is isotropic so that we may estimate the mean angular deviation $\langle\Delta\theta\rangle$ of the x-ray fluorescence after traversing of a carbon layer. It may be calculated considering the complex mechanism of the scattering in the inner cavity and from the walls of nanotubes $\langle\Delta\theta\rangle_{chan}$ as well as by the diffuse scattering in the space between separate nanotubes $\langle\Delta\theta\rangle_{diff}$. Hence, we can determine that

$$\langle\Delta\theta\rangle = \langle\Delta\theta\rangle_{diff} + \langle\Delta\theta\rangle_{chan} \tag{23}$$

The first term of the Eq.(23) depends on the texture of the nanotube's forest and on the mean distance between separate nanotubes or bundles of nanotubes tight attached to each other in the film, i.e. $\langle\Delta\theta\rangle_{diff} = \langle\Delta\theta\rangle_{text} + \langle\Delta\theta\rangle_{pass}$. The latter can be easily estimated from geometrical parameters while the microscopic pictures of the samples allow texture parameters to be defined. The analysis shows that for a layer thickness of 10 μm, the mean distance between separate nanotubes or bundles of nanotubes is about 100 Å. Moreover, we can evaluate also the mean angular deviation in the angular orientation of the forest of nanotubes of about $20° \div 30°$. Thus, for our samples the term $\langle\Delta\theta\rangle_{pass}$ corresponding to the straight ray propagation (without reflection "on a gleam", i.e., free propagation in a space between nanotubes) between the nanobundles in the carbon layer, is a few mrads (less than 1^0) order from below. This value is lower than the angular divergence of an x-ray beam channeling in a nanotube. However, this factor has a small influence, i.e. the density of rays following a "straight propagation" is quite small. It is also reasonable, that the angular width of the dispersion due to the nanolayer texture $\langle\Delta\theta\rangle_{text}$ can be estimated with the same method, e.g., analyzing the layer micropicture and the angular orientational disorder of nanotubes within the layer. Disorder may be estimated by the mean deviation value of the angular orientation of nanotubes that is about $20° \div 30°$.

The second term of the expression (23) is the term that describe the radiation scattering inside the oriented channels of the layer. Following the definition of the scattering in channeling phenomena (see [63]), with a simple model we can write the following equation for the average angle of scattering in a nanotube ($e = \hbar = c = m_0 = 1$)

$$\langle\Delta\theta\rangle_{chan} \sim \frac{1}{r_0 N_0} \int_{\omega_{\min}}^{\omega_{\max}} d\omega \ \frac{N_{ph}(\omega)}{\omega} \int_0^{r_0} dx \ \omega_0(x) \ , \tag{24}$$

where r_0 is an internal radius of the "hollow" part of nanotube, ω and $N_{ph}(\omega)$ correspond to the emitted photon energy and the number of photons, $\omega_0(x) = \sqrt{4\pi\rho(x)}$ is the energy of nanotube plasmon with a density of $\rho(x) = \frac{e^2}{m_0} N_e(x)$ [51,52]. From the expression (24) we can estimate that $\langle\Delta\theta\rangle_{chan} \leq \omega_{0\,\max} r_0/10 \approx 6° \div 10°$ at the given values of $\omega_{0\,\max} \simeq 100 \div 150$ eV and $r_0 \simeq 20 \div 30$ Å, that is in good agreement with the experimental data. Actually small angles of dispersion are affected by the interaction between x-ray quanta and electrons of the nanostructured media, which in nanotubes form axially symmetrical channel with smoothly varying density of the electronic cloud. Moreover it is also necessary to take into account the tunnel effect of quantum transition between channels in film formed by multiwall nanotubes. In the investigated samples the walls thickness has a range $\Delta d \sim 25 \div 65$ Å of the order or less than the transverse wavelength of the radiation $\lambda_\perp \sim 50 \div 100$ Å. Therefore, the probability of tunneling into the neighbour channels, $\propto \exp(-\Delta d \ / \ \lambda_\perp)$, is large enough. However, the process of tunneling is accompanied also by small-angle diffraction. The presence of a set of ordered walls introduces diffraction dispersion of x-ray and the first maximum mainly determined by diffraction, is observed at angles close to the angles of channeling [64].

A significant portion of the radiation mainly scattered at the condition of channeling by the hollow part of the nanotubes, is not guided due to the distributed disorder of the nanosystem orientation. In other words, a large angular width of the orientation distribution of the nanotubes leads to a fast dechanneling of the radiation due to the increase of the "reflection angles" from nanotube walls. Then the scattering of this dechanneled fraction of the beam is dominated by diffusion phenomena in the nanotube forest and explains also the weak peak in the angular distribution of the emerging radiation.

Modern technological manufacture processes addressed in the last years the scientific relevance of systems formed by layers of focusing nanotubes. The question about the possible use of hollow channels in nanostructures, discussed theoretically in the last years appears now feasible. However, the quality of these films is still far from ideal in order to be used as devices with defined crystal directions. Nevertheless, it is now possible start preliminary investigations about the structural effects of radiation dispersion in carbon nanotubes. In these systems the observed effects on the micron scale are due to the small sizes of the channels.

6. Acknowledgements

I would like to thank all our collaboration for studying coherent phenomena of radiation scattering in micro- and nanostructures of different periodicity, namely the SPARC team for the interest in organizing channeling projects within the FEL program of Frascati, Prof. N.F. Shul'ga and group of Kharkov PhTI for theoretical analysis of relativistic electron scattering in crystals and contribution in a proposal for new beamline construction, scientists of P.N. Lebedev Physical Institute and Institute for Röntgen Optics for their valuable research of X-ray transmission by capillary systems, Profs. W. Greiner, A. Soloviev, H. Backe and the Beam Test Facility team of LNF for their interest to study channeling undulator features for relativistic positrons. I am sincerely appreciated many fruitful discussions with Profs. X. Artru, M.A. Kumakhov, A Okotrub, and N. Nasonov, Drs. N. Zhevago and K. Ispirian and their promising proposals. I am particularly grateful to Profs. H. Überall, G. Diambrini Palazzi and Dr. G. Murtas for the invaluable historical and scientific information on pioneering studies of coherent radiation in crystals.

References

1. "Channeling 2004" - Workshop's Proceedings, Dabagov S.B., Ed., Proc. of SPIE **5974** (2005).

2. B. Ferretti, Nuovo Cimento **7** (1950) 118.
3. M.L. Ter-Mikaelian, Zh. Eksp. Teor. Fiz. **25** (1953) 296.
4. F.J. Dyson, and H. Überall, Phys. Rev. **99** (1955) 604.
5. E.L. Feinberg and I. Pomeranchuk, Nuovo Cimento Suppl. **3-X** (1956) 652.
6. H. Überall, Phys. Rev. **103** (1956) 1055.
7. H. Überall, Phys. Rev. **107** (1957) 223.
8. R. Frisch, and D. H. Olson, Phys. Rev. Lett. **3** (1959) 141.
9. W.R.H. Panofsky and A.N. Saxena, Phys. Rev. Lett. **2** (1959) 219.
10. G. Diambrini Palazzi, Rev. Mod. Phys. **40**(3) (1968) 611.
11. G. Bologna, G. Diambrini Palazzi, and G. P. Murtas, Phys. Rev. Lett. **4** (1960) 134.
12. G. Bologna, G. Diambrini Palazzi, and G. P. Murtas, Phys. Rev. Lett. **4** (1960) 572.
13. G. Diambrini Palazzi, Proc. Conf. Photon Interactions in the BeV Energy Range, Cambridge, Mass., **VI** (1963) 4.
14. G. Barbiellini, G. Bologna, G. Diambrini Palazzi, and G. P. Murtas, Phys. Rev. Lett. **8** (1962) 112.
15. G. Barbiellini, G. Bologna, G. Diambrini Palazzi, and G. P. Murtas, Phys. Rev. Lett. **8** (1962) 454.
16. G. Barbiellini, G. Bologna, G. Diambrini Palazzi, and G. P. Murtas, Phys. Rev. Lett. **9** (1962) 396.
17. G. Barbiellini, G. Bologna, G. Diambrini Palazzi, and G. P Murtas, Nuovo Cimento **28** (1963) 436.
18. G. Bologna, G. Lutz, H. D. Schultz, U. Timm, and W. Zimmermann, Nuovo Cimento **42** (1966) 844.
19. U. Timm, Fortschritte Der Physik **17** (1969) 12.
20. L. Criegee, G. Lutz, H. D. Schultz, U. Timm, and W. Zimmermann, Nuovo Cimento **16** (1966) 1031.
21. S. Kato, T. Kifrne, Y. Kimura, M. Kobayashi, K. Kondo, T. Nishikawa, H. Sasaki, K. Takanratsu, S. Kikuta, and K. Kohra, J. Phys. Soc. Japan **20** (1965) 303.
22. D. Alesini et al. "Status of the SPARC Project", presented at **PAC2005** 16-20/5/2005, Knoxville, Tennessee, USA.
23. http://www.sparx.it/
24. http://ilil.ipcf.cnr.it/plasmonx
25. Coherent Radiation Sources, A. W. Saenz and H. Überall, Eds., Springer, Berlin, 1985.
26. Proceedings of "Radiation from Relativistic Electrons in Periodic Structures", Nucl. Instr. Meth. **B145**(1,2) (1998).
27. M.Kh. Khokonov, and R.A. Carrigan, Jr., Nucl. Instr. Meth. **B145**(1,2) (1998)133.
28. N.F. Shul'ga, private communication.
29. V.A. Maisheev, private communication.
30. Radiation of Channeling Particles, Kumakhov M.A.,Ed., Energoatomizdat, Moscow, 1986.
31. "Kumakhov Optics and Application", Kumakhov M.A., Ed., Proc. of SPIE

4155 (2000).
32. S.B. Dabagov, Research Report of FIROS, 1992.
33. S.B. Dabagov, and M.A. Kumakhov, Proc. of SPIE **2515** (1995) 124.
34. S.B. Dabagov, M.A. Kumakhov, and S.V. Nikitin, Phys. Lett. **A203** (1995) 279.
35. G. Cappuccio, S.B. Dabagov, A. Pifferi, and C. Gramaccioni, Appl. Phys. Lett. **78** (2001) 2822.
36. 17th International Congress on X-Ray Optics and Microanalysis, Chamonix, France (September 22-26, 2003).
37. S.B. Dabagov, M.A. Kumakhov, S.V. Nikitina, et al., J. Synchrotron Rad. **2** (1995) 132.
38. N. Artemiev, A. Artemiev, V. Kohn, and N. Smolyakov, Phys. Scripta **57** (1998) 228.
39. S.B. Dabagov, and A. Marcelli, Appl. Opt. **38**, 7494 (1999).
40. S.V. Kukhlevsky, F. Flora, A. Marinai, et al., Nucl. Instr. Meth. **B168**, 276 (2000).
41. S.B. Dabagov, V.A. Murashova, N.L. Svyatoslavsky, et al., Proc. SPIE. **3444**, 486 (1998).
42. S.B. Dabagov, A. Marcelli, V.A. Murashova, et al., Appl. Opt. **39** (2000) 3338.
43. 5. S.B. Dabagov, A. Marcelli, G. Cappuccio, and E. Burattini, Nucl. Instr. Meth. **B187**(2) (2002) 169.
44. S.B. Dabagov, A. Marcelli, V.A. Murashova, et al., Proc. SPIE **4138** (2000) 79.
45. Chien Liu, and J.A. Golovchenko, Phys. Rev. Lett., **79** (1997) 788.
46. Yu.M. Alexandrov, S.B. Dabagov, M.A. Kumakhov, et al., Nucl. Instr. Meth. **B134** (1998) 174.
47. S. Iijima, Nature **354** (1991) 56.
48. Physical Properties of Carbon Nanotubes, R. Saito, G. Dresselhaus and M. S. Dresselhaus,Eds., Imperial College Press, London, 1998.
49. L. Langer, V. Bayot, E. Grivei et al. Phys. Rev. Lett. **76** (1996) 479.
50. V.V. Klimov, and V.S. Letokhov, Phys. Lett.**A222**(1996) 424.
51. G.V. Dedkov, Nucl. Instr. Meth. **B143**(1998)584.
52. N.K. Zhevago, and V.I. Glebov, Phys. Lett. A250(1998) 360.
53. E. Burattini, S.B. Dabagov and F. Monti, Nuovo Cimento **117B** (2002) 769.
54. S.B. Dabagov, Physics Uspekhi **46**(10) (2003)1053.
55. D. Gemmell, Rev. Mod. Phys. **46**(1974) 129.
56. W.Z. Li, S.S. Xie, L.X. Qian et al. Science **274** (1996) 1701.
57. J.I. Sohn, and S. Lee, Appl. Phys. **A74** (2002) 287.
58. M. Terrones, N. Grobert, J. P. Zhang et al. Chem. Phys. Letters**285** (1998) 299.
59. M. Mayne, N. Grobert, M. Terrones et al. Chem. Phys. Letters**338** (2001) 101.
60. N. Grobert, W.K. Hsu, Y.Q. Zhu et al. Appl. Phys. Lett.**75** (1999) 3363.
61. B.C. Satishkumar, A. Govindaraj, and C.R. Rao, Chem. Phys. Letters**307** (1999) 158.

62. Chr. Beyreuther, R. Hierl, and G. Wiech, Berlichte der Bunsen-Gesselschaft für Phys. Chem.**79** (1975) 1081.
63. S.B. Dabagov, Rad. Eff. Def. Solids **25** (1993) 103.
64. A.V. Okotrub, S.B. Dabagov, A.G. Kudashev et al. ZhETF Letters **81**(1) (2005) 34.

MONO-ENERGETIC ELECTRON GENERATION AND PLASMA DIAGNOSIS EXPERIMENTS IN A LASER PLASMA CATHODE

K. KINOSHITA, T. HOSOKAI, T. OHKUBO, A. MAEKAWA,
K. KOBAYASHI AND M. UESAKA

Nuclear Professional School, School of Engineering, University of Tokyo, 22-2 Shirane-Shirakata, Tokai, Naka, Ibaraki, 319-1188, Japan

A. ZHIDKOV

Department of Accelerator Physics and Engineering, National Institute of Radiological Sciences, 4-9-1, Anagawa, Inage, Chiba, 263-8555, Japan

Acceleration of plasma electrons by laser wake-filed is the most promising process to produce the next generation of compact accelerators because of its ultrahigh acceleration gradient. To achieve the efficient and stable mono-energetic acceleration of plasma electrons, we investigated the laser-plasma interaction and its correlation to electron acceleration through single shot measurements. We observed a thin laser channel when a mono-energetic spectrum was generated.

1. Introduction

Acceleration of plasma electrons by laser wake-field is the most promising process in order to produce compact accelerators because of its ultrahigh acceleration gradient. Recently, remarkable experimental results have been reported that a quasi-mono-energetic electron spectrum could be obtained through the self-injected laser plasma acceleration [1–3]. Moreover, the duration of such beams can be very short, the order of tens femtosecond. The feasibility of the mono-energetic acceleration and the femtosecond pulse generation has great significance for the future development of a high quality compact electron source via laser-plasma interaction. Complexity of the laser plasma interaction and variety of proposed acceleration schemes require detail measurements both plasma and accelerated parameters. Moreover, such measurements have to be simultaneous, though high stability and reproducibility of the laser pulse parameters is also crucial part of the research. Thus, we conducted single-shot measurement of the laser-produced plasma and the accelerated electrons to investigated the mono-energetic acceleration process.

2. Experimental setup

The laser system generates ultra-short intense laser pulses based on the chirped pulse amplification technique. The system consists of a laser oscillator, a pulse stretcher, a regenerative amplifier (REGEN), three multi-pass amplifiers, and a laser compressor. We use an acousto-optical crystal called as DAZZLER (Thales Laser Co.,Ltd.) that is installed between the stretcher and the REGEN to modify the shape of the laser spectrum and phase correlation to optimize the pulse duration after compression. The specifications of the system are as follows. The

pulse duration is 37 fs. The central wavelength is 790 nm. The maximum pulse energy is 600 mJ. The repetition rate is 10 Hz. The ASE level in this experiments is estimated with a third order crosscorrelator as 5×10^{-7} at 150 ps before the main pulse. The length of the laser pedestal can be controlled through a Pockels cell after the REGEN.

The experimental setup is shown in Fig. 1. Approximately 0.5 % energy of the main pulse is extracted through a splitter and is used as a probe pulse for the plasma diagnosis. The laser main pulse is focused on a gas jet with an off-axis parabolic mirror with the focal length of 178 mm. The gas jet assembles a shock-free gas nozzle with the Mach number 5 and a fast gas valve. The probe pulse passes through the gas jet perpendicularly to the direction of the main pulse. The time delay of the probe pulse can be shifted through the delay path on a motorized slide stage. The probe pulse is lead to a CCD camera through an imaging lens. Emitted electrons irradiate on a fluorescent screen (DRZ) placed downstream from the gas jet. Light emissions from the screen are measured with another CCD camera combined with an image intensifier. In front of the screen, Titanium foil with the thickness of 300 μm is placed in order to eliminate low energy electrons. Electron energy spectra can be obtained in case that a deflection magnet is placed between the gas jet and the screen. The integrated current transformer (ICT) can measure the charge of electron beams. These measurements can be done with one shot. A shadowgraph, interferogram, and Schlieren image can be taken with a femtosecond probe pulse. The polarization of the laser main pulse at the target can be rotated to be vertical or horizontal with a height shifter within the laser path. When the polarization is set vertical, it allows us to observe Thomson scattering, with the same CCD camera to measure the shadowgraph.

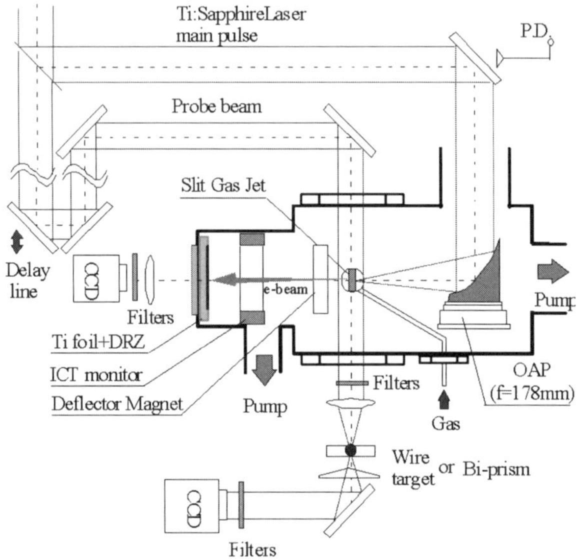

Figure 1. Experimental setup.

3. Experimental results

Figure 2 shows plasma interferograms and shadowgraphs, and corresponding electron distributions. In this figure, the probe pulse passes through the gas jet 2 ps before the main pulse. Thus, the image brought by the probe pulse reflects the preplasma distribution that is formed before the main pulse arrival. The existing prepulse consists of two components. One is the amplified spontaneous emission in the REGEN, and the other is the leakage of the main pulse from the REGEN during amplification. The nanosecond prepulse can be reduced through a Pockels cell after the REGEN that is called the pulse cleaner. The period of a round trip in the REGEN is 8 ns. It makes a small leakage pulse 8 ns before the main pulse. The contrast of the leakage pulse to the main pulse is about 5×10^{-5}. Furthermore, about 4 ns before the main pulse, there exists a satellite pulse, which is originated from the oscillator one pulse before the main. The contrast of the satellite pulse is about 10^{-4}. This pulse is also compressed and, then, has the almost same pulse duration as the main pulse. When these pulses are focused into the gas jet, ionization occurs due to the compressed prepulse and then the laser heating of the plasma with the nanosecond pedestal takes place. As shown in Fig.2, if the plasma cavity is as large as 800 μm, energetic electrons are not observed. On the other hand with the cavity size of 200 μm, high energy, low-emittance electrons appear.

Strong Thomson scattering at the edge of the plasma endorses formation of shockwave in the cavity. Corresponding shadowgraphs and Thomson scattering images taken at the same time are presented in Fig. 3. In these figures, polarization of the laser pulse is perpendicular to the direction of the probe pulse propagation, thus scattered light comes into the same direction of the probe pulse. The intensity of the Thomson scattering increase with the laser intensity and the plasma density, thus the bright spot in Fig. 3 indicates that the plasma at the edges of the plasma cavity is strongly compressed. As the laser pulse pedestal becomes longer, the size of the plasma cavity becomes larger and corresponding Thomson scattering shifts along with the cavity edge. The difference of the scattering intensities between both upstream and downstream edges is due to the ramp of the gas density; the laser pulse is focused on the entrance edge of the gas jet. As shown in Fig. 2, generation of energetic electrons is suppressed when a large plasma cavity is produced. It is known that plasma wave breaking at the edge of the shock front when the intense laser pulse

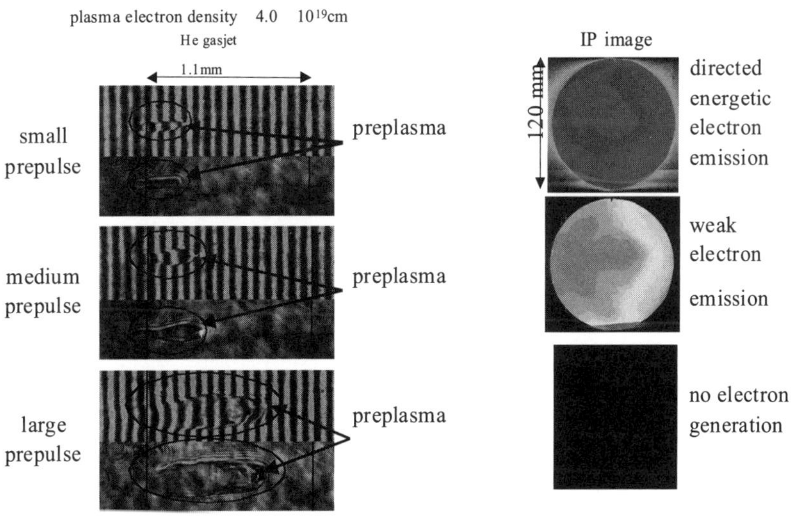

Figure 2. Observation of preplasmas in interferograms and shadowgraphs, and distributions of electron emission measured by imaging plates.

passes through it injects electrons in acceleration phase of laser wake [4–6]. If the plasma cavity is large, the shock expands and the front is far away from the focal point. If the size of the cavity is larger than the laser Rayleigh length, laser intensity at the shock front becomes low, thus consequent injection stimulated by wave breaking becomes weak. Moreover, plasma density in the cavity is lower and thus self-focusing is suppressed. It may shorten the acceleration length of the wake field and the energy gain by the injected electrons may become small. To achieve efficient injection and acceleration the matching between the shock front and the laser focal point is important.

A typical shadowgraph and interferogram images of the plasma are shown in Fig.4. The images are obtained for gas density of 4.0×10^{19} cm^{-3} (plasma density of 8.0×10^{19} cm^{-3}) and 11 TW peak power with 37 fs pulse duration of the main pulse with the smallest intensity of ASE. The laser intensity is 2.2×10^{19} W/cm^2. For these conditions, a peaked spot distribution of high-energy electrons is clearly observed on the center of the DRZ screen. The propagation direction of the laser pulses is from left to right in Fig.4. The polarization of the main laser pulse is parallel to the probe beam axis. Around the edge of the gasjet where the

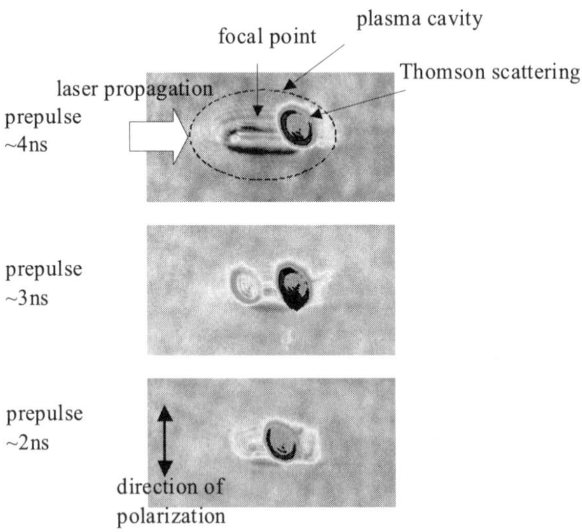

Figure 3. Observation of the Thomson scattering at the edge of the plasma cavity, superimposed on the shadowgraph image.

laser is focused, a faint channel-like structure on the center axis can be seen in Fig. 4. We attribute the appearance of channel-like structure to the focusing of the high intensity picosecond prepulse. Figure 5 shows an energy spectrum obtained in a single shot under the same conditions as in Fig. 4. It has a peaked distribution at 11.5 MeV with an energy spread of 10%(FWHM). The energy spectrum of the accelerated electrons in the plasma usually exhibits the 100% spread distribution but it exhibits the quasi mono-energetic distribution frequently. The energy spread fluctuates within the range of 35% time by time.

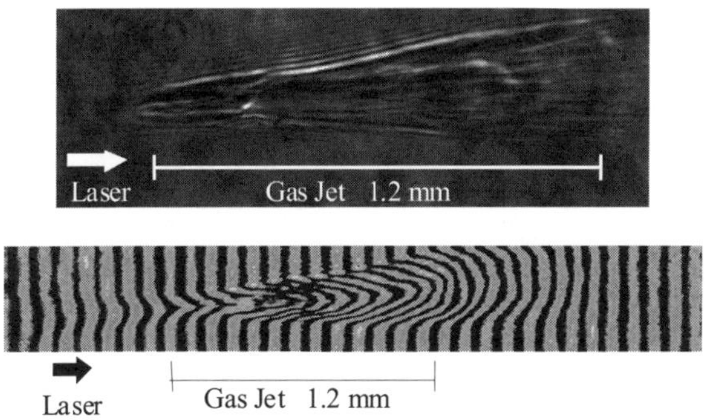

Figure 4. Single-shot plasma diagnosis; shadowgraph (up) and interferogram (down).

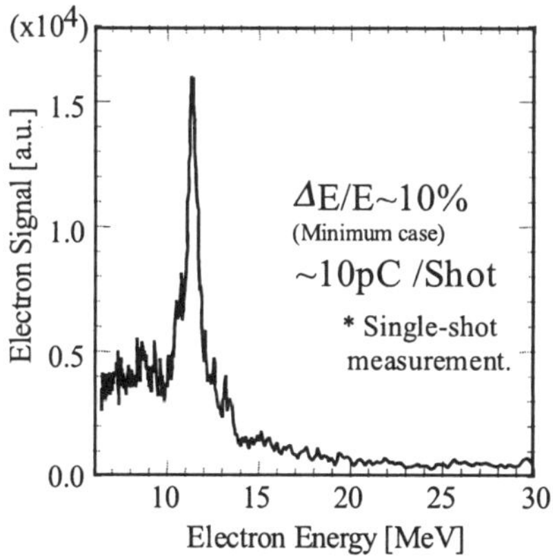

Figure 5. Electron energy spectrum.

4. Conclusion

We constructed the experimental setup for the single-shot measurements of the electron properties and the plasma distributions to study the laser plasma acceleration. Coupled with electron beam measurements, shadowgraphs and interferograms as the plasma diagnostics have been performed with a femtosecond probe pulse in a single shot. A quasi-monoenergetic electron bunch with energy peak at 11.5 MeV is detected at the plasma density of 8×10^{19} cm^{-3}.

Acknowledgments

This work was partially supported by the project of the Advanced Compact Accelerator Development, the Ministry of Education, Culture, Sports, Science, and Technology, Japan.

References

1. C. G. R. Geddes, Cs. Tóth, J. van Tilborg, E. Esarey, C. B. Schroeder, D. Bruhwiler, C. Nieter, J. Cary and W. P. Leemans, *Phys. Plasmas* **12**, 056709 (2005).
2. V. Malka, J. Faure, Y. Glinec, A. Pukhov and J.-P. Rousseau, *Phys. Plasmas* **12**, 056702 (2005).
3. E. Miura, K. Koyama, S. Kato, N. Saito, M. Adachi, Y. Kawada, T. Nakamura and M. Tanimoto, *Appl. Phys. Lett.* **86**, 251501 (2005).
4. T. Hosokai, K. Kinoshita, A. Zhidkov et al., *Phys. Rev.E* **67**, 6407 (2003).
5. T. Hosokai, K. Kinoshita, A. Zhidkov et al., *Phys. Plasmas* **11**, L57 (2004).
6. K. Kinoshita, T. Hosokai, A. Zhidkov et al, *Jpn. J. Appl. Phys.* **45** (2006) in press.

A HIGH-DENSITY ELECTRON BEAM AND QUAD-SCAN MEASUREMENTS AT PLEIADES THOMSON X-RAY SOURCE

J. K. LIM AND J. B. ROSENZWEIG *

Department of Physics and Astronomy,
University of California at Los Angeles,
Los Angeles, CA, 90095, USA
E-mail: jlim@physics.ucla.edu

S. G. ANDERSON AND A. M. TREMAINE†

Lawrence Livermore National Laboratory,
Livermore, CA, 94550, USA

A recent development of the photo-cathode injector technology has greatly enhanced the beam quality necessary for the creation of high density/high brightness electron beam sources. In the Thomson backscattering x-ray experiment, there is an immense need for under 20 micron electron beam spot at the interaction point with a high-intensity laser in order to produce a large x-ray flux. This has been demonstrated successfully at PLEIADES in Lawrence Livermore National Laboratory. For this Thomson backscattering experiment, we employed an asymmetric triplet, high remanence permanent-magnet quads to produce smaller electron beams. Utilizing highly efficient optical transition radiation (OTR) beam spot imaging technique and varying electron focal spot sizes enabled a quadrupole scan at the interaction zone. Comparisons between Twiss parameters obtained upstream to those parameter values deduced from PMQ scan will be presented in this report.

1. Introduction

Thomson backscattering between ultra-relativistic electron beam and chirped-pulse amplification (CPA) high-intensity IR laser has recently made

*This work was performed under the auspices of the U.S. Department of Energy under Contracts No. DE-FG-98ER45693 and No. DE-FG03-92ER40693.

†This work was performed under the auspices of the Lawrence Livermore National Laboratory under Contract No. W-7405-Eng-48 and the LLNL ILSA program under Contract No. LS04-001-B.

possible for the production of high-flux, ultra-short wavelength x-rays. Thomson radiations derived from this backscattering process is promised to be very useful for probing structural and dynamical arrangements of various different high-Z materials. At Lawrence Livermore National Laboratory, Thomson x-ray source called PLEIADES[1,2] has successfully demonstrated the production of 40-140 keV x-rays (according to $4\gamma^2$ Doppler frequency upshift) with sub-ps pulse length. An x-ray flux generated from PLEIADES has been measured at 10^7 photons per shot. This value is an order of magnitude improvement over the pre-PMQ final focus system operational period. This x-ray flux upgrade is resulted from focusing an electron beam to a tighter spot-size at the electron-laser interaction zone according to the proportionality of $N_x \propto 1/(\sigma_e^2 + \sigma_L^2)$ where σ_e and σ_L are electron beam and laser beam spot sizes.

Characterization of electron beam parameters at the interaction point can be used to accurately determine the outcome of x-ray properties such as its total divergence angle. Twiss parameters measurement performed at the interaction zone has been achieved via quad scan technique derived from the 'three-screen' quad scan method. In this measurement, electron beam spot sizes observed in a short beam waist region to be slightly larger than the normal expectation. In the current report, we report the modified Twiss parameter equations, which then are used in providing an improved fit with the PMQ scan measurement. The final electron beam focal spot size deduced from this modified analysis is consistent with beam size expectation determined from early design studies.

2. Electron Beam Transverse Dimension Measurements & PMQ scan analysis

2.1. *Focusing-Beam Profile*

A variation of quad-scan at the interaction point has been performed and compared with beam emittance obtained by standard quad-scan technique employed upstream. The method utilizes measurements of the transverse beam dimension in several longitudinal locations either by placing several beam monitoring screens or moving the beam (executed with moving the whole PMQ with linear translation motor) in z-direction with respect to a fixed screen. We employed the second approach, *i.e.*, varying the drift-space between the beam waist and a fixed screen. A data set obtained by this method has been fitted with the well-known beam profile equation which is a function of drift space z (a thin-lens approximation version):

$$\sigma^2(z) = \beta^* \varepsilon \left[1 + \left(\frac{\Delta z}{\beta^*} \right)^2 \right] + \sigma_{res}^2. \quad (1)$$

However, we noticed that the equation fits well if and only if an additional fitting parameter is added. This parameter is denoted by σ_{res} in Eq. (1). An apparent problem with the data is that near the beam waist, the transverse dimensions measured produced a flat feature over 5 mm longitudinal range in z essentially veiling the bottom saddle of the curve from the standard optical detection. Also, the interpretation of σ_{res} was uncertain at the time we reported in the *Phys. Rev. ST Accel. Beams* paper[3].

In this report, we investigate the possible electron beam "image shift" due to the image scanning process and the OTR angular divergence – $\Delta\theta_{OTR} \simeq 1/\gamma$ (γ is a relativistic factor)– contribution to "total emittance" at the focal point. First, we will calculate the value of the depth-of-field in the camera lens that we employed in the image-scanning measurement to relate to the analysis of the electron beam "image shift" effect. The derivation of the modified fitting equation will be shown after that.

2.2. *CCD Camera Depth-of-field Limitation*

In order to understand the problem of a depth-of-field at hand, we construct and use the information in the following diagram of a single thin-lens set up. In the Figure 1, an image of an object at s_o is clearly formed on the film plane located at s'_o, disregarding any lens aberration. When object is displaced from its optimal position, the image also displaces from the film plane forming a circle of blurry image. If the diameter of blurry image is small enough, the image resolution is still acceptable. Suppose the largest acceptable diameter is d such that all images are suitably in focus within a distance x, then object space conjugate to the interval x is the depth-of-field notated by $\alpha = s_2 - s_1$. Skipping detail derivation of depth-of-field parameter by quoting it from a book by Pedrotti[4], which is given by

$$\alpha = \frac{2Ads_o(s_o - f)f^2}{f^4 - (Ads_o)^2}, \quad (2)$$

where A is the *f*-number and f is the focal length of camera lens. For the PMQ scan experiment, a COHU 4910 equipped with a 1.4 *f*-number and 75 mm focal length lens has been used to grab the beam image. To understand how narrow this camera depth-of-field is for given actual experimental parameters, we compute the camera depth-of-field: $A = 1.4$, $s_o \cong 150$ mm,

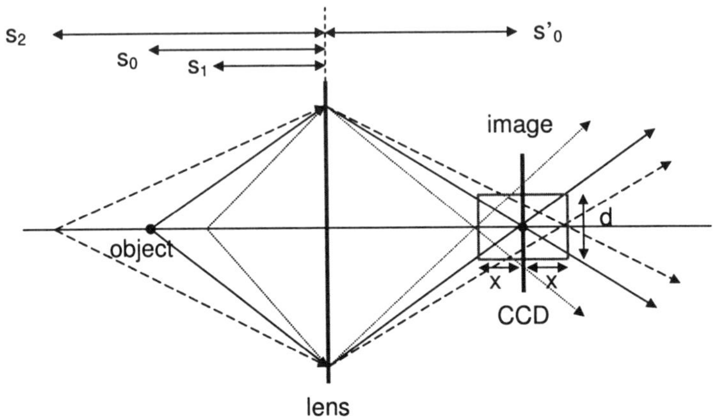

Figure 1. Idealized OTR Beam profile measurement schematic.

$f = 75$ mm and $d \cong 32$ μm (from the quad scan data). Using Eq. (2), a ballpark value of α_{CCD} ~100 μm is estimated, which is extremely shallow compared to the whole quad scan range.

2.3. *Modified Focusing-Beam Profile Equation*

It is then natural to postulate what effect might have on the PMQ scan curve in the event that the camera lens is not in focus at the exact beam object plane. In this case, we need to analyze how the beam image matrix is transported from the object to image plane, *i.e.*,

$$\sigma_f^c = M\sigma_o M^T, \tag{3}$$

where

$$\sigma_o = \begin{pmatrix} \sigma_{o,11} & \sigma_{o,21} \\ \sigma_{o,12} & \sigma_{o,22} \end{pmatrix} = \varepsilon \begin{pmatrix} \beta(s) & -\alpha(s) \\ -\alpha(s) & \beta(s) \end{pmatrix},$$

and the transport matrix for the camera optics system is

$$M = \begin{pmatrix} 1 - L_i/f & L_o + L_i - L_i L_o/f \\ -1/f & 1 - L_o/f \end{pmatrix}.$$

The transpose matrix is obtained by exchanging rows and columns of the matrix M as $(M_{ij})^T = M_{ji}$. Before going further with the derivation, we have made a simplification by assuming that the camera optics has a single lens with some effective focal length.

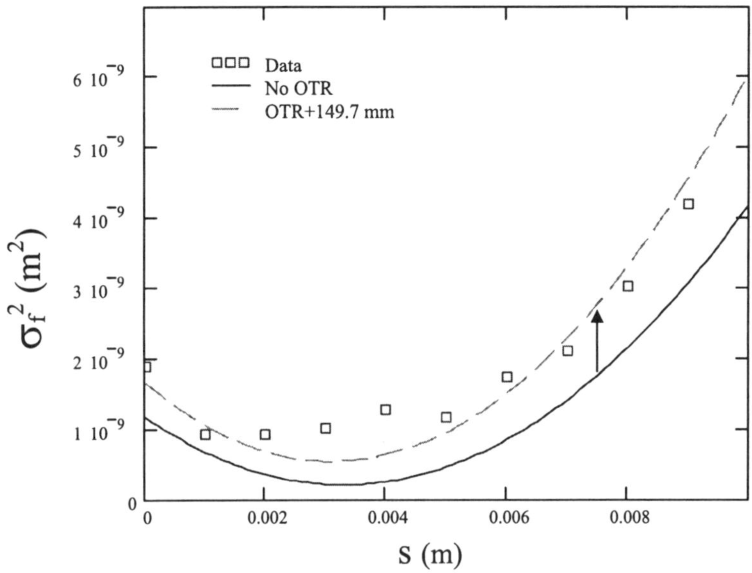

Figure 2.

From Eq. (3), we calculate the matrix element $\sigma^c_{f,11}$, which relates to the image being observed on the image plane:

$$\sigma^c_{f,11} = M_{11}(\sigma_{o,11}M_{11} + \sigma_{o,21}M_{21}) + M_{21}(\sigma_{o,12}M_{11} + \sigma_{o,22}M_{21}), \tag{4}$$

where $M_{11} = 1 - L_i/f$ and $M_{21} = L_o + L_i - L_oL_i/f$.

The Twiss parameters, $\beta(s)$, $\gamma(s)$ and $\alpha(s)$ of OTR image of focusing electron beam in the force-free drift are:

$$\beta(s) = \beta^* \left[1 + \left(\frac{s - s_w}{\beta^*}\right)^2 \left(1 + \frac{\beta^*}{\varepsilon}\Delta\theta^2_{OTR}\right)\right],$$

$$\gamma(s) = \frac{1}{\beta^*}\left(1 + \frac{\beta^*}{\varepsilon}\Delta\theta^2_{OTR}\right),$$

$$\alpha(s) = -\frac{s - s_w}{\beta^*}\left(1 + \frac{\beta^*}{\varepsilon}\Delta\theta^2_{OTR}\right), \tag{5}$$

where at the waist of a beam is $\gamma_o = 1/\beta^*$. The OTR light divergence is included as the term found in the second parentheses. The inclusion of OTR divergence angle as $\sigma^2_{x'} \rightarrow \sigma^2_{x'} + \Delta\theta^2_{OTR}$ effectively increases the "total emittance." The equation Eq. (4), then describes the OTR image of a focusing beam in a force-free drift formed on the image plane.

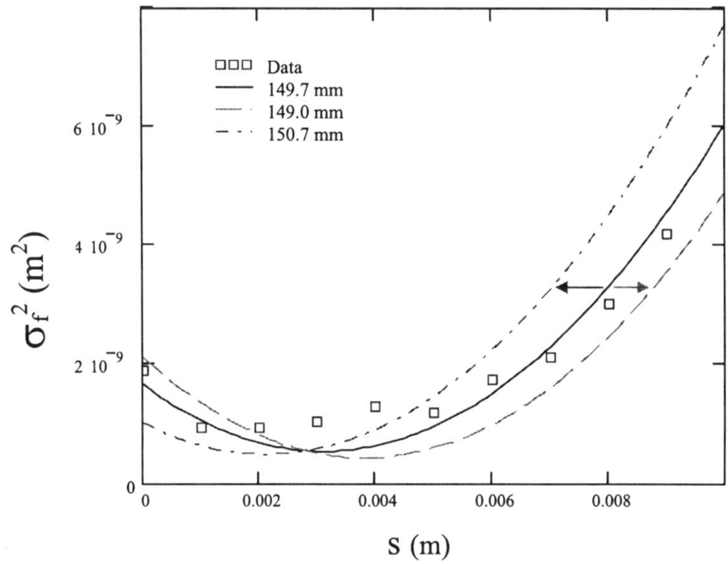

Figure 3.

In Eq. (4), two notable competing effects are shown, adding to the apparent beam image problem: depth-of-field and OTR divergence. First, we will start by examining what effects might OTR have on the final image being observed. In Figure 2, we plotted Eq. (4) with and without OTR effect. From the Figure, we notice that the OTR effect shifts the curve upward and fits somewhat better against the data points.

In Figure 3 as shown, we have isolated the case of a variation in the object distance. As before, a perfect fit against the data points is obtained when L_o =149.7 mm. Two curves with variations $L_o \pm \Delta L_o$ where the deviation value of ± 0.7 mm are considered. The positive sign shifts the curve to the left and the negative to the right as shown in Figure 3.

In Figure 4, we show both Eq. (1) and Eq. (4) plots for a comparison. Eq. (1) appears to have a better fit in the upper range of $s \geq 4$ mm, but

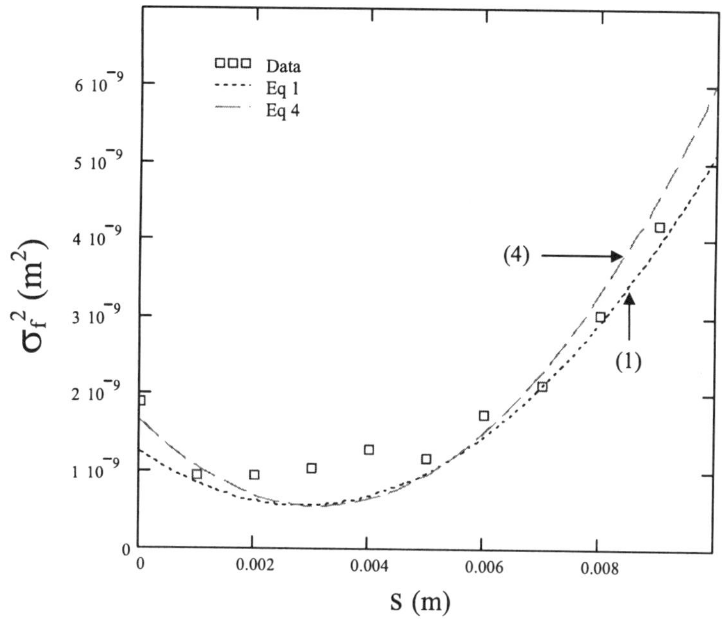

Figure 4.

Eq. (4) fits better in the lower range of $s \leq 4$ mm.

We note that in Eq. (4), an emittance from the standard quad scan data has been used. In the modified equation, we obtained the minimum β-function (β^*) to be 3.8 mm, slightly more than 1.62 mm obtained with Eq. (1). In the new analysis, we conclude that the minimum transverse beam size expected is 23.5 micron instead of 15.2 micron.

References

1. S. G. Anderson, *et. al.*, *Applied Physics B: Lasers and Optics* 10.1007/s00340-004-1455-0 (2004).
2. A. M. Tremaine, S. Anderson, S. Betts, J. Crane, D. J. Gibson, F. V. Hartemann, J. S. Jacob, P. Frigola, J. Lim, J. Rosenzweig and G. Travish, *Proceedings of the 2005 Particle Accelerator Conference* (2005).
3. J. K. Lim, *et. al.*, *Phys. Rev. ST Accel. Beams* **8**, 072401 (2005).
4. F. L. Pedrotti, S. J. and L. S. Pedrotti, *Introduction to Optics, Second Edition*, Prentice Hall, New Jersey (1993).

LASER PULSE CIRCULATION SYSTEM FOR COMPACT MONOCHROMATIC TUNABLE HARD X-RAY SOURCE

HARUYUKI OGINO [1], MENG DE [1], TOMOHIKO YAMAMOTO [1], FUMITO SAKAMOTO [1], KATSUHIRO DOBASHI [2], MITSURU UESAKA [1]

[1] *Nuclear Professional School, School of Engineering, University of Tokyo*
2-22 Shirakata-Shirane, Tokai-mura, Naka-gun, Ibaraki, 319-1188, JAPAN
[2] *National Institute of Radiological Sciences*
4-9-1 Anagawa, Inagi, Chiba, Chiba 263-8555, JAPAN
E-mail: ogino@nuclear.jp

We are construcing a laser electron Compton scattering monochromatic tunable hard X-ray source. It consists of the X-band (11.424 GHz) electron linear accelerator and Q-switch Nd:YAG laser. This work is a part of the JST(Japan Science and Technology Agency) project. The whole system is a part of the national project on the advanced compact medical accelerator development, hosted by NIRS(National Institute for Radiological Science). The University of Tokyo and KEK are working for the X-ray source. Main advantage of this X-ray source is monochromatic tunable hard X-rays(10-50keV) with the intensities of 10^8-10^9 photons/s. The table-top size X-ray source can generate dual energy monochromatic hard X-ray by turns and it takes about 40ms to chage the X-ray energy. It is calculated that the X-ray intensity is 10^7 photons/RF-pulse (10^8 photons/s in 10 pps) by the 35MeV electron and YAG laser(2J/pulse).The X-band beam line for the demonstration is under construction. We designed a laser pulse circulation system to increase the X-ray yield 10 times higer(up to 10^8 photons/RF-pulse,10^9photons/s). It can be proved that the laser total energy increases 10 times higher by the principle experiment with the lower energy laser(25mJ/pulse).

1. Introduction

Monochromatic hard X-ray(10-50 keV) are often used in medical science, biology and material science. For example, it is used in dynamic IVGAC[1,2], monochromatic X-ray CT, substruction imaging and dual energy X-ray CT[3]. Dual energy X-ray CT requires two monochromatic hard X-rays. The third generation light source can generate monochromatic tunable hard X-rays with high intensity ($10^{11}-10^{13}$photons/s).However, they are too large and to expensive to be applied and used widely for public use. Therefore, we are developing a compact X-ray source which can generate the monochromatic tunable hard X-ray. It is based on the laser-electron Compton scattering.It consits of the X-band (11.424 GHz) linac and Q-switch Nd:YAG laser.[4] It is shown in Figure 1. and the property of electron and laser beam is shown in Table.1. The scattered angle of photons are related to the energy, so we can generate the monochromatic X-ray by using the collimator.[6] We designed the laser pulse circulation system for the source. The pulse length of the Q-switch Nd:YAG laser is $10ns$(FWHM]) and it is shorter than the multiple bunch length of electron beam($1\mu s$). This means that most of the electron bunches do not collide with the laser pulse.[5] By the laser pulse circulation system, one laser pulse is blocked into the circuit and repeatedly collides with multiple electron bunch. Then, the X-ray intensity becomes higher.

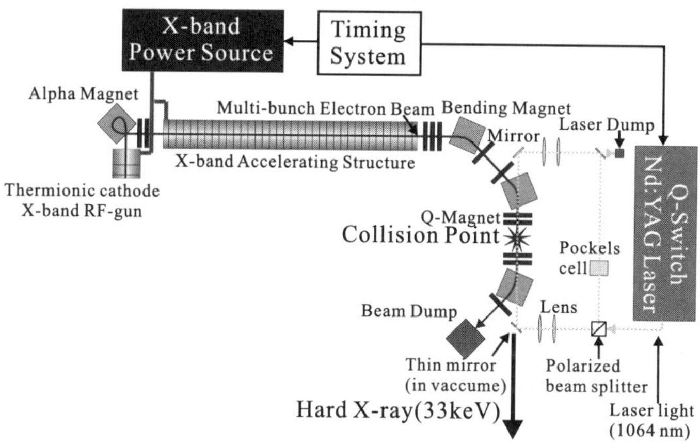

Figure 1. Laser electron Compton scattering compact monochromatic tunable hard X-ray source

2. Laser pulse circulation system

2.1. *Longitudinal direction*

There are two laser pulse circulation systems, one is about longitudinal direction and the other is about transverse direction. In the longitudinal direction, we change the laser polarization by using $\lambda/2$ waveplate and pockels cell. The laser pulse can be blocked into the circuit and collides with electron beams repeatedly. The bunch length of electron beam is $1\mu s$ and the laser revolving time is $17ns$, so one laser pulse can theoretically

Table 1. Properties of electron laser beam of the X-ray source

Electron	
Energy	: 35MeV
Charge	: 20pC/bunch
Number of bunches	: 10^5bunches/pulse
Beam size	: 100μm
Emittance	: 2.5πmm mrad
Nd:YAG Laser Fundamental	
Energy	: 2J/pulse
Pulse width	: 10ns(FWHM)
Repeating ratio	: 10pps
Wavelenght	: 1064nm
Beam size	: 100μm
Nd:YAG Laser Second harmonics	
Energy	: 1.4J/pulse
Pulse width	: 10ns(FWHM)
Repeating ratio	: 10pps
Wavelength	: 532nm
Beam size	: 100μm

collides with electron bunch more than 50times. It can theoreticaly increase the X-ray intensity more than 50times higher. Sketch of the longitudinal laser pulse circulation system is shown in Figure 2. Incident laser pulse passes through the polarizing beamsplitter with p polarization. After the laser electron Compton scattering, its polarization plane is changed from p polarization to s polatization by the $\lambda/2$ waveplate. The laser with p polarization goes through the polarizing beamsplitter and the laser with s polarization is reflected. So the laser pulse at the end of 1st lap is guided to the 2nd lap by the polarizing beamsplitter. In the 2nd lap, the laser pulse passes through the $\lambda/2$ waveplate again and its polarization changes from s polarization to p polarization. If the p polarization laser reaches to the polarizing beamsplitter again, the laser pulse go out the circuit. Then, we give high voltage to the pockels cell during the 2nd lap. The p polarization laser changes its polarization to s polarization during the 2nd lap. Therefore, the laser polarization plane at the polarization beamsplitter is always s polarization.The laser pulse can be blocked into the circuit and it collide with the electron beam repeatedly.

Total laser pulse energy in the laser circulation system I_n is,

$$I_n = \sum_{k=1}^{n} I_0 A^{k-1} = I_0 \frac{1 - A^n}{1 - A} \tag{1}$$

where I_0 is the initial pulse energy of the laser, A is the transmission efficiency per one circuit, and N is the total number of collisions. Since the revolving time is about 17ns, the laser pulse collides with electron bunch more than 50 times. With 90% transmission efficiency per one circulation(A=0.90), we expect that the X-ray intensity becomes 10 times higher. Therefore, the X-ray source can generate 10^9photons/s.

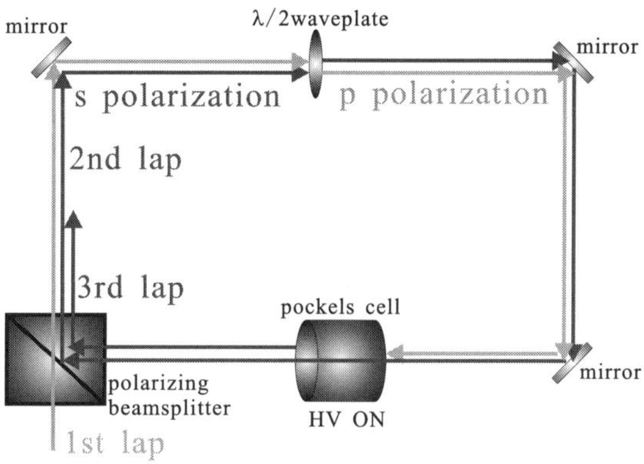

Figure 2. Longitudinal laser pulse circulation system

2.2. *Transverse direction*

In the transverse direction, we change the transverse laser positioning to match the electron beam positioning. For the laser electron Compton scattering experiment, we have to control the laser profile and electron beam profile at the collision point. The beam sizes of electron and laser beams are both $100\mu m$ at the collision point. The laser circulation system also requires high stability of the laser profile at the collision point. It is more difficult to change the electron beam positioning because we have to change the bending magnet and quadropole magnet parameter. So we change the laser positioning in our system. We change the transverse laser positioning and beam size by moving the four lenses shown in Figure 3. The transverse distance of the laser at the collision point is equal to the transverse distance of the four lenses. Then we can easily achieve the micro meter resolution. The injection lenses(see Figure 4) can changes the beam size at the collision point because it is inversely proportional to its beam size at the focusing lens.

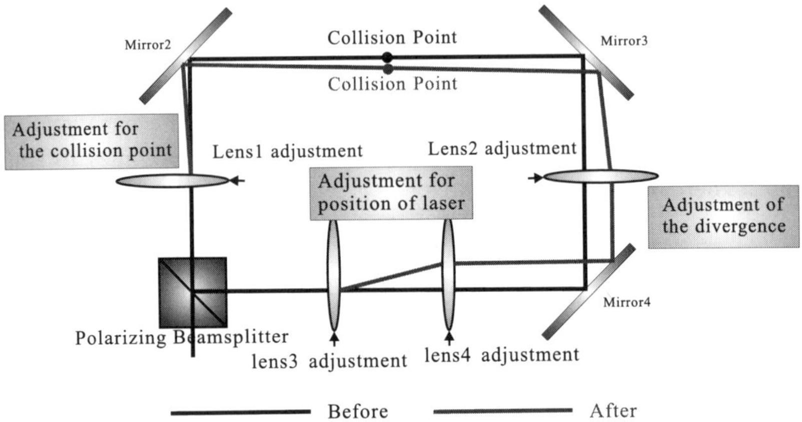

Figure 3. Transverse laser pulse circulation system

3. Experiment

We have done a proof of experiment for low power laser circulation system last year before adding the high power laser pulse circulation system to the X-ray source[7]. There is one problem about the air breakdown at the focusing point in the case that we install the same experimental setup for high power laser. We installed the concave lens at the focusing point to avoid the air breakdown.

3.1. *Setup*

The experimental setup is shown in Figure 4. The concave lens is installed between convex lens3 and convex lens5. We use a Nd:YAG Laser which is as same as the X-ray source, relatively lower energy(25mJ/pulse). We choosed the second harmonics wave($\lambda = 532nm$). The length of one circuit is 5m and the revolving time is about 17ns.
The aim of the experiment is to establish by proof the longitudinal and transverse laser pulse circulation system.

3.2. *Measurement*

We measured the each transmitted light from mirror. We can make certain of longitudinal laser pulse circulation by using the photoelectric tube shown in Figure 4. About the transverse direction laser pulse circulation, we also measure the transmitted light by using beam profiler. In both direction, we measure the transmitted light in order not to break the laser circulation condition.

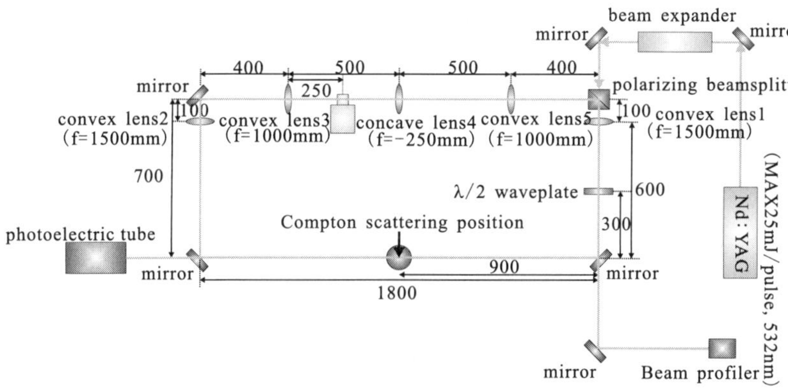

Figure 4. Low power laser circulation experimantal setup

4. Result

Figure 5 shows the output from the photoelectric tube and Figure 6 is the experimental picture. The revolving time is accurately equal to the theory and its signal indicates that laser pulse is bloked into the circuit more than 35times. We can calculate the transmission efficiency A and enhanced laser intensity by this result. The laser intensity of 1st lap is 0.75 and that of 15th lap is 0.19, then we obtain,

$$A = 0.90, \quad I_n = 10I_0. \tag{2}$$

Therefore, we have enhanced the laser intensity 10 times higer.

Figure 7 shows the laser transverse positioning when the lenses move horizontally with 10 μm steps. Figure 8 shows the transverse positioning when the lenses move vertically with 10 μm steps. In both figures, we measured laser centroid at the collison point. In Figure 7, it can be proved that the horizontal laser positioning can be changed by moving the lenses horizontally. The laser keeps its vertical positioning while changing its horizontal positioning. In Figure 8, it can be proved that the vertical laser positioning can be changed by moving the lenses vertically. The laser keeps its horizontal positioning while changing its verical positioning.

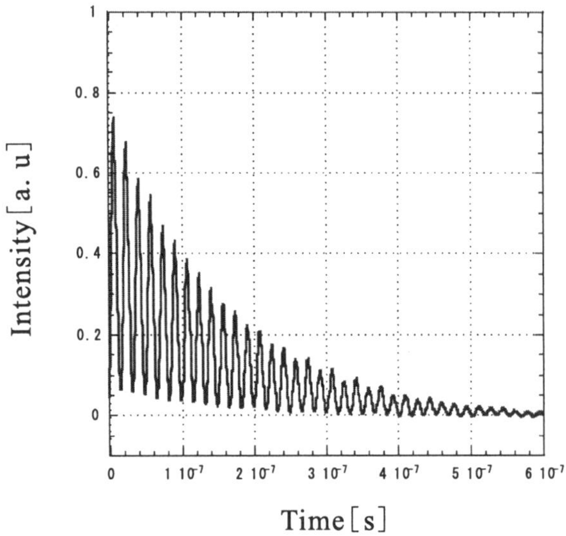

Figure 5. Laser intensity transition during the circulation

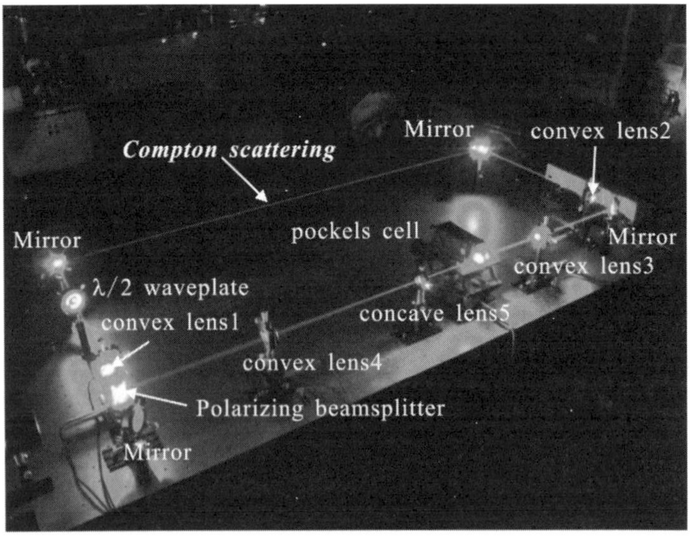

Figure 6. Laser pulse circulation experimental picture

Thatis, they proved that the transverse laser positioning can be changeed independently with 10μm steps.

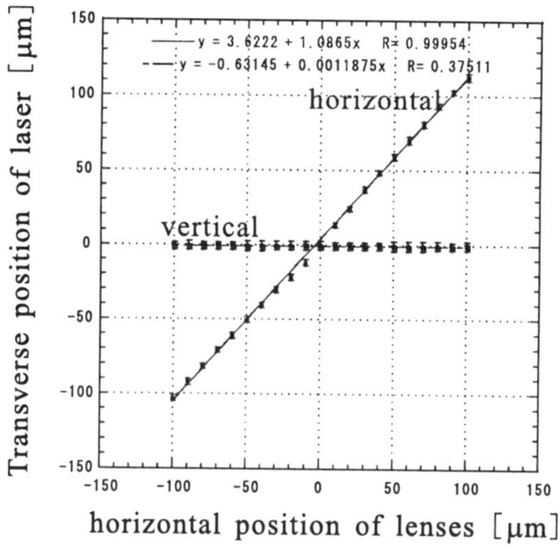

Figure 7. Transverse laser positioning during the horizontal lenses shift

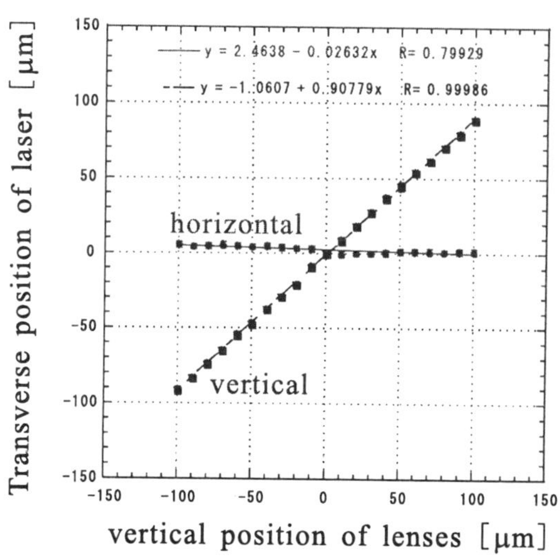

Figure 8. Transverse laser positioning during the vertical lenses shift

5. Conclusion

We are developing the compact monochromatic tunable hard X-ray source. It is based on laser electron Compton scattering. It consists of X-band linac and Q-switch Nd:YAG laser. We designd the laser pulse circulation system to generate the X-ray with higer intensity. By the proof of experiment with the second harmonics wave(532nm), We have enhanced the laser intensity 10 times higher and established transverse laser positioning with 10μm steps. We can expect that the X-ray intensity will be enhanced 10 times higer in the laser electron Compton scattering experiment.
We will do futher studies and experiments for the fundamental wave circulation.

6. Acknowledgements

This study is performed in the national project of "Development of Advanced Compact Accelerators" in Japan, partially supported by Research Program on Development of Innovative Technology (#0494) in Japan Science and Technology Agency.

References

1. E. Rubenstein, et al. *Proc. National Academy Sci. USA* 83(1986)
2. S. Ohtsuka, et al. *British Journal of Radiology* 72(1999)
3. M Torikoshi, et al. *Opt.* 371(2000).
4. M. Uesaka, et al. *Proc. XXI International LINAC Conference in Korea,* Gyeongju, Korea, 626(2002)
5. K. Dobashi, et al. *Jpn. J. Appi. Phys.* 4A 44(2005)
6. K. Dobashi, et al. *Proc. 8th European Particle Accelerator Conference in Paris, France* 667(2002)
7. F.Ebina, et al.Nuclear Instruments and Methods in Physics Research B241(2005)

LIMITS ON PRODUCTION OF NARROW BAND PHOTONS FROM INVERSE COMPTON SCATTERING*

J. ROSENZWEIG AND O. WILLIAMS

UCLA Dept. of Physics and Astronomy, 405 Hilgard Ave., Los Angeles, CA 90034, USA

In using the inverse Compton scattering (ICS) interaction as a high brilliance, short wavelength radiation source, one collides two beams, one an intense laser, and the other a high charge, short pulse electron beam. In order to maximize the flux of photons from ICS, one must focus both beams strongly, which implies both use of short beams and the existence of large angles in the interaction. One aspect of brilliance is the narrowness of the wavelength band emitted by the source. This paper explores the limits of ICS-based source brilliance based on inherent wavelength broadening effects that arise due to focal angles, laser energy density, and finite laser pulse length effects. It is shown that for a nominal 1% desired bandwidth, that one obtains approximately one scattered photon per electron in a head-on collision geometry.

1. Introduction

The use of inverse Compton scattering (ICS) in future light sources [1] spans many different fields of application and an extremely wide range of wavelengths. In the case of lowest scattered energy, in the angstrom spectral region (10's of keV) one finds potentially revolutionary applications in medicine and materials characterization [2,3]. At the higher energy photons (few MeV), one finds application to examination of nuclear resonances for non-invasive nuclear materials detection; at the ~100 MeV range one may use ICS as an integral step in obtaining polarized positrons4 for linear colliders; finally, ICS is proposed for converting electron energy at the hundreds of GeV scale to photon form in γ - γ colliders [5].

A common thread in the vast majority of applications is that one aims to maximize the number of scattered photons while simultaneously limiting the bandwidth of the scattered photons. For example, in medical imaging and therapy applications [6], a typical demand would be for $10^9 - 10^{10}$ X-ray photons in a 1% bandwidth These two goals are at odds, however, as many of the techniques used in generating higher scattered flux cause either a range of red-shifts (e.g. strong focusing), or higher harmonics (nonlinear scattering). The purpose of this paper is to quantify the interplay between these competing design criteria, and identify the physical limitations on narrow-band ICS photon production, with the goal of informing both current ICS experiments [7,8] and planning for advanced ICS applications.

---

* This work was performed under the auspices of the U.S. Department of Energy under contract number DE-FG03-92ER40693.

The basic geometry for ICS collisions that are nominally conducted head-on is shown in Fig. 1. The spread in incident angles resulting from the divergence of the laser beam is indicated by θ_i, and the final state scattering angle by θ_s. The central frequency of the laser is indicated by ω_L, and the nominal velocity of the electron is taken to be $v = \beta c$, where $\beta \cong 1$ and thus $\gamma = U_e / m_e c^2 = \left(1 - \beta^2\right)^{-1/2} >> 1$, i.e. the beam is ultra-relativistic.

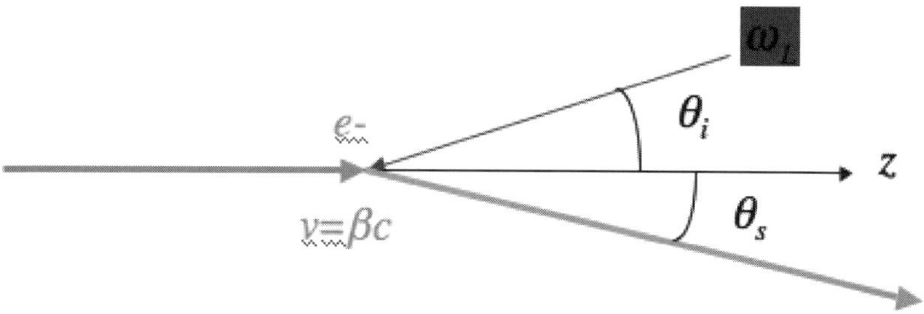

Figure 1. Geometry for nominal head-on (0°) ICS collisions, with spread in incident angle θ_i due to laser divergence shown, as well as final state scattering angle θ_s.

At this stage, we assume for the sake of analysis that the electrons in the beam are all directed perfectly along the z-axis. This assumption allows us to concentrate on the angles due to the spread in laser photon direction, which is generally much larger than the spread in electron angles. As the spot sizes of the electron and laser beams are equal, this assertion is quantified by comparing the emittance of the electron beam with the effective "emittance" of the laser, and requiring $\varepsilon \cong \varepsilon_n / \gamma << \lambda_L / 4\pi$. For example, in the PLEIADES experiment at LLNL [9], even though ε_n was quite large due to nonideal fields in the linac, this requirement is satisfied, as ε_n, while $\lambda_L / 4\pi = 6.4 \times 10^{-8}\,\text{m}$. In fact, at PLEIADES, the electron beam was straightforwardly focused to a smaller spot than the laser [10] due in part to this disparity in "emittances".

We will thus find that the estimate on ICS photons found within a certain bandwidth and angular spread is valid under this assumption. We must then add to the analysis a realization that the spread in electron beam angles leads to a breakdown in the direct correlation between lab scattering angle and wavelength. In order to obtain a certain bandwidth, a simple angular cut may not be sufficient. This point is discussed at the end of our analysis.

2. ICS analysis in the thomson Limit

Our ICS spectral analysis is performed in the Thomson limit, where the scattered photons as observed in the nominal beam rest frame do not have an appreciable wave-length shift (Compton shift) upon scattering. In this frame, the

Lorentz transformation of the $\left(\omega,\vec{k}c\right)$ 4-vector from the lab frame yields the frequency ω', and the transverse and longitudinal components of the vector potential $k'_{\perp}$ and k'_z:

$$\begin{aligned}
\omega' &= \gamma\left(\omega_L - \beta c k_L \cos(\theta_i)\right) = \gamma\omega_L\left(1+\beta\cos(\theta_i)\right) \\
k'_{\perp} &= k_L \sin(\theta_i) \\
k'_z &= \gamma\left(k_L \cos(\theta_i) - \beta\omega_0 / c\right) = \gamma k_L\left(\cos(\theta_i)+\beta\right).
\end{aligned} \tag{1}$$

The Thomson limit is obtained when $\hbar\omega' << m_e c^2$, or explicitly $\hbar\omega_L\gamma\left(1+\beta\cos(\theta_i)\right) << m_e c^2$. For example, at the Neptune ICS experiment (which has $\theta_i = \pi/2$), the rest frame photon energy is ($\lambda_L = 10.6$ μm, $\gamma = 30$) 3.6 eV. Thus we make the Thomson approximation that the frequency is independent of scattering angle, and $\omega'_s = \omega' = \gamma\omega_L\left(1+\beta\cos(\theta_i)\right)$, and the wave-number components of the scattered photons are

$$\begin{aligned}
k'_{\perp s} &= \frac{\omega'_s}{c}\sin(\theta') = k_L\gamma\left(1+\beta\cos(\theta_i)\right)\sin(\theta') \\
k'_{zs} &= k_L\gamma\left(1+\beta\cos(\theta_i)\right)\cos(\theta')
\end{aligned} \tag{2}$$

where θ' is the rest frame scattering angle. We note in the Thomson limit one may simply derive the emitted angular intensity (photon distribution) from the expression

$$\frac{dP_s}{d\Omega'} = \frac{e^2\dot{v}^2}{4\pi c^3}\cos^2(\theta'). \tag{3}$$

We will make use of this distribution in our calculation of the final scattered spectrum. In this regard, note also that the total Thomson scattering cross-section $\sigma_T = 8\pi r_e^2/3$ is Lorentz invariant.

The frequency of the scattered radiation is found by a final Lorentz transformation to the laboratory frame,

$$\omega_s = \gamma^2\omega_L\left(1+\beta\cos(\theta_i)\right)\left(1+\beta\cos(\theta')\right). \tag{4}$$

Equation 4 contains all needed information about the red-shifting of radiation which is emitted at finite angles. In order to interpret it, one needs to substitute for the rest frame angle θ', however, in terms of the lab angle, which can be found by finding the wave vector components,

$$\begin{aligned}
k_{\perp s} &= k_L\gamma\left(1+\beta\cos(\theta_i)\right)\sin(\theta') \quad \text{and} \\
k_{zs} &= k_L\gamma^2\left(1+\beta\cos(\theta_i)\right)\left(\cos(\theta')+\beta\right),
\end{aligned} \tag{5}$$

to give the expression

$$\tan(\theta_s) = \frac{k_{\perp s}}{k_{zs}} = \frac{1}{\gamma}\left(\frac{\sin(\theta')}{\cos(\theta') + \beta}\right). \tag{6}$$

In the small-angle approximation Eq. 6 yields, simply, $\theta_s \cong \theta'/2\gamma$.

The final, small-angle expression for the scattered frequency in the Thomson limit is thus

$$\omega_s = 4\gamma^2 \omega_L \left(1 - \frac{\theta_i^2}{2}\right)\left(1 - \frac{(\gamma\theta_s)^2}{2}\right). \tag{7}$$

The leading factor in Eq. 7 is the maximum Doppler shift given by the system kinematics, while the first and second correction terms give the effect of incident and scattering angles being nonvanishing. The final angle-induced red-shift is familiar from the FEL condition, where resonant emission occurs when the when emitted wave-front overtakes electron by λ_r in an undulator wavelength λ_u, or $\sim \lambda_L/2$ in the Thomson system. In all systems where a relativistic electron is emitting radiation, the relative small-angle red shift is given approximately by $(\gamma\theta)^2/2$.

2. Photon production and bandwidth

With these results in hand, we can parameterize the wavelength through the final angle. This approach has the benefit of also relating to the photon angular distribution, and thus the number of photons produced within a given bandwidth. In other words, the emission angle may serve as a proxy for the bandwidth. However, because of the finite spread in incident angles, and the fact that the red-shift due to this type of error observed on-axis is Doppler shifted as $(\gamma\theta_s)^2/2$, one may not simply limit the acceptance angle to clean the spectrum of emitted radiation. One must instead use other standard techniques, such as use of crystal-based monochromators [11] to pass the desired bandwidth. With this in mind, we can proceed to discuss the number of photons produced within a bandwidth using its functional dependence on the emission angle, leaving the extraction of only that bandwidth (if needed) for further discussion.

The efficiency of the photon production as a function of maximum emission angle (relative to the electron direction) is

$$\eta_{acc}(\theta'_{\max}) \equiv \frac{N_{acc}(\theta'_{\max})}{N_{total}} = \tfrac{1}{2}\left(1 - \cos^3(\theta'_{\max})\right) \cong \frac{3}{4}\theta'^2_{\max} \tag{8}$$

or in terms of lab angle, $\eta_{acc}(\theta_s) \cong 3(\gamma\theta_s)^2$. Written as a function of rms bandwidth, this *acceptance* efficiency is simply

$$\eta_{acc}(\theta_{\max}) \cong 6\sqrt{3}(BW_{rms})_{acc}. \tag{9}$$

There is also an expansion of the bandwidth of scattered light which is not apparent from the geometry shown in Fig. 1 — the spread in initial wavelength due to the finite character of the pulse train,

$$\left(BW_{rms}\right)_{FT} \cong \left(\sqrt{2}k_L\sigma_z\right)^{-1} = \frac{\lambda_L}{\sqrt{8}\pi\sigma_z}. \tag{10}$$

This can be mitigated by choosing a longer laser pulse. However, there is a practical limit on the laser pulse length, in that it may not exceed the Rayleigh length,

$$\sigma_z \leq Z_R = \frac{4\pi\sigma_x^2}{\lambda_L}, \tag{11}$$

due to the free-space expansion of the laser over lengths exceeding this limit.

We must also examine the spread in incident angles due to the focusing of the laser which, as stated above, dominate the angles due to the electron beam emittance. Away from the laser focus, the rms spread in laser angle due to diffraction is $\theta_{L,rms} = \frac{\sigma_x}{Z_R} = \frac{\lambda_L}{4\pi\sigma_x}$, giving bandwidth

$$\left(BW_{rms}\right)_{foc} = \frac{\left(\theta_L^2\right)_{rms}}{2} = \left(\frac{\lambda_L}{4\pi\sigma_x}\right)^2. \tag{12}$$

A question arises as to whether this spread in angles actually displays itself *inside* of the laser focus, where the phase fronts are nearly flat, and do not show the angles seen outside of the focus. The answer is affirmative; one needs to consider the effect of the Guoy phase shift, which is a function of z, as $\tan(\varphi(z)) = z/Z_R$. This dynamic phase shift gives a spread in longitudinal wave-number,

$$BW_{\text{Guoy}} = \frac{\Delta k_z}{k_z} = \frac{1}{k_z}\frac{d\varphi}{dz} \cong \frac{\lambda_L}{2\pi Z_R} = \frac{\lambda_L^2}{8\pi^2\sigma_x^2}, \tag{13}$$

giving the same result as Eq. 12, when one considers the rms spread. This increase in laser bandwidth can be seen as arising from the free-space dispersion relation, $k_z = \sqrt{(\omega_L/c)^2 - k_\perp^2}$, with $k_z \cong \sqrt{k_L^2 - \sigma_x^{-2}} \cong k_L\left(1 - \frac{1}{2}\left(\lambda_L/2\pi\sigma_x\right)^2\right)$.

We now adopt the restriction that, in order to maximize photon production, we choose the shortest Rayleigh range consistent with avoiding the "hour-glass" effect, $\sigma_z = Z_R = 4\pi\sigma_x^2/\lambda_L$. Under this condition, we have $\theta_{L,rms} = \frac{\lambda_L}{4\pi\sigma_x}$ $= \frac{\lambda_L}{4\pi}\sqrt{\frac{4\pi}{\lambda_L\sigma_z}} = \sqrt{\frac{\lambda_L}{4\pi\sigma_z}}$, and so

$$\left(BW_{rms}\right)_{foc} = \theta_{L,rms}^2 = \sqrt{2}\left(BW_{rms}\right)_{FT}, \tag{14}$$

and, remarkably, the bandwidth due to the Fourier transform of the laser pulse is nearly the same as that due to the spread in angles in the laser focus.

A final, critical, effect which induces spread in the scattered radiation spectrum is the nonlinearity of the motion due essentially to relativistic effects caused by the size of the laser field. A way to see this is that the electron has motion in the laser field with maximum induced angle of

$$\theta \approx \frac{p_\perp}{p_0} \cong \frac{a_L}{\gamma}, \text{ where } a_L = \frac{eE_L}{m_e c\omega_L}, \tag{15}$$

giving a relative red-shift of $(\gamma\theta)^2 = a_L^2/2$. This shift is again familiar from FEL theory, if one substitutes the normalized undulator vector potential a_u for its counterpart corresponding to the laser, a_L. Thus the spread in angles, due to the fact that the nonlinear motion-inducing electric field is not uniform, yields an rms contribution to the scattered bandwidth of

$$\left(BW_{rms}\right)_{NL} \cong \frac{a_L^2}{7.7}. \tag{16}$$

This result is valid only for $a_L \ll 1$; for higher fields the motion takes on a "figure-8" character, and higher harmonics appear in the motion and the emitted spectrum [12].

3. Constraints on luminosity and photon flux

We are now in a position to calculate the photon luminosity $N_\gamma = \mathcal{L}\sigma_T$, which gives the total scattered number of photons as $N_\gamma = \mathcal{L}\sigma_T$. Maximizing the luminosity by our optimum choice of Rayleigh range, we have

$$\mathcal{L} = \frac{N_L N_{e-}}{4\pi\sigma_x^2} = \frac{N_L N_{e-}}{\lambda_L \sigma_z} = \frac{U_L N_{e-}}{hc\sigma_z}, \tag{17}$$

where U_L is the total laser energy and h is Planck's constant.

Use of U_L allows us to connect the luminosity to the laser energy density

$$u_{\text{EM}} = \frac{N_L h\nu_L}{(2\pi)^{3/2}\sigma_z\sigma_x^2} = \frac{U_L}{\sqrt{\pi/2}\lambda_L^2\sigma_z^2}, \tag{18}$$

and thus the peak electric field value, as

$$E_0 = \sqrt{\frac{2u_{EM}}{\varepsilon_0}} = \sqrt{\frac{2U_L}{\sqrt{2/\pi}\varepsilon_0\lambda_L\sigma_z^2}}. \tag{19}$$

A constraint on the maximum tolerable a_L (and therefore the bandwidth due to nonlinear effects) thus also implies a limit on the total laser energy

$$U_{L,\max} = \frac{\varepsilon_0 E_L^2 \sigma_z^2 \lambda_L}{(2\pi)^{1/2}} = \frac{k_L \sigma_z^2 a_{L,\max}^2}{(8\pi)^{1/2} r_e} m_e c^2 . \tag{20}$$

We now arrive at an expression for the maximum achievable luminosity in terms of only the laser pulse length, wavelength, and normalized field,

$$\mathcal{L}_{\max} = \frac{U_{L,\max} N_{e-}}{hc\sigma_z} = \frac{k_L a_{L,\max}^2 \sigma_z m_e c^2}{\sqrt{8\pi} r_e hc} N_{e-} = \frac{\alpha a_{L,\max}^2 k_L \sigma_z}{\sqrt{8\pi} r_e^2} N_{e-}, \tag{21}$$

with associated maximum number of photons per collision

$$N_\gamma = \mathcal{L}_{\max} \sigma_T = \frac{\sqrt{8\pi}\alpha (k_L \sigma_z) a_{L,\max}^2}{6} N_{e-}. \tag{22}$$

This result, intriguingly, can be put in terms of the desired bandwidths, as

$$N_\gamma \cong 0.76 \frac{(BW_{rms})_{acc} (BW_{rms})_{NL}}{(BW_{rms})_{foc}} N_{e-}. \tag{23}$$

Equation 23 is a powerful expression that can be used to guide design. The most obvious point that is deduced immediately from it is that one may use a *smaller* $(BW_{rms})_{foc}$ to obtain more scattered photons. This is because the only way to have more laser photons in the interaction given the assumed limit on energy density is to have a larger beam, and accompanying smaller $(BW_{rms})_{foc}$.

As an illustration of Eq. 23's use, in many applications one desires a bandwidth on the order of 1%, so both $(BW_{rms})_{acc}$ and $(BW_{rms})_{NL}$ can be chosen maximally to be near 0.01. We shall see that a reasonable U_L is obtained when $(BW_{rms})_{foc} \sim 10^{-4}$. With these assumptions we have the remarkable result that $N_\gamma \approx N_{e-}$! In order to evade this limit, one must be able to relax some assumptions concerning the laser, e.g. one may consider guiding it using a plasma channel [13].

4. A short design exercise

The emphasis in our analysis has been to constrain the laser, and so one must ask whether our constraints produce reasonable demands on the laser design. The first choice one must make in specifying the laser is in wavelength. From our analysis it is straightforward to see that the total laser energy scales in this choice as $U_L \propto \lambda_L$, due to the limits on maximum tolerable field strength, and the simple fact that $a_L \propto \lambda_L$. Thus, given our assumptions, one finds that contrary to the argument that large λ_L gives proportionally more photons per pulse for fixed U_L, one is not permitted to use this laser with the same field value (intensity). We therefore assert that one should use a *short* wavelength laser, perhaps an 800 nm Ti:Sapphire or even a harmonic.

With the choice of λ_L, one then specifies $(BW_{rms})_{foc}$, which sets the dimensions of the beam unambiguously. Finally, one then chooses $(BW_{rms})_{NL}$, which then gives the energy density, and one arrives at the laser energy,

$$U_L = 0.1\frac{(BW_{rms})_{NL}}{(BW_{rms})_{foc}^2}\frac{\lambda_L}{r_e}m_e c^2. \tag{24}$$

Equation 24 shows explicitly that it is hard to exploit a larger beam size, as, and one is constrained by laser design and cost.

We have put together a full list of parameters to explore the utility of our approach in Table 1. We also add the parameters associated with an electron beam in order to specify the scattered wavelength (choosing a medical X-ray type example) and the emittance needed for the design. One can see that the beam spot size is rather small, with a Rayleigh range of under 0.5 mm. This type of focus may be obtained using permanent magnet quadrupoles, as done at PLEIADES [10]. Note that even in this, perhaps lowest electron beam energy $\varepsilon << \lambda_L / 4\pi$.

Table 1. Example of ICS interaction design for medical X-rays.

Laser wavelength λ_L	800 nm	Laser energy U_L	1 J
$(BW_{rms})_{foc}$	1.5×10^{-4}	Pulse length $\sigma_z(=Z_R)$	450 μm
$(BW_{rms})_{acc}$	0.01	Spot size σ_x	5.3 μm
$(BW_{rms})_{NL}$	0.01	Electrons N_{e-}	6×10^9
Electron beam energy	38 MeV	Scat. photon energy	35 keV
Emittance $\varepsilon = \varepsilon_n / \gamma$	2×10^{-8} m-rad	Scatt. photons in band	3×10^9

4. Comments on monochromatization

Depending on the application, the angular spread and bandwidth growth of the scattered photons, induced by the finite laser pulse length, nonlinear scattering effects and intrinsic angular spread at the laser focus, necessitates the use of a monochromator before delivery of the ICS photons to the user end station.

The emission spectrum of ICS photons is similar to that of insertion devices (particularly wigglers) in synchrotron radiation facilities. Typically, the electron beam energy of these rings is 0.8-10 GeV, thus producing very low divergence photon beams following undulators and wigglers [14]. In the case of the lower energy beams which might be used in ICS photon facilities (1-150

MeV), the divergence angle is on the order of mrad. As with undulator radiation, ICS photons are red-shifted when observed off the central axis. Also, depending on the interaction angle and laser polarization, the higher harmonics produced during nonlinear scattering have similar angular positions (e.g. two off-axis lobes of 2nd harmonic) to undulator radiation.

Because the double differential spectrum of the nonlinear scattering is quite broad, the ability to monochromatize this source is of concern. However, with modern ray tracing programs, EUV/soft x-ray diffraction gratings and grazing incidence mirrors of specific geometry (e.g. ellipsoidal), monochromators can be developed where beams of relatively high divergence (≥30 mrad) and large bandwidth can be collimated and diffracted [15], subsequently being sent through a pinhole for wavelength selection. The reflection coefficient at each mirror varies depending on the incidence angle of a photon of given energy. Other energies can also be reflected following the same ray traces, but with lower efficiency [16]. This can actually be utilized in multi-mirror systems to help filter out undesirable, higher energy photons.

Once the photon beam is collimated it can be incident on a diffraction grating without concern for the variance in frequency with angle. All the photons have the same incidence angle on the grating; therefore they are diffracted into their respective angles based on wavelength and follow the grating equation $mN\lambda = \sin\alpha + \sin\beta$ where m is the order number, N is the line density of the grating, and α and β are the incidence and diffraction angles, respectively (opposite sign if on different sides of grating normal where $\beta<0$) [17]. For EUV/soft x-rays, typically the 2nd order is used to maximize energy resolution.

Imaging spectrometers for the analysis of soft x-ray emission high-order harmonics in laser-plasma experiments have already been developed and have been shown to have wide wavelength and angular acceptance (5-75 nm and 10 mrad, respectively) [18].

Acknowledgments

The authors would like to thank K.J. Kim and S. Reiche for comments concerning the role of electron angle in ICS spectrum.

References

1. R. W. Schoenlein, et al., Science 274, 236 (1996).
2. C. Rose-Petruck, et al., Nature 398, 310 (1999).
3. C. W. Siders, et al., Science 286, 1340 (1999).
4. T. Omori, *et al., Nucl. Instrum. Methods A* 500, 232 (2003).
5. I.F. Ginzburg *et al.,* Pis'ma v ZhETF **34** (1981) 514.
6. S. Ohtsuka, et al. *British Journal of Radiology* 72(1999)
7. M. Babzien, et al., *Phys Rev Letters* **96**, 054802 (2006)
8. O. Williams, et al., these proceedings.
9. David J. Gibson, *et al., Phys. Plasmas*, 11 2857 (2004)

10. J. K. Lim, et al., *Phys. Rev. ST Accel. Beams* **8**, 072401 (2005)
11. J. S. Reid and G. J. Milne, J. Appl. Cryst. 21, 992 (1988).
12. G. A. Krafft, A. Doyuran and J. B. Rosenzweig, *Phys. Rev. E* **72**, 056502 (2005)
13. R. Yoder and J.B. Rosenzweig, these proceedings.
14. Attwood, David, "Soft X-ray and Extreme Ultraviolet Radiation: Principles and Applications"
15. O'Hara, David, *Private communication*, Parallax Research, Inc., February 2006.
16. Center for X-ray Optics website, http://www-cxro.lbl.gov/ .
17. Peatman, W.B., "Gratings, Mirrors, and Slits: Beamline Design for Soft X-ray Synchrotron Radiation Sources", Gordon and Breach Science Publishers, 1997.

18. Poletto, Luca, et al., "Optical design of a spectrometer-monochromator for the extreme-ultraviolet and soft x-ray emission of high-order harmonics", Applied Optics vol. 42, No. 31, November 2003.

PRELIMINARY RESULTS FROM THE UCLA/SLAC ULTRA-HIGH GRADIENT CERENKOV WAKEFIELD ACCELERATOR EXPERIMENT

M. C. THOMPSON*, H. BADAKOV, J. B. ROSENZWEIG, G. TRAVISH†

Department of Physics and Astronomy,
University of California, Los Angeles
Los Angeles, CA 90095
E-mail: thompson93@llnl.gov

M. HOGAN‡, R. ISCHEBECK‡, N. KIRBY‡, P. MUGGLI§, A. SCOTT◇, R. SIEMANN‡, D. WALZ‡, R. YODER♦

‡*Stanford Linear Accelerator Center, Stanford, CA, 94309*
§*University of Southern California Los Angeles, CA, 90089*
◇*University of California, Santa Barbara Santa Barbara, CA, 93106*
♦*Manhattan College Riverdale, NY, 10471*

The first phase of an experiment to study the performance of dielectric Cerenkov wakefield accelerating structures at extremely high gradients in the GV/m range has been completed. This experiment takes advantage of the unique SLAC FFTB electron beam and its demonstrated ultra-short pulse lengths and high currents (e.g., σ_z = 20 μm at Q = 3 nC). The FFTB electron beam has been successfully focused down and sent through varying lengths of fused silica capillary tubing with two different sizes: ID = 200 μm / OD = 325 μm and ID = 100 μm / OD = 325 μm. The pulse length of the electron beam was varied in the range 20 μm < σ_z < 100 μm which produced a range of electric fields between 2 and 20 GV/m at the inner surface of the dielectric tubes. We observed a sharp increase in optical emissions from the capillaries in the middle part of this surface field range which we believe indicates the transition between sustainable field levels and breakdown. If this initial interpretation is correct, the surfaced fields that were sustained equate to on axis accelerating field of several GV/m. In future experiments we plan to collect and measure coherent Cerenkov radiation emitted from the capillary tube to gain more information about the strength of the accelerating fields.

---

*Current Affiliation: Lawrence Livermore National Laboratory
†Spokesman for the collaboration at Erice.

1. INTRODUCTION

The accelerating field achievable in conventional radio frequency accelerators is ultimately limited by the breakdown of the metallic accelerating structure. It is well understood that the field sustainable in a cavity increases with the frequency of the cavity [1,2]. The empirical relation for the surface breakdown field of modern metallic cavities is given by

$$E_s = 220(f[\text{GHz}])^{1/3}\ \text{MV/m}, \tag{1}$$

where f is the RF frequency and E_s is the maximum sustainable surface electric field, which is about 2.5 times greater than the maximum accelerating field [2]. It is difficult to simply follow this scaling by using ever higher frequency cavities due to what could be called the "THz gap"; the current lack of practical high power source between X-band microwaves ($\sim$ 11 GHz) and infrared lasers ($\sim$ 28 THz).

There is considerable interest in the optical limit of break down scaling and the building of dielectric structures powered by high intensity laser beams. Various studies have indicated that GV/m accelerating fields should be possible in dielectric laser accelerators [3] as long as the driving laser pulses are very short [4]. The difficulty with laser accelerators, however, is that they operate at very short wavelengths (e.g. 10.6 μm at 28 THz), which greatly complicates beam injection and phasing.

Dielectric accelerators can also be powered directly by high energy charged particle beams via wakefield excitation [5,6], eliminating the need to generate high power electromagnetic waves. Additionally, a particle beam driven dielectric wakefield accelerator can operate at essentially any wavelength since the accelerating wakefield wavelength is determine by the dielectric geometry. This type of accelerator has been studied in depth over the last several years, but with the maximum fields limited to 10's of MeV/m by the lack of ultra-short drive beams. According to the scaling laws which govern dielectric wakefield accelerators, however, the recently achieved 20 μm pulse-length beams obtained at the Stanford Linear Accelerator Center (SLAC) Final Focus Test Beam (FFTB) facility may be sufficient to generate longitudinal fields in excess of 1 GV/m.

2. THEORY AND SIMULATION

An electron beam driven dielectric wakefield accelerator is essentially an unform dielectric tube coated on the outside with a conductor, see Fig. 1. When an intense electron beam passes through center of the tube its

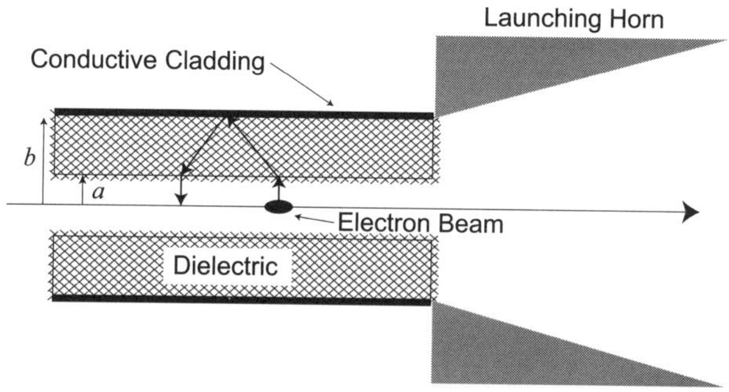

Figure 1. Schematic of the dielectric wake experiment.

electric field bends at the Cerenkov angle within the dielectric, reflects off the outside conducting layer, and returns to the axis where it can be used to accelerate other particles. For an ultra-relativistic beam ($\gamma \gg 1$) it is fairly straightforward to calculate the radial field on the surface of the dielectric $E_{r,surface}$, and the decelerating field within the beam $E_{z,decel}$, to first approximation through simple application of Gauss' law. Such an analysis yields the expression

$$E_{r,surface} = \frac{1}{\epsilon_0} \frac{Q}{(2\pi)^{3/2} a\sigma_z + \frac{\sqrt{\varepsilon-1}}{\varepsilon}\pi a^2} \left[\frac{\text{V}}{\text{m}}\right], \tag{2}$$

where all quantities are in SI units and Q is the beam charge, a is the inner radius of the dielectric, σ_z is the rms length of the beam, $\epsilon_0 = 8.854 \text{ x } 10^{-12}\ \text{Fm}^{-1}$ is the permittivity of free space, and ε is the dielectric constant. Furthermore, $E_{r,surface}$ and $E_{z,decel}$ are related by the equation

$$E_{z,decel} = \frac{\sqrt{\varepsilon - 1}}{\varepsilon} E_{r,surface}. \tag{3}$$

Table 1. Experiment Design Parameters

Bunch Charge (Q)	3 nC
Bunch Energy	30 GeV
Beam Radius (σ_r)	10 μm
Beam Length (σ_z)	100 - 20 μm
Inner Dielectric Radius (a)	50 and 100 μm
Outer Dielectric Radius (b)	162 μm
Dielectric Relative Permittivity (ε)	~ 3
Peak Decelerating Field	8 GV/m
Peak Accelerating Field	12 GV/m
Peak Field at Dielectric	22 GV/m

By symmetry, it is clear that the accelerating wake behind the driving electron beam $E_{z,accel}$ must be on the order of $E_{z,decel}$, although a more sophisticated analysis is required to predict $E_{z,accel}$ precisely. It is clear from these equations that maximizing the decelerating wakefield, and hence the accelerating wakefield, requires maximizing Q while minimizing σ_z and a. It should be noted that a must be chosen several times larger than σ_r to prevent damage to the dielectric from the beam halo.

The parameters for our experiment, which is based upon the unprecedented combination of high charge, short pulse duration, and small spot size available in the SLAC FFTB beam, are listed in Table 1. The electric field predictions of Eqs. 2 and 3 compare well to those of more rigorous an-

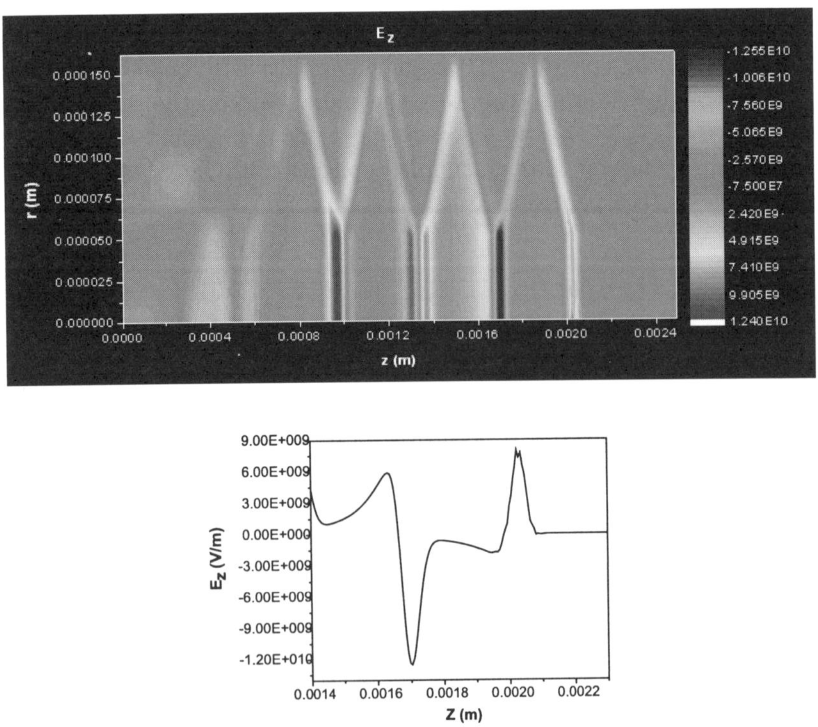

Figure 2. OOPIC simulation results for parameters $Q = 3$ nC, beam energy 30 GeV, $\sigma_r = 10$ μm, $\sigma_z = 20$ μm, $a = 50$ μm, $b = 162$ μm, and $\varepsilon = 3$. At top is a contour plot showing the z component of the electric field (E_z) throughout the simulation region. At bottom is a plot of E_z along the z axis at a radius of 10 μm showing an enlargement of the first wake oscillation. The electron pulse centroid coincides with the peak of E_z on the right.

alytical calculations [8,9] and numerical simulations. Using the values listed in Table 1 and choosing $\sigma_z = 20\ \mu$m and $a = 50\ \mu$m, Eq. 3 predicts a decelerating wakefield within the electron bunch of 8.2 GV/m. OOPIC simulations [10] using the same parameters give a value of 8 GV/m for the peak decelerating field, see Fig. 2. This simulation also indicates a peak accelerating field of 12.5 GV/m, as shown on the right in Fig. 2, and a peak field of 22 GV/m at the surface of the dielectric.

3. BREAKDOWN THRESHOLD OBSERVATIONS

Our experiment is the first in the GV/m regime of electron beam driven dielectric wakefield accelerators. The primary goal of the first phase of the experiment was to assess dielectric material survivability. Since the experiment uses dielectric tubes with 100 and 200 μm IDs, we limited the tube length to about a cm in order to mitigate alignment issues. The energy gained or lost by beam particles was therefore only on the order of 10 MeV, while the full energy spread of the highly compressed FFTB beam is several GeV. Consequently, energy changes were not resolvable in this experiment. The primary signature of high field wakes in this first phase of the experiment was the detection of dielectric breakdown.

This dielectric wakefield experiment is unique in that it produced fields comparable to those in earlier laser breakdown studies. We are particularly interested in differentiating material breakdown in wake-excited, relatively long wavelength ($> 50\ \mu$m photon) systems, as compared with optically (laser) excited systems. Quantum absorption of photons to produce free electrons in the material is known to be an initiator of avalanche ionization [4]. This mechanism should be mitigated in the wake experiment, since the individual photon energies are lower than in the laser case, perhaps allowing higher fields to be tolerated. In the case of optically excited fields, the breakdown limit for a 70 fs pulse is less than 2 GV/m in fused silica [4]. As discussed below, it seems that surface fields higher than 2 GV/m where sustained in our dielectric wake experiment at similar pulse lengths.

The dielectric tubes for this experiment were modified from off the shelf synthetic fused silica capillary tubing ($\varepsilon \cong 3$), which we procured in 100 and 200 μm ID sizes. Samples were prepared from this bulk product by cutting pieces to length, removing the outer polyimide jacket, polishing the cut ends, and coating the outside of the tube with vapor deposited aluminum. Multiple samples were packaged together for the experiment in a modular mounting block, Figs. 3 and 4. The mounting block is about 1 cm wide, by

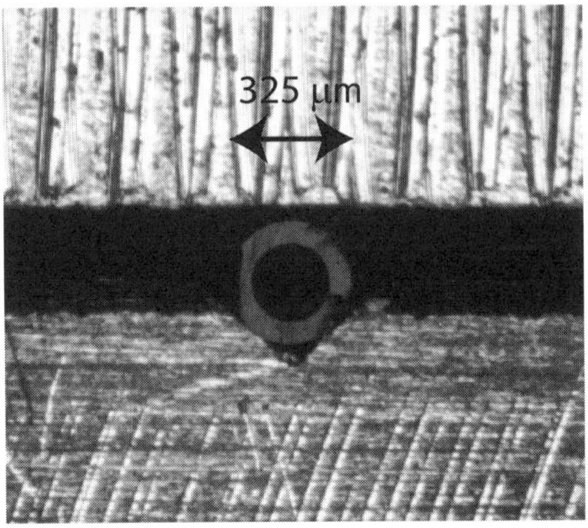

Figure 3. End on view of a 200 μm ID, 325 μm OD capillary tube (unpolished) held in the mounting block.

5 cm long, by 0.5 cm high and is designed to hold 10 samples with 2.5 mm center to center spacing. The mounting block also has an optical transition radiation (OTR) screen for locating the beam orbit and a series of blank alignment holes, as shown in Fig. 4.

Our methodology for aligning the experiment was to move the dielectric tubes onto the established beam orbit rather than trying to steer the beam through the tubes. To accomplish this, the tube sample mounting block was placed on optics mounts and motorized stages to provide all the necessary degrees of freedom. An initial rough alignment was done during the experiment installation with an alignment laser matched to the electron beam orbit. Final alignment began by observing the electron beam on the fiber holder OTR screen with a telescopic camera and an insertable pelicle mirror, see Fig. 4. Once the location of the beam center in the camera field of view was established, the fiber holder could be moved through the fixed camera frame and properly aligned with the established beam position. Before trying to propagate the beam through the fibers, we sent the beam through a series of successively smaller blank alignment holes. By passing the beam through these metal walled "dummy" holes and monitoring x-ray emissions in the accelerator enclosure we planned to ensure that the beam is well aligned and its halo is small enough to fit through the dielectric tube. In practice, this alignment technique seemed to work very well, although we could not rule out beam halo scrapping in the blank holes or dielectric tubes since our only x-ray detector was part of the machine protection sys-

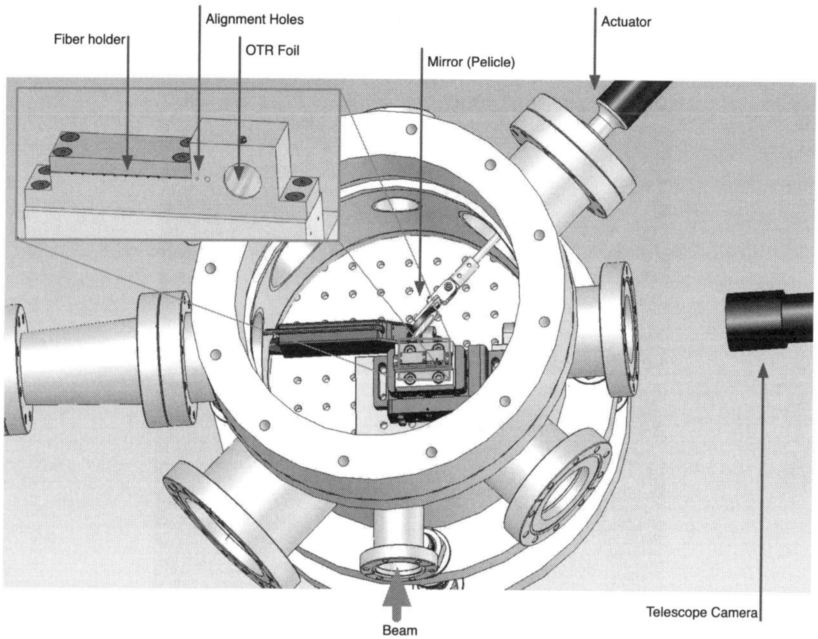

Figure 4. CAD illustration of the experimental setup.

tem and was relatively insensitive. Alignment procedures will be further refined and use better x-ray diagnostics in future runs of the dielectric wake experiment.

Once beam propagation through the dielectric tubes was established, we began to vary the wakefield magnitude by adjusting the beam pulse length and search for an optical signal of breakdown on a side view telescopic camera (not shown in Fig. 4). This experiment was repeated for samples with different tube IDs. Individual tube samples were systematically exposed to different field levels for varying periods of time for a later study of differences in surface damage.

While analysis of the collected data is still preliminary, the observations seem to indicate a clear breakdown threshold. Most of the observations were of 200 μm ID fibers and the general impression is that the visible light output of the fibers jumped up sharply in the middle of the beam pulse length range, Fig. 5. This signal was clearly visible on both the side view telescopic camera and a wide-angle top-view general observation camera. The initiation of breakdown discharges are most likely responsible for the optical radiation increase. The visible wavelength light recorded at

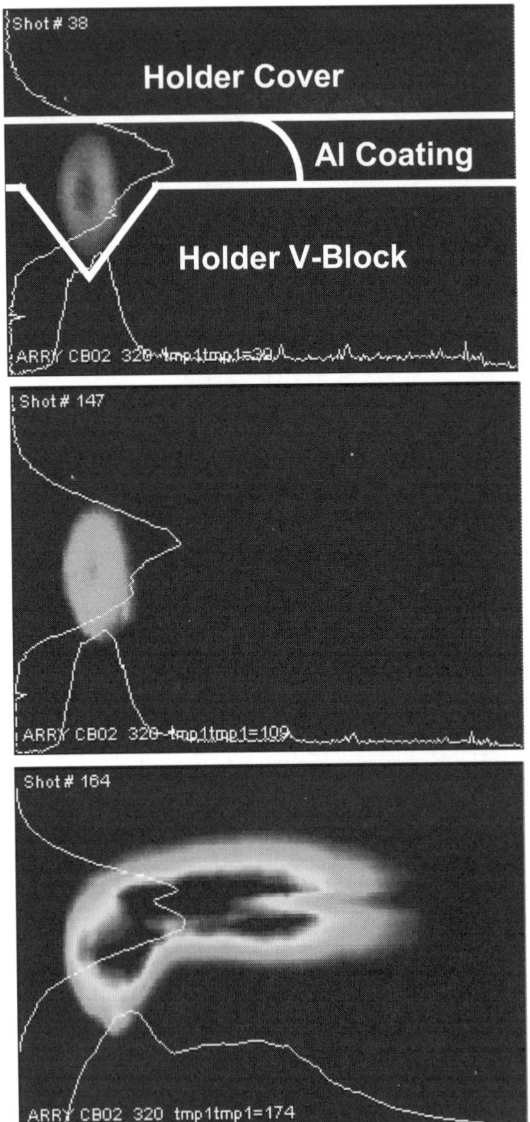

Figure 5. Visible light images (displayed with the same scaled-to-intensity color map) of a 200 μm ID fiber tip as the electron beam pulse current is ramped up. In the top image, the electron beam pulse length is long ($\sim$ 100 μm σ_z) and the fiber end glows faintly with incoherent optical wavelength emissions. The various boundaries of the fiber holder are also illustrated in the top frame, along with the area in which the fiber's aluminum coating blocks optical emissions. In the middle frame the electron beam is shorter and incoherent emission has increased. In the bottom frame the electron beam is near its minimum pulse length ($\sim$ 20 μm σ_z) and a breakdown event has occurred producing a large flash of visible light.

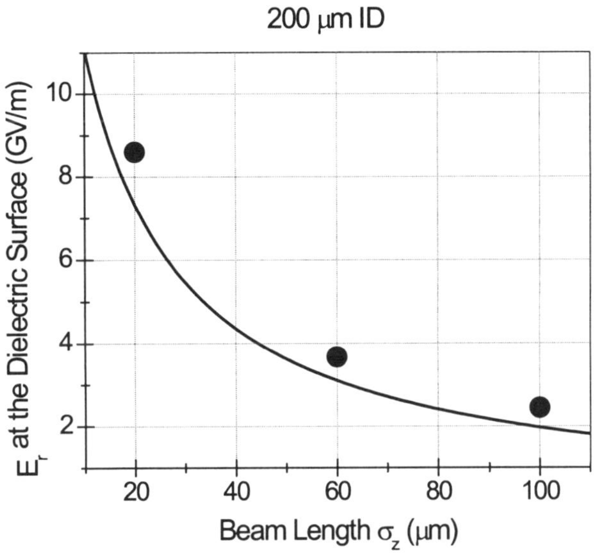

Figure 6. Radial electric field at the inner surface of a 200 μm ID fused silica capillary tube as a function of electron beam pulse length. The line is the prediction of Eq. 2 and the dots are the results of OOPIC simulations. Note that the slight offset arises because the OOPIC results give the peak field while Eq. 2 give the average field over the effective length of the beam. It is not yet clear which method of evaluating the surface fields produces a better figure of merit.

long pulse length / low fields is probably due to a combination of incoherent cerenkov radiation, incoherent transition radiation, and scintillation. Once a breakdown occurs, however, the light from these mechanisms is small compared to the visible emissions from the plasma formed during the breakdown event. It is clear from the observations that breakdown events start to occur after the beam has been compressed below the midpoint of its range. If we plot the surface fields at the dielectric verses the beam pulse length, Fig. 6, we find that at the midpoint ($\sigma_z \sim 60$ μm) the surface fields at the dielectric are in the range 3 - 4 GV/m. OOPIC simulations indicated that this range of surface fields corresponds to 1.5 - 2 GV/m on-axis accelerating fields in the wake.

Further analysis of the data will allow us to make a more accurate determination of the fused silica capillary tube breakdown threshold. In addition to a more careful analysis of the visible light observations and associated beam parameters, we also intend to examine the dielectric tubes with various diagnostics to detect structural damage from the breakdown events. We hope to be able to correlated the number of breakdown events a particular tube experienced to the degree of pitting or other forms of damage that are evident in the postmortem analysis.

One form of damage which was immediately obvious from cursory examination of the fibers was the vaporization of their aluminum cladding. This type of damage does not impact the study of breakdown, but would be unacceptable in acceleration experiments. When preparing the samples, we assumed that the aluminum coating thickness was unimportant as long as it was substantially thicker than the skin depth at the frequencies of interest. Consequently, we deposited a layer of aluminum about 1 μm thick on the outside of the fibers. After the experiment we discovered that as the number of electron pulses sent through a particular fiber increased, the amount of aluminum remaining on the tube's outer surfaces decreased. We believe that the vaporization of the aluminum resulted from resistive heating of the aluminum cladding due to the electron beam image current. This heating may result in the lose of aluminum through either ablation, long time scale heating of the coating, or a combination of both. In the ablative scenario, each high current beam pulse, which can range from 3 – 18 kA depending on the pulse length, heats the thin aluminum coating enough to evaporate a small amount of the metal. The temperature of the aluminum cladding would not increase over the long term, however, since residual heat would dissipate between beam pulses. A bulk heating scenario is also possible due to the relative thermal isolation and low mass of the aluminum coating. In this case, as beam pulses travel through the fiber at 1 Hz, the heat deposited by each pulse builds up in the cladding until it is hot enough to sublimate. Initial calculations indicate that significant energy is deposited into the aluminum through resistive heating, but do not resolve which material displacement mechanism dominates. While further work is needed, the vaporization problem can probably be solved in either case by increasing the thickness of the cladding and increasing its thermal contact with the holding block.

4. COHERENT CERENKOV RADIATION

In the second phase of the experiment we plan to measure the quantity and spectral distribution of the coherent Cerenkov radiation (CCR) produced in the dielectric as part of the wake excitation process. An independent measurement of the wakefield strength can be extracted from the CCR properties. The total amount of energy radiated as CCR is approximately

$QE_{z,decel}L_d/2$, where L_d is the dielectric length. This energy is distributed over narrow spectral lines of wavelength $\lambda_n \cong 4(b-a)\sqrt{\varepsilon - 1}/n$, where n is an integer, and the energy in each line is approximately proportional to n up to the roll off wavelength $\lambda_n \cong \sqrt{2\pi}\sigma_z$ [7]. For example, using the peak field parameters from Table 1 we find that 150 mJ of CCR will be produced under these conditions with spectral lines at 633, 316, 211, 158, 126, and 106 μm. This THz CCR will have about 5 times the intensity of coherent transition radiation emitted from the surface of the tube end [7]. The CCR will also have a very different spectrum from the coherent transition radiation, which allows for easy differentiation. The CCR exiting the end of the dielectric tube will be collected by an off-axis paraboloid and transported through a vacuum window to a detector. The amount of CCR delivered to the detector will be increased by using a launching horn, similar to those used in microwave antennas, which is just a $\sim 10°$ metal cone placed after the tube exit, see Fig. 1.

5. CONCLUSIONS

We have completed the first phase of a two phase experiment which explores the feasibility of GV/m class Cerenkov wakefield accelerators. The observations of breakdown events made in the first phase of the experiment seem to indicate that a dielectric wakefield accelerating structure based on fused silica can support surface fields of at least 3 - 4 GV/m, which corresponds to accelerating fields of 1.5 - 2 GV/m, without breakdown. These values indicate that a beam driven dielectric accelerator can sustain higher fields, due to its lower photon energy, than a laser driven dielectric accelerator. Further analysis of the data will enable us to refine this initial assessment of the breakdown fields. In the second phase of the experiment we will resolve the problem of aluminum cladding vaporization and measure the coherent Cerenkov radiation emitted by the interaction of the electron beam with the dielectric structure. Analysis of the coherent Cerenkov radiation will give an independent measurement of the strength of the fields within the dielectric.

6. ACKNOWLEDGEMENTS

This work was performed under the auspices of the US Department of Energy under Contract No. DE-FG03-92ER40693 and W-7405-ENG-48.

References

1. W. D. Kilpatrick, *Rev. Sci. Instr.*, **28**, 824 (1957).
2. G. A. Loew and J. W. Wang, "RF Breakdown and Feild Emission", SLAC-PUB-4845 (1989).
3. J. Rosenzweig, A. Murokh, and C. Pellegrini, *Phys. Rev. Lett.*, **74**, 2467 (1995).
4. D. Du, *et al.*, *Appl. Phys. Lett.*, **64**, 3071 (1994).
5. W. Gai, *et al.*, *Phys. Rev. Lett.*, **61**, 2756 (1988).
6. J. G. Power, *et al.*, *Phys. Rev. ST Accel. Beams*, **3**, 101302 (2000).
7. J. B. Rosenzweig, *et al.*, "An Ultra-High Gradient Cerenkov Wakefield Acceleration Experiment at SLAC FFTB," In *Advanced Accelerator Concepts: Eleventh Workshop*, AIP Conf. Proc. No. 737, 811 (2004).
8. R. Siemann and A. Chao, "Wakefields in a Dielectric Tube with Frequency Dependent Dielectric Constant", SLAC ARDB report 368 (2004).
9. R. Siemann, "Decelerating Gradient for FFTB Dielectric Tube Experiment", SLAC ARDB report 371 (2004).
10. D. L. Bruhwiler, *et al.*, *Phys. Rev. ST Accel. Beams*, **4**, 101302 (2001).

STATUS OF THE POLARIZED NONLINEAR INVERSE COMPTON SCATTERING EXPERIMENT AT UCLA

O. WILLIAMS, A. DOYURAN, R.J. ENGLAND, J. LIM, J.B. ROSENZWEIG, G. TRAVISH

Department of Physics, University of California, Los Angeles
Los Angeles, CA 90095, USA

C. JOSHI, S. TOCHITSKY

Department of Electrical Engineering, University of California, Los Angeles
Los Angeles, CA 90095, USA

An Inverse Compton Scattering (ICS) experiment investigating the polarized harmonic production in the nonlinear regime has begun which will utilize the existing terawatt CO_2 laser system and 15 MeV photoinjector in the Neptune Laboratory at UCLA. A major motivation for a source of high brightness polarized x-rays is the production of polarized positrons for use in future linear collider experiments. Analytical calculations have been performed to predict the angular and frequency spectrums for various polarizations and different scattering angles. We report on the experimental set-up and status.

1. Introduction

The production of third-generation light source quality x-rays using low energy electron beams as well as the possibility of using it as a source of polarized positrons in high energy physics experiments has spurred much interest in inverse Compton scattering (ICS) [1,2]. Such applications require an understanding of the nonlinear and polarization properties of the scattering process during high laser intensities (where the laser vector potential, $a_o \geq 1$).

The spectral and angular distribution of the nonlinear scattering process has not been determined experimentally and it has yet to be shown whether the laser polarization is transferred to the scattered photons during the nonlinear interaction. The physics behind this scattering mechanism should give more insight into the limitations of using a laser or other electromagnetic field as an undulator for the production of x-rays.

The Neptune Laboratory at UCLA houses a terawatt-class CO_2 laser system capable of achieving $a_o=1$ [3], the necessary condition to obtain nonlinear scattering and hence higher harmonic generation. Very strong focusing is therefore required at the interaction point (IP) for both the laser and electron beam. In order to focus the 14 MeV, 300 pC, low emittance beam produced by the 1.6 cell S-Band photoinjector and linac, the development of an array of strong focusing permanent magnet quadrupoles (PMQ's) was necessary. The strong diffraction due to the short laser focal length and small aperture of the PMQ's did not permit counter-propagation and necessitated a transverse (90°) scattering geometry between the laser and electron beam.

2. Experimental Set-up

The layout of the interaction region is shown in Figure 1. In order to obtain rms beam sizes near 25 μm, strong focusing is required. Conventional electromagnet quads were not employed due to the dispersion, chromatic aberrations, and strong space charge forces associated with focusing the lower energy electron bunch [4]. To achieve the small beam sizes, PMQ's were developed with a gradient of ~115 T/m. Five total PMQ's are arranged in a modified triplet configuration (F-DD-FF) with only 1.5 cm to the minimum beam waist (IP), followed by a symmetric triplet for recollimation. Both PMQ assemblies are manually adjustable with respect to the IP and remotely adjustable as a unit to allow scanning of the waist along the beam axis.

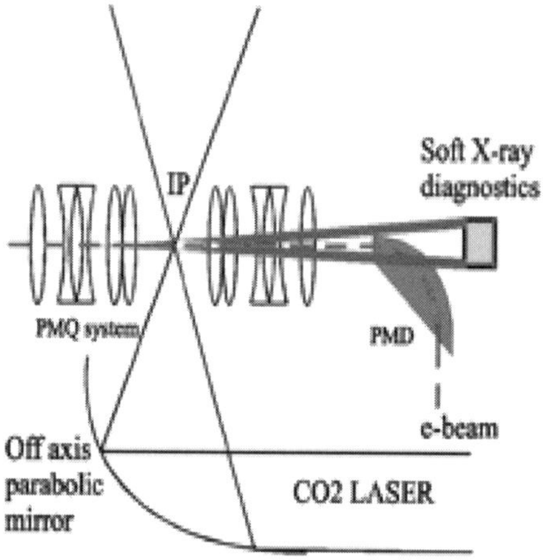

Figure 1: Experimental drawing of the interaction region

It can be seen in the expression for the laser vector potential, $a_0 = \lambda_L e E_L / 2\pi m_e c^2$, where e is the charge of an electron, E_L is the energy of the laser pulse, and m_e is the mass of an electron, that longer wavelength light, λ_L, makes a_o large, giving an advantage to CO_2 lasers.

However, a high intensity spot is still required at the IP and for 500 GW of 10.6 μm light, this necessitates a 25 μm spot (RMS). This strong focusing is accomplished by a 5" copper off-axis parabolic mirror with an F# of 1.5 oriented 90° with respect to the beam axis, resulting in orthogonal scattering with the electron beam. The design parameters for the electron and laser beams are shown in Table 1.

Table 1: Design Electron and Laser Beam Parameters

Parameter	Value
Electron Beam Energy	14 MeV
Beam Emittance (norm. RMS)	5 μm
Electron Beam Spot size (RMS)	25 μm
Beam Charge	300 pC
Bunch Length (RMS)	4 ps
Laser Beam radius at IP (RMS)	25 μm
CO_2 Laser Wavelength	10.6 μm
CO_2 Laser Rayleigh Range	0.75 mm
CO_2 Laser Power	500 GW
CO_2 Laser Pulse Length	200 ps

Due to the low energy electron beam, the scattered photons are emitted into a rather large cone of half angle ($\theta \sim 1/2\gamma$) θ=15 mrad (see Table 2 for more scattered photon parameters). This large angle imposed strong space constraints on the experimental layout. In order to place the x-ray diagnostics (i.e. x-ray CCD camera) as close to the source as possible to fully image any off-axis harmonics, the electron beam needed to be bent and dumped immediately following recollimation, requiring a compact permanent magnet dipole.

Table 2: Calculated Scattered Photon Parameters

Parameter	Head-on	Transverse
Scattered photon wavelength	5.3 nm	10.7 nm
Scattered photon energy	235 eV	118 eV
Scattered photon pulse duration (FWHM)	10 ps	10 ps
Interaction time	5 ps	0.33 ps
# of periods electrons see	283	10
# of photons emitted per electron	3.34	0.11
Total # of photons	$6.3x10^9$	$2x10^8$
Half Opening Angle (θ)	2.7 mrad	15 mrad
Bandwidth	0.35%	10 %

2.1. *Permanent Magnet Quadrupole (PMQ)*

The PMQ's are constructed using commercially available, NdFeB, M=1.2 T cubes supported in a hexagonal iron yoke with hyperbolic iron tips about the beam axis. An aluminum spacer is used to constrain the individual cubes within the yoke, as shown in Figure 2. To assure a uniform, magnetic-to-mechanically matched field about the beam axis, the hyperbolic tips, yoke, and aluminum spacer are electric discharge machined (EDM) for better precision and the cubes measured with a hall probe and sorted. Using the pulse wire method [5], the magnetic axis was found to be within 25 μm of the geometric axis.

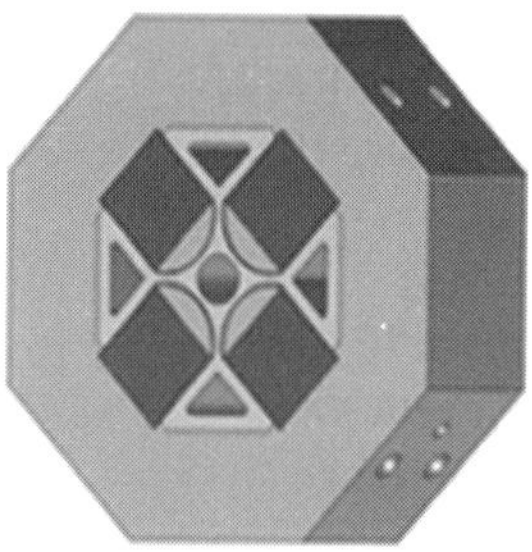

Figure 2: PMQ design using 4 cube magnets, hyperbolic pole tips and aluminum spacers.

2.2. *Permanent Magnet Dipole (PMD)*

The PMD is shown in Figure 3. The size of the gap had to be considered not only for desired field strength, but also because too small of a gap could clip the scattered photons. The resultant design included NdFeB, M=1.32 T magnets cut circularly to optimize scattered photon transport and a 16 mm gap, yielding a dipole field of ~0.78 T/m. This produces a bend radius of 60 mm for 14 MeV electrons. Field measurements are in excellent agreement with RADIA 3D simulations. To accommodate off-energy and off-trajectory particles, the PMD has been placed on a heavy load in-vacuum actuator which translates vertically. Changing the position at which the electrons enter the PMD correspondingly changes their exit position and angle. This allows for acceptance of 12-14 MeV beams with transverse offsets and entrance angles.

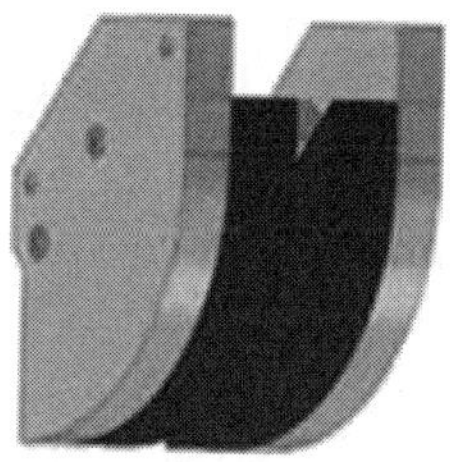

Figure 3: Compact dipole using permanent magnets. The beam enters from the left and is bent 90° upwards.

2.3. *IP Diagnostics*

The spatial overlap and synchronicity of arrival at the IP between the electron and laser beam is critical to obtain sufficient flux of scattered photons. The high density of charge in the 25 μm electron spot at the IP makes traditional YAG or phosphor imaging less desirable due to limited spatial resolution [6]. Therefore, we chose to image the electron beam using optical transition radiation (OTR). The difficulty of imaging a 90° interaction is also complicated by the compactness of the experiment. The distance between the focusing and recollimating quads is only 3 cm and must accommodate the diffracting laser beam. The transport of light at the IP intended for alignment diagnostics therefore required an unconventional layout. This prompted the development of a highly reflective, sharp-edged, aluminum pyramid used to simultaneously image the two beams as seen in Figure 4.

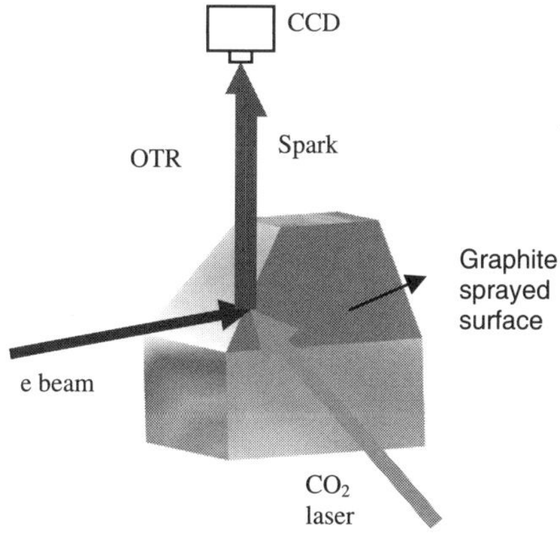

Figure 4: The IP imaging diagnostic "pyramid".

The 10.6 μm light, not lying in the visible spectrum, requires the application of a thin layer of micron-sized graphite on the laser imaging surface of the pyramid which "sparks" when hit by the laser, allowing for both beams to be overlapped along the pyramid's edge as imaged by a vertically mounted CCD camera. The two beams must also be synchronized in time at the IP. Temporal overlap is accomplished through a Germanium crystal acting as a "gate" [7]. When the beam electrons are incident on the crystal oriented at 45°, a semiconductor plasma is formed which is opaque to the 10.6 μm laser light. By measuring the transmitted intensity of the laser, the relative timing between the beams is revealed and allows for tuning of the arrival time at the IP.

2.4. *X-ray Imaging*

Two types of x-ray imaging devices are planned for use in the experiment, a soft x-ray CCD camera (on loan from the ANL APS) and a microchannel plate (MCP) detector. The x-ray camera will be used to image primarily the beam axis with good resolution. Because the x-ray source is diverging so strongly and the x-ray camera has an imaging area of only 12.3x12.3 mm^2, it resides in a reentrant housing allowing it to be inserted in the vacuum chamber, closer to the source.

The MCP detector has a 40 mm diameter imaging area and is mounted further from the source than the x-ray camera, yet still has 75% more field of view. After being cross-calibrated against the x-ray camera, the MCP will be used to better image the beam off-axis, particularly the expected higher harmonics.

In addition to the angular distribution imaged by the two detector types, future plans include concurrently measuring the frequency spectrum using a soft x-ray imaging spectrometer in order to obtain the double differential spectrum (DDS). The effects of high laser intensity on red-shifting the fundamental and spectral broadening of the harmonics are expected to be revealed with this diagnostic.

Polarization measurements in the soft x-ray region are not easy and in order to adequately show that the laser polarization is imprinted onto the scattered photons during the nonlinear interaction it has been proposed to use transmissive multilayer optics. Thus far this appears to be the simplest method for soft x-ray polarimetry although obtaining such optics has yet to be discussed.

3. Calculation Tool

The experiment is well-coupled with a strong theoretical framework on the expected DDS. A Mathematica routine has been developed which predicts the angular and frequency spectra for the scattered photons at different scattering angles and laser polarizations and pulse shapes. A more detailed discussion on the expected distributions of higher harmonics can be found in Ref. [8].

4. Status of the experiment

Rough alignment was performed using a graphite-coated phosphor to simultaneously image the two beams. Upon achieving spatial overlap we moved the Germanium crystal to the IP. Adjusting the delay line of the seed for the 10.6 μm regen allowed for relative timing shifts between the two beams until an obvious crossing point was observed. It can be seen in the plot below that before the e-beam has arrived (100-275 ps) the ratio of the reference signal reflected off the input window of the chamber and the transmitted signal detected on the other side of the crystal is nearly unity (normalized). Upon arrival of the e-beam (~275-290 ps) the semiconductor plasma is formed, greatly reducing the transmitted 10.6 μm intensity. The recombination time of the plasma is on the order of nanoseconds, therefore this method is only effective when the (few ps) e-beam arrives late and is thereafter delayed appropriately.

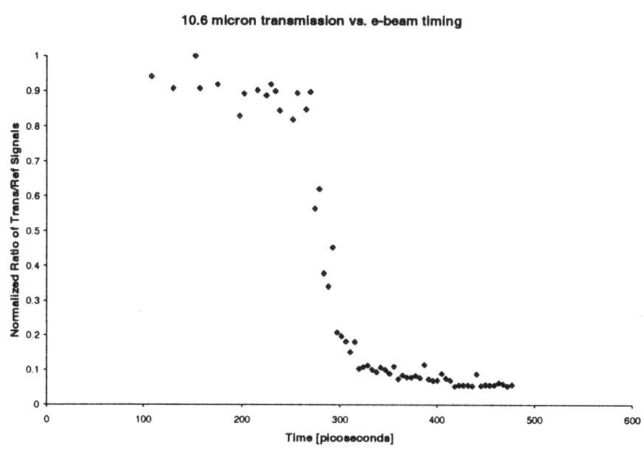

Figure 6: Effect of e-beam timing on transmission of 10.6 μm light through a germanium crystal.

Because this method of synchronization necessitates spatial alignment, we are confident the beams are "seeing" each other. Upon installation of the x-ray

CCD camera and familiarization with image acquisition software, the search for first x-rays will begin.

References

1. W. Thomlinson, *Nucl. Inst. Methods A* **319**, 295 (1992).
2. T. Omori et al., *Nucl. Inst. Methods A* **500**, 232 (2003).
3. S. Tochitsky, et al., AAC 2004 Proceedings.
4. J. Lim, et al., *Phys Rev STAB* **8**, p. 072401 (2005).
5. G. Rakowsky, et al., PAC 1999 Proceedings.
6. A. Murokh, et al., PAC 2001 Proceedings.
7. S. Tochitsky, et al., *Physics of Plasmas* **11**, No. 5, 2875 (2004).
8. G.A. Krafft, et al., *Phys Rev E* **72**, 056502 (2005).

COUPLING LASER POWER INTO A SLAB-SYMMETRIC ACCELERATOR STRUCTURE

R. B. YODER

Dept. of Physics
Manhattan College
Riverdale, NY 10471, USA
E-mail: ryoder@mailaps.org

J. B. ROSENZWEIG

Dept. of Physics and Astronomy
University of California—Los Angeles
Los Angeles, CA 90095, USA

We present a detailed analysis of the coupling of external laser fields into a resonant acceleration structure. The structure is slab-symmetric and consists of metal boundaries lined with dielectric; coupling is accomplished via an array of narrow, equally-spaced slots in the upper structure boundary. Effects of the couplers on the structure fields are investigated using both theory and 2-dimensional time-dependent simulation. Details of the resonance are found to be highly sensitive to the coupling slot geometry: the presence of the couplers can lead to frequency detuning, changes in the field breakdown limits and overall Q-factor, and distortions of the field pattern. Beam wakefields are enhanced by the presence of the slots, but found to have no significant effect on the beam transport. The resonant accelerating fields in this case are nearly constant along the short transverse direction and are found to have between 10 and 15 times the amplitude of the driving radiation.

1. Introduction

Particle acceleration using lasers continues to be of great interest as a potential method for producing ultrahigh field gradients on the order of GV/m. Among the strategies that have been considered for laser acceleration is the use of a non-traditional, translationally-symmetric geometry for the resonant structure, rather than the conventional cylindrical symmetry.[1] Such a 'slab geometry' avoids many experimental difficulties inherent in other common approaches, such as plasma accelerators. The advantageous properties of slab symmetry have been discussed in Refs. 1 and 2, and include the strong suppression of transverse beam wakefields, lack of higher-order synchronous modes, mitigation of field breakdown via the use of dielectric materials, and accleration of high-charge beams without significant beam loading. With a resonant structure, peak accelerating fields may be much higher than those of the drive laser.

The basic theory of acceleration and wakefields in a slab-symmetric resonant structure has been presented and discussed in several papers over the last decade,[2,3,4] together with some preliminary versions of an accelerator at $\lambda = 10.6\ \mu$m. In this paper, we examine the details of coupling power into a version of this device resonant at terahertz frequencies ($\lambda = 340\ \mu$m), using a novel coupler design which relies on a series of transverse slots. To our knowledge, only one previous paper[5] has attempted a detailed coupling analysis of a laser-driven accelerator; the results presented here indicate that failure to consider coupling methods may be a serious omission.

Most examples of laser-powered structures in the literature have considered some variation of end-coupling: a laser beam is focused into one end of a slow-wave structure, which is able to contain and guide a speed-of-light mode.[6,7,8] However, in practice it is nearly impossible to inject both laser and electron beams collinearly into any such device. Furthermore, the total pump power which can be injected into such a structure is severely limited by the damage threshold of its end face. For these reasons, a side-coupling scheme appears to be the most practical way to design a laser-driven acceleration structure. In that case, a series of couplers (or a partially transmitting boundary) can be used, allowing the incident laser power to be distributed over a reasonably large area and minimizing the laser flux as much as desired. A successful side-coupling scheme must also involve mode conversion, however, since the structure is not being powered in the desired accelerating mode. Hence, coupler analysis and efficiency must be included in any realistic structure design.

2. Structure Geometry and Field Theory

Fundamentally, this slab-symmetric accelerator consists of a pair of parallel dielectric planes, separated by a narrow vacuum gap and bounded above and below by a thin conductive layer. Periodic slots in the conductor provide a means for coupling radiation into the gap and also enforce longitudinal periodicity in the structure fields; this modulation prevents the structure fields from consisting of 'zero-mode' Fabry-Perot modes. Instead, when resonance is achieved, the mode pattern is dominated by a longitudinal standing wave having a phase velocity exactly equal to the speed of light. The array of slots extends throughout the interaction length (defined as, roughly, the diameter of the laser spot impinging on the upper surface). The concept is illustrated in Fig. 1.

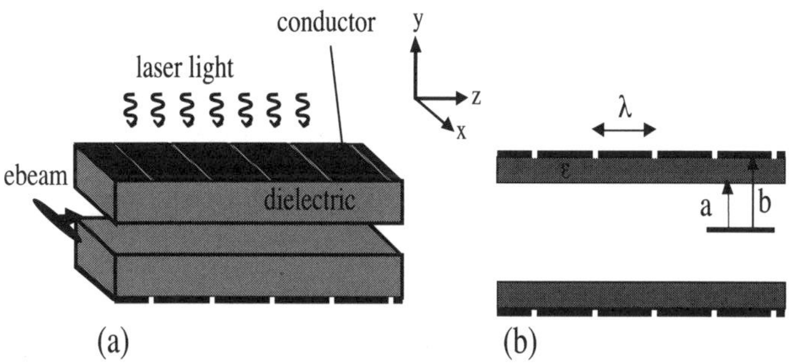

Figure 1. (a) Schematic drawing of the structure geometry. Two layers of dielectric-lined conductor surround a vacuum gap; a very wide electron beam is injected into the gap and travels in the $+z$ direction, while radiation (polarized in z) is coupled in from above through transverse slots in the conductor. (b) A cross-section in x.

Any periodic modulation in the coupling or dielectric properties of the structure would serve to generate usable accelerating fields, and previous approaches to this design have each taken a different approach to generating this periodicity.[1,3,9] (A related idea is used in a recently-proposed Bragg-reflector accelerator.[5]) The periodic slots used here represent the optimum choice for coupling power preferentially into the accelerating mode, albeit at the cost of perturbing the field. Coupling slots also appear on the lower surface, as shown in the figure; these serve to symmetrize the design and also allow the use of transmitted radiation as a diagnostic.

2.1. *Summary of analytic mode theory*

The accelerating-mode fields for a structure with infinite transverse extent are straightforward to derive and have been presented in e.g. Refs. 3 and 10; the effects of a finite transverse dimension on beam stability have been considered in Ref. 2. We will simply quote relevant results here. Using the axes and structure parameters defined in Fig. 1, the vacuum gap has width $2a$ and the structure itself width $2b$, with dielectric (of relative permittivity ϵ) occupying the regions from $|y| = a$ to $|y| = b$. The structure is illuminated from the $+y$ direction by a laser polarized in the z direction, with beam propagation along the z-axis. If one assumes infinite extent in x and phase velocity c for the accelerating mode, one obtains

$$E_z = E_0 \cos(k_z z) e^{i\omega t} \tag{1}$$

$$E_y = E_0 k_z y \sin(k_z z) e^{i\omega t} \tag{2}$$

in the vacuum region and

$$E_z = AE_0 \sin[k_z \sqrt{\epsilon - 1}(b - y)] \cos(k_z z) e^{i\omega t} \tag{3}$$

$$E_y = -\frac{AE_0}{\sqrt{\epsilon - 1}} \cos[k_z \sqrt{\epsilon - 1}(b - y)] \sin(k_z z) e^{i\omega t} \tag{4}$$

within the dielectric. The allowed wavenumbers $k_{z,n}$ are the roots of

$$k_{z,n} a \frac{\sqrt{\epsilon - 1}}{\epsilon} = \cot[k_{z,n} \sqrt{\epsilon - 1}(b - a)] \tag{5}$$

and the relative amplitude of the electric field in the dielectric is given by

$$A = \csc[\sqrt{\epsilon - 1}(b - a) k_{z,n}]. \tag{6}$$

In practice, the $n = 1$ fundamental mode gives the optimal field configuration.

The accelerating modes of a physically realizable structure, one which is very wide in the horizontal dimension but *not* infinite, are described in Ref. 2. While the field pattern is no longer independent of y, the variation with y is on the order of a few percent, assuming a plausible laser spot size.

Longitudinal and transverse wakefields generated by a ribbon beam in a finite structure were calculated analytically in the same paper,[2] showing good agreement with numerical simulation. Instability growth due to wakefields of a finite-width beam is also considered in detail. It is shown that, though both the flute and multi-bunch beam break-up instabilities may occur, they grow far more slowly than in comparable cylindrical geometries, and may be controlled by relatively straightforward means, including vertical focusing and detuning methods.

3. Slot Field: Analysis

As a way to understand the behavior of the slot fields and estimate their magnitude, we can consider each slot to represent a parallel-plate transmission line of length Λ equal to the thickness of the structure's metal wall (as illustrated in Fig. 2). The line is driven by wall currents on the inner surface of the structure boundaries and terminates in an open circuit at the outer surface. This strategy assumes that the slot width w is small compared to the radiation wavelength λ_0, implying that the z-directed field within the slot can be taken as approximately constant.

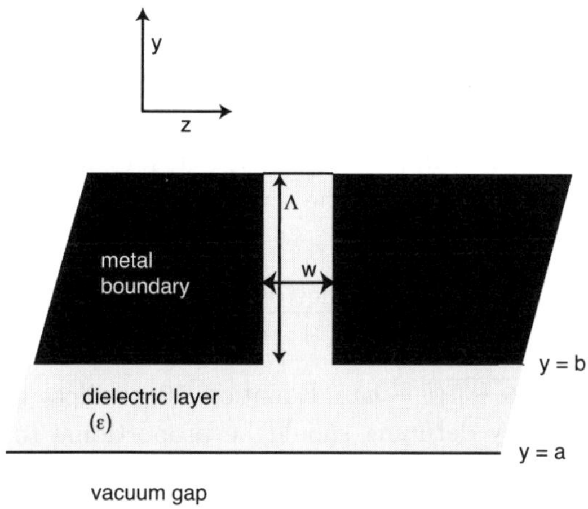

Figure 2. Detail of coupling slot geometry, showing parameters used in the analysis.

Fields within the slots then have the form

$$E_z(y,t) = 2E_0 \cos[k_g(\Lambda - y + b)]\, e^{i\omega t - ik_g(\Lambda+b)} \tag{7}$$

$$H_x(y,t) = 2iE_0(\sqrt{\epsilon}/\eta_0) \sin[k_g(\Lambda - y + b)]\, e^{i\omega t - ik_g(\Lambda+b)} \tag{8}$$

where the slot extends from $y = b$ to $y = b + \Lambda$, η_0 is the impedance of free space, and $k_g = 2\pi/\lambda_g$, with the guide wavelength $\lambda_g = \lambda_0/\sqrt{\epsilon}$. E_0, the excitation amplitude of the propagating waves, can be estimated from the wall current J_x associated with the accelerating field in the structure; we note that E_0 will be approximately inversely proportional to ϵ, since it effectively arises from the field generated by a time-varying charge distribution at $y = b$. Note that setting $\Lambda = \lambda_g/4$ results in a quarter-wave matching condition in which the slot fields vanish at $y = b$, the inner surface; the structure fields are then unperturbed by the presence of the slots. We will show below that simulation of the structure filling under these conditions confirms this behavior, with very long filling times, i.e. large values for the coupling Q-factor.

To estimate the perturbation of the resonant frequency by the presence of the slot fields, we can employ a version of the Slater perturbation theorem. In this case, we have

$$\frac{\Delta\omega}{\omega_0} = \frac{\int_{\text{slots}}(\mu|H|^2 - \epsilon|E|^2)\,dV}{\int_{y<b}(\mu|H|^2 + \epsilon|E|^2)\,dV}, \tag{9}$$

where the fields in the numerator are given by Eqns. 7 and 8 and in the denominator by the unperturbed structure fields, Eqns. 1-4. We implicitly assume that the field energy in the slots is small compared to the total field energy of the structure, which will be true for sufficiently narrow slots, and ignore the perturbation of the structure fields by the slots. Explicit calculation gives the result

$$\frac{\Delta\omega}{\omega_0} = \frac{-4\lambda_0(\sin 2k_g\Lambda)/k_g w}{\pi^4[a\sin^2\zeta + \frac{2}{3}\epsilon a\cos\zeta + (b-a)\epsilon]}, \tag{10}$$

in which $\zeta = k_z\sqrt{\epsilon - 1}(b-a)$. Equation 10 predicts that for small w and Λ the frequency detuning should be proportional to Λ/w, and that there should be no detuning effect for $\Lambda = \pi/2k_g$, as mentioned previously. Reasonable agreement with simulation is found for small values of w.

4. Simulation Results

4.1. *Time-dependent Filling*

Simulation of the time-dependent filling of the structure through the coupling slots has been demonstrated using both GdfidL and a custom two-dimensional finite-difference code.[3] The theoretical dependence of the frequency perturbation on slot dimensions agrees approximately with simulation for small slot depths, though frequency resolution is limited by the simulation grid size. Further, the use of quarter-wavelength matching in the slot length to give unperturbed structure fields was verified, as shown in Fig. 3(a), where a two-dimensional calculation shows strong fields in the slots that have no effect on acceleration fields when the slot length Λ is approximately $\lambda_g/4$. In the same figure, these fields are compared with those of a structure with 5-μm coupling slots, much shorter than $\lambda_g/4$. For this second case, the accelerating field is deformed in the vicinity of the slot and the resonant frequency is increased.

Figure 4 compares the relative amplitudes of the accelerating and Fabry-Perot modes for these two structures (detailed parameters of which are presented in Table 1). Clearly, the quarter-wavelength slots give a structure with high Q-factor—the accelerating field amplitude is more than 15 times that of the drive laser—with a concomitant long filling time (more than

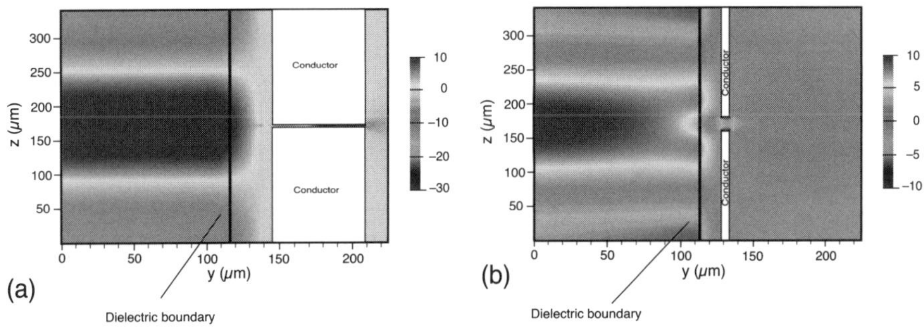

Figure 3. Map of E_z over one structure period in z, for slab-symmetric structures with (a) quarter-wave slots, $\Lambda = 69\ \mu\text{m} \approx \lambda_g/4$, and (b) short slots ($\Lambda = 5\ \mu\text{m} \ll \lambda_g/4$), with dimensions given in Table 1. Only the half-plane above the symmetry axis at $y = 0$ is shown. Field strengths are in relative units.

half a nanosecond). Likewise, the structure shows excellent mode quality, as the competing Fabry-Perot mode has less than one-half percent the amplitude of the accelerating mode. However, the field strength within the slots is also very large—at the outer end, more than three times the peak accelerating field. The 5-μm slots lead to a significantly reduced Q-factor and field enhancement, with noticeably poorer mode quality, but the field in the slot is in general *less* than the peak accelerating fields. (There will, however, be relatively large surface fields at the inner and outer corners of the slot. While the field magnitude in the slot center is at most 0.4 times the accelerating field amplitude E_{max}, there are also field spikes at the outer slot aperture with magnitudes that are sensitive to the simulation grid but are on the order of E_{max}. The slot corners will hence remain the most likely breakdown location.) Furthermore, the filling time for the 5-μm slots (150–200 ps) is much more realistic for an experiment driven by a short-pulse source such as would be obtained by frequency conversion of a CO_2 laser. The shorter pulse duration also reduces the risk of breakdown. Both structures show accelerating fields on the order of 100 MV/m when driven with a 100 MW laser source focused to a 2 cm spot.

4.2. *Wakefields*

The vanishing of transverse wakefields in a smooth slab-symmetric structure, using an infinitely wide beam, has been verified using simulation in e.g. Refs. 3 and 4. Longitudinal wakefields under those conditions are neg-

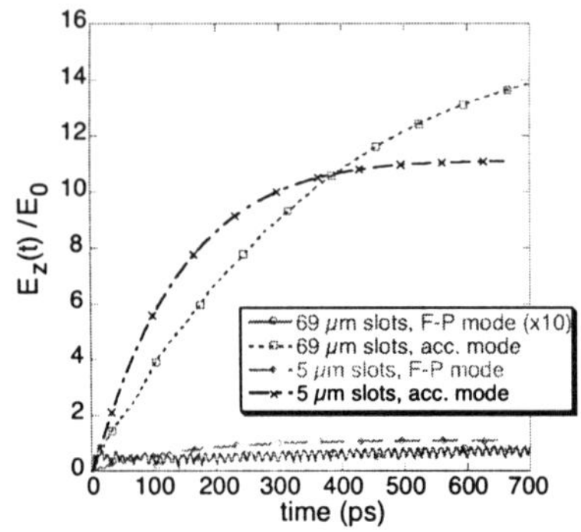

Figure 4. Time history of electric field amplitudes for the accelerating (speed-of-light) mode and the Fabry-Perot (zero) mode, comparing the two structure designs from Fig. 3. All fields are normalized to the amplitude of the driving radiation.

Table 1. Structure dimensions and filling parameters from time-dependent 2D simulation for the two versions of the 340-μm slab-symmetric structure presented in Figs. 3 and 4.

Parameter	Quarter-wave Slots	Short Slots
Vacuum gap ($2a$)	230 μm	237 μm
Dielectric thickness ($b - a$)	30 μm	17 μm
Dielectric relative permittivity ϵ	3.0	11.69
Slot width w	5 μm	20 μm
Slot depth Λ	69 μm	5 μm
Accelerating field $E_{\max}/E_0$	15.7	11.3
Peak accelerating field (per 100 MW drive power)	122 MV/m	89 MV/m
Fabry-Perot mode amplitude	$0.0038E_{\max}$	$0.099E_{\max}$
$1/e$ filling time τ	335 ps	137 ps
$Q = \omega\tau$	1850	750
Maximum field in slots	$3E_{\max}$	1.3–1.5 $E_{\max}$
Frequency detuning $\Delta\omega/\omega_0$	0	+0.033

ligible for an electron bunch length on the order of λ_0. However, one might reasonably expect the presence of the coupling slots to modify (and possibly strengthen) the wakefields, due to the passage of the electron beam near what is essentially a diffraction grating. Under these conditions, coherent radiation can be produced via the Smith-Purcell effect,[11,12] with the relation

$$\lambda = \frac{\ell}{|n|}(1/\beta - \cos\theta) \qquad (11)$$

connecting radiation wavelength λ, grating period ℓ, diffraction order n, electron velocity β, and emission angle θ. As $\beta \to 1$ for a relativistic beam, the emission angle $\theta \to 1/\gamma$, as usual, and λ becomes much less than ℓ. We therefore expect the Smith-Purcell radiation to be dominated by wavelengths in the nanometer range; further, the angular power distribution of forward radiation for a relativistic beam (discussed in Ref. 12) implies that such radiation is exponentially suppressed unless $4\pi x_0/\ell \approx 1/\gamma$, where x_0 is the offset of the electron beam from the grating. For the 340-μm slot-coupled accelerator discussed here, the left and right sides of this relation differ by nearly two orders of magnitude. Suppression of forward radiation implies a negligible interaction of the beam with its own wakefields; nevertheless, the form of the wakefields behind the bunch will in general be altered by the presence of the slots, due to the presence of a confined (non-synchronous) grating mode near the surface.

To quantify these additional effects, we have simulated the passage of a ribbon electron beam through 17 periods of a slotted structure using the 2D particle-in-cell code OOPIC.[13,14] Comparing the longitudinal (retarding) wakefields with and without slots, there is only a small change in the peak field magnitude, though the field pattern becomes less regular, as shown in Fig. 5. A structure resonant at $\lambda_0 = 340$ μm driven by a beam having $\sigma_z = 30$ μm has a retarding wakefield of only 10 kV/m, as shown in Fig. 6. The wake pattern clearly shows multimode components, but it is still true that the wake radiation has group velocities significantly less than c.

The transverse wakefield cancellation, while exact in the unslotted case, becomes a partial suppression in the slotted structure. Transverse fields of up to 20 kV/m exist within the structure, whereas fields in the 1 MV/m range are found within the dielectric; details of the field shape depend on the slot geometry. Figure 6 compares the resulting transverse wakes for several slot widths. There is no clear trend with slot width, but the narrowest slots (10 μm) appear to be the most physically similar to a grating, with the largest wakefields as a result.

5. Summary

The results presented here show clearly that any analysis of a laser-driven acceleration structure must include consideration of the coupling, or risk considerable inaccuracy. While distributed side-coupling is advantageous for this structure, in view of the damage limits on end-coupling, important consequences of this choice must be taken into account.

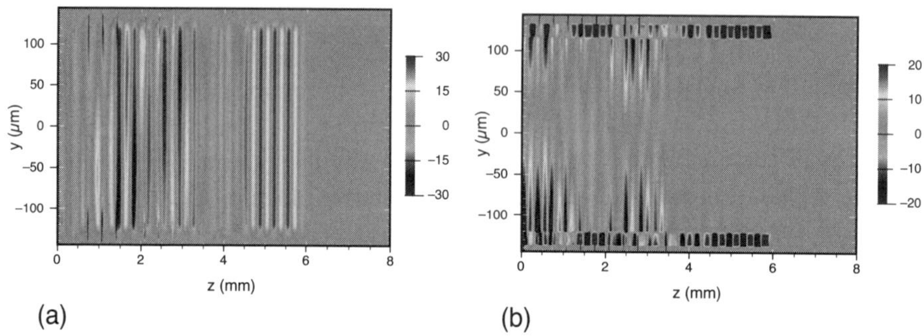

Figure 5. Wakefields (from OOPIC simulation) behind an infinitely-wide electron beam passing through a slab-symmetric structure with an array of coupling slots extending the length of the structure. Structure dimensions are as given in the right-hand column of Table 1; the (cold) electron beam has $\sigma_z = 100$ μm, $\sigma_r = 30$ μm, is located at $z = 5.8$ mm and is moving in the $+z$ direction. (a) Longitudinal wakefields (E_z), in kV/m. (b) Transverse wakefields, $E_y - cB_z$, in kV/m. Note that wakefields within the dielectric are saturated; they are generally at least an order of magnitude larger than those in the vacuum gap.

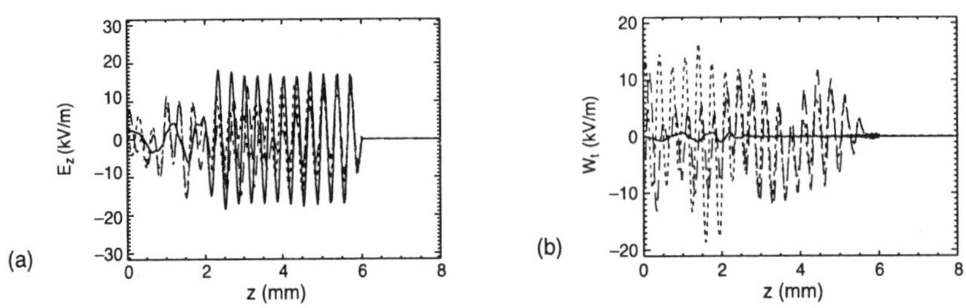

Figure 6. Lineouts of beam wakefields as a function of axial position, for the same fixed structure geometry and electron beam parameters as in Fig. 5 but with slots, as follows: solid, no slots; short dashes, 10 μm-wide slots; long dashes, 20 μm-wide slots. (a) Longitudinal wakefields E_z, along the z-axis. (b) Transverse wakefields $W_t = E_y - cB_z$, along the line $x = 0, y = +33$ μm.

We find that the time-dependent filling of this accelerator structure—including the final axial field value—depends sensitively on the parameters of the coupling slots. Hence, the overall efficiency of the structure and its acceleration gradient depend on the details of the coupling slot dimensions. Sensitive dependence of the gradient on the slot size implies, in turn, that simulations of structure behavior are limited in their accuracy and predictive power, and that manufacturing tolerances for the slots will be tight. Other consequences of the coupling slots include the perturbation of the resonant structure frequency by a few percent. This requires that the design be modified to compensate, and/or that the vacuum gap be adjustable in practice. While the slots do add a multimode component to the beam wakefields, and perturb the vanishing of the transverse wakes even for an infinitely-wide structure, they do not cause significant beam radiation, energy loss, or emittance growth.

The slot-coupled, slab-symmetric design presented here is competitive, with acceleration gradients on the order of several hundred MV/m, and appears to avoid serious limitations from material breakdown or damage. Other versions of the design, based on the same principles but without metallic boundaries, could allow scaling the device to much lower wavelength if desired.

References

1. J. Rosenzweig, A. Murokh, and C. Pellegrini, *Phys. Rev. Lett.* **74**, 2467 (1995).
2. A. Tremaine, J. Rosenzweig, and P. Schoessow, *Phys. Rev. E* **56**, 7204 (1997).
3. J. B. Rosenzweig and P. V. Schoessow, in *Advanced Accelerator Concepts, Eighth Workshop, 1998*, edited by W. Lawson *et al.*, AIP Conf. Proc. 472 (1999), p. 693.
4. R. B. Yoder and J. B. Rosenzweig, in *Advanced Accelerator Concepts, Eleventh Workshop, 2004*, edited by V. Yakimenko, AIP Conf. Proc. 737 (2004), p. 288.
5. Z. Zhang, S. G. Tantawi, and R. D. Ruth, *Phys. Rev. ST Accel. Beams* **8**, 071302 (2005).
6. A. Mizrahi and L. Schächter, *Phys. Rev. E* **70**, 016505 (2004).
7. X. E. Lin, *Phys. Rev. ST Accel. Beams* **4**, 051301 (2001).
8. B. M. Cowan, *Phys. Rev. ST Accel. Beams* **6**, 101301 (2003).
9. P. V. Schoessow and J. B. Rosenzweig, in *Proceedings of the 1999 Particle Accelerator Conference*, ed. A. Luccio and W. MacKay (IEEE, 2000), p. 3624.
10. R. B. Yoder and J. B. Rosenzweig, in *The Physics and Applications of High Brightness Electron Beams*, proc. of the 2002 ICFA Workshop (World Scientific, 2003), p. 525.
11. S. J. Smith and E. M. Purcell, *Phys. Rev.* **92**, 1069 (1953).
12. K. J. Woods *et al.*, *Phys. Rev. Lett.* **74**, 3808 (1995).
13. J. P. Verboncoeur, A. B. Langdon, and N. T. Gladd, *Comput. Phys. Commun.* **87**, 199 (1995).
14. D. L. Bruhwiler *et al.*, *Phys. Rev. ST Accel. Beams* **4**, 101302 (2001).

List of Participants

List of Participants

David Alesini
INFN-LNF
Via E. Fermi 40, 00044 Frascati (Roma) ITALY
david.alesini@lnf.infn.it

Scott Anderson
LLNL, Lawrence Livermore National Laboratory,
Livermore, CA, 94550, USA
anderson131@llnl.gov

Gerard Andonian
UCLA, Department of Physics & Astronomy
405 Hilgard Ave., Los Angeles, CA 90095
gerard@physics.ucla.edu

K.L.F. Bane
SLAC, Stanford Linear Accelerator Center,
2575 Sand Hill Road, Menlo Park, CA 94025, USA
kbane@slac.stanford.edu

Llan Ben-zvi
BNL, Brookhaven National Laboratory,
Upton, New York, USA
benzvi@bnl.gov

Carlo Joseph Bocchetta
Sincrotone Trieste S.C.p.A.
34012 Trieste, ITALY
carlo.bocchetta@elettra.trieste.it.

Rodolfo Bonifacio
INFN Sezione di Milano, ITALY
e Centro Brasileiro de Pesquisas Fìsicas, Rio de Janeiro, BRASIL
rodolfob@cbpf.br

Manuela Boscolo
INFN-LNF
Via E. Fermi 40, 00044 Frascati (Roma), ITALY
manuela.boscolo@lnf.infn.it

Charles A. Brau
Vanderbilt University, Department of Physics,
Nashville, TN 37235, USA
charles.a.brau@vanderbilt.edu

Francesco Broggi
INFN Sezione di Milano, Laboratorio LASA,

Via Fratelli Cervi 201, 20090 Segrate, Milano, ITALY
francesco.broggi@mi.infn.it

Luciano Catani
INFN-Roma2
Via della Ricerca Scientifica 1, 00133 Roma , ITALY
luciano.catani@roma2.infn.it

Rossana Centioni
INFN-LNF Via E. Fermi 40, 00044 Frascati (Roma), ITALY
rossana.centioni@lnf.infn.it

Yu-Jiuan Chen
LLNL, Lawrence Livermore National Laboratory,
Livermore, CA, 94550, USA
yjchen@llnl.gov

Alan Cook
UCLA, Department of Physics & Astronomy
405 Hilgard Ave., Los Angeles, CA 90095, USA
alancook@physics.ucla.edu

Paolo Craievich
Sincrotone Trieste S.C.p.A.
34012 Trieste, ITALY
paolo.craievich@elettra.trieste.it

Sultan Dabagov
INFN-LNF
Via E. Fermi 40, 00044 Frascati (Roma) ITALY
sultan.dabagov@lnf.infn.it

Les Dallin
Canadian Light Source Inc. University of Saskatchewan
101 Perimeter Road, Saskatoon, SK, Canada. S7N 0X4
les.dallin@lightsource.ca

Gerardo D'Auria
Sincrotone Trieste S.C.p.A.
34012 Trieste, ITALY
gerardo.dauria@elettra.trieste.it

Simone Dimitri
Sincrotone Trieste S.C.p.A.
34012 Trieste, ITALY
simone.dimitri@elettra.trieste.it

David Dowell
SLAC, Stanford University,
2275 Sand Hill Road, MS18, Menlo Park, Ca, 94025, USA

dowell@slac.stanford.edu

Mike Dunning
UCLA, Department of Physics and Astronomy,
University of California, Los Angeles, CA 90095-1547, USA
mdunning@physics.ucla.edu

R.Joel England
UCLA, Department of Physics and Astronomy,
University of California, Los Angeles, CA 90095-1547, USA
england@physics.ucla.edu

Mario Ferianis
Sincrotone Trieste S.C.p.A.
34012 Trieste, ITALY
mario.ferianis@elettra.trieste.it

Massimo Ferrario
INFN-LNF
Via E. Fermi 40, 00044 Frascati (Roma), ITALY
massimo.ferrario@lnf.infn.it

Daniele Filippetto
INFN-LNF
Via E. Fermi 40, 00044 Frascati (Roma) ITALY
daniele.filippetto@lnf.infn.it

Klaus Floettmann
DESY, Deutsches Elektronen-Synchrotron,
Notkestrasse 85, 22603 Hamburg, GERMANY
klaus.floettmann@desy.de

Valeria Fusco
INFN-LNF
Via E. Fermi 40, 00044 Frascati (Roma), ITALY
valeria.fusco@lnf.infn.it

Romain Ganter
PSI, Paul Scherrer Institute,
5232 Villigen, SWITZERLAND
romain.ganter@psi.ch

Giancarlo Gatti
INFN-LNF
Via E. Fermi 40, 00044 Frascati (Roma), ITALY
giancarlo.gatti@lnf.infn.it

Luca Giannessi
ENEA
Via Enrico Fermi 45, 00044 Frascati (Roma), ITALY

giannessi@frascati.enea.it

Dario Giove
INFN Sezione di Milano - Laboratorio LASA,
Via Fratelli Cervi 201, 20090 Segrate, Milano, ITALY
dario.giove@mi.infn.it

Silvia Giromini Vannucci
INFN-LNF
Via E. Fermi 40, 00044 Frascati (Roma), ITALY
silvia.giromini@lnf.infn.it

Danilo Giulietti
Dipartimento di Fisica "E. Fermi", Università di Pisa & INFN
Largo B. Pontecorvo, n.3, 56127, Pisa, ITALY
danilo.giulietti@df.unipi.it

William Graves
MIT, Massachusset Institute of Technology
77 Massachusetts avenue Cambridge, MA 02139-4307, USA
wsgraves@mit.edu

Jang-hui Han
DESY, Deutsches Elektronen-Synchrotron,
Notkestrasse 85, 22603 Hamburg, GERMANY
jang.hui.han@desy.de.

Mark Hogan
SLAC, Stanford University,
2275 Sand Hill Road, MS18, Menlo Park, Ca, 94025, USA
hogan@slac.stanford.edu

Katja Honkavaara
DESY, Deutsches Elektronen-Synchrotron,
Platanenallee 6, 15738 Zeuthen, GERMANY
katja.honkavaara@desy.de

Ben Ismail
LLR, École polytechnique,
91128 PALAISEAU Cedex, FRANCE
a_benismail@yahoo.co.uk

Dino Jaroszynski
Department of Physics□University of Strathclyde□
John Anderson Building□107 Rottenrow,□Glasgow G4 0NG
dino@phys.strath.ac.uk

Mark Jong
Canadian Light Source Inc. University of Saskatchewan
101 Perimeter Road, Saskatoon, SK, Canada S7N 0X4

mark.dejong@lightsource.ca

Thorsten Kamps
BESSY
Berliner Elektronenspeicherring-Gesellschaft fur Synchrotronstrahlung
D-12489 Berlin, GERMANY
kamps@bessy.de

Emanuel Karantzoulis
Sincrotone Trieste S.C.p.A.
34012 Trieste, ITALY
karantzoulis@elettra.trieste.it

Sergiy Khodyachykh
DESY, Deutsches Elektronen-Synchrotron,
Platanenallee 6, 15738 Zeuthen, GERMANY
sergiy.khodyachykh@desy.de

Kwang-Je Kim
University of Chicago & Argonne National Laboratory
5640 South Ellis Avenue, Chicago, IL 60637
kwangje@aps.anl.gov

Kenichi Kinoshita
Nuclear Professional School, School of Engineering, University of Tokyo,
22-2 Shirane - Shirakata, Tokay, Naka, Ibaraki, 319-1188, JAPAN
kino@nuclear.jp

Rami Kishek
Institute for Research in Electronics and Applied Physics, University of Maryland,
College Park, MD 20742, USA
ramiak@ebte.umd.edu

Peter Kneisel
Center for Advanced Studies of Accelerators, Jefferson Lab,
12000 Jefferson Ave. Newport News, VA 23606, USA
kneisel@jlab.org

Kazuyoshi Koyama
AIST, National Institute of Advanced Industrial Science and Technology,
Tsukuba, Ibaraki 305-8568, JAPAN
k.koyama@aist.go.jp

Geoffrey A. Krafft
Center for Advanced Studies of Accelerators, Jefferson Lab,
12000 Jefferson Ave. Newport News, VA 23606, USA
krafft@jlab.org

Patrick Krejcik
SLAC, Stanford University,

2275 Sand Hill Road, MS18, Menlo Park, Ca, 94025, USA
pkr@slac.stanford.edu

Joseph Lidestri
Columbia University - Biochemistry & Molecular Biophysics
630 West 168th Street, New York, NY 10032, USA
jpl27@columbia.edu

Steve Lidia
LBNL, Lawrence Berkeley National Laboratory
Berkeley, CA 94720, USA
smlidia@lbl.gov

Jae Lim
UCLA, Department of Physics & Astronomy
405 Hilgard Ave., Los Angeles, CA 90095, USA
jlim@physics.ucla.edu

Torsten Limberg
DESY, Deutsches Elektronen-Synchrotron,
Notkestrasse 85, 22603 Hamburg, GERMANY
torsten.limberg@desy.de

Cecile Limborg- Deprey
SLAC, Stanford University,
2275 Sand Hill Road, MS18, Menlo Park, Ca, 94025, USA
limborg@slac.stanford.edu

Vladimir Litvinienko
BNL, Brookhaven National Laboratory,
Upton, New York, USA
vl@bnl.gov

Jom Luiten
Eindhoven University of Technology
Den Dolech 2, P.O. Box 513, Eindhoven 5600 MB, THE NETHERLANDS
o.j.luiten@tue.nl

Mauro Migliorati
INFN-LNF & Univ. Roma, Accelerator Division
via Enrico Fermi, 40, 00044 Frascati (Roma), ITALY
mauro.migliorati@uniroma1.it

B.L. Militsyn
A.F.G. van der Meer, FOM Institute of Plasma Physics ìRijnhuizenî,
P.O. Box 1207, 3430 BE Nieuwegein, THE NETHERLANDS
b.l.militsyn@rijnh.nl

Andrea Mostacci
INFN-LNF & Univ. Roma, Accelerator Division

via Enrico Fermi, 40, 00044 Frascati (Roma), ITALY
andrea.mostacci@uniroma1.it

Anke-Susanne Mueller
Forschunhszentrum Karlsruhe
Hermann-von-Helmholtz-Platz 1 76344 Eggenstein-Leopoldshafen
anke-susanne.mueller@iss.fzk.de

Haruyuky Ogino
Nuclear Professional School, School of Engineering, University of Tokyo,
22-2 Shirane - Shirakata, Tokay, Naka, Ibaraki, 319-1188, JAPAN
ogino@nuclear.jp

Anatoliy Opanasenko
NSC KIPT, National Science Center Kharkov Institute of Physics & Technology
Akademicheskaya St., Kharkov, 61108, UKRAINE
opanasenko@kipt.kharkov.ua

Anne Oppelt
DESY, Deutsches Elektronen-Synchrotron,
Platanenallee 6, 15738 Zeuthen, GERMANY
anne.oppelt@desy.de

Hywel Owen
CCLRC, Daresbury Laboratory
Daresbury Warrington Cheshire, UK WA4 4AD
h.owen@dl.ac.uk

Dennis Palmer
Titan Pulse Sciences Division
2700 Merced Street, San Leandro, CA 94577, USA
dtpalmer@titan.com

Luigi Palumbo
INFN-LNF & Univ. Roma, Accelerator Division
via Enrico Fermi, 40, 00044 Frascati (Roma), ITALY
luigi.palumbo@uniroma1.it

Marco Pedrozzi
PSI, Paul Scherrer Institute,
5232 Villigen, SWITZERLAND
marco.pedrozzi@psi.ch

Claudio Pellegrini
UCLA, Department of Physics & Astronomy
405 Hilgard Ave., Los Angeles, CA 90095, USA
pellegrini@physics.ucla.edu

Giuseppe Penco
Sincrotone Trieste S.C.p.A.

34012 Trieste, ITALY
giuseppe.penco@elettra.trieste.it

Vittoria Petrillo
INFN Sez. Milan, University of Milan
via Celoria 16, 20133 Milano, ITALY
petrillo@mi.infn.it

Triveni Rao
BNL, Brookhaven National Laboratory,
Upton, New York, USA
triveni@bnl.gov

Sven Reiche
UCLA, Department of Physics & Astronomy
405 Hilgard Ave., Los Angeles, CA 90095, USA
reiche@physics.ucla.edu

Albert Reitsma
University of Strathclyde□
John Anderson Building□107 Rottenrow□Glasgow G4 ONG, UK
a.reitsma@phys.strath.ac.uk

Alberto Renieri
ENEA
Via Enrico Fermi 45, 00044 Frascati (Roma), ITALY
renieri@frascati.enea.it

Lenny Rivkin
PSI, Paul Scherrer Institute,
5232 Villigen, SWITZERLAND
leonid.rivkin@psi.ch

Mary Jo Robertson
UCLA, Department of Physics & Astronomy
405 Hilgard Ave., Los Angeles, CA 90095, USA
mjrobert@physics.ucla.edu

James Rosenzweig
UCLA, Department of Physics & Astronomy
405 Hilgard Ave., Los Angeles, CA 90095, USA
rosen@physics.ucla.edu

Jorg Rossbach
DESY, Deutsches Elektronen-Synchrotron,
Notkestrasse 85, 22603 Hamburg, GERMANY
joerg.rossbach@desy.de

Andrea Renato Rossi
INFN Sez. Milan, University of Milan

via Celoria 16, 20133 Milano, ITALY
andrea.rossi@mi.infn.it

Raphael Roux
Laboratoire de L'accelerateur Linèaire, IN2P3-CNRS, Universitè de Paris-Sud,
B.P 34, 91898 Orsay, FRANCE
roux@lal.in2p3.fr

Robert Ryne
LBNL, Lawrence Berkeley National Laboratory
Berkeley, CA 94720, USA
RDRyne@lbl.gov

Akira Sakumi
Nuclear Professional School, School of Engineering, University of Tokyo,
22-2 Shirane - Shirakata, Tokay, Naka, Ibaraki, 319-1188, JAPAN
asakumi@nuclear.jp

Fernando Sannibale
LBNL, Lawrence Berkeley National Laboratory
Berkeley, CA 94720, USA
FSannibale@lbl.gov

Levi Schachter
Department of Electrical Engineering, Technion - Israel Institute of Technology,
Haifa 32000, ISRAEL
levi@ee.thecnion.ac.il

Angelo Schiavi
University of Rome "La Sapienza", Dip. di Energetica,
Via A. Scarpa 14, 00161, Roma, ITALY
angelo.schiavi@uniroma1.it

John Schmerge
SLAC, Stanford University,
2275 Sand Hill Road, MS18, Menlo Park, Ca, 94025, USA
schmerge@slac.stanford.edu

Siegfried Schreiber
DESY, Deutsches Elektronen-Synchrotron,
Platanenallee 6, 15738 Zeuthen, GERMANY
siegfried.schreiber@desy.de

Jacek Sekutovicz
DESY, Deutsches Elektronen-Synchrotron,
Notkestrasse 85, 22603 Hamburg, GERMANY
jacek.sekutowicz@desy.de

Luca Serafini
INFN Sez. Milan, University of Milan

via Celoria 16, 20133 Milano, ITALY
luca.serafini@mi.infn.it

Timur Shaftan
BNL, Brookhaven National Laboratory,
Upton, New York, USA
shaftan@bnl.gov

John Smedley
BNL, Brookhaven National Laboratory,
Upton, New York, USA
smedley@bnl.gov

Todd Smith
HEPL 122C Stanford University
Stanford, CA 94305-4085, USA
Todd.Smith@Stanford.edu

Bruno Spataro
INFN-LNF
Via E. Fermi 40, 00044 Frascati (Roma), ITALY
bruno.spataro@lnf.infn.it

Friedrich Staufenbiel
FZR, Forschungszentrum Dresden Rossendorf
POB 510119, 01314 Dresden, GERMANY
F.Staufenbiel@fzd.de

Lazar Staykov
DESY, Deutsches Elektronen-Synchrotron,
Platanenallee 6, 15738 Zeuthen, GERMANY
lazar.staykov@desy.de

Frank Stephan
DESY, Deutsches Elektronen-Synchrotron,
Platanenallee 6, 15738 Zeuthen, GERMANY
frank.stephan@desy.de

Gennady Stupakov
SLAC, Stanford University,
2275 Sand Hill Road, MS18, Menlo Park, Ca, 94025, USA
stupakov@slac.stanford.edu.

Franco Tazzioli
INFN-LNF
Via E. Fermi 40, 00044 Frascati (Roma), ITALY
franco.tazzioli@lnf.infn

M.C. Thompson
UCLA, Department of Physics & Astronomy

405 Hilgard Ave., Los Angeles, CA 90095, USA
thompson@physics.ucla.edu

Paolo Tomassini
IPCF- CNR & INFN Edificio C - Polo Fibonacci
Largo B. Pontecorvo, 3 - 56127 Pisa, ITALY
tomassini@ipcf.cnr.it

Limberg Torsten
DESY, Deutsches Elektronen-Synchrotron,
Notkestrasse 85, 22603 Hamburg, GERMANY
torsten.limberg@desy.de

Gil Travish
UCLA, Department of Physics & Astronomy
405 Hilgard Ave., Los Angeles, CA 90095, USA
travish@physics.ucla.edu

Srinivasan-Rao Triveni
BNL, Brookhaven National Laboratory,
Upton, New York, USA
triveni @bnl.gov

Alexander Tron
Lebedev Physical Institute
Leninsky Pr. 53, 119991 Moscow, RUSSIA
atron@sci.lebedev.ru

Mitsuru Uesaka
Nuclear Professional School, School of Engineering, University of Tokyo,
22-2 Shirane - Shirakata, Tokay, Naka, Ibaraki, 319-1188, JAPAN
uesaka@nuclear.jp

Cristina Vaccarezza
INFN-LNF
Via E. Fermi 40, 00044 Frascati (Roma), ITALY
cristina.vaccarezza@lnf.infn.it

S.B. Van Der Geer
Eindhoven University of Technology
Den Dolech 2, P.O. Box 513, Eindhoven 5600 MB,THE NETHERLANDS
s.b.van.der.geer@tue.nl

Chun-xi Wang
Argonne National Laboratory,
9700 South Cass Avenue, Argonne, Illinois 60439, USA
wangcx@aps.anl.gov

Takahiro Watanabe
BNL, Brookhaven National Laboratory, Upton, New York, USA

twatanabe@bnl.gov

Oliver Williams
UCLA Department of Physics & Astronomy
405 Hilgard Ave., Los Angeles, CA 90095, USA
obw@ucla.edu

Vitaly Yakimenko
BNL, Brookhaven National Laboratory,
Upton, New York, USA
yakimenk@bnl.gov.

Rodney Yoder
Department of Physics Manhattan College Riverdale, NY 10471, USA
ryoder@mailaps.org

Max Zolotorev
LBNL, Lawrence Berkeley National Laboratory
Berkeley, CA 94720, USA
Max_Zolotorev@lbl.gov